Lecture Notes in Intelligent Transportation and Infrastructure

The series "Lecture Notes in Intelligent Transportation and Infrastructure" (LNITI) publishes new developments and advances in the various areas of intelligent transportation and infrastructure. Merging theoretical foundations, practical applications, and forward-looking insights, LNITI provides a comprehensive understanding of both the state-of-the-art and the future prospects within this dynamic field.

LNITI is designed to be an inclusive platform that covers an extensive array of topics including, but not limited to intelligent transportation systems, smart mobility, intelligent logistics, critical infrastructure, smart architecture, smart cities, intelligent governance, construction design, data security, operational analysis, optimal route planning, digitalization, autonomous vehicles, the evolution of transport systems as well as green and sustainable urban structures. The series contains monographs, conference proceedings, edited volumes, lecture notes and textbooks. Of particular value to both the contributors and the readership are the short publication timeframe and the world-wide distribution, which enable wide and rapid dissemination of high-quality research output.

Proceedings published in the series are indexed by INSPEC.

All books, including proceedings, published in the series are evaluated by Web of Science

Olegas Prentkovskis · Irina Yatskiv (Jackiva) ·
Paulius Skačkauskas · Mykola Karpenko ·
Michał Stosiak
Editors

TRANSBALTICA XIV: Transportation Science and Technology

Proceedings of the 14th International Conference TRANSBALTICA, September 14–15, 2023, Vilnius, Lithuania

Springer

Editors
Olegas Prentkovskis
Vilnius Gediminas Technical University
Vilnius, Lithuania

Irina Yatskiv (Jackiva)
Transport and Telecommunication Institute
Riga, Latvia

Paulius Skačkauskas
Vilnius Gediminas Technical University
Vilnius, Lithuania

Mykola Karpenko
Vilnius Gediminas Technical University
Vilnius, Lithuania

Michał Stosiak
Wrocław University of Science
and Technology
Wrocław, Poland

ISSN 2523-3440 ISSN 2523-3459 (electronic)
Lecture Notes in Intelligent Transportation and Infrastructure
ISBN 978-3-031-52651-0 ISBN 978-3-031-52652-7 (eBook)
https://doi.org/10.1007/978-3-031-52652-7

This Springer imprint is published by the registered company Springer Nature Switzerland AG
The registered company address is: Gewerbestrasse 11, 6330 Cham, Switzerland

Paper in this product is recyclable.

Preface

The international scientific conference "Transbaltica: Transportation Science and Technology" is a traditional, annual event of the Faculty of Transport Engineering in Vilnius Gediminas Technical University (VILNIUS TECH), organized in cooperation with various partners since 2001. The fourteenth conference "Transbaltica: Transportation Science and Technology" was held on 14–15 September 2023 in Vilnius, Lithuania. The conference was organized in partnership with the Ministry of Transport and Communications of the Republic of Lithuania, JSC Girteka Logistics and Transport Innovation Association.

"Transbaltica 2023" received 100 contributions from 12 countries around the world. After a careful single-blind peer-review process, which also involved reviewers from various parts of the world, 61 papers were accepted. Each paper underwent evaluation by two reviewers: a member of the scientific committee and an external reviewer. During the conference, researchers presented their works, covering a number of scientific problems within the research fields of transport engineering, transportation and logistics, as well as other disciplines and interdisciplinary areas related to transport system.

Current proceedings of "Transbaltica 2023" consists of five parts, featuring these main areas of interest:

1) **Intelligent Vehicles and Infrastructure** cover highly automated and autonomous driving, and infrastructure for autonomous and connected vehicles.
2) **Combustion in Engines, Alternative Technologies, Energy Management and Emissions** provides useful information regarding alternative fuels, fuel mixtures, combustion process control methods and efficient use of energy.
3) **Vehicle Engineering and Dynamics** discusses timely issues in vehicle modelling and simulations, vehicle safety systems and railway transport, and relating technologies and applications
4) **Logistics and Transportation** describes the most recent research trends regarding green logistics, supply chain connectivity, computational logistics, as well as new circumstances affecting the carriage of goods and passengers, and possible solutions to them.
5) **Railway Transport** covers various rail-based transport systems, dynamics and mechanics of rail vehicles, rail electrification, rail transport infrastructure, planning and design and other advanced rail technologies.

The organizers of the international scientific conference "Transbaltica: Transportation Science and Technology" and the editors of these proceedings would like to acknowledge all reviewers who helped evaluate conference submissions and refine contents of this volume. Organizers of the conference also acknowledge all the authors who have chosen "Transbaltica: Transportation Science and Technology" as the publication platform for their research and would like to express our hope that their papers will foster further developments in the design and analysis of complex transport systems, offering a valuable and timely resource for scientists, researchers, practitioners, and students who work in all the areas mentioned above.

Olegas Prentkovskis
Irina Yatskiv (Jackiva)
Paulius Skačkauskas
Mykola Karpenko
Michał Stosiak

Organization

Organizing Committee

Chairman

Olegas Prentkovskis — Vilnius Gediminas Technical University, Lithuania

Secretary

Darius Aleksonis — Vilnius Gediminas Technical University, Lithuania

Members

Darius Bazaras — Vilnius Gediminas Technical University, Lithuania

Edgar Sokolovskij — Vilnius Gediminas Technical University, Lithuania

Giedrius Garbinčius — Vilnius Gediminas Technical University, Lithuania

Saugirdas Pukalskas — Vilnius Gediminas Technical University, Lithuania

Scientific Committee

Chairman

Paulius Skačkauskas — Vilnius Gediminas Technical University, Lithuania

Secretary

Mykola Karpenko — Vilnius Gediminas Technical University, Lithuania

Members

Gintautas Bureika	Vilnius Gediminas Technical University, Lithuania
Adam Deptuła	Opole University of Technology, Poland
Marianna Jacyna	Warsaw University of Technology, Poland
Irina Jackiva	Transport and Telecommunication Institute, Latvia
Jolanta Janutėnienė	Klaipeda University, Lithuania
Raimundas Junevičius	Vilnius Gediminas Technical University, Lithuania
Michał Stosiak	Wroclaw University of Science and Technology, Poland
Helena M. Ramos	University of Lisbon, Portugal
Pavlo Maruschak	Ternopil Ivan Pul'uj National Technical University, Ukraine
Artur Kierzkowski	Wroclaw University of Science and Technology, Poland
Vidas Žuraulis	Vilnius Gediminas Technical University, Lithuania
Quirino Estrada	Autonomous University of Ciudad Juárez, Mexico
Tomislav Rožić	University of Zagreb, Croatia
Matej Babic	University of Novo Mesto, Slovenia
Paweł Śliwiński	Gdańsk University of Technology, Poland
Nicholas Langlois	EISGELEC, France
Iyad Alomar	Transport and Telecommunication Institute, Latvia
Metin Mutlu Aydin	Ondokuz Mayıs University, Turkey
Anna Borucka	Military University of Technology, Poland
Ivona Bajor	University of Zagreb, Croatia
Mohsen Besharat	University of Leeds, UK
Kristina Čižiūnienė	Vilnius Gediminas Technical University, Lithuania
Olja Čokorilo	University of Belgrade, Serbia
Maksym Delembovskyi	Kyiv National University of Civil Engineering and Architecture, Ukraine
Cyril Fischer	Institute of Theoretical and Applied Mechanics, Czech Republic
Piotr Gorzelańczyk	Stanisław Staszic State University of Applied Sciences in Piła, Poland
Viktoriia Ivannikova	Dublin City University Business School, Ireland
Ilya Jackson	Massachusetts Institute of Technology, USA
Sylwester Kudźma	Rotary Power, UK

Contents

Logistics and Transportation

Railway Transport

Intelligent Vehicles and Infrastructure

Application of Laser Technologies for Scanning Communication Routes While Restoring the Infrastructure of Ukraine

Sergii Panchenko, Yevgeniia Ugnenko, Elena Uzhviieva, Yevhen Korostelov, and Nataliia Sorochuk(✉)

Ukrainian State University of Railway Transport, Kharkiv, Ukraine
nat.sorochuk50@gmail.com

Abstract. The design work in modern software systems with laser technology for scanning the surface of communication routes is carried out in a unified digital model. It allows for the introduction of software systems into the information environment along with the adaptation and support of information modelling technologies. The unified digital surface model includes digital models and objects of relief, situation, geology and communications that require various methods of collecting the initial data and their processing. Modern software systems provide the initial data quality and the possibility of forming a unified digital model. Owing to the flexible multi-organizational structure of data and the possibility of their conversion, a unified digital model can be transferred to any software product and integrated into the technological chains of the life cycle of objects built on laser scanning platforms. The research part of the software complexes can be integrated and implemented in the technological chain of the entire life cycle of objects. The use of modern software systems with laser technologies for scanning the surface of communication routes allows specialists to build an effective integrated manufacturing and technological environment starting with the preparation of initial data for designing and ending with the transfer of the design data to the construction site. Geoinformation science not only integrates the solutions developed by geodetic science and cartography in this field but also implements them in the system information environment. Geoinformation systems contain solutions implemented in the software product, which are related to the establishment of the reference system, the transformation of data from one coordinate system to another, the creation of topological models, the analysis of data, provided that these data are presented in a unified coordinate system, etc. Solutions to many problems in GIS are based on the key process called Geo-referencing, i.e. establishing the relationships of the model of geographical objects with the Earth coordinate system and the cartographic projection. Therefore, geoinformation systems require the knowledge of the basic positions of coordinate systems on the Earth's surface and the cartographic projections.

Keywords: Laser scanning · Unified digital model · Initial data · Information modelling · Software complex · Information environment

O. Prentkovskis et al. (Eds.): TRANSBALTICA 2023, LNITI, pp. 3–11, 2024.
https://doi.org/10.1007/978-3-031-52652-7_1

1 Introduction

The interest in laser scanning and geoinformation technologies has been constantly growing due to their effective application for engineering research [1].

They are used for solutions of local, regional and global problems associated with sustainable development of territories. Modern geoinformation systems and laser scanning technologies are expanding their research methods and providing digital tools for managing and operating spatial data, modelling processes, visualizing data, models and processes by applying advanced computer techniques and specialized tools for processing and analysing geodata [2].

The technology of processing laser-scanning data using the 3D SCAN system [3, 4] allows solving the following tasks:

- to load point clouds in different formats;
- to display point clouds in 3D and 2D shapes;
- to upload and display a photo image with the geospatial reference in the *.kml format together with the point cloud;
- to filter 'noise' in the point cloud;
- to recognize point and linear objects of the situation and to create 3D and 2D topographic objects;
- to select a terrain or an area with specified slope parameters;
- to carry out an adaptive thinning of the point cloud and construct a digital terrain model (DTM);
- to create and edit topographic objects for the preparation of topographic plans when implementing projects on the reconstruction of communication routes; and
- to export data in convenient formats for further creation of a digital surface model (DSM) for engineering purposes.

Modern software systems allow creating a digital terrain model in a semi-automatic mode. This requires the following:

- to select a terrain and specify the parameters suitable for this type of terrain. As a result, a database with the points related only to the terrain will be created; and
- to adjust the obtained point cloud according to the requirements for the digital terrain model (maximum distance between points on flat areas, minimum size of terrain microforms). Consequently, the database with the number of points according to the number of pickets in the instrumental topographic survey will be created.

Thus, using laser scanning technologies and taking into account the features of engineering surveys for linear objects, it is possible to analyse the project management on the basis of the information available.

2 Analysis of Recent Research and Publications

The operation of modern laser rangefinders used in scanners is based on pulse and phase distance measuring methods. In the process of scanning, the direction of laser beam propagation and the distance to the object's points are recorded. The scanner creates a point array (cloud) of laser reflections from the objects in the scanner's field of view with five characteristics, namely spatial coordinates (x, y, z), intensity and real colour. Usually, the characteristics of the real colour for each point are recognized using a digital camera. Leica Geosystems (Switzerland) is a recognized world leader in the development of laser technologies and production of laser scanners [6].

The main advantages of laser scanning are:

- lower costs of topographic surveying;
- fewer or no additional surveys of the object;
- more accurate and comprehensive surveys and, consequently, elimination of ambiguities inherent in desktop studies; and
- minimal field work time.

Added advantages:

- rapid result delivery;
- shortened overall project cycle;
- high level of detail and improved quality of the result;
- work safety during surveying;
- non-destructive survey technique; and
- possibility of using point clouds by different experts, which increases the technique effectiveness.

Areas of application for laser scanning:

- creating 3D models of complex engineering structures and technological equipment with a high degree of detail and accuracy;
- surveying facades of historical buildings, monuments and unique objects to be reconstructed;
- surveying communication routes;
- surveying tunnels;
- monitoring buildings and structures;
- determining the scope of earthworks and/or technological capacities; and
- keeping records of the consequences of emergencies [7, 8].

The laser scanning of objects is carried out by modern 3D scanners [9]. These devices not only simplify the process of creating 3D models, but also allow solving problems with maximum accuracy and reliability.

The advantages of 3D scanning are instant 3D object visualization, high accuracy, safety, portability, convenience, great variety of objects, accurate colour rendering, and scanning of moving objects.

The 3D scanner can greatly facilitate the human activity in many fields. The technology of volumetric scanning is dynamically developing and provides unique opportunities from planning medical operations and creating a design layout to controlling the product quality. The 3D scanner is needed for determining the object's shape with high accuracy in the shortest possible time.

There are special features of modern computer-aided design systems (CAD) fitted with the laser scanning technology for projects on the reconstruction of communication routes while creating a digital surface model. They are:

- 3D SCAN is intended for creating DSM for engineering purposes according to point clouds. The software complex uses tools that make it possible to quickly and conveniently create a digital model of the situation in the form of point, linear and area conventional signs, to select relief points and to create DTM. The resulting DSM can be arranged as a topographic plan, and then transferred for further processing in modern software systems or exported in exchange formats;
- 3D SCAN created on the DAT 4 platform has a similar interface (workspaces, properties window, and history window). The general modules are classifier editor, geodetic library, template editor, drawing model, drawing editor, and web maps;
- 3D SCAN can import:
- point clouds in the *.las format or text formats;
- geolocation of photo images with the *.kml reference;
- raster images with or without references, as well as data of cartographic web services.

When importing point clouds, the information about the colour of points, intensity of the reflected signal, as well as point classifications performed by the software, are read when forming a point cloud file.

The point cloud contains about 50 million points, while there is practically no delay in their displaying. The number of points displayed does not affect the DSM quality and all methods use the initial data of the point cloud located in the *.cpc file.

To use web services and work with geolocation photos, it is necessary to set a coordinate system in the project on the reconstruction of communication routes, in which a point cloud is created. If the information on the coordinate system in the *.las file is available, these data will be read and the coordinate system will be installed in the project on the reconstruction of communication routes.

The 3D SCAN software complex has a database of EPSG coordinate systems (CS) which can be used to quickly obtain the parameters of many CSs by code, in particular all zones of USK2000 [10], CS42, some areas of CS63, most of the current CSs in foreign territories [11].

Using web maps, there is no need to use other software systems, and 3D SCAN can immediately solve the issue of general analysis of the terrain in the scanning area by clarifying the direction of communication routes and measuring distances to the objects. When working with mobile scanning data if scanning and photographing have been carried out in different times, it can be difficult to compare the structure of folders with photos of scanned route sections.

By applying the gradient fill for the height, it is possible to clearly highlight the terrain features, which is particularly important for unrecognized point clouds. Thus, details at different heights can be displayed more clearly by controlling the gradient. This makes it possible to achieve visual clarity of curbs and other objects with a low height on the communication route sections.

The 3D scanning has a number of functions with which it is possible to create quickly and conveniently vector objects according to point clouds in semi-automatic mode, and to refer these objects to classifiers. Certainly, it is possible to manually display the objects according to the point cloud. In order to correctly and unambiguously determine the mark for the points of created objects (first and foremost it relates to the 'Plan' window where there is no third coordinate), it is advisable to select the terrain with the 'Highlighting the terrain' command. This command creates a new point cloud containing only relief points.

It should be noted that the mathematical filter does not ensure the rejection of all non-relief points and the storage of relief ones. Similarly, if the classification of relief and vegetation is available, it is possible to specify the relief points in the initial data with the 'Extract layers' command and to select the *ground layer.

According to the algorithm, the active point cloud will be sequentially processed by the so-called 'Filter windows' with different sizes. The minimum size of the 'Filter window' is one meter; the maximum is 20 m [12].

Each 'Filter window' will get its threshold of excess depending on the specified maximum and minimum threshold of excess, as well as the average slope of the terrain. If the height of the object in the 'Filter window' exceeds the specified threshold, such an object will be deleted.

3 Determination of the Purpose and Objectives of the Study

The purpose and objectives of the study have been formulated on the basis of the analysis of the latest research and publications as well as by taking into account the interest in laser scanning and geoinformation technologies.

The purpose of the study is to determine digital tools for spatial data management.

The main objectives of the study are:

- to model the operation processes for communication routes;
- to define the source data management when solving applied problems of designing communication routes; and
- to analyse existing models for recognizing linear and point objects.

4 Main Part of the Study

Let us consider methods for recognizing objects in the point cloud. Some objects, for example, the edge of asphalt and concrete surfaces on sections of communication routes, cannot not be recognized due to the lack of a clear boundary (because some areas of asphalt and concrete surface are in poor condition or badly destroyed). In this case, it is possible to visually recognize the edge of asphalt-and-concrete coating and to outline it using the point cloud. Topographers experienced in field of encoding will be able to appreciate the function of quick search for an object using its code (corresponds to the basic DAT codes) in the classification. Here, the reliability of delineation of the asphalt-and-concrete coating edge will correspond to the real fieldwork.

If clear objects are available, automatic recognition can be applied. A typical example of such an object is power line wires (Fig. 1).

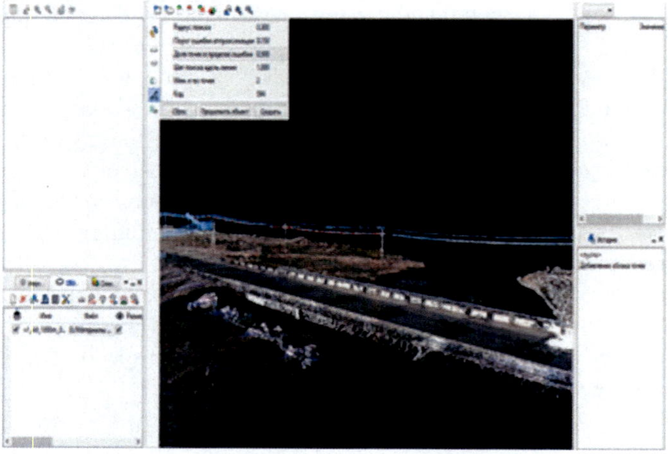

Fig. 1. Recognition of linear topic objects.

The same tool is available for recognizing point objects. For example, while putting manually poles and road signs, the operator can accidentally capture a point quite far from the sign base (due to projections). In such cases, it is convenient to use the tool for recognizing point topic objects (Fig. 2).

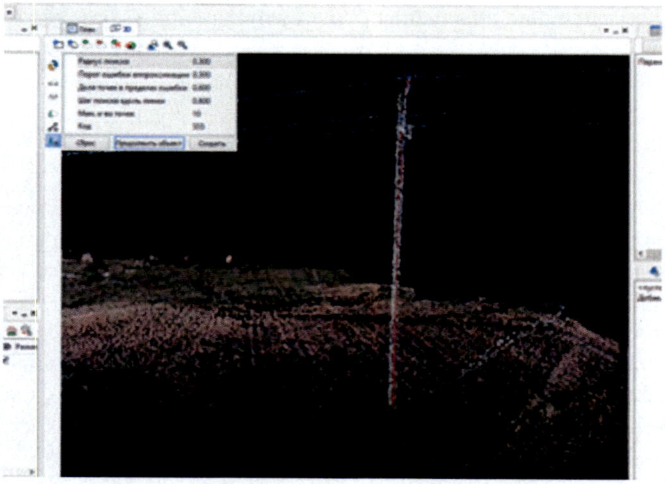

Fig. 2. Recognition of point topic objects.

The objects created are immediately displayed with the corresponding conventional sign in the 'Plan' window as those created in the 'Plan' window, similarly to point topic objects recognized and marked in the designated terrain model.

The pre-determined cloud of relief points can be used to create DTM. The initial data transmit information about the terrain, but this information is redundant, as a large number of points complicate the modelling of the surface and further work with the project on the reconstruction of communication routes.

The minimum approximation area corresponds to the database in the 'Relief microforms' window.

Parameters of the mean and minimum deviation depend on the terrain nature [13]. After filtering, the number of points is close to the initial data of a full-scale survey. Using the model points, it is possible to build the surface of the terrain by taking into account the structural lines. Structural lines are linear topic objects on relief points with a sign of relief as presented in the classification of objects.

In many cases, the correct displaying of the terrain surface while designing the topographic plan requires the creation of a linear topic object at the top and bottom of the slopes. The slopes are almost unreadable in the 'Plan' window, while in the 3D window in manual mode, it is possible to delineate slopes, however the experience indicates that those surveying make many mistakes in long objects, missing, for example, breaks in the slopes where secondary roads adjoin.

The resulting cloud also clearly shows slopes along the highway, turns and breaks of the slopes on the adjoining roads. If necessary, the terrain and slope maps can be displayed together, and this mode can be used for drawing the structural lines of the top and bottom of a slope (Fig. 3).

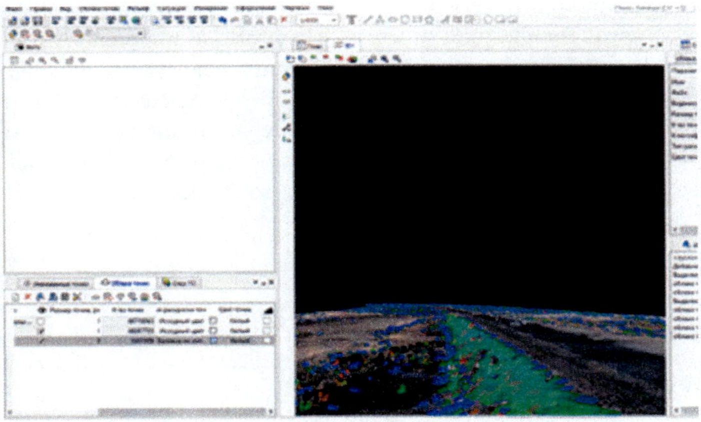

Fig. 3. Recognition of roadbed slopes.

The result can be saved in the *.gds4 project for transferring to software packages and exported to *dxf,* mif or *mid.

The gds4 format saves points and the topic object (TO).

In the *dxf format, according to the configured correspondence scheme of point topic objects, the initial data can be transmitted by blocks of linear topic objects by means of configured line types; DSM is also transmitted in the form of horizontals and triangulation (3D Face).

5 Conclusion

The laser scanning technology for the surface of communication routes provides powerful digital tools for managing spatial data and modelling operation processes of communication routes.

The data collected can be stored in the form of sets or files. In addition, these data can be organized in related groups called models. In order for heterogeneous data and models to be processed in the same system, they must be arranged and reduced to a single information model in which they complement each other. The result of data management is the creation of such an information model that can effectively store the data in the database and efficiently process them in information systems using different technologies.

The data management provides qualitatively new properties for geodata. It is the data management that makes it possible to use geographical data for solving a wide range of applied tasks of control, analysis, logistics, planning, design, forecasting, resource use, monitoring, etc.

The initial and pre-processed information includes a plethora of parameters, some of which may duplicate each other. Here special models can be used to reduce the amount of information on real objects; these models can store information on the basic properties and secondary properties of the objects of study. One of the features of data collection in geoinformatics is that the initial data can have not only different dimensions, but also be measured using different measurement scales. The data management in geoinformatics creates conditions for editing data of different dimensions and measurement scales in a single environment for the joint analysis. Therefore, a comprehensive analysis of heterogeneous data measured using different measurement scales is possible [1, 2, 14].

While managing various initial data on objects, their characteristics, forms and interrelations, different descriptive information turns into sets of models which are used for processing in geoinformation technologies. An integrated information model includes a set of simpler models, which requires optimization so that to be efficient.

References

1. Zatserkovny, V.I., Burachek, V.G., Zheleznyak, O.O., Tereshchenko, A.A.: Geoinformation systems and databases: monograph. Book. 2 (Nizhyn Mykola Gogol State University, Nizhyn), p. 237 (2017)
2. Shipulin, V.D.: Basic principles of geoinformation systems: textbook. manual (Kharkiv National Academy of Municipal Economy, Kharkov, 2010, p. 313 (2010)
3. Dorozhynskyi, O.L.: Geoinformation systems and databases: monograph. Terrestrial laser scanning in photogrammetry. Publishing House of Lviv Polytechnic (2014)
4. PromScan3D. https://www.promscan3d.com
5. AERO3D engineering. https://aero3d.ua
6. Shevchenko, T.G., Moroz, O.I., Alarm, I.S.: Geodetic instruments (Polytechnic, Lviv, 2006), pp. 455–459 (2006)
7. Leica Geosystems AG. https://leica-geosystems.com/en-us/
8. Principle of laser scanning. https://ngc.com.ua/ua/info/whats_hds.html

9. Wu, C., Yuan, Y., Tang, Y., Tian, B.: Application of terrestrial laser scanning (TLS) in the architecture, engineering and construction (AEC) Industry. Sensors **22**(1), 265 (2022). https://doi.org/10.3390/s22010265
10. On the approval of the Procedure for the use of the USK-2000 State Geodetic Reference System of coordinates in the implementation of land management works Order 02.12.2016 No. 509 Document z1646-16, valid, current edition – Adoption dated 02.12.2016. [Electronic resource]. https://zakon.rada.gov.ua/laws/show/z1646-16
11. Cherniaha, P., Kubakh, S.: Advantages and disadvantages of different coordinate systems and geodetic projections during land cadastre management. Modern achievements of geodetic science and production, 2010. issue. II (20), pp. 62–66 (2010)
12. Instructions for topographic surveying at scales of 1:5000, 1:2000, 1:1000 and 1:500. Document z0393-98, current edition – Edition dated 09/28/1999, basis – z0653-99. [Electronic resource]. https://zakon.rada.gov.ua/laws/show/z0393-98
13. Law of Ukraine On topographical, geodetic and cartographic activities. Document 353-XIV, valid, current edition – Edition dated 07/27/2013, basis – 367-VII: [Electronic resource]. https://zakon.rada.gov.ua/laws/show/353-14
14. Ischuk, O.O., Korzhnev, M.M., Koshlyakov, O.E.: Spatial analysis and modeling in GIS: Tutorial. Under the editorship Acad. D.M. Grodzinsky. K.: Kyiv University Publishing and Printing Center, p. 200 (2003)

The Current State of Using Drones for Property Protection in Slovakia

Andrej Vel'as(✉), Jakub Ďurica, and Martin Boroš

University of Žilina, Univerzitná 8215/1, 010 26 Žilina, Slovakia
Andrej.Velas@uniza.sk

Abstract. The protection of individuals and property can be defined as procedures and processes aimed at achieving a state of security for individuals and property using available resources, including technical, personnel, information, and others. State institutions such as police forces and armed forces, as well as non-state institutions, primarily private security services and guards, are dedicated to ensuring the protection of individuals and property. The level and method of this protection are determined by the security situation (crime rate) in a specific country and the legal regulations (laws, standards, and decrees) that govern possible methods of protection. The financial situation of a particular company and the available resources/finances for safeguarding individuals and property significantly impact the level of protection provided.

This article is primarily focused on the physical protection of objects by drones. The issue at hand pertains to security management. The aim of this article is to describe the possible methods of property protection using drones or robots, along with their advantages, disadvantages, and potential barriers to their implementation. During the COVID-19 pandemic, security managers had to contend with a lack of personnel due to COVID-19-related impacts or their involvement in testing and monitoring individuals entering facilities. This deficiency could not be compensated for by employing regular staff. Given the increasing cost of labour, security managers today are seeking effective ways to safeguard the perimeters of objects or large open areas within companies. The use of drones presents an interesting and promising solution, although it is not without its complexities.

Keywords: Security · Protection · Drones

1 Introduction

Based on practical experience in Central Europe, the current state of property protection is as follows: Property protection is a combination of technical components of physical protection, organizational and regulatory measures, and personnel resources (guards). All these elements create the Physical Protection System or physical security. The Physical Protection System (PPS) consists of system elements combined and designed to achieve the required level of protection. PPS is rarely the same in different locations or objects due to the numerous differences in objects (business sector), environment,

O. Prentkovskis et al. (Eds.): TRANSBALTICA 2023, LNITI, pp. 12–21, 2024.
https://doi.org/10.1007/978-3-031-52652-7_2

objectives, and threats. PPS is a term used to define the integration of people, procedures, and equipment to protect assets from threats and risks. Fischer defines PPS as a way to protect objects against theft, vandalism, sabotage, and unauthorized access [1]. According to Loveček [2] and Zou [3], the main aims of PPS are to protect assets or facilities from theft, sabotage, or other anthropogenic attacks and to equally integrate people, processes, and facilities. Smith writes that physical security comprises measures designed to protect people, prevent unauthorized access to property, and safeguard property from sabotage, damage, and theft [4].

The basic functions of PPS are detection, deceleration and response. Some authors write about deterrence as the fourth basic function of PPS, but this can be assigned to elements that have the task of slowing down entry [5, 6]. Deterrence is one of the basic tasks of mechanical barriers [7].

2 Drones in the Security Field

The use of drones for property protection has been discussed since 2017 during workshops for security managers organized by the Association of Security Managers in Slovakia. The association comprises approximately 25 managers (with the number of managers and associate members growing), representing various industries ranging from automotive (Volkswagen, Brose, Škoda, Jaguar Land Rover, and Continental) to services, food, energy, communications, and education. They face a significant challenge - the lack of human resources for ensuring security and conducting patrols across large areas within company perimeters.

For instance, in Slovakia, which, according to Eurostat [8], is one of the countries with the lowest average wages in Europe, the Slovak Chamber of Private Security reports that the wage costs per employee in private security services amount to 65,780 Euros per year (including employer's taxes and state fees) [9]. Security managers suggest that with this budget when considering multiple employees, it becomes feasible to acquire professional drones. In the context of cost reduction, any automated solution becomes appealing.

We asked security managers in Slovakia via email in August of 2023 to answer the following questions:

1. Do you currently use drones for property protection?
2. Are you interested in using drones for property protection in the future?
3. Which parts of the organization do you consider the most important for asset protection using drones? Which parts have the potential to be protected by drones? Select or add:

   - extensive parking lots,
   - production halls,
   - storage areas,
   - perimeter,
   - other ... (please specify)?

4. What types of data should drones provide for property protection? Please select or add:

- video/images of corresponding quality,
- thermal cameras (night vision),
- fire detection,
- air quality sensors (usable in case of gas leaks),
- other ... (please specify)?

5. What advantages do you see in the potential use of drones for property protection compared to traditional detection technologies?
6. What limitations do you associate with the use of drones for property protection?

Five out of the seven members of the Security Managers Association Board of Directors responded.

The responses to the first question indicated that companies are not currently utilizing drones for asset protection.

In response to the second question, they expressed an interest in using drones for property protection, particularly for outdoor locations. They also mentioned that, at present, this solution is seen as rather innovative.

Regarding the answers to the third question, potential areas for drone-assisted protection were identified as rooftops, construction sites, and the monitoring of large perimeters. They also highlighted the importance of using drones for responding to extraordinary and crisis events, serious industrial accidents, ecological incidents, and fires. Drones have the potential to protect lives and enhance safety during the organization of large-scale events in secure areas.

Regarding the desired outputs from drones, the managers concurred with the provided response options and added the following points. Image quality must ensure real-time and uninterrupted signal transmission. Night flights require special permission from aviation authorities, making them unrealistic for commercial applications. Drones should be equipped with voice output capabilities to communicate with intruders.

In response to question number 5, which focused on the advantages of using drones, they emphasized that drones offer a flexible alternative for rapidly adapting to changing security situations. Drones are particularly valuable for monitoring challenging-to-access locations, providing an overview of complex situations at the scene, and facilitating swift operational decision-making. Additionally, drones can contribute to cost savings and enable more efficient personnel deployment, ultimately leading to heightened security levels.

The primary limitations were addressed in the sixth question. According to the managers, the main challenges include legislative conditions and technological complexities when compared to camera and perimeter detection systems. The need for a trained operator pilot was highlighted. Autonomy in drone operation is currently limited, and operators must maintain visual contact with the drone, precluding flights behind obstacles.

Security managers pointed out that they currently lack solutions for protecting their assets from unauthorized drones.

Patterson [10] published the utilization of drones in various industry sectors. The graph (see Fig. 1) also illustrates the percentage of drones used in the security sector. In the same article, he highlights the significant surge in the drone industry and the substantial increase in the drone market, reaching billions of dollars.

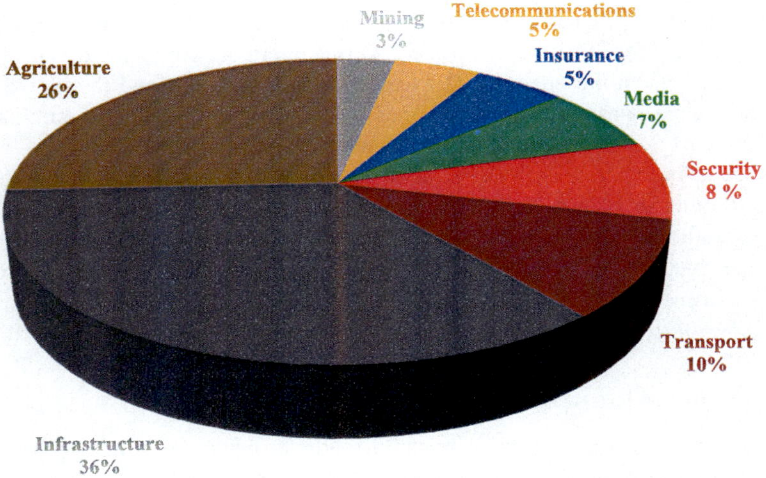

Fig. 1. Drone usage by industry sector [10].

According to security managers in Slovakia [11], drones are considered a cost-saving solution by substituting personnel (guards) with drones that can autonomously monitor the internal or external perimeter of objects or companies. Matt Sloan also addresses the topic of drones and their role in property protection in Security magazine [12]. Additionally, the Commercial Property magazine [13] asserts that the return on investment for drones falls within a 6-month timeframe, and drones have the potential to replace 3–4 security personnel with a single drone.

2.1 The Current Legislation for the Use of Drones

The utilization of drones for property protection entails specific limitations that restrict their application. The existing legislation governing drone use throughout the EU has been effective since 2021, established by Commission Regulation (EU) No. 2019/947. This standardized framework outlines operational constraints for drones, drone categorization, requirements for marking and registration, and the training of drone operators/pilots, culminating in a subsequent examination [14]. Commission Regulation (EU) No. 2019/947 is binding across the EU, excluding Slovakia, which has not yet implemented this regulation. Slovakia continues to govern drone flight rules through the decree issued by the Transport Authority (2/2019), which closely mirrors the EU regulations.

The overarching rules for drone flights (except for hobby drones) specify a maximum altitude of 120 m above ground level, adequately accommodating the requirements for safeguarding individuals and property. The lateral range is capped at 1,000 m, or visual observation (pilots are prohibited from using DJI Googles or other FPV glasses). Glasses can be used by co-pilots or camera operators. The requisite presence of an operator constitutes a significant hurdle for deploying aerial drones. Night flights are possible with explicit permission from the Transport Authority. Navigating in conditions of reduced visibility, such as fog, heavy snow, and strong winds, remains a challenging predicament.

Ensuring a minimum distance of 50 m from people is intricate, as closer proximity is at times necessary for effective monitoring. Drone operators are mandated to complete training specific to drone operation. Private training companies and the University of Zilina offer training programs for drone operators. In the future, with the anticipation of legislative changes, the utilization of autonomous drones is expected.

For ground-based drones (UGV - unmanned ground vehicles), the regulatory requirements for training are comparatively less stringent, and they are not subject to the aviation authority's authorization process. Ground drones possess the advantage of accommodating larger batteries, resulting in an extended operational range. Given the projected increase in drone deployment [15] the regulatory landscape governing their usage is expected to evolve.

2.2 Visions of the Use of Drones in Security Areas

In the field of physical protection, drones will most probably be utilized primarily for monitoring large-scale objects. The vision for the near future involves employing clusters of flying or ground drones that can return to their base for recharging or maintenance. Anticipated legislative changes will further enable drone intervention.

Aliter Technologies is developing compelling solutions in the form of the Virtual Mast (VIMA). This entails perimeter monitoring utilizing a drone connected to a power supply device via a cable. The drone can consistently monitor the object for 3–4 months and can take off when weather conditions change. Another offering from Aliter is an airship designed for object monitoring. Both systems come equipped with their own software for defining perimeters, along with support from thermal imaging cameras. Customers have the option to select the desired camera type and communication format [16].

The advantage of flying drones lies in their capacity to cover extensive areas without the need for installing fixed cameras. Drones contribute to significant cost and time-saving for monitoring vast spaces, such as parking lots in the automotive industry or pipeline installations in sectors like oil, gas, energy, and transportation. Their speed and ability to promptly reach the location of security incidents outpace those of security guards.

Furthermore, the integration of artificial intelligence into drones enhances their capability to detect security incidents from a distance. Intelligent algorithms grounded in deep learning, a machine learning method, offer advantages for recognizing security-related events.

Advantages of Using Drones in Property Protection:

- Lower costs compared to personnel-based protection.
- Higher speed than the movement speed of individuals or guards (beneficial for pursuing intruders).
- Coverage of large areas inaccessible to the human eye from ground level.
- Rapid repositioning and changing of the field of view.
- Capability to monitor and track a single object.
- Ability to employ thermal imaging systems (nighttime motion detection).
- Applicability in hard-to-reach areas (energy, overhead lines, gas industry, oil industry – oil pipelines).

- Potential for retrospective analysis of records.
- Ability to deter intruders.
- Quiet operation.
- Ability to detect environmental changes (gas leaks, fires, technological failures, and more).

Disadvantages of Using Drones in Property Protection:

- Current legal regulations mandate constant monitoring of the drone's movement by the operator.
- Limited possibilities for intervention or provision of first aid.
- Higher operating costs compared to physical protection methods.
- Limited flight time (approximately 30–40 min, with the option to replace the battery after landing).
- Necessity to use multiple drones for continuous coverage (substituting during charging).
- Usage range is affected by weather conditions (aeronautical limitations, reduced battery life in winter).
- Dependence on radio communication (susceptible to interference).
- Additional expenses for constructing a Drone Monitoring Center.
- Compliance with the General Data Protection Regulation (GDPR) [17].

So, what is the possible vision for business protection using drones? In the case of employing flying drones: Drones should traverse the perimeter of the enterprise over land owned by the company responsible for safeguarding the premises. A sufficient number of drones can be calculated to ensure continuous 24/7 business monitoring. This necessitates the establishment of drone nests where drones can be recharged alternately. Presently, the average battery life of a drone is 30–40 min. It is imperative to revise legal regulations and standards to permit the autonomous movement of drones or management through artificial intelligence, enabling drones to search for security incidents. The system must be integrated with a weather station capable of "grounding" drones in case of abrupt weather changes.

It is necessary to establish a control and monitoring centre (see Fig. 2) according to EN 50518 Monitoring and Alarm Receiving Center standards, facilitating the transmission of images/videos from drones. In cases of security outsourcing, it becomes possible to relay images beyond the protected premises, as long as compliance with GDPR regulations is maintained. Given that each drone must have an operator, the current suitability of drone usage lies in crisis/emergencies or expansive areas where a "bird's eye" perspective is essential.

A straightforward equation to calculate drone endurance (total flight time) [18] is as follows:

$$\text{Endurance (hrs)} = (\text{Battery Capacity (Ah)})/(\text{Current (Amps)}) \tag{1}$$

Endurance also depends on the drone's weight, speed, size and payload.

The formula can be used to determine the drone's range:

$$\text{Range(miles)} = (\text{RPM.V.60.Pitch})/12.5260.\text{Endurance (hrs)} \tag{2}$$

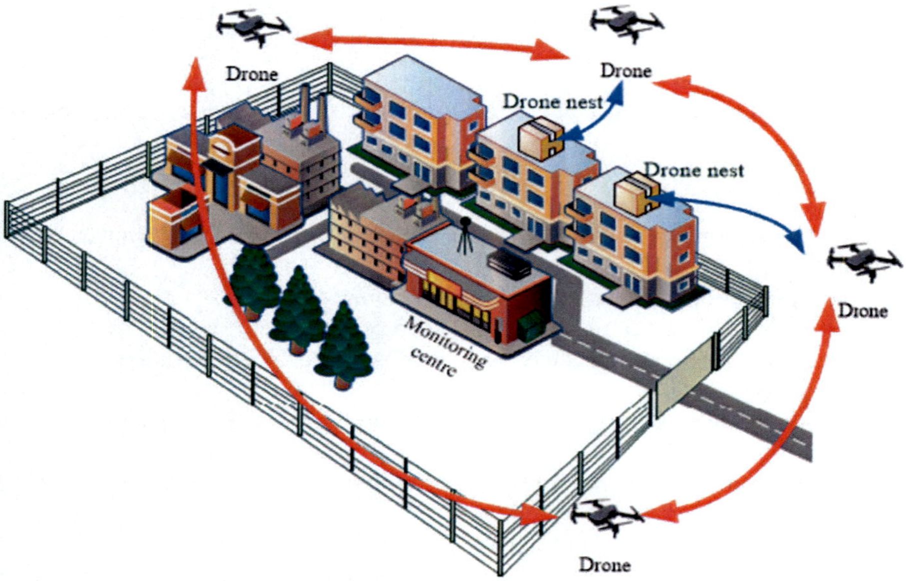

Fig. 2. Visualization of Drones Usage.

where RPM (revolutions per minute) represents the rotor RPM, pitch signifies the pitch of the drone's propeller in inches, Endurance denotes the duration the drone remains airborne in hours, and V stands for the supply voltage in volts (V) [18].

A simpler alternative, considering battery life and legal regulation requirements, involves ground drones or drones moving along the object's perimeter fence, as illustrated in Fig. 3. This approach would enable the drone to draw power from a rail along its path, facilitating relatively straightforward data transmission simultaneously. A drawback lies in the use of a fence perimeter detection system based on vibration sensing, while winter icing could pose a challenge.

Other researched possibilities include spherical robots, which, thanks to their spherical shape, can move swiftly on the ground and monitor the object. Their limitations encompass reduced obstacle traversal ability, restricted adaptability to stairs, and relatively low camera mounting height. Additionally, autonomous ground vehicles equipped with camera mounting stands and other sensors can be utilized for perimeter protection [19]. Prototypes have already been tested for object protection, with the main constraint being snow accumulation during winter. However, the advantage lies in the potential for sensor placement, such as monitoring gas leaks and fires.

A key constraint in employing drones for property protection is the safeguarding of personal data. If the drone's monitoring extends beyond the protected area, adherence to GDPR rules is imperative.

Fig. 3. Visualization of Fence Drone Usage.

3 Conclusion

This article highlights the evolving landscape of property protection in Central Europe, specifically focusing on the integration of drones into physical security systems. Property protection is a multifaceted concept that combines technical components, organizational measures, and personnel resources to create a Physical Protection System. PPS aims to safeguard assets and facilities from theft, sabotage, and unauthorized access.

The use of drones for property protection has gained attention since 2017, particularly as a response to challenges posed by limited human resources for security patrolling across large areas. The article presents insights gathered from security managers in Slovakia, who were surveyed about their perspectives on using drones for asset protection. While companies are not currently using drones for protection, there is a growing interest in their potential. Security managers identify various areas where drones could be valuable, including monitoring rooftops, construction sites, and large perimeters.

The advantages of using drones for property protection are numerous, including cost savings, rapid response capabilities, coverage of vast and challenging-to-access areas, and integration with advanced technologies like thermal imaging and AI. However, the adoption of drones in security also comes with limitations such as legal regulations requiring operator supervision, operating costs, limited flight time, weather dependency, and data protection concerns under GDPR.

The article underlines the need for legislative changes to permit more autonomous drone operations and artificial intelligence-based management. It envisions a future where drones play a crucial role in enterprise protection, patrolling perimeters and responding to security incidents. A control and monitoring centre, compliant with relevant standards, would facilitate data transmission from drones. Ground-based drones and other robotic systems, such as spherical robots and autonomous ground vehicles, are also explored as potential options for perimeter protection.

Overall, the integration of drones into property protection systems represents a promising avenue for enhancing security measures, addressing challenges of resource limitations, and adapting to changing security dynamics. However, the successful implementation of drones in this context requires careful consideration of technological advances, legal frameworks, and data privacy concerns.

Acknowledgement. This research was funded by Scientific Grant Agency of the Slovak Republic, grant number 1/0173/21 Research of measures implemented by security managers in the organizations related to the occurrence and spread of COVID-19 and in other emergency situations and by grant of Slovak Research and Development agency nr. APVV-20-20676 Monitoring and tracing of movement and contact of persons in medical facilities.

References

1. Fischer, R., Halibozek, E., Green, D.: Introduction to Security, 8 edn. Burtterworth-Heinemann (2008). ISBN 978-1493303250
2. Loveček, T., Reitšpís, J.: Projektovanie a hodnotenie systémov ochrany objektov, 1 edn. EDIS – vydavateľstvo UNIZA, Žilina (2011). ISBN 978-80-554-0457-8
3. Zou, B., et al.: Insider threats of physical protection systems in nuclear power plants: prevention and evaluation. In: Progress in Nuclear Energy, vol. 104 (2018). https://doi.org/10.1016/j.pnucene. https://www.sciencedirect.com/science/article/pii/S01491970173020. Accessed 21 May 2023
4. Smith, C.L.: Trends in the development of security technology. In: Gill, M. (Ed.), The Handbook of Security. Palgrave Macmillian Ltd, Basingstoke, Great Britain (2006)
5. Fennelly, L.J., Perry, M.A.: Physical Security: 150 Things You Should Know, 2 edn. Butterworth-Heinemann (2017). ISBN 978-0-12-809487-7
6. Houbing, S., et al.: Cyber-Physical Systems. Foundations, Principles and Applications. Acamedic Press, Cambridge (2017). https://doi.org/10.1016/C2015-0-00708-0. ISBN: 978-0-12-803801-7
7. Benson, M.: Overview of Physical Protection Systems Design and Evaluation. In: Content from Training Module 3 from Integrated Security Design Workshop [SAND2013-5143P] (2013). https://www.osti.gov/servlets/purl/1115144. Accessed 7 June 2023
8. Minimum wage statistics. (2023). https://ec.europa.eu/eurostat/statistics-explained/index.php?title=Minimum_wage_statistics. Accessed 3 July 2023
9. Minimum wage costs for the performance of the security employee's work (2023). https://www.sksb.sk. Accessed 4 July 2023
10. Patterson, J.: An Aerial View of the Future – Drones in Construction (2018). https://www.geospatialworld.net/blogs/an-aerial-view-of-the-future-drones-in-construction. Accessed 4 July 2023
11. Association of Security managers in Slovakia, homepage (2023). Workshop findings. www.abm.sk. Accessed 6 July 2023
12. Sloan, M.: Drones as Security tools. SECURITY magazine (2023). https://www.securitymagazine.com/articles/98699-drones-as-security-tools. Accessed 22 June 2023
13. Физическая охрана vs сторожевой дрон. Commercial property Nr. 9/2019. Accessed 22 June 2023

14. TOUCHIT magazine Homepage, Popularity of drones growth. https://touchit.sk/drony-rastu-na-popularite-aj-medzi-slovakmi-ako-je-to-s-pokutami-pri-poruseni-pravidiel/348465. Accessed 5 June 2023
15. Scylla Homepage, How Drones Are Used to Optimize Physical Security. https://www.scylla.ai/how-drones-are-used-to-optimize-physical-security. Accessed 7 June 2023
16. Aliter Technologies Homepage. https://www.aliter.com/. Accessed 7 June 2023
17. Prasanth, A. T.: Pros and cons of using drones for perimeter security, asmag.com Security &IoT. https://www.asmag.com/showpost/33352.aspx. Accessed 4 July 2023
18. Drone Capabilities - Endurance and Range. https://www.911security.com/learn/airspace-security/drone-fundamentals/drone-capabilities-endurance-range. Accessed 2 July 2023
19. Autonomous vehicle to patrol perimeter at Eastern Goldfields Regional Prison. Accessed 6 July 2023

User and Operator Friendly Outdoor Car Parking Lot Occupancy Detection (OCPLOD) System Design: Ondokuz Mayıs University Example

Metin Mutlu Aydın[1](✉), Zahid Enes Genç[1], Recep Arslan[1], and Dimitris Potoglou[2]

[1] Ondokuz Mayıs University, 55139 Samsun, Turkey
metinmutluaydin@gmail.com
[2] Cardiff University, Cardiff, NJ CF10 3AT, UK
potogloud@cardiff.ac.uk

Abstract. The increase in global population has brought about significant traffic problems in recent years. There has been a substantial rise in vehicle production and ownership with improving economic conditions and production capacity. People's desire to go anywhere with their private vehicles leads to significant traffic problems in finding parking spaces. Transportation authorities have implemented various applications to reduce parking problems, but the issue still remains serious. This problem leads to unnecessary fuel consumption and time loss for drivers looking for vacant parking spaces within parking facilities. While sensors are used in multi-level parking lots to address this issue, there are limited applications available for solving this problem in outdoor parking lots. Among these applications, free space detection can also be achieved using the cameras. However, existing software solutions are often expensive and not user-friendly. The also require dependency on specific companies for maintenance and calibration. Thus, users don't want to use it actively. In this study, user-friendly software has been developed at Ondokuz Mayıs University, which utilizes one or more cameras installed in three selected pilot outdoor parking lots to detect empty spaces based on the captured images. The software allows the parking lot operator to define vacant spaces, determine parking lots, and calibrate them according to their preferences. The developed software also offers flexibility to the user, cost-effectiveness after commercialization, and the ability for the user to make any necessary adjustments after installation. By using the software, drivers' concerns and time spent in searching for free parking spaces within the parking lots have decreased. Consequently, traffic congestion and delays in parking lots have also decreased, leading to increased fuel efficiency and providing environmental benefits.

Keywords: Parking lot management · Camera-based monitoring · Intelligent transportation systems · Fuel savings · Delay reduction

O. Prentkovskis et al. (Eds.): TRANSBALTICA 2023, LNITI, pp. 22–36, 2024.
https://doi.org/10.1007/978-3-031-52652-7_3

1 Introduction

Smart cities offer various opportunities for technological advancements and improved accessibility. As modernization progresses, people are experiencing a significant increase in vehicle ownership. Many towns and cities around the world, especially urban areas, are encountering challenges in locating available parking spaces in all parking areas. This scarcity of parking spaces has resulted in inconvenience for vehicle users during their trips and may negative effect on driver behaviours. Additionally, it has given rise to various issues such as traffic congestion, traffic accidents, stress and difficulties in finding suitable parking lots, and limited parking accessibility [1]. According to a United Nations [2] report, the global population was evenly divided between urban and rural areas, primarily due to an unprecedented surge in urbanization. However, this rapid urban growth in the last century has led to several challenges such as increased traffic congestion, traffic chaos in cities, air pollution, and a lack of parking spaces. In the last two decades, private car utilization has significantly increased, yet the availability of parking spaces has not kept pace [3]. This imbalance results in citizens wasting a considerable amount of time searching for available parking spaces, exacerbating traffic congestion and air pollution. Current technological systems focus on to optimize parking space utilization to estimate the utilization status of parking areas, such as determining the vehicle numbers and available lots.

In recent years, numerous authors have put forth computer vision-based approaches to tackle parking lot management challenges. These approaches primarily involve processing images obtained from parking lots and addressing various objectives. These objectives encompass: a) Automatic detection of parking space positions; b) Classification of individual parking spaces: and (c) Vehicle detection and counting in images. All these tasks serve as fundamental components of smart parking solutions, which aim to provide automated parking lot management. Examples of such management include dynamic pricing based on the number of cars in the parking lot and guiding drivers to the nearest available parking space. Smart Parking solutions are crucial as they not only optimize the utilization of parking spaces but also save both time and fuel for drivers [4, 5]. Nowadays, smart parking systems have also begun to be integrated into the concept of smart cities. Thus, smart parking solutions promote the application of cutting-edge technologies such as the Internet of Things, computer vision, machine learning, and 5g networks to enhance the effective parking areas.

This study developed software that detects available free parking spaces in three different outdoor parking lots located within Ondokuz Mayıs University (OMU) Central Campus, using three different cameras placed in those parking lots. The information about the available spaces is then displayed on digital screens at the entrances of the parking lots, allowing drivers to see the vacant spots. With this user-friendly software, both time and fuel can be saved as the confusion in the outdoor parking area and its surroundings decreases. Along with reduced fuel consumption, an environmental gain can also be achieved. Additionally, drivers will experience less stress due to improved operational performance of the outdoor parking lots, leading to indirect health benefits. The developed empty parking space detection software is unique compared to its existing examples. The developed new software works as completely user-friendly and it allows the operator using the software to perform all maintenance and calibration tasks after installation. This software enables parking lot operators to define the boundaries of all parking slots themselves compared to existing samples. The software also supplies opportunity to make necessary adjustments and modifications (if needed) and connects the desired camera footage to the software. Thus, users can make edits using the tools provided by the software in case of errors. Additionally, it has a distinctive structure compared to its current examples which area actively used in the fields This software also supplies full access and editing rights to the users and it is a big difference from the current other examples. Moreover, there will be no maintenance or update requirements from the developer after the initial installation that is generally expected to be preferred by outdoor parking lot operators.

2 Literature Review

Numerous studies have been conducted to develop efficient parking systems to estimate parking occupancy and count cars in car parking areas. In their study, Almeida et al. [6] introduced a new parking lot dataset, revealing the need for different classification techniques to address primary challenges. They provided a parking lot dataset consisting of 105,837 images captured from various angles in different parking areas. Another parking lot dataset with 480,000 images under different climatic conditions was proposed using convolutional neural networks [7]. These datasets highlighted the importance of employing different classification techniques to address primary challenges. In another study, Ahrnbom et al. [8] proposed machine learning algorithms SVM and Linear Regression with six quantized values. They also developed an algorithm capable of obtaining a set of canonical parking spaces and estimating their structure based on a pre-trained model generated from the system. Huang and Wang [9] utilized image-based methods to classify parking spaces and cars, while Lee et al. [10] developed a monitoring approach using videos for outdoor parking lots. Horprasert et al. [11] presented background subtraction for static images obtained from cameras, enabling statistical estimation of background and calculation of parking occupancy. Huang et al. [12] proposed a Bayesian structure for formulating vacant parking spaces that worked effectively during both day and night. Masmoudi et al. [13] focused on the issue of missed spaces between parked cars, which can affect system accuracy. Jermsurawong et al. [14] presented a customized method for determining parking space occupancy based on specific objectives.

Other studies have also explored various features and techniques in parking systems. Tschentscher et al. [15] and Ahrnbom et al. [8] utilized colour histograms and colour space analysis, respectively. Delibaltov et al. [16] developed a unique 3D model-based structure that effectively addressed occlusion issues encountered during occupied parking space calculation. Wu et al. [17] employed a Bayesian classifier based on edges, corners, and wavelet features to detect cars and overcome challenges posed by illumination changes. Seo and Urmson [18] proposed a method utilizing aerial images as input for detecting empty spaces in parking lots. They also introduced an algorithm capable of obtaining canonical parking spaces and estimating their structure using a pre-trained model generated by the system. Several authors, including Lu et al. [19], Tang et al. [20], Badiiet et al. [21], and Kotb et al. [22], have contributed to this line of research. For example, Huang and Wang [12] employed image-based methods to classify both parking spaces and cars. Lee et al. [10] developed a monitoring approach for outdoor parking lots using videos. Horprasert et al. [11] presented background subtraction techniques to estimate background information at specific time intervals and statistically analyze the data for calculating parking occupancy. In a study by Schneiderman and Kanade [23], Gabor filters were used to train a classifier on images captured under different lighting conditions, specifically focusing on unoccupied parking slots. This approach enabled the system to detect whether a vehicle was present in the slot or not. Bayesian structures have been found effective for formulating vacant parking spaces, functioning well in both day and night situations. Manase et al. [24] emphasized the occlusion challenges in parking lots, particularly the spaces between parked cars that can be missed and impact system accuracy. Saharan et al. [25] proposed a sensor-based solution for collecting real-time parking availability data. Jermsurawong et al. [14] presented a customized method for determining parking space occupancy using neural networks and visual-based feature extraction. Furthermore, recent studies have explored emerging concepts such as IoT-based techniques, image processing, and fog removal elimination [26–29]. These advancements contribute to the development of smarter parking systems in Smart Cities. Currently, In the scope of smart cities, the detection of free parking spaces in outdoor parking lots primarily relies on two different techniques and technologies:

- Sensor-Based Technologies: These technologies involve the installation of sensors beneath the asphalt to detect the presence or absence of vehicles. The sensors can accurately identify whether a parking space is occupied or vacant. By utilizing this approach, cities can efficiently manage parking spaces, mitigating challenges such as traffic congestion, reducing air pollution, minimizing waiting times for parking, and enabling a more self-governing parking system.
- Vision-Based Technologies: Image-based technologies utilize cameras to observe parking areas and employ computer vision methods to determine capacity utilization. By analyzing the images captured by the cameras, the system can determine the availability of parking spaces. This approach provides a cost-effective and flexible solution for parking management. Vision-based technologies offer advantages such as easier maintenance and lower installation costs compared to sensor-based systems.

The adoption of these technologies in smart cities facilitates effective parking space management, helping to address associated challenges and improving the overall urban experience. By accurately detecting parking space availability, cities can optimize parking utilization, reduce congestion, and enhance the convenience and satisfaction of residents and visitors.

Sensor-Based technologies are widely implemented in smart parking systems for smart cities, offering various benefits such as minimizing traffic jams, reducing air pollution, and decreasing waiting times for parking vehicles. This technology enables self-governing parking systems that enhance overall efficiency. One implementation of Sensor-Based technology utilizes Bluetooth, allowing users to identify vacant parking slots through their smartphones [30]. This system finds application in diverse settings like shopping malls, cinema halls, and universities. Nath et al. [31] discussed two different variations of the parking system, concluding that IoT (Internet of Things) technology, particularly at level 1, is well-suited for implementing smart parking solutions. Users can identify available parking slots using their smartphones or laptops [32]. Wireless sensors, computer vision, and Android technologies are employed to identify parking spaces [33]. While sensor-based networks and vision-based systems have been proposed for Parking Detection Systems (PDS), sensor-based networks have limitations due to the high cost and complexity of installing individual sensors. Consequently, large parking areas with numerous spaces cannot be feasibly monitored with individual sensors. In contrast, vision-based systems offer advantages in terms of cost and maintenance compared to sensor-based systems. As a result, a framework has been suggested that determines the optimum parking systems through advanced feature extraction and machine learning techniques, leveraging the benefits of vision-based approaches.

Another popular and effective technique is vision-based technology utilization. Vision-based technologies have gained popularity as an effective technique in parking systems, with several researchers making significant contributions in this field. For instance, Al-Kharusi and Al-Bahadly [34] proposed an innovative system that utilizes image processing techniques to detect parking spaces. Their system captures and processes images of the parking lot, providing drivers with access to this processed information to locate available parking spots. This groundbreaking approach has paved the way for a unique parking management system based on License Plate Recognition [35]. Not only does this system provide accurate vehicle information, but it also records entry and exit times. Additionally, it offers real-time video streaming, ensuring the availability of the most up-to-date and relevant information about vehicles, even at high speeds. The LPR model has demonstrated impressive results, achieving a remarkable 95% success rate in real-time implementation. Another noteworthy advancement is the introduction of an innovative parking space detection system based on image processing techniques, which offers improved efficiency and accuracy compared to moving objects [36]. This system relies on image processing instead of sensor-based technology, as it is considered more effective and cost-efficient. The study not only sheds light on existing parking services but also discusses their economic viability, highlighting the limitations of current parking methods in terms of reliability, modernity, and efficiency [37]. Furthermore, a noteworthy approach involves the implementation of a secure parking system that utilizes video analysis to determine human behaviours [38]. An image registration algorithm is employed to register incoming and outgoing vehicles, maintaining a record of available parking slots. Support vector machine classification is also utilized to monitor activities within the parking area.

Additionally, an effective car parking detection system based on Wireless Network Sensors (WSIV) has been established, where nodes detect and monitor the availability of parking lots [39]. The parking management team collects detailed information on vacant parking spaces, security, and statistical data using CrossBow Motes to implement the system. In another study by Postigo et al. [40], the authors propose an innovative method that utilizes a static surveillance camera to estimate the size of non-measured free parking areas. Various systems described in the literature have their own advantages and limitations. However, careful observation has highlighted the need for a new intelligent outdoor parking system to effectively address the existing problems. Currently, many car parking areas lack proper management systems and rely on manual handling without the support of scientific equipment or systems. This often leads to significant time wastage for people searching for suitable parking spaces, particularly in metropolitan cities where the availability of free parking spaces is scarcer compared to rural areas. The main reason behind this scarcity is the lack of modern technologies installed in parking current outdoor parking areas. The aim of this study is to provide parking lot operators with the possibility of monitoring and management of empty spaces with simple equipment at their disposal. During the study, image processing algorithms will be used to first detect the empty space. Then, with the help of the developed algorithm, the system will be able to instantly display the results to the user and operator. Users will be able to achieve maximum performance by changing the parking limits at any time using the developed algorithm and the prepared software. In contrast to the existing studies in the literature, the algorithm developed within the framework of this study and the software that provides full open access to the users will be presented.

3 Study Site and Datasets

In the study, an image processing-based car parking slot detection analysis was performed on a dataset of a total peak 6 h in examined three car park areas which are located at the main campus of the Ondokuz Mayıs University in Türkiye (Fig. 1).

a)

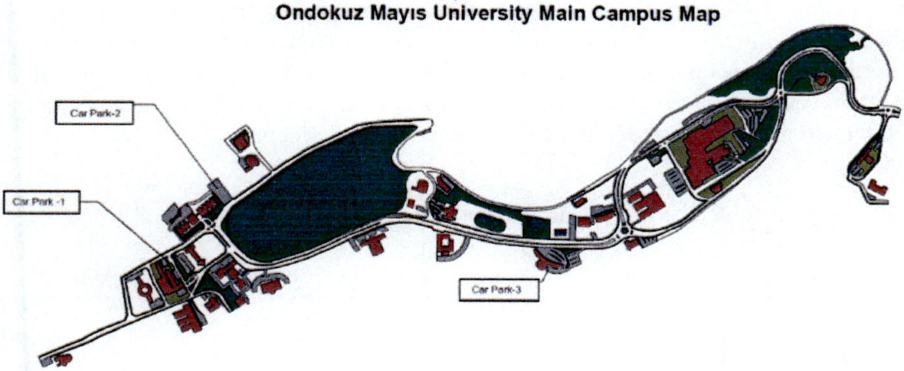

b)

Fig. 1. a) A satellite image and b) schematic location of the Ondokuz Mayıs University main campus.

The study dataset has been obtained on different days and shows different parking behaviors at the examined three car parks. The dataset has been collected using three cameras that are oriented at different outdoor parking areas. All three cameras are named as "Cam-1", "Cam-2" and "Cam-3" during the data collection process. "Cam-1" observes Car Park 1 (Fig. 2a), "Cam-2" monitors Car Park-2 (Fig. 2b), and "Cam-3" observes the Car Park-3 (Fig. 2c).

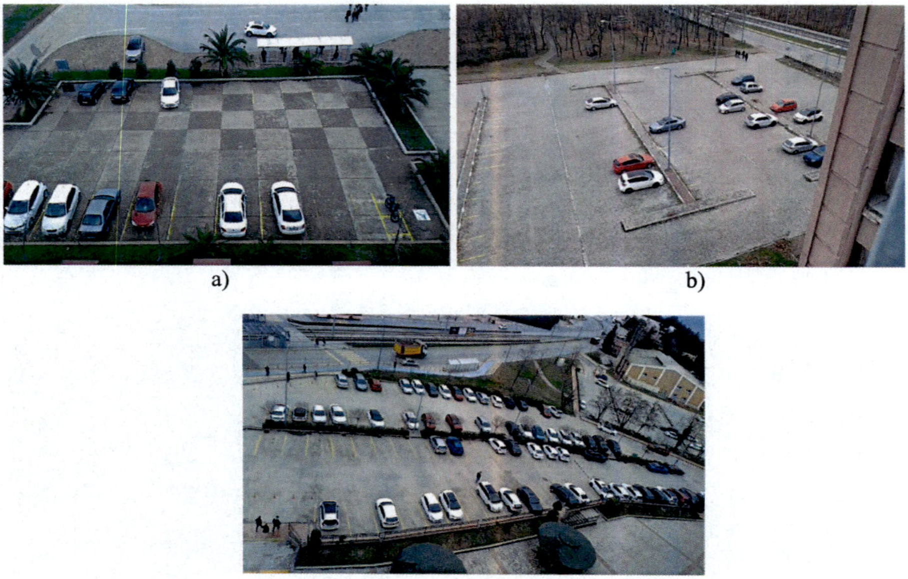

a) b) c)

Fig. 2. Examined car parks (a) Car Park-1 (b) Car Park-2 and (c) Car Park-3.

Considering the varying perspectives of the cameras in the three car parks, the collected video footage exhibits different geometries. To achieve this, a total of 6 h of video was recorded for each camera across the three car parks, at a frame rate of 1 fps. The videos, amounting to two hours per car park, were captured on different days, ensuring coverage of various parking space configurations, including both "empty" and "fully" occupied scenarios. Table 1 provides a summary of the recorded videos included in the dataset, along with the observed vehicle counts for each video.

Table 1. Some information for the data collection and analysis.

Camera No	Car Park No	Video Length (min)	Total Observed Veh.
Cam-1	Car Park-1	120	69
Cam-2	Car Park-2	120	93
Cam-3	Car Park-3	120	116

4 Methods and Analysis

In the study, a benchmark approach was used to determine the occupancy status of parking areas from the video recordings. Specifically, it was considered solutions based on image processing from a video recording which is a popular computer vision. For this purpose, used working scheme for the study is summarized in the given Workflow in Fig. 3.

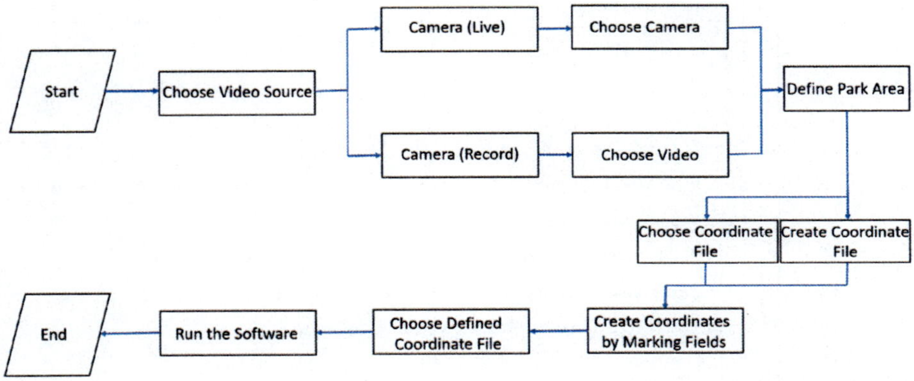

Fig. 3. Flowchart of the used methods during the occupied area determination.

The study, is focused on developing automatic full and empty parking spaces detection by detecting the coordinates of each parking space, manually. This process is shown in Fig. 4 where the operator defines each parking coordinate in the beginning of the system installation process.

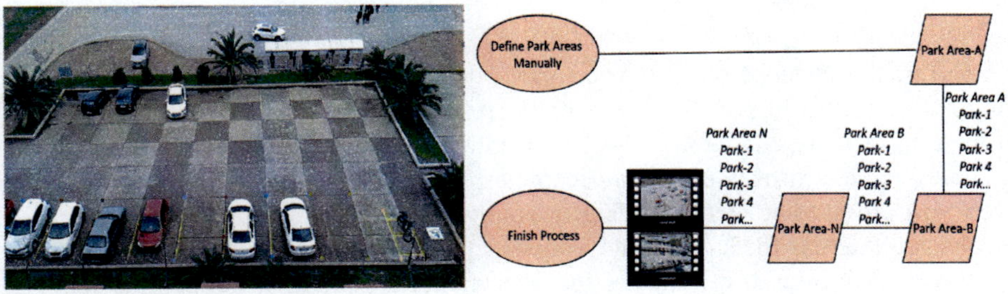

Fig. 4. Determination of the coordinates of slots to develop software that detects occupied or empty park slots.

4.1 Car Park Occupancy Detection

In the study, a software is aimed to develop to detect occupied car parks. In accordance with this objective, a user interface design will be created using the Python programming language and the Tkinter library as given in Fig. 5.

In the software development process, the main purpose is to design a user-friendly interface to supply an easily understandable appearance. Thus, it may allow users to easily use it and make adjustments and calibrations suitable for their operations.

```
def checkParkingSpace(self, imgCrop):
    try:
        count = cv2.countNonZero(imgCrop)
        if count < self.samplingImageDotControl:
            cellState = True
            stateColor = (0, 150, 0)
        else:
            cellState = False
            stateColor = (0, 0, 150)
        if self.coordinate[1] != cellState:
            cv2.fillPoly(self.imageCoordDrawed, [np.array(self.coordinate[0])], (stateColor))
            self.coordinates[self.coordinate_Area][self.coordinate_parkCell][1] = cellState
        with open(self.coordinateFileAdress, "w+") as coordfileWrite:
            coordfileWrite = json.dump(self.coordinates, coordfileWrite, indent=2)
    except Exception as exc:
        self.errorLog(f"park alanı taranamadı : {exc}")
def checkParkingSpacesParallel(self, imgCrops):
    with concurrent.futures.ThreadPoolExecutor(max_workers=4) as executor:
        futures = [executor.submit(self.checkParkingSpace, imgCrop) for imgCrop in imgCrops]
        concurrent.futures.wait(futures)
```

Fig. 5. An example coding of the developed algorithm.

During the development of the occupancy detection system, it is crucial to highlight the changes occurring in the region of interest compared to the image background. To determine these changes, the region of interest needs to be determined within the frame. Each area corresponds to an occupied spot, and the pixel-wise differences within each area are accumulated. To identify empty lots in the image, the technique used involves subtracting the background from the current frame. If the pixel weight exceeds a predefined threshold, it indicates the presence of a vehicle in the parking lot. The differences are then scaled, stored, and utilized to determine the changes in the examined area. By comparing the real pixel value with previous values, the system can filter out values that exceed the threshold. If night vision cameras or the necessary lighting infrastructure are available, the developed algorithm can work smoothly even at night. The slightest change in the pixel values in the parking lots can be used to determine whether the parking lot is empty or occupied. Again, empty parking spaces can be easily detected in rainy or cloudy weather using the developed algorithm. If the parking lot is covered with snow, the algorithm can detect the car by the change of the pixel value when the cars stop in the parking lot, but if the car is completely covered with snow, i.e.

if the pixel value is the same in the entire lot, the algorithm unfortunately cannot detect the change until the car moves and cannot detect empty.

In the developed system, A Convolutional Neural Network (CNN) algorithm is used. This algorithm is a type of deep learning algorithm that is particularly well-suited for image recognition and processing tasks. The algorithm trains using a large dataset of labelled images, where the network learns to recognize patterns and features that are associated with specific objects or classes. In the study, it is worth noting that the used cameras are fixed, and the positioning of the parking lots is part of a calibration process that only needs to be performed once during the camera installation process. Therefore, the position of each lot is known, it is normal to determine the number of available parking lots by evaluating the status of each lot (free or occupied). For this scenario, three potential approaches are explored: image classification, object detection, and semantic segmentation. The image classification approach operates as given: an input image for the analysis, a small rectangular image patch is sampled around each lot to ensure it covers the entire lot. These image examples are sampled during both the training and testing process, relying on the specified coordinates of the parking lots during camera installation using CNN algorithm. Each extracted lot is labelled as either 'empty' or 'full' based on the occupancy status of the corresponding stall. Subsequently, a classifier is trained to differentiate between 'empty' and 'full' stalls. During analysis, the trained data is employed to determine the late status of each parking lot, enabling the system to get the count of occupied parking spaces. This situation is visualized in Fig. 6.

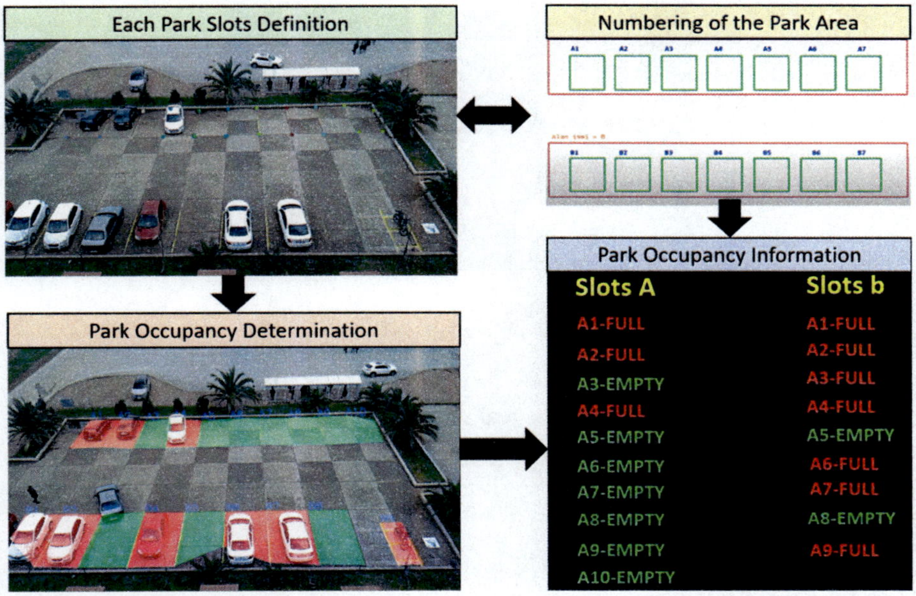

Fig. 6. Working infrastructure of the developed user-friendly car park area detection software.

In the system design, area examination limits are generated and assessed for car park area detection in the three examined car park areas, as depicted in Fig. 7. The obtained dataset has been manually named to provide the following information:

- Park area: This identifies the region where the parking spaces are located (Fig. 4).
- Total number of cars: This indicates the count of cars which is observed in parking area (Fig. 6).
- Bounding boxes around each car: A bounding box is provided for each car within the observed parking area (Fig. 7).
- Coordinates of the four corners for each parking lot: The coordinates are specified for each lot as shown in Fig. 8.

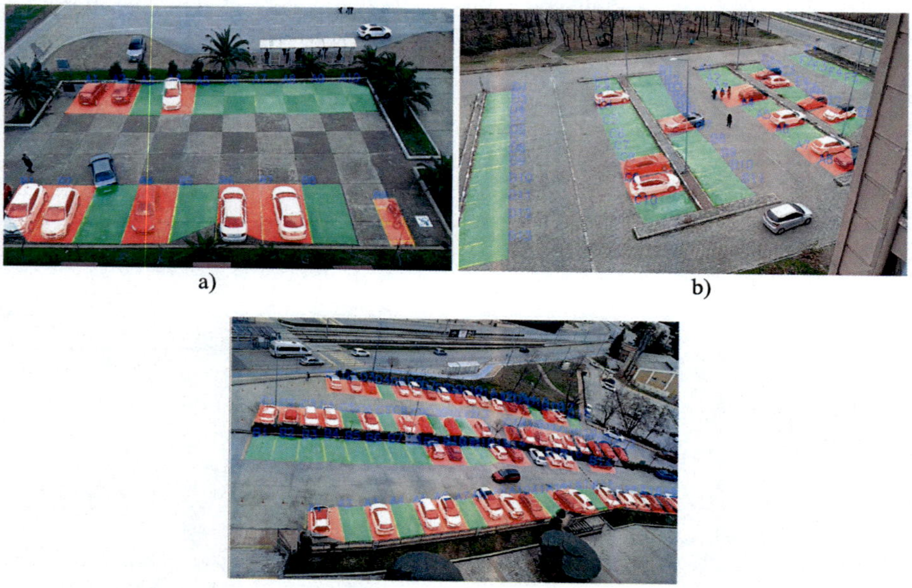

a) b) c)

Fig. 7. Bounding boxes annotated for the full park and red for the empty park lot.

During the development process, a real screen image was captured in a real parking lot using an installed fixed camera. In the image, parking lots are indicated by green colour when they are empty and red colour when they are fully occupied (see Fig. 8).

Fig. 8. Parking lots for empty and fully occupied parking slots.

5 Conclusion

In the study, the occupancy status of three different parking areas (lots-based) in Ondokuz Mayıs University Campus has been investigated by using an image-based method. The study results were obtained from a real-site data from the examined areas. An algorithm has been developed using parking utilization videos obtained from real sites (a data source) and the data has been processed into areal processed data to make it ready for programming. In the next step, user-friendly software has been developed for outdoor parking areas. With the help of this software, users can define parking area boundaries, and the developed algorithm within the software can detect free and occupied parking spaces in sample parking lots. The results obtained are displayed using green colour for free parking spaces and red colour for occupied parking spaces. This process has resulted in obtaining a detailed dataset that can be used for analysis and planning related to parking utilization. With this dataset, errors in the software have been identified and resolved, resulting in the development of software capable of detecting free and occupied spaces in all areas.

By using the software, drivers' concerns and time spent searching for free parking spaces within the parking area have decreased. Consequently, traffic congestion and delays in parking areas have also decreased, leading to increased fuel efficiency and providing environmental benefits. Additionally, the software records entries and exits in all parking areas, enabling the digital collection of daily parking data. It is foreseen that these accurate and real-time parking utilization and occupancy data can be effectively used in future parking planning efforts.

Overall, the software has proven to be beneficial in reducing parking search time for drivers, minimizing traffic congestion and delays in parking areas, increasing fuel efficiency, and contributing positively to the environment. Moreover, the collection of daily parking data through the software is expected to be valuable for future parking planning activities by researchers, planners and road authorities.

Acknowledgements. This study was conducted under a research project titled "i-gCar4ITS: Innovative and Green Carrier Development for Intelligent Transportation System Applications" which was supported by British Council. The authors would like to thank British Council for this support.

References

1. Amato, G., Carrara, F., Falchi, F., Gennaro, C., Meghini, C., Vairo, C.: Deep learning for decentralized parking lot occupancy detection. Expert Syst. Appl. **72**, 327–334 (2017)
2. United Nations. World Urbanization Prospects: The 2007 Revision. United Nations Publications (2022)
3. Di Mauro, D., Moltisanti, M., Patanè, G., Battiato, S., Farinella, G.M.: Park smart. In: International Workshop on Traffic and Street Surveillance for Safety and Security, pp. 1–2. IEEE (2017)
4. Paidi, V., Fleyeh, H., Håkansson, J., Nyberg, R.G.: Smart parking sensors, technologies and applications for outdoor parking lots: a review. IET Intell. Transp. Syst. **12**(8), 735–741 (2018)
5. Polycarpou, E., Lambrinos, L., Protopapadakis, E.: Smart parking solutions for urban areas. In: 2013 IEEE 14th International Symposium on A World of Wireless, Mobile and Multimedia Networks (WoWMoM), pp. 1–6. IEEE (2013)
6. Almeida, P., Oliveira, L.S., Silva, E., et al.: Parking space detection using textural descriptors. In: Systems, Man, and Cybernetics (SMC), IEEE International Conference, pp. 3603–3608. IEEE. (2013)
7. Do, H., Kim, J., Lee, K., et al.: Implementation of CNN based parking slot type classification using around view image. In: Proceedings of the 2020 IEEE International Conference on Consumer Electronics (ICCE) (2020)
8. Ahrnbom, M., Astrom, K., Nilsson, M.: Fast classification of empty and occupied parking spaces using integral channel features. In: IEEE Computer Society Conference on Computer Vision and Pattern Recognition Workshops, pp. 1609–1615 (2016)
9. Huang, C.C., Wang, S.J.: A hierarchical Bayesian generation framework for vacant parking space detection. IEEE Trans. Circuits Syst. Video Technol. **20**(12), 1770–1785 (2011)
10. Lee, C.H., Wen, M.G., Han, C.C., et al.: An automatic monitoring approach for unsupervised parking lots in outdoors. In: Proceedings of International Carnahan Conference on Security Technology, pp. 271–274 (2005)
11. Horprasert, T., Harwood, D., Davis, L.S.: A statistical approach for real-time robust background subtraction and shadow detection. In: IEEE ICCV (1999)
12. Huang, C.C., Tai, Y.S., Wang, S.J.: Vacant parking space detection based on plane-based Bayesian hierarchical framework. IEEE Trans. Circuits Syst. Video Technol. **23**(9), 1598–1610 (2013)
13. Masmoudi, I., Wali, A., Alimi, A.M.: Parking spaces modelling for inter spaces occlusion handling. In: International Conference in Central Europe on Computer Graphics, Visualization and Computer Vision, pp. 119–124 (2014)
14. Jermsurawong, J., Ahsan, U., Haidar, A., et al.: One-day long statistical analysis of parking demand by using single-camera vacancy detection. J. Transp. Syst. Eng. Inf. Technol. **14**(2), 33–44 (2014)
15. Tschentscher, M., Koch, C., König, M., et al.: Scalable real-time parking lot classification: an evaluation of image features and supervised learning algorithms. In: Proceedings of the International Joint Conference on Neural Networks, p. 15 (2015)
16. Delibaltov, D., Wu, W., Loce, R.P., et al.: Parking lot occupancy determination from lamp-post camera images. In: IEEE Conference on Intelligent Transportation Systems, Proceedings, ITSC, pp. 2387–2392 (2013)
17. Wu, Q., Huang, C., Wang, S., et al.: Robust parking space detection considering inter-space correlation. In: IEEE International Conference on Multimedia and Expo, pp. 659–662 (2007)
18. Seo, Y.W., Urmson, C.: Utilizing prior information to enhance self-supervised aerial image analysis for extracting parking lot structures. In: Proceedings of IEEE/RSJ International Conference on Intelligent Robots and Systems, pp. 339–344 (2009)

19. Lu, R., Lin, X., Zhu, H., et al.: SPARK: a new VANET-based smart parking scheme for large parking lots. In: Proceedings – IEEE INFOCOM, pp. 1413–1421 (2009)
20. Tang, C., Wei, X., Zhu, C., et al.: Towards smart parking based on fog computing. IEEE Access **6**, 70172–70185 (2018)
21. Badii, C., Nesi, P., Paoli, I.: Predicting available parking slots on critical and regular services by exploiting a range of open data. IEEE Access **6**, 44059–44071 (2018)
22. Kotb, A.O., Shen, Y.C., Huang, Y.: Smart parking guidance, monitoring and reservations: a review. IEEE Intell. Transp. Syst. Mag. **9**(2), 6–16 (2017)
23. Schneiderman, H., Kanade, T.: Object detection using the statistics of parts. Int. J. Comput. Vis. **56**, 151–177 (2004)
24. Manase, D.K., Zainuddin, Z., Syarif, S., et al.: Car detection in roadside parking for smart parking system based on image processing. In: 2020 International Seminar on Intelligent Technology and Its Applications (ISITIA), pp. 194–198 (2020)
25. Saharan, S., Kumar, N., Bawa, S.: An efficient smart parking pricing system for smart city environment: a machine learning-based approach. Futur. Gener. Comput. Syst. **106**(1), 622–640 (2020)
26. Kaur, S., Bansal, R.K., Mittal, M., et al.: Mixed pixel decomposition based on extended fuzzy clustering for single spectral value remote sensing images. J. Indian Soc. Remote Sens. **47**(3), 427–437 (2019). https://doi.org/10.1016/j.compeleceng.2018.08.018
27. Mittal, M., Kumar, M., Verma, A., et al.: FEMT: a computational approach for fog elimination using multiple thresholds. Multimed. Tools Appl. **80**(1), 227–241
28. Sanjeewa, E.D.G., Herath, K.K.L., Madhusanka, B.G.D.A., et al.: Visual attention model for mobile robot navigation in domestic environment. Glob. Sci. J. **8**(7), 1960–1965 (2021)
29. Ghansiyal, A., Mittal, M., Kar, A.K.: Information management challenges in autonomous vehicles: a systematic literature review. J. Cases Inf. Technol. **23**(3), 58–77 (2021)
30. Kumar, I., Manuja, P., Soni, Y., Yadav, N.S.: An integrated approach toward smart parking implementation for smart cities in India. In: Kolhe, M., Tiwari, S., Trivedi, M., Mishra, K. (eds.) Advances in Data and Information Sciences. LNNS, vol. 94, pp. 343–349. Springer, Singapore (2020). https://doi.org/10.1007/978-981-15-0694-9_32
31. Nath, A., Konwar, H.N., Kumar, K., Islam, M.R.: Technology enabled smart efficiency parking system (TESEPS). In: Sarma, H., Bhuyan, B., Borah, S., Dutta, N. (eds.) Trends in Communication, Cloud, and Big Data. LNNS, vol. 99, pp. 141–150. Springer, Singapore (2020). https://doi.org/10.1007/978-981-15-1624-5_14
32. da Cruz, M.A.A., et al.: An IoT-Based Solution for Smart Parking. In: Singh, P., Pawłowski, W., Tanwar, S., Kumar, N., Rodrigues, J., Obaidat, M. (eds.) Proceedings of First International Conference on Computing, Communications, and Cyber-Security (IC4S 2019). LNNS, vol. 121, pp. 213–224. Springer, Singapore (2020). https://doi.org/10.1007/978-981-15-3369-3_16
33. Sarangi, M., Mohapatra, S., Tirunagiri, S.V., Das, S.K., Babu, K.S.: IoT aware automatic smart parking system for smart city. In: Mallick, P., Balas, V., Bhoi, A., Chae, G.S. (eds.) Cognitive Informatics and Soft Computing. AISC, vol. 1040, pp. 469–481. Springer, Singapore (2020). https://doi.org/10.1007/978-981-15-1451-7_50
34. Al-Kharusi, H., Al-Bahadly, I.: Intelligent parking management system based on image processing. World J. Eng. Technol. (2014)
35. Tian, Q., Guo, T., Qiao, S.: Design of intelligent parking management system based on license plate recognition. J. Multimed. **9**(6), 774–780 (2014)
36. Yusnita, R., Norbaya, F., Basharuddin, N.: Intelligent parking space detection system based on image processing. Int. J. Innov. Manag. Technol. **3**(3), 232–235 (2012)
37. Faheem, F., Mahmud, S.A., Khan, G.M., et al.: A survey of intelligent car parking system. J. Appl. Res. Technol. **11**(5), 714–726 (2013)

38. Rizwan, M., Habib, H.A.: Video analytics framework for automated parking. Tech. J. Univ. Eng. Technol. (UET) Taxila, Pakistan **20**(4), 87–94 (2015)
39. Alharbi, A., Ahlkias, G., Yamin, M., et al.: Web-based framework for a smart parking system. Int. J. Inf. Technol. **13**, 1495–1502 (2021)
40. Postigo, C.G., Torres, J., Menendez, J.M.: Vacant parking area estimation through background subtraction and transience map analysis. IET Intell. Transp. Syst. **9**(9), 835–841 (2015)

Exploring Concrete Scrap as a Promising Building Material for Restoration Ukraine's Transport Infrastructure

Katerina Krayushkina(✉), Oleksandr Lapenko, and Nataliia Makhinko

National Aviation University, Liubomyra Huzara 1, 03058 Kyiv, Ukraine
{kateryna.kraiushkina,lapenko.oleksandr, nataliia.makhinko}@npp.nau.edu.ua

Abstract. This paper investigates the possibility of using concrete scrap as a filler in the manufacture of concrete and reinforced concrete products as promising alternative building materials. These products will be used to restore the transport infrastructure of Ukraine as cost-effective substitutes for traditional road building materials. To study the issue of physical, mechanical and technical properties of concrete scrap, a number of tests were carried out. The study aimed to determine the possibility of using concrete scrap as an aggregate in the manufacture of concrete products without deteriorating the physical mechanical, construction, and technical properties. The article analyses the existing experience in the use of secondary crushed stone. Based on this, it can be seen that concrete scrap is actively used in construction. The advantage of using it is also a reduction in the cost of work and profitability. The authors carried out experimental studies to study the physical and mechanical properties of crushed stone-sand mixture and concrete based on aggregate from concrete scrap.

Keywords: Gravel-sand mixture · Concrete scrap · Construction waste · Recycling · Scrap-based concrete

1 Introduction

Early morning of February 24, 2022, Russia launched a full-scale invasion of Ukraine without declaring war. Russian troops began intensive shelling of the Armed Forces of Ukraine and also launched rocket and bomb attacks on airfields, and industrial, residential and infrastructure facilities throughout Ukraine.

To date, in Ukraine, due to Russian military aggression, the volume of destruction waste has reached 10–12 million tons per year, according to preliminary estimates, which is comparable to the amount of annual municipal solid waste.

The problem is essentially threefold. It includes the following issues: social (rehabilitation of housing and engineering infrastructure), technical and economic (construction materials and products and financial resources), and environmental (alienation and pollution of the territory by landfills for the storage of construction waste) [1–3].

O. Prentkovskis et al. (Eds.): TRANSBALTICA 2023, LNITI, pp. 37–47, 2024.
https://doi.org/10.1007/978-3-031-52652-7_4

The most obvious solution to the problem would be to involve this waste in the economic circulation as much as possible. Especially considering that the cleanup of territories from war waste is taking place in all de-occupied territories.

The Cabinet of Ministers of Ukraine has established a clear algorithm for dealing with destruction waste, which is set out in the "Procedure for handling waste from the destruction of buildings and structures" formed in connection with damage as a result of hostilities, terrorist acts, sabotage or liquidation of their consequences.

In 2022, Ukrainian standart DSTU 9171:2021 "Guidelines for ensuring the balanced use of natural resources in the design of structures" was issued. It approves the reuse and recycling of materials and products at a level of at least 70%, which is part of the implementation of the agreement with the European Union.

To make this process as efficient as possible, the first stage is the arrangement of sites for the temporary placement of destruction waste.

Currently, about 144,000 tons of war waste are located at 48 sites organised in Ukraine. Destruction waste is taken to temporary storage sites, where it is sorted, namely, hazardous, construction and other waste are separated. Construction waste is shredded for reuse, in particular, the post-war restoration of infrastructure facilities in Ukraine.

At the same time, the Russian invasion caused more than one hundred thousand billion hryvnias of damage to Ukraine's road network. After all, Ukrainian infrastructure is the second target of the enemy after military facilities. However, unfortunately, this figure is increasing virtually every day, as rocket and artillery attacks continue.

According to the latest data, more than 25,000 km of roads, bridges and overpasses, as well as airfields have been damaged in Ukraine. All these infrastructure facilities need to be restored as soon as possible, also using economical and environmentally friendly technologies [26].

This article shows the results of the conducted studies to determine the possibility of using concrete scrap for to restoration of infrastructure objects in Ukraine.

2 Analysis of Existing Research Results

A review of the scientific literature showed that research in the field of the use of secondary aggregates based on concrete scrap has been going on not only in our country but also abroad for about 70 years [4]. Demolition waste was first used in Germany after World War II, but it is only recently that this practice has spread worldwide as a promising method of disposal [5].

The crushed stones of various fractions and screenings of crushing are the products of crushing concrete scrap. The fractional composition of coarse aggregate and crushing screenings, as well as the shape of the grains, depends on the crushing technology and the corresponding installation.

The presence of particles of cement mortar on the grains of the original crushed stone is a distinctive feature of crushed concrete aggregates from natural ones of natural origin. The amount of cement mortar in various fractions of crushed stone varies widely and can reach 36–39%. Depends on the size of the aggregate and has a significant impact on the final properties of concrete based on it [6–8].

The presence of a solution component on crushed stone grains significantly worsens its basic properties: it significantly increases its water absorption, crushability, and abrasion, and reduces frost resistance.

The properties of secondary fillers are significantly inferior to natural ones, they have a lower density, higher porosity and water absorption, and are 70% less resistant to abrasion [10, 14, 15]. These properties of secondary aggregates influence the properties of concrete.

An increase in the strength and other characteristics of secondary crushed stone is possible when crushed according to modes that ensure the destruction of mainly cement stone. To ensure this mode of grinding, special equipment is created, for example, vibro-jaw or cone inertial crushers [9, 10]. Improving the characteristics of secondary crushed stone is also possible due to various technological methods, in particular, mechanical processing in a concrete mixer, and impregnation with reinforcing polymer solutions [11–15].

Another way to achieve high strength values in comparison with the initial strength of crushed concrete is the use of water-reducing additives with a simultaneous increase in the cement content.

The works [16–21] provide data on the same compressive strength of concrete with a 30% content of secondary aggregate in a mixture of natural aggregate and concrete with the use of natural aggregates.

It is noted that the complete replacement of natural coarse aggregate with crushed stone from crushed concrete, when manufactured at a lower water-cement ratio compared to concrete on traditional aggregates, contributes to an increase in its strength characteristics. However, with equivalent water-cement ratios, the strength characteristics of concretes using aggregate from crushed concrete are inferior to traditional ones.

After analysing the experience of using concrete scrap in construction, we can conclude that already in our time, due to organisational measures, the use of rational technological schemes for processing waste concrete and reinforced concrete, the use of more modern equipment and the improvement in the quality of aggregate from crushed concrete ensured its competitiveness with natural crushed stone [22–25].

The concrete scrap obtained after processing can be used to restore Ukraine's infrastructure, namely: when laying layers of road and airfield structures, when laying the foundation or covering footpaths, parking lots, walking alleys, slopes along rivers and canals; for reclamation, improvement and planning of territories; for the manufacture of concrete and reinforced concrete products, foundations for storage, industrial premises and small mechanisms.

To confirm the findings, a number of laboratory studies of concrete scrap were carried out. The purpose of which is to analyse the possibility of using concrete scrap to replace crushed stone materials in the manufacture of building concrete and reinforced concrete structures, crushed stone and sand mixtures that will be used in the restoration of the transport infrastructure of Ukraine.

3 Scrap Concrete Research

When performing the experimental part of the work, 2 batches of secondary aggregate (large and small) were made from concrete scrap, strength class B30. According to the established crushing mode in the laboratory jaw crusher. The products of this processing were sorted and were mixtures of non-cohesive materials from crushed stone with a maximum particle size of 40 mm (70%) and screenings with a particle size of less than 10 mm (30%).

The properties of crushed stone materials obtained in this way were examined for compliance with the requirements of DSTU B V.2.7–30.

The requirements for crushed stone materials for the construction of pavements and bases of road surfaces from non-cohesive materials are given in Table 1 (according to DSTU B V.2.7–30).

Categories of roads are accepted according to the Ukrainian standard DSTU B V.2.3–4.

Table 1. Requirements for crushed stone materials.

No.	Name indicator	Indicator values for pavement layers			
		coatings		base	
		road category			
		IV (bottom layer)	V	I–III	IV–V
1	Strength grade (by crushing capacity), not lower than, MPa:				
	- crushed stone from igneous metamorphic rocks	1,000	800	800	600
	- crushed stone from sedimentary rocks and metallurgical slags	800	600	600	300
	- crushed gravel	800	600	600	400
2	Abrasion grade, no lower than	C-II	C-III	C-III	C-III – C-IV
3	Grade of crushed stone in terms of frost resistance for areas with an average air temperature of the coldest month of the year, ° C, not lower than:				
	- from 0 to minus 5	F15	F15	F15	–
	- from minus 5 to minus 10	F25	F25	F25	F25
4	The content of lamellar (flaky) and needle-shaped grains,% by weight, not more than	15	15	35	–

The test results of crushed stone-sand mixtures of various types (C1, C7, C8), most used in road construction, selected from crushed concrete scrap, are given in Table 2. Tests were carried out for compliance with the requirements of DSTU B V.2.7–30.

Table 2. Results of determining the grain composition of crushed stone from concrete scrap.

Maximum grain size, mm	Index	Total residue on the control sieve with hole size, mm							
		40	20	10	5	2.5	0.63	0.16	0.05
40	Actual results	3.5	28.4	42.3	57.2	67.2	77.1	90.0	93.0
	Requirements to C1	0–15	20–40	35–60	45–70	55–80	65–90	75–92	80–93
	In accordance with the requirements	+	+	+	+	+	+	+	+
	Requirements to C7	0–15	20–40	40–65	55–80	65–90	85–95	95–100	95–100
	In accordance with the requirements	+	+	+	+	+	–	–	+
	Requirements to C8	0–15	15–30	30–55	40–70	55–80	75–90	85–95	95–100
	In accordance with the requirements	+	+	+	+	+	+	+	+

Compliance of the grain composition of the concrete scrap mixture with the requirements of DSTU B V.2.7–30 for a mixture of type C7 is graphically shown in Fig. 1 (Table 3 and Table 6).

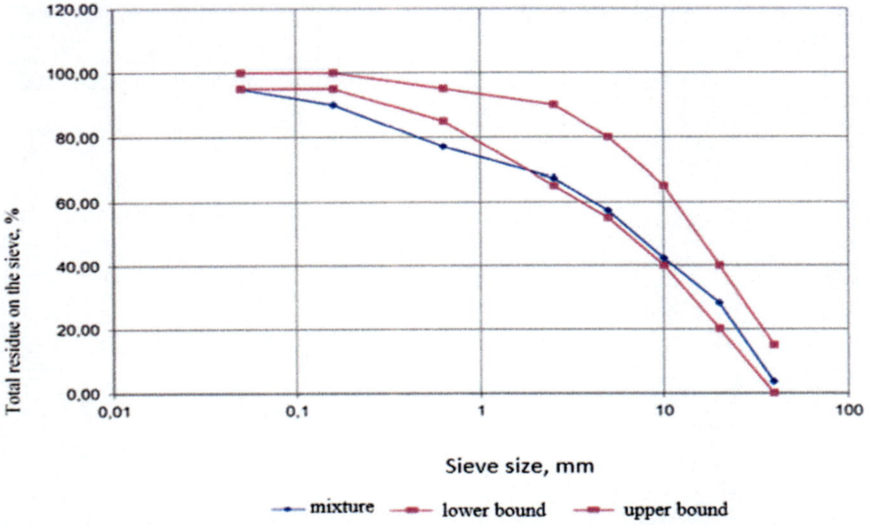

Fig. 1. Graphical representation of the compliance of the grain composition of the mixture of concrete scrap with the requirements for a mixture of type C7.

Table 3. Test results of crushed stone mix from concrete scrap.

No	The name of the indicators	Requirements DSTU B V.2.7–30	Test results
1	The content of dusty and clay particles, % by mass	not more than 5.0	4.23
2	Moisture, % by weight	–	2.6

Table 4. Results of determining the grain composition of screenings from concrete scrap.

Index	Total residue on the control sieve with hole size, mm								
	10	5	2.5	1.25	0.63	0.315	0.14	0.071	Pallet
Actual grain composition	0.12	23.82	53.20	63.80	72.17	80.34	90.81	95.97	100.00
Requirements DSTU B V.2.7–30	0–15	15–30	30–55	40–70	55–80	75–90	85–95	95–100	100

Table 5. Results of determining the grain composition of crushed stone from concrete scrap.

Index	Total residue on the control sieve with hole size, mm								
	25	20	10	5	2.5	0.63	0.14	0.071	Pallet
Actual grain composition	0.23	19.16	32.06	55.95	76.22	86.34	94.57	96.77	100.00
Requirements DSTU B V.2.7–30	0–15	10–20	20–40	40–65	55–80	65–90	85–95	95–100	100

Table 6. Results of testing crushed stone and crushed stone-sand mixture from concrete scrap.

No	The name of the indicators	Requirements from DSTU B V.2.7–30	Actual values of indicators for factions		
			5–10	10–20	20–40
1	True density, g/cm^3	–	2.51	2.50	2.50
2	Content of flaky and needle-shaped grains, % by weight	not more than 35	2.3	6.4	5.1
3	Grain shape group	–	rounded	rounded	rounded
4	Content of weak grains, %	not more than 10	4.8	3.6	2.9
5	The content of clay in the breasts, %	till 0.25	0.0	0.0	0.0
6	Water absorption, %	–	0.17	0.25	0.22
7	Weight loss during crushability strength test, % by weight	–	16.44	16.05	15.85
8	Weight loss when testing crushed stone for abrasion, % by weight	over 25 to 35 inclusive	28.1	28.5	32.2
9	Brand of crushed stone by abrasion	C-II	C-II	C-II	C-II
10	Weight loss after frost resistance test (accelerated method), % by mass	<10.0	9.2	8.4	8.5
11	Brand of crushed stone for frost resistance	F35	F35	F35	F35

The test results showed that, according to the grain composition, the crushed stone-sand mixture from concrete scrap can be attributed to the crushed stone-sand mixture of type C8 according to DSTU B V.2.7–30.

The studied crushed stone from concrete scrap according to the grain composition can be attributed to a fraction of 5–20 mm, however, according to the residues on a sieve with a hole size of 20 mm, this crushed stone does not meet the requirements of Table 4 DSTU B V.2.7–30.

The results of testing the components of crushed stone and crushed stone-sand mixture from concrete scrap showed that the content of flaky and needle-shaped grains, % by weight, in crushed stone of a fraction of 5–10 mm is 2.3%, crushed stone of a fraction of 10–20 mm is 6.4%, and for crushed stone of a fraction of 20–40 mm – 5.1%. According to the requirements of Table 5 of DSTU B V.2.7–30, according to the grain shape, crushed stones of a fraction of 5–10 mm, 10–20 mm and 20–40 mm can be classified as a rounded group.

The content of weak grains, % by weight, is 4.8% for crushed stone of a fraction of 5–10 mm, crushed stone of a fraction of 10–20 mm is 3.6%, and crushed stone of a fraction of 20–40 mm is 2.9%.

The mass loss of crushed stone during the crushing strength test, % by weight, for a crushed stone of a fraction of 5–10 mm is 16.44%, crushed stone of a fraction of 10–20 mm is 16.05%, and crushed stone of a fraction of 20–40 mm is 15.85%. Since the requirements for weight loss when testing crushed stone from crushed cement concrete have not been established, it is impossible to establish the brand of crushed stone by crushability.

The mass loss of crushed stone during the test for abrasion, % by weight, for a crushed stone of a fraction of 5–10 mm is 28.1%, crushed stone of a fraction of 10–20 mm is 28.5%, and crushed stone of a fraction of 20–40 mm is 32.2%. According to the classification in DSTU B V.2.7–74, the grade for the abrasion of crushed stone of all fractions is C-II.

The weight loss after the frost resistance test, % by weight, by the accelerated method by alternately holding the crushed stone in a sodium sulfate solution and drying showed that the investigated crushed stone of a fraction of 5–10 mm, 10–20 mm and 20–40 mm, respectively, Table 4 DSTU B V. 2.7–75 can be attributed to the F35 grade, since after eight test cycles, the mass loss of crushed stone of each fraction was no more than 10.0%. With an increase in the number of test cycles to 10, the mass loss of crushed stone of each fraction was more than 10.0%.

Next, concrete sample tubes 15 × 15 × 15 mm in size were made for comparison. Samples of compositions No. 1 and No. 3 (river sand with Mk = 1.2, granite crushed stone of a fraction of 5–40 mm). Samples of compositions No. 2 and No. 4 (river quartz sand with Mk = 1.2, concrete scrap fraction 5–40 mm). Portland cement class PC 400 was used in all compositions. The samples were kept under normal conditions until the standard age. The compositions are shown in Table 7. The test results of the obtained concrete are presented in Table 8.

Table 7. Concrete compositions.

No. of composition	Granite crushed stone, kg	Crushed stone from crushed concrete, kg	Cement, kg	Sand, kg	Water/Cement
1	1,200		260	690	0.65
2	1,180		345	570	0.65
3		1,200	260	690	0.65
4		1,180	345	570	0.65

Table 8. Concrete test results.

No of composition	Density, kg/m^3	Strength, MPa
1	2,318	29.6
2	2,304	29.3
3	2,318	30.4
4	2,304	28.8
Requirements DSTU B V.2.7–176:2008	–	25–30

Analyzing the data in Table 4, it can be seen that when replacing a large natural aggregate (from granite crushed stone of a mixed fraction of 10–40 mm) with a large aggregate from crushed concrete, the strength of heavy concrete with the same water-cement ratio practically does not decrease. This indicates the possibility of replacing natural crushed stone aggregate with concrete scrap in low-grade concrete.

4 Conclusions

According to the results of the studies of concrete scrap from crushed cement concrete, in terms of grain composition and physical and mechanical properties, they comply with the requirements of DSTU B V.2.7–30. The crushed stone-sand mixture corresponds to types C1 and C8 according to DSTU B V.2.7–30.

Taking into account that the studied concrete scrap in the form of crushed stone and crushed stone-sand mixture has on its surface microparticles of cement grade no lower than PC 400, it can be used for:

- arrangement of base layers (upper and lower) using the wedge method without adding binders;
- arrangement of base layers by the method of dense mixtures;
- strengthening the upper (working layer) subgrade to increase the bearing capacity of the road structure
- devices for the foundation of road and airfield structures with the addition of an inorganic binder;
- when carrying out repair work using cold recycling technology, a mixture of mineral materials, is added to the milled mixture.
- pavement from concrete scrap is possible on roads IV – V categories, and base layers I – V categories.

It has been established that concrete scrap as a substitute for crushed stone materials meets the regulatory and technical requirements for physical and mechanical characteristics and can be used in concrete mixtures of strength class B 30 and below.

The using of concrete scrap for the restoration of infrastructure facilities in Ukraine is expedient and meets the principles of the concept of "sustainable development". The main provisions provide for saving materials and energy, increasing the durability of structures and reducing the negative impact on the environment, including the preservation of irreplaceable sources of natural resources.

References

1. Burak, M.P., Ryshchenko, T.D.: Construction materials science. Tutorial. KhNAMG, Kharkiv (2007)
2. Gusev, B.V., Zagursky, V.A.: Secondary use of concrete. Stroyizdat (2008)
3. Smal, M.V., Dzyubynska, O.V., Shelkovich, O.B.: Word experience of reproductive use of concrete in construction production. Modern technologies and calculation methods in construction, p. 7 (2017)
4. Valovoy, O. I., Eremenko, O.Yu., Valovoy, M.O. Corrosion resistance of concrete on aggregates from metallurgical industry waste. Resource-saving materials, constructions, buildings and structures: coll. of science works, vol. 32, pp. 15–22 (2016)
5. Valovoy, M.O.: Manufacturing technology and analysis of crack resistance of test samples of beams on the waste of mining and beneficiation plants. Resource-saving materials, constructions, buildings and structures: collection. of science works, vol. 25, pp. 233–239 (2013)
6. Valovoi, A., Koval, P., Eremenko, A., Valovoi, M., Volkov, S.: Durability of beams with hybrid reinforcement from metal and basalt fiber reinforced polymer (BFRP) armature. MATEC. In: Web of Conferences, vol. 230, p. 2035 (2018)
7. Efimenko, A.Z.: Concrete waste - raw material for the production of effective building materials. Concr. Technol. **2**, 17–24 (2014)
8. Denisevich, D., Dymakova, A., Schneider, A., Ibe, E.: Physical and mechanical features of materials based on concrete scrap. Herald Eurasian Sci. **3** (2020)
9. Tolypin, A., Tolypina, N.: An effective way to recycle 3d printing concrete scrap. Constr. Mater. Prod. **4**(2), 12–18 (2021)
10. Makhinko, A., Makhinko. N.: To the Calculation of the Optimal Level of Reliability by Using Economic Indicators, vol. 47, pp. 251–259 (2020)
11. Ounis, A., Ebead, U., Judd, S.: Life cycle cost analysis of structural concrete usingseawater, recycled concrete aggregate, and GFRP reinforcement. Constr. Build. Mater. **175**, 152–160 (2018)
12. Eong, C., Keum, S., Jin-Kyu, S., Bang Yeon, L.: Composite properties of high-strengthpolyethylene fiber-reinforced cement and cementless composites. Compos. Struct. **138**(15), 116–121 (2016)
13. Carrascal, A. Polanco, T.: The Use of Construction Waste to Remediate a Thermally Active Spoil Heap Applied Sciences, vol. 13, no. 12 (2023)
14. Pertile, E., Dvorský, T., Václavík, V.: Design and construction of foundations for industrial-facilities in the areas of former post-mining waste dumps. Energies **15**, 57–66 (2022)
15. Charvat, J., Svoboda, J., Macalova, K.: Use of construction waste and recycled rubber for the preparation of cement composites August 2021 Selected Scientific Papers. J. Civ. Eng. **16**(1) (2021)
16. Stevulova, N., Estokova, M. Holub, E., Singovszka, K.: Characterization of demolition construction waste containing asbestos. Release of fibrous dust particles. Appl. Sci. **10**(11) (2022)
17. Sainz-Aja, A., Carrascal, A. Polanco, T., Thomas, J.: Determination of the optimum amount of superplasticizer additive for self-compacting concrete. Appl. Sci. **10**(9) (2020)
18. Sormunen, P., Karki, T.: Recycled construction and demolition waste as a possible source of materials for composite manufacturing. J. Build. Eng. **24**, 100742 (2019)
19. Bakhrakh, A., Solodov, A., Naruts, V., Larsen, O., Alimov, L., Voronin, V.: High-performance self-compacting concrete with the use of coal burning waste. In: IOP Conference. Series: Earth and Environmental Science, vol. 90, pp. 012–033 (2017)

20. Rao, A., Jha, K., Misra, S.: Use of aggregates from recycled construction and demolition waste in concrete. Resour. Conserv. Recycl. **50**, 71–81 (2020)
21. Tam, V., Soomro, A., Evangelista, T.: A review of recycled aggregate in concrete applications. Constr. Build. Mater. **172**, 272–292 (2020)
22. Wahlstr, M., Bach, R., Smeets, A.: Construction and demolition waste: challenges and opportunities in a circular economyttinen. In: European Topic Centre Waste and Materials in a Green Economy, p. 10 (2020)
23. Hoffmann, A.L.C.: Recycled concrete and mixed rubble as aggregates: influence of variations in composition on the concrete properties and their use as structural material. Sustainable construction materials and technologies. Coventry (2017)
24. Marinkovic, S., Ignjatovic, I., Radonjanin, V.: Life-cycle assessment of concrete with recycled aggregates. In: Handbook of Recycled Concrete and Demolition Waste, pp. 569–604 (2013)
25. Akbarnezhad, A., Ong, C., Tam, M., Zhang, M.: Effects of the parent concrete properties and crushing procedure on the properties of coarse recycled concrete aggregates. Mater. Civ. Eng. **25**, 1795–1802 (2022)
26. Akbarnezhad, A., Ong, M., Zhang, C., Tam, T.: Microwave - assisted beneficiation of recycled concrete aggregates. Constr. Build. Mater. **25**, 3469–3479 (2021)
27. https://minfin.com.ua/realty/articles/na-vosstanovlenie-dorog-ponadobyatsya-sotni-milliardov-chast-sredstv-mozh

Resilience of Life Support Systems for Crewed Autonomous Transport Systems for Extended Space Missions in Isolated Environment

Igor Kabashkin(✉) and Sergey Glukhikh

Transport and Telecommunication Institute, Lomonosov 1, Riga 1019, Latvia
kiv@tsi.lv

Abstract. Sustaining human life during the space missions necessitates highly reliable and resilient life support systems (LSS) aboard space stations. This paper presents a methodology for modelling of LSS reliability and resilience using Markov chain. A three-tier redundant LSS architecture is analysed, comprising: 1) redundant regenerative life support reactors, 2) backup storage systems with catalysts and reagents to restore physicochemical reactors and microbial strains to restore biotechnological reactor, and 3) emergency life support recourses. Markov models are developed to simulate the architecture's reliability by modelling possible system states and transition probabilities. The model assesses the LSS architecture's resilience through the ability to absorb failures while still providing life support. The described approach presented may be useful for optimizing life support systems configurations at the early stage of LSS development for maximum its resilience.

Keywords: Autonomous transport · Space missions · Life support systems · Reliability · Redundancy · Fault tolerance

1 Introduction

The advent of space travel has brought a new category of transport systems - autonomous spacecraft designed to traverse the vast expanse of space.

The sphere of crewed space missions presents a unique type of autonomous transport system, one that operates in an extraordinarily isolated environment. Unlike conventional transport systems, space missions demand a self-sustaining ecosystem that can support human life, along with providing autonomous navigation and communication capabilities. The crewed space missions can be delved as an independent class of autonomous transport systems, showcasing their unique attributes, challenges, and overarching significance.

A critical component of these missions is the Life Support Systems (LSS), designed to autonomously maintain a habitable environment within the spacecraft. LSS handles air purification, temperature control, waste management, and provision of food and water. The operation of LSS underpins the success of any crewed mission, making it a vital aspect of the spacecraft's autonomy.

O. Prentkovskis et al. (Eds.): TRANSBALTICA 2023, LNITI, pp. 48–58, 2024.
https://doi.org/10.1007/978-3-031-52652-7_5

The efficiency and reliability of the LSS directly impact the feasibility of mission objectives. The development of models for analysing the resilience and reliability of LSS poses a significant challenge, but one that carries profound importance.

This paper analyses a proposed three-level life support system architecture to provide the needed reliability and autonomy for long-duration deep space missions. The architecture includes:

Level 1 - Redundant physicochemical and biotechnological reactors to regenerate essential resources in closed-loop fashion, including water, oxygen, methane, carbon dioxide, and food.

Level 2 - Backup storage systems with expendable catalysts, reagents, and microbial strains to restore impaired first-level systems in case of failure.

Level 3 - Emergency cache of reserve consumable supplies of water, oxygen, and food as a temporary haven.

The model, designed to evaluate, and predict the reliability of LSS under various conditions encountered in space missions is proposed in the paper.

The structure of the article contains 5 sections: the present introduction, a review of publications on the topic (Sect. 2), a description of the architecture of LSS under study (Sect. 3), a description of the reliability model of the LSS (Sect. 4) and conclusions (Sect. 5).

2 Related Works

Life Support Systems on modern spacecraft are highly sophisticated and designed to meet the unique challenges of supporting human life in space. There are two key types of life support - physicochemical and bioregenerative. Physicochemical systems are the ones primarily in use today, while bioregenerative systems are more experimental and aim to integrate biological elements to create a self-sustaining ecosystem.

The International Space Station (ISS) utilises one of the most advanced physicochemical LSS, known as the Environmental Control and Life Support System (ECLSS) [1].

The paper [2] reviews the development of systems to utilize natural resources on Mars to enable advanced missions. It focuses on relatively mature NASA technologies for in-situ resource utilisation (ISRU), an important but very immature space technology.

Looking to the future, research is being done on bioregenerative life support systems, which incorporate biological components like plants and microorganisms to recycle air, water, and waste [3]. The objective is to create a more sustainable and self-sufficient system, particularly for long-duration space missions or settlements on other celestial bodies [4].

For instance, NASA's Advanced Life Support (ALS) project [5] and the European Space Agency's MELiSSA (Micro-Ecological Life Support System Alternative) [6] are some of the notable programs working towards the development of efficient bioregenerative systems.

The reliability and resilience of LSS are of paramount importance in ensuring mission success and crew safety.

The paper [7] discusses considerations in developing highly reliable life support systems for a human mission to Mars, where spare parts and crew return are not options as on the ISS. A plan focused on demonstrated reliability within constraints is needed.

The designing flexible, reliable life support systems for deep space using storage, recycling, or both to minimise mass while meeting reliability requirements are discussed in [8]. The goal is ultra-reliable, minimum mass life support for deep space.

The literature review on reliability modelling and analysis of environmental control and life support systems for space stations, where spare parts or quick crew recovery may not be feasible is provided in [9].

The definition and approach for understanding of reliability, robustness, and resilience for dependable LSS space systems are discussed in [10]. It is shown that reliability is increasingly important for space systems facing changing conditions and unknowns.

While previous studies have analysed the reliability of individual life support system components, subsystems, and end-to-end architectures, they do not fully consider the capabilities for restoration and reconfiguration enabled by heterogeneous redundant systems with functional layers. Existing models also lack the probabilistic analysis of reliability and process interconnectivity offered by a comprehensive framework approach. This paper puts forth a new model to fill the gap, capturing the dynamic mechanisms of failure, recovery, and redundancy management possible in a multi-tiered life support system architecture. By accounting for both the risks of component failures and the mitigating benefits of standby reserves, reconfigurability, and repairability, a more complete picture of overall system reliability emerges. The proposed Markov modelling approach provides a pathway to demonstrate the resilience possible for critical deep space life support applications and identifies areas needing further technology maturation.

3 Architecture of LSS

Ensuring the resilience of LSS is critical for the success of long-term crewed space missions. To enable autonomous habitation far from Earth, the onboard LSS must be capable of reliably regenerating essential resources like water, oxygen, and food even in the event of equipment failures. A three-tiered approach can provide this needed resilience (Fig. 1).

The first level consists of the baseline physicochemical and biotechnological reactors that recycle resources and sustain the environment. This includes systems like the water processor that filters urine and condenses humidity to produce potable water, the oxygen generator that splits water into oxygen and hydrogen, and the plant growth chambers that fix carbon dioxide while producing food. Regular operation of these systems enables self-sufficiency.

The second level is a backup system containing expendable catalysts, reagents, and microbial strains that can temporarily replace impaired first-level systems. For example, if the water processor malfunctions, chemical disinfection systems can purify water until repairs are made. Oxygen candles and tanks provide breathing air if the oxygen generator fails. Stored algae and cyanobacteria cultures can supplement food production if plant growth is disrupted. These emergency backups buy time for repairs while avoiding catastrophic failure.

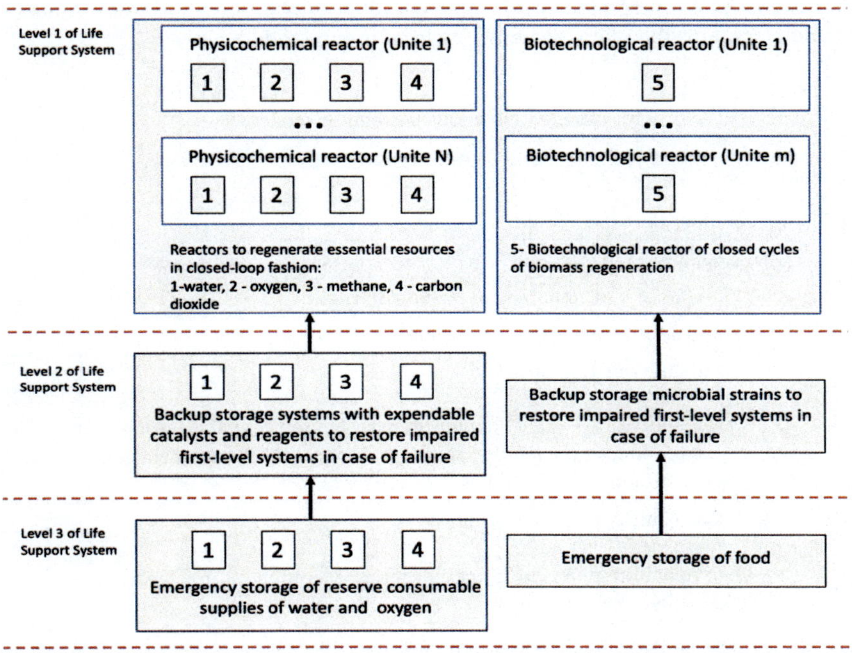

Fig. 1. Three-level life support system architecture (developed by the authors).

The third and final level consists of reserve consumable supplies like water, oxygen, and food that the crew can use if regeneration systems in levels one and two are exhausted. This provides a buffer against prolonged outages. The reserves must be sized to match the crew metabolism and planned mission duration.

By modelling the reliability of each LSS component using real-world failure data, the overall resilience of the multi-tier architecture can be analysed quantitatively.

This paper analyses a proposed three-level life support system architecture to provide the needed reliability and autonomy for long-duration deep space missions.

This layered approach provides defence-in-depth to ensure uninterrupted functioning of all life support processes even given individual component degradations or failures.

4 Model Formulation and Solution

The development of models for analysing the resilience and reliability of LSS poses a significant challenge, but one that carries profound importance.

Resilience modelling allows us to quantify the ability of an LSS to absorb disturbances and still maintain its function. A resilient system can handle a certain degree of fault or malfunction and continue to operate, albeit potentially at a reduced capacity. In long-duration space missions, where the availability of spares and the feasibility of maintenance are limited, understanding, and enhancing system resilience is of paramount importance.

In this study, the authors used a LSS reliability model based on Markov chains. Table 1 contains the main designations used in the model.

Table 1. Notations used in the model.

Notations	Parameters
λ	Failure Rate of LSS unit
μ	Repair Rate of LSS unit
m	Number of biotechnological reactors in LSS
N	Number of physicochemical reactors in LSS
A	Availability of LSS at whole
A_i	Availability of $i(i = \overline{1, 5})$ circuit for regeneration essential resources in closed-loop fashion: $i = 1$ - water, $i = 2$ - oxygen, $i = 3$ - methane, $i = 4$ - carbon dioxide, $i = 5$ - biomass regeneration
$U = 1 - A$	Unavailability of LSS at whole
$U_i = 1 - A_i$	Unavailability of $i(i = \overline{1, 5})$ circuit for regeneration essential resources in closed-loop fashion
t_s	Average switching time of redundant LSS unit at first level of LSS with parameter of Poisson's flow $\nu = 1/t_s$
t_φ	The average launch time of the second level backup storage system in the case of LSS unit failure at first level of LSS with parameter of Poisson's flow $\varphi = 1/t_\varphi$
p_i	Stationary probability of the state S_i with i failed elements of the first level of LSS
p_{Bi}	Stationary probability of the turn-on state of the LSS second level B_i of backup storage systems operation with expendable catalysts and reagents to restore impaired first-level systems in case of i failure LSS unit at first level for hot redundancy of units
p_{Ei}	Stationary probability of the turn-on state of the LSS third level E_i of emergency storage of reserve consumable supplies of water, oxygen or food in case of i failure LSS unit at first level for cold redundancy of units
p_F	Stationary probability of the state with LSS full failure

LSS in space missions comprise several integral subsystems, each playing a vital role in maintaining a habitable environment for the crew in an isolated setting. In assessing the overall reliability and resilience of the LSS, a crucial metric often considered is availability, defined as the proportion of time a system is in a functioning state.

The multiplicative relationship between the availabilities of the subsystems arises from the dependencies between them. The LSS can only function correctly and ensure crew survival if all its subsystems are operational because a failure in any subsystem can lead to the failure of the entire LSS. Therefore, the overall availability of the LSS can be represented by the availabilities of its subsystems:

$$A = \prod_{i=1}^{5} A_i$$

LSS comprises multiple subsystems, each serving a crucial role in maintaining habitability within the spacecraft. While these subsystems vary considerably in terms of their functions and physical implementation principles, they share a common trait: a three-level redundancy architecture.

The implementation of a standardised three-level redundancy architecture in all LSS subsystems allows for the use of a standardized reliability model across all subsystems.

Let us construct a unified Markov model of the reliability for all subsystems under study with the LSS architecture, shown in the Fig. 1. Let's consider two possibilities for organising redundancy of the main components (units) of the architecture at the first level - a hot standby and a cold standby.

1. *The hot redundancy of the main units of the LSS architecture at the first level.*

The graph of states transitions for the Markov chain with hot standby of the main components of the LSS architecture at the first level is shown in the Fig. 2.

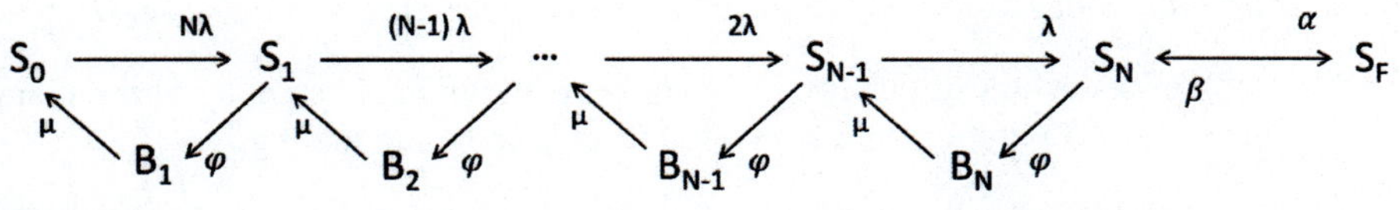

Fig. 2. The graph of states transitions for the Markov chain with hot standby of the main components of the LSS architecture.

The Markov chain was used to describe the examined system with next types of states: S_i - state where i LSS units have failed at first LSS level; B_i - state where i LSS units have failed at first LSS level and backup storage systems with expendable catalysts, reagents, or microbial strains of LSS second level are in operation to restore impaired first-level systems; S_F – state of LSS full failure.

The availability analysis of one single LSS subsystem for three-level redundancy architecture can be done using Markov models with classical Kolmogorov–Chapman differential equations for the system in a steady-state conditions.

For the transition graph of the Markov chain, which describes the use of the hot reserve of the main units of the LSS subsystems at the first level (Fig. 2), we can write the following system of equations for stationary state probabilities in accordance with the general rules [11]:

$$2\lambda p_0 = \mu p_{B1} \tag{1}$$

$$[(n - k\lambda + \varphi)]p_k = (n - k + 1)\lambda p_{k-1} + \mu p_{Bk+1}$$
$$(k = 1, \ldots n - 1) \tag{2}$$

$$\mu p_k = \varphi p_k (k = 1, \ldots n - 1) \tag{3}$$

$$(\alpha + \varphi)p_n = \lambda p_{n-1} + \beta p_F \tag{4}$$

$$\mu p_{Bn} = \varphi p_n \tag{5}$$

$$\beta p_F = \alpha p_n \tag{6}$$

This system of equations should be supplemented with the normalising equation:

$$p_0 + p_F + \sum_{1}^{n} (p_i + p_{Bi}) = 1 \tag{7}$$

For each of i LSS subsystems ($i = 1, \ldots 5$) under consideration the availability A_i can be determined by the probability of a failure state p_F, which can be determined by solving the above system of equations:

$$A_i = 1 - U_i = 1 - p_F \tag{8}$$

2. *The cold redundancy of the main units of the LSS architecture at the first level.*

The graph of states transitions for the Markov chain with cold standby of the main components of the LSS architecture at the first level is shown in the Fig. 3.

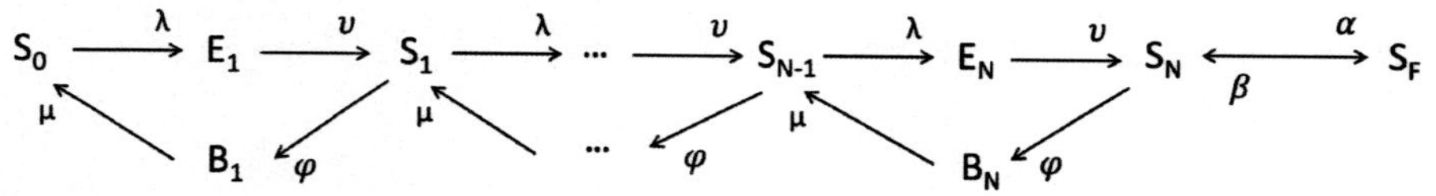

Fig. 3. The graph of states transitions for the Markov chain with cold standby of the main components of the LSS architecture.

The Markov chain was used to describe the examined system with next types of states: S_i - state where i LSS units have failed at first LSS level; B_i - state where i LSS units have failed at first LSS level and backup storage systems with expendable catalysts, reagents, or microbial strains of LSS second level are in operation to restore impaired first-level systems; E_i - state where i LSS units have failed at first LSS level and emergency storage of reserve consumable supplies provides the necessary needs of the crew for life support resources; S_F – state of LSS full failure.

For the transition graph of the Markov chain, which describes the use of the cold reserve of the main units of the LSS subsystems at the first level (Fig. 3), we can write the following system of equations for stationary state probabilities in accordance with the general rules [26]:

$$\lambda p_0 = \mu p_{B1} \tag{9}$$

$$\upsilon p_{E1} = \lambda p_0 \tag{10}$$

$$\mu p_{B1} = \lambda p_0 \tag{11}$$

$$(\lambda + \varphi)p_{k-1} = \nu p_{Ek-1} + \mu p_{Bk}(k = 2, \ldots n - 1) \tag{12}$$

$$\nu p_{Ek} = \lambda p_{k-1}(k = 2, \ldots n - 1) \tag{13}$$

$$\mu p_{Bk} = \varphi p_k(k = 2, \ldots n - 1) \tag{14}$$

$$(\alpha + \varphi)p_n = \nu p_{En} + \beta p_F \tag{15}$$

$$\beta p_F = \alpha p_n \tag{16}$$

This system of equations should be supplemented with the normalising equation:

$$p_0 + p_F + \sum\nolimits_1^n (p_i + p_{Bi} + p_{Ei}) = 1 \tag{17}$$

For each of i LSS subsystems ($i = 1, \ldots 5$) under consideration the availability A_i can be determined by the probability of a failure state p_F, which can be determined by solving the above system of equations:

$$A_i = 1 - U_i = 1 - p_F \tag{18}$$

Numerical Example

Let's determine the readiness of one of the LSS subsystems, the main set of equipment of which is duplicated and is in hot standby.

The graph of transitions of states of the Markov chain with cold standby of the main components of the LSS architecture at the first level is shown in the Fig. 4.

Using the system of Eqs. (1)–(7) describing the hot standby, for our case n = 2, we can obtain the following expressions to determine the stationary probabilities of finding the system under study in different states:

$$p_1 = \frac{\lambda}{\varphi}p_0,\ p_2 = \frac{\lambda^2}{\varphi^2}p_0,\ p_{E1} = \frac{\lambda}{\nu}p_0,\ p_{E2} = \frac{\lambda^2}{\nu\varphi}p_0,$$

$$p_{B1} = \frac{\lambda}{\mu}p_0,\ p_{B2} = \frac{\lambda^2}{\mu\varphi},\ p_F = \frac{\alpha\lambda^2}{\beta\varphi^2}p_0$$

Using above equations and normalising condition (7) we can define an expression for the probability p_0:

$$p_0 = \left[1 + \lambda\left(\frac{1}{\varphi} + \frac{1}{\nu} + \frac{1}{\mu}\right) + \lambda^2\left(\frac{1}{\nu\varphi} + \frac{1}{\nu\mu} + \frac{1}{\varphi^2} + \frac{\alpha}{\beta\varphi^2}\right)\right]^{-1}$$

In this case the availability of LSS i subsystem is described by Eq. (8) when substituting the obtained values of p_F and p_0 in them:

$$A_i = 1 - U_i = 1 - p_F$$

$$A_i = 1 - \frac{\alpha\lambda^2}{\beta\varphi^2}\left[1 + \lambda\left(\frac{1}{\varphi} + \frac{1}{\nu} + \frac{1}{\mu}\right) + \lambda^2\left(\frac{1}{\nu\varphi} + \frac{1}{\nu\mu} + \frac{1}{\varphi^2} + \frac{\alpha}{\beta\varphi^2}\right)\right]^{-1}$$

Figure 4 shows the dependences of the unavailability U of the three-level LSS subsystem as a function of the mean time between failures T_0 of the basic LSS units at the first level of redundancy and various values of the time t_φ for the typical values of the remaining LSS reliability indicators [7, 8].

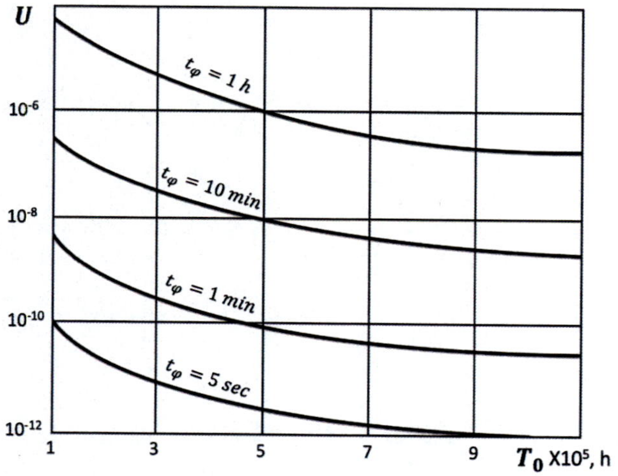

Fig. 4. Function of the unavailability U of the three-level LSS subsystem.

Figure 4 depicts the unavailability function for a hypothetical three-level life support subsystem architecture based on the developed Markov model. Several key insights can be drawn from analysing the curves.

The unavailability shows an inverse exponential relationship with the mean time between failures (MTBF) of the Level 1 components. As the MTBF increases, subsystem unavailability drops rapidly at first before gradually levelling off. This illustrates the substantial reliability gains achievable by using highly reliable Level 1 technologies. Even modest improvements to increase MTBF can dramatically boost overall subsystem availability.

Reducing the average activation time for the Level 2 backups also decreases unavailability, as shown by the three curves. The faster switching to Level 2 on Level 1 failure, the less likelihood of complete failure during the transition. This quantifies the importance of rapid fault response capabilities in the architecture.

While the gains diminish, continued reduction in unavailability is possible by increasing MTBF and lowering Level 2 activation time. This highlights the benefit of pushing both metrics to practical limits, as the residual unreliability has exponential dependence on each. Very high MTBF values for Level 1 paired with near-instantaneous Level 2 switching would minimise unavailability.

5 Conclusion

This paper presented a methodology for evaluating the reliability and resilience of multi-level redundant life support architectures using Markov modelling techniques. The proposed three-tier design provides built-in redundancy and autonomy to ensure continuous functioning even with individual component failures.

The Markov models developed enable quantitative assessment of failure tolerance over long-duration missions. The numerical example highlighted how subsystem availability depends directly on the reliability of Level 1 components and response times of backups. This reveals that high inherent reliability and rapid fault response are dual necessities for minimizing overall system unreliability.

The analysis shows that unavailability diminishes exponentially but asymptotically as MTBF increases and Level 2 activation time decreases. While substantial gains are possible initially, residual unreliability persists due to propagation of uncertainties in the complex interconnected architecture. This highlights the need to push reliabilities towards practical limits while managing failure risks through all three tiers of redundancy.

An integrated modular architecture is implied, allowing repairs and replacements of failed components without complete system shutdown. This maintainability is essential over multi-year missions were returning to Earth for major servicing is impossible. Intelligent diagnostics and crew training will play key roles in enabling autonomous maintenance and repairs in deep space.

The modelling approach presented provides a powerful methodology for optimizing life support architecture resilience. It enables probabilistic analysis during design phase to balance reliability, redundancy, and other performance factors like mass under constraints. This can help map technology development priorities for achieving fault tolerance goals.

The paper focuses on quantitative modelling of LSS reliability. Further work should address qualitative elements like failure modes and human factors in operating and maintaining the life support architecture over demanding multi-year missions. Overall, the paper takes steps toward a comprehensive methodology for designing maximally resilient life support capabilities essential for humanity's deep space voyages.

References

1. Schneider, W.F., Perry, J.L., Broyan, J.L., et al.: NASA Environmental Control and Life Support Technology Development and Maturation for Exploration: 2019 to 2020. Overview. In: International Conference on Environmental Systems ICES-2020. TTU-IR, [location not given] (2020). https://ttu-ir.tdl.org/bitstream/handle/2346/86395/ICES-2020-200.pdf. Accessed 01 Aug 2023
2. Starr, S.O., Muscatello, A.C.: Mars in situ resource utilization: a review. Planetary Space Sci. **182** (2020). https://doi.org/10.1016/j.pss.2019.104824
3. Ellery, A.: Supplementing closed ecological life support systems with in-situ resources on the moon. Life **11**, 770 (2021). https://doi.org/10.3390/life11080770
4. Keller, R., Goli, K., Porter, W., Alrabaa, A., Jones, J.A.: Cyanobacteria and algal-based biological life support system (BLSS) and planetary surface atmospheric revitalizing bioreactor brief concept review. Life **13**, 816 (2023). https://doi.org/10.3390/life13030816
5. Rapp, D.: Mars life support systems. Int. J. Mars Sci. Explor. **2**, 72–82 (2006). http://www.marsjournal.org/contents/2006/0005/files/rapp_mars_2006_0005.pdf. Accessed 01 Aug 2023
6. Hendrickx, L., et al.: Microbial ecology of the closed artificial ecosystem MELiSSA (Micro-Ecological Life Support System Alternative): reinventing and compartmentalizing the Earth's food and oxygen regeneration system for long-haul space exploration missions. Res. Microbiol. **157**(1), 77–86 (2006). https://doi.org/10.1016/j.resmic.2005.06.014
7. Jones, H.W.: Developing reliable life support for mars. In: 47th International Conference on Environmental Systems ICES-2017, pp. 16–20. NASA, Charleston, South Carolina (2017). https://ntrs.nasa.gov/api/citations/20170010347/downloads/20170010347.pdf. Accessed 01 Aug 2023
8. Jones, H.W.: Design and Analysis of a Flexible, Reliable Deep Space Life Support System. NASA (2016). https://ntrs.nasa.gov/api/citations/20160005782/downloads/20160005782.pdf. Accessed 01 Aug 2023
9. Sharma, G., Rai, R.N.: Reliability modelling and analysis of environmental control and life support systems of space stations: a literature survey. Acta Astronautica **155**, 238–246 (2019). https://doi.org/10.1016/j.actaastro.2018.12.010
10. Jones, H.W.: Going beyond reliability to robustness and resilience in space life support systems. In: 50th International Conference on Environmental Systems ICES-2021, pp. 12–15. TTU-IR (2021). https://ttu-ir.tdl.org/bitstream/handle/2346/87122/ICES-2021-140.pdf?sequence=1&isAllowed=y. Accessed 01 Aug 2023
11. Trivedi, K.S., Bobbio, A.: Reliability and Availability Engineering Modeling, Analysis, and Applications, 2nd edn. Cambridge University Press, Cambridge (2017)

Design of a Vehicle Monitoring System for the Needs of Security Managers

Martin Boros(✉), Andrej Velas, Zuzana Zvakova, and Jozef Svetlik

University of Žilina, Univerzitná 8215/1, 010 26 Žilina, Slovakia
{martin.boros,andrej.velas,zuzana.zvakova,
jozef.svetlik}@uniza.sk

Abstract. Due to the impact of the COVID-19 pandemic, several needs of the company have been created, which include, for example, knowing the current location and route of the car from the point of view of safety. The current market offers several options for online monitoring, but as part of our scientific and research activities, we focused on creating our own system, using a programmable microcontroller, and then comparing the reliability with the commonly used ones.

The aim of the contribution is to point out the reliability of the created system during experimental measurements compared to the middle-class module used in practice. Testing took place simultaneously while driving a car along a pre-selected route. The results of the experimental testing are surprising.

Keywords: Security managers · Localization · COVID-19

1 Introduction

The current era provides great opportunities for the use of modern technologies, often referred to as smart. Means and systems designed to protect property are no exception, such as controlling an electric security system using a mobile phone, guarding an object using drones, autonomous access systems and many other solutions. In the case of vehicle monitoring, there has been no significant change and there is still a choice of monitoring using the global system for mobile communication, hereinafter referred to as GSM, or the global positioning system, hereinafter referred to as GPS. We consider GSM to be the second generation of mobile systems, the so-called 2G network, the main idea of which was the transition to a digital system. In this period, we begin to encounter data transmission, so-called GPRS transmissions. GSM transmission has been used commercially since 1991, currently, there are several frequency variants of this system that are used for different types of transmission - voice transmission, data transmission, etc. [1–3].

The development of GPS technology dates to 1973, when the development of the NAVSTAR GPS system was started, which followed the American GNSS Transit system. Originally, these systems were used as naval systems whose task was to locate submarines and were intended primarily for military purposes. After the downing of a civilian airliner

O. Prentkovskis et al. (Eds.): TRANSBALTICA 2023, LNITI, pp. 59–66, 2024.
https://doi.org/10.1007/978-3-031-52652-7_6

in 1983, US President Ronald Regan decided that the system would be made fully available to the public after development was completed [1].

The entire transmission of information in GPS works in such a way that the satellites send coded messages containing information about their position as well as about the position of other satellites together with the expression of the exact time, which tells about the sending of the message from the satellite. The given codes are subsequently received by the GPS receiver, which calculates the time difference between the sending of the message by the satellite and the reception of the message by the satellite. Calculation algorithms are used to synchronize and express time, as there are atomic clocks in satellites and ordinary clocks in receivers. To determine the position using GPS, it is necessary for the receiver to communicate with several satellites at the same time. With the influence of the largest possible number of satellites, it is possible to determine the most accurate position, as shown in Fig. 1 [4, 5].

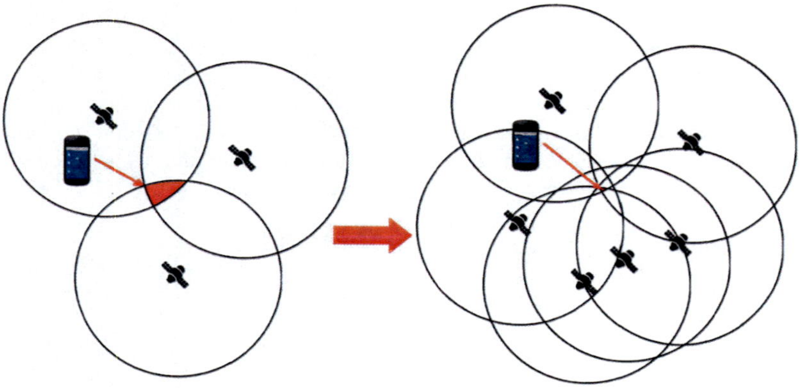

Fig. 1. An example of determining the exact location [4].

There are several types of GPS receivers in the Slovak or foreign market, either standalone or as part of some other device (mobile phone, smart watch, etc.). In addition to the classic use of GPS receivers for navigation systems, we also know specific uses, for example for monitoring company cars or guarding work machines on construction sites.

Company car monitoring has been a very popular option in recent years for management surveillance of employees for security reasons. All cars designed to transport money must be equipped with such a system [6, 7]. With the help of GPS, a superior employee can monitor whether the vehicle is moving along a pre-agreed route, whether it is going at an appropriate speed or whether it has accidentally stopped [8, 9].

Analyzing the market for monitoring devices for cars was carried out by Ing. Matej Klocáň and Bc. Erik Lettrich, who focused on testing selected GPS systems. From the results of the works as well as from the available sources of professional literature, there is an assumption that GPS systems are not sufficiently reliable in densely built-up dwellings, underground parking lots and the like. As part of the scientific research activities of the Department of Security Management, Faculty of Security Engineering, we decided to create our own GPS device with which we will be able to focus on a more detailed investigation and identification of places in the selected region with the probability of poor or insufficient GPS localization. In addition to the tests, various practical tests of both electrical and mechanical systems are continuously implemented at the department [10].

The basis of the monitoring device created by us is the programmable Arduino microcontroller. We decided to use Arduino because of its complexity of use and acquired knowledge from previous test devices. With the help of Arduino, as part of the scientific research activities of the department, we managed to build a test table in which 9 PIR detectors were simultaneously tested, a test device for alarm transmission systems and we created a proposal for an electrical security system [11, 12].

Arduino is comparable in appearance and functionality to classic simple programmable microcontrollers. Although only in appearance, Arduino is an open electronic platform that uses a simple type of hardware and software that can read different types of input requests and convert them into an output pulse. Through the output pulse, it is possible to activate the motor, turn on the relay, activate the lighting and much more.

Arduino is used by both scientists and the public, and even due to its simplicity and efficiency, a community has been created that publicly publishes its projects with a complete description of the program [9, 13].

2 Methodology

As part of the first phase, it was necessary to create our own test device, the basis of which was the Arduino MEGA microcontroller. The other parts were a GPS module for obtaining data on latitude and longitude, altitude, speed of movement, number of connected satellites, a GSM module for controlling user modes via SMS, an LCD for displaying data from the GPS module, a real-time RTC module for determining the current time, a microSD card reader, for saving the record of the position and the distance travelled, a thermometer and a hygrometer and a battery for power supply. The overall connection of the system is shown in Fig. 2.

The second phase was the procurement of a commonly used module, which was a monitoring unit from the NAM Slovakia company marked NCL 21. This is a device that can be used, for example, to guard vehicles and things, monitor the movement of people, track leisure activities or guard packages and containers. It is not intended for fixed installation, it can be freely carried and, if necessary, attached to any metal surface thanks to the magnetic system. It allows a choice between six operating modes, which have a significant impact on the maximum operating time of the device. Control of the monitoring unit as well as display of the current position or the traveled path works through a mobile and web application. The device consists of an internal GPS and GSM antenna, a SIM card, a thermometer, an accelerometer, an accumulator, a contactless charging coil, magnetic contacts and a waterproof, mechanically resistant case.

The last phase was the creation of comparative scenarios, which were a road between two villages, crossing a tunnel and a densely populated part. For the needs of the article, we decided to evaluate only the tests focused on the transition between two municipalities and the tunnel transition.

As part of GPS connectivity testing, it is possible to encounter several errors that are caused by ionospheric effects, the shift in the orbit of satellites, and the like. In the past,

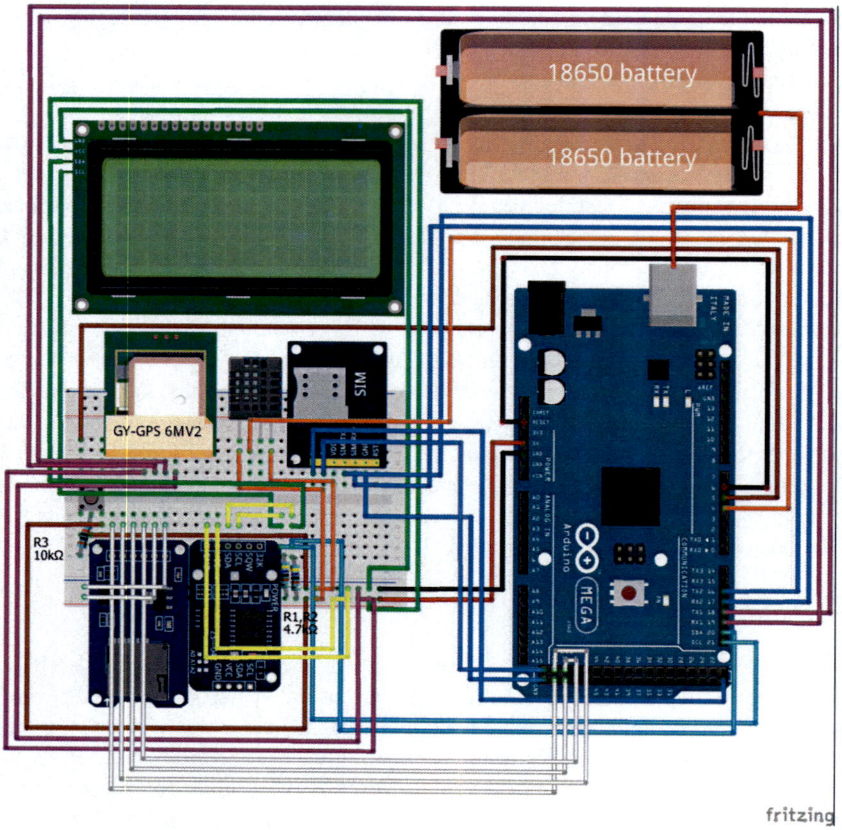

Fig. 2. Graphical connection of the system in the Fritzing program.

the main factor in the error rate of the GPS signal was the Selective Availability (SA) error. Another significant error is the geometric distribution of satellites, which can be affected by the number of visible satellites and thus the quality of the signal, because the more satellites the receiver has in view, the higher the quality of accuracy [14].

3 Results

The first scenario tested was the transition between two municipalities, Martin and Ondrášová in Slovakia. The resulting rendering of the route of both systems is shown in Fig. 3.

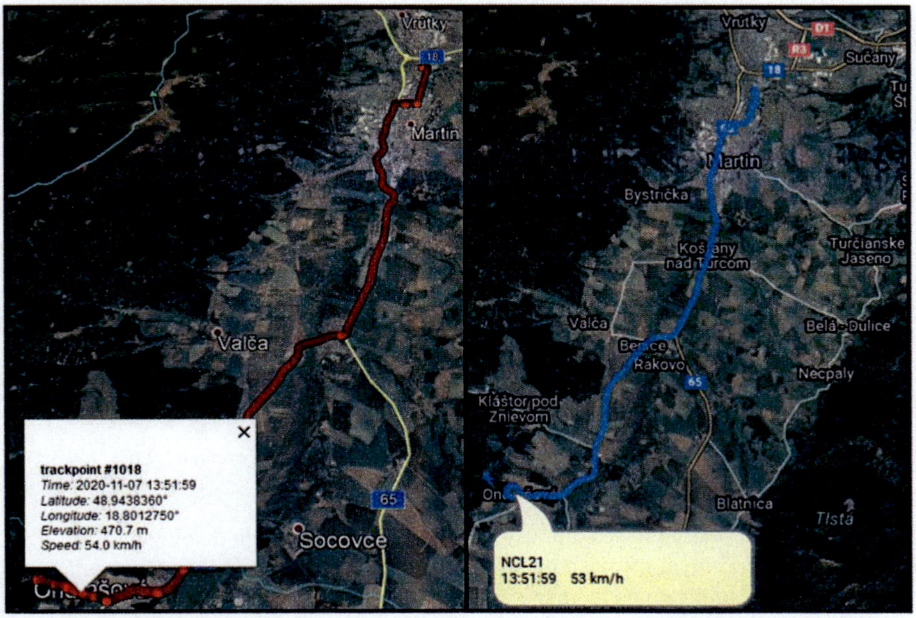

Fig. 3. Illustration of the comparison of determining the accuracy of the location of both devices between two municipalities.

On the left side of picture no. 3 shows the output generated on the basis of the collected positional data by the designed monitoring device, and on the right side is the monitoring result of the borrowed monitoring unit NCL 21. Looking at the figure, it can be seen that the plotted paths are very similar, which implies that the recording of the traveled path, where the emphasis is not placed on the close details of the environment, is practically on the same level. This claim is of course supported by several tests and not just this one demonstrated. The figure also shows the way in which it is possible to learn additional information about the movement at a specific point of the track by marking it with the mouse on the computer. Outputs from the designed device are enriched with information such as time, latitude and longitude, altitude and speed of movement. The web application belonging to the borrowed device provides information about the time and speed of movement. Additional information about the traveled path is intentionally displayed at the same time point of the path so that it is possible to compare the current detected movement speed of both devices. As can be seen in the figure, the speed of movement varies by only one kilometre per hour, which may also be due to a slight deviation in the setting of the time parameters of the devices.

The second testing was focused on places where it was assumed to have a lower signal value, namely tunnels, specifically at their entrances and exits. The Považský Chlmec tunnel was chosen for this testing and the individual results of the testing are shown in Fig. 4 and Fig. 5.

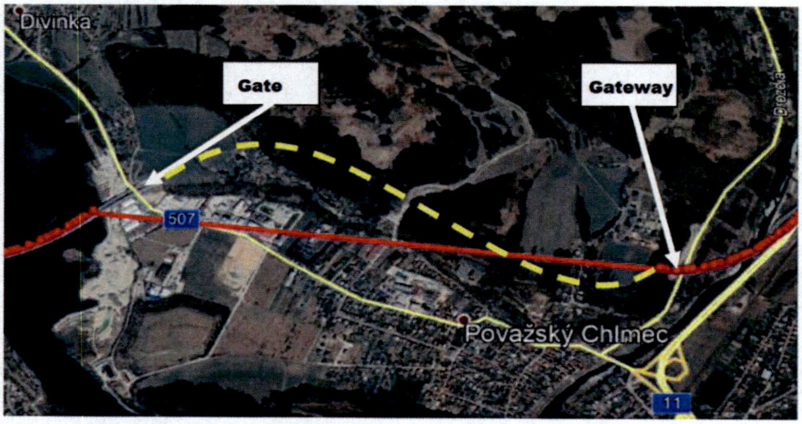

Fig. 4. Representation of the travelled route for the system created by us.

In Fig. 4 it is possible to see the way the software works when displaying positional data. The moments when the designed monitoring device was able to receive data from GPS satellites are represented by many red points on the map. When entering the tunnel, the GPS signal was interrupted, and it was restored approximately 200 m after exiting the tunnel. A signal interruption is represented by a long straight red line connecting two red points representing the last recorded position before entering the tunnel and the first one after exiting it. The area where the tunnel goes underground is represented by a dashed yellow line.

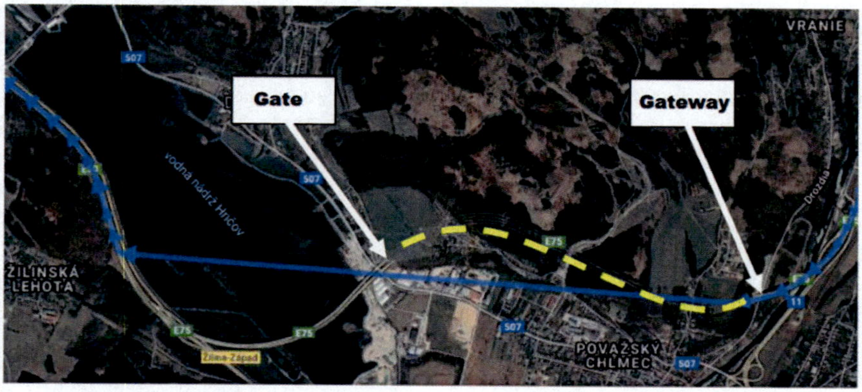

Fig. 5. Representation of the travelled route for the NCL 21 monitoring unit.

A very similar principle, which was described when using the "Google Earth" software, is also used in the web application intended for the borrowed device. Recorded locations are represented by blue arrows and a long blue line indicates a GPS signal failure. The big difference between the two devices lies in re-establishing contact with the GPS. While the designed device was able to do this after approximately 200 m, the borrowed monitoring unit needed approximately 2 kms to do so.

This example is clearly visible when comparing Fig. 4 and 5. In these cases, the connection between the receiving unit and the satellites is restored. Whereas in the case of our device, the connection was realized as we write after about 200 m, practically immediately after the exit from the tunnel. In the case of the borrowed unit, it was only after 2 km to some part beyond the bridge. During this time, the unit was probably looking for free satellites from which it could get location information. The disadvantage is that the manufacturer of the borrowed unit does not indicate the type of designation of the GPS antenna and therefore it is not possible to determine whether it is a difference in the type of antenna, or whether the delay is caused by the slowing down of other parts of the unit.

As we can see based on the results of the tests, we can say that it is possible to compete with the commonly used system using the modular elements of the Arduino platform.

4 Conclusion

The article focuses on the issue of the possibility of monitoring cars as well as trucks using a GPS locator. The research took place in two phases, namely the production of the own localizer and experimental testing, which was supplemented with a monitoring unit commonly used in practice.

With the help of experimental tests, it was shown that the system created by us, operating with the Arduino MEGA microcontroller, has a better GPS antenna. We can say this assessment based on the drawn map, which we travelled according to the device, as well as the assessment of the travelled route from the point of view of speed, number of connected satellites, etc. It is also interesting that in the sum of the costs, the system created by us is cheaper than the commonly offered monitoring unit. The conclusions of the research offer further questions, or options for the direction of research, which could be oriented towards the reliability of GPS elements in logistics centers, and densely populated areas, or to think about a GPS monitoring system in tunnels or other buildings.

Acknowledgement. This article was supported by the Scientific Grant Agency as part of project no. 1/0173/21 Research on measures implemented by security managers in organizations in connection with the spread of COVID-19 and in other emergency situations.

References

1. Lettrich, E.: Car monitoring options. Bachelor thesis, University of Žilina (2019)
2. Lovecek, T., Strakova, L., Kampova, K.: Modeling and simulation as tools to increase the protection of critical infrastructure and the sustainability of the provision of essential needs of citizens. Sustainability **13**(11) (2021)
3. Siser, A., Lovecek, T., Maris, L.: Simulation of possible assault vectors in an attack using a real-life waterworks object as a use case. In: 12th International Scientific Conference of Young Scientists on Sustainable, Modern and Safe Transport (2017)

4. Dolejš, J.: Jak funguje zaměření polohy pomocí GPS?. https://www.svetandroida.cz/gps-princip/. Accessed 28 July 2023
5. Cooper, M., Sornalingam, S., Heath, J.: Battle to retain GPS. Battle to retain GPs: defining "generalism" is key. BMJ-Br. Med. J. (381)
6. Sventekova, E., Cicmancova, S.: Risk assessment of rail transport. In: Transport Means - Proceedings of the International Conference (2013)
7. Klucka, J., Gruendichler, R., Ristvej, J.: Relations of COVID-19 and the risk management framework. Sustainability **13**(21) (2021)
8. Vandlickova, M., Markova, I., Holla, K., Gaspercova, S.: Evaluation of marblewood dust's (Marmaroxylon racemosum) effect on ignition risk. Appl. Sci.-Basel **11**(15) (2021)
9. Bera, B., Chattaraj, D., Das, A.K.: Designing secure blockchain-based access control scheme in IoT-enabled internet of drones deployment. Comput. Commun. **153**, 229–249 (2020)
10. Rehak, D., Hromada, M., Lovecek, T.: Personnel threats in the electric power critical infrastructure sector and their effect on dependent sectors: overview in the Czech Republic. Saf. Sci. **127** (2020)
11. Rybak, A., Joostberens, J., Manowska, A., Pielot, J.: The impact of environmental taxes on the level of greenhouse gas emissions in Poland and Sweden. Energies **15**(12) (2022)
12. Bluszcz, A.: Conditions for maintaining the sustainable development level of EU member states. Social indicators research (2018)
13. Alecksen, T., Whicker, R.: Retrospective detection sensitivity for gps-based gamma radiation surveys. Health Phys. (2023)
14. Jurišica, L., Vitko, A., Duchoň, F., Kaštan, D.: Systém GPS – princíp, prednosti a nedostatky, In: AUTOMA (2011). https://automa.cz/Aton/FileRepository/pdf_articles/42982.pdf

Employing Digital Twins in Operation and Maintenance Management of Transportation Systems

Robert Giel(✉), Sylwia Werbińska-Wojciechowska, and Klaudia Winiarska

Wroclaw University of Science and Technology, Wybrzeze Wyspianskiego 27, 50-370 Wroclaw, Poland

{robert.giel,sylwia.werbinska,klaudia.winiarska}@pwr.edu.pl

Abstract. The chapter is focused on the digital twin concept use as a supporting tool for operation and maintenance management decision-making processes. A digital twin is a virtual representation of a connected physical asset and gives the capability to predict its conditions in the future. Indeed, it allows for decreasing the resources needed to design, produce, and keep technical assets in a good operational state. Therefore, the main goal is to present the digital twin concept implementation possibilities in the area of operation and maintenance management activities performance. As a result, a paper presents a short literature review on the defined research area. Later, the digital twin concept implementation possibilities in relation to management processes in the operation and maintenance area are presented. The work ends with conclusions and directions for further research.

Keywords: Digital Twin · Operation and maintenance management · Transportation system · Technical system

1 Introduction

The importance of maintenance management functions has increased over the past five years. According to the standard BS EN 13306:2010 [34], it is defined as *all management activities that determine the objectives and strategies of maintenance and their implementation through appropriate tools, such as maintenance planning, control, and supervision, improvement of methods in the organization, together with consideration of the economics of operation.* Indeed, it refers to the process of effectively managing and overseeing the activities and resources required to ensure effective and efficient maintenance tasks in an organization.

The main goals of maintenance management include maximizing efficiency and optimizing the safety of the environment and employees by ensuring the reliability and availability of physical assets when required [26]. This is made possible by keeping technical equipment in an acceptable condition or restoring that equipment to acceptable working conditions [23]. Specific assignments are performed within this process to get closer to the assumed goals. These assignments are shown in Fig. 1.

O. Prentkovskis et al. (Eds.): TRANSBALTICA 2023, LNITI, pp. 67–76, 2024.
https://doi.org/10.1007/978-3-031-52652-7_7

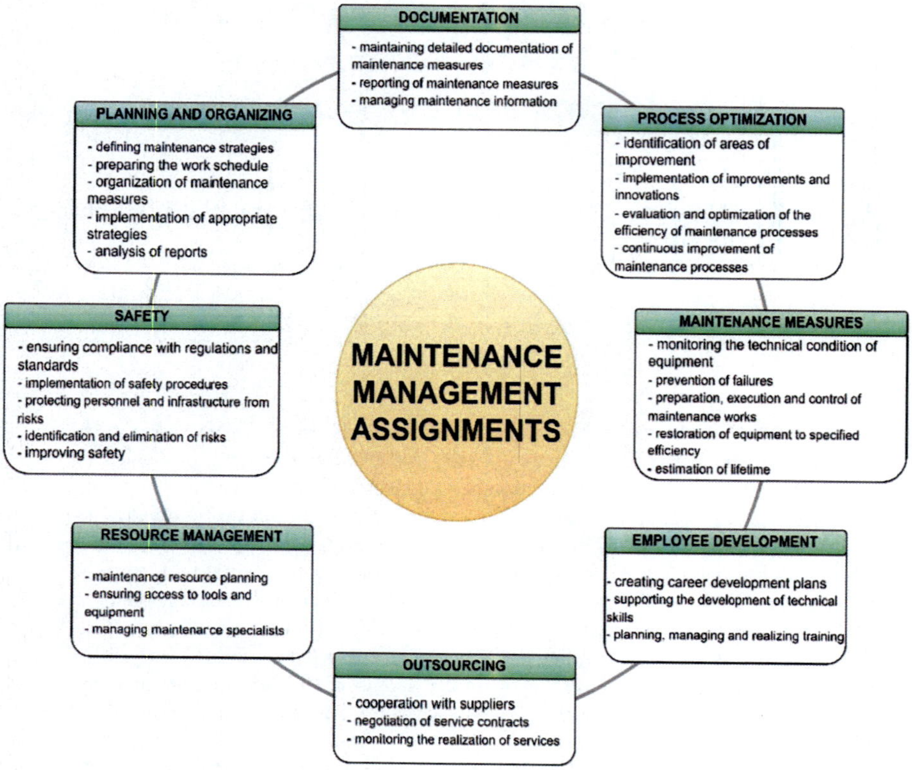

Fig. 1. Assignments within maintenance management.

Effective maintenance management is crucial in achieving operational reliability and minimizing downtime of assets. This is possible by maximizing resource availability [27] and proper decision-making process performance [28]. According to the standard BS EN 16646:2014 [5], the management of maintenance processes includes five sub-processes, which are:

- setting maintenance goals and strategies,
- planning maintenance activities,
- resource management and development,
- maintenance activities execution,
- post-maintenance actions and continuous improvement.

Proper maintenance management requires a specified hierarchy of decision-making. Indeed, the maintenance decision pyramid can be used to organize management activities [24]. At the top of the pyramid is the level of long-term strategic maintenance management activities. It is, for example, the selection of an overall maintenance strategy or the definition of maintenance goals. The next level is tactical maintenance management, which includes selecting concrete methods for conducting maintenance activities. At the same time, at the bottom of the pyramid is operational maintenance management, which provides for decisions related to daily task performance. In this pyramid, the decision level from top to bottom is increasingly organized, and the risks associated with making a wrong decision differ according to the scale of the consequences.

One of the key decision areas in maintenance management is selecting a suitable maintenance strategy. This decision depends on an organization's specific needs and goals, available resources, or the complexity of the dilemma. In the literature, one may identify the main five maintenance strategies:

- *reactive strategy*, which involves repairing technical equipment after failure occurrence to restore it to the required condition [16],
- *preventive strategy*, which is based on regular maintenance activities performance of equipment regardless of their present condition and wear level [4],
- *condition-based strategy*, which involves taking maintenance actions when the monitored parameters determining the technical condition of equipment reach limit values [3],
- *predictive strategy*, which consists of predicting the occurrence of technical equipment failures before they actually occur by monitoring system operating conditions and using prognosis models [13],
- *proactive strategy* applies to planned maintenance activities that are carried out before an emergency occurs [22].

The modern approach to maintenance management is based on the use of data, their analysis, and advanced tools that enable precise decision-making. Modern technologies such as Big Data analytics, Machine Learning, Artificial Intelligence, Internet of Things, and Digital Twin (DT) allow better data collection, analysis, and interpretation. This leads to more accurate decision-making. As a result, many authors underline the possibility of using the Digital Twin approach for effective operation and maintenance management activities performance (see, e.g. [10, 21]). Therefore, the next Section presents a summary of DT use in the area of technical system maintenance.

2 Digital Twin for Maintenance Management of Technical Systems, Including Transportation Systems – Literature Review

Digital Twin approach, has an important role in the context of maintenance management. DT supports operational and maintenance decisions, thereby contributing to improved efficiency of maintenance activities [1]. Using DT for maintenance management makes it possible to control resources in an enterprise better. This causes increased reliability and durability of the organization's technical equipment. In addition, DT contributes to the evolution of maintenance and the shift to more advanced maintenance strategies, such as proactive maintenance.

A literature review about the use of DT in the maintenance area can be found, e.g., in [8, 12, 30]. In [10], the authors wrote about applying DT in various maintenance strategies performance. In addition, there can be underlined review articles focusing on the use of DT for condition monitoring and predictive maintenance [2, 9, 18, 31].

The authors [17] described the possibilities of applying DT concepts in Industry 4.0. At the same time, the authors of [6] and [20] presented its implementation challenges. Review articles on the application of DT maintenance in smart and sustainable manufacturing were also created to increase its efficiency [14, 25].

DT is applied throughout the life cycle of a physical object, starting from the design phase, through the production and operation phases, and ending with its withdrawal from operation.

Therefore, the paper [19] presents the types of DT applications in different phases of the life cycle of a technical object (Fig. 2).

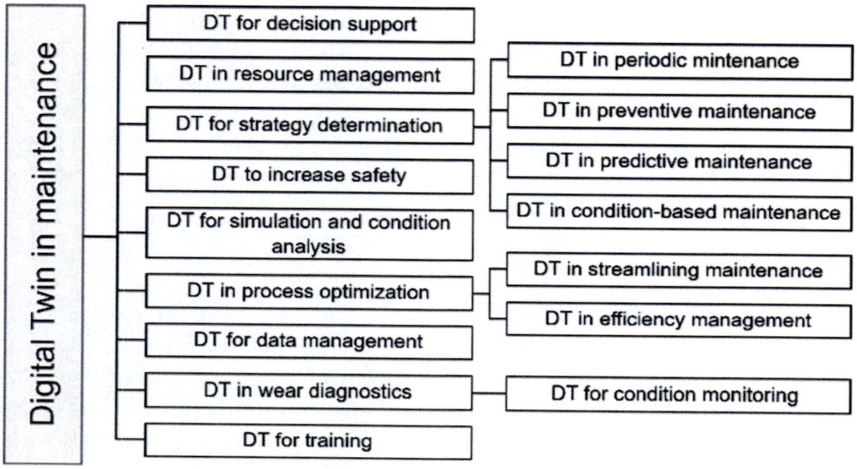

Fig. 2. The main research areas defined in the literature for DT in maintenance.

DT provides intelligent maintenance management by automatically analysing collected data related to a technical facility's operation, technical state, or workload. On their basis, it is possible to predict failures, anticipate maintenance, and plan corrective actions by responding to certain anomalies in real time. As a result, using DT in maintenance increases technical systems' reliability, efficiency, and safety. A detailed indication of the differences between traditional predictive maintenance and DT-based predictive maintenance was described by the authors in [31]. In addition, a comprehensive review of DT use in the transportation area is provided, e.g., in [11] The next Section presents the framework concept for DT in the area of maintenance management activities performance in the context of the asset management approach.

3 Framework for DT as a Supporting Tool for Operational and Maintenance Management Activities Performance

In the context of maintenance management, DT plays an essential role because, among other things, it offers the possibility to evolve the way operation and maintenance are carried out but also contributes to improving reliability, efficiency, and safety of technical systems. Therefore, this article proposes the Digital Twin conceptual framework as a supporting tool in performing the main physical asset management activities.

The operational and maintenance management activities should be perceived from a strategic dimension. Indeed, they should cover those decisions taken to shape an organization's future maintenance requirements and business goals [15]. This broader perspective makes it necessary to consider the scope of operational and maintenance management in relation to technical systems life cycle stages: specification, acquisition, panning, operation, performance evaluation, improvement, and disposal [15]. In this broader context, we may define operational and maintenance functions as physical asset management.

Asset management is defined in the ISO 55000 standard as *the coordinated activity of an organization to realize value from assets* [32]. Based on this, the main objective of operation and maintenance, which is linked to the overall organizational goals, should be to maximize the profitability of the organization by performing activities by providing the highest level of organization's operational availability by retaining working equipment in an acceptable condition or returning the equipment to an acceptable working condition [29]. Thus, following [29], we may present the scope of technical systems operation and maintenance system (Fig. 3).

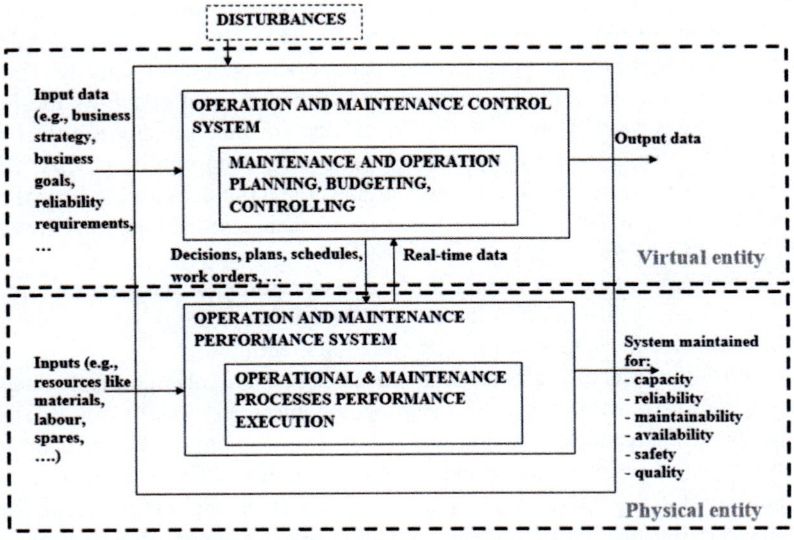

Fig. 3. Operation and maintenance of technical systems with the main entities. [29] Source: own contribution based on.

Although there are no simple solutions for implementing an integral asset management model, the literature proposes a set of management models (see e.g. [7]). Another solution can be the implementation of the digital twin concept. DT framework includes physical and virtual layers (Fig. 4). Their discussion on the example of internal transport systems maintenance is presented in [11].

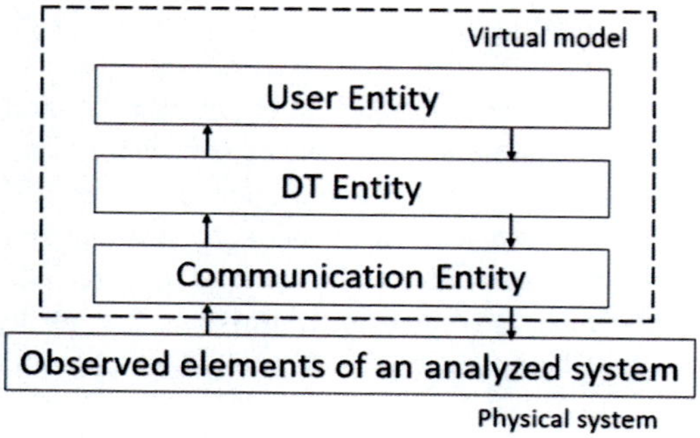

Fig. 4. The main four layers of digital twin reference model based on ISO 23247 standards.

Taking one step further, according to the ISO 55000 family of standards [32, 33] and the ISO 23247 standard [35–38], we may define the digital twin framework in relation to the main performed asset management activities. It contains a set of layers representing the solution's physical and virtual parts (Fig. 5).

According to Fig. 5, we may define five main operational and maintenance management phases: strategic phase, identification phase, control phase, execution phase, and improvement phase. The first phase is responsible for defining the main operational and maintenance management strategy in relation to the defined business goals or reliability requirements stated for the technical devices being operated in the company. The identification phase starts when the main priorities for management activities are defined. This phase contains two DT layers – the physical one and the virtual one, as well as the communication entity. The physical layer includes system decomposition, sensors selection, and necessary data collection. Indeed, it gives the input data for the virtual model of assets. The communication entity provides the exchange of information between the physical layer and the virtual one.

The identification phase is responsible for a work-order generation as a primary output. The stages for information integration, decision-making processes, and work-order generation are performed in the user and digital entities.

The control phase is also a part of the virtual layer. The planning and scheduling process allows for transforming the strategic goals into operational plans. This is the starting point for the execution phase, which is part of the physical layer. During this phase, two types of actions may be performed – operational and maintenance. When a maintenance task is scheduled, the post-maintenance testing should be performed. The results from this phase are the inputs for the improvement phase. Due to continuous improvement requirements, the asset's technical condition is updated (based on its virtual model use), the main condition state prognosis is conducted, and, when necessary, the maintenance program is updated.

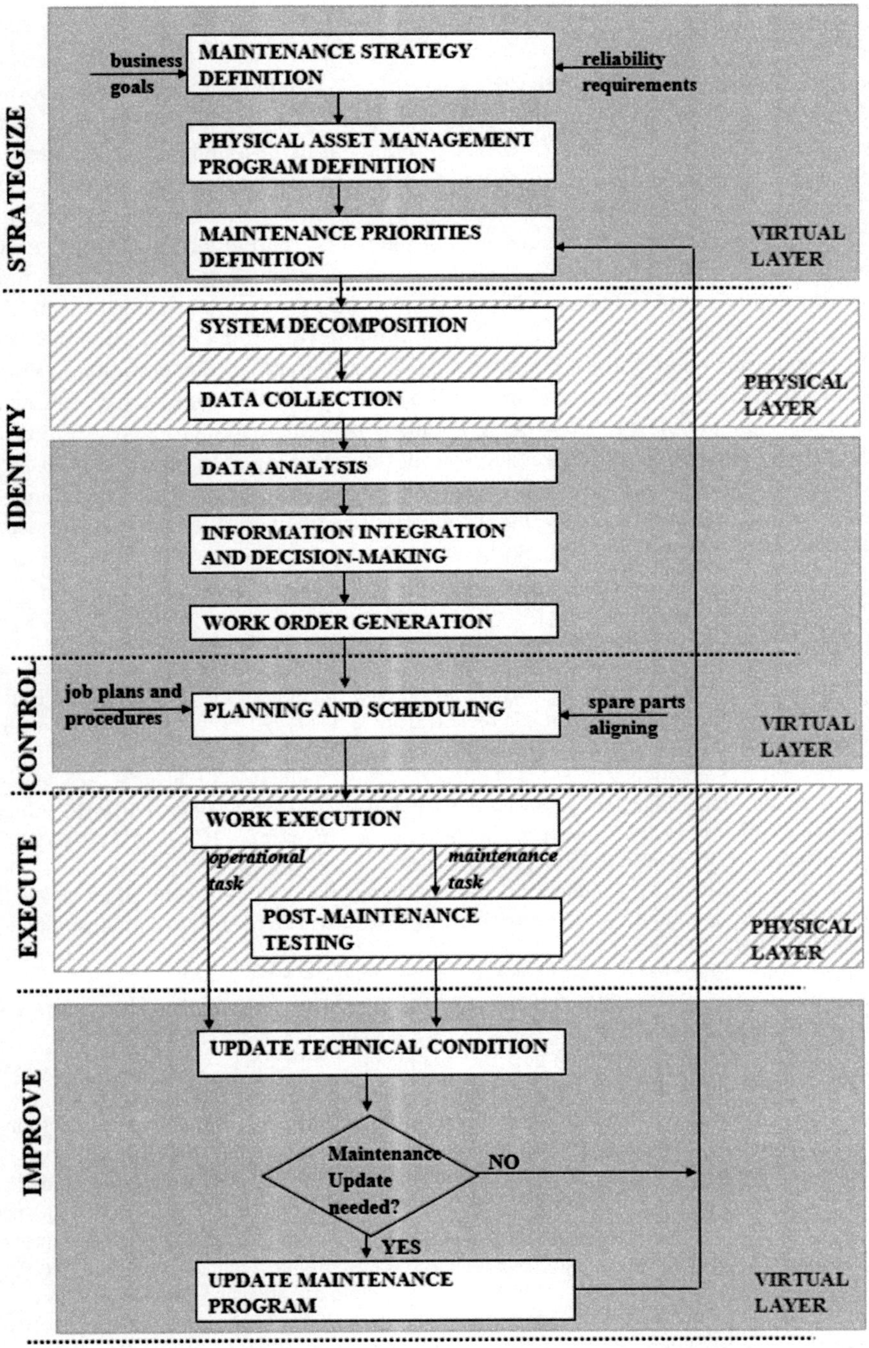

Fig. 5. Framework for DT concept based on ISO 23247 standards and ISO 55000 family of standards.

4 Conclusions

The article presents the conceptual framework for DT implementation for supporting technical system operation and maintenance management activities. According to the conducted literature review, implementing digital twins in the maintenance management for technical systems, including transportation systems, still needs development, especially at the system level. The main known solutions mostly focus on the development of DTs for technical devices (single ones or networks), particularly focusing on fault diagnosis, system elements monitoring and control, or supporting the main operational decisions to provide effective and safe operation of the system. At the same time, the literature review indicates the growing importance of implementing DT in the maintenance management of in-house logistic systems, particularly internal transportation systems. The use of DT in this area brings a number of benefits to maintenance, such as optimized maintenance operations, proactive facility management, increased productivity, reduced downtime of logistic infrastructure, and overall improved logistics services.

However, there is still a need to develop an approach that focuses on maintenance management activities holistically. This problem is especially visible for logistic systems. The authors will develop such a complex approach in their future research works.

References

1. Agnusdei, G.P., Elia, V., Gnoni, M.G.: Is digital twin technology supporting safety management? A bibliometric and systematic review. Appl. Sci. **11**, 1–17 (2021). https://doi.org/10.3390/app11062767
2. Aivaliotis, P., Georgoulias, K., Alexopoulos, K.: Using digital twin for maintenance applications in manufacturing: state of the art and gap analysis. In: Proceedings - 2019 IEEE International Conference on Engineering, Technology and Innovation, ICE/ITMC 2019 (2019). https://doi.org/10.1109/ICE.2019.8792613
3. Alaswad, S., Xiang, Y.: A review on condition-based maintenance optimization models for stochastically deteriorating system. Reliab. Eng. Syst. Saf. **157**, 54–63 (2017). https://doi.org/10.1016/j.ress.2016.08.009
4. Basri, E.I., Razak, I.H.A., Ab-Samat, H., Kamaruddin, S.: Preventive maintenance (PM) planning: a review. J. Qual. Maint. Eng. **23**, 114–143 (2017). https://doi.org/10.1108/JQME-04-2016-0014
5. BSI EN 16646. BSI Standards Publication - Maintenance—Maintenance within physical asset management. 40 (2014)
6. Cimino, C., Negri, E., Fumagalli, L.: Review of digital twin applications in manufacturing. Comput. Ind. **113**, 103130 (2019). https://doi.org/10.1016/j.compind.2019.103130
7. Crespo Marquez, F.A., Diaz, V.G.-P., Fernandez, J.F.G.: Advanced Maintenance Modelling for Asset Management. Springer, Cham (2018). https://doi.org/10.1007/978-3-319-58045-6
8. D'Amico, R.D., Erkoyuncu, J.A., Addepalli, S., Penver, S.: Cognitive digital twin: an approach to improve the maintenance management. CIRP J. Manuf. Sci. Technol. **38**, 613–630 (2022)
9. van Dinter, R., Tekinerdogan, B., Catal, C.: Predictive maintenance using digital twins: a systematic literature review. Inf. Softw. Technol. **151**, 107008 (2022). https://doi.org/10.1016/j.infsof.2022.107008
10. Errandonea, I., Beltrán, S., Arrizabalaga, S.: Digital twin for maintenance: a literature review. Comput. Ind. **123** (2020). https://doi.org/10.1016/j.compind.2020.103316

11. Giel, R., Werbińska-Wojciechowska, S., Winiarska, K.: Framework for digital twins concept implementation in internal transportation systems. In: Proceedings of the RelStat 2023 Conference (in review)
12. Gosavi, A., Le, V.K.: Maintenance optimization in a digital twin for industry 4.0. Ann. Oper. Res. (2022). https://doi.org/10.1007/s10479-022-05089-1
13. Haarman, M., Mulders, M., Vassiliadis, C.: Predictive maintenance 4.0. Predict the unpredictable (2017)
14. He, B., Bai, K.J.: Digital twin-based sustainable intelligent manufacturing: a review. Adv. Manuf. **9**, 1–21 (2021). https://doi.org/10.1007/s40436-020-00302-5
15. Jardine Andrew, K.S., Tsang, A.H.C.: Maintenance, Replacement, and Reliability. Theory and Applications. CRC Press (2021)
16. de Jonge, B., Scarf, P.A.: A review on maintenance optimization. Eur. J. Oper. Res. **285**, 805–824 (2020). https://doi.org/10.1016/j.ejor.2019.09.047
17. Kenett, R.S., Bortman, J.: The digital twin in industry 4 0 a wide-angle perspective. Qual. Reliab. Eng. Int. **38**, 1357–1366 (2022)
18. Liu, H., Xia, M., Williams, D., Sun, J., Yan, H.: Digital twin-driven machine condition monitoring: a literature review. J. Sens. **2022** (2022). https://doi.org/10.1155/2022/6129995
19. Liu, M., Fang, S., Dong, H., Xu, C.: Review of digital twin about concepts, technologies, and industrial applications. J. Manuf. Syst. **58**, 346–361 (2021). https://doi.org/10.1016/j.jmsy.2020.06.017
20. Melesse, T.Y., Di Pasquale, V., Riemma, S.: Digital twin models in industrial operations: a systematic literature review. Procedia Manuf. **42**, 267–272 (2020). https://doi.org/10.1016/j.promfg.2020.02.084
21. Menegon, J., Isatto, E.L.: Digital twins as enablers of structure inspection and maintenance. Gestão Prod. **30**, 1–13 (2023). https://doi.org/10.1590/1806-9649-2022v30e4922
22. Muganyi, P., Mbohwa, C.: Proactive maintenance strategic application to advance equipment reliability. In: Proceedings of the International Conference on Industrial Engineering and Operations Management 2018, pp. 3300–3309 (2018)
23. Nowakowski, T., Tubis, A., Werbińska-Wojciechowska, S.: Evolution of technical systems maintenance approaches – review and a case study. In: Burduk, A., Chlebus, E., Nowakowski, T., Tubis, A. (eds.) ISPEM 2018. AISC, vol. 835, pp. 161–174. Springer, Cham (2019). https://doi.org/10.1007/978-3-319-97490-3_16
24. Pintelon, L., Van Puyvelde, F.: Maintenance management defined. In: Maintenance Decision Making. Acco, Leuven (Belgium), pp. 3–13 (2006)
25. del Real Torres, A., Andreiana, D.S., Ojeda Roldán, Á., Hernández Bustos, A., Acevedo Galicia, L.E.: A review of deep reinforcement learning approaches for smart manufacturing in industry 4.0 and 5.0 framework. Appl. Sci. **12** (2022). https://doi.org/10.3390/app122312377
26. Sabaei, D., Erkoyuncu, J., Roy, R.: A review of multi-criteria decision making methods for enhanced maintenance delivery. Procedia CIRP **37**, 30–35 (2015). https://doi.org/10.1016/j.procir.2015.08.086
27. Shenoy, D., Bhadury, B.: Maintenance Resources Management: Adapting MRP. Taylor & Francis (2005)
28. Wang, H., Ye, X., Yin, M.: Study on Predictive Maintenance Strategy. Int. J. u- e- Serv. Sci. Technol. **9**, 295–300 (2016). https://doi.org/10.14257/ijunesst.2016.9.4.29
29. Werbinska-Wojciechowska, S.: Technical System Maintenance. Delay-Time-Based Modelling. Springer, Cham (2019). https://doi.org/10.1007/978-3-030-10788-8
30. Xia, J., Zou, G.: Operation and maintenance optimization of offshore wind farms based on digital twin: a review. Ocean Eng. **268**, 113322 (2023). https://doi.org/10.1016/j.oceaneng.2022.113322
31. Zhong, D., Xia, Z., Zhu, Y., Duan, J.: Overview of predictive maintenance based on digital twin technology. Heliyon **9**, e14534 (2023). https://doi.org/10.1016/j.heliyon.2023.e14534

32. ISO 55000: Asset management - overview, principles and terminology (2014)
33. ISO 55001: Asset management—Management systems—Requirements (2014)
34. BS EN 13306:2017: Maintenance. Maintenance terminology (2018)
35. ISO/DIS 23247-1 Automation systems and integration—Digital Twin framework for manufacturing—Part 1: Overview and general principles (2020)
36. ISO/DIS 23247-3 Automation systems and integration—Digital Twin framework for manufacturing—Part 3: Digital representation of manufacturing elements (2020)
37. ISO/DIS 23247-2 Automation systems and integration—Digital Twin framework for manufacturing—Part 2: Reference architecture (2021)
38. ISO/DIS 23247-4 Automation systems and integration—Digital Twin framework for manufacturing—Part 4: Information exchange (2021)

Combustion in Engines, Alternative Technologies, Energy Management and Emissions

Vibrations of Micro-hydraulic Pipes Induced by Pulsatile Fluid Flow

Michal Stosiak[1(✉)], Mykola Karpenko[2], Paulius Skačkauskas[2], Adam Deptuła[3], and Justyna Krawczyk[1]

[1] Faculty of Mechanical Engineering, Wroclaw University of Science and Technology, Łukasiewicza 7/9, 50-371 Wroclaw, Poland
michal.stosiak@pwr.edu.pl, justyna.krawczyk@pwr.wroc.pl

[2] Faculty of Transport Engineering, Vilnius Gediminas Technical University, Plytinės g. 27, 10105 Vilnius, Lithuania
{mykola.karpenko,paulius.skackauskas}@vilniustech.lt

[3] Faculty of Production Engineering and Logistics, Opole University of Technology, Prószkowska 76, 45-758 Opole, Poland
a.deptula@po.edu.pl

Abstract. The paper points out that micro-hydraulic pipes are induced to flex vibrations by the pulsating flow of the working fluid. Some sources of pressure pulsation in the micro-hydraulic system are indicated. Among the sources of performance pulsations, the performance pulsation generated by a positive displacement micropump is given. Capacity pulsations and impedance of the hydraulic system result in pressure pulsations. The pressure pulsation results in mechanical vibrations of the periodic micro-hydraulic lines. It is noted that under certain conditions, pressure pulsation amplification caused by hydraulic resonance can occur in the micro-hydraulic pipe. The conditions under which such a phenomenon can occur are given, and the resonant lengths of micro-hydraulic pipes are determined for a given pressure pulsation frequency.

Keywords: Micro-hydraulics · Vibrations · Flow pulsation · Tests

1 Introduction

Despite the fact that microfluidic systems are not yet widely used, this field of technology is very popular among researchers and innovative companies that are trying to introduce this type of products to the market. Hydraulic microsystems are systems in which the flowing medium is fluid. The elements that give this fluid the required flow and pressure have dimensions from several hundred nanometers to several centimeters. Micro-hydraulic elements are adapted to small flows (from 2 to 50 cm^3/s) or very small flows (below 2 cm^3/s). Hydrostatic microsystems can be used wherever the use of pneumatic or electromechanical systems is limited due to the requirements regarding the possibility of developing large forces or moments while maintaining precise movement of the executive elements and easy control of this movement. However, micro-hydraulic

O. Prentkovskis et al. (Eds.): TRANSBALTICA 2023, LNITI, pp. 79–90, 2024.
https://doi.org/10.1007/978-3-031-52652-7_8

systems are also required to be silent in operation. Although micro-hydraulic systems are much less noisy than conventional hydraulic systems, the standards regarding permissible noise levels relate their values to the installed power. The power installed in micro-hydraulics is an order of magnitude or two lower than in conventional hydraulics, so the permissible noise levels are lower than for conventional hydraulics. The problem of noise in the operation of micro-hydraulic systems is inextricably linked to the pulsation of pressure in these systems and the mechanical vibrations of micro-hydraulic elements, which may have different sources. Micro-hydraulic pipes can be induced to vibrate by the pulsating flow of the working fluid or by vibrations of the structure on which they are installed. In micro-hydraulic systems, the pulsating flow of the working fluid results mainly from the design of the displacement pump and the work kinematics of its displacement elements (teeth, pistons, etc.) [1, 2].

Figure 1 shows the time course of the instantaneous capacity of a typical displacement pump.

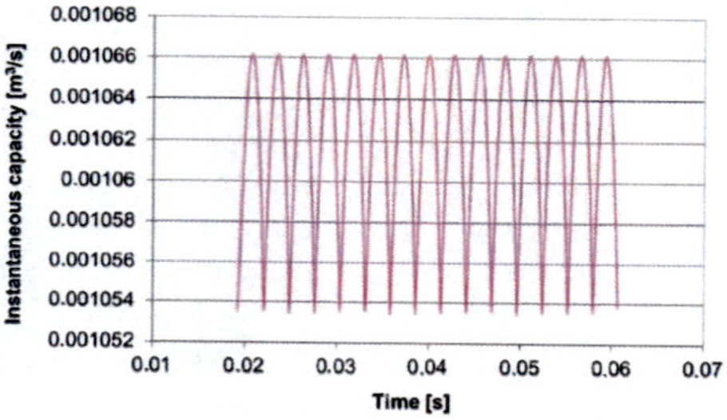

Fig. 1. Theoretical course of instantaneous capacity as a function of time; rotational speed 1450 rpm.

A working machine or device equipped with a micro-hydraulic system is a source of mechanical vibrations with a wide frequency spectrum [3, 4]. These vibrations, transmitted to the bodies of hydraulic valves, may induce vibrations of valve control elements (sliders, poppets) [5]. Figure 2 shows the amplitude-frequency spectrum of the directional control valve slider vibrations, excited by the vibrations of the ground (e.g. the frame of the machine on which the valve is installed).

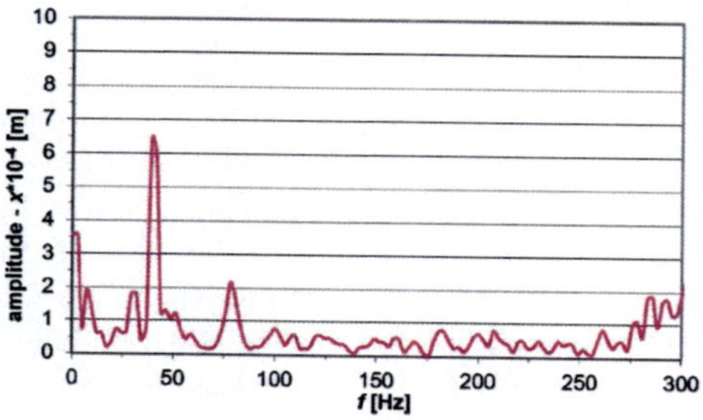

Fig. 2. Amplitude-frequency spectrum of hydraulic directional control valve slider vibrations forced by ground vibrations with frequency $f = 40$ Hz.

The oscillating spool of the manifold causes pressure pulsations with frequencies corresponding to the oscillations of the spool. The resulting pressure pulsation may induce additional vibrations in the hydraulic pipes.

The problem of vibrations of pipes transporting fluids is of interest to researchers around the world. The authors of paper [6] analytically describe non-linear free vibrations of a pipe in a viscous, damping environment. Using non-linear Von-Kármán strain relationships, strain gradient theory and Hamilton's principle, non-linear equations of motion were derived in the presence of a viscous damping medium. The non-linear equations of motion are discretized and transformed into a set of non-linear ordinary differential equations using Galerkin's method. Closed expressions for the non-linear frequency, critical fluid velocity and damped time response of a functionally graded porous fluid-carrying micropipe were obtained for the first time using a homotopy analysis method. The authors showed that the vibration period increases with increasing fluid velocity, the damping factor. The authors of the paper [7] also use a non-linear equation of motion derived using Hamilton's principle to describe the dynamics of the pipe. The partial differential equation is discretized using the Galerkin method. In work [8], the non-linear, size-dependent resonance characteristics of flexible fluid-carrying micropipes subjected to harmonic loading were investigated. The non-linear equation and the corresponding boundary conditions of the system are developed from a modified stress-pair continuum theory combined with Euler-Bernoulli beam theory and von Kármán geometric non-linearity. The authors analyzed the influence of various parameters, i.e. the bending stiffness factor, which represents the size-dependent effect, the slenderness factor and the dimensionless mean flow velocity, on the non-linear, size-dependent forced vibration characteristics of the system. In work [9], a new theoretical model was developed to analyze non-linear vibrations of three-dimensional (3D) cantilever micropipes transporting fluid. Particular attention was paid to the derivation and analysis of reduced equations and the influence of small scale on periodic motions. The material length scale parameter and geometric non-linearities due to large deflection, given by the Lagrang strain tensor, were included in the principal equations.

2 Mathematical Description of Pipe Vibrations Excited by Pulsating Flow

The fluid flowing in the micropipe may cause its bending vibrations when a particular speed value is exceeded, which leads to the loss of pipe stability. The value of the flow velocity at which this occurs is usually called the critical flow velocity. The critical velocity depends on the assumed boundary conditions, i.e. the method of supporting the pipe.

The description of bending vibrations of a hydraulic pipe with flowing fluid assumes treating the pipe as a Bernoulli-Euler beam [10, 12]:

$$EJ\frac{\partial^4 w}{\partial x^4} - \frac{\partial}{\partial x}\left[(S - pA)\frac{\partial w}{\partial x}\right] + b\frac{\partial w}{\partial t} + m_c\frac{D^2 w}{Dt^2} + m_r\frac{\partial^2 w}{\partial t^2} + (m_c + m_r)gcos\alpha = 0, \tag{1}$$

where the value of the equation component (S-pA) depends on the pressure p and the type of constraints limiting the freedom of axial displacement of the pipe.

After further transformations and assuming that the pipe is located horizontally ($\alpha = 0$), the flow is uniform, there is no external tensile force, and there is no pressure other than that resulting from the flow resistance. Also, there is no external resistance of the medium in which the pipe is located ($b = 0$), the equation takes the form:

$$EJ\frac{\partial^4 w}{\partial x^4} + m_c U^2 \frac{\partial^2 w}{\partial x^2} + 2m_c U \frac{\partial^2}{\partial x \partial t} + (m_r + m_c)\frac{\partial^2 w}{\partial t^2} = -(m_r + m_c)g. \quad (2)$$

It should be noted that the pipe deflection function to satisfy Eq. (2) should have two terms: one corresponding to the static deflection, which is the result of the weight of the pipe and fluid, and the other corresponding to the dynamic deflection. This can be written as follows:

$$w(x, t) = w_s(x, t) + w_d(x, t), \quad (3)$$

moreover, it is required that these functions satisfy the equations respectively:

$$EJ\frac{d^4 w_s}{dx^4} + m_c U^2 \frac{d^2 w_s}{dx^2} = -(m_r + m_c)g, \quad (4)$$

$$EJ\frac{\partial^4 w_d}{\partial x^4} + m_c U^2 \frac{\partial^2 w_d}{\partial x^2} + 2m_c U \frac{\partial^2 w_d}{\partial x \partial t} + (m_r + m_c)\frac{\partial^2 w_d}{\partial t^2} = 0. \quad (5)$$

In order to obtain the final solution of the above equations (treating the pipe as a Bernoulli-Euler beam), it is necessary to formulate the boundary conditions - the method of supporting the pipe. Obtaining a more accurate description requires the consideration of wave phenomena in the pipe and fluid during non-stationary processes, and the pipe must be treated as a Timoshenko beam. This leads to partial differential equations of the hyperbolic type, of which the solution methods were given earlier.

Equation (5) can be used to analyze the flexural vibrations of a pipe with flowing fluid supported by an articulated joint. The first part of the Eq. (5) corresponds to the elastic force of the material of the bent pipe, the second part of this equation describes the force needed to change the stream direction and adjust it to the current curvature of the axis, and the Coriolis inertia force corresponds to the third part. This force arises when the stream section rotates with an angular velocity of $\partial\theta/\partial t$, where $\theta = \partial w/\partial x$. The last, fourth part of this equation corresponds to the transverse inertia force of the pipe with the fluid during its bending movement. The solution of Eq. (5) requires specifying the boundary conditions. By solving this equation, it is possible to determine, for the assumed boundary conditions, the basic angular frequency of bending vibrations of a pipe filled with a stationary fluid:

$$\omega_0 = \left(\frac{\pi}{l}\right)^2 \sqrt{\frac{EJ}{m}}. \quad (6)$$

On the other hand, the basic and second angular frequencies of a simply supported pipe filled with flowing fluid can be written as [12]:

$$\left(\frac{\omega_j}{\omega_0}\right)^2 = \beta \pm \left\{\beta^2 - 4\left[1 - \left(\frac{U}{U_d}\right)^2\right] \cdot \left[4 - \left(\frac{U}{U_d}\right)^2\right]\right\}^{1/2}, \quad (7)$$

$j = 1, 2,$

where

$$\beta = 8,5 - \left(\frac{U}{U_d}\right)^2 \left[2,5 + \left(\frac{16}{3\pi}\right)^2 \cdot \frac{m_c}{2m}\right]. \tag{8}$$

The analysis of the formula (7) shows that when the flow velocity increases, the natural frequency of the pipe decreases, and when it reaches the critical value U_d, it decreases to zero. The critical value of the velocity for the pressure flow and the pipe stretched axially by an external force S is determined by the equation [12]:

$$U_d = \left[\frac{l}{m_c} \cdot \left(\frac{\pi^2}{l^2} \cdot EJ - pA + S\right)\right]^{1/2}. \tag{9}$$

3 Micro-hydraulic Pipe Vibration Research

The subject of experimental research was a steel hydraulic micropipe with an internal diameter of 4 mm, an external diameter of 6 mm, a length of 194 cm, and a circular cross-section. The micropipe was supported at the ends. The schema of the hydraulic system with the marked measurement points is shown in Fig. 3. The pressure pulsation in the tested micro-hydraulic system (points p_1 and p_2) and mechanical vibrations of the pipe in two mutually perpendicular directions X and Z (points a_1 and a_2) were measured. In the tested system, a gear micropump with external gearing was used, with a unitary capacity of 0.8 cm^3/rev, driven by an electric motor with a rotational speed of 1,380 rpm. The static value of the pressure in the system was determined by a throttling microvalve.

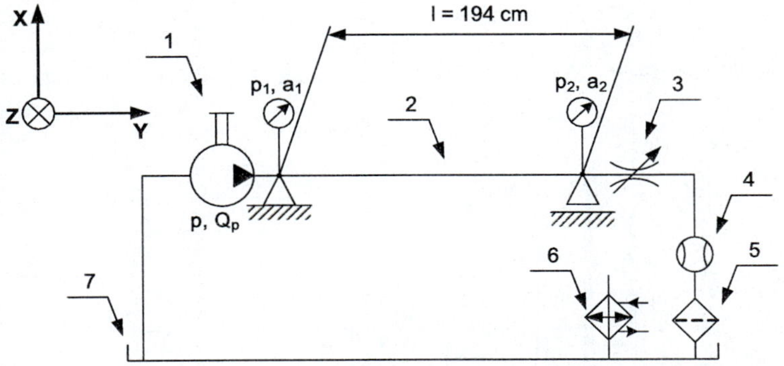

Fig. 3. Schema of the research system: p_1 – piezoelectric sensor for pressure pulsation measurement at the pump, p_2 – piezoelectric sensor for pressure pulsation measurement at the throttling valve, a_1 – 3-axis accelerometer, a_2 – 1-axis accelerometer, a_3 - 3-axis accelerometer, 1 – micropump, q_p = 0.8 cm^3/rev, 2 – hydraulic micropipe, inner diameter 4 mm, 3 – adjustable throttle microvalve, 4 – flowmeter, 5 – hydraulic oil filter, 6 – hydraulic oil cooler, 7 – hydraulic oil tank.

Figure 4 shows the details of the location of the measuring sensors in the system.

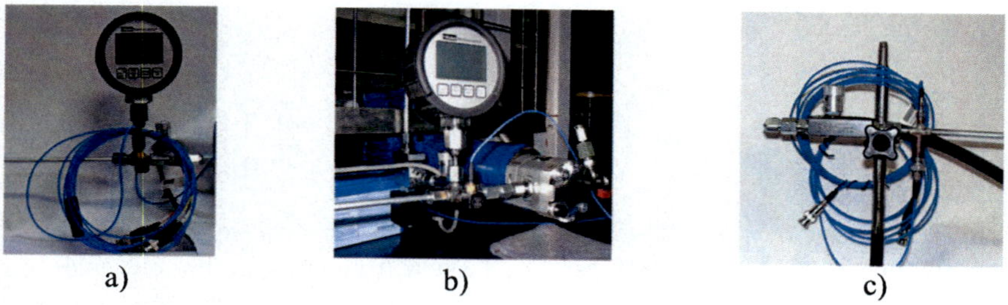

Fig. 4. Measuring point of mechanical vibration acceleration of the hydraulic micropipe and pressure pulsation in the micro-hydraulic system: a), b) at the micropump, c) at the throttle microvalve.

Figures 5, 6 and 7 present the amplitude-frequency spectra of pressure pulsations in the tested system and the mechanical vibrations of the hydraulic micropipe. Horizontal (Z direction) and vertical (X direction) vibrations of the pipe in the direction perpendicular to the pipe axis (according to Fig. 3).

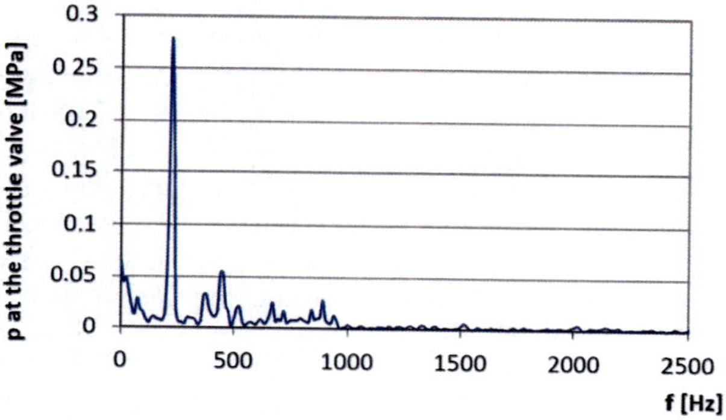

Fig. 5. Pressure pulsation at the throttle valve; $p_{\text{average}} = 5$ MPa, $Q = 1{,}16$ dm^3/min.

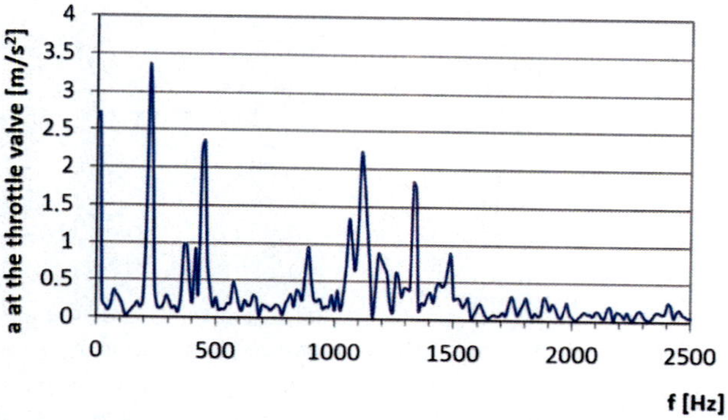

Fig. 6. Horizontal vibrations of the pipe at the throttle valve in the direction perpendicular to the pipe axis; $p_{\text{average}} = 5$ MPa, $Q = 1.16$ dm^3/min. Z direction, perpendicular to the axis of the pipe.

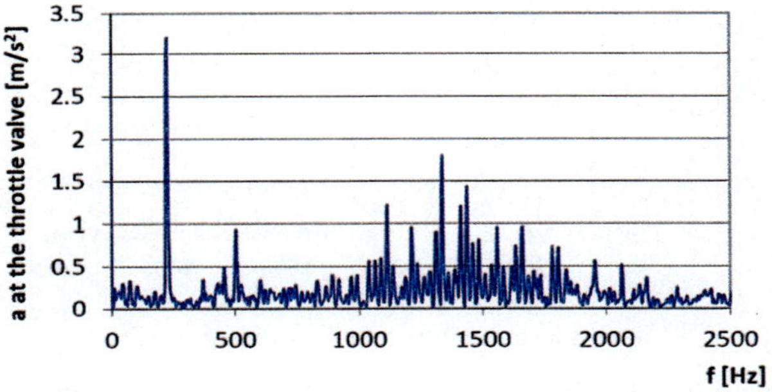

Fig. 7. Vertical vibrations of the pipe at the throttle valve in the direction perpendicular to the pipe axis; $p_{average} = 5$ MPa, $Q = 1.16$ dm^3/min. X direction, perpendicular to the axis of the pipe.

In order to reduce the bending vibrations of the micropipe, its rigid restraint was introduced in the middle of its length, Fig. 8.

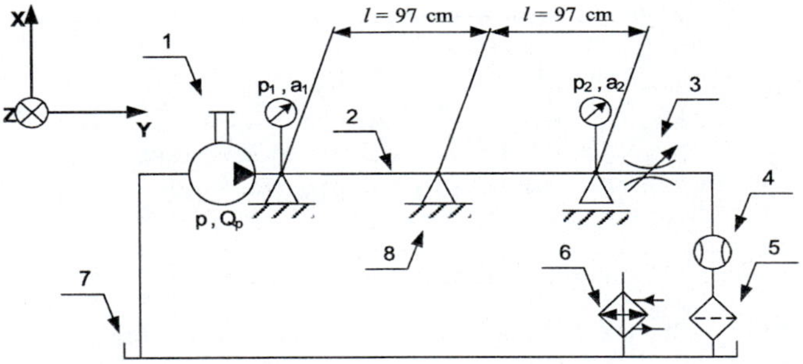

Fig. 8. Schema of the research system: p_1 – piezoelectric sensor for pressure pulsation measurement at the pump, p_2 – piezoelectric sensor for pressure pulsation measurement at the throttling valve, a_1 – 3-axis accelerometer, a_2 – 1-axis accelerometer, a_3 – 3-axis accelerometer, 1 – micropump, $q_p = 0.8$ cm^3/rev, 2 – hydraulic micropipe, inner diameter 4 mm, 3 – throttle hydraulic microvalve, 4 – flowmeter, 5 – hydraulic oil filter, 6 – hydraulic oil cooler, 7 – hydraulic oil tank, 8 – rigid restraint of the micropipe in the middle of its length.

Figure 9 and 10 present the results of the comparison of the pipe vibration acceleration at point a_2 (according to Fig. 8) in the vertical (X) and horizontal (Z) directions. This comparison includes the case in which the hydraulic micropipe was supported at its ends (at the pump and the throttle valve) and the case in which the micropipe was additionally restrained in the middle of its length - as shown in Fig. 8 point 8.

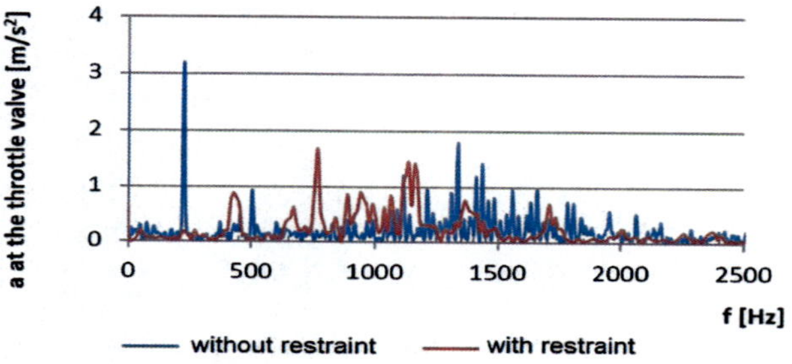

Fig. 9. Comparison of the spectrum of acceleration of vertical vibrations at the end of the micropipe for the case of its restraint in the middle of its length and the absence of this restraint; $p_{average} = 5$ MPa, $Q = 1.16$ dm^3/min.

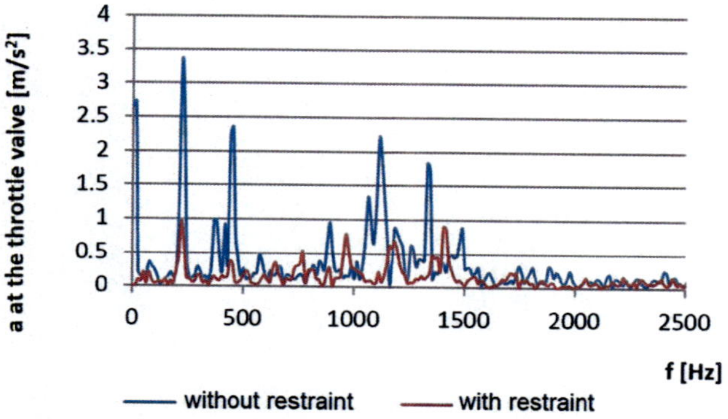

Fig. 10. Comparison of the spectrum of acceleration of horizontal vibrations at the end of the micropipe for the case of its restraint in the middle of its length and the absence of this restraint; $p_{\text{average}} = 5$ MPa, $Q = 1.16$ dm^3/min.

The analysis of the presented results of the excitation of bending vibrations of micropipes indicates that there is a coincidence between the components of the pressure pulsation spectrum (excitation) and the components of mechanical vibration acceleration (system response), which manifests itself in the formation of mechanical vibrations with frequencies corresponding to the frequency of pressure pulsations in the considered system.

For the unit used during the tests (micro-gear pump with external gearing), the basic component of the pulsation spectrum is 230 Hz, for the number of teeth z = 10 and the rotational speed on its shaft n = 1,380 rpm. In addition, it is observed in Fig. 9 and Fig. 10 that the introduction of a micropipe restraint in the middle of its length (Fig. 8) results in a significant reduction in the amplitude of bending vibration accelerations resulting from the pressure pulsation of the hydraulic micropump (in both tested directions).

4 Hydraulic Resonance in a Hydraulic Long Line

In hydraulic systems, under certain conditions, due to wave phenomena, we can deal with a hydraulic long line (HLL). We deal with HLL when the length of the pipe is equal to or greater than the length of the pressure wave propagated in it. It is even recommended to treat the hydraulic pipe as a long line already when the condition.

$\lambda_f \leq \frac{c_0}{10 f_{max}}$ is met (λ_f – wavelength, c_0 – speed of pressure wave propagation in the pipe, f_{max} – maximum excitation frequency) [11]. Then, such a pipe is treated as an element of a system with distributed parameters, so it should be taken into account that changes in pressure and flow rate propagate along the axes of pipes with a finite speed in the form of traveling and reflected waves. The system with a long hydraulic line can be presented in the illustrative way shown in Fig. 11. The HLL is connected to the P flow generator (displacement pump) using the V1 element and to the O receiver (e.g. motor, hydraulic cylinder, throttle valve, Etc.) using the V2 element.

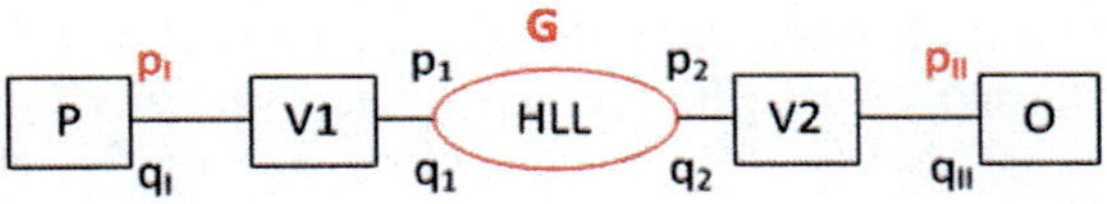

Fig. 11. Schematic representation of the hydraulic system with a hydraulic long line.

The hydraulic system presented in this way can also be presented in the form of a complex one-port, as shown in Fig. 12.

Fig. 12. Complex one-port system with a hydraulic long line.

The elements V1, HLL and V2 are treated as two-ports described by the transmittances V1, HLL and V2, while the receiver O represents a one-port element of the system. The above representation of the hydraulic system in the form of a complex one-port, as in Fig. 12, can be found in work [11]. The matrix transfer function of the entire system has the following form [11, 13]:

$$\begin{bmatrix} p_I \\ q_I \end{bmatrix} = \boldsymbol{\Pi} \cdot \begin{bmatrix} p_{II} \\ q_{II} \end{bmatrix}, \tag{10}$$

where $\boldsymbol{\Pi} = V1 \cdot HLL \cdot \mathrm{V2}$.

Individual terms of the transfer function matrix Π of the hydraulic system are marked as $\pi_{11}, \pi_{12}, \pi_{21}, \pi_{22}$.

Knowing the load impedance, defined as [11, 13]:

$$Z_k = \frac{p_{II}}{q_{II}}, \tag{11}$$

the transmittance of the system with a long hydraulic line is obtained, indicating the amplification of the value of the pressure pulsation amplitude at the end of the pipe to the value of the pressure pulsation amplitude at the beginning of the pipe [11, 13]:

$$G_{p_{II}p_I} = \frac{p_{II}}{p_I} = \frac{Z_k}{\pi_{11} \cdot Z_k + \pi_{12}}. \tag{12}$$

When parameterizing the hydraulic system, it is possible to graphically present the transmittance of the pipe as a function of the excitation frequency (pressure pulsations) and its length. This form is convenient because it allows you to determine the resonant lengths of the pipe for a known pressure pulsation value. It was assumed that the pipe is powered by a pump with capacity $Q_0 = 2.33 * 10^{-4}$ [m³/s], pressure at the end of the pipe $p_0 = 5$ MPa, radius of the pipe R = 0.006 m, fluid density $\rho = 860$ km/m³, fluid viscosity $\nu = 50 * 10^{-6}$ [m²/s], wave propagation speed $v = 1{,}288$ m/s. The pipe ends with a throttle valve, so the final impedance can be expressed as $Z_k = \frac{2 \cdot \Delta p_0}{Q_0}$, where Δp_0 is the pressure difference across the valve. It was adopted $Z_k = 4.8 * 10^{10}$ [Ns/m⁵].

Figure 13a shows the transmittance of the pipe as a function of its length and excitation frequency. On the other hand, Fig. 13b shows the transmission function of the pipe as a function of its length for the given excitation frequency f = 230 Hz.

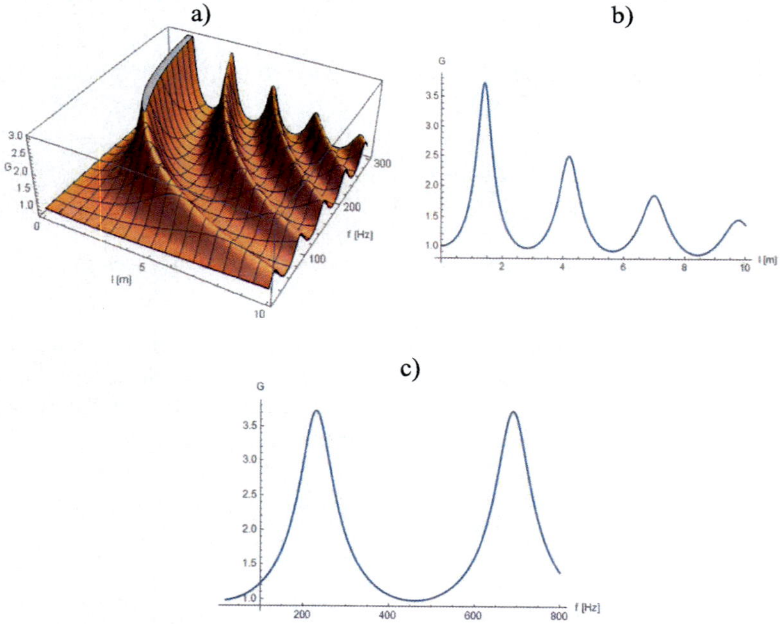

Fig. 13. Pipe transmittance $G_{p2,p1}$ as a function of: a) its length and frequency of excitation, b) of its length for the excitation frequency 230 Hz, c) excitation frequency for the pipe length 1.4 m.

As can be seen from Fig. 13, there are hydraulic resonances in the hydraulic pipe. For a given excitation frequency, there is such a length of the hydraulic pipe that will result in the amplification of the pressure pulsation amplitude. And vice versa: for a given length of the pipe, there is such a frequency of excitation that will lead to an increase in the amplitude of pressure pulsations.

If we assume that the wavelength λ_f is related to its propagation speed c_0 and frequency in the equation:

$$\lambda_f = \frac{2\pi c_0}{\omega}. \tag{13}$$

The condition of the first hydraulic resonance of the pipe can be specified [11]:

$$\frac{L}{\lambda_f} = \frac{1}{4}. \tag{14}$$

This is represented by the maximum value of the pressure pulsation amplitude at the receiver in relation to the pulsation at the beginning of the pipe. For the propagating frequency of the wave in the pipe equal to 230 Hz and its propagation speed in the steel pipe equal to 1,288 m/s, the wavelength will be:

$$\lambda_f = 5.6\,\text{m}. \tag{15}$$

And the resonant length of the pipe L_R (Fig. 13b):

$$L_R = \frac{\lambda_f}{4} = 1.4\,\text{m}. \tag{16}$$

This means that for a pressure pulsation frequency of 230 Hz and a pipe length of 1.4 m at the end of the pipe, the amplitude of the pressure pulsation will amplify in relation to the amplitude of the pressure pulsation at the beginning of the pipe.

5 Conclusions

The paper indicates the problem of the excitation of micro-hydraulic pipe vibrations. The pressure pulsation caused by the pulsating operation of the displacement pump was considered as the source of excitation. Experimental research shows that there is a coincidence between pipe vibration and pressure pulsation. As it has been shown, the introduction of the pipe restraint in its half-length effectively reduces its vibrations. An additional source of pressure pulsation may be the vibrating control element of the hydraulic valve. The paper shows that under certain conditions the amplitudes of pressure pulsations in the hydraulic pipe may be strengthened, and thus the excitation acting on the pipe may be strengthened. A condition has been given that must be met in order for wave phenomena to occur in a hydraulic pipe - overlapping of the traveling and reflected waves and amplification of the pressure pulsation amplitude. The designer of the hydraulic system, bearing in mind the reduction of pressure pulsations and knowing the frequency and speed of the wave propagating in the pipe, should design the length of the hydraulic pipes in such a way that there is no amplification of the pressure pulsation amplitudes. On the other hand, knowing the length of the hydraulic pipe and the wave propagation speed, it is possible to determine the resonant frequencies for which the pressure pulsation amplitude will increase in the pipe.

References

1. Xu, W., et al.: An advanced pressure pulsation model for external gear pump. Mech. Syst. Signal Process. **187**, 109943 (2023). https://doi.org/10.1016/j.ymssp.2022.109943. ISSN 0888-3270
2. Huang, S., Song, Y., Yin, J., Xu, R., Wang, D.: Experimental and numerical investigation on the pressure pulsation in reactor coolant pumps under different inflow conditions. Nucl. Eng. Technol. **55**(4), 1310–1323 (2023). https://doi.org/10.1016/j.net.2022.12.023. ISSN 1738-5733
3. Skliros, C.: A case study of vibration fault diagnosis applied at Rolls-Royce T-56 turboprop engine. Aviation **23**(3), 78–82 (2019). https://doi.org/10.3846/aviation.2019.11900
4. Miao, H., Wang, C., Li, C., Song, W., Zhang, X., Xu, M.: Nonlinear dynamic modeling and vibration analysis of whole machine tool. Int. J. Mech. Sci. **245**, 108122 (2023). https://doi.org/10.1016/j.ijmecsci.2023.108122. ISSN 0020-7403
5. Stosiak, M., et al.: Research of vibration effects on a hydraulic valve in the pressure pulsation spectrum analysis. J. Mar. Sci. Eng. **11**, 301 (2023). https://doi.org/10.3390/jmse11020301
6. Sabahi, M.A., Saidi, A.R., Khodabakhsh, R.: An analytical solution for nonlinear vibration of functionally graded porous micropipes conveying fluid in damping medium. Ocean Eng. **245**, 110482 (2022). https://doi.org/10.1016/j.oceaneng.2021.110482. ISSN 0029-8018
7. Hu, K., Wang, Y.K., Dai, H.L., Wang, L., Qian, Q.: Nonlinear and chaotic vibrations of cantilevered micropipes conveying fluid based on modified couple stress theory. Int. J. Eng. Sci. **105**, 93–107 (2016). https://doi.org/10.1016/j.ijengsci.2016.04.014. ISSN 0020-7225
8. Dehrouyeh-Semnani, A.M., Nikkhah-Bahrami, M., Yazdi, M.R.H.: On nonlinear vibrations of micropipes conveying fluid. Int. J. Eng. Sci. **117**, 20–33 (2017). https://doi.org/10.1016/j.ijengsci.2017.02.006. ISSN 0020-7225
9. Guo, Y., Xie, J., Wang, L.: Three-dimensional vibration of cantilevered fluid-conveying micropipes—types of periodic motions and small-scale effect. Int. J. Non-Linear Mech. **102**, 112–135 (2018). https://doi.org/10.1016/j.ijnonlinmec.2018.04.001. ISSN 0020-7462
10. Maciel, V.S.F., Kheiri, M., Franzini, G.R.: Passive suppression of flow-induced vibrations of a cantilevered pipe discharging fluid using non-linear vibration absorbers. Int. J. Non-Linear Mech. **144**, 104053 (2022). https://doi.org/10.1016/j.ijnonlinmec.2022.104053. ISSN 0020-7462
11. Kudźma, Z.: Tłumienie pulsacji ciśnienia i hałasu w układach hydraulicznych w stanach przejściowych i ustalonych. Oficyna Wydawnicza Politechniki Wrocławskiej. Wrocław (2012)
12. Gryboś, R.: Drgania konstrukcji wzbudzone przepływem. Wydawnictwo Politechniki Śląskiej, Gliwice (2005). ISBN 83-7335-188-4
13. Krus, P.: Dynamic models for transmission lines and hoses. In: Fleury, A.D.T., Rade, D.A., Kurka, P.R.G. (eds.) DINAME 2017. LNME, pp. 531–547. Springer, Cham (2019). https://doi.org/10.1007/978-3-319-91217-2_38

Experimental Study of the Multi-disc Negative Brake for a Hydraulic Motor

Pawel Sliwinski(✉) and Ryszard Jasinski

Gdansk University of Technology, Gdansk, Poland
{pawel.sliwinski,ryszard.jasinski}@pg.edu.pl

Abstract. This paper describes the methodology for experimental testing of a multi-disc brake. The construction of this brake was also present. The brake is dedicated to hydraulic motors with a small working volume. Experimental tests were carried out on a brake with plates immersed in oil and, for comparison, tests were carried out on a dry brake. As a result of the tests, the permissible torque (load) that is able to transfer the brake and the pressure required to release the brake were determined. The temperature rise in the chamber of the uncoupled brake (both wet and dry brake), whose shaft with active discs was rotated at speed of 1,500 rpm, was also examined. The influence of the deformation of the brake discs on the characteristics of the torque of friction in these discs was also observed.

Keywords: Negative brake · Multi-disc brake · Disc · Hydraulic motor · Oil

1 Introduction

In hydraulic drive systems of machines and devices hydraulic motor is an executive element. In some drive systems, due to safety requirements, it is necessary to use an additional hydraulic motor shaft lock. An example of such a drive system is the winch drive or the drive of the transport and assembly system [1]. In such systems, the shutdown or loss of hydraulic power must cause the brake to engage and lock the motor shaft. Thus, in a mechanical and rigid manner, uncontrolled descent of the load is secured. The hydraulic motor only starts when the brake is released. Having a hydraulic system available, it is desirable to use a hydraulically released brake in conjunction with a hydraulic motor. Then the pressure signal to release the brake is taken from the high-pressure line supplying the hydraulic motor. An example of such a motor-brake assembly can be found e.g. in [1]. The working assembly of such a lock is a multi-disc brake, the idea of which is similar to the principle of operation of a multi-disc clutch. The disadvantage of such clutches, and thus multi-disc brakes, may be deformation of friction discs and pressure plates. This leads to an uneven distribution of pressures on the contact surfaces [2, 3], a decrease in the value of the transmitted torque and to uneven wear of the friction discs [4–6]. In a multidisc clutch the distributions of the contact pressure on the surfaces near the pressure plate are more equable than those the distributions of the contact pressure on the surfaces near the opposite plate [2]. Furthermore, in the multidisc clutch is the contact pressure concentration on small regions of the nominal frictional

O. Prentkovskis et al. (Eds.): TRANSBALTICA 2023, LNITI, pp. 91–102, 2024.
https://doi.org/10.1007/978-3-031-52652-7_9

interface [3, 4, 7]. The deformations of the friction plates and thus the distribution of pressures on these plates are influenced by thermal phenomena occurring in the working clutch or brake [8–10].

It should be added that the brake discs are immersed in the working fluid, i.e. in hydraulic oil. Oil significantly reduces the coefficient of friction and thus there is a risk of brake slipping, which is unacceptable. Therefore, it is advisable to conduct research aimed at gaining knowledge about the phenomena occurring in such blockades.

If the brake hydraulic motor is operated for short periods (e.g. a few tens of seconds max.), the hydraulic system is simplified and the brake chamber is not supplied with working fluid. The brake then runs dry. Therefore, it is advisable to examine the operating parameters of the dry brake. It is especially advisable to check the temperature increase in the brake chamber and thus determine the permissible working time of this brake.

For the needs of drive systems with satellite motors with a very small working volume (which can be loaded with a torque of up to 65 Nm), a prototype of a negative-acting hydraulic multi-disc brake was developed from scratch in the hydraulics laboratory of the Gdansk University of Technology. Negative action means that in the normal state (the system is not working) the brake is locked. Only when the pressure signal is applied, the brake is released. The brake pads are immersed in mineral oil. So it's a wet brake. A general view of the hydraulic motor coupled with the brake is shown in Fig. 1

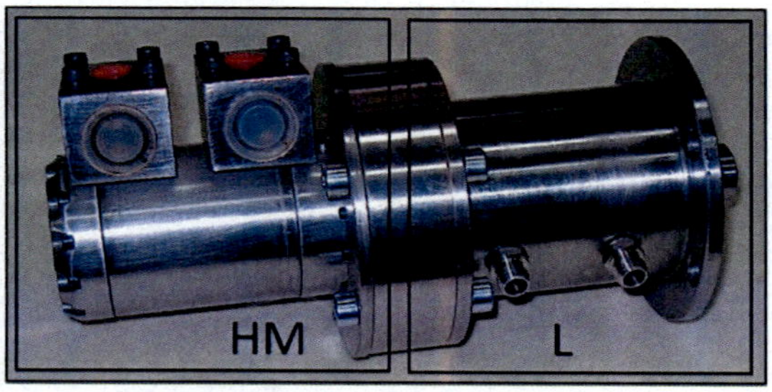

Fig. 1. General view of the hydraulic motor (HL) with the negative brake (L).

When designing the brake, there were problems with the proper selection of the friction coefficient and the number of brake discs. The friction discs were imposed from above by the target manufacturer of this brake. Therefore, it was not possible to interfere with the geometric dimensions of the tiles. The value of the coefficient of friction for a steel-steel friction pair immersed in oil ranges from 0.08 to 0.17, and the permissible pressures for such a pair range from 0.8 to 1.4 MPa [4, 5, 7]. Moreover, it is recommended that the total number of friction discs in clutches and brakes be less than 30 [5, 6]. This limitation results from the fact that the moment of friction in the clutch or brake is not directly proportional to the number of discs, because there is friction in the splines of the discs [4–7, 11]. Therefore, in the calculation of multi-disc clutches and brakes, the coefficient k is introduced to reduce the frictional moment. The value of this coefficient is $0.76 \div 1$, with the value 1 for two discs and 0.76 for ten discs [4–6, 12].

The main aim of the research was to determine the allowable torque (load) that the brake can transfer and the pressure required to release the brake. Due to the fact that the hydraulic motor can work for a long time at high rotational speeds (even 1,500 rpm), therefore with the brake disconnected, it is advisable to check the temperature in the brake chamber. The tests were carried out when the space of the discs chamber was filled with Total Azolla 46 mineral oil ("wet" brake) and, for comparison, tests were carried out on the brake without lubricating liquid ("dry" brake).

2 The Object of Research – Multi-disc Negative Brake

Figure 2 shows the construction of the multi-disc brake prototype intended for testing. The moment (torque) is transferred by the set of friction discs 1 and 2. The value of this torque should be at least 65 Nm.

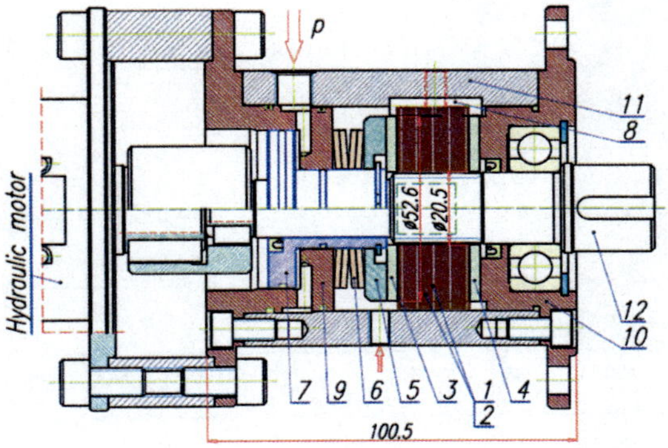

Fig. 2. Structure of the multi-disc negative brake: 1,2 – friction discs (active and passive); 3,4 – washer; 5 – clamp; 6 – disc springs; 7 – piston, 8 – pin; 9,10,11 – body elements, 12 – shaft.

The friction discs are made of NIMAX tool steel [13]. The design of these discs is shown in Fig. 3. The roughness of the discs surfaces is Ra = 0.10 ÷ 0.26.

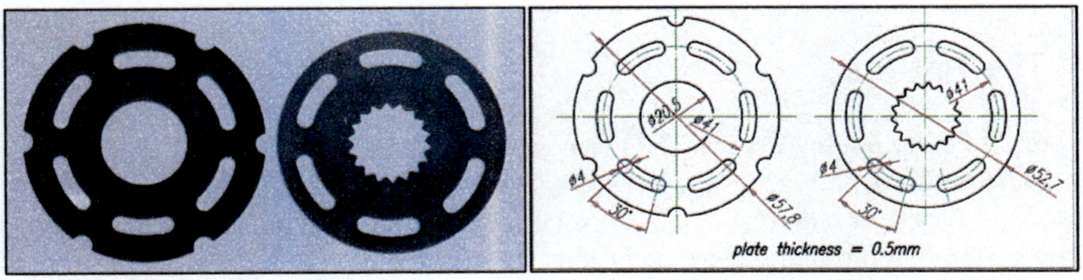

Fig. 3. Friction discs and their dimensions: passive disc (left) and active disc (right).

To calculate the amount of friction surface, the following was adopted:

- torque transmitted by the brake M = 65 Nm;
- the contact area of active and passive disc A = 1,513 mm^2 (according to 0);

- allowable surface pressure for the friction disc material $p_{all} = 0.96$ MPa [6];
- coefficient of friction $\mu = 0.08$ [6].

For the data adopted in this way, it was calculated that the number of friction surfaces should be at least 38 and the clamping force of the disc package should be F = 1,452 N. Thus, 19 active and 20 passive discs are used in the brake.

The set of discs is compressed by a package of four disc springs arranged in an 'X" (Fig. 2). The brake uses four HENNLICH disc springs with dimensions 45x22.4x1.75 (in mm), made of steel 1.1231 (DIN 17222) [14]. The characteristics of the spring package are shown in Fig. 4.

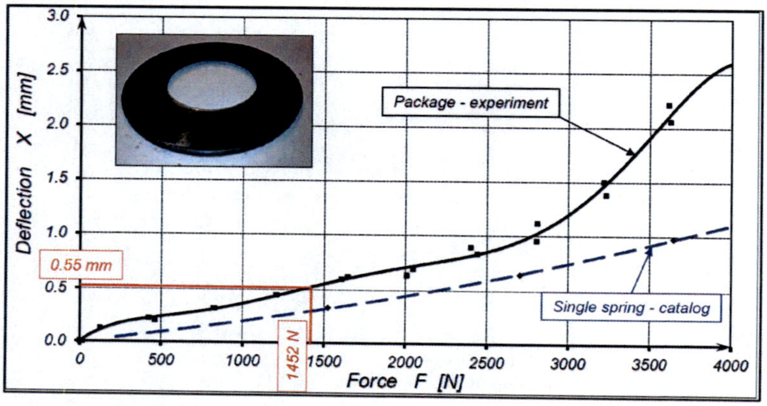

Fig. 4. Experimentally determined characteristics of a package of four disc springs arranged in X (as in 0) and characteristics of a single spring according to catalog data.

The brake is released when the piston 7 moves to the left (Fig. 2). Then the spring pack 6 is compressed and the friction discs 1 and 2 are loosened. The piston is moved after a certain pressure value is applied to the cylinder chamber through the P channel (Fig. 2).

3 The Test Rig

In order to carry out the experimental tests, the test stand was configured according to the scheme shown in Fig. 5. The tested brake 2 was coupled with the shaft of worm gear 12 (with a gear ratio of 1:120). The rotation of the brake shaft is thus forced by the electric motor 11 through this transmission.

During the test of the brake the following parameters were registered:

- pressure p in the port P of the brake (Fig. 2) – strain gauge pressure transducer, range $0 \div 10$ MPa and class 0.3;
- torque M – strain gauge force transducer FT mounted on the arm 0.5 m (the arm attached to the brake body), range $0 \div 500$ N, class 0.1;

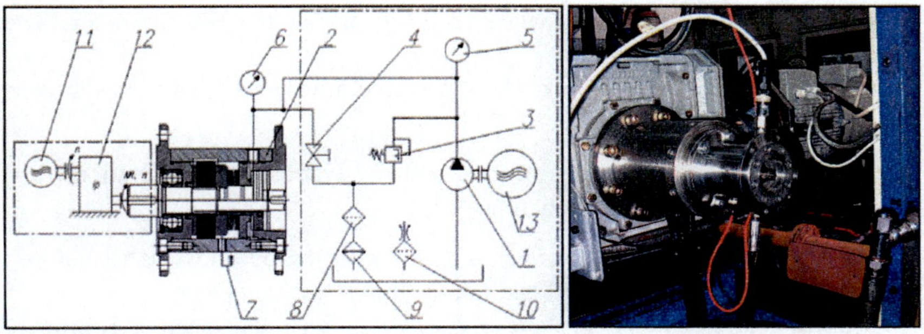

Fig. 5. Scheme of the measuring system (left) and general view of the brake on the test stand: 1 – feeding pump, 2 – tested brake, 3 – overflow valve, 4 – ball valve, 5 – pressure gauge (in the pump discharge port), 6 – pressure sensor (in the brake piston chamber), 7 – temperature sensor, 8 – drain filter, 9 – cooler, 10 – filler and breathing filter, 11 – electric motor with frequency converter, 12 – worm gear, 13 – electric motor.

- rotational speed of the brake shaft – inductive sensor, accuracy of measurement ±0.01 rpm;
- temperature T of liquid in the brake chamber – RTD temperature sensor, class A, max. measurement error 0.5 °C.

4 Method of Carrying Out the Research

Due to the fact that the brake can be used in a very simple system, where the brake chamber will not be flushed with the working liquid, the working parameters of this brake should be examined for the case of dry brake discs. Therefore tests of a dry brake (with dry discs) and a wet brake (with discs immersed in mineral oil) were carried out. Each time a set of new discs was used. So after dry testing, the discs were no longer used for oil testing. Before each measurement, the brake was disassembled, thoroughly washed and dried.

For tests of wet brake, each disc was rinsed with oil during assembly to ensure proper oil penetration between these discs while the brake was in operation.

Brake tests were carried out in the following stages:

1) stage 1: measurement of the brake release pressure. Sequence of actions:
   - the pressure in the brake chamber was set at about 60 bar;
   - the worm gear drive was started and the rotational speed of the brake shaft was set to 1 rpm;
   - the recording of pressure and torque measurements was started and the pressure was reduced to zero with valve 3 (Fig. 5);
2) stage 2: measurement of the torque during starting and stopping the drive with the brake locked. Sequence of actions:
   - locked brake, i.e. pressure in the brake piston chamber equal to zero;
   - the recording of torque measurement results was started;

  - the rotation of the brake shaft was forced (by starting the electric motor 11) at a constant speed of 1 rpm;
  - after one full rotation of the brake shaft, the drive and recording of the measurement results were turned off;

3) stage 3: measurement of the torque during brake activation and deactivation with the gear drive engaged. Sequence of actions:
  - the pressure in the brake chamber was set with the overflow valve before the measurement at the level of 40 bar, i.e. the brake was unlocked;
  - the worm gear, i.e. the brake shaft drive, was started – at a speed of 1 rpm;
  - the brake was blocked (by releasing the pressure as a result of opening the ball valve 4 - pump 1 connected directly to the tank);
  - after time Δt, the brake was unlocked by increasing the pressure in the system as a result of closing the ball valve;

4) stage 4: measurement of the brake torque and temperature with the brake released and the rotational speed of 1,500 rpm. This measurement was aimed at:
  - determination of the temperature increase in the body during brake operation in order to determine whether it is necessary to cool the brake discs. The measurement was ended when the temperature exceeded 80 °C;
  - determination of the moment of friction, on the basis of which it is possible to determine the power lost in the brake during operation, resulting from the viscosity of the factor filling the space of the brake discs.

The parameters recorded during the tests were:

- brake friction torque (braking torque) – measured with a force sensor placed on a 0.5 m long arm – torque measurement on the brake body and not on the shaft (Fig. 5);
- temperature of the working medium in the brake body,
- pressure in the brake piston chamber,
- rotational speed (and angle of rotation) of the brake shaft.

5 Test Results

5.1 Stage 1 - Brake Release Pressure

The characteristics of the pressure in the piston chamber and the corresponding torque in the brake are shown in Fig. 6. These characteristics show that the pressure required to release the brake is at least 30 bar.

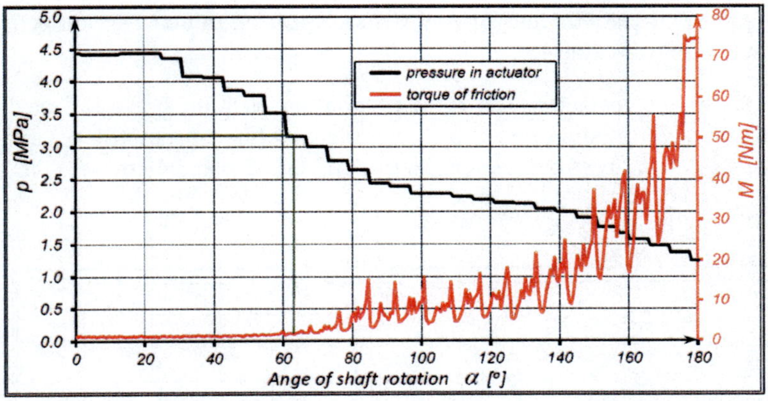

Fig. 6. Characteristics of friction torque and pressure in the brake actuator chamber during brake application with the shaft rotating.

5.2 Stage 2 - Friction Torque During Starting and Stopping the Drive with the Brake Locked

In the clamped brake, after setting the shaft in slow rotation (n = 1 rpm), it was observed that the friction torque in the brake discs does not assume a constant value (Fig. 7). The torque characteristic takes the shape of a sine wave reaching the minimum and maximum of its value.

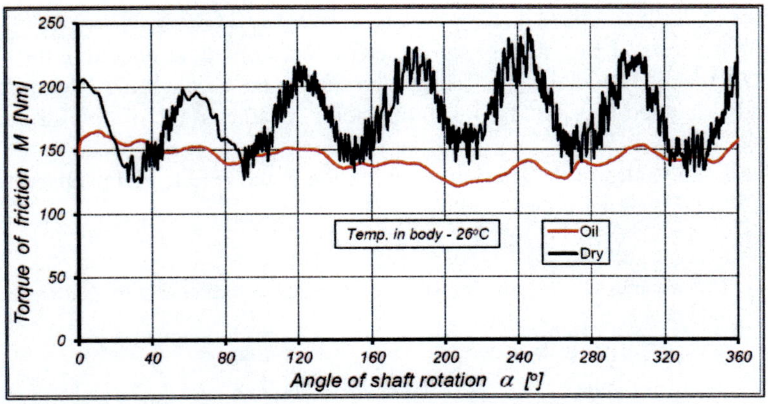

Fig. 7. Characteristics of the friction moment in the clamped brake as a function of the angle of rotation of the brake shaft. Dry brake and wet brake (oil-wetted discs and oil-filled chamber).

5.3 Stage 3 - Friction Torque During Brake Application and Disconnection with the Drive Running

Figure 8 shows the characteristics of the friction torque in the brake activated during shaft rotation (shaft rotation forced by a worm gear with a speed of n = 1 rpm). The characteristics of the friction moment have a similar, sinusoidal course as in stage 3 (Fig. 7).

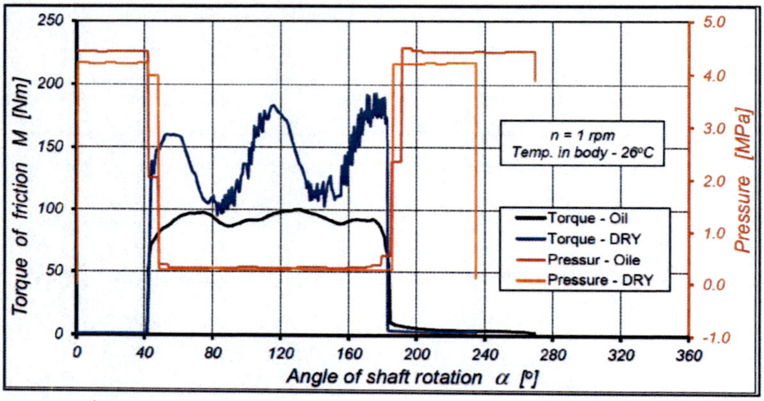

Fig. 8. Characteristics of the friction torque as a function of the angle of rotation of the brake shaft. The brake is activated during the rotation of the shaft at a constant speed n = 1 rpm. Dry brake and wet brake (oil-wetted discs and oil-filled chamber).

During the tests, it was observed that the friction moment after setting the shaft of the clamped brake in rotation (Fig. 7) is greater than in the case when the rotating brake was already clamped (Fig. 8). This phenomenon is characteristic of both dry and wet brakes.

The extreme values of the friction torque in the brake read from the characteristics as in Fig. 8 and Fig. 9 are presented in Table 1.

Table 1. The minimum and maximum values of the friction torque in a dry and wet brake.

	Stage 2		Stage 3	
	min	max	min	max
Dry	135	245	98	190
Wet (OIL)	127	158	87	100

The summary presented in Table 1 shows that the dry brake is able to transfer the greatest load. The smallest load is transferred by the brake with discs immersed in oil.

5.4 Stage 4 - Heating Time of the Released Brake and Power Losses

Figure 9 shows the temperature characteristics of fluids in the released brake body rotating at 1,500 rpm.

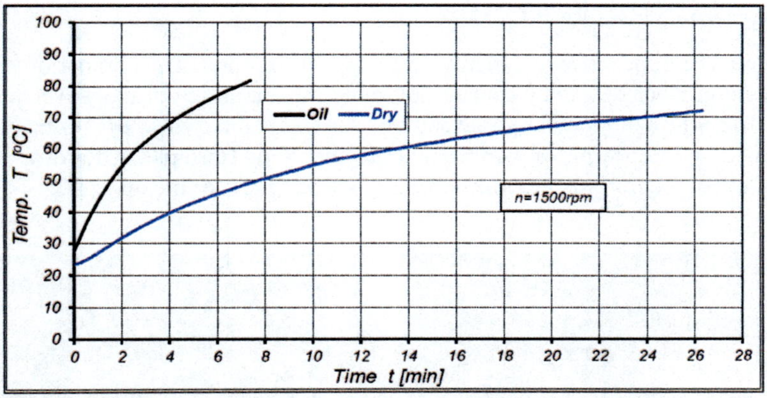

Fig. 9. Temperature characteristics in the brake body. Brakes released. Brake shaft rotational speed n = 1,500 rpm.

Figure 10 show the characteristics of the friction torque in a released brake rotating at a speed of n = 1,500 rpm and whose discs were washed with various fluids. The torque value decreases rapidly over time when the discs are washed with oils. This is related to the decrease in viscosity caused by the increase in temperature.

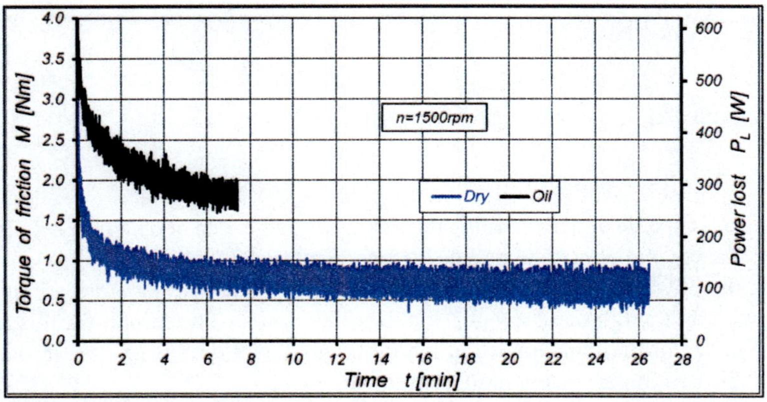

Fig. 10. Characteristics of friction torque in the wet and dry brake. Brake released. Brake shaft rotational speed n = 1,500 rpm.

5.5 Discs After Tests

After the tests, the brake was disassembled. The roughness of the surfaces of the discs was measured and is:

- for discs working in oil Ra = 0.10 ÷ 0.28;
- for dry-running discs Ra = 0.08 ÷ 0.30.

Waviness of both active and passive discs was also found. The number of waves is equal to the number of kidney-shaped holes and the amplitude of the wave was max. 0.2 mm. To confirm the waviness, one pair of discs was painted with red ink and installed in the brake. A short-term test was carried out according to step 3. After disassembly, uneven wear of the ink was found, confirming the waviness of the discs (Fig. 11).

Fig. 11. The condition of the active disc after a dry test (left) and after a test with a pair of discs coloured with red ink (right).

6 Conclusions

The tests of the brake have proven that:

a) the brake is capable of withstanding the required engine load of 65 Nm. The minimum load value beyond which the brake slip may occur is 87 Nm;
b) during the brake tests, the friction discs were deformed. Wavy deformations of the discs with an amplitude of about 0.2 mm were formed. The reason for this may be the influence of the temperature in the rubbing contact on the stresses in the discs, which resulted in permanent deformation;
c) the friction torque in the brake does not maintain a constant value as a function of the shaft rotation angle (0 and 0), which is the effect of the waviness of the discs. The maximum value of the friction torque is even 125% of the minimum value for oil and 194% for a dry brake (0);

d) there are lower values of the friction torque in the brake for the case when the brake is locked while the shaft is rotating (stage 3). This can be explained by the fact that there is a very good distribution of oil in the space between the brake discs. And in the case of a dry brake, the surfaces of the discs may have worn in slightly;
e) the brake can work both dry and wet;
f) in the oil-filled brake chamber the temperature reaches 80 °C after 7 min of brake operation at the brake shaft rotational speed of 1,500 rpm! On the other hand, the dry brake chamber heats up to 75 °C only after 25 min. Such a rapid increase in temperature in the chamber filled with oil is caused by energy loss in the oil moving between the brake plates (influence of the oil viscosity);

Therefore, the brake should not work if its chamber is only filled with oil. The brake chamber should be flushed with a continuous stream of oil. This will cool the brake plates.

Analysing the test results, it is concluded that the corrugation of the discs is beneficial from the point of view of brake operation. Overloading the system, and therefore overloading the brake, will cause the brake to rotate by a small angle, or an angle for which there is a state of equilibrium between the friction torque in the brake and the external load. Thus, the wet brake can be overloaded even up to 100 Nm (0).

References

1. SM-Hydro homepage. https://smhydro.com.pl/silniki-hydrauliczne/. Accessed 30 June 2023
2. Yang, L., Ma, B., Ahmadian, M., Li, H., Vick, B.: Pressure distribution of a multidisc clutch suffering frictionally induced thermal load. Tribol. Trans. **59**(6), 983–992 (2016). https://doi.org/10.1080/10402004.2015.1126877
3. Abdullah, O., Rasham, A., Jobair, H.: The influence of frictional facing thickness on the contact pressure distribution of multi-disc dry clutches. FME Trans. **46**, 33–38 (2018). https://doi.org/10.5937/fmet1801033A
4. Bąk, M.: Wpływ parametrów konstrukcyjnych sprzęgieł mokrych i ich sterowania na zmiany wybranych wielkości fizycznych w trakcie zmiany przełożenia przekładni mechanicznej (Influence of design parameters and control of wet clutches on selected physical quantities' changes during gear change of a gearbox). Ph.D. thesis, Gdansk University of Technology, Gdansk (2022)
5. Osiński, Z.: Sprzęgła i Hamulce (Clutches and Brakes), PWN (2000)
6. Dietrich, M.: Podstawy konstrukcji maszyn (eng. Fundamentals of Machine Design), vol. 2. PWN Publishing House, Warszawa (2022)
7. Bąk, M., Patrosz, P., Sliwinski, P.: Torque transmitted by multi-plate wet clutches in relations to number of friction plates and their dimensions. In: Stryczek, J., Warzynska, U. (eds.) NSHP 2020. LNME, pp. 367–376. Springer, Cham (2021). https://doi.org/10.1007/978-3-030-59509-8_33
8. Zagrodzki, P.: Numerical analysis of temperature fields and thermal stresses in the friction discs of a multidisc wet clutch. Wear **101**, 255–271 (1985). https://doi.org/10.1016/0043-1648(85)90080-8
9. Holgerson, M.: Wet clutch engagement characteristics. Ph.D. thesis, Lulea University of Technology (1999)
10. Davis, C.L., Sadeghi, F., Krousgrill, C.M.: A simplified approach to modeling thermal effects in wet clutch engagement: analytical and experimental comparison. J. Tribol. **122**, 110–118 (2000)

11. Bąk, M.: Torque capacity of multidisc wet clutch with reference to friction occurrence on its spline connections. Sci. Rep. **11**, 21305 (2021). https://doi.org/10.1038/s41598-021-00786-6
12. Bąk, M.: Comparison of mathematical models of torque transmitted by multi-disc wet clutch with experimental results. In: Prentkovskis, O., Yatskiv, I., Skačkauskas, P., Maruschak, P., Karpenko, M. (eds.) TRANSBALTICA 2022. LNITI, pp. 383–392. Springer, Cham (2023). https://doi.org/10.1007/978-3-031-25863-3_36
13. Uddeholm homepage. https://www.uddeholm.com/poland/pl/products/uddeholm-nimax/. Accessed 30 June 2023
14. Hennlich homepage. https://www.hennlich.pl/produkty/sprezyny-sprezyny-talerzowe-157.html. Accessed 30 June 2023

Using of the Trucks with Electrical Drive on the Farm Enterprises

Valerii Dembitskyi[1](✉), Volodymyr Sakhno[2], Igor Murovanyi[1], and Mykola Maiak[1]

[1] Lutsk National Technical University, Lutsk, Ukraine
dvm2@meta.ua
[2] National Transport University, Kyiv, Ukraine

Abstract. In present research work there was investigated a problem of using of the electric drive trucks on the farm enterprises. According to the results of the literature sources analysis there was found a great interest of scientists and users concerning the usage of the commercial electric vehicles in the agriculture. There was made a mathematical modeling of the power consumptions by trucks of gross weight 4,000, 8,000, 10,000 and 12,000 kg taking into account their different operating conditions, a movement by an asphalt concrete, road-metal, ground pavement. There was determined that a vehicle movement by a pavement with a high value of rolling resistance coefficient leads to an abrupt increasing of the power consumption, up to 2…2.5 times. There was made a comparison of the values of received power consumptions with the results that were received by another scientists using another research methods. A deviation of received results of the specific power consumptions consists within 5…20%, that is accessible and says about their sufficient convergence. A perspective of further investigations should be directed to a possibility of quick replenishment of the power supply, a determination of rational correlation between different power sources, and to a receiving of detailed characteristics of the vehicles operating conditions that are used in the agriculture.

Keywords: Vehicle · Electric drive · Agriculture · Power · Power consumptions · Operating conditions

1 Introduction

Nowadays an active using of the electrical drive vehicles is not a news anymore. The cars have completely entered the market and are actively operated by both private owners and business structures, especially in taxis transportation. A development of the trucks market in Ukraine can be characterized by an effort to decrease the fuel consumptions, the level of pollutions by using the alternative fuel types, such as biofuel, hydrogen etc. [1, 2]. Under the influence of scientific and technical advance, there are gradually introduced new technologies and a commercial transport with electric drive. But one charge vehicle run prevents a wider introduction of the electrical vehicles in commercial

O. Prentkovskis et al. (Eds.): TRANSBALTICA 2023, LNITI, pp. 103–112, 2024.
https://doi.org/10.1007/978-3-031-52652-7_10

transport. Thus, according to the researches results [3] an extraordinary majority of the vehicles, in fact independently from their carrying capacity, have a cruising range on an electric drive within 100…500 km.

It is worth saying that an opinion and a perception of such vehicles by public significantly influences the development of the trucks market. Thus, in a research work [4] the researchers concentrate on the differences in perception of the experienced and unexperienced users of the electric vehicles. They have chosen 344 farmers among which a part of respondents that have an electric vehicle operating experience is 30%, and their age is mainly 35…50 years, along with that the age of people who have not any electric vehicle operating experience is more than 53 years. The respondents were asked to evaluate an effectiveness of the electric vehicles and the agricultural electric vehicles by such parameters as charge time, power, availability of charging point, power consumption etc. Though none of chosen characteristics was recognized unimportant, there existed the strict differences in perception between those who had and had not an electric vehicle operating experience. These researches results show a necessity of making some changes in popularization of the electric vehicles in order to increase their popularity in agricultural sector.

During the electric drive trucks operating in agriculture, there are present some factors that complicate a possibility of their full usage, namely a significant diversity of the driving conditions (traffic modes, condition and type of pavement, dependence of a vehicle operating intensity from a season, type of field work etc.).

The aim of given research work is to determine the perspectives of the electric drive trucks using on the farm enterprises by an investigation of their specific power consumptions.

Literature sources analysis. A modern state of the researches concerning the using of the electric drive vehicles in agriculture demonstrates a considerable interest of scientists in solving this problem. In a research work [5] there were investigated the tendencies and perspectives of development of the off-road electric vehicles and robots based on green energy in agricultural sector. According to the results of conducted re-searches, it was offered to pay attention on the next questions:

- improvement of a power accumulation technology and recharge for the agricultural electric vehicles;
- finding of new standards for the electric agricultural vehicles, robots and machines;
- using of light little agricultural vehicles in a group inspire of heavy vehicles, in order to reduce the environmental influence, to lower the expenses and to ensure a flexibility;
- development of the advanced technologies, management systems, intellectual units, such as common and assigned vehicles, Internet data bases analysis, artificial intellect, block chain, Internet of things, new sensors and computer algorithms.

In an article [6] there was examined a new type of electric trucks with limited distance, where the electric vehicles are used to plan a route in an inverse flow logistics. A model is made basing on a real problem in agriculture, that joins both a determination of some stores location and finding the routes with a restriction for limited distance.

In a research work [7] the scientists from India have made an electric vehicle on solar panels. The offered vehicle can be used for many purposes in agriculture, for ex-ample for sowing the seeds, levelling, air-spraying by pesticides, hoeing and plans gathering.

Taking into consideration a necessity to assure a branch of Brazilian agricultural sector by the new technologies, a research [8] purposed to find the construction specifications of autonomous electric vehicle that serves as a multi-purpose platform that was firstly developed to remove the wild grass and to satisfy the farmer's needs. In order to get and examine these needs, there was used a methodology of product development in its information phase. As a result there were found 19 project specifications with corresponding specific denominations that allows to find the physical and economical parameters that will be a base for developing an autonomous electric vehicle on its conceptual base.

In order to improve the electric vehicles characteristics, in an article [9], basing on the stand tests, there was examined an effectiveness of a power plant of two different power engines. In this research work there was found, basing on a comparison of three methods of modeling of motor effectiveness, that when the measured values in tests of motor effectiveness are insufficient, so the function of neural networks has a low capacity of generalization in the full operating conditions, and an exactness of regression polynomial is limited. Basing on this fact, there was offered a new modeling method, due to which it was proved that the tractors with double-engine drive system have the optimal economic parameters.

In a research work [10] there was described a light, module and energy efficient planform of a robotic vehicle for wide using in agriculture – a small robotic agricultural vehicle. The researches results of the characteristics of the agricultural electric vehicles are given in the research works [11–16].

The conducted researches analysis demonstrates a considerable interest of population on a passage for the ecologically clean types of transport and on their usage in agriculture. Along with that, the problems of vehicles universalization, that is their us age for both transportation and as a tractor vehicles, are important and actual.

2 Researches Results

To determine a necessary quantity of electric power within the given research, it is worth to take into consideration three vehicle traffic modes: acceleration, braking and constant speed movement. These modes can be determined from a power balance equation [17].

To overcome the movement resistance forces, there should be found a necessary quantity of electric power:

$$N_{\Sigma} = P_{\Sigma} \cdot V, \tag{1}$$

where V – vehicle speed, m/s; P_{Σ} – amount of the vehicle movement resistance forces.

By the chosen output data and by the values of rolling resistance coefficient in a function of vehicle speed one can find the power consumptions during a vehicle acceleration, for the vehicles of gross weight 4,000, 8,000, 10,000 and 12,000 kg (see Fig. 1).

From the received results there was established that a specific power consumption is decreasing with a vehicle mass increasing and a constant movement speed increasing. Such difference can be from 2% on a ground road up to 15% on an asphalt concrete pavement. Along with that, on the speeds up to 6 m/s such difference is less than 1.0%.

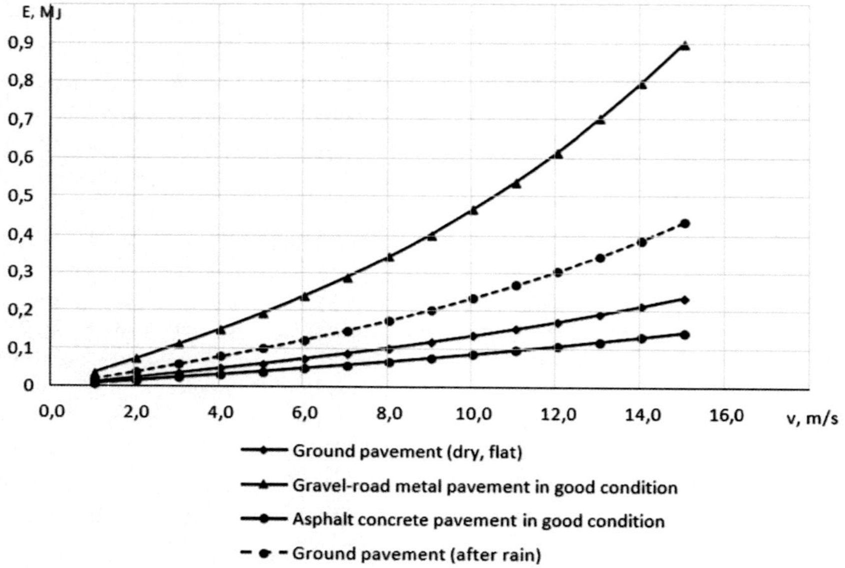

Fig. 1. Power consumptions during a vehicle acceleration of gross weight 12,000 kg.

An important part of every vehicle with electric drive is a system of recuperative braking which, depending on the operating conditions, allows to give back to system up to 30% of consumed power. Along with that, taking into consideration the difficult operating conditions, a low speed movement with a high rolling resistance coefficient, a part of recuperative braking process would be rather low, that is why it was decided to make the researches without taking into account the recuperative braking system.

In order to evaluate in detail the power consumptions during the movement on different speeds, there was found a specific index of power consumption per one kg of vehicle mass. The calculations results are given in a graphic on Fig. 2 for a vehicle that is moving by a wet ground pavement.

In a research work [18], during the researches of the tractor technics, there were analyzed the existed driving cycles, as a result of what there was taken a decision to make the researches using an elementary part of driving cycle.

An elementary part of driving cycle is characterized by three movement modes: acceleration, constant speed movement, braking.

A calculation of power consumption on an elementary part of driving cycle would give a possibility to compare the indices of different vehicles and to determine the advantages and/or disadvantages of using the electric drive vehicles. A duration of acceleration mode is a time necessary to get a speed 8 m/s (28.8 km/h) with an acceleration 0.5 m/s^2. A duration of braking mode is a time necessary to get a speed 0 m/s with a deceleration 0.5 m/s^2. A duration of a constant speed movement, according to the results of the researches shown in [19] is approximately equal to an acceleration time and is 16 s.

On Fig. 3 there are given the power consumptions during a movement of a vehicle of gross weight 12,000 kg on an elementary part of driving cycle.

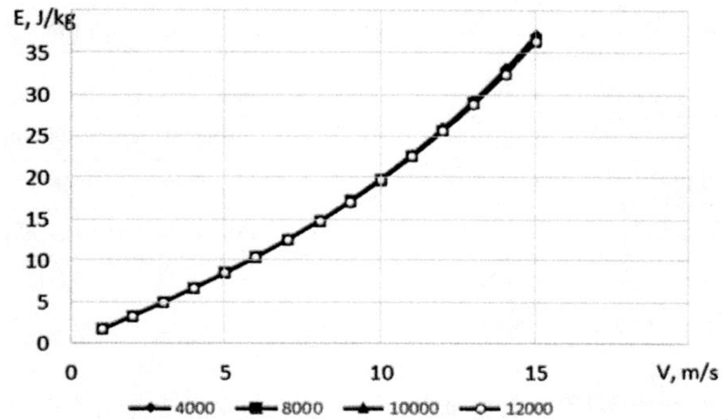

Fig. 2. Specific power consumptions during a vehicle movement by a wet ground pavement.

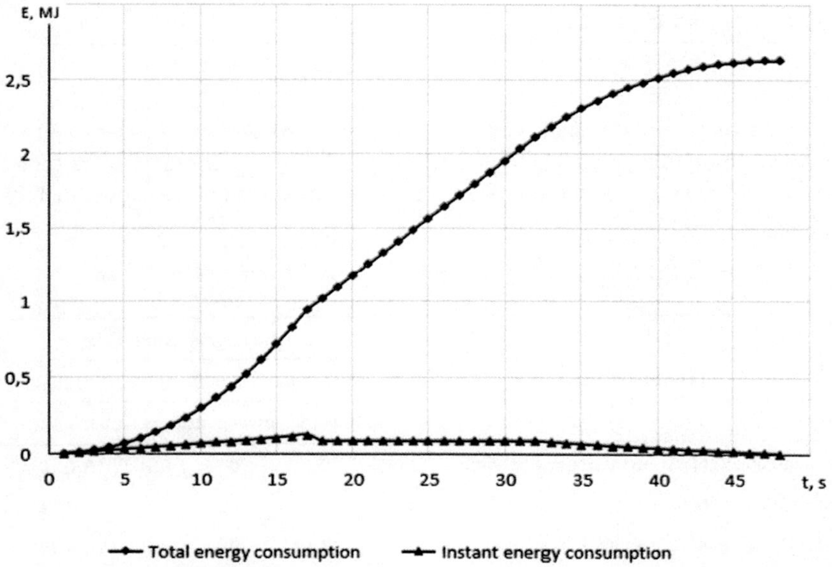

Fig. 3. Power consumptions during a movement of a vehicle of gross weight 12,000 kg on an elementary part of driving cycle by a dry ground road.

The results of calculation of the power consumptions during a vehicle movement are given in Table 1.

Table 1. Power consumptions during the vehicles movement on an elementary part of driving cycle, MJ (kWh).

Type of road pavement	Vehicle gross weight, kg			
	4,000	8,000	10,000	12,000
	Power consumptions, kWh			
Asphalt concrete pavement in good condition	0.11	0.21	0.26	0.31
Asphalt concrete pavement in satisfactory condition	0.11	0.22	0.27	0.33
Gravel-road metal pavement in good condition	0.12	0.24	0.30	0.36
Gravel-road metal pavement in satisfactory condition	0.13	0.27	0.33	0.40
Ground pavement (dry, flat)	0.17	0.34	0.42	0.50
Ground pavement (after rain)	0.37	0.73	0.91	1.10

A length of elementary part of driving cycle consists 248 m. The received values of power consumptions are taken to the specific indices, namely there were found the power consumptions per 1 ton-kilometer. The calculation results are shown in Table 2.

Table 2. Specific power consumptions during the vehicle's movement.

Type of road pavement	Vehicle gross weight, kg			
	4,000	8,000	10,000	12,000
	Specific power consumptions, kWh/t km			
Asphalt concrete pavement in good condition	0.11	0.11	0.10	0.10
Asphalt concrete pavement in satisfactory condition	0.11	0.11	0.11	0.11
Gravel-road metal pavement in good condition	0.12	0.12	0.12	0.12
Gravel-road metal pavement in satisfactory condition	0.13	0.14	0.13	0.13
Ground pavement (dry, flat)	0.17	0.17	0.17	0.17
Ground pavement (after rain)	0.37	0.37	0.37	0.37

In a research work [20] there were conducted the researches concerning a power consumption by the trucks. During the researches the authors have divided the vehicles by a weight onto eight categories, according to Table 3.

During the researches in this research work there were chosen the vehicles of gross weight 4,000, 8,000, 10,000, 12,000kg, that, according to a classification [20], belong to 2b, 5, 6, 7.

Table 3. Trucks categories (according to [20]).

Vehicle category	2b	3	4	5	6	7	8
Gross weight, pound	8501 - 10000	10000-14000	14000 - 16000	16000 - 19500	19501 - 26000	26001 - 33000	more than 33000
Gross weight, kg	3856 - 4536	4537 - 6350	6351 - 7256	7257 - 8845	8846 - 11793	11794 - 14968	more than 14969

The authors of the researches [20] have determined the specific power consumptions for different vehicles categories. The specific power consumptions, received during the calculations (Table 2) are confronted with the power consumptions given in [20]. Along with that, during this confrontation there was made a conversion from an American system of units to a system of units SI, there was taken that 1 mile = 1.6093 km (see Table 4). In order to compare there was chosen an asphalt concrete pavement in good condition. The results are shown in Table 5.

Table 4. Specific power consumptions by different weight categories of the vehicles.

Weight category of a vehicle	Specific power consumptions			
	kW h/mile		kW h/mile	
	min	max	min	max
2b	0.4	0.8	0.249	0.497
5	0.55	1.5	0.342	0.932
6	0.6	1.6	0.373	0.994
7	1.35	2.5	0.839	1.553

Table 5. Results of comparison of the specific power consumptions.

Weight category of a vehicle	Gross weight of a vehicle, kg	Estimated data, kW h/km	Values, according to [20], kW h/km			Deviation from average, %
			min	max	Average	
2b	4,000	0.44	0.249	0.497	0.373	18.0
5	8,000	0.85	0.342	0.932	0.808	5.2
6	10,000	1.05	0.373	0.994	0.684	53.6
7	12,000	1.25	0.839	1.553	1.196	4.5

While analyzing the data given in Table 5, there were noticed the considerable deviations from average from 4,5 to 53%, that is rather significant, that is why there was made a comparison with upper and lower limit of specific power consumptions (mini-mum and maximum). The results are shown in Table 6.

According to the data shown in Table 6, there was found that the calculated specific power consumptions of the vehicles of weight categories 2b, 5 and 7 are within the range of the specific power consumptions values given in [20]. For the vehicles of weight category 6 (10,000 kg) there was noticed an exceeding of maximum values limit on 5,6%, that can be also estimated admissible (Fig. 4).

Table 6. Results of comparison of the specific power consumptions regarding to limit values.

Weight category/Gross weight of a vehicle, kg	Estimated data, kW h/km	Values, according to [20], kW h/km			
		min	Deviation, %	max	Deviation, %
2b/4000	0.44	0.249	76.7	0.497	−11.5
5/8000	0.85	0.342	148.5	0.932	−8.8
6/10000	1.05	0.373	181.5	0.994	5.6
7/12000	1.25	0.839	49.0	1.553	−19.5

Basing on the received values, it was stated that the results of calculation of the power consumptions by the vehicles are adequate and are within the data received in another ways by another scientists.

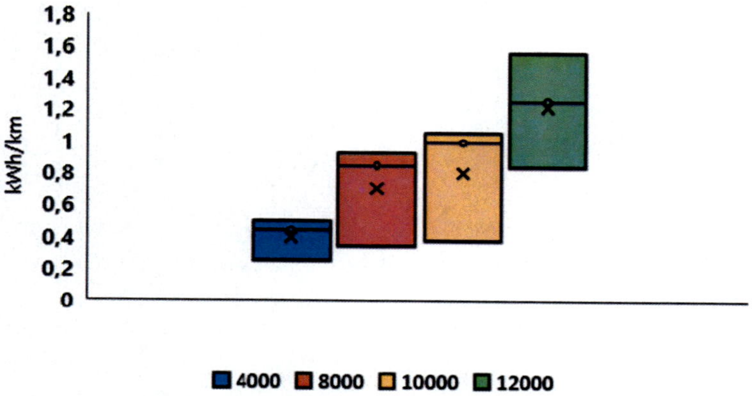

Fig. 4. Results of comparison of the specific power consumptions regarding to limit values.

There is no doubt that during a vehicle movement by a road pavement with another rolling resistance coefficient, a gravel, road metal, ground pavement, the power consumptions would strongly differ from the average one, especially given in [20].

3 Conclusions

After have analyzing the market of commercial vehicles with electric drive, there was noticed an increase tendency of a segment of the average carrying capacity trucks. Irrespective of the vehicles carrying capacity, their cruising range is within 100…500 km.

While making the theoretical researches, there were determined the power consumptions by the vehicles of gross weight 4,000, 8,000, 10,000 and 12,000 kg and specific power consumptions by the vehicles during their movement by asphalt concrete, gravel-road metal and ground pavement of different condition and there was found an increasing of these specific consumptions up to 2 times, dependently on the type and condition of base pavement. By the results of analysis of the received specific power consumptions with the data given in the researches results of another scientists, there were found the deviations that consist from 5 to 20%, and in fact are within the determined ranges that are considered accessible for present researches.

By the results of modeling the power consumptions in different operating conditions, there was found a decreasing of cruising range during the vehicles operation on the ground pavement. For dry flat pavement a cruising range is decreasing up to 38%, for wet pavement (after rain) such decreasing is up to 72%.

Further investigations should concentrate on a possibility to quickly replenish a power reserve, a determination of rational correlation between different power sources, and also a receiving of detailed characteristics of the operating conditions of the vehicles used in agriculture.

References

1. Yemets, B.V., Melnychuk, S.V., Rudzinskyi, V.V., Lomakin, V.O.: Modeling of the dynamics of agricultural vehicles during operation on alternative fuel. In: Proceedings of the 9th International Scientific and Technical Internet Conference “Problems and prospects for the development of road transport”, 14–15 April 2021: Collection of Scientific Papers, pp. 93–97. VNTU, Vinnytsia (2021)
2. Kolodnytska, R.V.: Problems and prospects of using diesel biofuel and hydrogen in automobile transport. In: Materials of the 9th International Scientific and Technical Internet Conference “Problems and Prospects of the Development of Automobile Transport”, 14–15 April 2021: Collection of Scientific Works, pp. 139–143. VNTU, Vinnytsia (2021)
3. IEA. Global EV Outlook 2021, IEA, Париж (2021). https://www.iea.org/reports/global-ev-outlook-2021
4. Riedner, L., Mair, C., Zimek, M., et al.: E-mobility in agriculture: differences in perception between experienced and non-experienced electric vehicle users. Clean Tech. Environ. Policy **21**, 55–67 (2019). https://doi.org/10.1007/s10098-018-1615-2
5. Ghobadpour, A., Monsalve, G., Cardenas, A., Mousazadeh, H.: Off-road electric vehicles and autonomous robots in agricultural sector: trends, challenges, and opportunities. Vehicles **4**(3), 843–864 (2022). https://doi.org/10.3390/vehi-cles4030047
6. Theeraviriya, C., Sirirak, W., Praseeratasang, N.: Location and routing planning considering electric vehicles with restricted distance in agriculture. World Electr. Veh. J. **11**(40), 61 (2020). https://doi.org/10.3390/wevj11040061
7. Karthikeyan, P., Vinothini, S., Lakshmanan, M., Selvam, N., Jenopaul, P., Kumaresan V.: Design and implementation of solar powered electric vehicle for multitasking application in agriculture. Ann. Roman. Soc. Cell Biol. **25**(2), 923–930. (2021). https://www.annalsofrscb.ro/index.php/journal/article/view/1042
8. Rosales, J.C., et al.: Design specifications of an autonomous electric vehicle for use in family farm units. Engenharia Agrícola **42** (2022). https://doi.org/10.1590/1809-4430-Eng.Agric.v42n3e20210220/2022

9. Cheng, Z., Zhou, H., Lu, Z.: A novel 10-parameter motor efficiency model based on I-SA and its comparative application of energy utilization efficiency in different driving modes for electric tractor. Agriculture **12**(3), 362 (2022). https://doi.org/10.3390/agriculture12030362
10. Bawden, O,, Ball, D., Kulk, J., Perez, T., Russell, R.: A lightweight, modular robotic vehicle for the sustainable intensification of agriculture. In Chen, C. (ed.) Proceedings of the 16th Australasian Conference on Robotics and Automation 2014, pp. 1–9. Australian Robotics and Automation Association (ARAA), Australia (2014)
11. Joo, D., Cho, J.-H., Woo, K., Kim, B.-T., Kim, D.-K.: Electromagnetic field and thermal linked analysis of interior permanent-magnet synchronous motor for agricultural electric vehicle. IEEE Trans. Magn. **47**(10), 4242–4245 (2011). https://doi.org/10.1109/TMAG.2011.2149504
12. Gorjian, S., Ebadi, H., Trommsdorff, M., Sharon, H., Demant, M., Schindele, S.: The advent of modern solar-powered electric agricultural machinery: a solution for sustainable farm operations. J. Clean. Prod. **292**, 126030 (2021). https://doi.org/10.1016/j.jclepro.2021.126030. ISSN 0959-6526
13. Talwariya, A., Singh, P., Kolhe, M.: Non-cooperative game theory based stepwise power tariff model using Monte-Carle simulation for agricultural consumers. Open Agric. **4**(1), 418–425 (2019). https://doi.org/10.1515/opag-2019-0041
14. Schmidt, J.R., Cheein, F.A.: Assessment of power consumption of electric machinery in agricultural tasks for enhancing the route planning problem. Comput. Electron. Agric. **163**, 104868 (2019). https://doi.org/10.1016/j.compag.2019.104868. ISSN 0168-1699
15. Grigore, A.I., et al.: Study on the conversion of a conventional tractor into an electric tractor. Ann. Univ. Craiova-Agric. Montanol. Cadastre Ser. **50**(2), 308–315 (2020)
16. Cristea, M., et al.: Types of electric motor controllers that can be used on agricultural electric tractors. Ann. Univ. Craiova-Agric. Montanol. Cadastre Ser. **51**(2), 235–250 (2021)
17. Sakhno, V.P., Bezborodova, G.B., Mayak, M.M., Sharay, S.M.: Cars: traction-speed properties and fuel economy. Scientific Guide, 174 p. KVITS, Kyiv (2004)
18. Zakharchuk, O.V.: The use of gas fuel in auto-tractor technology: a monograph. Oleh Viktorovych Zakharchuk, 164 p. IVV Lutsk NTU, Lutsk (2016)
19. Dembitsky, V.M., Mazylyuk, P.V., Pavlyashyk, S.M.: Adaptation of the driving cycle to real traffic conditions of city buses. In: Dembitskyi, V.M., Mazilyuk, P.V., Pavlyashik, S.M. (eds.) Scientific Notes. Interuniversity Collection, vol. 62, pp. 98–102. Lutsk National Technical University, Lutsk (2018)
20. Smith, D., et al.: Medium and heavy-duty vehicle electrification: an assessment of technology and knowledge gaps. United States (2020). https://doi.org/10.2172/1615213

Simulation of Thermoelectric Coolers for Automotive Temperature Stabilization Systems

Andrii Bukaros[1(✉)] and Oleg Onishchenko[2]

[1] Odesa Military Academy, Odesa 65009, Ukraine
andrey.bucaros@gmail.com
[2] National University - Odessa Maritime Academy, Odesa 65052, Ukraine

Abstract. The analysis of literary sources showed that thermoelectric devices are increasingly used in road transport as generators of electrical energy and as a part of temperature stabilization systems. Both directions provide for the control of thermoelement parameters, which requires the determination of their dynamic properties.

This paper considers the method of creating a simple dynamic model of the Peltier element, which describes the main electrothermal processes occurring in it. The synthesis of the model was carried out by solving the one-dimensional thermoelement thermal conductivity equation with given initial and boundary conditions using the integral Laplace transform. On the basis of the obtained operator equation, transfer functions were synthesized and a simulation model was built.

Verification of the obtained model was carried out in the Matlab/Simulink software environment by comparing the time dependences of the cold junction temperature at different values of direct current with the results of numerical simulation of Peltier elements obtained by other researchers. The verification showed that the discrepancy between the simulation results obtained in this paper and the known ones does not exceed 5%, which indicates the efficiency of the considered method. At the same time, the proposed model is much simpler and clearer.

The indicated advantages of the developed model make it possible to easily calculate setting parameters and build automotive temperature stabilization systems using thermoelectric coolers. Further improvement of the model is possible by taking into account the temperature dependence of thermoelectric parameters and the heat exchange of the thermoelement hot surface with the environment.

Keywords: Thermoelectric cooler · Peltier element · Dynamic model · Simulation

1 Introduction

In recent decades, the problem of shortage and rational use of energy resources has become acute all over the world and, in particular, in road transport. One of the promising ways to solve this problem is the use of economical and reliable thermoelectric

O. Prentkovskis et al. (Eds.): TRANSBALTICA 2023, LNITI, pp. 113–120, 2024.
https://doi.org/10.1007/978-3-031-52652-7_11

devices. The principle of thermoelectric devices operation can be described by two inverse processes: based on the Seebeck thermoelectric effect, the conversion of thermal energy into electrical energy, which is used in thermoelectric generators, and based on the Peltier effect, the conversion of electrical energy into thermal energy, which is used in thermoelectric converters.

The use of thermoelectric generators to convert waste heat of exhaust gases into electricity is of great interest to researchers. The authors of papers [1, 2] claim that about 40% of the energy produced by an internal combustion engine is released into the atmosphere by exhaust gases in the form of heat. This leads to significant environmental pollution and increases the negative impact of the greenhouse effect. Compared to other methods of exhaust gas energy recovery, thermoelectric generators have a number of advantages [3, 4]: environmental friendliness, absence of noise, vibration, working fluids, high reliability, low operating costs, etc.

Another promising direction of research is the use of thermoelectric converters (Peltier elements), in particular, to maintain comfortable climatic conditions in the car interior and control the temperature of car batteries. In contrast to widespread vapor compression refrigeration units [5], thermoelectric coolers contribute to reducing the emission of refrigerants into the atmosphere and reducing fuel consumption [6, 7]. Thermoelectric cooling of car batteries allows to increase the efficiency and service life of batteries due to relatively low inertia and a wide range of operating temperatures of Peltier elements [8, 9].

The given directions of thermoelectric effects application in road transport involve controlling the performance of thermoelectric devices, especially when they are used as coolers. It is obvious that to build effective thermoelectric temperature stabilization systems, it is necessary to know the Peltier elements dynamic properties. It should be noted that known dynamic models of thermoelements based on electrothermal analogies [10] or discretization of nonlinear heat conduction equations [11] are too overloaded and inexpedient for the synthesis of classic proportional–integral–derivative temperature controllers [12].

To overcome this contradiction, this paper proposes a method of Peltier element simple dynamic model synthesis based on elementary transfer functions.

2 Synthesis of the Peltier Element Mathematical Model

The object of research is a Peltier thermoelement, which consists of two n- and p-type semiconductor branches (Fig. 1). Voltage is applied to one ends of these branches (hot junction), the other ends are electrically connected (cold junction).

Taking into account all factors affecting the thermoelectric properties of the thermoelement is extremely difficult. For a qualitative assessment of dynamic parameters, we will consider a one-dimensional model of the thermoelement, which is a uniform rod. We assume that there is no heat removal from the side surfaces of the thermoelement, and the temperature of the hot surface (junction) of the Peltier element is unchanged and equal to the ambient temperature. In the paper we will consider the heat exchange of a cold surface (junction) with the environment according to Newton's law without an additional heat load q_l (Fig. 1) from the side of the cooled object.

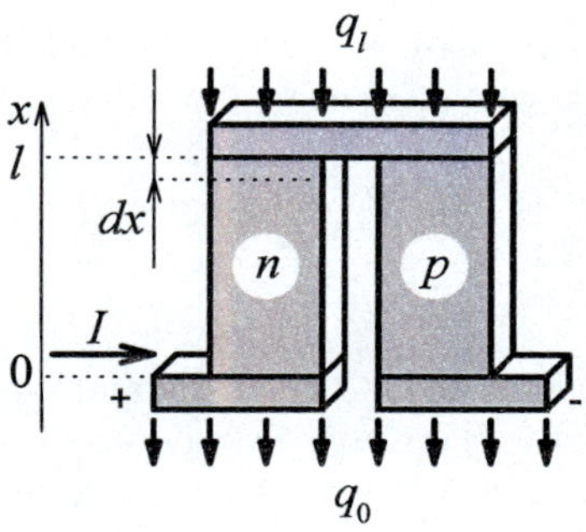

Fig. 1. Peltier thermoelement.

Connection points of thermoelements with working surfaces and insulating plates are considered objects with concentrated parameters, therefore, their temperature does not depend on the x coordinate. The temperature dependences of the thermoelement thermoelectric parameters are neglected. We also neglect the Thompson effect, since, as research [10] shows, its influence is quite small compared to other thermoelectric effects.

According to the assumptions made, the thermal conductivity equation of the Peltier thermoelement will be written as [13]:

$$c \cdot \frac{\partial T(x,t)}{\partial t} = k \cdot \frac{\partial^2 T(x,t)}{\partial x^2} + \rho \cdot \left(\frac{I(t)}{A}\right)^2, \tag{1}$$

where c is the volume heat capacity of the thermoelement rod, J/(cm^3 · K); ρ is the thermoelement rod resistivity, Ohm · cm; k is the thermal conductivity, W/(cm · K); A is the cross-sectional area, cm^2; T is the temperature, K; t is the time, s; x is the coordinate along thermoelement length, cm; I is the current flowing through the thermoelement, A. The initial condition:

$$T(x,0) = T_a, \tag{2}$$

where T_a is the ambient temperature. Boundary conditions for the given task will have the following form:

– for a hot surface:

$$T(0,t) = T_a, \tag{3}$$

– for a cold surface, considering the heat balance in the volume $A \cdot dx$ (Fig. 1):

$$\begin{aligned} c \cdot A \cdot dx \cdot \tfrac{\partial T(x,t)}{\partial t} + c_l \cdot \tfrac{\partial T(l,t)}{\partial t} &= k \cdot A \cdot \tfrac{\partial T(x,t)}{\partial x} + \tfrac{\rho \cdot dx}{A} \cdot I^2(t) \\ -\alpha \cdot T(l,t) \cdot I(t) &+ \tfrac{r_l}{A} \cdot I^2(t) + h \cdot A \cdot [T_a - T(l,t)], \end{aligned} \tag{4}$$

where l is the thermoelement length, cm; c_l is the heat capacity of the insulating plate, J/K; α is the Seebeck coefficient, V/K; h is the heat transfer coefficient, W/(cm^2 · K); r_l is the contact resistance of the thermoelement junction with the cold surface,

Ohm · cm^2. If we direct dx → 0 and go to the boundary, then after elementary transformations the second boundary condition will take the form [14]:

$$-\frac{\partial T(l,t)}{\partial x} - \frac{1}{k}\left[\frac{c_l}{A}\cdot\frac{\partial T(l,t)}{\partial t} + \frac{\alpha}{A}\cdot T(l,t)\cdot I(t) - h\cdot[T_a - T(l,t)] - \frac{r_l}{A^2}\cdot I^2(t)\right] = 0. \tag{5}$$

To find the transfer function of the Peltier thermoelement, we apply the integral Laplace transform for Eq. (1) with boundary conditions (3), (5). At the same time, as it is customary in the classical automatic control theory, we consider the initial conditions to be zero. As a result, we get:

$$\frac{\partial^2 T(x,s)}{\partial x^2} - \frac{c}{k}\cdot s\cdot T(x,s) + \frac{\rho}{k\cdot A^2}\cdot I^2(s) = 0, \tag{6}$$

$$T(0,s) = 0, \tag{7}$$

$$\frac{\partial T(l,s)}{\partial x} + \frac{1}{k}\left[\frac{c_l}{A}\cdot s\cdot T(l,s) + \frac{\alpha}{A}\cdot T(l,s)\cdot I(s) - h\cdot[T_a - T(l,s)] - \frac{r_l}{A^2}\cdot I^2(s)\right] = 0, \tag{8}$$

where s is the Laplace variable.

We look for the general solution of Eq. (6) in the form:

$$T(x,s) = C_1\cdot e^{x\sqrt{\frac{c}{k}\cdot s}} + C_2\cdot e^{-x\sqrt{\frac{c}{k}\cdot s}} + \frac{\rho}{s\cdot c\cdot A^2}\cdot I^2(s). \tag{9}$$

To determine the constants C_1 and C_2, we substitute the expression (9) into the boundary conditions (7), (8). As a result, after elementary transformations, we get the following system of linear equations:

$$\begin{cases} C_1 + C_2 = -\frac{\rho}{s\cdot c\cdot A^2}\cdot I^2(s) \\ C_1\cdot e^{l\sqrt{\frac{c}{k}\cdot s}}\cdot\sqrt{\frac{c}{k}\cdot s} - C_2\cdot e^{-l\sqrt{\frac{c}{k}\cdot s}}\cdot\sqrt{\frac{c}{k}\cdot s} \\ = -\frac{1}{k}\left[\frac{c_l}{A}\cdot s\cdot T(l,s) + \frac{\alpha}{A}\cdot T(l,s)\cdot I(s) - h\cdot[T_a - T(l,s)] - \frac{r_l}{A^2}\cdot I^2(s)\right]. \end{cases} \tag{10}$$

Solving the system of Eqs. (10) relative to the constants C_1 and C_2, substituting the obtained values into the general solution (9), and introducing hyperbolic functions, we obtain the following solution of the operator Eq. (6) regarding the image of the cold surface temperature:

$$\begin{aligned} T(l,s) = &\left[\rho - \frac{\rho}{cosh\left(l\sqrt{\frac{c\cdot s}{k}}\right)} + \frac{\sqrt{\frac{c\cdot s}{k}}\cdot r_l}{coth\left(l\sqrt{\frac{c\cdot s}{k}}\right)}\right]\frac{I^2(s)}{s\cdot c\cdot A^2} \\ &- \frac{\frac{c_l}{A}\cdot s\cdot T(l,s) + \frac{\alpha}{A}\cdot T(l,s)\cdot I(s) - h\cdot[T_a - T(l,s)]}{k\cdot\sqrt{\frac{c\cdot s}{k}}\cdot coth\left(l\sqrt{\frac{c\cdot s}{k}}\right)}. \end{aligned} \tag{11}$$

For the transition from hyperbolic functions to fractional-rational expressions, we will apply the expansion of the hyperbolic cosine and cotangent into power series and limit ourselves to the first two terms, while the error will not exceed 4%. Thus, by substituting the decomposed hyperbolic functions into expression (11) after elementary transformations, we obtain the final solution of Eq. (6):

$$T(l,s) = \frac{K_{J1}\cdot s + K_J}{(\tau_J\cdot s+1)(\tau_P\cdot s+1)}\cdot I^2(s) - \frac{K_P\cdot T(l,s)}{\tau_P\cdot s+1}\cdot I(s) + \frac{K_T\cdot[T_a - T(l,s)]}{\tau_P\cdot s+1}, \tag{12}$$

where the transfer coefficients and the time constants:

$$K_J = \frac{l\cdot(\rho\cdot l + 2r_l)}{2\cdot k\cdot A^2} = \frac{R+2R_l}{2\cdot K},\ K_{J1} = \frac{c\cdot l^3\cdot(\rho\cdot l+3r_l)}{6\cdot(k\cdot A)^2} = \frac{\tau_J\cdot(R+3R_l)}{3\cdot K},$$
$$\tau_J = \frac{c\cdot l^2}{2\cdot k} = \frac{C}{2\cdot K},\ \tau_P = \frac{A\cdot c\cdot l^2 + 3\cdot c_l\cdot l}{3\cdot k\cdot A} = \frac{C+3c_l}{3\cdot K},\ K_P = \frac{\alpha\cdot l}{k\cdot A} = \frac{\alpha}{K},\ K_T = \frac{h\cdot l}{k}, \tag{13}$$

where R is the electrical resistance of the thermoelement branch, Ohm, R_l is the electrical resistance of the thermoelement junction with the cold surface, Ohm, C is the heat capacity of the thermoelement branch, J/K, K is the thermal conductance of the thermoelement branch, W/K.

In expression (12), the first term models the release of Joule heat in the volume of the Peltier thermoelement branch and in the contact resistance of the cold junction, the second term models the absorption of Peltier heat, the third models the heat exchange of the cold surface with the external environment.

3 Researching the Peltier Element Mathematical Model by Means of Simulation

On the basis of the operator Eq. (12), it is possible to build a simulation model of the Peltier thermoelement, the structural diagram of which is shown in Fig. 2. As can be seen, the obtained model consists of three elementary transfer functions, each of which describes a separate thermal process occurring in the thermoelement in accordance with (12).

We will verify the model by comparing the results of simulation in the Matlab/Simulink environment with the results of research [11]. To do this, we will use the parameters of widely used Peltier modules based on Bi-Te alloys [11]: module height $l = 0.14$ cm, volume heat capacity of the branches $c = 1.4$ J/(cm$^3\cdot$ K), thermal conductivity of the branches $k = 0.014$ W/(cm $\cdot$ K), resistivity $\rho = 0.0015\ \Omega\cdot$ cm, Seebeck coefficient $\alpha = 240\cdot 10^{-6}$ V/K, cross-sectional area $A = 0.01$ cm^2, initial temperature $T_a = 300$ K, heat capacity of the insulating plate $c_l = 0.0012$ J/K; value of contact resistance $r_l = 5\cdot 10^{-6}\ \Omega\cdot$ cm^2, of heat transfer coefficient of the cold surface $h = 10^{-3}$ W/(cm$^2\cdot$ K).

As a result of the model study, the time dependences of the cold surface temperature (shown in Fig. 3 by solid lines) were obtained when feeding the thermoelement with constant currents of the following values: 1 A, 2 A, 3 A. For comparison, dashed lines in Fig. 3 show similar dependencies obtained in [11].

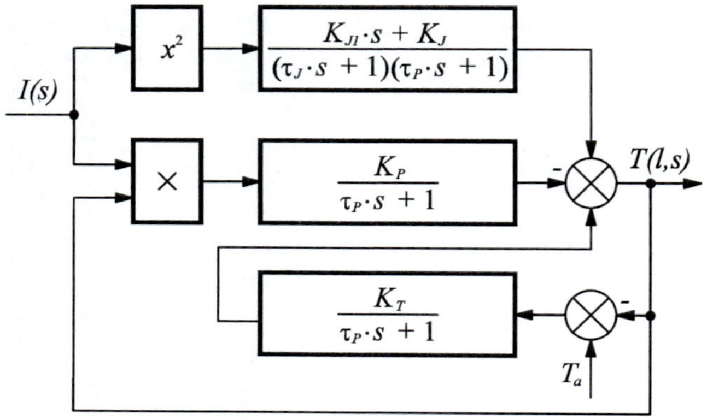

Fig. 2. Structural diagram of the Peltier thermoelement simulation model.

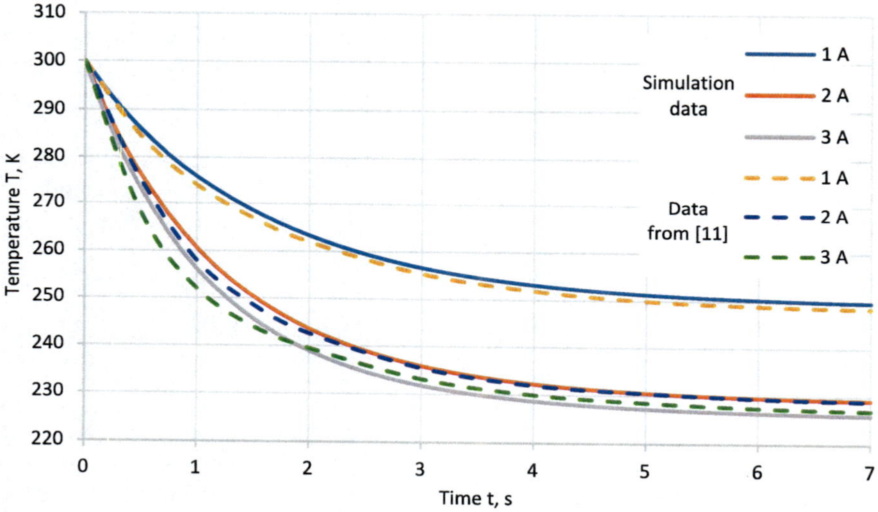

Fig. 3. Time temperature dependences of the thermoelement cold surface.

Analysis of Fig. 3 allows us to assert a relatively high level of coincidence of the time temperature dependences curves, obtained according to the model in Fig. 2 and in paper [11], in static and dynamic mode. The static error for all curves does not exceed 2%. The maximum deviation in dynamics is observed for the 3 A curve and does not exceed 5%.

The discrepancy between the time dependences of the thermoelement cold surface temperature obtained by the developed and known models can be explained by the neglect of the temperature dependence of thermoelectric parameters, in particular, the Seebeck coefficient α. As shown in the paper [15], with a decrease in the temperature of the thermoelement based on Bi-Te alloys, the value of the Seebeck coefficient α decreases almost linearly. Thus, at thermoelement temperature values close to 300 K

$\alpha > 240 \cdot 10^{-6}$ V/K, Peltier heat absorption is more intense, the process of lowering the cold junction temperature occurs faster (dashed lines in Fig. 3 are located below the solid lines). At thermoelement temperature values close to 200 K $\alpha < 240 \cdot 10^{-6}$ V/K, Peltier heat absorption is less intense, the process of lowering the temperature of the cold junction occurs more slowly (dashed lines in Fig. 3 are located above solid lines).

It should be noted that the obtained simulation model (Fig. 2) is much simpler and clearer in comparison with known analogues [10, 11]. The models developed by other authors in the Matlab/Simulink environment [16–18] do not take into account the phenomenon of thermal conductivity and represent the Peltier element as an inertialess thermoelectric converter. In contrast to the mentioned analogues, the developed model (Fig. 2) takes into account not only the phenomenon of thermal conductivity, but also the heat capacity of insulating plates and heat release in the contact resistance. The simplicity of the model makes it quite easy to adjust the automatic performance controllers of Peltier elements as part of automotive temperature stabilization systems using well-known methods [12].

4 Conclusion

The obtained results and the maximum dynamic error value of 5% in comparison with the known simulation results testify to the efficiency of the proposed method of Peltier thermoelement dynamic model synthesis.

The obtained model consists only of three elementary transfer functions and makes it possible to easily calculate and optimize the setting parameters of automotive temperature stabilization systems based on Peltier thermoelements.

For future experimental researches, this model must be improved by taking into account the heat exchange of the thermoelement hot surface with the environment and the temperature dependence of the thermoelectric parameters.

References

1. Burnete, N.V., Mariasiu, F., Depcik, C., Barabas, I., Moldovanu, D.: Review of thermoelectric generation for internal combustion engine waste heat recovery. Prog. Energy Combust. Sci. **91**, 101009 (2022)
2. Lu, H., et al.: Experiment on thermal uniformity and pressure drop of exhaust heat exchanger for automotive thermoelectric generator. Energy **54**, 372–377 (2013)
3. Alghoul, M.A., et al.: A review of thermoelectric power generation systems: roles of existing test rigs/prototypes and their associated cooling units on output performance. Energy Convers. Manag. **174**, 138–156 (2018)
4. Patil, D.S., Arakerimath, R.R., Walke, P.V.: Thermoelectric materials and heat exchangers for power generation – a review. Renew. Sustain. Energy Rev. **95**, 1–22 (2018)
5. Bukaros, A., Onishchenko, O., Herega, A., Trushkov, H., Konkov, K.: Simulation modeling of vapor compression refrigeration unit temperature modes. Stud. Syst. Decis. Control **454**, 253–266 (2023)
6. Yogesh, N., Omkar, S., Gorakhnath, G., Omkar, G.: Air conditioning system in car using thermoelectric effect. Int. J. Eng. Res. Technol. **9**(6), 374–377 (2020)

7. Raut, M.S., Walke, P.V.: Thermoelectric air cooling for cars. Int. J. Eng. Sci. Technol. **4**(5), 2381–2394 (2012)
8. Lyu, Y., Siddique, A.R.M., Majid, S.H., Biglarbegian, M., Gadsden, S.A., Mahmud, S.: Electric vehicle battery thermal management system with thermoelectric cooling. Energy Rep. **5**, 822–827 (2019)
9. Lu, M., Zhang, X., Ji, J., Xu, X., Zhang, Y.: Research progress on power battery cooling technology for electric vehicles. J. Energy Storage **27**, 101155 (2020)
10. Piggott, A.: Detailed transient multiphysics model for fast and accurate design, simulation and optimization of a thermoelectric generator (TEG) or thermal energy harvesting device. J. Electron. Mater. **48**(9), 5442–5452 (2019). https://doi.org/10.1007/s11664-019-06952-x
11. Kotsur, M.: Optimal control of distributed parameter systems with application to transient thermoelectric cooling. Adv. Electr. Comput. Eng. **15**(2), 117–122 (2015)
12. Ang, K.H., Chong, G.C.Y., Li, Y.: PID control system analysis, design, and technology. IEEE Trans. Control Syst. Technol. **13**(4), 559–576 (2005)
13. Yang, R., Chen, G., Kumar, R.A., Snyder, J.G., Fleurial, J.-P.: Transient cooling of thermoelectric coolers and its applications for microdevices. Energy Convers. Manag. **46**(9–10), 1407–1421 (2005)
14. Anatychuk, L.I., Vikhor, L.M., Kotsur, M.P., Romaniuk, I.F., Soroka, A.V.: Optimal control of transient thermoelectric cooling process in the mode of minimum power consumption. J. Thermoelectricity **1**, 74–87 (2020)
15. Witting, I.T., et al.: The thermoelectric properties of bismuth telluride. Adv. Electron. Mater. **5**, 1800904 (2019)
16. Electrothermal converter - MATLAB. https://www.mathworks.com/help/sps/ref/peltierdevice.html. Accessed 30 July 2023
17. Simplified TEG Model or Peltier Device using Power Library - File Exchange - MATLAB Central. https://www.mathworks.com/matlabcentral/fileexchange/115400-simplified-teg-model-or-peltier-device-using-power-library. Accessed 30 July 2023
18. Simulink Model of TEG module - File Exchange - MATLAB Central. https://www.mathworks.com/matlabcentral/fileexchange/74694-simulink-model-of-teg-module. Accessed 30 July 2023

Emissions of Petroleum Products from Roads into Roadside Soils as Part of Exhaust Gas Emissions and Surface Wastewater

Valentina Iurchenko[1](✉), Oksana Melnikova[1], and Larysa Mykhailova[2]

[1] O.M. Beketova National University of Urban Economy in Kharkiv, 17, Marshal Bazhanov Street, Kharkiv, Ukraine
yurchenko.valentina@gmail.com
[2] Brandenburg University of Technology at Cottbus, Konrad-Wachsmann-Allee, 6, Cottbus, Germany

Abstract. The highway causes an intensive emission of petroleum products into the roadside space and their active deposition by soils to ecologically dangerous concentrations. We considered two main ways of petroleum product transfer from the road: air (dispersion in the atmospheric air of gaseous emissions of vehicles) and water (with surface wastewater formed on roads, as a result of flooding of roadside territories and splashdown). Based on the data of the physical and chemical analysis of air, water and soil environments, a tentative assessment of the intensity of petroleum products transfer from the highway to the roadside soils by each of these ways was made using the ratio of concentrations of aliphatic and aromatic hydrocarbons in the exhaust gases, surface waste water formed on roads and in soils of the nearest roadside space as an indicator. According to the results of determining this indicator, it was found that in the emission of petroleum products from the highway and their accumulation in the soils of the roadside space, the entry of these contaminants in the exhaust gases by air predominates.

Keywords: Highway · Environmental hazard · Petroleum product emissions · Gaseous emissions · Surface wastewater · Aliphatic hydrocarbons · Aromatic hydrocarbons

1 Introduction

Traffic on highways is caused an intensive technogenic load on the atmospheric air, soils of roadside territories and hydrosphere objects. The flow of pollutants from the highway into the roadside space occurs in two ways: by air (dispersion in the atmospheric air of gaseous emissions of vehicles) and water (with surface wastewater formed on roads, as a result of flooding of roadside areas and splashdown) [1–4].

The result of such impacts is intensive contamination of soils of the roadside space, among which the greatest environmental hazard in terms of the level of exceeding maximum permissible concentration (MPC) is pollution by petroleum products (PP), including polycyclic aromatic hydrocarbons (PAHs). The concentration of PPs in soils

O. Prentkovskis et al. (Eds.): TRANSBALTICA 2023, LNITI, pp. 121–129, 2024.
https://doi.org/10.1007/978-3-031-52652-7_12

immediately adjacent to roads may exceed the MPC (200 mg/kg) by more than 10 times. And PP concentrations in soils in dynamics of distance from a road testifies that intensive pollution (especially from the windward side of a road) occurs on distance about 10 m from a road, then PP concentration in soils sharply decreases (practically in 10 times) which allowed to assume the impact of intensive runoff from the road surface on this character of pollution [5, 6].

Monitoring of surface wastewater generated in urban areas has shown that road traffic is one of the main sources of their pollution in urban areas [7]. In addition to the most volatile substances, PP emitted from non-point sources can be deposited on the surfaces of urban areas and subsequently washed by stormwater into sewage systems and natural water bodies. Recently, the presence of volatile organic compounds (VOCs), mainly products of partial degradation of PPs, in dissolved and colloidal forms in stormwater has been demonstrated. This indicates that the mass transfer of these compounds, and consequently surface water effluent toxicity and associated risks, is potentially higher than previously assumed [8].

There are many known possible sources of traffic-related roadway washout contamination. These include materials used in vehicles (tires, vehicle bodywork, engine components, and care products); exhaust fumes and particulates; pavement and markings; and road equipment (road signs, separators, and barriers). The amount of pollutants generated and accumulated on the surface of highways depends on traffic-related factors such as speed, composition, and number of traffic vehicles [5, 9].

After a comprehensive screening, including a search in scientific literature and databases, expert opinions, ranking method and chemical hazard identification, the highest priority environmental hazards were selected from a list of 1,100 volatile organic compounds found in highway runoff. The results showed the following order of priority: polycyclic aromatic hydrocarbons (PAHs) > C_{20}–C_{40} alkanes > alkylphenols > phthalates > aldehydes > phenolic antioxidants > bisphenol A > oxygen-containing PAHs > C_5–C_{12} naphtha > amides > amines. The use of modern methods of removal of surface wastewater from roads and methods of their deep cleaning will reduce the uncontrolled emission of PPs and products of their oxidation from roads on the ecosystems of roadside space and increase the level of environmental safety of these areas [10–12].

Determining the proportion of aqueous and airborne PP transport from highways to roadside soils is not an easy task, since it is difficult to accurately attribute the identified compounds to pollution specifically by motor vehicle emissions or runoff from roads based on their chemical composition. The composition of PPs in automobile exhaust gases has been studied in some detail [2, 13]. Aliphatic and aromatic hydrocarbons have been identified in it, including particularly hazardous compounds – PAHs. The composition of runoff from roads includes most compounds from automobile emissions, as well as a number of other hydrocarbons, including aromatic ones - products of abrasion of tires, road surface, fuel leaks, etc. [3, 5, 9, 14].

The purpose of the work is to evaluate the contribution of water and atmospheric ways of PP transfer from highways to soil contamination of the adjacent roadside space, as well as to establish the features of pollution of rain washouts from roads by sorbed on particles and emulsified PP.

2 Objects and Methods

The gaseous emissions, run-offs from highway and soil in the areas adjacent to urban and suburban roads and road facilities (RF) were investigated. Gaseous emissions were studied by direct method, i.e., by chromatography of gaseous samples, and indirectly, by analysis of PPs sorbed by snow, the samples which were taken at a distance of 1 m from the edge line of the highway. The samples of the run-off from the roads, including RF territories were obtained by piece wash-off method (the period without rains before sampling was 10–15 days.) The wash-off was performed with cotton tampon in the definite volume of distilled water (500 cm^3) from the definite coat area (~1,500 cm^2), limited by the wooden grid [15].

Based on the area of the run-off and the volume of water used for the run-off, according to the formulas for calculating the characteristics of surface run-off developed by various regulatory documents [15], it is possible to recalculate the obtained hydrochemical parameters of the run-off to characteristics adequate to those obtained when analyzing real rain run-off from the road. The water level created by the artificial run-off was 3.3 mm. According to regulatory documents, in the absence of long-term observations, the water level created by rain may be assumed to be within 5–10 mm. Thus, the studied artificial flushing from the roads was approximately 1.5–3.0 times more concentrated than the real one formed in the first 10 min of rain.

To analyze the composition of gaseous substances, carburetor engine exhaust gases and roadside air samples were taken using Auer sorption tubes (type B) with subsequent desorption and gas chromatographic analysis using a mass spectrometric detector [16]. Identification and quantification of the concentrations of individual HCs included in the PP of the roadside soil of the roads under study were performed by gas-liquid chromatography (chromatograph Fisons 8065, HT8 column) using a mass spectrometer (Fisons MD800) as the detector. Commercial diesel fuel and Hewlett Packard alkane standard Part. No. 18710-60170 were used for quantification and identification of total hydrocarbons and n-alkanes, respectively.

The total content of aliphatic and aromatic hydrocarbons in soils, in melted snow and artificial run-off from highways were estimated by gravimetric method [17] using different extractors for analysis – chloroform and hexane. This made it possible to fractionate PPs polluting water environments or soil PPs by dividing them into a fraction of conditionally aliphatic PPs – gasoline, kerosene, diesel fuels (hexane extract), and a fraction of conditionally complex PPs – high-molecular-weight, aromatic and heterorganic (the difference between the PP content extracted successively with chloroform and then hexane and the content of PPs extracted only with hexane) [18].

3 Results and Discussion

3.1 Pollution of Gaseous Emissions from Vehicles, Rain Run-Off from Roads and Roadside Soils by Aliphatic and Aromatic Hydrocarbons

In order to make an approximate assessment of the impact of the air and water ways of PP transport on the soils of roadside space, we considered and compared the ratio of concentrations of aliphatic and aromatic hydrocarbons in automobile exhaust emissions, in gaseous emissions that reach the ground in roadside space (snow cover), in surface run-off water that is generated on roads, and in soils of the roadside space closest to the road.

The absolute concentrations of aliphatic and aromatic hydrocarbons in the studied media are influenced by many factors, and the ratio of their concentrations is a more stable indicator. For example, the concentration of pollutants in stormwater is extremely dependent on the time after the onset of rain (which is not always indicated in scientific publications). And the ratio of concentrations of various pollutants in surface wastewater is more stable. According to scientific and technical literature (Table 1), the ratio of aliphatic to aromatic hydrocarbons in stormwater generated on roads differs significantly (more than 10 times) from this indicator in gaseous emissions.

Table 1. Ratio of concentrations of aliphatic and aromatic hydrocarbons in gaseous and aqueous emission of PP created by the road (data from scientific and technical literature).

Objects that create the PP emission	Aliphatic/aromatic hydrocarbon ratio
Emissions of exhaust gases from internal combustion engines running on:	
Petrol	0.03–0.52 [13]
Diesel	2.35–1.78 [13]
Petrol fuel standard (Germany)	0.50 [19]
Surface wastewater generated on the road	27.28–61.38 [20]

The data obtained during the study of gaseous and aqueous media formed on roads in the Kharkiv region are presented in Table 2.

As can be seen, the established ratio of the concentrations of aliphatic and aromatic hydrocarbons in the emissions of internal combustion engines is very close to the data presented in the scientific and technical literature (Table 1). Moreover, in the roadside atmosphere, which was sampled at a distance of 1 m from the edge line at a height of 0.5–1.0 m, this ratio increases. This is probably due to a decrease in the proportion of aromatic compounds that quickly condense and settle from the air to the ground surface. The ratios of concentrations of aliphatic and aromatic hydrocarbons obtained in the study of road run-off from the Kharkiv region differ significantly (3–9 times lower) from the data in the scientific and technical literature, which is possibly due to the different composition of fuel, road construction materials, traffic speed, traffic flow, etc.

Table 2. Ratio of concentrations of aliphatic and aromatic hydrocarbons in gaseous and aqueous emission of PP created by the road and its RF, established in the experimental studies.

Objects that create the PP emission	Ratio aliphatic/aromatic hydrocarbons
Emissions of exhaust gases of internal combustion engines running on gasoline	1.28*
Atmosphere of the roadside space of an urban highway	3.0*
Vehicle emissions (adsorbed by snow cover)	
urban roads	1.19–1.25**
suburban highways	3.93**
gas station	0.95–0.90**
parking lot	0.78**
Summer season surface wastewater**	
suburban highways	1.5–6.9**
gas station	0.33–1.0**
parking lot	0.9**

* – data from gas chromatographic analysis
** – data on the results of petroleum product fractionation using different extractors

The data in Table 1 and 2 were compared with the ratio of concentrations of aliphatic and aromatic hydrocarbons in the soils of the territories adjacent to the roads (Table 3). And only the surface soil layer of the areas closest to the road (1 m from the edge line of the highway) was taken into consideration, because with increasing distance from the road and layer depth, these factors have an increasingly significant impact on the selected indicator.

Table 3. The ratio of concentrations of aliphatic and aromatic hydrocarbons in the soils of areas directly adjacent to highways.

Objects that create the PP emission	Aliphatic/aromatic hydrocarbon ratio
Urban roads	7.93*–8.26*
Suburban highways	1.30*; (0.42–1.08)**
Suburban parking lot	(0.3–1.35)**
Suburban gas station	(1.0–1.35)**

* – data from gas chromatographic analysis
** – data on the results of petroleum product fractionation using different extractors

According to the scientific and technical literature (Table 1), the ratio of aliphatic and aromatic hydrocarbon concentrations ranges from an average of 1.0 in gaseous emissions to 44.4 in road run-offs. As can be seen (Table 3), the ratio of concentrations of aliphatic and aromatic hydrocarbons in soils adjacent to roads is closer to the value of this indicator in gaseous emissions generated on roads. The particularly low value of this indicator in the soils adjacent to suburban highways, where there are no curbs and the soils are directly affected by road run-offs, is noteworthy. According to the results obtained, even on these roads, the composition of PP pollution of soils directly adjacent to the road is formed mainly by gaseous emissions. This is probably due to the fact that gaseous emissions from cars are a permanent factor of soil pollution, and road run-offs are episodic phenomena.

3.2 Pollution of Rain Run-Offs from Roads by Sorbed on Particles and Emulsified PP

The pollutants that contaminate road run-off reach the road surface in the form of exhaust gases, aerosols and fine dust, leaks from standing vehicles, leaks and spills of fuel and lubricants, water used to wash contaminated parts of a vehicle (wheels, underbody) during repair work, gas station operator errors, and large and small "breathers" at gas stations.

The concentration of PPs in road run-offs depends on the duration of the rain-free period (Fig. 1). As can be seen, a sharp increase in the concentration of PPs in surface run-off from the highway of the studied gas station was observed after 6 days, mainly due to a sharp increase in the fraction of conditionally aliphatic PPs.

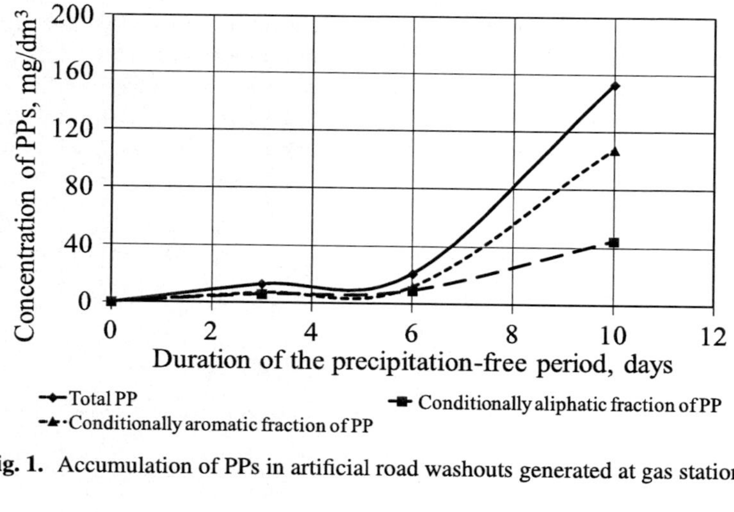

Fig. 1. Accumulation of PPs in artificial road washouts generated at gas stations.

The increase in the concentration of PPs in road and industrial run-off is due to a number of factors: the concentration of suspended solids that actively adsorb PPs, traffic flow, vehicle speed, technical condition of equipment, qualification level of industrial runoff workers, etc.

Suspended solids (SS) are also environmentally hazardous substances that pollute surface wastewater generated on roads, and they are solid particles of exhaust gases from cars – insoluble (solid carbon, metal oxides, silicon dioxide, sulfates, nitrates, asphaltenes) and soluble in organic solvents (resins, varnish, soot, aromatic fractions contained in fuel and oil), abrasion products of tires and brake pads, road dust with road surface particles, etc. These insoluble pollutants are present in surface wastewater in the form of a coarse suspension with a particle size of more than 100 μ and in the form of a fine suspension or emulsion with a particle size of 100–0.1 μ. Colloidal substances in wastewater have a particle size of 0.1–0.001 μ.

It is known that the sludge adsorbs various pollutants, including PPs. The content of PPs in the ash filtered from the surface wastewater of the studied objects is shown in Fig. 2.

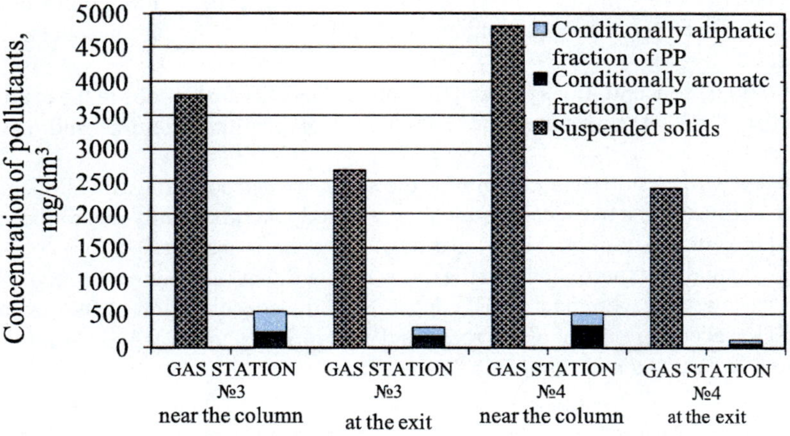

Fig. 2. Content of PPs in the SS from surface run-off from different suburban RF.

As can be seen from Fig. 2, the accumulation of PPs in road run-offs at road infrastructure facilities is influenced by the concentration of SS, which actively adsorb PPs.

The accumulation of PP on the surface of roads is influenced by various man-made factors (presence of bulk cargo transportation, traffic intensity, vehicle condition, road condition, etc.) and natural factors (wind, precipitation). For example, the direction and strength of the wind, which carries the PP from the road surface to the adjacent territories and vice versa, has a major impact. Suburban highways, as a rule, do not have engineering structures in the form of curbs and are located in open areas, and therefore are well blown by winds, which carry the ash.

4 Conclusions

Vehicular traffic on roads is caused an intense flow of PPs to the areas adjacent to the road and the accumulation of PPs in roadside soils. The main factors affecting the intensity of the flow of PP from roads to adjacent areas are the presence of a curb, traffic intensity, engine operation, composition and speed of vehicles, etc.

The transfer of PP from the road to the roadside space occurs in two ways: by air (dispersion in the atmospheric air of gaseous emissions of vehicles) and by water (with surface run-off water formed on roads, as a result of flooding of roadside areas and splashdown). The ratio of the concentrations of aliphatic and aromatic hydrocarbons was chosen as a criterion for identifying the source of PPs in roadside soils.

According to the literature, this ratio in car exhaust gases was 0.03–2.35, and in surface wastewater generated as a result of road run-offs, 27.28–61.38. According to our own research, the ratio of concentrations of aliphatic and aromatic hydrocarbons in the emissions of gasoline-powered internal combustion engines is close to the data of scientific and technical literature, it is 1.28 on a highway; and in the atmosphere of the roadside of a city road 3.0. An indirect study of motor vehicle emissions based on the content of PPs in the snow cover (analysis of PP extracts into chloroform and hexane) revealed a very close ratio of aliphatic and aromatic hydrocarbon concentrations of 0.8–1.3.

The ratio of concentrations of aliphatic and aromatic hydrocarbons in the artificially obtained road run-off was much lower than according to the scientific and technical literature: 1.5–5.9.

Based on the results of determining the ratio of concentrations of aliphatic and aromatic hydrocarbons in roadside soils, it was found that it is mainly due to the supply of PPs in the composition of exhaust gases by atmospheric means.

In the course of experimental studies, it was found that the concentration of PPs in surface wastewater generated on roads depends on the concentration of SS that adsorb PPs. And the accumulation of SS on roads and road infrastructure facilities is influenced by various natural and anthropogenic factors.

References

1. Siddiqui, E., Pandey, J.: Atmospheric deposition: an important determinant of nutrients and heavy metal levels in urban surface runoff reaching to the Ganga River. Arch. Environ. Contam. Toxicol. **82**(2), 159–161 (2022). http://surl.li/jaowc
2. Xinyue, C., Zhiliang, Y., Xianbao, S., Yu, Y., Xi, J.: On-road emission characteristics of VOCs from light-duty gasoline vehicles in Beijing. China Atmos. Environ. **124**(B), 146–155 (2016). https://doi.org/10.1016/j.atmosenv.2015.06.019
3. Polkowska, Ż, Skarżyńska, K., Dubiella-Jackowska, A., Staszek, W., Namieśnik, J.: Evaluation of pollutant loading in the runoff waters from a major urban highway (Gdansk Beltway, Poland). Glob. NEST J. **9**(3), 269–275 (2007)
4. Gillis, P., Parrott, J., Helm, P.: Environmental fate and effects of road run-off. Arch. Environ. Contam. Toxicol. **82**, 159–161 (2022). https://doi.org/10.1007/s00244-021-00906-3
5. Baensch-Baltruschat, B., Kocher, B., Stock, F., Reifferscheid, G.: Tyre and road wear particles (TRWP)—a review of generation, properties, emissions, human health risk, ecotoxicity, and fate in the environment. Sci. Total. Environ. **733**, 137823 (2020). https://doi.org/10.1016/j.scitotenv.2020.137823
6. Mykhailova, L., Fischer, T., Iurchenko, V.: Distribution and fractional composition of petroleum hydrocarbons in roadside soils. Appl. Environ. Soil Sci. Article ID 938703, 6 (2013). https://doi.org/10.1155/2013/938703
7. Iurchenko, V., Melnikova, O., Mykhailova, L., Lebedeva, E., Mikhalevich, N.: Supporting of ecological safety of run-off from the territory of objects of road infrastructure, contaminated

by petroleum products. In: Gopalakrishnan, K., Prentkovskis, O., Jackiva, I., Junevičius, R. (eds.) TRANSBALTICA 2021. LNITI, pp. 10–17. Springer, Cham (2020). https://doi.org/10.1007/978-3-030-38666-5_2

8. Leroy, M., et al.: Performance of vegetated swales for improving road runoff quality in a moderate traffic urban area. Sci. Total. Environ. **566–567**, 113–121 (2016). https://doi.org/10.1016/j.scitotenv.2016.05.027
9. Trenouth, W., Gharabaghi, B.: Highway runoff quality models for the protection of environmentally sensitive areas. J. Hydrol. **542**, 143–155 (2016). https://doi.org/10.1016/j.jhydrol.2016.08.058
10. Markiewicz, A., Björklund, K., Eriksson, E., Kalmykova, Y., Strömvall, A.-M., Siopi, A.: Emissions of organic pollutants from traffic and roads: priority pollutants selection and substance flow analysis. Sci. Total. Environ. **580**, 1162–1174 (2017). https://doi.org/10.1016/j.scitotenv.2016.12.074
11. Björklund, K., Malmqvist, P.A., Strömvall, A.M.: Simulating organic pollutant flows in urban stormwater: development and evaluation of a model for nonylphenols and phthalates. Water Sci. Technol. **63**, 508–515 (2011). https://doi.org/10.2166/wst.2011.251
12. Kalmykova, Y., Björklund, K., Strömvall, A.M., Blom, L.: Partitioning of polycyclic aromatic hydrocarbons, alkylphenols, bisphenol A and phthalates in landfill leachates and stormwater. Water Res. **47**, 1317–1328 (2013). https://doi.org/10.1016/j.watres.2012.11.054
13. Kalmykova, Y., Moona, N., Strömvall, A.M., Björklund, K.: Sorption and degradation of petroleum hydrocarbons, polycyclic aromatic hydrocarbons, alkylphenols, bisphenol A and phthalates in landfill leachate using sand, activated carbon and peat filters. Water Res. **56**, 246–257 (2014). https://doi.org/10.1016/j.watres.2014.03.011
14. DBN B.2.5. – 75:2013 Sewage Systems. External networks and building. Substantive provisions of planning
15. A system of standards in the field of environmental protection and rational use of resources. Atmosphere. Norms and methods for measuring the content of carbon monoxide and hydrocarbons in exhaust gases of vehicles with gasoline or gasoline-fueled engines: DSTU 4277:2004. - [Effective as of July 01, 2004]. - SE "DERZHAVTOTRANSNSNDIPROEKT" (State Automobile Research and Design Institute), 2004 - (National Standard of Ukraine)
16. Gloria, J., Jacob, M., Reiko, H., Jennifer, C.: GS-MS determination of volatile organic compounds in gasoline and diesel emissions. Dartmouth Undergraduate J. Sci. 47–53 (2006). http://surl.li/izumo
17. List of methods for performing measurements (determinations) of the composition and properties of samples of environmental objects, emissions, wastes and discharges provisionally allowed for use by the State Inspectorate of Ukraine, approved by the Head of the State Environmental Inspectorate of Ukraine 2013 (Chief State Inspector of Ukraine for Environmental Protection on 01.03.2013) (in Ukraine)
18. Glyaznetsova, Y., Zuyeva, I., Chalaya, O., et al.: IV the International internet is a symposium on a persorption and extraction of April, 25 - on September, 30, pp. 212–218 (2011). http://www.ich.dvo.ru/~isse/2011/images/stories/files/extraction.pdf
19. Iurchenko, V., Ugnenko, I., Melnikova, O., Mykhailova, L., Lebedeva, E., Viselga, G.: Emissions of polycyclic aromatic hydrocarbons by road transport into roadside areas. In: Prentkovskis, O., Yatskiv, I., Skačkauskas, P., Junevičius, R., Maruschak, P. (eds.) TRANSBALTICA 2021. LNITI, pp. 263–269. Springer, Cham (2022). https://doi.org/10.1007/978-3-030-94774-3_26
20. Polkowska, Ż., Grynkiewicz, M., Zabiegała, B., Namieśnik, J.: Levels of pollutants in runoff water from roads with high traffic intensity in the city of Gdańsk, Poland. Polish J. Environ. Stud. **10**(5) 351–363 (2001). http://surl.li/izuog

Parameter Analysis of the Series Hybrid Vehicle Propulsion System

Andrius Macutkevičius and Raimundas Junevičius(✉)

Vilnius Gediminas Technical University, Vilnius, Lithuania
{andrius.macutkevicius,raimundas.junevicius}@vilniustech.lt

Abstract. A hybrid electric vehicle propulsion system in an off-road application can benefit not only for better mileage, but can also increase motor traction and vehicle power at the wheels. In some specific cases, the battery pack becomes a bottleneck in the system. Not all vehicle batteries can provide the required peak current and voltage to the drivetrain. This depends on battery size, terrain, driving speed, battery temperature and many other parameters. A generator can provide additional power when needed. In this paper the interaction between battery and generator is presented to show the advantages and limitations of the system.

Keywords: Hybrid vehicle · Range extender · Sensitivity analysis

1 Introduction

Electric vehicle (EV) drive architecture become different compared to vehicles with internal combustion engines. Main weight of EV states from energy storage system. There is energy density of electric vehicle battery generally from 30 to 190 Wh/kg [16]. When liquid energy source for combustion engines is 12,1–13,9 kWh/kg [4]. Battery also can make big influence on EV price. Economically battery price is main barrier to their deployment [9]. In ecological state of view electric vehicles usage on roads make less environmental affect. The main ecological problems of EV are related to the manufacturing of vehicle and electricity, maintenance and vehicle utilization.

A component that can extend electric vehicle trip distance using well known technology is electric generator which transform liquid energy source to electricity. Using series hybrid vehicle design, generator usage could be fitted to vehicle as battery charger. Battery charging system increases the overall weight of the vehicle. Given the high fuel energy potential and the high weight of the batteries, a battery charging system makes sense at the moment. By using a small battery and charging from the mains, can be ensure ecological driving over short distances. However, the generator is used when the distance travelled increases or the driving mode increases the energy consumption beyond what the battery system can provide.

O. Prentkovskis et al. (Eds.): TRANSBALTICA 2023, LNITI, pp. 130–139, 2024.
https://doi.org/10.1007/978-3-031-52652-7_13

1.1 Generator in Hybrid Vehicle

The integration of generators in electric vehicles (EVs) has been the primary focus of studies revolves around improving the efficiency, range, and overall performance of electric vehicles by incorporating generators into their design (Fig. 1).

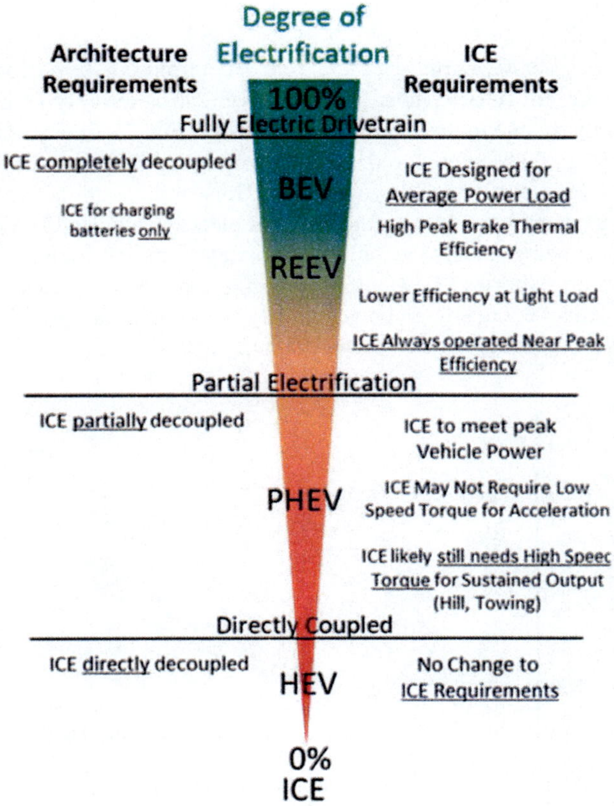

Fig. 1. Degree of electrification for different types of electric vehicles [18].

Researchers have explored various generator types, including internal combustion engines, fuel cells, and micro gas turbines [6] to augment the power supply and extend the driving range of EVs. These generators serve as range extenders, providing additional electrical energy to recharge the batteries or directly power the vehicle's electric motor. Advanced control algorithms, hybridization strategies, and energy management systems are used for optimal utilization of the generator and minimize energy losses. Efficiency optimization has been a major objective. There also can be mentioned possibility to use electric motors as generators then vehicle is braking [15].

The impact of generator usage on environmental sustainability has been a critical consideration. Alternative fuels, such as biofuels or hydrogen, researchers have strived to reduce greenhouse gas emissions and promote a greener transportation ecosystem [14]. It should also be stressed that the focus must be on reducing overall fuel consumption [10]. Furthermore, researchers have investigated the challenges related to the integration of generators in EVs, such as packaging constraints [19], weight considerations [1], and system complexity.

1.2 Battery

Electric vehicle (EV) batteries play a crucial role in determining the performance, range, and overall efficiency of electric vehicles. Electric vehicle batteries typically use lithium-ion (Li-ion) chemistry due to their high energy density (Fig. 2). Energy density refers to the amount of energy that can be stored in a given volume or weight of the battery. Higher energy density allows for longer driving ranges. Researchers are actively working on improving energy density to increase the range of electric vehicles [3]. Other emerging battery chemistries being researched include solid-state batteries, lithium-sulfur batteries [20] and lithium-air batteries [7]. Another important parameter relates to how quickly the battery can deliver energy. High power density is essential for rapid acceleration and efficient regenerative energy when vehicle braking [13].

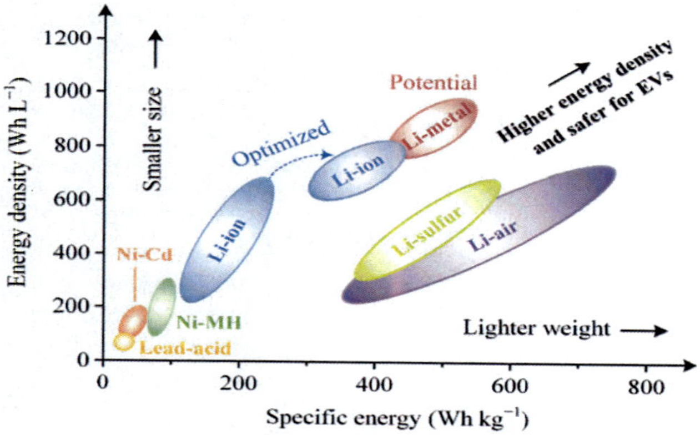

Fig. 2. Different battery types compare chart of energy density and specific energy [3].

State of charge (SOC) indicates the current stored energy level of the battery, usually represented as a percentage. It helps drivers or algorithms estimate the remaining range and plan charging accordingly. Accurate SOC estimation is crucial for optimizing battery performance and vehicle range [5]. The actual and usable capacity of a battery is different because the SOC of each battery has to be within a certain range. This is reason of Battery management system (BMS) creation. BMS monitors and controls various aspects of the battery, including SOC, state of health (SOH), temperature, voltage, and current [12]. It ensures safe and efficient operation, prevents overcharging or over-discharging, and balances cell voltages within the battery pack.

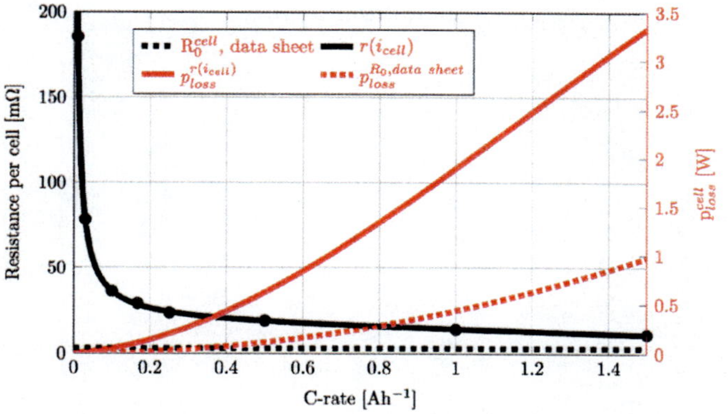

Fig. 3. Battery internal resistance influence on power loss [11].

Effective thermal management involves cooling or heating the battery to maintain it within the desired temperature range, preventing overheating or excessive cold. Temperature affects chemical processes in battery [2]. At low temperatures, the capacity of the battery decreases and the internal resistance increases. The internal resistance affects the internal temperature of the battery. Higher resistance increases the battery's energy dissipation (Fig. 3). Black solid curve shows battery cells measured internal resistance in different discharge rates [11]. But in the main case internal resistance must be constant as dashed black line. Red lines are relevant to power loss, which is calculated.

2 Modelling of Hybrid Vehicle

Modelling based on model described in previous publication [8]. This model describes electric vehicle drive energy distribution. It designed in general form to investigate general energy management parameters. Simple architecture makes calculation faster. This is important parameter for next stage of analysis. Vehicle itself is 4-by-4 off-road vehicle with two electric motors. One on front and other on rear axle. Power to the motors is supplied from battery pack and from diesel generator. Vehicle drivetrain do not have direct link from internal combustion engine to the wheels. Engine is used just to power the battery pack and electric motors.

New model has fixed battery parameters. Model estimate battery voltage according to capacity and system current. Also, model power loss is calculated according to internal resistance:

$$P_{Batt.Loss} = I^2 R_{Batt}. \tag{1}$$

where I – amperage from battery, A; R_{Batt} – battery internal resistance, Ω (This parameter shown in Fig. 4).

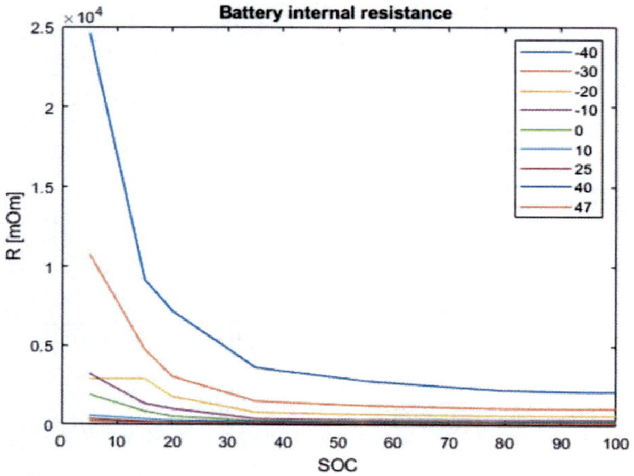

Fig. 4. Battery internal resistance, different temperature curve.

The model also includes a power generator. It switches on and off depending on the state of charge of the battery. The energy generated by the generator is fed into the battery. This generates a positive amperage $I_{Gen.}$ and charges the battery or compensates the energy resulting from the vehicle loads:

$$I_{Gen.} = \frac{P_{Gen.}}{U_{Batt.}}, \tag{2}$$

where $I_{Gen.}$ – current from generator, A; $P_{Gen.}$ – generator power, W; $U_{Batt.}$ – system voltage, V.

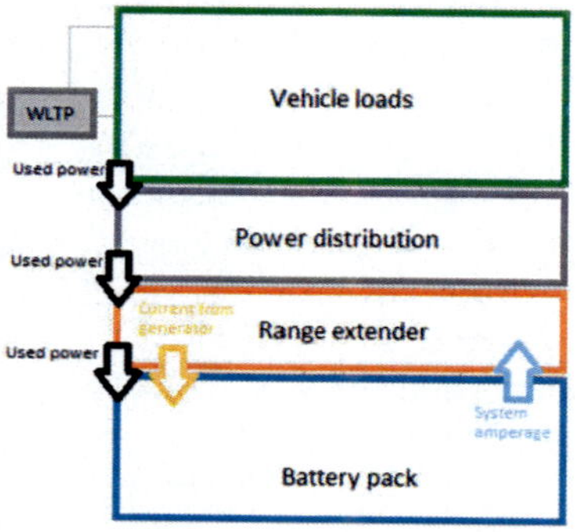

Fig. 5. Electric vehicle model Simulink scheme: vehicle loads (green block), power distribution (grey block), generator (orange block), battery parameters calculation (blue block).

The model created in MATLAB (Fig. 5) uses the specified parameters framework to simulate a vehicle's driving cycle. The input used for the model is the WLTP driving cycle vehicle speed. The WLTP cycle is divided into different parts of driving scenarios from city to highway, making the synthetic drive cycle close to real driving [17].

3 Results

Sensitivity analysis is a crucial method employed in system analysis. By conducting sensitivity analysis, it becomes possible to assess the degree and specific system parameters that exert influence on performance. In the present study, selected five parameters to analyse the system. These parameters encompass the battery energy loss, total energy consumption, generator-generated energy, as well as the average and final levels of battery charge. Parameter range shown in the Table 1. The selected driving mode for analysis assumes that the vehicle's battery initially holds a 40% charge. Charging of the battery commences when the state of charge (SOC) drops to 30%. Once the SOC reaches 80%, the charging process by the generator is halted. The battery pack utilized in the system comprises two 35 kWh batteries. Additionally, the car's air resistance coefficient is measured at 0.56, while the frontal surface area is determined to be 3.5 m^2.

Table 1. Electric Trips used for sensitivity analysis.

Parameter	Value	
	Min	Max
Temperature, °C	10	35
Rolling resistance, –	0.01	0.3
Slope angle, rad	0	0.1745
Vehicle mass, kg	3000	5500
Generator power, kW	60	123,2

Upon analysing the sensitivity of system characteristics to various parameters, it is evident that the selection of the slope angle holds the highest influence on the overall system performance in most cases (Fig. 6). The generator exhibits maximum energy generation when it the power output is elevated and when the batteries operate at higher temperatures. Notably, the battery charge level experiences a significant impact from temperature variations, as lower temperatures result in increased resistance and subsequently affect the charge level.

By examining the hypothetical characteristic surfaces, it becomes apparent that as the vehicle mass and slope angle escalate, both energy consumption and energy generation by the generator increase. When considering the average state of charge (SOC) of the battery, which is influenced by the generator power and temperature, a critical threshold is identified wherein the generator's energy output declines, and the average SOC remains above 54%.

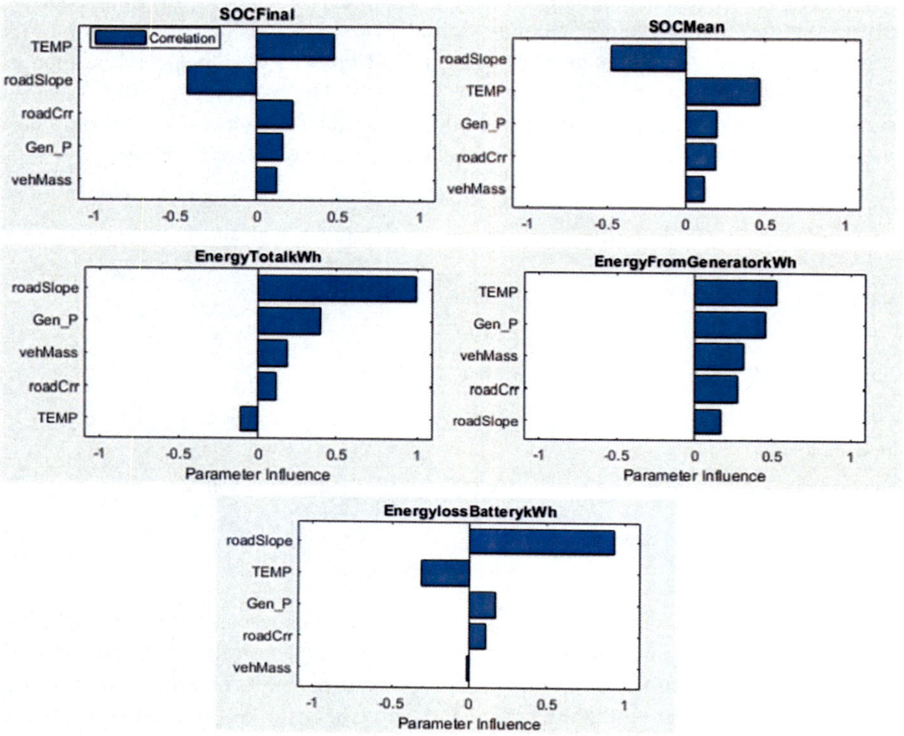

Fig. 6. Parameters influence on characteristics chart.

Furthermore, noteworthy observations are made concerning the vehicle generator. It is observed that the generator generates a significantly higher amount of energy within the power output range of 85 to 110 kW. Similar observations arise when comparing the average SOC value with the car's mass in relation to the generator's power output.

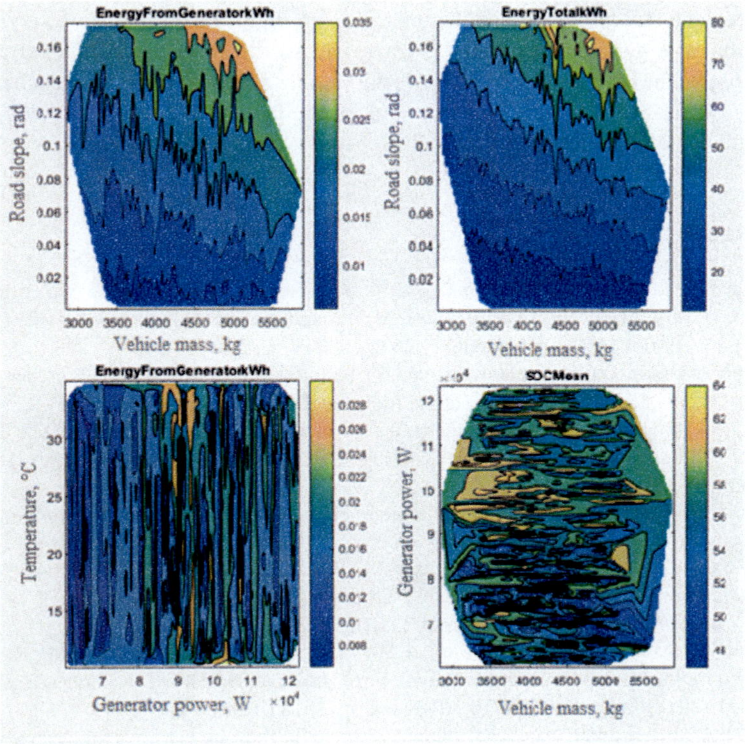

Fig. 7. Hypothetical characteristic surfaces.

4 Conclusion

1. The internal parameters of the vehicle battery are the key factor in vehicle performance in off-road applications. Temperature, internal resistance and SOC affect the discharge current and this has a direct impact on the power and torque at the wheels. Figure 7 shows the balance between energy consumption and energy production. Steep slopes close to 32° require huge amounts of energy, and depending on the vehicle weight, the battery power may not be sufficient to drive at the required speed profile. For this reason, for a vehicle with a total weight of around 5 tonnes, it is necessary to have an 80–110 kW generator.
2. A heavy vehicle drains the battery several times faster than a light one, if we want to have the same output parameters in terms of speed or acceleration. In real life, these values will vary and the heavier vehicle will have lower accelerations, but it will still have a several times higher impact on battery life then the lighter vehicle, as the charge-discharge periods become very fast. This also heats the battery itself and therefore the internal battery resistance varies. This can be seen clearly in Fig. 7 and 6. Extra power from the generator can help to save battery life.

3. Based on the two previous conclusions, it is enough to state that in heavy driving conditions the generator provides extra power to the system and this extra power helps to save the battery pack and allows the vehicle's electric motors to reach nominal parameters at different diving speeds and loads. This results in better electric drive performance on different terrains.

References

1. Alcázar-García, D., Romeral Martínez, J.L.: Model-based design validation and optimization of drive systems in electric, hybrid, plug-in hybrid and fuel cell vehicles. Energy **254**, 123719 (2022). https://doi.org/10.1016/j.energy.2022.123719
2. Ali, H.M.: Thermal management systems for batteries in electric vehicles: a recent review. Energy Rep. **9**, 5545–5564 (2023). https://doi.org/10.1016/j.egyr.2023.04.359
3. Block, A., Song, C.H.: Exploring the potential of material information in patent data: the case of solid-state batteries. J. Energy Storage **71**, 108123 (2023). https://doi.org/10.1016/j.est.2023.108123
4. BU-1007: Net Calorific Value. (2010, September 7). Battery University. https://batteryuniversity.com/article/bu-1007-net-calorific-value
5. Chen, Y., Li, R., Sun, Z., Zhao, L., Guo, X.: SOC estimation of retired lithium-ion batteries for electric vehicle with improved particle filter by H-infinity filter. Energy Rep. **9**, 1937–1947 (2023). https://doi.org/10.1016/j.egyr.2023.01.018
6. Karvountzis-Kontakiotis, A., Andwari, A.M., Pesyridis, A., Russo, S., Tuccillo, R., Esfahanian, V.: Application of micro gas turbine in range-extended electric vehicles. Energy **147**, 351–361 (2018). https://doi.org/10.1016/j.energy.2018.01.051
7. Lee, Y.I., et al.: Mechanical balance of plant design of lithium-air batteries for electric vehicles. J. Energy Storage **70**, 107969 (2023). https://doi.org/10.1016/j.est.2023.107969
8. Macutkevičius, A., Junevičius, R.: Identification of specific system parameter space in initial research stage. In: Prentkovskis, O., Yatskiv, I., Skačkauskas, P., Maruschak, P., Karpenko, M. (eds.) TRANSBALTICA 2022. LNITI, pp. 375–382. Springer, Cham (2023). https://doi.org/10.1007/978-3-031-25863-3_35
9. Mohammadi, F., Saif, M.: A comprehensive overview of electric vehicle batteries market. E-Prime – Adv. Electr. Eng. Electron. Energy **3**, 100127 (2023). https://doi.org/10.1016/j.prime.2023.100127
10. Norouzi, N., Majedi, S.: Dynamic modeling of the effect of Vehicle Hybridization policy on carbon emission and energy consumption (2022)
11. Ollas, P., Thiringer, T., Persson, M., Markusson, C.: Battery loss prediction using various loss models: a case study for a residential building. J. Energy Storage **70**, 108048 (2023). https://doi.org/10.1016/j.est.2023.108048
12. Pannerselvam, S., Narayanan, V., Gireesh Kumar, T.: Energy efficient machine learning based SMART-A-BLE implemented wireless battery management system for both hybrid electric vehicles and battery electric vehicles. Procedia Comput. Sci. **218**, 235–248 (2023). https://doi.org/10.1016/j.procs.2023.01.006
13. Prasanthi, A., Shareef, H., Errouissi, R., Asna, M., Mohamed, A.: Hybridization of battery and ultracapacitor for electric vehicle application with dynamic energy management and non-linear state feedback controller. Energy Convers. Manag. X **15**, 100266 (2022). https://doi.org/10.1016/j.ecmx.2022.100266
14. Puricelli, S., Cardellini, G., Casadei, S., Faedo, D., van den Oever, A.E.M., Grosso, M.: A review on biofuels for light-duty vehicles in Europe. Renew. Sustain. Energy Rev. **137**, 110398 (2021). https://doi.org/10.1016/j.rser.2020.110398

15. Rizzo, G., Tiano, F.A., Mariani, V., Marino, M.: Optimal modulation of regenerative braking in through-the-road hybridized vehicles. Energies **14**(20), Article no. 20 (2021). https://doi.org/10.3390/en14206835
16. Summary and Comparison of Battery Characteristics (2021). https://www.pveducation.org/ko/태양광/summary-and-comparison-of-battery-characteristics
17. Teoh, J., Morris, S., Chew, K.: Performance analysis of electric vehicle in worldwide harmonized light vehicles test procedure via vehicle simulation models in ADVISOR, pp. 215–220 (2019). https://doi.org/10.1109/ICSEngT.2019.8906356
18. Tran, M.-K., et al.: A review of range extenders in battery electric vehicles: current progress and future perspectives. World Electr. Veh. J. **12**(2), Article no. 2 (2021). https://doi.org/10.3390/wevj12020054
19. Widyantara, R.D., Zulaikah, S., Juangsa, F.B., Budiman, B.A., Aziz, M.: Review on battery packing design strategies for superior thermal management in electric vehicles. Batteries **8**(12), Article no. 12 (2022). https://doi.org/10.3390/batteries8120287
20. Xiong, Z., Chen, Q., Qin, J., You, J., Cheng, S.: Tubular NiCo2S4 hierarchical architectures as sulfur hosts for advanced rechargeable lithium sulfur batteries. Int. J. Electrochem. Sci. **18**(6), 100159 (2023). https://doi.org/10.1016/j.ijoes.2023.100159

Vehicle Engineering and Dynamics

Rational Choice of Powers Ration of Engines of Tractor Vehicle and Active Trailer Link

Mykhailo Podryhalo[1] (✉), Ruslan Kaidalov[2], Igor Gritsuk[3], and Vasyl Omelchenko[2]

[1] Kharkiv National Automobile and Highway University, Kharkiv, Ukraine
pmikhab@gmail.com
[2] National Academy of the National Guard of Ukraine, Kharkiv, Ukraine
[3] Kherson State Maritime Academy, Kherson, Ukraine

Abstract. The article proves that the use of automobile trains, in comparison with single automobiles, allows to increase productivity and reduce the cost of automobile transportation. The use of an active trailed link allows to increase the dynamic properties of automobile trains. In addition, the use of an active trailed link makes it possible to increase the energy efficiency of automobile trains by increasing the efficiency of the wheel drive. It was established that increasing the efficiency of the wheel drive is possible due to the rational distribution of torque (traction moments) and power between the wheels of the tractor and the trailed link. According to the criterion of the maximum efficiency of the wheel drive, a method of determining the rational (optimal) distribution of traction moments and, accordingly, the power of the electric motors of the tractor and trailer (semi-trailer) is proposed. It was determined that the rational ratio of traction moments (power) is proportional to the ratio of the total circular stiffness of the tires of the tractor and trailer (semi-trailer). Various options for the distribution of traction moments (powers) for the combination of tractor vehicles and trailers (semi-trailers) with different combinations of the driving wheels number are considered. The presented results show that the ratio of the number of driving wheels of the tractor vehicle and the active trailed link has a significant influence on the choice of a rational ratio of traction moments on the specified links of road trains.

Keywords: Automobile train · Tractor vehicle · Active trailed link · Wheel drive

1 Introduction

The use of automobile trains, in comparison with single automobiles, allows to increase productivity and reduce the cost of automobile transportation. The use of an active trailed link allows to increase the dynamic properties of automobile trains, namely, the ability to move in given road conditions with certain average technical speed, as well as to move uphill. In addition, the use of an active trailed link allows to increase the energy efficiency of automobile trains by increasing the efficiency of the wheel drive. Increasing the efficiency of the wheel drive is possible due to the rational distribution of torques (traction) and power between the wheels of a tractor and a trailed link.

The appearance of automobiles with electromechanical power units greatly simplified the construction of road trains with an active trailed link. However, the issue of

O. Prentkovskis et al. (Eds.): TRANSBALTICA 2023, LNITI, pp. 143–148, 2024.
https://doi.org/10.1007/978-3-031-52652-7_14

rational distribution of traction moments and, accordingly, engine power between the tractor and trailer (semi-trailer) is not considered in the known literature.

2 Analysis of Previous Achievements and Publications

Analysis of the used structures of trailers and semi-trailers, carried out by the authors earlier in the work [1]. Various schemes of drive for driving wheels of the tractor and the trailed link are considered. The scheme of the combined electromechanical drive of trailer (semi-trailer) wheels is also considered. However, the specified scientific work did not consider the issues of determining the efficiency of the wheel drive and the distribution of traction moments between the wheels of the tractor and the trailer (semi-trailer).

The use of a new approach to considering the chassis of an automobile [2] as a single mechanism with a road (link-stand) allowed the authors of the works [3, 4] to determine the rolling resistance of the wheels as a component of the total energy losses in the transmission during power transmission from the engine to the drive wheels. Analytical expressions for determining the efficiency of a single drive wheel of a wheel drive of a two-axle automobile and a multi-axle automobile containing both driving and driven wheels were obtained in [3, 4] works.

The study [4] determined the rational distribution of torques between the front and back wheels of a two-axle electric vehicle:

$$\begin{aligned}(\beta_M)_{opt} &= \frac{M_{\text{tf}}}{M_{\text{tf}}+M_{tb}} = 0{,}5\left(1+\frac{M_{ff}-M_{fb}}{M_e\cdot U_{tr}\cdot\eta_{tr}^{mom}}\right)\\ &= 0{,}5\left[1+\frac{mgfR_d(b-a)}{L\cdot M_e\cdot U_{tr}\cdot\eta_{tr}^{mom}}\right],\end{aligned} \tag{1}$$

where M_{tf}, M_{tb} – total torques on the wheels of the front and rear axles, respectively; M_{ff}, M_{fb} – total torques of rolling resistance of the front and back wheels, respectively,

$$M_{ff} = mgf\cdot R_d\frac{b}{L}; \tag{2}$$

$$M_{fb} = mgf\cdot R_d\frac{a}{L}; \tag{3}$$

where m – mass of an automobile; g – acceleration of gravity, $g = 9.81\ \frac{m}{s^2}$; f – coefficient of rolling resistance of wheels; R_d – dynamic wheel radius; L – longitudinal wheelbase of an automobile; a, b – the distance between the front and back axles of the mass center of an automobile on the horizontal plane passing through the indicated axles; M_e – effective torque of an automobile engine; U_{tr}; η_{tr}^{mom} – gear ratio and transmission efficiency; η_{tr}^{mom} is determined on the part of the transmission to the wheels.

Product $M_e\cdot U_{tr}\cdot\eta_{tr}^{mom}$ for an electric automobile is equal to the total torque $M_{t\Sigma}$ on all wheels,

$$M_eU_{tr}\cdot\eta_{tr}^{mom} = M_{t\Sigma} = M_{\text{tf}} + M_{tb} \tag{4}$$

It should be noted that when talking about the rational distribution of torques between the axes, we mean the optimal distribution for an acceptable objective optimization function. In works [3, 4], as it has already mentioned, the function of the instantaneous efficiency of the wheel drive of a two-axle automobile was adopted as the objective function.

In the above-mentioned works [3, 4], as well as in other well-known works [5–9, 10], when considering the possibility of increasing the energy efficiency of automobile trains, the rational (optimal) distribution of torques (power) between the wheels of tractors and trailer links was not determined.

3 Presentation of the Main Material

The distribution of torques between the front and back wheels of the tractor is rational, chosen in accordance with the ratio (1).

Assuming that the active trailed link has two driving axes, the distribution of torques between which is subject to ratio (1). In this case, the maximum efficiency coefficients of the wheel drive are separately realized on each link of the road train. In this case, the total efficiency of the road train will be:

$$\left(\eta_{dr}^{mom}\right)_{\Sigma} = \beta_I \left(\eta_{dr}^{mom}\right)_{\max I} + (1 - \beta_I)\left(\eta_{dr}^{mom}\right)_{\max II}, \tag{5}$$

where β_I – coefficient of distribution of traction moments on the wheels of the tractor vehicle.

$$\beta_I = \frac{M_{tI}}{M_{tI} + M_{tII}} = \frac{N_{tI}}{N_{tI} + N_{tII}}; \tag{6}$$

where M_{tI}, M_{tII} – total traction moments on the wheels of the tractor vehicle and the trailed link, respectively; N_{tI}, N_{tII} – total moments on the wheels of the tractor and trailer; $\left(\eta_{dr}^{mom}\right)_{\max I}$; $\left(\eta_{dr}^{mom}\right)_{\max II}$ – maximum (extreme) values of the efficiency of the wheel drive of the tractor vehicle and the trailed link, respectively.

The indicated values of the maximum efficiency of the wheel drive of a two-axle automobile are determined in work [4] and are found by the formula:

$$\begin{aligned}\left(\eta_{dr}^{mom}\right)_{\max} &= 1 - \left(\frac{\frac{M_{ff}^2}{C_{cir}} + M_{ff} + M_{fb}}{M_e \cdot U_{tr} \cdot \eta_{tr}^{mom}} + \frac{M_e \cdot U_{tr} \cdot \eta_{tr}^{mom} - 2M_{fb}}{C_{cir}}\right) \\ &= 1 - \left[\frac{\frac{m^2 g^2 f^2 \cdot R_d^2}{L^2 \cdot C_{cir}}\left(a^2 + b^2\right) + mgf \cdot R_d}{M_e \cdot U_{tr} \cdot \eta_{tr}^{mom}} + \frac{M_e \cdot U_{tr} \cdot \eta_{tr}^{mom} - 2mgfR_d \frac{a}{L}}{C_{cir}}\right].\end{aligned} \tag{7}$$

Expression (5) for the total efficiency of the wheel drive of the road train taking into account ratios (4), (7) will take the form:

$$\begin{aligned}\left(\eta_{dr}^{mom}\right)_{\Sigma} = \beta_I &\left\{1 - \left[\frac{m_I^2 g^2 f_I^2 R_{dI}^2}{L_I^2 C_{cirI}}\left(a_I^2 + b_I^2\right) + m_I g f_I R_{dI} + \frac{\beta_I \cdot M_{t\Sigma} - 2m_I g f_I \frac{a_I}{L_I} R_{dI}}{C_{cirI}}\right]\right\} + (1 - \beta_I) \\ &\left\{1 - \left[\frac{\frac{m_{II}^2 g^2 f_{II}^2 R_{dII}}{L_{II}^2 C_{cirII}}\left(a_{II}^2 + b_{II}^2\right) + m_{II} g f_{II} R_{dII}}{(1 - \beta_I) M_{\kappa\Sigma}} + \frac{(1 - \beta_I) M_{t\Sigma} - 2m_{II} g f_{II} \frac{a_{II}}{L_{II}} R_{dII}}{C_{cirII}}\right]\right\}.\end{aligned} \tag{8}$$

In expression (8), all indices "I" refer to the tractor vehicle, and all indices "II" refer to the trailed link.

The rational (optimal) value of the coefficient of distribution of traction moments on the wheels of the tractor vehicle is found using the well-known optimization method:

$$\frac{d}{d\beta_I}\left[\left(\eta_{dr}^{mom}\right)_\Sigma\right] = 0. \tag{9}$$

For maximum function $\left(\eta_{dr}^{mom}\right)_{\max\Sigma}$ for β_I it is necessary for the inequality to hold:

$$\frac{d^2}{d\beta_I^2}\left[\left(\eta_{dr}^{mom}\right)_\Sigma\right] < 0, \tag{10}$$

at the value $\beta_I = (\beta_I)_{opt}$.

As a result of these actions, it was determined that:

$$(\beta_I)_{opt} = \frac{C_{cirI}}{C_{cirI} + C_{cirII}} = \left(1 + C_{cirII}/C_{cirI}\right)^{-1}. \tag{11}$$

This result corresponds to $\left(\eta_{dr}^{mom}\right)_\Sigma = \left(\eta_{dr}^{mom}\right)_{\max\Sigma}$, as inequality (10) holds.

Figure 1 shows the dependence of the ratio of the total circular stiffness of the tires of the wheels of the trailed link and the tractor.

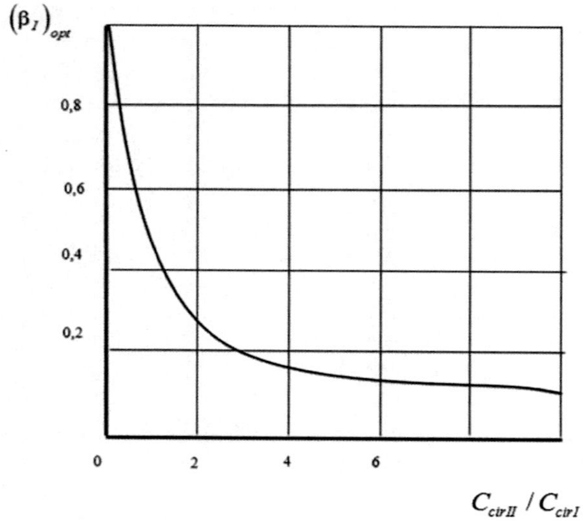

Fig. 1. Dependence $(\beta_I)_{opt} = F(C_{cirII}/C_{cirI})$.

It can be seen from Fig. 1 that with an increase in the ratio C_{cirII}/C_{cirI} the rational value of the coefficient decreases $(\beta_I)_{opt}$ distribution of traction moments (tractive power) on the wheels of the tractor vehicle and, accordingly, an increase of the part on the wheels of the active trailed link.

Let's consider various options for the distribution of traction moments (powers) for the combination of tractor automobiles and trailers (semi-trailers) with different combinations of the number of driving wheels (Table 1). Let's assume that all wheels of the road train have the same circular stiffness.

Taking into account the accepted assumption, formula (11) can be presented in the form:

$$(\beta_I)_{opt} = (1 + n_{II}/n_I)^{-1}, \tag{12}$$

where n_I/n_{II} – the number of wheels of the tractor and trailer, respectively.

The results presented in Table 1 show that the ratio of the number of driving wheels of the tractor and the active trailed link has a significant impact on the choice of a rational ratio of traction moments on the specified links of road trains. The use of the proposed recommendations will improve the energy efficiency of road trains in the established traffic modes.

Table 1. The influence of the scheme of the automobile chassis on the coefficient $(\beta_I)_{opt}$ distribution of traction moments (power) on the wheels of the tractor vehicle.

Scheme of the automobile chassis	$(\beta_I)_{opt}$
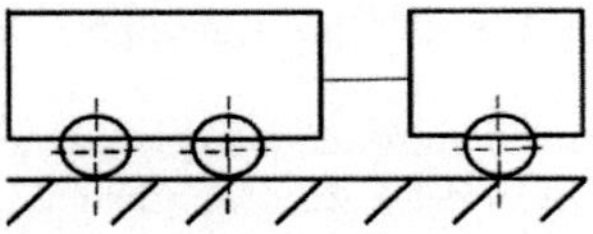	0.67
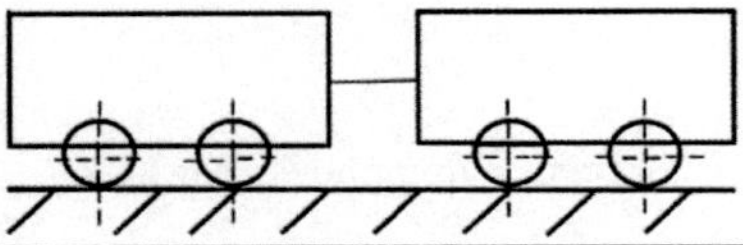	0.50
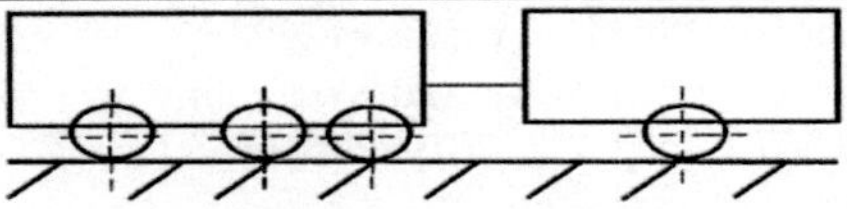	0.75
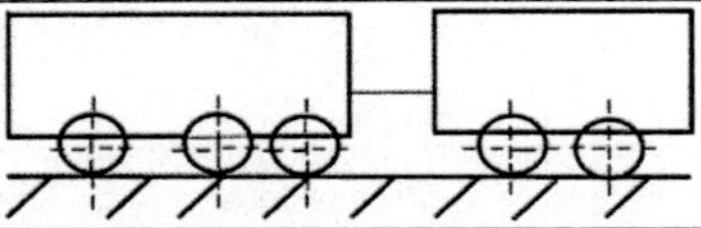	0.60
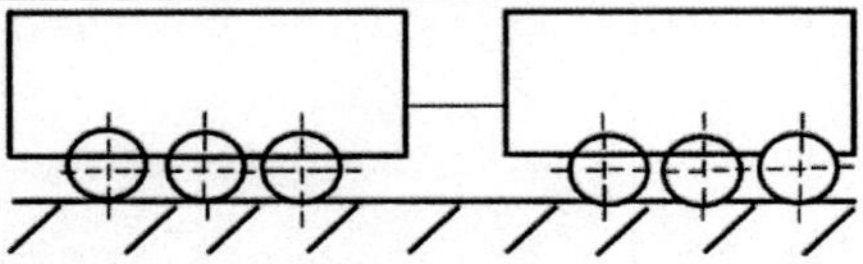	0.50
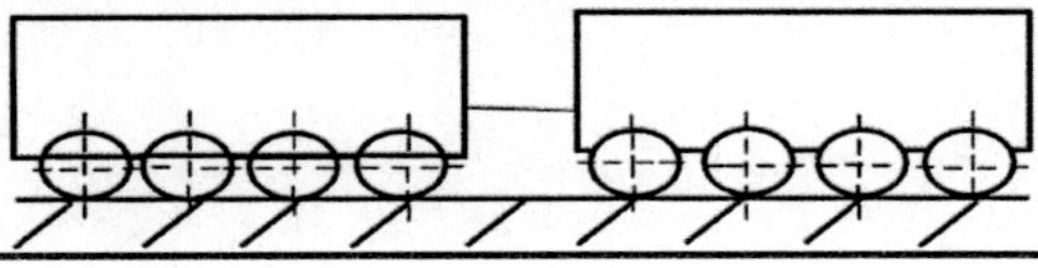	0.50

4 Conclusions

As a result of the conducted research, the total coefficient of efficiency of the drive is proposed as an indicator of energy efficiency of an automobile train with an active semi-trailer.

For road trains with electric and combined electromechanical drive of the driving wheels of the tractor vehicle and an active trailed link, a rational ratio of the distribution of traction moments (tractive power) to the wheels of the tractor vehicle is determined. The specified coefficient depends on the ratio of the total circular stiffness of the tires of the wheels of the tractor vehicle and the trailed link.

References

1. Mazin, O.: Increasing the energy efficiency of automobiles during maneuvering by reducing unproductive energy consumption. Sci. J. **2**(15), 10–15 (2020)
2. Tarasov, Y.: Scientific bases of ensuring the technical level of motor vehicles during design and modernization. Sci. J. **3**(17), 7–13 (2021)
3. Artemov, N., Lebedev, A.: The rational accelerated method and its application in dynamics of automobiles. Miskdruk **4**, 51–56 (2012)
4. Poliakov, V., Sakhno, V.: Three-Track Road Trains. Maneuverability. National Transport University, Kyiv (2013)
5. Sakhno, V., Gumeniuk, P., Marchuk, R.: Maneuverability and safety of movement of three-link road trains of various layout schemes. Visnyk of Donetsk Acad. Automobile Transp.: Sci. J. **4**, 12–17 (2011)
6. Kutkov, H.: Tractors and automobiles: theory and technological properties. INFRA, Kharkiv (2016)
7. Shevchuk, R.: Operating indicators of tractors and automobiles: workshop on calculating indicators. Lviv National Agrarian University, Lviv (2018)
8. Volkov, V., Boboshko, A.: Vehicle dynamics. KNAHU, Kharkiv (2008)
9. Sakhno, V., Glinchuk, R., Marchuk, R., Onyshchiuk, V.: Maneuverability of three-link road trains of the "B-Double" type. Probl. Road Transp. **7**, 187–198 (2010)

Progressive Tool Modernization Using Sensor Technology in Automotive Parts Manufacturing

Juras Skardžius, Saulius Nagurnas(✉), and Vidas Žuraulis

Vilnius Gediminas Technical University, Plytinės g. 25, 10105 Vilnius, Lithuania
{juras.skardzius,saulius.nagurnas,vidas.zuraulis}@vilniustech.lt

Abstract. In this research presented an experimental study, which extends usage of eddy current sensors in outdated progressive stamping tool and overall press system in automotive parts manufacture. During the test various tool sections were modified in order to implement sensor's and keep optimal design principle. Experiments were carried out using 4 mm thickness, cold rolled steel material. The purpose of the research was to automatically detect material scrap before it leaves imprints on part and stop the press in order not to damage stamped part of tool itself. Scrap thickness that needs to be detected was established by experimental way and was 0.1 mm. Research results were consist of separate detection steps and conclusion taken only after serial type production with simulated more severe conditions than actual production. Overall results were influenced by the design of the existing tool; however, scrap detection on any place of strip was achieved and additional scrap detection map was created. Determinate that detection is influenced by tool dimensions, existing design and used sensor's type.

Keywords: Progressive tool · Stamping · Sensor · Quality · Eddy current sensor · Performance · Optimization · Automotive · Parts manufacture

1 Introduction

In progressive stamping, the key element of the process is the tool used to produce the parts. Other important elements of the process are the press, the strip feeding and straightening line, and the part quality inspection devices. Product tool is a mechanical device, so its modernization has the greatest added value for the company performance. This requires sensors that monitor the work of the tool, a controller that processes the incoming data and is able to make decisions. According to the general process correlation, the controller should evaluate the quality of the part in real time and inform the operator in case of instabilities.

Plant machinery and prearranged maintenance are the fey factors for effective manufacturing in automotive manufacturing [1]. Historically, the function of the first sensor's was to protect the press and tool from overloading [2]. Overloads can be caused by an incorrectly set press slide, material feed pitch or raised scrap metal. In order to avoid overload, all dies are equipped with mechanical or hydraulic safety systems [3]. Sensor systems can be used to monitor the applied force and track the stability of the process [4].

O. Prentkovskis et al. (Eds.): TRANSBALTICA 2023, LNITI, pp. 149–161, 2024.
https://doi.org/10.1007/978-3-031-52652-7_15

A detailed review is provided by Su and Ravindran [5], where authors discuss various types of sensors, their application, subtleties of installation, principles of operation and summarize the possibilities of use. An important point is the material feed misalignment. This problem is probably the most common, apart from wear of parts, in progressive stamping, so every tool has a automatic feed sensor called diesaver. It is an optical or inductive type sensor standing at the end of the tool, which monitors the position strip by press cycle so that strip stops in the specified range every time. Both types of sensors have their advantages, but according to Reghem [6], their reliability is adversely affected by the lubricants used in the process and the resulting metal dust. Considering these factors, inductive sensors are more resistant, but their efficiency is reduced by losses due to the magnetic field (eddy currents) that can form in the strip. Reghem mentions eddy currents as a side effect, although this principle is based on a widely used sensor. An article published by Kevin Kuang [7] compares inductive, proximity and eddy current sensors. All these sensors work well in aggressive environments and allow non-contact measurements. Also, their operating principle is similar, only inductive and proximity sensor's measure voltage drop, while eddy current sensor's measure the change of complex resistance in their coil. According to article authors, the shortcomings of inductive and proximity sensors are related to the ferromagnetic core in them and the so-called iron loss due to the magnetic field generated by it. This affects: 1) an uneven output signal due to uneven absorbed of the magnetic field, as a result of which calibration is required; 2) high temperature expansion coefficient of the core, as a result, the accuracy of the sensor's also varies at different temperatures; 3) so-called iron loss increases with frequency, so the upper frequency limit is 50 Hz.

Eddy current sensor's do not have a ferromagnetic core, so the issues listed above can be avoided. Advantages of eddy current sensor's: 1) high measurement frequency, which can reach 100 kHz; 2) high accuracy - available a resolution is up to 0.5 μm; 3) temperature stability.

Important to note that in order to obtain accurate data for the discussed application, the distance to the measured object cannot be greater than 4 mm. This is the biggest disadvantage of eddy current sensors. Due to the measured distance, the possibilities of their use are narrowed, however they are suitable for monitoring scrap metal or shavings that have fallen on the moving strip. When scrap appears on the strip, the pressure plate (lifter) tilts and the distance to the sensor changes. Due to the high accuracy, the sensor allows detecting the anomaly of the product, but does not indicate where it occurred.

The change in load caused by the raised scrap changes is very small compared to the total load on the press, so their detection is complicated. Sophisticated signal processing methods are used to solve this problem. For example, Zhou et al. [8] analyzed the possibility of using strain sensors installed to monitor the press load to detect errors then part moving to the next stamping station. Using a method based on reoccurrence plots, the researchers successfully detected such events, although at first glance the load signal was virtually indistinguishable from the one obtained when the part was correctly moved. Another example is the study by Li and Bassiuny [9]. These authors draw attention to the low strain signal-to-noise ratio and the transient nature of this signal, which makes it difficult to extract useful information from it using traditional methods. As an alternative, the authors test the latent model, wavelet and Hilbert–Huang transforms. The effectiveness of these methods is tested in the progressive molding of an automobile engine hood, during which piezoelectric strain sensors are used to detect incorrect tape feeding. Based on the test results, the authors conclude that the Hilbert–Huang transform is the most suitable for this.

2 Research Methodology

Referring to the proposed research hypotheses and the planned practical use, it was decided to install two eddy current sensors on stop blocks of the tool. The technical data of the eddy current sensors are presented in Table 1, and their installation locations are shown in Fig. 1.

Table 1. Technical specification of eddy current sensor.

Characteristic	Value
Manufacturer	TRsystems GmbH, Division Unidor
Model	WSD S2/10MF
Measurement range	0–2 mm
Linearity	±0.12 mm
Repeatability	0.02 mm
Resolution	<0.001 mm
Response time	<1 ms

The purpose of these sensors operation is based on distance measurement: EC1 and EC2 indicate the position of the stripper plate relative to its supports. Work area dimensions of tested tool are 1,250 mm in length and 870 mm in wide.

All sensors were connected to the data processing and control system compactPRESS smartLINE. The key element is an intelligent controller that continuously monitors and analyzes the signals of the digital and analog sensors connected to the system to protect the press with the tool installed in it and ensure an optimal and controlled

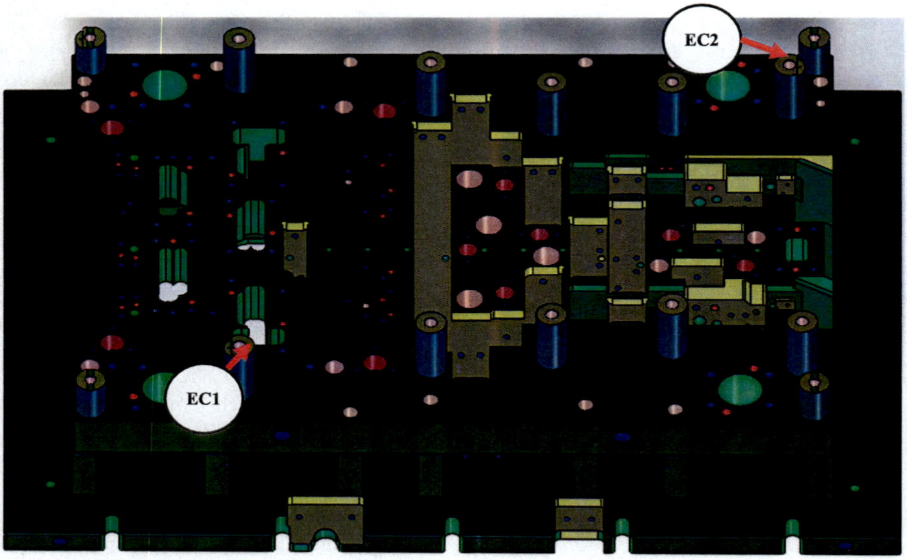

Fig. 1. Sensor's EC1 and EC2 position.

Table 2. Technical data of data processing system.

Characteristic	Value
Manufacturer	TRsystems GmbH, Division Unidor
Model	UNiDOR compactPRESS smartLINE
Controller	Ether CAT Master RT
Press angle setting module	CEVV58S-00117
Analog signal input module	Ether CAT Box

production process. Other elements of this system are described in Table 2, while their interconnections and practical application in a stamping line are shown in Fig. 2.

The task of the research was to determine the capabilities of the eddy current sensors installed in our experimental tool and check whether they are sufficient to detect the moments when unauthorized imprints are formed during stamping process. In order to find out if hypothesis is effective, the following steps were done:

1. After analyzing the real production data, it was determined what high scrap leaves intolerable imprints in the part and consider this height as the limit height h_l;
2. Two eddy current sensors were installed in the stop blocks of the tool bottom plate in the opposite corners (see Fig. 2);
3. Eddy current sensors were calibrated so that they react only when the difference in the distances measured by them is greater than h_l. It is important to avoid unnecessary stops of the stamping process;

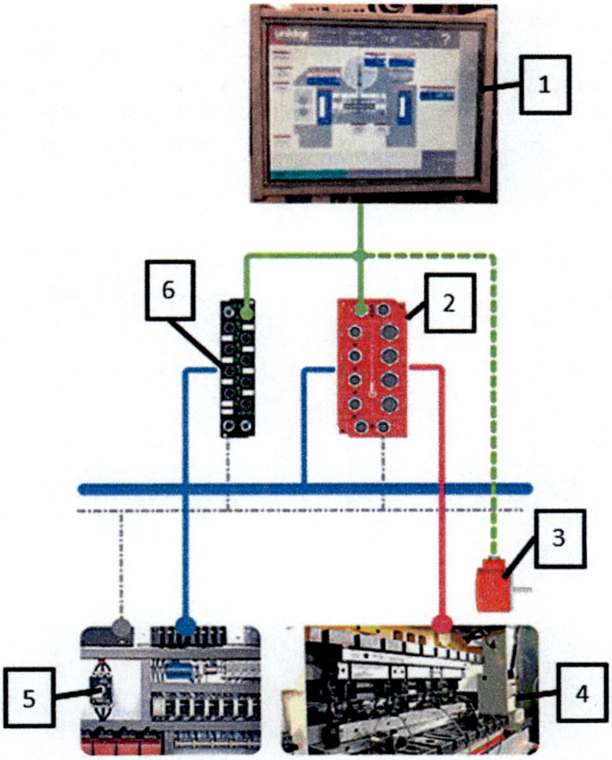

Fig. 2. Sensor's Communication scheme of sensor's system: 1) display screen; 2) press sensor's connection adapter; 3) additional input; 4) press sensor's; 5) tool sensor's; 6) tool sensor's connection adapter.

4. Placed fragments of calibrated metal sheet of different thicknesses on the strip in various places of the tool and made strokes. In order for the experiment to be successful, the system needed to detect all fragments with a thickness greater than h_l and not react to thinner fragments;
5. Next system tested using several fragments at once. In this way, checked if there can be a situation where scrap falls in several places of the strip, so pressure divided equally and system could not detect it;
6. Finally tested the system under real conditions. To speed up the experiment, used the maximum allowable lubrication and removed the waste repellers from the punches, which increase the probability that the waste stick to the punch and therefore be lifted and fall onto the belt.

3 Results and Their Analysis

The experiment was started with a nominal 4 mm thickness, S355MC material at 25 cycles per minute speed, using the level of lubrication specified in the tool's instruction manual and 50 parts produced at this rate. During the production, readings of sensor's

EC1 and EC2 were taken. Based on the obtained data, the signal curves of both sensor's, presented in Fig. 3 and Fig. 4 were drawn, and calculated how much the measured distance varies at each point. After evaluating the obtained results, conclusion was taken that the "noise" of the sensor capture is 50 μm amplitude. Due to the noise, an inactive zone can mean that small foreign objects will be "invisible" and the reliability of the system will decrease.

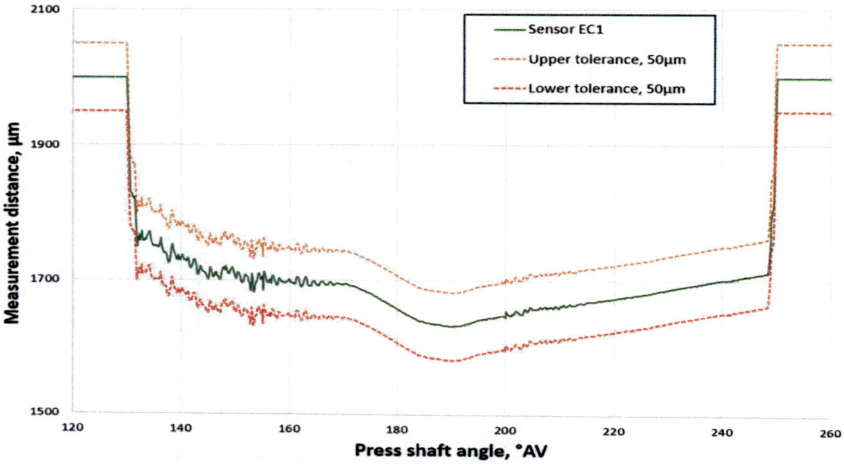

Fig. 3. Upper and lower limits of sensor EC1.

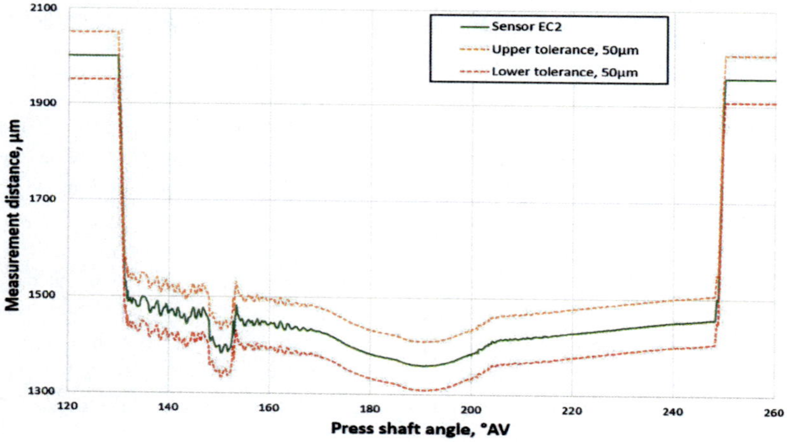

Fig. 4. Upper and lower limits of sensor EC2.

To determine an acceptable imprint dept, 5 × 10 mm hardened steel (HRC – 65) plate on the strip were stuck, imitating scrap waste. After completing the full cycle with different thickness steel plates, parts were assembled and inspected. Results of the research shown in Fig. 5. Decision was taken that imprints left by 50 μm plate are acceptable, by 100 μm plate are marginal, and by 200 μm plate are not allowed.

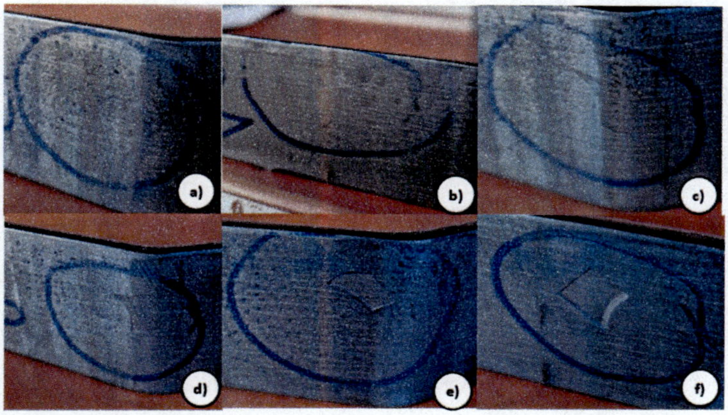

Fig. 5. Imprints on the part caused by different thickness steel plate: a) - 50 μm; b) - 100 μm; c) - 150 μm; d) - 200 μm; e) - 300 μm; f) - 400 μm.

To make sure that the sensor's detect waste metal of the same thickness anywhere on the strip, "scrap waste" map were tried to create. After training the system in serial production mode, at a constant speed of 25 strokes per minute, on stopped tool a plate of calibrated thickness in specially selected places were glued (Fig. 6) and tried to restart it, observing whether the sensors would detect a scrap waste or not. The position and thickness of the fragment during the tests are presented in Table 3, and the results of foreign body detection are presented in Table 4 ("+" means that the test was successful and the sensor's detected the foreign body, and "-" means that the test was unsuccessful).

As shown in test results, sensors were not able to detect smaller scrap waste than 0.3 mm, even when sensor's median was lower than foreign body thickness. This could be explained by testing material thickness itself (thick material compensates scrap thickness by deformation) and design of the tool (comparatively, scrap plate thickness to stripper plate dimensions). The decision to carry out an experiment to examine production stability with implemented sensor's was taken.

Press was left to work in automatic - serial mode, 25 cycles per minute speed and all process stops were recorded. Their summary is presented in Table 5.

As shown in the log, few unplanned stops occurred in first hour, but is not unusual in early production and stalls occurred unrelated to the sensor's being tested. Scrap detection curves are shown in Fig. 7.

Scrap plates were stuck to the oiled coil and carried deep into the tool. Sensors EC1 and EC2 immediately ordered to stop the process, so the protection worked properly. Later, 0.3 mm plate on coil were stuck. In this case, the sensors did not stop the process, although they were not supposed to as determinate in the earlier test. Test lasted around 5 h and none of production stalls occurred related to the sensors.

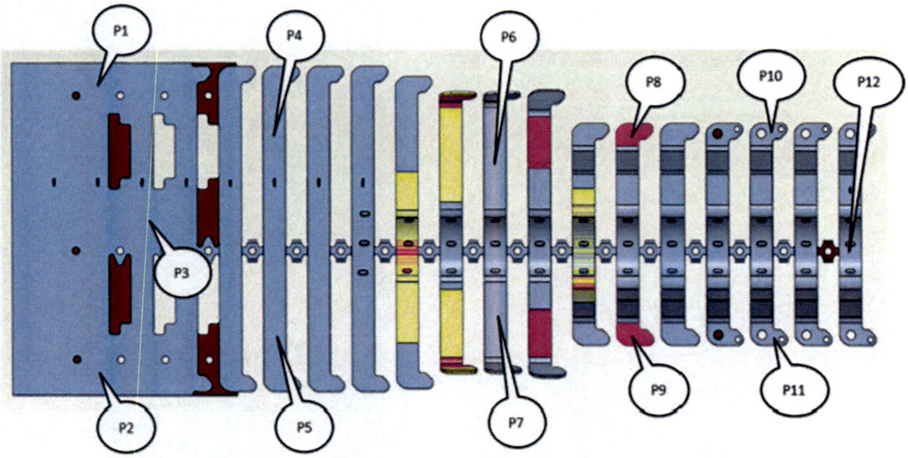

Fig. 6. Positions of steel fragments used in research.

Table 3. Place and scrap thickness combinations used in research.

Test number	Fragment position and thickness, mm											
	P1	P2	P3	P4	P5	P6	P7	P8	P9	P10	P11	P12
1	0.4	–	–	–	–	–	–	–	–	–	–	–
2	–	0.4	–	–	–	–	–	–	–	–	–	–
3	–	–	0.4	–	–	–	–	–	–	–	–	–
4	–	–	–	0.4	–	–	–	–	–	–	–	–
5	–	–	–	–	0.4	–	–	–	–	–	–	–
6	–	–	–	–	–	0.4	–	–	–	–	–	–
7	–	–	–	–	–	–	0.4	–	–	–	–	–
8	–	–	–	–	–	–	–	0.4	–	–	–	–
9	–	–	–	–	–	–	–	–	0.4	–	–	–
10	–	–	–	–	–	–	–	–	–	0.4	–	–
11	–	–	–	–	–	–	–	–	–	–	0.4	–
12	–	–	–	–	–	–	–	–	–	–	–	0.4
13	0.3	–	–	–	–	–	–	–	–	–	–	–
14	–	0.3	–	–	–	–	–	–	–	–	–	–
15	–	–	0.3	–	–	–	–	–	–	–	–	–
16	–	–	–	–	–	0.3	–	–	–	–	–	–
17	–	–	–	–	–	–	0.3	–	–	–	–	–
18	–	–	–	–	–	0.2	–	–	–	–	–	–
19	–	–	–	–	–	–	0.2	–	–	–	–	–
20	–	–	–	–	–	0.1	0.1	–	–	–	–	–
21	–	–	–	–	–	–	–	–	–	0.3	–	–
22	–	–	–	–	–	–	–	–	–	–	0.3	–
23	–	–	–	–	–	–	–	–	–	0.1	0.1	–
24	–	–	–	–	–	–	–	0.3	0.3	–	–	–
25	–	–	–	–	–	–	–	0.2	0.2	–	–	–
26	–	–	–	–	–	–	–	0.1	0.1	–	–	–
27	–	–	–	–	–	–	–	–	–	–	–	0.3
28	–	–	–	–	–	–	–	–	–	–	–	0.2
29	–	–	–	–	–	–	–	–	–	–	–	0.1
30	–	–	–	0.3	0.3	–	–	–	–	–	–	–
31	0.3	–	0.3	–	–	–	–	–	–	–	–	–

(continued)

Table 3. (*continued*)

Test number	Fragment position and thickness, mm											
	P1	P2	P3	P4	P5	P6	P7	P8	P9	P10	P11	P12
32	–	0.3	–	0.3	0.3	–	–	–	–	–	–	–
33	0.3	0.3	0.3	–	–	–	–	–	–	–	–	–

Table 4. Test results.

Test number	Result	Test number	Result	Test number	Result
1	+	12	+	23	+
2	+	13	–	24	+
3	+	14	–	25	+
4	+	15	–	26	+
5	+	16	+	27	+
6	+	17	+	28	+
7	+	18	+	29	–
8	+	19	+	30	–
9	+	20	–	31	–
10	+	21	+	32	–
11	+	22	+	33	–

Table 5. Serial production mode test results.

Cause No	Time	Description
1	08:15	Start
2	08:21	Stoppage due to part quality
3	08:43	Stoppage due to part quality
4	08:58	Stoppage due to part quality
5	09:25	Stoppage due to manually added 0.4 mm scrap plate sensors EC1 and EC2 detection
6	09:45	Manual press stoppage by operator
7	10:06	Stoppage due to part quality
8	10:25	Stoppage due to manually added 0.3 mm scrap plate which sensors EC1 and EC2 did not notice
9	10:35	Stoppage due to manually added 0.4 mm scrap plate sensors EC1 and EC2 detection
10	10:39	Manual press stoppage by operator
11	10:51	Stoppage due to manually added 0.4 mm scrap plate sensors EC1 and EC2 detection
12	11:35	Stoppage due to manually added 0.3 mm scrap plate which sensors EC1 and EC2 did not notice
13	12:40	Stoppage due to manually added 0.4 mm scrap plate sensors EC1 and EC2 detection
14	13:00	End of production

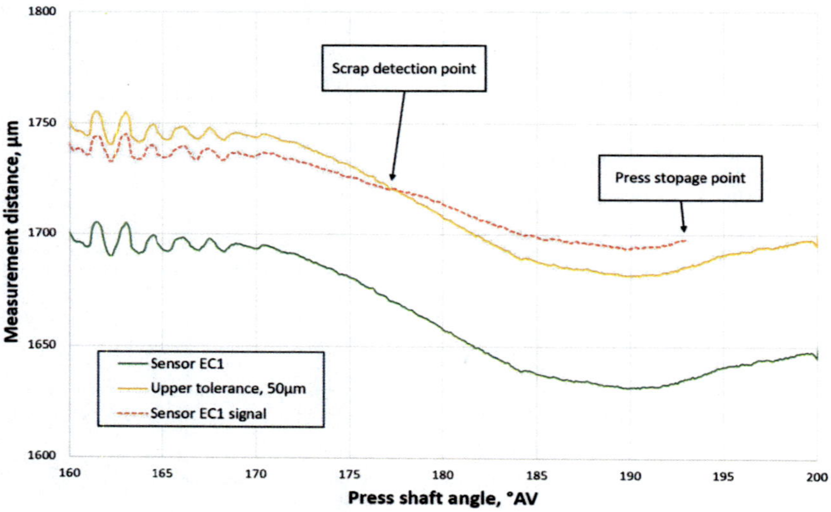

Fig. 7. 0.4 mm thickness scrap detection by EC1 sensor.

4 Conclusion

Experiment of eddy current sensors implementation into progressive tool for the scrap detection considering successful, although further development is still expected. The result demonstrated a high reliability of scrap detection as well as a low frequency of unplanned triggers. Unfortunately, it was not possible to reach the desired limit of 0.1 mm thick metal body, but this can be explained by the ratio of material and scrap thickness. 0.4 mm scrap body detection was achieved with implemented sensor's, which is only 10% of used material thickness. It is likely that this system would be more effective in smaller progressive tools due to smaller diagonal distance.

References

1. Oke, S.A., Bamigbaiye, A., Oyedokun, O.I.: A mathematical model for evaluating accelerated wear due to inefficient maintenance. Transport **21**(2), 105–111 (2006)
2. Dickerson, G.: Demystifying die protection. Sensors and controls: eyes and brain of die protection. Stamping J. (2010)
3. Zhou, C., et al.: An automatic process monitoring method using recurrence plot in progressive stamping processes. IEEE Trans. Autom. Sci. Eng. **13**, 1102–1111 (2016)
4. Overly, S.: Recovering from press overload: protection systems help to sense overload and minimize machine, tool damage. The Fabricator. https://www.thefabricator.com/thefabricator. Accessed 30 May 2023
5. Ravindran, D., Su, Y.-C.: Sensor's for sheet metal forming, in sheet metal forming - processes and applications. ASM Int. Mater. Park 301–315 (2012)
6. Reghem, P., et al.: Precision of an eddy current sensor with magnetic circuit losses (1994)
7. Kuang, K.: Magnetic Sensor's - Principles and Applications (2012)

8. Zhou, C., et al.: An automatic process monitoring method using recurrence plot in progressive stamping processes. IEEE Trans. Autom. Sci. Eng. **13**(2), 1102–1111 (2016)
9. Li, X., Bassiuny, A.M.: Transient dynamical analysis of strain signals in sheet metal stamping processes. Int. J. Mach. Tools Manuf **48**(5), 576–588 (2008)

The Conceptual Model for Increasing Wear Resistance and Lubrication Efficiency for Non-conformal and Conformal Friction Units from the Standpoint of Micro-EHD Theory

Mykola Dmitrichenko, Oleksandr Milanenko(✉), Anatoliy Savchuk, Yuliia Turytsia, Maksim Pavlovskiy, Oleksiy Kushch, and Andriy Bobro

National Transport University, Kyiv, Ukraine
milanmasla@gmail.com

Abstract. A conceptual model for improving wear resistance and lubrication efficiency for non-conformal and conformal friction units operating under extreme operating conditions with insufficient lubrication, which causes micro-EHD breaks in the lubricating layer during the transition to different modes from fluid to boundary friction, taking into account the multifactorial nature of the task of assessing contact-mechanical rheological and physicochemical aspects according to the criterion evaluation for the realization of stable micro-EHD lubrication according to the criterion of lubrication mode, selection of the optimal microgeometry for the contact-mechanical properties of the stress-strain state of friction surfaces in the actual plane according to the criterion of microgeometry; increasing the share of hydrodynamic pressure of lubricating layers relative to the contact pressure on the protrusions of micro irregularities by the pressure criterion; taking into account rheological parameters for changing the piezoelectric viscosity coefficient with increasing pressure by the rheological criterion; assessing the appearance of a plastic-strain state by the plasticity criterion; modifying and adjusting the qualitative and quantitative chemical composition of components in the lubricating medium to reduce the average local temperature of friction surfaces by the temperature criterion.

Keywords: Micro-EHD lubrication · Wear resistance · Lubrication efficiency

1 Introduction

Studies for non-conformal friction units having concentrated local (point) contact with a small actual contact plane, which are characteristic of various friction bearing units (ball and tubular contact roller bearings) and non-conformal ICE units (cam mechanisms and some gearboxes), timing units and some gears) and for conformal ICE units with linear contact (between the upper compression ring and the inner wall of the ICE cylinder liner), must be carried out within the framework of contact-mechanical, rheological and physicochemical aspects.

O. Prentkovskis et al. (Eds.): TRANSBALTICA 2023, LNITI, pp. 162–166, 2024.
https://doi.org/10.1007/978-3-031-52652-7_16

At the present stage of development of tribology, there are insufficiently solved applied problems of determining the maximum contact stresses and strains generated in discrete areas in the actual plane in the near-surface deformed contact zone and the level and location of the center of occurrence and the occurrence of the zone of localization of the tangential stress concentrator in the subsurface contact zone, and of generalizing the stress-strain state of real contact friction surfaces in actual areas of frequent fracture in depth and along the rolling direction. An appropriate assessment of the stress-strain state of friction surfaces will make it possible to select the optimal microgeometry (shape) of the contact according to the corresponding equivalent mechanical properties. The problem is that existing calculations for assessing the stress-strain state of friction surfaces are based on solving the Hertzian contact problem, in which the contact microgeometry is taken from a hypothetical nominal contact plane. But, in lubricated nonconformal friction units operating under extreme operating conditions, the actual contact plane differs significantly from the nominal plane in the zone of discrete contact areas along the vertices of micro-bumps, in terms of the necessary conditions for the breakdown of the lubricating layer's EHD. Thus, the corresponding calculations on the nominal plane of the Hertz problem can significantly distort the results, for example, for conditions of insufficient lubrication (mixed friction), when the microgeometry along the actual plane (Fig. 1a) differs significantly from the microgeometry (Fig. 1b) under conditions of abundant lubrication (liquid friction).

At the present stage of development of tribology, the boundary transition from fluid to mixed friction, which occurs under conditions of insufficient lubrication, is best explained from the standpoint of the micro-EHD theory of lubrication. The characteristic features of micro-EHD lubrication are clearly shown in Fig. 2a, b.

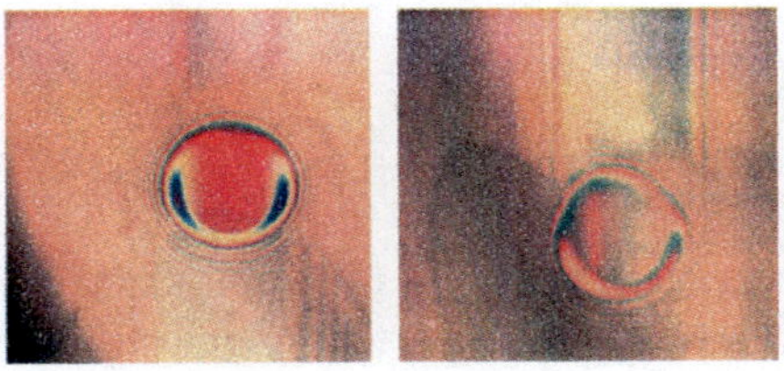

a) b)

Fig. 1. a, b. Micro-interferograms of the actual point contact plane of a non-conformal assembly under conditions of abundant lubrication (a) and insufficient lubrication (b).

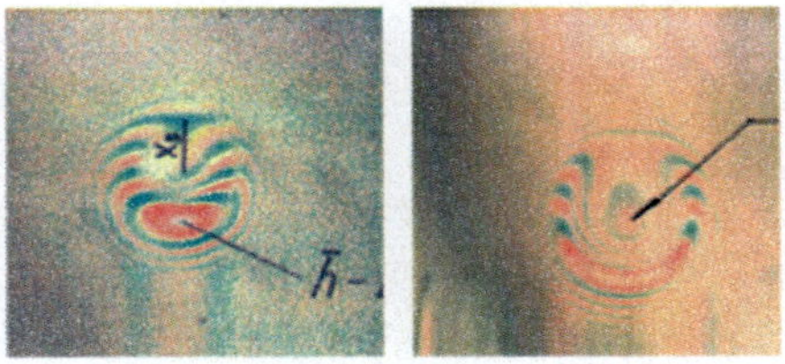

a) b)

Fig. 2. a, b. Micro-interferograms of the micro-EHD of the point contact zone of a non-conformal assembly with a characteristic displacement (a) and formation of a depression (b) of the film thickness in the central contact region in the low-temperature regime.

Under extreme conditions of operation of non-conformal friction units, which is associated with the occurrence of insufficient lubrication at low temperatures, when the lubricant does not have time to enter the contact zone and relax in time in the zone at the contact exit due to the manifestation of non-Newtonian properties of lubricating layers, the zone (cavity) of minimum thickness begins to move to the central contact area (see Fig. 2a) and form there, forming a second depression already in the central contact area (see Fig. 2b), as shown in the micro-interferograms made by optical interferometry. The corresponding depressions in the film thickness affect the patterns of hydrodynamic and contact pressures. The formation of micro-EHD of the lubricating layer occurs when the lubrication efficiency is characterized by the predominance of the contact pressure $P_{a'}$ in discrete contact areas over the hydrodynamic pressure $P_{f'}$ of the lubricating layer, that is, when $P_{a'} > P_{f'}$, but the total pressure P does not exceed the critical pressure value $P_{кр}$. Exceeding the total pressure P by ΔP (the value of the bearing capacity) can lead to ruptures in the lubricating layer and the appearance of burr. The thickness of the lubricating layer in the micro-EHD contact zone of non-conformal friction units (Fig. 3) has certain features that distinguish it from the thickness of the EHD lubricating layer, namely, a characteristic depression in the central contact area according to the micro-interferogram (see Fig. 2, b), which occurs under conditions of insufficient lubrication at low temperature.

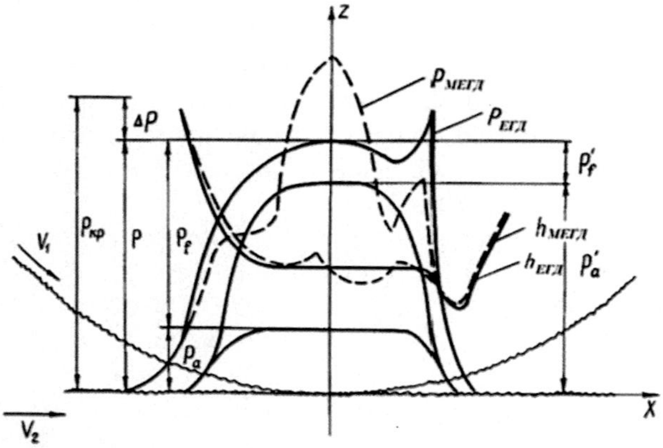

Fig. 3. The macro button chooses the correct format automatically [1].

Thus, in the extreme operation of non-conformal friction units of bearings and internal combustion engines, there is an interaction between the protrusions of micro-bumps at a certain thickness of the lubricating layer, which is associated with insufficient lubrication due to the non-stationary friction mode and/or the manifestation of non-Newtonian pseudoplastic properties of the lubricating layer in the contact zone on a rheological basis. Under these conditions, micro-wedges will arise, for which it is necessary to form optimal lubricating layers in terms of thickness, which will cause hydrodynamic pressure P_f to increase the lifting force by increasing the rolling speed or using a lubricant with optimal rheological properties to bring the lubrication regime to a stable micro-EGD regime in compliance with the condition $P_{MEGD} < P_{кр}$.

Under non-stationary stop-and-go friction conditions of non-conformal and conformal assemblies, the transition from boundary to mixed friction and from mixed to fluid friction, and vice versa, is an inevitable process that is associated with the timely and regular supply of lubricant to the contact zone. The timely and regular supply of lubricant to the contact zone is very difficult to realize in practice. In this regard, accepting the corresponding inevitable process of insufficient lubrication under extreme operating conditions, the issue of implementing sustainable micro-EGD lubrication is considered using an integrated approach to improve the lubrication efficiency and wear resistance of friction pairs in terms of contact-mechanical, rheological, and physicochemical aspects.

2 Criteria of the Conceptual Model

Considering the multifactorial nature of the task of improving the lubrication efficiency and wear resistance of bearing assemblies and conformal and non-conformal ICE units, an integrated approach was used to evaluate the contact-mechanical, rheological, and physicochemical aspects, and on its basis.

Based on the above considerations, a conceptual model for increasing wear resistance and lubrication efficiency for non-conformal and conformal friction units was developed from the standpoint of micro-EHD theory in the transition from fluid to boundary friction by criterion evaluation (Table 1).

Table 1. Criteria of the conceptual model for increasing wear resistance and lubrication efficiency for non-conformal and conformal friction units from the standpoint of micro-EHD theory for different friction modes.

No.	Name	Definition
1	Criterion for lubrication mode [2]	$\lambda = \frac{h}{\sqrt{R_{a1}^2 + R_{a2}^2}}$
2	Microgeometry criterion [3]	$k\prime = 1,03 \cdot \beta^{0,64}.$
3	Pressure criterion [4]	$\frac{P_h}{P_\Sigma} = \frac{1}{\lambda} \cdot \left(\frac{h_{min}}{h_0}\right)^{6,3}$
4	Rheological criterion [5]	$\alpha = \frac{\ln(\eta_1/\eta_2)}{P_2 - P_1}$
5	Plasticity criterion [4]	$\mu = \frac{E\prime}{H_\mu} \cdot \left(\frac{\sqrt{R_{a1}^2 + R_{a2}^2}}{r}\right)^{\frac{1}{2}}$
6	Temperature-mechanical criterion [4]	$T = T_{\mathcal{M}} + \Delta T$

3 Conclusions

According to the developed conceptual model, the conditions for preventing the rupture of the micro-EHD lubrication layer are established, which are achieved by a set of measures to improve tribotechnical performance according to the following criteria: realization of stable micro-EHD lubrication according to the criterion of lubrication mode *(λ)*, selection of optimal microgeometry with respect to the contact-mechanical properties of the stress-strain state of friction surfaces along the actual plane according to the criterion of microgeometry *(k)*; increase in the share of hydrodynamic pressure of lubricating layers relative to the contact pressure on the protrusions of micro-irregularities according to the pressure criterion (P_h/P_Σ) taking into account rheological parameters to change the piezoelectric viscosity coefficient with increasing pressure according to the rheological criterion (α); assessment of the appearance of a plastic-strain state according to the plasticity criterion (μ); modification and adjustment of the qualitative and quantitative chemical composition of components in the lubricating medium to reduce the average local temperature of friction surfaces according to the temperature criterion (T).

References

1. Houpert, L., Hamrock, B.: EHD lubrication calculation used as a tool to study scuffing. In: Houpert, L., Hamrock, B. (eds.) Leeds-Lyons Symposium: Mechanism and Surface Distress, pp. 146–155 (1985)
2. Dmytrychenko, N.F., Milanenko, O.A.: Lubricating action of oils in conditions of elastohydrodynamic lubrication: Monography. Informavtodor, Kiev (2009)
3. Milanenko, O.A.: Influence of microgeometry in the point contact zone on fatigue life under resting friction conditions for friction bearing assemblies. Probl. Tribol. **28–1**(107), 6–12 (2023)
4. Khebda, M., Chichinadze A.V.: Handbook of tribotechnics. In: Edited Khebda, M., Chichinadze, A.V. (eds.) in 3 parts. Mechanical engineering (Part 2: Lubricants, lubrication technique, sliding and rolling supports (1990)
5. Dmitrichenko, N.F., Bilyakovich, O.N., Savchuk, A.N., Milanenko, A.A., Turitsya, Y.A.: The effect of rheological parameters on the tribotechnical characteristics of modified I-40A oil. J. Friction Wear **39**(2), 164–168 (2018)

The Influence of Track Vertical Irregularities on the Crane Dynamic Behaviour

Viačeslav Petrenko(✉)

Vilnius Gediminas Technical University, Plytinės St. 25, 10105 Vilnius, Lithuania
viaceslav.petrenko@vilniustech.lt

Abstract. Cranes, such as gantry, beam, overhead cranes, that move along longitudinal rails are used at low speeds and those cranes do not have suspension system. It is believed that at such speeds there are no complex dynamic forces. The aim of this study is to simulate the movement of overhead crane with a different combination of various vertical irregularities and determine the level of dynamic forces.

Keywords: Overhead crane · Vertical irregularities · Dynamic behaviour

1 Introduction

One of the popular types of rail transport is various cranes (gantry, beam, overhead) that move along longitudinal rails. Classical railway transport rolling stock is moving at high speeds (100–400 km/h) and is characterized by the appearance of complex dynamic processes. Railway cranes are used at low speeds, for example, the maximum speed of gantry cranes should not exceed 1 m/s. Therefore, cranes practically do not have suspension system and it is believed that at such speeds there are no complex dynamic processes, such as in classical railway transport rolling stock.

Cranes have complex systems for moving weights, controlling the position of the weight, reducing the swing of the weight. The dynamic processes of these systems have been well studied and described in numerous publications. For example, the dynamic behaviour of supporting structure of an overhead crane during the operation of a hoisting mechanism analysed in [1]. Or analysis of trolley or wheel–rail beam system impact on the dynamic behaviour of the container crane analysed in [2, 3]. Dynamic reaction forces of an overhead crane during lifting process described in [4]. Weight oscillation reduction for overhead cranes with different payload and different rope length described in [5].

In some cases, for example, when increasing the length of crane tracks, it is required to increase the efficiency of the crane. Obviously, the simplest method to increase efficiency is to increase the speed of the crane, which increases the level of dynamic forces. The main aim of this study is to simulate the movement of overhead crane with a different combination of various vertical irregularities and determine the level of dynamic forces.

O. Prentkovskis et al. (Eds.): TRANSBALTICA 2023, LNITI, pp. 167–177, 2024.
https://doi.org/10.1007/978-3-031-52652-7_17

2 Overhead Crane Computer Simulation

2.1 Overhead Crane Model

The Universal Mechanism (UM) program package was used for simulation of kinematics and dynamics of overhead crane. The crane model (Fig. 1) was created as a hybrid model, consisting of:

- a crane frame, which consists of several elements that simulate the elastic deformations of the longitudinal beam;
- integrated and adapted UM software of railway wheel sets with axle boxes.

Fig. 1. Overhead crane model.

Since it is planned to simulate the dynamics of an overhead crane at a constant speed with range from 0.8 m/s to 5.0 m/s, the systems for generating tractive force or braking force are not used in the model. A more detailed description of the model can be found in [6] article.

2.2 Crane Tracks Model

Crane track model presented as an inertial rail track model. That model considers crane rails as rigid bodies under crane wheels. Inertial rails has three degree of freedom. Equation of motion for inertial rail are given in crane coordinate system and under rail base is modelled as a Special force of Bushing type [7].

Crane track model uses new R65 type rails, with height of rail 0.18 m and dissipative properties: vertical stiffness 50 MN/m, lateral stiffness 43 MN/m, roll stiffness 5.0 MN/m, lateral damping 2.4 MN s/m, vertical damping 0.2 MN s/m, roll damping 2.0 MN s/m.

Contact forces are computed using the models by W. Kik and J. Piotrowski. In the model by W. Kik and J. Piotrowski FASTSIM algorithm, modified for non-elliptical contact patches, is used to compute creep forces [7].

To create irregularities in the crane track, only vertical irregularities was generated. For vertical irregularities creation the deterministic function was used:

$$y = \frac{H}{2(1 - cosx)}. \tag{1}$$

In that irregularities model, wave length 20 m, wave height 50 mm, displacement of the first wave 10 m, number of waves 100 (for 2,000 m length).

3 Crane Software Simulation Process

3.1 Main Stages of the Simulation Process

Experimental part was divided into four consistent steps. On the first step crane track without any irregularities is used. To get out of a symmetrical dynamic situation, the crane load is attached to a beam with a slight asymmetry. This step allows us to understand the behaviour of the crane model and the conditions of the experiment.

Next two steps took place when the main geometrical parameters of irregularities where changed:

- vertical irregularities height (40 mm, 50 mm and 60 mm);
- vertical irregularities length (19 m, 20 m and 21 m).

Changing these parameters allows you to understand the influence of the shape of irregularities on the dynamic behaviour of the crane.

In all three steps left and right rails vertical irregularities was symmetrical, so the last step represents an asymmetrical arrangement of rails irregularities. In this series of experiments, the vertical irregularities waves are arranged unsymmetrically, with a steps: 1, 2, 3, 4, 5, 6, 7, 8, 9 and 10 m. This range of asymmetry guarantees the maximum elevation of one rail relative to the other.

All four simulation steps were carried out at different speeds of the crane model. A wide range of speed changes was determined from 0.8 m/s to 5.0 m/s with a speed change step of 0.3 m/s (in total 15 speed steps). The simulation process ends when the crane travels 200 m and due to the different speed of the crane, the simulation time varies from 250 s to 40 s (Fig. 2).

3.2 Crane Dynamic Behaviour Simulation Results

The first step simulation uses crane track without any irregularities. Due to the asymmetric arrangement of the attached load, it is possible to observe a gradual displacement of the crane wheelsets. At low speeds, 0.8 or 1.7 m/s the wheels are pressed against the rails, but at high speeds, for example at 5.0 m/s, the wheel sets rebound (Fig. 2).

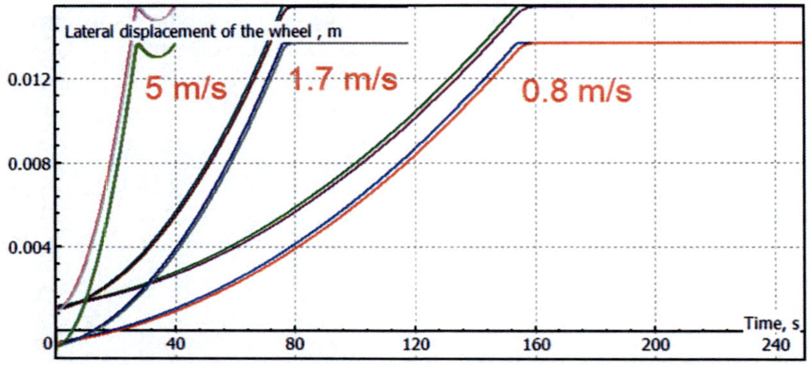

Fig. 2. Lateral displacement of the crane wheels relative to the rail at a speed of 0.8 m/s, 1.7 m/s and 5.0 m/s (chart lines from the right to the left).

The dynamic behaviour of the crane can be determined by the level of total lateral forces acting on the wheels along the Y-axes of the track system. The graph shows (Fig. 3) that at a speed of 0.8 m/s, the wheels are pressed against the rails and slides along the rails with a constant maximal force of 12 kN. At higher speeds, impact loads occur between the wheelc and the railc, for example, at a speed of 1.7 m/s, multiple 8–12 kN lateral forces strikes occurs, but at a speed of 5.0 m/s, single strikes with a force of about 6 kN occur.

The results of the first simulation step, in the absence of rails irregularities, show the logical and predictable dynamic behavior of the crane. This means that the created parametric model of the crane, the model of the crane rails and the simulation conditions are created correctly and further simulation is possible.

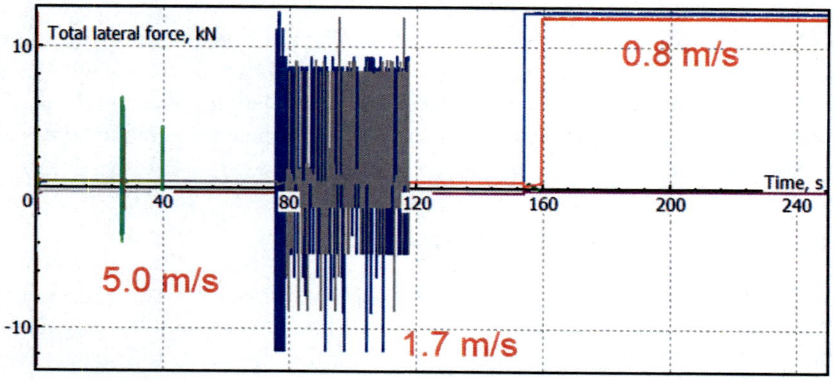

Fig. 3. Total lateral force acting on the wheel along the Y-axes of the track system at a speed of 0.8 m/s, 1.7 m/s and 5.0 m/s.

At the next stage of simulation, the influence of the height of symmetrical vertical irregularities (waves) on the dynamic behaviour of the crane was analysed. The simulation was carried out for the entire range of speed with change in the wave height.

Simulation result shows, that even for lowest crane speed 0.8 m/s, after pressing the crane wheels sets to the rails, a wobbling movement of the crane first wheel occurs (Fig. 4).

With an increase in the speed of the crane, the pattern of the lateral displacement of the first wheel practically does not change, only the wobbling amplitude slightly decreases (Fig. 5).

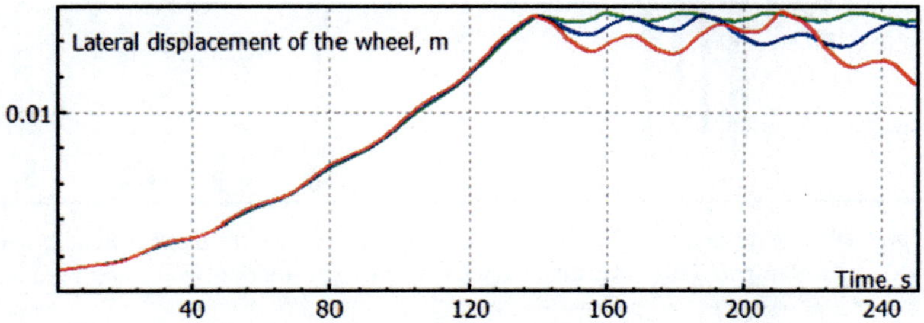

Fig. 4. Lateral displacement of the crane first wheel relative to the rail at a speed of 0.8 m/s of 40 mm (green), 50 mm (blue) and 60 mm (red) vertical irregularities height.

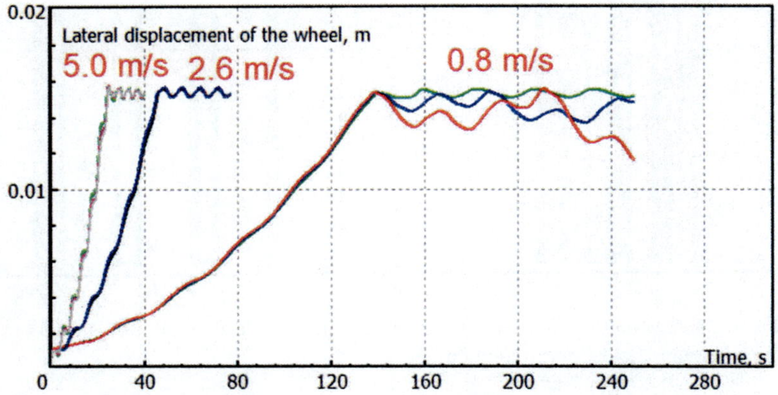

Fig. 5. Lateral displacement of the crane first wheel relative to the rail at a different speed of 40 mm, 50 mm and 60 mm vertical irregularities height (marked in different colours).

The dynamic behavior of the crane can be estimated from the change in the magnitude of the total lateral force. At minimum speed, the lateral force is almost constant, but as the speed increases, side impacts occur, reaching 12 kN at a speed of 5 m/s, the impact force decreases to 6 kN (Fig. 6).

Due to the conical shape of the profile of the flange of the crane wheels, in the moments, when wheels strikes the rails, a decrease in the normal force occurs up to 2 kN (Fig. 7). The distribution of the normal force in the entire range of speeds from 0.8 m/s till 5.0 m/s, remains practically identical.

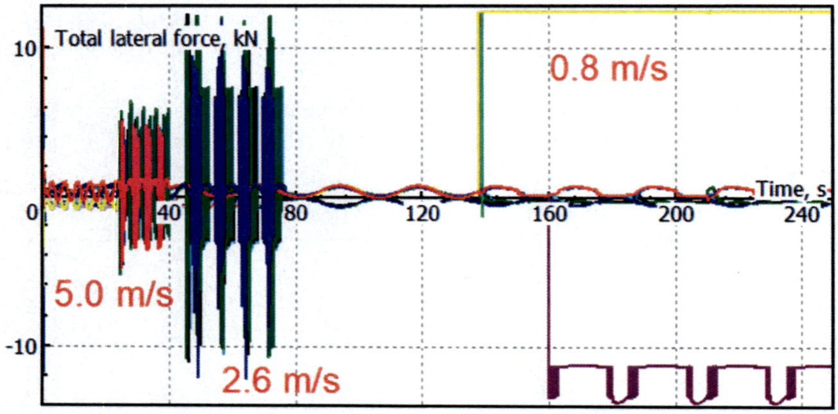

Fig. 6. Total lateral force acting on the all wheels along the Y-axes of the track system at a different speed of 40 mm, 50 mm and 60 m vertical irregularities height (marked in different colours).

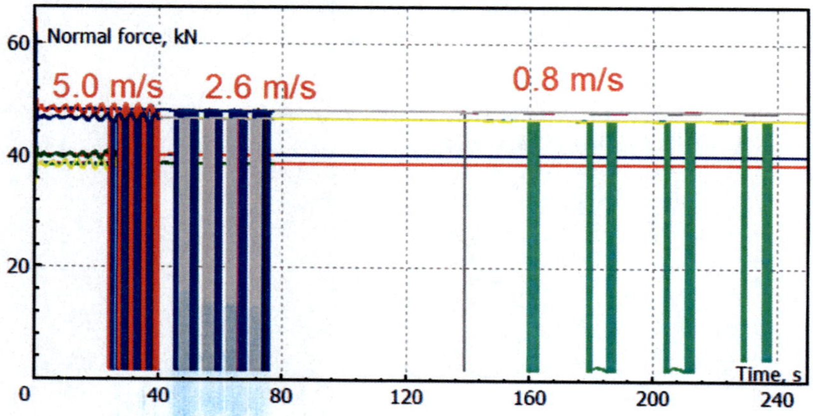

Fig. 7. Normal forces on the all wheels at a different speed of 40 mm, 50 mm and 60 m vertical irregularities height (marked in different colours).

The influence of the length of symmetrical vertical irregularities on the dynamics behavior of the crane was simulated in the next step. During the simulation, the initial wavelength of 20 m increased and decreased by 1 m, while simulating all 15 crane speed steps from 0.8 m/s till 5.0 m/s.

At first, the dynamic behaviour of the crane was evaluated by the lateral displacement of the crane wheels. The Fig. 8 graphs shows that with an increase in the wavelength, the amplitude of the oscillations of the second wheel increases till 0.04 m for speed 1.1 m/s.

Analysing the results obtained for the entire speed diapason, it can be seen that at low and high speeds, lateral oscillations of the crane wheels occur, but at a speed of 2.6 m/s, the wheels are pressed against the rails, practically without lateral play (Fig. 9).

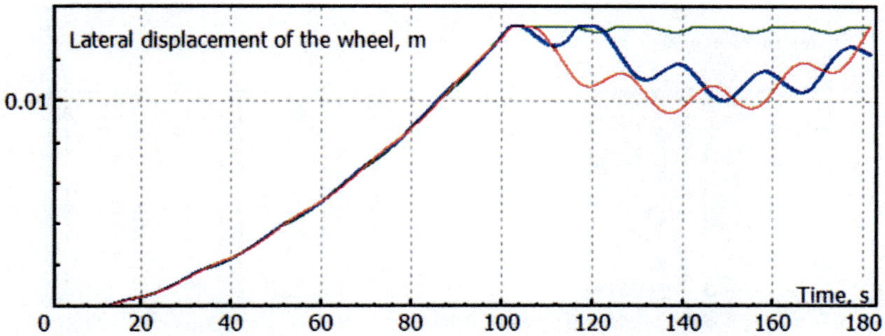

Fig. 8. Lateral displacement of the crane second wheel relative to the rail at a speed of 1.1 m/s of 19 m (green), 20 m (blue) and 21 m (red) vertical irregularities length.

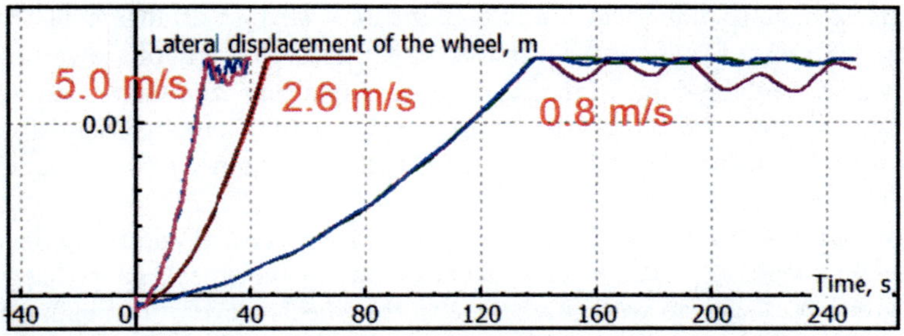

Fig. 9. Lateral displacement of the crane second wheel relative to the rail at a different speed of 19 m, 20 m and 21 m vertical irregularities length (marked in different colours).

Dynamic behavior of the crane when the wavelength of irregularities changes according to the level of total lateral and normal forces acting on the crane wheels. At this step, the distribution and level of forces are similar to the results of the previous experiment (Fig. 10 and 11). At low speeds, for example 0.8 m/s, the wheels are pressed against the rails, and as the speed increases, periodical side impacts occur with simultaneous unloading of the crane wheels.

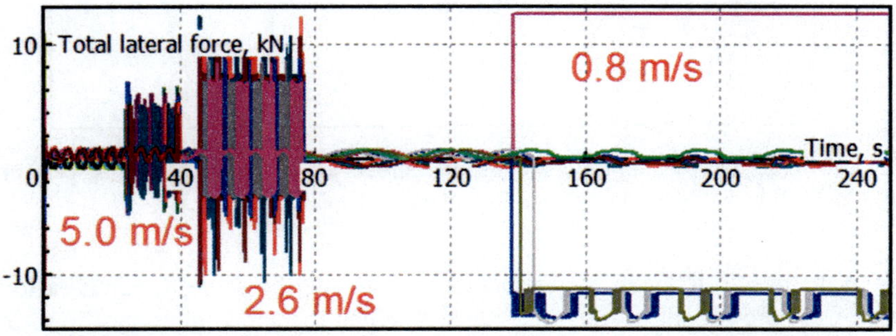

Fig. 10. Total lateral force acting on the wheel along the Y-axes of the track system at a different speed of 19 m, 20 m and 21 m vertical irregularities length (marked in different colours).

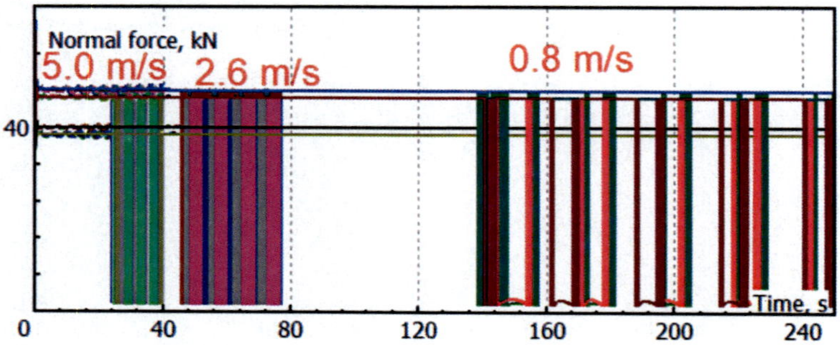

Fig. 11. Normal forces at a different speed of 19 m, 20 m and 21 m vertical irregularities length (marked in different colours).

At the last stage of numerical simulation, the most adverse arrangement of vertical irregularities is created - asymmetric. In this case, the wave length of the vertical irregularities is 20 m, and the range of the asymmetric displacement of the vertical waves of the left and right rail is from 1 m to 10 m (in 1 m steps). In this multivariant simulation, calculations were made for 15 different speeds with 10 asymmetrical displacements of vertical irregularities.

With an increase in the asymmetry of vertical irregularities, the movement of the crane wheels changes (Fig. 12). With a small asymmetry, the wheels are pressed against the rails (green line), with an average asymmetry, the wheels move with a small amplitude (red line), and with a maximum asymmetry, the crane constantly wobbles from one rail to another (blue line).

With the maximum asymmetry of irregularities, an impact of the wheel on the rail occurs (red circle) and it is obvious that this leads to significant dynamic impact loads.

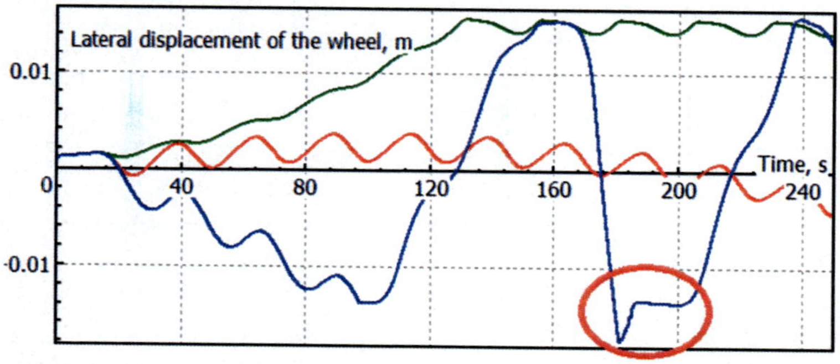

Fig. 12. Lateral displacement of the crane first wheel relative to the rail at a 0.8 m/s speed of 1 m (green), 4 m (red) and 10 m (blue) vertical irregularities displacement.

When the speed of the crane increases, we can see the same three characteristic variants of the movement of the crane wheels. When the speed increases up to 5 m/s (Fig. 13), the wheels either press against the rails (green line), or move along a chaotic curve (red line), or the wheels wobbles from one rail to another (blue line).

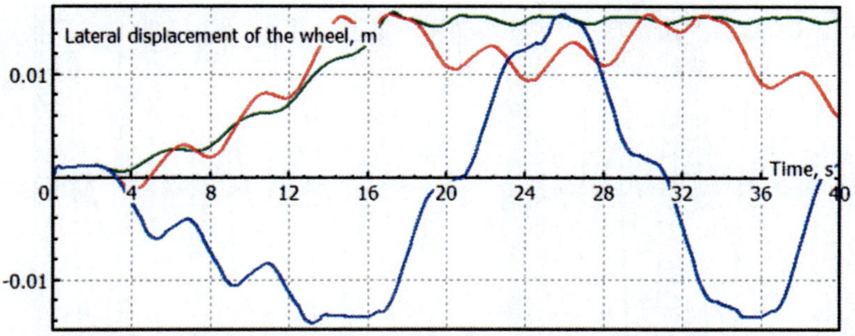

Fig. 13. Lateral displacement of the crane first wheel relative to the rail at a 5.0 m/s speed of 1 m (green), 4 m (red) and 10 m (blue) vertical irregularities displacement.

Analyzing the total lateral forces (Fig. 14), it can be seen that the maximum force level of 12 kN occurs at the maximum asymmetry of vertical irregularities, when the wheels wobbles from one rail to the other.

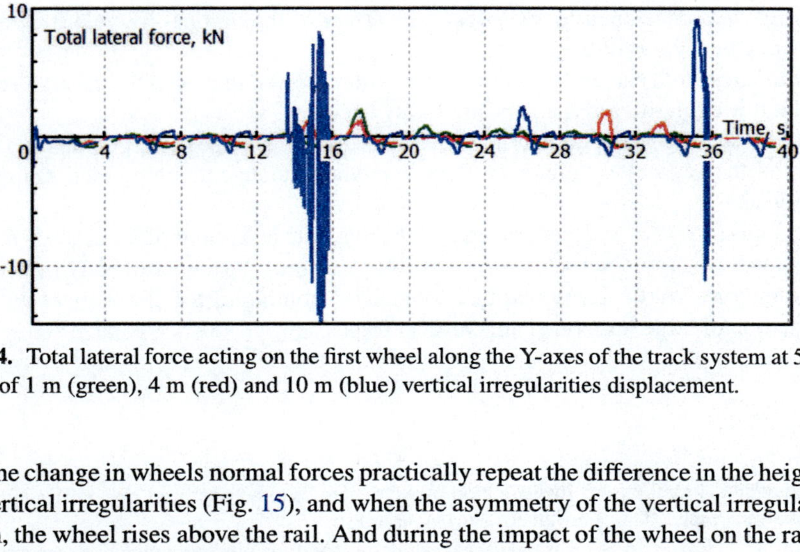

Fig. 14. Total lateral force acting on the first wheel along the Y-axes of the track system at 5.0 m/s speed of 1 m (green), 4 m (red) and 10 m (blue) vertical irregularities displacement.

The change in wheels normal forces practically repeat the difference in the heights of the vertical irregularities (Fig. 15), and when the asymmetry of the vertical irregularities is 4 m, the wheel rises above the rail. And during the impact of the wheel on the rail, the frequency of the change in force increases significantly.

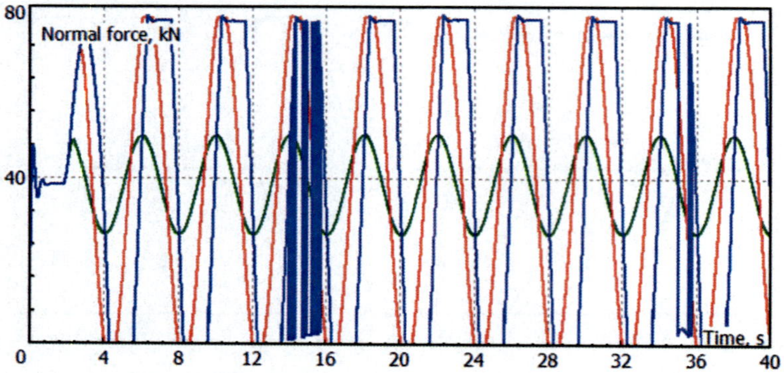

Fig. 15. First wheel normal forces at a 5.0 m/s speed of 1 m (green), 4 m (red) and 10 m (blue) vertical irregularities displacement.

4 Conclusions

1. In this study, an overhead crane model, crane tracks and numerical experiment conditions were developed to investigate the effect of the vertical irregularities on the dynamic behaviour of the crane.
2. The first research stage of numerical simulation showed that even for the crane tracks without irregularities, dynamic forces may occur.
3. At the second stage of simulation, with a symmetrical change in the geometric parameters of the vertical irregularities, the crane wheels slide and are pressed against the rail head.
4. The final stage of modelling proved that the dynamic behaviour of the crane is mainly affected by the asymmetrical displacement of the vertical irregularities. With a certain asymmetry of vertical irregularities, intensive wobbling of the crane arises with the formation of large horizontal and vertical forces.

References

1. Bogdevičius, M., Vika, A.: Investigation of the dynamics of an overhead crane lifting process in a vertical plane. Transport **20**(5), 176–180 (2005)
2. Zrnić, N.Đ, Hoffmann, K., Bošnjak, S.M.: Modelling of dynamic interaction between structure and trolley for mega container cranes. Math. Comput. Model. Dyn. Syst. **15**(3), 295–311 (2009)
3. Tong, M., Shen, Y.: Research on coupled vibration of wheel–rail beam system in the high-speed quayside container crane. Mech. Adv. Mater. Struct. **27**(17), 1473–1482 (2020)
4. Spruogis, B., Jakštas, A., Turla, V., Iljin, I., Šešok, N.: Dynamic reaction forces of an overhead crane of lifting. Transport **26**(3), 279–283 (2011)
5. Chen, Y.-F., Huang, A.-C.: Oscillation reduction for overhead cranes with time-varying payload and rope length. J. Chin. Inst. Eng. **37**(2), 259–267 (2014)

6. Petrenko, V., Ževžikov, P.: Investigation of dynamic behaviour of an overhead crane using multivariant calculations. In: Prentkovskis, O., Yatskiv (Jackiva), I., Skačkauskas, P., Junevičius, R., Maruschak, P. (eds.) TRANSBALTICA 2021. LNITI, pp. 58–67. Springer, Cham (2022). https://doi.org/10.1007/978-3-030-94774-3_6
7. Laboratory of Computational Mechanics. Simulation of Rail Vehicle Dynamics Homepage. http://www.universalmechanism.com/download/80/eng/08_um_loco.pdf. Accessed 05 May 2021

Determining the Growth of Energy Consumption and Power to Increase the Speed of the Car

Sergiy Shuklinov(✉), Mykola Mykhalevych, and Yuliia Umantseva

Kharkiv National Automobile and Highway University, Kharkiv, Ukraine
schuklinovsn@gmail.com

Abstract. In city conditions, the car moves in an unstable mode. This causes frequent acceleration and worsens the efficiency and environmental friendliness of the internal combustion engine as it operates in transient mode. The use of "mild hybrid" technology can improve these performance properties. At the same time, the issue of rational parameters of components for hybridizing a car arises, as they increase its cost. Traditionally, an electric motor helps with acceleration and provides energy recovery during braking. The power of the electric motor and traction battery can be justified based on the analysis of analytical dependencies that describe the movement of the vehicle. To solve this problem, it is necessary to determine the level of additional energy to overcome the work of the vehicle's drag forces when its speed changes. The theoretical foundations are written taking into account the assumption of a change in speed within specified limits with constant acceleration, which is possible when considering changes in vehicle speed over short periods of time. The developed analytical dependencies are the theoretical basis for determining the rational value of the electric motor power when using the "mild hybrid" technology and allow us to determine the target task of controlling the "mild hybrid" technology when accelerating a car to improve its economic and environmental characteristics.

Keywords: Mild hybrid · Car acceleration · Motor power · Energy consumption · Traction force · Resistance forces

1 Introduction

In city conditions, the car moves in an unstable mode [1, 2]. During acceleration of the car, the engine works in transient mode [3]. The economy and environmental friendliness of the internal combustion engine in the transitional mode deteriorate significantly [4]. Application of "mild hybrid" technology allows improving these operational properties [5–7, 9]. In this case, it is rational to ensure the acceleration of the car by connecting the electric motor to the transmission, which makes it possible to perform a slow transition process of the internal combustion engine and improve its environmental friendliness [6]. During deceleration, it is rational to switch this electric motor to generator mode to recover the car's kinetic energy. [10–12] Determining the power level of the electric

O. Prentkovskis et al. (Eds.): TRANSBALTICA 2023, LNITI, pp. 178–186, 2024.
https://doi.org/10.1007/978-3-031-52652-7_18

motor and battery capacity to ensure the desired acceleration of the car when the speed of movement increases within the given limits is of scientific interest to many scientists [13–17]. To solve this problem, it is necessary to determine the level of additional energy to overcome the work done by the forces of resistance to the movement of the car during the change in the speed of the car.

2 Theoretical Foundations

The level of additional energy determines the minimum capacity of the battery. The additional power that must be supplied to the driving wheels when changing the speed with the desired acceleration determines the minimum power of the electric motor.

Figure 1 shows the dependences of the total traction force on the driving wheels P_k, which is created by the internal combustion engine, the total force of drag and road resistance $P_\psi + P_w$, as well as the additional traction force on the driving wheels that needs to be created by the electric motor ΔP_{kE} on the distance covered when accelerating from speed v_1 to speed v_2. The figure also shows the dependence of the total traction force created by the internal combustion engine and the electric motor $\Sigma P_k = f(s)$. On the section $o - a$ the car moves in a steady state with a speed of v_1.

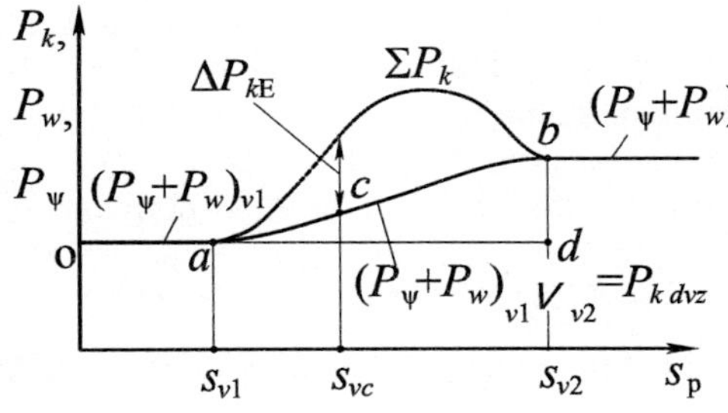

Fig. 1. Car forces vs drive distance dependency in different driving modes.

As it is known, the steady mode of movement with speed v_1 occurs under the condition $P_{k\ dvz} = (P_\psi + P_w)_{v_1}$. On the section $a - b - c$, the car moves with acceleration from speed $v1$ to speed v_2. Acceleration of the car occurs because, on this section, the total traction force on the driving wheels ΣP_k is greater than the total resistance force $(P_\psi + P_w)_{v_1 \to v_2}$. Because of the use of the "mild hybrid" technology, a slow increase in the traction force on the driving wheels $P_{k\ dvz}$, which is created by the internal combustion engine in accordance with the condition $P_{k\ dvz} = (P_\psi + P_w)_{v_1 \to v_2}$, where $(P_\psi + P_w)_{v_1 \to v_2}$ – the value of the total resistance force for the corresponding speed of movement as it increases from v_1 to v_2. In this case, the amount of acceleration of the car is determined by the level of additional force ΔP_{kE} created by the electric motor.

To determine the value of the minimum battery capacity required to accelerate the car from speed v_1 to speed v_2, it is necessary to find out what work is done by the resistance forces during acceleration of the car on this section of the road. To do this, we will perform an analysis of the balance of forces that act on the car during its movement.

Given the action of the driving force and resistance forces, the force balance equation can be written in the form (1).

$$\Sigma P = \Sigma P_k - (P_\psi + P_w), \tag{1}$$

where $\Sigma P_k = P_{k\,dvz} + \Delta P_{kE} = \frac{(M_e + M_{eE})\cdot u_{tr}\cdot \eta_{tr}}{r_d}$ – full traction force of the driving wheels (ΔP_{kE} – additional traction force on the drive wheels created by the electric motor; M_e – internal combustion engine torque; M_{eE} – torque of the electric motor; u_{tr}, η_{tr} – gear ratio and transmission efficiency; r_d – dynamic radius of the driving wheels); P_ψ, P_w – road resistance and drag forces.

When the car is in steady motion, the sum of all forces acting on the car is zero. When the car accelerates, this amount differs from zero and is called the acceleration resistance force P_j, i.e. $\Sigma P = P_j$. In the theory of the car, equations in the form of a balance of driving forces and forces of movement resistance are used to analyze the modes of its movement.

$$\Sigma P_k = P_\psi + P_w + P_j, \tag{2}$$

where $P_j = j_a \cdot \delta_{vr} \cdot {}^{G_a}/_{g}$ – the force of resistance to acceleration of the car (G_a – the force of gravity of the car; g – the acceleration of free fall; δ_{vr} – the coefficient of consideration of the rotating masses of the car; j_a – the acceleration of the car).

It is obvious from (1) and (2) that to accelerate the car, it is necessary to increase the total traction force, i.e. supply additional energy from the engine. The amount of additional energy is determined by the increase in energy of the forces of resistance to the movement of the car. The elementary energy dE_o of the forces of movement resistance during acceleration of the car is determined at the elementary increment of the path ds.

$$dE_o = P_\psi \cdot ds + P_w \cdot ds + P_j \cdot ds. \tag{3}$$

Taking into account the differential determination of the speed of movement, it is possible to write.

$$ds = v \cdot dt, \tag{4}$$

where v – the speed of the car; dt – time differential of the car.

The differential of the car's movement time can be expressed from the differential determination of the car's acceleration.

$$dt = \frac{dv}{j_a}. \tag{5}$$

Taking into account (4) and (5), Eq. (3) will take the form.

$$dE_o = P_\psi \cdot \frac{v}{j_a} \cdot dv + P_w \cdot \frac{v}{j_a} \cdot dv + P_j \cdot \frac{v}{j_a} \cdot dv. \tag{6}$$

Let's write down the resistance forces in Eq. (6) in expanded form and after transformations we get.

$$dE_o = \frac{G_a \cdot \psi \cdot v}{j_a} \cdot dv + \frac{k_w \cdot F_a \cdot v^3}{j_a} \cdot dv + \frac{G_a}{g} \cdot \delta_{vr} \cdot v \cdot dv. \tag{7}$$

To determine the increase in energy to overcome resistance forces when the speed of the car increases from speed v_1 to speed v_2, we integrate Eq. (7).

$$\int_{v_1}^{v_2} dE_o = \int_{v_1}^{v_2} \frac{G_a \cdot \psi \cdot v}{j_a} \cdot dv + \int_{v_1}^{v_2} \frac{k_w \cdot F \cdot v^3}{j_a} \cdot dv + \int_{v_1}^{v_2} \frac{G}{g} \cdot \delta_{vr} \cdot v \cdot dv. \quad (8)$$

To solve Eq. (7), we assume that the acceleration of the car from speed v_1 to speed v_2 occurs with a constant acceleration j_{sr}. Taking into account the accepted assumption, we will determine the increase in energy on the driving wheels, necessary to overcome the resistance forces when the speed of the car increases from v_1 to v_2.

$$E_o = \frac{G_a \cdot \psi}{2j_{sr}} \cdot (v_2^2 - v_1^2) + \frac{k_w \cdot F_a}{4j_{sr}} \cdot (v_2^4 - v_1^4) + \frac{G_a \cdot \delta_{vr}}{2g} \cdot (v_2^2 - v_1^2). \quad (9)$$

Equation (9) represents the balance equation of the car's energy gain during acceleration. The right-hand side of (9) characterizes the energy expenditure for changing the speed of the car from v_1 to v_2 with constant acceleration j_{sr}. The first two components characterize the energy losses during the acceleration of the car to overcome the road resistance E_ψ and drag E_w. It is obvious that the greater the acceleration with which the increase in speed occurs, the less energy is spent on overcoming the forces of road resistance and drag. The third component in the right-hand side of formula (9) is equal to the increase in kinetic energy of the car E_v as a result of the change in the value of the speed of steady motion from v_1 to v_2. It should be noted that energy losses for increasing the car's kinetic energy do not depend on its acceleration. The left part of Eq. (9) characterizes the necessary increase in energy, which must be supplied from the engine to the driving wheels to increase the speed of the car with a given acceleration. Energy losses during the acceleration of the car to overcome the forces of road resistance and drag correspond to the area of the figure $a - b - c - d$ in Fig. 1. The area of the figure bounded by the curves $a - b - c$ and ΣP_k characterizes the energy losses to overcome the inertial forces of the car during acceleration.

The minimum power of the electric motor, which must be brought to the driving wheels when changing the speed with the desired acceleration, is determined from the balance of forces of the car (Fig. 2). At steady motion with a speed of v_1, the torque of the internal combustion engine is supplied to the drive wheels, which creates a traction force $P_{k\ dvz}$ according to the partial characteristic $P_{k\ \alpha 1}$ with fuel supply α_1. At the same time, the traction force $P_{k\ dvz}$ is equal to the sum of the resistance forces $(P_\psi + P_w)_{v_1}$ at the speed v_1. Steady motion of the car at a speed v_2 after acceleration is characterized by the traction force $P_{k\ dvz}$ according to the partial characteristic $P_{k\ \alpha 2}$ with fuel supply α_2. At the same time, the traction force $P_{k\ dvz}$ is equal to the sum of the resistance forces $(P_\psi + P_w)_{v_2}$ at the speed v_2.

To calculate the power that must be developed by the engine to accelerate the car, we will determine the elementary increase in power.

$$dN_o = P_\psi \cdot dv + P_w \cdot dv + P_j \cdot dv, \quad (10)$$

where dv is the elementary acceleration of the car's speed.

When the movement resistance forces are expanded, Eq. (10) takes the form.

$$dNo = G_a \cdot \psi \cdot dv + k_w \cdot F_a \cdot v^2 \cdot dv + \frac{G_a}{g} \cdot \delta_{vr} \cdot j \cdot dv. \tag{11}$$

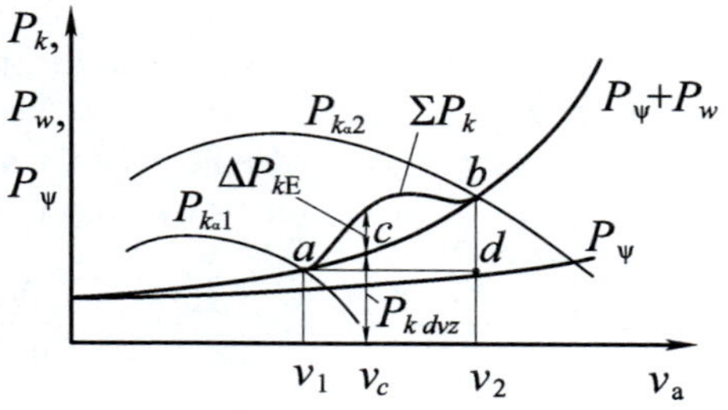

Fig. 2. The graph of the balance of forces of the car.

To determine the increase in power to increase the speed of the car from speed v_1 to speed v_2 with acceleration j_a, let's integrate Eq. (11) by the elementary speed increase.

$$\int_{v_1}^{v_2} dN_o = \int_{v_1}^{v_2} G_a \cdot \psi \cdot dv + \int_{v_1}^{v_2} k_w \cdot F_a \cdot v^2 \cdot dv + \int_{v_1}^{v_2} \frac{G_a}{g} \cdot \delta_{vr} \cdot j_a \cdot dv. \tag{12}$$

To solve Eq. (12), we assume that the acceleration of the car from speed v_1 to speed v_2 occurs with a constant acceleration j_{sr}. Taking into account the accepted assumption, we will determine the additional power that must be supplied to the driving wheels to overcome the resistance forces during acceleration from speed v_1 to speed v_2.

$$N_o = G_a \cdot \psi \cdot (v_2 - v_1) + \frac{k_w \cdot F_a}{3} \cdot \left(v_2^3 - v_1^3\right) + \frac{G_a}{g} \cdot \delta_{vr} \cdot j_{sr} \cdot (v_2 - v_1). \tag{13}$$

The first two components on the right-hand side of formula (13) characterize the increase in power that must be supplied to the drive wheels when the value of the steady-state speed changes from v_1 to v_2. This increase in power on the driving wheels characterizes the increase in the power of the road resistance and drag and is determined by the area of the figure $a - b - c - d$ in Fig. 2. The third component of expression (13) characterizes the power spent on acceleration of the car and in Fig. 3 is determined by the area of the figure bounded by the lines $a - b - c$ and ΣP_k.

3 Analysis of the Results

Figure 3 shows the dependence of energy consumption during car acceleration with different average accelerations. Energy consumption E_o during acceleration of the car with different average accelerations is calculated according to formula (9).

As can be seen from Fig. 2, the loss of energy to increase the kinetic energy of the car does not depend on its acceleration. At the same time, the time and path of movement

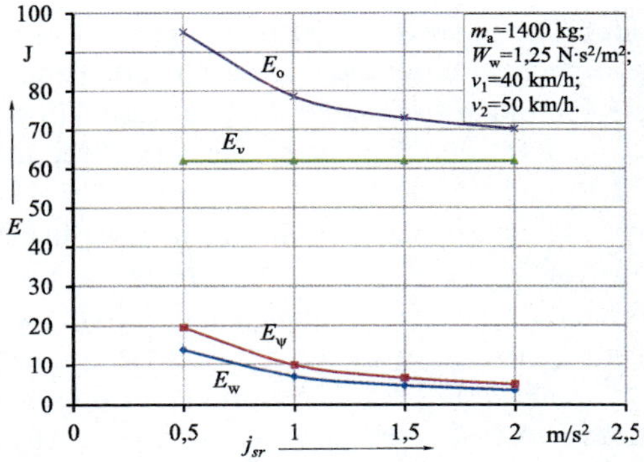

Fig. 3. The graph of the balance of energy consumption during acceleration of the car with different average accelerations: m_a is the mass of the car; W_w is the car's smoothness factor.

from speed v_1 to v_2 decrease when accelerating the car with greater acceleration, which leads to a decrease in energy consumption E_w and E_ψ to overcome the forces of air and road resistance.

Figure 3 also shows the dependence of energy expenditure E_w and E_ψ on overcoming the forces of drag and road resistance during acceleration with different average accelerations. The dependence of E_v reflects the energy expenditure to increase the kinetic energy of the car during acceleration from v_1 to v_2 with different values of constant acceleration j_{sr}.

Figure 4 shows the power dependence during acceleration of the car with different average accelerations. The necessary increase in power N_o delivered to the driving wheels during acceleration of the car with different average accelerations is calculated according to formula (13). Figure 4 also shows the dependence of the required increase in power N_w and N_ψ to overcome the forces of drag and road resistance during acceleration with different average accelerations. The dependence of N_v reflects the increase in power to increase the kinetic energy of the vehicle when accelerating from v_1 to v_2 with different values of constant acceleration j_{sr}.

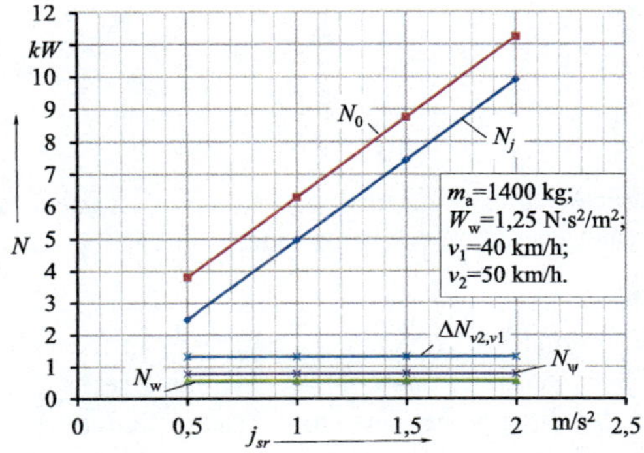

Fig. 4. Dependence of powers for overcoming resistance to movement with constant acceleration.

If the time constant of the internal combustion engine changes the mode in which environmental indicators are ensured, exceeds the acceleration time of the car, the energy consumption of the electric motor for acceleration increases (Fig. 5). The energy consumed by the electric motor is characterized by the area of the figure bounded by the curves $a - c - b$ and ΣP_k and $(P_\psi + P_w)_{v_2}$. That is, the internal combustion engine is loaded more slowly when the speed increases than the increase in the forces of resistance to the movement of the car $(P_\psi + P_w)_{v_1 \to v_2}$. And the electric motor must be able to increase the traction force ΔP_{kE}, which is supplied to the driving wheels. After reaching the speed of movement v_2, the electric motor supports the internal combustion engine during the movement to point b (Fig. 5). Accordingly, the power of the electric motor should increase. This power of the electric motor is characterized by the area of the figure bounded by the curves $a - c - b$ and ΣP_k (Fig. 6). It can be seen from Fig. 6 that the traction force on the driving wheels, which is created by the internal combustion engine $P_{k\ dvz}$, grows more slowly than the increase in resistance forces $(P_\psi + P_w)_{v_1 \to v_2}$.

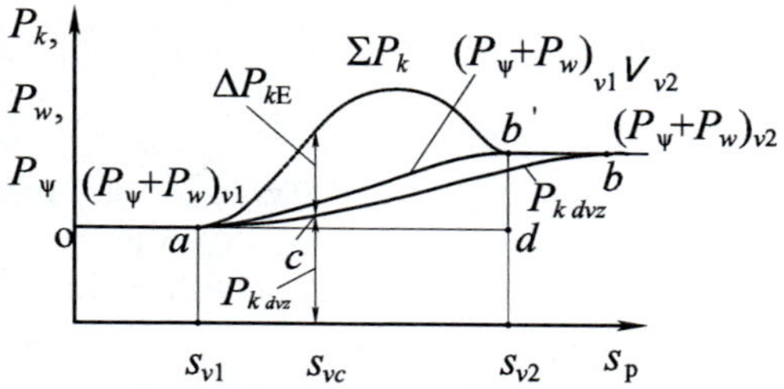

Fig. 5. Dependencies of $P_k = f(s)$, $P_w = f(s)$, $P_\psi = f(s)$, $\Delta P_{kE} = f(s)$, $\Sigma P_k = f(s)$ if the time of the transient process of the internal combustion engine is greater than the acceleration time.

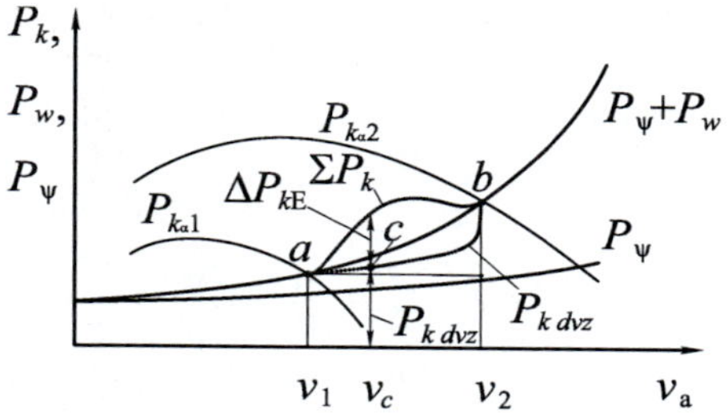

Fig. 6. The graph of the car‘s force‘s balance, if the time of transient process of the internal combustion engine is greater than the acceleration time.

At the same time, the traction force $P_{k\ dvz}$ increases to the value of the sum of the resistance forces $(P_\psi + P_w)_{v_2}$ at a constant value of the speed v_2, and the traction force of the electric motor ΔP_{kE} disappears. The slower the growth of the traction force $P_{k\ dvz}$ in the transient process, the more the power of the electric motor is required to ensure the desired acceleration from speed v_1 to speed v_2.

When driving a car, the driver determines the level of desired acceleration by setting a control signal on the accelerator pedal. To form the desired acceleration of the car on its driving wheels, it is necessary to create the corresponding total traction force. The application of the "mild hybrid" technology allows you to create a rational balance between the component forces of the electric motor and the internal combustion engine. At the same time, regardless of the level of the control signal set by the driver, the increase in the power of the internal combustion engine should occur in a mode that provides pleasant indicators of its economy and environmental friendliness.

The specified level of acceleration of the car is provided by the corresponding increase in the traction force on the driving wheels, which is created by the electric motor. The increase in this component of the traction force occurs due to the increase in power developed by the electric motor depending on the level of the desired acceleration of the car and the time of movement with this acceleration. It should be noted that the consumption of electrical energy during slow acceleration from speed v_1 to speed v_2 increases.

4 Conclusions

The developed analytical dependencies are the theoretical basis for determining:

- the rational value of the power of the electric motor when applying the "mild hybrid" technology, which will ensure the improvement of the economic and environmental characteristics of the car during its acceleration;
- obtained analytical dependencies allow to determine the consumption of electricity during the acceleration of the car when applying the "mild hybrid" technology;
- parameters of the target task of controlling the "mild hybrid" technology when accelerating the car to improve its economic and environmental characteristics.

Acknowledgement. This work was conducted under the Scientific research "Development of a system that improves the energy efficiency and environmental friendliness of motor transport based on the "mild hybrid" technology. 0122U000654, funded by the Ministry of Education and Science of Ukraine.

References

1. Kashkanov, A., Kashkanova, A., Podrigalo, M., Klets, D. et al.: Estimation Parameters of Braking of Vehicles Category M1 at Definition of Circumstances Road Accidents. SAE Technical Paper 2022-01-1166 (2022). https://doi.org/10.4271/2022-01-1166
2. Mykhalevych, M., Shuklinov, S., Dvadnenko, V., Yaryta, O.: The perspectives of mild hybrid technology for creation of vehicle hybridization system. Automob. Transp. (50), 29–39 (2022). https://doi.org/10.30977/AT.2019-8342.2022.50.0.04
3. Liu, R., Zhao, X., Zhu, X., Ma, J.: Statistical characteristics of driver acceleration behaviour and its probability model. Proc. Inst. Mech. Eng. Part D J. Automob. Eng. **236**(2–3), 395–406 (2021). https://doi.org/10.1177/09544070211018039

4. Al-Samari, A.: Study of emissions and fuel economy for parallel hybrid versus conventional vehicles on real world and standard driving cycles. Alex. Eng. J. **56**(4), 721–726 (2017). https://doi.org/10.1016/j.aej.2017.04.010
5. Nissan. Welcome to the Nissan Intelligent Mobility Technology Tour Nissan's Advanced Technology Development, https://www.nissan-global.com/PDF/NIMTT2019_Overall_Presentation_e.pdf. Accessed 02 May 2022
6. Liu, Y., Liao, Y.G., Lai, M-C.: Fuel economy improvement and emission reduction of 48 V mild hybrid electric vehicles with P0, P1, and P2 architectures with lithium battery cell experimental data. Adv. Mech. Eng. (2021). https://doi.org/10.1177/16878140211036022
7. Foust, T., Jones, R., Graves, E., McCoskey, J., Yoon, H.-S.: Effect of an electric vehicle mode in a plug-in hybrid electric vehicle with a post-transmission electric motor. Int. J. Electr. Hybrid Veh. **8**(4), 289–301 (2016). https://doi.org/10.1504/IJEHV.2016.080728
8. Mitropoulos, L.K., Prevedouros, P.D., Kopelias, P.: Total cost of ownership and externalities of conventional, hybrid and electric vehicle. Transp. Res. Procedia **24**, 267–274 (2017). https://doi.org/10.1016/j.trpro.2017.05.117
9. Taoudi, A., Haque, M.S., Luo, C., Strzelec, A., Follett, R.F.: Design and Optimization of a Mild Hybrid Electric Vehicle with Energy-Efficient Longitudinal Control (2021). https://doi.org/10.4271/14%2D10%2D01%2D0005
10. Walsh, J., Muneer, T., Celik, A.N.: Design and analysis of kinetic energy recovery system for automobiles: case study for commuters in Edinburgh. J. Renew. Sustain. Energy **3**, 013105 (2011). https://doi.org/10.1063/1.3549152
11. Armenta-Déu, C., Cortés, H.: Analysis of kinetic energy recovery systems in electric vehicles. Vehicles **5**, 387–403 (2023). https://doi.org/10.3390/vehicles5020022
12. Leontiev, D., Klimenko, V., Mykhalevych, M., Don, Y., Frolov, A.: Simulation of working process of the electronic brake system of the heavy vehicle. In: Palagin, A., Anisimov, A., Morozov, A., Shkarlet, S. (eds.) MODS 2019. AISC, vol. 1019, pp. 50–61. Springer, Cham (2020). https://doi.org/10.1007/978-3-030-25741-5_6
13. Wager, G., Whale, J., Braunl, N.: Battery cell balance of electric vehicles under fast-DC charging. Int. J. Electr. Hybrid Veh. **8**(4), 351–361 (2016). https://doi.org/10.1504/IJEHV.2016.080732
14. Mamala, J., Śmieja, M., Prażnowski, K.: Analysis of the total unit energy consumption of a car with a hybrid drive system in real operating conditions. Energies **14**, 3966 (2021). https://doi.org/10.3390/en14133966
15. Mikhalevich, M., Yarita, A., Turenko, A., Leontiev, D., et al.: Assessment of Operation Speed and Precision of Electropneumatic Actuator of Mechanical Transmission Clutch Control System. SAE Technical Paper 2018-01-1295 (2018). https://doi.org/10.4271/2018-01-1295
16. Calderaro, V., Galdi, V., Graber, G., Piccolo, A., Cogliano, D.: An algorithm to optimize speed profiles of the metro vehicles for minimizing energy consumption. In: Proceedings of the 2014 International Symposium on Power Electronics, Electrical Drives, Automation and Motion (SPEEDAM), Ischia, Italy, 18–20 June 2014, vol. 6, pp. 813–819 (2014)
17. Kim, Y., Lee, I., Kang, S.: Eco assist techniques through real-time monitoring of BEV energy usage efficiency. Sensors **15**, 14946–14959 (2015). https://doi.org/10.3390/s150714946

Optimization of the Power Drive of a Mild Hybrid Vehicle

Volodymyr Dvadnenko(✉) and Oleksandr Dziubenko

Kharkiv National Automobile and Highway University, Kharkiv, Ukraine
dvadnenkovladimir@gmail.com

Abstract. The prospects of hybrid cars in modern conditions are shown. However, hybrid cars are more complex and more expensive than conventional cars. To provide a competitive price for hybrid vehicles, Mild Hybrid Electric Vehicles (MHEVs) have been developed that use smaller electric motors and smaller capacity batteries. To improve the economic and environmental parameters of a mild hybrid vehicle, a new algorithm for the operation of a traction electric motor is proposed and substantiated, which makes the regenerative braking of a mild hybrid vehicle more efficient. In the proposed vehicle in hybrid mode, in order to minimize the power of the traction electric drive, the operation mode of the internal combustion engine with a low load, but with a high specific fuel consumption, is replaced by electric drive. This is a uniform movement with a relatively low speed (40–60 km/h) and low accelerations on an asphalt road without a noticeable rise. The idle mode of the internal combustion engine is also excluded. The performed calculations and experimental studies show that the uniform movement of such a hybrid car with a low-power electric motor in the electric drive mode is possible in urban conditions, for which a special algorithm has been developed. It is shown that the most suitable for such a car is an asynchronous electric motor. A technique for optimizing the control of a traction asynchronous motor based on the frequency scalar method, taking into account the load of the electric motor, is proposed and justified. The expediency and efficiency of the proposed methods of driving a car and a traction asynchronous electric motor are shown.

Keywords: Mild hybrid vehicle · Algorithm of operation of mild hybrid Vehicle motors · Hybrid vehicle control algorithm · Asynchronous power motor · Power electric drive

1 Introduction

An effective solution to the problem of reducing harmful emissions is the widespread transition to electric vehicles. But there are many reasons that do not allow this to be done quickly [1]. These are the low mileage of an electric vehicle without recharging a traction battery, a long charging time, and an underdeveloped infrastructure of charging stations. An electric car currently has a high cost and a lot of weight, primarily due to the battery with a high energy capacity. To charge a large number of electric vehicles, it will be necessary to significantly increase the capacity of power plants and electrical networks,

O. Prentkovskis et al. (Eds.): TRANSBALTICA 2023, LNITI, pp. 187–194, 2024.
https://doi.org/10.1007/978-3-031-52652-7_19

in addition, most power plants do not meet modern environmental requirements. In modern conditions, therefore, hybrid cars are more convenient to operate [2]. However, hybrid cars have a higher price tag than ICE-only cars.

It is possible to reduce the cost as well as the weight of the hybrid vehicle traction motor by reducing the power of the traction motor. This will reduce the current consumption and allow the use of a traction battery with a lower capacity. Therefore, the battery will be cheaper, lighter and take up less space. This is done in modern - mild-hybrid-electric-vehicle - (MHEV) [3]. Such cars have an internal combustion engine, and the electric drive is used to start the internal combustion engine in a start-stop system and to increase the torque of a running internal combustion engine. There are MHEV, in which the electric motor is also used to drive the drive wheels [3, 4]. However, due to the low power of the electric drive, such an MHEV drive is used only for slow movement. If we want to use an electric drive in an MHEV without a running internal combustion engine, then the operation algorithm of a traditional hybrid car, in which starting and accelerating on an electric motor, becomes unusable due to insufficient acceleration when accelerating. For the new MHEV algorithm, we use the fact that for uniform electric movement at relatively low speeds in urban mode, a relatively low electric motor power is required. For such a movement, it is necessary to choose the optimal gear ratio from the motor shaft to the drive wheels. In addition, it is necessary to develop an algorithm that describes under what driving conditions an internal combustion engine should be used and under what electric drive. This algorithm can be applied by the driver who drives the car, or it can be implemented by the automatic transmission control system of the car.

2 Description of the Proposed Mild Hybrid Vehicle

Modern cars must have a powerful engine to provide the required dynamic qualities in city traffic when starting from a standstill and picking up speed. In a mild hybrid car, the internal combustion engine is quite powerful, but not the traction motor.

In [5, 6], an algorithm for the operation of mild hybrid car engines is proposed, in which starting from a standstill and accelerating is performed using only an internal combustion engine with manual or automatic gear shifting. In populated areas, when the speed is limited to 50 km/h, uniform movement or movement with slight acceleration in the speed range of 40–60 km/h is possible on an electric motor with the internal combustion engine turned off. At the same time, in a mild hybrid vehicle, the electric motor shaft must have a kinematic connection with the drive wheels with a gear ratio optimized for this speed. The algorithm is based on the fact that for uniform movement at a specified speed in the city, a relatively small electric motor power is required, provided that the vehicle speed is such that the electric motor operates in the high-speed zone, when its power is close to maximum. For this, the gear ratio from the electric motor to the wheels is chosen so that at a vehicle speed of 50 km/h, the speed of the electric motor corresponds to the speed at which it has maximum power. Then the electric motor operates in a relatively narrow speed range, high torque is not required, and therefore a large current from the battery is not required. This allows the use of a traction battery of relatively small capacity with small dimensions, low weight and low cost. The main advantage of this algorithm is manifested in city traffic, because the consumption of electric energy from the battery during uniform movement alternates with recharging the battery during regenerative braking. Therefore, the recovery efficiency is increased. In this electric mode, a significant contribution to fuel economy is made not only by recuperation, but also by the convenient possibility of using the acceleration-coasting mode, in which the kinetic and potential energy accumulated by the vehicle is directly used for transport work without transformations. It is these possibilities that can be most effectively implemented using an asynchronous traction motor with frequency control, which has a minimum braking torque in a de-energized mode. This is due to the fact that in the electric drive mode we have a simpler and more convenient way for the driver to use the overrun modes when the accelerator and brake pedals are released; acceleration when the accelerator pedal is pressed; regenerative braking, when the brake pedal is only retracted from the stop and emergency braking, when the brake pedal is strongly depressed.

In the proposed mild hybrid car, a relatively inexpensive and light electric motor and a traction battery are used, therefore, the weight of the car and the price increase slightly compared to a conventional car. This, as well as the relatively low fuel consumption, increases its competitiveness.

The internal combustion engine of the proposed car works more often during acceleration. This, unlike a traditional hybrid car or a plug-in hybrid car, provides better internal combustion engine thermal conditions and provides better interior heating.

3 Calculation of an Asynchronous Traction Electric Drive

As is known, the full controllability of the electric drive can be achieved if the electromagnetic torque of the motor is controlled. In all types of electric motors, the torque is formed as a result of the interaction of the magnetic fields of the stator and rotor. The maximum torque occurs at orthogonal (at 90°) position of the vectors of the magnetic fields of the stator and rotor. The rotating magnetic field of the stator is formed by sinusoidal currents in its windings. The position of the rotating vector of the stator magnetic field is determined by the phase relationships of the currents in the stator windings, so we can talk about the stator current vector, both in a three-phase coordinate system and in a two-phase (orthogonal) coordinate system. To obtain unambiguous control functions, the value of the magnetic flux of the rotor and the vector of the stator flux linkage must be independent of each other, and then one of them can be maintained constant, and with the help of the other, regulation can be carried out. In DC motors and synchronous valve motors with electromagnetic excitation, there are separate electrical circuits for controlling the magnetic flux and armature current. In squirrel-cage asynchronous motors, there is only one channel in which both current components are combined and the task of the control system is to separate them. Mathematically, this problem is solved using the equations of a generalized electric machine in vector form. Hence the name of the method - "vector control".

Thus, the principle of vector control of an asynchronous motor consists in separate control of the rotor magnetic flux and stator flux linkage using independent components of the stator current, corresponding to the projections of the current vector. In fact, this happens by controlling the voltage and frequency of the phases of the induction motor. The vectors that determine the electromagnetic moment can be represented both in a rotating synchronous coordinate system and in a fixed coordinate system oriented along the field of the stator poles. The concept of vector control is fully suitable for DPT, if the excitation vector and the armature vector are combined into vectors presented in a coordinate system oriented along the axis of the main poles. In DPT, the orthogonality of these vectors at any position of the rotor is provided by the brush-collector assembly. The difference between IM and DPT lies only in the fact that in IM both vectors are compared in a coordinate system that rotates with the flow, while in DPT such a coordinate system is motionless. The real currents of the IM stator flow in the fixed windings and correspond to the projections of the current vector onto the fixed system of phase coordinate axes. Therefore, in the vector control of IM, it is necessary to solve the problems of coordinate transformations between the following coordinate systems: between rotating and stationary, between orthogonal and three-phase.

As is known, DC motors are well suited for a traction electric drive, therefore, AM with vector control is also well suited for a traction electric drive [7].

There is a method for frequency control of an asynchronous motor in the scalar mode [8]. At present, despite the competition with vector control methods, it is quite widespread, because allows solving many technical problems of a mass electric drive easier and more efficiently. In the case of scalar control with maintaining a constant ratio of voltage to frequency, it is necessary to increase the stator supply voltage with an increase in the rotation speed. However, the synchronous frequency of the induction motor is not equal to the shaft speed, and the slip of the induction motor depends on the load. Thus, the control system with scalar control without additional feedback cannot accurately control the operation of the motor under varying load.

The scalar control method is relatively simple to implement, but has significant drawbacks:

- has poor dynamics when starting the engine from a stationary state,
- has a relatively narrow range of rotation speed adjustment (no more than 1:10),
- poorly adapted to load changes.

A distinctive feature of the vector control system for an asynchronous traction electric drive is the need to use an additional computing unit in which the modulus of the rotor flux vector and its current angular position are estimated. This is done by solving in real time a system of differential equations compiled in accordance with the mathematical model of an asynchronous electric motor.

Compared to the scalar, the vector control method has the following advantages:

- wide range of speed control;
- smooth regulation of the engine rotation speed in the entire frequency range;
- the ability to maintain a constant speed when the load of the electric drive changes;
- reduction of losses during transient processes in the drive (in connection with this, some increased efficiency of the engine).

A vector frequency converter is needed where a wide control range is required, especially at low speeds. Despite a number of significant advantages, it should be noted that the computational complexity of the vector control method is high, and when calculating the optimal drive operating modes, it is necessary to take into account a large number of drive parameters. Briefly, the disadvantages of vector control can be said as follows: it is expensive to develop, manufacture and maintain. Therefore, where a large depth of regulation is not needed and there is no requirement to provide high torque at low speed, the scalar method is used. It is precisely these operating conditions of the induction motor that take place in the proposed algorithm for a mild hybrid vehicle, since such a mild hybrid vehicle does not use the operation of an electric motor at low speeds and operates in a relatively narrow speed range. This makes it possible to further reduce the cost of the electric traction drive system. To this end, in the scalar method of frequency control of an asynchronous traction motor in the proposed mild hybrid vehicle, the electric motor load is taken into account by introducing additional feedback to control the changing load. To do this, with a uniform movement of a car on an electric drive-in urban condition, the technical solution proposed in [9] is used. According to this technical solution, the frequency control channel selects a frequency proportional to the vehicle speed, and the phase voltage control channel selects such a three-phase alternating voltage, which, when applied to all three phases, provides a given power factor on one of them. Consider the implementation and justification of this method.

In the proposed mild hybrid vehicle, the asynchronous traction motor is kinematically connected to the secondary shaft of the mechanical gearbox. Induction motor control uses a frequency control channel and a phase voltage control channel. The frequency control channel uses the vehicle speed sensor, since its signal is proportional to the speed of the axis of the induction motor. Knowing the number of pairs of stator poles and the coefficient of proportionality between the frequency of the vehicle speed sensor signal and the shaft speed of the asynchronous motor, we obtain by adding the nominal absolute slip the frequency of the three-phase alternating voltage that must be supplied to the asynchronous motor.

To determine the optimal voltage, which in the general case depends not only on the frequency, but also on the load of the electric motor, we use the law of M.P. Kostenko. According to the law of M.P. Kostenko [10], the normalized voltage is equal to the product of the normalized frequency and the normalized torque to the power of 1/2, i.e.

$$\frac{U}{U_{nom}} = \frac{f}{f_{nom}}\sqrt{\frac{M}{M_{nom}}}, \tag{1}$$

where U_{nom}, f_{nom} and M_{nom} are rated voltage, frequency and torque, respectively. In other words, if the saturation in the engine is weak, then for optimal operating conditions it is necessary to change the relative voltage in proportion to the product of the relative frequency and the square root of the relative torque. In this case, the operation of an asynchronous motor takes place at an almost constant power factor, a constant margin of stability and a constant absolute slip. Since the voltage and frequency can be set and controlled by the control system, there remains an unpredictable load torque that varies within certain limits. However, if voltage control is introduced, focusing not only on the change in frequency, but also on maintaining the optimal power factor, then, as follows from Kostenko's law (1), the change in the load of the asynchronous motor can also be taken into account. Consequently, the servo control system for slip and traction torque can be closed in phase current (through phase voltages) taking into account the power factor, since the power factor is easily calculated by a specialized microcontroller from the signals of the phase current and phase voltage sensors. Power factor control is performed on one phase because all phases are symmetrical and behave the same if the same voltage is applied to them.

Power factor can be determined using low-cost specialized microcircuits (microcontrollers for electronic kilowatt-hour meters, such as ADE7755). The voltage control channel uses the signals of the current and voltage sensors of the motor phase, which are processed by a specialized controller and then fed to the microcontroller of the control unit. The task of the control unit is to change the voltage of the motor phases in accordance with the frequency, but, in addition, correct it so as to maintain the set value of the power factor. Why given, and not maximum, because for an asynchronous motor, a small reactive power is needed to form the magnetic field of the rotor. As a result, it is possible to use a relatively small power scalar frequency-controlled induction motor in a mild hybrid vehicle, which reduces the cost of the hybrid vehicle.

On Fig. 1 shows a block diagram of the system for implementing the proposed method for controlling a traction asynchronous electric motor in a mild hybrid vehicle, where 1 is a traction battery, 2 is a power controlled three-phase bridge, 3 is an asynchronous electric motor, 4 is a voltage control channel, 5 is a frequency control channel, 6 – vehicle speed sensor.

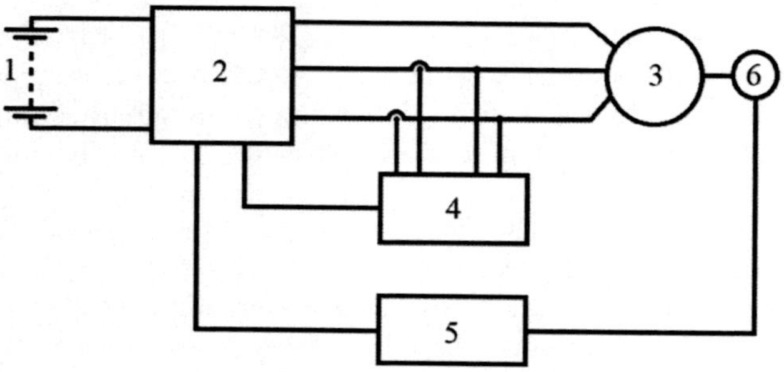

Fig. 1. Block diagram of the AM control system.

By minimizing the stator current at the operating point of the mechanical characteristic, taking into account the AM stability margin, we obtain an adaptive scalar control system. Such minimization can be done using tabular microcontroller frequency control. The induction motor is designed to have maximum efficiency and maximum power factor at rated load. By adjusting the frequency and voltage of the stator supply, it is possible to change the slip and thereby change the power factor, therefore, to achieve high efficiency, as well as the ability to use the control principle to minimize electrical power losses in the inverter-motor system.

4 Calculation of the Transmission Coefficient from the Shaft of the Electric Motor to the Driving Wheels of the Vehicle

Speed range from 40 to 60 km/h provides the possibility of driving only on the electric motor of the mild hybrid vehicle [6, 11].

Let's write the road resistance force in total with the air resistance force [12]

$$P_r = \psi G + \frac{1}{2} C_x A \rho_a V^2, \tag{2}$$

where G - vehicle weight, $\psi = f_r \cos\alpha + \sin\alpha$ - total road resistance coefficient, α - road inclination angle, f_r - rolling resistance coefficient, η_{tr} - transmission efficiency, δ - coefficient of accounting for rotating masses, V is the speed of the car, ρ_a - the air density, C_x is the streamlining coefficient, g is the acceleration of gravity, A is the frontal area of the car.

This force in uniform motion is equal to the traction force of the electric motor

$$P_T = \frac{M_e(V) i_{eg} i_{mg} \eta_{Tp}}{r_w}, \tag{3}$$

where M_e (V) - speed-dependent torque of the electric motor, η_{tr} - transmission efficiency, i_{mg} - main gear ratio, i_{eg} - transmission ratio between the electric motor and the secondary shaft of the gearbox.

Based on this equality, we obtain the power of the electric motor

$$N = M_W \omega_W, \tag{4}$$

where N - electric motor power, M_w and ω_W, respectively, are the moment and angular velocity of the driving wheel, and $\omega_W = 2\pi\ V/r_W$.

Let's evaluate the dynamics in the selected speed range [13] The acceleration time t_a from V1 to V2 can be calculated by integrating the reciprocal of the vehicle's instantaneous acceleration over the required speed interval over the speed:

$$t_a = \int_{V1}^{V2} \frac{1}{j} dV = \int_{V1}^{V2} \frac{\delta G}{g(P_T - P_a - \psi G)} dV, \tag{5}$$

where P_a is the force of air resistance, j is the acceleration of the car.

5 Conclusions

Considering the above, it can be concluded that the implementation of the developed algorithm makes it possible to create a hybrid power plant of a mild hybrid vehicle, which improves economic and environmental parameters, and also makes it possible to achieve low cost compared to hybrid vehicles with electric motors based on a synchronous electric machine with permanent magnets. This algorithm of work will also reduce not only the cost, but also the weight of a mild hybrid car. Therefore, a mild hybrid vehicle with a relatively small power induction motor and a small TAB can be made efficient and cost effective. The results obtained can be useful for the transport industry.

Acknowledgement. This work was conducted under the Scientific research "Development of a system that improves the energy efficiency and environmental friendliness of motor transport based on the "mild hybrid" technology", 0122U000654, funded by the Ministry of Education and Science of Ukraine.

References

1. Ning, D., Prasad, K., Lie, T.: The electric vehicle: a review. Int. J. Electric Hybrid Veh. **9**(1), 49–66 (2017)
2. Travis, F., Ryan, J., Evan, G., Joshua, C., Hwan-Sik, Y.: Effect of an electric vehicle mode in a plug-in hybrid electric vehicle with a post-transmission electric motor. Int. J. Electr. Hybrid Veh. **8**(4), 289–301 (2016)
3. Taoudi, A., Haque, M.S., Luo, C., Strzelec, A., Follett, R.F.: Design and optimization of a mild hybrid electric vehicle with energy-efficient longitudinal control. SAE Int. J. Electr. Veh. **10**, 55–78 (2021)
4. Mild Hybrid Electric Vehicle (MHEV) – examples. https://x-engineer.org/mild-hybrid-electric-vehicles-mhev-examples/. Accessed 29 Aug 2020
5. Dvadnenko, V.Ya., Pushkar, O.B.: Improving the economic and environmental characteristics of a microhybrid car. Motor Transport: Sat. scientific tr. (45), 12–22 (2019)
6. Mykhalevych, M., Shuklinov, S., Dvadnenko, V., Yaryta, O.: Prospects of "mild hybrid" technology for creating a hybridization system of vehicles. Automob. Transp. (50), 29–39 (2022). https://doi.org/10.30977/AT.2019-8342.2022.50.0.04
7. Pankratov, A.I.: Controlling systems of electric drives, 2nd edn. DDMA, Kramatorsk (2018)
8. Lavrinenko, Y.M., Marchenko, O.S., Savchenko, P.I., Sinyavskyi, O.Yu., Voytiuk, D.G., Lysenko, V.P.: Electric drive. Lir-K, Kyiv (2009)
9. Utility model patent №151236 Method of frequency control of an asynchronous traction electric motor/ Application №u2O2107659/Dvadnenko, V.Ya., Dzyubenko, O.A. statement 28.12.202k published 06/22/2022, Bui. №25
10. Vidmish, A.A., Yaroshenko, L.V.: Fundamentals of the electric drive. Theory and practice. Part I. VNAU. Vinnitsa (2020)
11. Bazhinov, O.V., Dvadnenko, V.Ya.: Automotive hybrid power drive. Kharkiv. FOP Brovin O.V. (2016)
12. Volkov, V.P.: The theory of exploitation properties of the car. Kharkiv, KhNAHU (2003)
13. Dvadnenko, V.Ya.: Mathematical model of a conversion hybrid car in the electric drive mode. Bull. KhNADU (72), 41–48 (2016)

The Hazards of Batteries Used in Electric Vehicles and Ensuring Their Safety

Tomas Pasaulis(✉) and Robertas Pečeliūnas

Vilnius Gediminas Technical University, Saulėtekis Av. 11, 10303 Vilnius, Lithuania
{tomas.pasaulis,robertas.peceliunas}@vilniustech.lt

Abstract. Due to ecological reasons, the number of electric vehicles in the world is increasing every year, so it is important to ensure that electric vehicles on the road are safe. This article analyzes the principle of operation of lithium-ion (li-ion) batteries used in electric vehicles, the main components, materials, and the main requirements for batteries. The types of cells used in li-ion batteries, their advantages and disadvantages are briefly reviewed. The dangers that can be caused by damaged or malfunctioning batteries and the reasons of these hazards are also discussed. Electric vehicle battery installation strategies and tools used to ensure safety are described.

Keywords: Electric vehicles · Battery · Crash · Hazard · Safety

1 Introduction

The number of electric vehicles is growing for environmental reasons, and in the near future, electric vehicles are expected to make up a large part of the total car park. The advantage of electric vehicles is that they do not emit any emissions and helps reduce the greenhouse effect. Rechargeable lithium-ion (li-ion) batteries are generally used in electronic devices and are considered to be potentially the best energy source for electric vehicles due to the high energy density of these batteries and the high number of charging cycles [1]. However, the safety and reliability of electric vehicle batteries present major challenges for the electrification of public and private transport [2]. It is electric car batteries that are the most important component of the car, which unfortunately can cause danger to the driver and passengers of the vehicle for various reasons, so it is very important to analyze the problems that arise and take all the necessary steps to prevent accidents related to electric vehicle batteries. The purpose of this article is to review the working principle of li-ion batteries used in electric vehicles, the main batteries components, the possible batteries hazards and their safety.

2 Structure, Properties and Working Principle of Li-ion Batteries

In order to find out the safety and possible threats of li-ion batteries, it is necessary to analyze the structure and working principle of li-ion batteries. Li-ion batteries generally consist of four main components: cathode, anode, separator and electrolyte. As the battery discharges, lithium ions travel from the anode to the cathode through the electrolyte,

O. Prentkovskis et al. (Eds.): TRANSBALTICA 2023, LNITI, pp. 195–202, 2024.
https://doi.org/10.1007/978-3-031-52652-7_20

generating the electrical current that is needed to keep the electric vehicle moving. As the battery charges, the cathode releases lithium ions, which return to the anode (Fig. 1) [1].

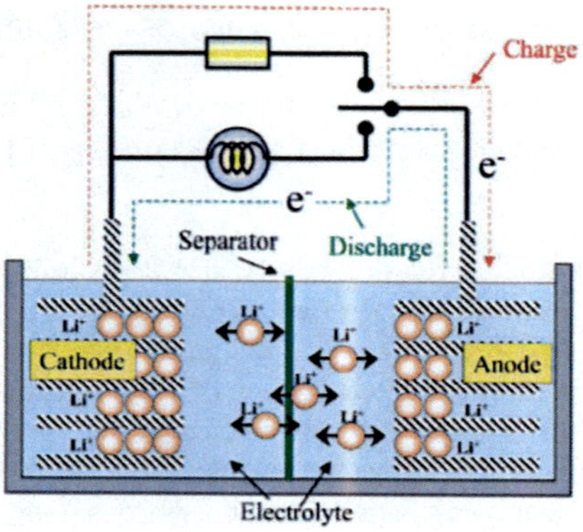

Fig. 1. Working principle of li-ion battery [3].

Generally, the technological properties of electric batteries depend on the materials from which the battery components are made. Cathodes in batteries can be made using manganese oxide, lithium cobalt oxide, lithium nickel cobalt oxide, or lithium iron phosphate. The anode is usually made using carbon, and recently lithium titanium has gained a lot of interest due to its thermal stability and potential safety. The electrolytes used in li-ion batteries are mainly based on aprotic organic solvents. The most commonly used electrolytes are mixtures of various carbonates and a dissolved salt [4]. Li-ion batteries with liquid electrolytes have good performance, the electrolyte solutions offer high conductivity and excellent wetting of the electrode surface [5]. However, the liquid electrolytes based on aprotic organic solvents which usually used in li-ion batteries are highly flammable [4]. Solving this problem aims to replace liquid organic electrolytes with non-flammable solid-state electrolytes [6]. Solid-state electrolytes can be grouped into inorganic ceramic electrolytes and organic polymeric electrolytes [5]. The separator is the main element that prevents electrical contact between electrodes and allows the ions to move. Currently used separators consist of a thin microporous polyolefin membranes made of polyethylene, polypropylene, or laminates of both. The polyethylene of this type of separator begins to melt when it reaches a temperature of 135 °C, while the polypropylene maintains its integrit. As the polymer melts, the pores become closed, creating an isolating layer that provides some protection against short circuit and overcharge. Separators made of ceramic materials can also be used, such separators are characterized by high temperature stability and good chemical stability [4].

Li-ion batteries are very widely used because they have high energy density, high power density, long cycle life, low self-discharge rate and no memory effect [7]. The energy density of the battery is very important parameters that describes the ability of the battery to hold energy, the higher energy density, the more energy the battery can store [8]. Ensuring the proper battery temperature and operating environment can ensure maximum battery energy capacity [2]. In a general sense, a good battery for an electric vehicle should be characterized by lightness, affordability, safety and durability [8].

3 Electrical Vehicle Structure and Types of Cells Used in Li-ion Batteries

An electric car is powered by an electric motor and a battery. These types of cars use an external electrical charging socket or can charge the battery through regenerative braking [9]. The main components of an electric vehicle may be located in different places, but the principle is based on a single platform. Figure 2 shows how the components of an electric car can be layout [10].

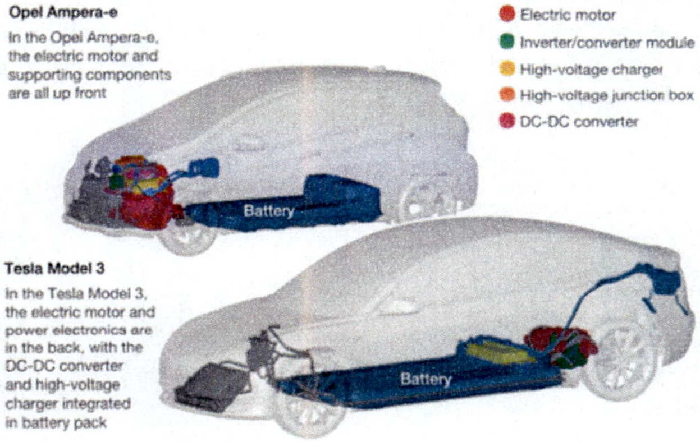

Fig. 2. Electric vehicle components and their layout [10].

Li-ion battery packs use different types of battery cells. The cells are different from each other in design and characteristics. There are generally three types of cells used in li-ion batteries. The first type is cylindrical cells, which are characterised by their high mechanical stability and easy manufacture, which results in a lower price. Cylindrical cells cannot fully fill the space of the battery pack, but they allow the air to flow more easily around the group of cylindrical cells, which improves the thermal control of the battery [10, 11]. Another type is prismatic cells, which offer an advantage in terms of battery area utilisation, but raises challenges for thermal control as there is no clear space between the cells [10]. The third type of cell used in batteries is the pouch cell. These cells have a flat shape which helps to achieve a very high battery space efficiency, but like prismatic cells, this type of cell is more difficult to dissipate heat and therefore temperature control is very important in batteries using this type of cell [11]. The types of cells used in batteries and their structure are shown in Fig. 3.

Shape	Cylindrical	Prismatic	Pouch
Diagram	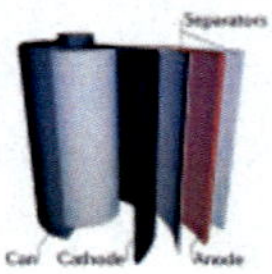	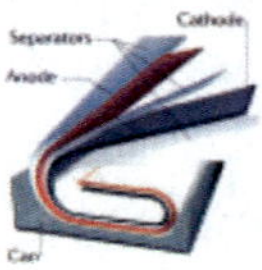	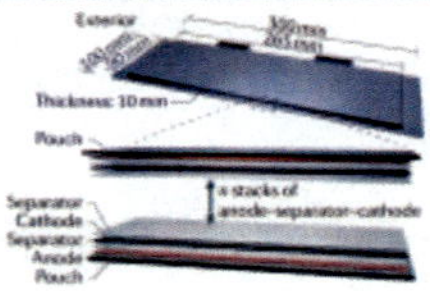
Electrode Arrangement	Wound	Wound	Stacked
Mechanical Strength	++	+	–
Heat Management	–	+	+
Specific Energy	+	+	++
Energy Density	+	++	+

Fig. 3. Types of cells used in li-ion batteries [10].

Typically, a battery module consists of connected cells and then several modules are combined to create a battery pack (Fig. 4) [12]. Hundreds or thousands of cells can be connected in parallel and in series to form one large battery pack to get the specific voltage and capacity required. However, battery management in a large battery pack is complex enough to ensure their safety [13].

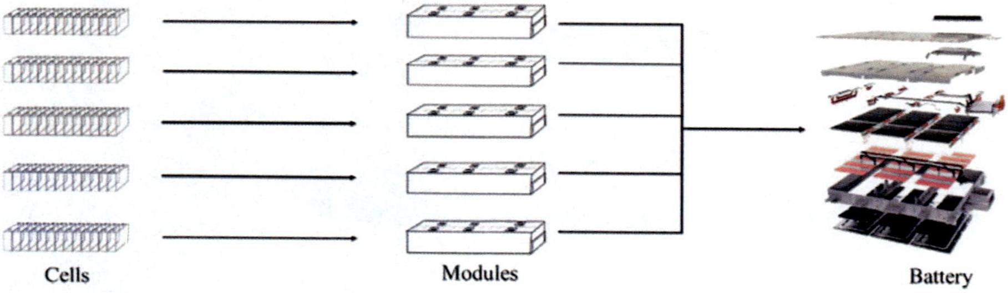

Fig. 4. Schematic connection between cell, module and battery [13].

When choosing the type of battery cell, car manufacturers mainly consider their battery integration technology and battery performance. As an example, the Tesla S3 electric car model uses cylindrical cells [13].

4 Hazards of Batteries Used in Electric Vehicles

The hazards of electric vehicles can be classified into electrical, chemical and thermal [14]. These types of hazards can be caused by poor control systems, battery defects, disasters or traffic accidents [13].

Electrical hazards - one of the possible fears is that it may be unsafe and dangerous to touch the car after the accident because of the high-voltage components that are integrated in the car. In this case, the electric vehicle must be designed to be completely safe from high-voltage elements after a serious accident. Normally, when a car airbags are triggered by impact, the contacts of the high-voltage circuit open quickly and the high-voltage system is shut down [14]. High-voltage components must be located outside the passenger compartment in a safe place to avoid danger to passengers [2]. The purpose of high voltage protection is to secure that the passengers of a car do not get an electric shock during or after an accident. The concept of high voltage protection for electric vehicle is based on protection against direct and indirect contact with high voltage components [15].

Damaged electric vehicle batteries can also cause chemical hazards. Lithium-ion batteries contain many potentially hazardous materials [4]. If an electric vehicle is on fire it is crucial to identify whether the fire has spread to the battery or not. In the event of a fire, the exothermic processes in the battery may release toxic substances due to self-reinforcing exothermic processes. In this case, hydrocarbons, CO and especially hydrogen fluoride could be released. Hydrogen fluoride is highly toxic and can cause health damage or deadly consequences. The hydrocarbons released can ignite or worse, get collected in free space and ignite later onwards. If the mixture of gases released is above the explosive threshold, a spark may cause a flame burst or possible explosion [14].

In a traffic accident, batteries can be mechanically damaged by the loads applied, leading to internal short circuits or damage to the casing, which can cause thermal runaway of batteries [13]. Typically, lithium-ion cells have a restricted temperature operating limits within which they operate normally. This temperature range is normally not higher than 60 °C. If the battery cells are exposed to higher temperatures, side reactions called exothermic reactions may occur. Exothermic reactions reinforcing the already ongoing heat generation, leading to the thermal runaway [14]. Thermal runaway is an exothermic circuit reaction in which the battery cells begin to heat up by themselves at a rate higher than 0.2 °C/min [2]. This process normally starts by the decomposition of the solid electrolyte interface at temperature as low as 90–120 °C. If thermal runaway occurs in one cell, neighboring cells or modules can be affected by the temperature increase of the first cell, so the other cell as well initiates this process. This process is difficult to stop because it is self-sustaining process as the cathode releases its own oxygen for an exothermic reaction [14]. The deformation caused by the impact can lead to thermal runaway of the battery, which is why it is important to measure the safety of batteries in collision [16]. In this case, the nail penetration test is used by many institutions and companies to investigate thermal runaway [17]. Considering the importance of heat, the battery pack should be installed where there is good air flow to increase heat dissipation and protect the batteries [2].

Also, for electric vehicles, unwanted vibrations can lead to hazards. An electric vehicle moves on various road surfaces, which can cause vibrations that can affect the electric vehicles batteries. Lack of proper vibration isolation has been cited as one of the main reasons for battery pack durability failures in the market. When vibrations reach the frequency range of 100 Hz, resonance occurs and battery cells starts disassociate from each other, which is one of the main reasons for the reduced the battery life [2]. As vibration can cause damage to the battery structure, the design of an electric vehicle battery pack requires a reliability analysis and optimisation of the external structures, as well as the installation of protective elements with vibration isolation functions, thus avoiding the failure of electrical and mechanical components due to fatigue caused by vibrations [18, 19]. Vibration durability tests are used to better understand the behaviour of a given system when a vehicle is subjected to mechanically induced vibrations appropriate to the operating environment. The tests are performed to determine the desired maximum lifetime of the vehicle and to determine the lifetime of li-ion batteries, as well as other types of batteries for use in cars [20].

5 Safety of Electric Vehicles in Traffic Accidents

The safety of electric cars is very important. It is essential to have good protection for the components of electric vehicles in the event of a vehicle collision. The biggest attention should be given to the car's battery, which should not be damaged even in the most serious accidents. Frontal collisions are not so dangerous for the battery of an electric car because there is a large deformation space in the front of the car and special deformation paths have been designed in all cars (Fig. 5) [21].

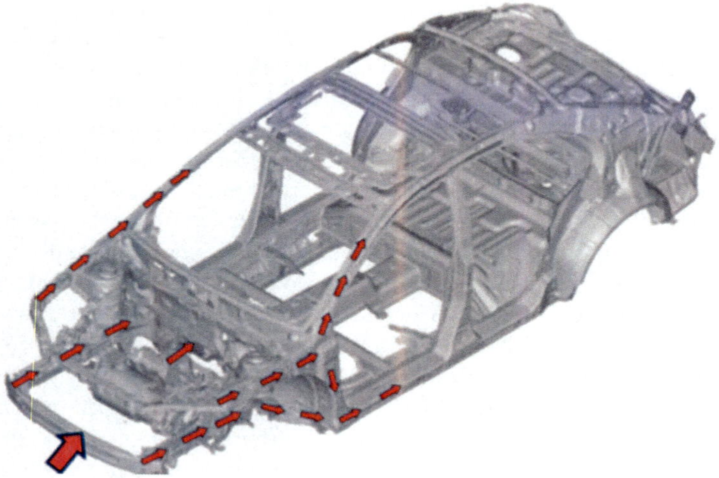

Fig. 5. Deformation paths for frontal impact [22].

The deformation paths absorb the loads caused by the impact, so in this case the car's batteries are sufficiently well protected against possible damage in a traffic accident [21]. Unlike in a frontal collision, batteries can be seriously damaged in the event of a side impact of a car, because in this case there is a very short deformation path [23]. It should also be noted that an impact to the rear of the car with a small overlap can cause the batteries to be crushed and deformed [24]. It is important to consider all possible impact directions when installing batteries. The passenger area of the car is also known as the primary safe zone and is considered to be the main safe zone that is protected from impact loads. In many current electric vehicles, the batteries are placed in the primary safe zone. As both batteries and passengers are in the same space, the batteries must be strongly protected against mechanical failures in order not to be a hazard to passengers [25]. Batteries should be installed in areas of the car that are least affected in the event of a crash [13]. Knowing the most crucial loading condition can significantly help in the design of the protective construction of the battery module and the battery pack [17]. Taking into account the results of the deformation model safety protection zones, it is indicated that the floor area of the car is considered to be the most suitable location for the placement of the batteries, in order to ensure and minimise the probability of deformation of the battery in different types of crash scenarios [26]. The correct placement and protection of the batteries is very important, as otherwise the batteries could be damaged in an accident, which could lead to various hazards that could be dangerous for the driver and passengers.

6 Conclusions

1. The main reason why li-ion batteries are most commonly used in electric vehicles is their high energy density, which describes the battery's ability to store energy.
2. An electric vehicle battery pack consists of a number of modules and the modules consist of a number of cells, which can be cylindrical, prismatic and pouch cells. The type of cells affects the temperature control of the battery and the space efficiency of the battery.
3. There are three major hazards of electric vehicle batteries: electrical hazards, chemical hazards and thermal hazards. The safety of batteries is also affected by various vibrations.
4. Electric vehicle batteries are the most difficult to protect in the event of a side impact due to the small deformation area. In the event of a frontal impact, batteries are normally sufficiently well protected.
5. It is very important that the batteries are protected from possible damage in the event of an accident, which could cause danger to the driver or passengers of the electric vehicle. In order to avoid hazards, a proper battery installation strategy must be in used.

References

1. Chawla, N., Singh, S., Bharti, N.: Recent advances in non-flammable electrolytes for safer lithium-ion batteries. Batteries **5**(1), 19 (2019)
2. Arora, S., Shen, W., Kapoor, A.: Review of mechanical design and strategic placement technique of a robust battery pack for electric vehicles. Renew. Sustain. Energy Rev. **60**, 1319–1331 (2016)
3. Huang, P., Ping, P., Wang, Q., Sun, J.: The combustion behavior of large scale lithium titanate battery. Sci. Rep. **5**(1), 7788 (2015)
4. Ruiz, V., Pfrang, A., Kriston, A., Omar, N., Van den Bossche, P., Boon-Brett, L.: A review of international abuse testing standards and regulations for lithium ion batteries in electric and hybrid electric vehicles. Renew. Sustain. Energy Rev. **81**, 1427–1452 (2018)
5. Zheng, Y., et al.: A review of composite solid-state electrolytes for lithium batteries: fundamentals, key materials and advanced structures. R. Soc. Chem. (2020)
6. Zhao, W., Yi, J., He, P., Zhou, H.: Solid-state electrolytes for lithium-ion batteries: fundamentals, challenges and perspectives. Electrochem. Energy Rev. **2**, 574–605 (2019)

7. Xiong, R., Ma, S., Li, H., Sun, F., Li, J.: Toward a safer battery management system: a critical review on diagnosis and prognosis of battery short circuit. iScience (2020)
8. Alanazi, F.: Electric vehicles: benefits, challenges, and potential solutions for widespread adaptation. Appl. Sci. **13**(10), 6016 (2023)
9. Angeline, P.S., Rajkumar, M.N.: Evolution of electric vehicle and its future scope. Mater. Today Proc. **33**, 3930–3936 (2020)
10. Pham, X.M., Bui, G.V., Pham, H., Pham, L.H.P.: Design process of electric vehicle power system. Appl. Mech. Mater. **907**, 101–114 (2022)
11. Un, C., Aydin, K.: Thermal runaway and fire suppression for electric vehicle batteries. In: International Conference on Progresses in Automotive Technologies, Istanbul (2020)
12. Linse, C., Kuhn, R.: Design of high-voltage battery packs for electric vehicle. In: Advance in Battery Technologies for Electric Vehicle, pp. 245–263. Woodhead Publishing Series in Energy (2015)
13. Chen, Y., et al.: A review of lithium-ion battery safety concerns: the issues, strategies, and testing standards. J. Energy Chem. **59**, 83–99 (2021)
14. Geisbauer, C., et al.: Scenarios involving accident-damage electric vehicles. Transp. Res. Procedia **55**, 1484–1489 (2021). 14th International scientific conference on sustainable, modern and safe transport. (2021)
15. Chombo, P.V., Laoonual, Y., Wongwises, S.: Lessons from the electric vehicle crashworthiness leading to battery fire. Energies **14**(16), 4802 (2021)
16. Kotter, P., Kisters, T., Schleicher, A.: Dynamic impact tests to characterize the crashworthiness of large-format lithium-ion cells. J. Energy Storage **26**, 100948 (2019)
17. Li, W., Zhu, J., Xia, Y., Gorji, M.B., Wierzbicki, T.: Data-driven safety envelope of lithium-ion batteries for electric vehicles. Joule **3**, 2703–2715 (2019)
18. Hua, X., Thomas, A.: Effect of dynamic loads and vibrations on lithium-ion batteries. J. Low Freq. Noise Vib. Act. Control **40**(4), 1927–1934 (2021)
19. Zhang, J., Zhang, L., Sun, F., Wang, Z.: An overview on thermal safety issues of lithium-ion batteries for electric vehicle application. Special section on battery energy storage and management systems. IEEE (2018)
20. Somerville, L., Hooper, J.M., Marco, J., McGordon, A., Lyness, C., Walker, M., Jennings, P.: Impact of vibration on the surface film of lithium-ion cells. Energies **10**(6), 741 (2017)
21. Gu, X., Wang, W., Xia, L., Jiang, P.: A system optimisation design approach to vehicle structure under frontal impact based on SVR of optimised hybrid kernel function. Int. J. Crashworthiness (2019)
22. Yu, L., Gu, X., Qian, L., Jiang, P., Wang, W., Yu, M.: Application of tailor rolled blanks in optimum design of pure electric vehicle crashworthiness and lightweight. Thin-Walled Struktures **161**, 107410 (2021)
23. Uerlich, R., Sanalkumar, K.A., Bokelmann, T., Vietor, T.: Finite element analysis considering packaging efficiency of innovative battery pack designs. Int. J. Crashworthiness (2019)
24. Biqiang, Z., Xiaodang, L., Yajun, Z.: The structure optimization analysis of electric vehicle in small offset rear end collision. Procedia Eng. **137**, 103–108 (2016)
25. Kukreja, J., Nguyen, T., Siegmund, T., Chen, W., Tsutsui, W., Balakrishnan, K., Liao, H., Parab, N.: Crash analysis of a conceptual electric vehicle with a damage tolerant battery pack. Extreme Mech. Lett. **9**, 371–378 (2016)
26. Navale, A.B., Chippa, S.P., Chougule, D.A., Raut, P.M.: Crashworthiness aspects of electric vehicle design. Int. J. Crashworthiness **26**(4), 368–387 (2020)

Logistics and Transportation

Distribution Optimization for Connected Autonomous Vehicles (CAV) Considering Fuel Consumption Optimization

Guangyuan Yang[1](✉), Hui Ma[1], Keqi Chen[2], and Aoran Zhou[3]

[1] College of Transportation Engineering, Chang'an University, Xi'an 710064, China
2020134002@chd.edu.cn

[2] School of Automobile, Chang'an University, Xi'an 710064, China

[3] Chang'an Dublin International College of Transportation at Chang'an University, Xi'an 710064, China

Abstract. With the maturity of autonomous driving and Internet of Vehicles, as well as the increasingly serious climate change problem, the research and application of smart logistics and green logistics have attracted much attention. With urban distribution as the research object, a distribution optimization model considering connected autonomous vehicles (CAV) for fuel consumption optimization at intersections is constructed, and an enhanced differential evolutionary algorithm (EDEA) is designed to find the optimal solution. The model considers intersection fuel consumption, time window constraint and penalty cost, and aims at minimizing the sum of distribution cost, carbon tax cost and penalty cost, focusing on optimizing the speed strategy of CAV at intersections to reduce fuel cost and carbon emission. EDEA generates the initial population by backward learning strategy and random strategy, which effectively improves the quality of the initial population. The intersection operator and variation operator in the differential evolution algorithm are tested, and the operator with the best adaptation degree is selected from them. The distribution optimization model and EDEA are applied to verify the distribution problem of autonomous vehicles in a courier distribution center in Xi'an. The results show that the constructed model considers fuel consumption at intersections and can decrease the carbon tax cost, distribution cost and penalty cost arising from the speed change of autonomous vehicles at the intersection. The EDEA proposed in this paper has a better performance in finding the optimal value, and the optimization target value is 18.3% lower than the Differential Evolutionary Algorithm (DEA) and 35.9% lower than the Standard Genetic Algorithm (SGA).

Keywords: Distribution optimization · Connected autonomous vehicle (CAV) · Fuel consumption optimization · Enhanced Differential Evolution Algorithm (EDEA) · Intersection speed strategy

O. Prentkovskis et al. (Eds.): TRANSBALTICA 2023, LNITI, pp. 205–222, 2024.
https://doi.org/10.1007/978-3-031-52652-7_21

1 Introduction

In recent years, more and more logistics and distribution vehicles are shuttling in the city every day, which brings high labor costs and carbon emission problems while facilitating the life of citizens. How to reduce logistics and distribution costs, reduce carbon emissions and improve customer satisfaction has become an urgent problem to be solved.

Dantzig et al. [1] first proposed the vehicle routing problem (VRP) in 1959. This problem usually takes the minimization of distribution cost or time as the optimization objective, and the constraints usually include supply and demand matching, vehicle loading capacity, vehicle mileage and customer time window [2]. Zhang [3] studied simultaneous delivery VRP with soft time window constraint under single-center static road network and multi-center semi-open VRP with soft time window constraint under multi-center dynamic road network, respectively. Ye [4] conducted a path optimization study on backhaul idling, which is widespread in road freight, and the study integrated customer service priority, time constraint, vehicle load capacity, toll fee, and fuel cost constraints. Liu et al. [5] constructed a joint distribution path problem model for energy efficient vehicles considering the needs of cold chain logistics in terms of distribution cost and energy saving. Zhang et al. [6] proposed a hybrid multi-objective evolutionary algorithm based on combined timing to solve the multi-vehicle path problem with time windows with the criterion of minimizing the number of vehicles and time wastage at the same time. Most of the current research on VRP focuses on the route selection and optimization problem, by optimizing the route to reduce costs, while ignoring the optimization of vehicle fuel consumption.

The road conditions and vehicle conditions at intersections are very complex, and vehicles usually need to slow down or stop at intersections, which can significantly in-crease fuel consumption and carbon emissions. By optimizing the frequency and magnitude of vehicle speed changes at intersections, it not only helps to reduce traffic congestion, but also helps to lower fuel cost and exhaust emissions. Based on the vehicle-road cooperative system, Zhang [7] studied the guided vehicle speed curve planning method with the minimum energy consumption for the passage characteristic signal and vehicle passage of vehicles at signal intersections. Zhao [8] analyzed the passage trajectory of a vehicle through an intersection by means of a follower model, and simulated the fuel consumption generated by a vehicle through a signal intersection and the pollution caused., Zhu et al. [9] investigated the driving strategy of signalized intersections with the goal of energy saving by automatically obtaining real-time vehicle position and speed, traffic light status and countdown information using V2I communication system at the intersection section. Lakouari et al. [10] investigated the CO2 emission at roundabout intersections using a metacellular automata model. Wei et al. [11] proposed a joint optimization method to improve the fuel saving performance of a fuel cell vehicle (FCV) by addressing the speed planning and energy management problems when passing through multiple consecutive signal intersections. Ma et al. [12] proposed a vehicle trajectory planning method that can significantly reduce the average fuel consumption and the number of lane changes of a connected autonomous vehicle (CAV) at independent signal intersections, and validated it in a mixed traffic environment consisting of an artificially driven vehicle and a CAV. Most of the current studies on the optimization of vehicle efficiency and fuel consumption at intersections only focus on intersections, but ignore the optimization of intersections, vehicles, and vehicle path planning as a whole in an intelligent Internet of Vehicles environment.

In the context of the development of intelligent and connected transportation systems, CAV distribution has become a popular topic for improving distribution efficiency [13–17]. Autonomous vehicles equipped with intelligent vehicle networking technology can largely reduce urban traffic congestion, traffic accidents, and carbon emissions. Technology companies such as Mobileye, Waymo, Baidu and Cruise, as well as major car companies, have all made significant achievements in the development and application of autonomous driving technology. The order volume of Baidu's self-driving travel service has exceeded 2 million orders, and the autonomous delivery vehicles of a number of take-out and delivery companies have been used in semi-closed environments such as campuses and communities. Self-driving technology has entered the critical stage of commercial application. In the available research, Lu et al. [18] proposed a smart Telematics intersection signal timing method using V2X network to reach real-time monitoring and control of intersection traffic conditions. Yang [19] proposed a two-layer control framework that enables CAVs to approach signalized intersections with good driving comfort based on the characteristics of networked vehicles that can communicate with traffic control centers in both directions in a connected vehicle environment to induce the speed of vehicles and achieve fine control of intersections to improve intersection passage speed and avoid traffic accidents. Shen et al. [20] proposed a two-layer control framework that enables CAVs to approach signalized intersections with good driving comfort. From the above studies, it is clear that CAVs and vehicular networks show great potential for application in existing studies and are very suitable for applications in areas such as smart logistics.

Considering the current status that few studies incorporate fuel consumption optimization, self-driving technology, and Internet of Vehicles into the overall optimization of VRP, we propose a CAV distribution optimization model considering intersection fuel consumption optimization in the environment of Internet of Vehicles. Since most of the speed changes of CAVs occur in the intersection area, the model focuses on the fuel consumption optimization of CAVs in the intersection area, establishes the relationship between speed changes and carbon emissions, reduces fuel consumption and carbon emissions by controlling the frequency and magnitude of speed changes, and integrates the speed optimization function with the cost optimization function in the form of segmentation function to achieve optimization of the total cost of distribution. The Enhanced Differential Evolutionary Algorithm (EDEA) is designed to solve the model. EDEA uses a backward learning strategy to generate the original population, which can greatly improve the efficiency of the algorithm without increasing the original search space. Finally, the feasibility and performance of the model and algorithm proposed in this paper are verified by example studies.

2 CAV Distribution Optimization Model Considering Total Cost Optimization

2.1 Problem Statement

In this section, a CAV distribution optimization model is established, and the cost optimization is taken as the optimization target of the model. First, this model screens the available paths based on the locations of distribution centers and customers. Then, based on the screened feasible paths, relevant features of intelligent vehicle networking technology are introduced, and the signal information at the intersection is transmitted to the CAV by the cooperative vehicle infrastructure system (CVIS) in real time. When the CAV enters the intersection area, the real-time timing status of the intersection is obtained through the vehicle network. After that, based on the relationship between vehicle speed variation and fuel usage, fuel consumption is calculated in real time, real-time intersection speed planning and decision making is performed, and carbon emission in the distribution process is calculated by mass balance equation. Finally, the above decisions are integrated, so that the CAV path optimization tends to the optimal sum of distribution cost, carbon tax cost and time window penalty cost.

This model is based on the following assumptions.

1. In the intelligent vehicle networking environment, information such as CAV speed, CAV location, and intersection signal timing can be exchanged and shared among CAVs and between CAVs and CVIS;
2. The optimization objective of this model is to minimize the sum of CAV distribution cost, carbon tax cost and penalty cost;
3. Each CAV has the same loading capacity and its maximum load capacity is not exceeded in a single transport;
4. There is a linear relationship between transportation cost and transportation distance;
5. All CAVs depart from the distribution center and return to the distribution center after completing their tasks;
6. The model adopts soft time window constraint, if the CAV fails to deliver the goods to the customer within the specified time window, the customer will not reject the goods, but the delivery service provider will be penalized according to the length of the default time to compensate for the loss of the customer and improve customer satisfaction.

2.2 Intersection Fuel Consumption Optimization Model

Model Parameters and Variables. The intersection fuel consumption optimization model is mainly supported by the intersection timing status, location information and other related information, and the CAV autonomously decides the speed strategy to reduce fuel consumption by reducing the frequency and magnitude of speed changes. According to the different states of the signal when CAV enters the intersection range, this model is divided into two states, green and red, and discusses the speed strategies that CAV can adopt according to the remaining time length of the current state respectively. The logic structure of this model is shown in Fig. 1.

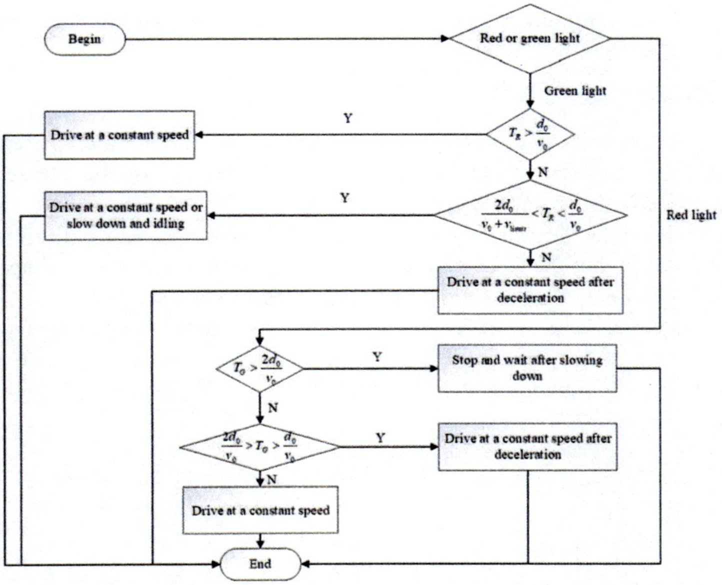

Fig. 1. Logical structure diagram of intersection speed optimization model.

The meanings of the parameters and variables in the model are as follows: T_R is the remaining time for the signal to change from green to red; T_G is the remaining time for the signal to change from red to green; V_{limit} is the speed limit of the road; V_0 is the initial speed of the CAV when it enters the intersection range; t_a is the acceleration time; t_d is the deceleration time; t_c is the time to travel at a uniform speed; t_s is the idle waiting time; t_0 is the time for the CAV to receive the CVIS signal and make decision; D_U is the upstream range of the intersection; D_D is the downstream range of the intersection; d_0 is the distance from CAV to the intersection; a_a is the acceleration of CAV; a_d is the deceleration of CAV; $a_{a\text{-}max}$ is the maximum value of a_a; $a_{d\text{-}max}$ is the maximum value

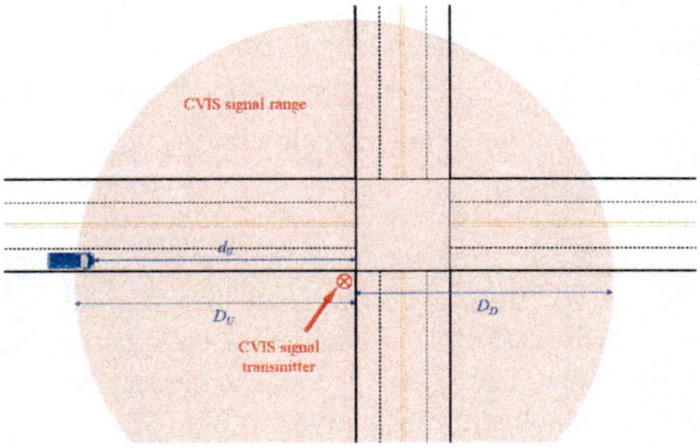

Fig. 2. Schematic diagram of CAV crossing intersection.

of a_d; $a_{d\text{-}min}$ is the minimum value of a_d. The schematic diagram of CAV through the intersection is shown in Fig. 2.

When the Current Signal is Green.

1. When T_R is greater than the time required for the CAV to pass the upstream of the intersection at a uniform speed, i.e., $\frac{d_0}{v_0} \le T_R$, the CAV can pass the intersection at a uniform speed with V_0 without changing its speed;
2. When T_R is less than the time required for the CAV to pass upstream of the intersection at uniform speed, but the CAV can pass the intersection within the road speed limit, i.e., $\frac{2d_0}{v_0+v_{\lim it}} \le T_R < \frac{d_0}{v_0}$, the following 2 strategies can be used.
   a) Accelerated passage. At the upstream of the intersection, the quantitative expression for the CAV speed strategy is given in Eq. (1). At the downstream of the intersection, the quantitative expression for the CAV speed strategy is given in Eq. (2).

$$\begin{cases} \int_0^{t_a}(v_0 + a_a t)dt + \int_0^{t_c}(v_0 + a_a t_a)dt + \int_0^{t_0} v_0 dt = d_0 \\ t_a + t_c = T_R \\ a_a \le a_{a-max} \\ v_0 + a_a t_a \le v_{limit} \end{cases} \tag{1}$$

$$\begin{cases} \int_0^{t_d}(v_0 + a_a t_a - a_d t)dt + \int_0^{t_c} v_0 dt = D_D - d_0 \\ a_d \le a_{d-max} \end{cases} \tag{2}$$

   b) Stop and wait. At the upstream of the intersection, the quantitative expression for the CAV speed strategy is given in Eq. (3). At the downstream of the intersection, the quantitative expression for the CAV speed strategy is given in Eq. (4).

$$\begin{cases} \int_0^{t_d}(v_0 - a_a t)dt + \int_0^{t_0} v_0 dt = d_0 \\ t_d + t_s = T_G \\ a_d \le a_{d-max} \end{cases} \tag{3}$$

$$\begin{cases} \int_0^{t_a}(v_0 + a_a t - a_d t_d)dt + \int_0^{t_c} v_0 dt = D_D - d_0 \\ a_a \le a_{a-max} \end{cases} \tag{4}$$

3. When T_R is much less than the time required for the CAV to pass the upper reaches of the intersection at uniform speed, and the CAV cannot pass the intersection within the speed limit, i.e., $0 \le T_R < \frac{2d_0}{v_0+v_{\lim it}}$, then the CAV should stop and wait, and its speed strategy is the same as the stop and wait strategy when $\frac{2d_0}{v_0+v_{\lim it}} \le T_R < \frac{d_0}{v_0}$.

When the Current Signal is Red.

1. When T_G is greater than the time required for the CAV to maintain a low speed through the upstream of the intersection without stopping, i.e., $\frac{2d_0}{v_0} < T_G$, the CAV must stop and wait at the intersection, at which time the quantitative expression of the CAV speed strategy is shown in Eq. (5) upstream of the intersection. The quantitative expression for the CAV speed strategy is given in Eq. (6).

$$\begin{cases} \int_0^{t_d}(v_0 - a_d t)dt + \int_0^{t_0} v_0 dt = d_0 \\ t_d + t_s = T_G \\ a_d \le a_{d-max} \end{cases} \tag{5}$$

$$\begin{cases} \int_0^{t_a}(v_0 + a_a t - a_d t_d)dt + \int_0^{t_c} v_0 dt = D_D - d_0 \\ a_a \le a_{a-max} \end{cases} \tag{6}$$

2. When T_G is greater than the time required for the CAV to maintain a uniform speed upstream of the intersection, but can maintain a low speed through without stopping when passing through the intersection, i.e., $\frac{d_0}{v_0} < T_G \le \frac{2d_0}{v_0}$. At the upstream of the intersection, the quantitative expression for the CAV speed strategy is given in Eq. (7). At the downstream of the intersection, the quantitative expression for the CAV speed strategy is given in Eq. (8).

$$\begin{cases} \int_0^{t_d}(v_0 - a_d t)dt + \int_0^{t_c}(v_0 - a_d t_d)dt + \int_0^{t_0} v_0 dt = d_0 \\ t_d + t_c = T_G \\ a_d \le a_{d-max} \end{cases} \tag{7}$$

$$\begin{cases} \int_0^{t_a}(v_0 + a_a t - a_d t_d)dt + \int_0^{t_c} v_0 dt = D_D - d_0 \\ a_a \le a_{a-max} \end{cases} \tag{8}$$

3. The CAV can pass through the intersection with V_0 when T_G is less than the time required for the CAV to maintain the original speed upstream of the intersection, i.e., $0 \le T_G \le \frac{d_0}{v_0}$.

2.3 CAV Distribution Total Cost Optimization Model

In this section, based on the intersection fuel consumption optimization model pro-posed in Sect. 2.2, the fuel cost at the intersection is allocated to the whole distribution process in the form of segmentation function, and the CAV distribution optimization model is established with the optimization objective of minimizing the overall CAV distribution cost. The meanings of the parameters and variables of the model are as follows: i,j,p are customer sequence numbers; K is the number of CAVs owned by the distribution center; k is the CAV sequence number, $k \in \{1,2,\ldots,K\}$; L is the number of customers served by this distribution center; b_k is the initial load of each CAV; d_i is the amount of goods received by customer i; $[e_i,l_i]$ is the time window of customer i; e_i is the left time window; l_i is the right time window; $x_{ijk} = 1$ means vehicle k drives from i to j, otherwise 0; $y_{ik} = 1$ means customer i is served by vehicle k, otherwise 0; s_{ij} is the distance from customer i to customer j; t_{ij} is the travel time of the CAV from customer i to customer j; a_i is the service time of customer i; w_i is the waiting time of the CAV at customer i; g_i is the arrival time of CAV at customer i; V_{ijk} is the speed of CAV k from i to j; E is the fuel consumption; A is the unit fuel cost; u_e is the early arrival penalty weight; u_l is the delayed arrival penalty weight; $h_n(x)$ is the fuel consumption versus speed, $n \in \{1,2\}$, where, when $n = 1$, represents the fuel consumption of CAV at the intersection. When $n = 2$, represents the fuel consumption relationship of CAV at non-intersection; μ is the CO2 emission factor; θ is the emission factor of other pollutants; and $F(x)$ is the carbon tax function.

The CAV distribution cost is

$$c_{trans} = AE = \sum Ah_n(x) = \sum_{n=1,2} A \sum_{i=0}^{L}\sum_{j=0}^{L}\sum_{k=1}^{K} h_n(V_{ijk}) \tag{9}$$

The cost of the carbon tax is

$$c_{co_2} = f(E) = \sum E(\mu + \theta) \tag{10}$$

The penalty cost is

$$c_2 = \sum_{i=0}^{L} \sum_{k=1}^{K} y_{ik}(d_i + q_i) \times \left[u_e \max((e_i - g_i), 0) + u_l \max((g_i - l_i), 0))\right] \tag{11}$$

According to the data and simulation steps in the literature [21], the relationship between CAV speed, acceleration and fuel consumption at the intersection is

$$E = (28.03 - 5.6lnV)\frac{|a|}{2.001} \tag{12}$$

The objective function is

$$\min \sum_{n=1,2} A \sum_{i=0}^{L} \sum_{j=0}^{L} \sum_{k=1}^{K} h_n\left(V_{ijk}\right) + \sum_{i=0}^{L} \sum_{k=1}^{K} y_{ik}(d_i + q_i) \times \left[u_e \max((e_i - g_i), 0) + u_l \max((g_i - l_i), 0))\right] + f(E) \tag{13}$$

The constraints are

$$\sum_{k=1}^{K} y_{ik} = 1, i \in \{1, 2, \cdots, L\} \tag{14}$$

$$\sum_{j=0}^{L} \sum_{k=1}^{K} x_{ijk} = 1, i \in \{1, 2, \cdots, L\} \tag{15}$$

$$\sum_{i=1}^{L} x_{ijk} = 1, \forall j, \forall k \tag{16}$$

$$\sum_{i=1}^{L} \sum_{k=1}^{L} x_{ijk} = \sum_{p=1}^{L} \sum_{k=1}^{K} x_{jpk}, \forall p \tag{17}$$

$$\sum_{k=1}^{K} \sum_{i=0}^{L} x_{ijk}(g_i + t_{ij} + w_i + a_i) = g_j, j \in \{1, 2, \cdots, L\} \tag{18}$$

$$w_j = max\{e_i - g_i, 0\} \tag{19}$$

$$e_i \leq g_i + w_i \leq l_i, i \in \{1, 2, \cdots, L\} \tag{20}$$

$$\sum_{i,j \in S \times S} x_{ijk} \leq |s| - 1, s \in \{1, 2, \cdots, L\} \tag{21}$$

$$\sum_{k=1}^{K} b_k = \sum_{i=1}^{L} d_i \tag{22}$$

$$0 \leq V_{ijk} \leq V_{max} \tag{23}$$

Equation (9) denotes the transportation cost of CAV, Eq. (10) denotes the carbon tax cost, and Eq. (11) denotes the penalty cost. Equation (12) is the relationship between

CAV speed, acceleration and fuel consumption at the intersection. The objective function in Eq. (13) aims to minimize the sum of the distribution cost, carbon tax cost, and penalty cost of CAV in Eqs. (9)–(11). Constraints (14)–(17) indicate that each customer is served only once, constraint (18) indicates the time from customer i to customer j, constraint (19) indicates the waiting time of customer j after the CAV departs from customer i, constraint (20) indicates that the arrival of the CAV and the waiting time at customer i must be within the time window, constraint (21) eliminates the subloop, and constraint (22) indicates that the CAV's initial load is equal to the sum of the number of shipments received by each customer, and constraint (23) is a CAV speed constraint.

3 Algorithm Design

Differential Evolutionary Algorithm (DEA) is an optimization algorithm evolved from genetic algorithm, in which the generation of offspring population consists of two steps: (i) generating variance vector from parent differential vector; (ii) crossover the variance vector generated in the first step with the parent population to generate offspring. As an algorithm that evolves through a population, DEA's excellent global search strategy, low-complexity variation operation, and independent competitive survival strategy are its most significant advantages, so it can improve the simplicity of genetic operation to a certain extent and is more superiority-preserving [22]. However, the algorithm has the disadvantages of poor global search capability, slow convergence, and easy to fall into local optimization.

To address the shortcomings of DEA, an enhanced differential evolutionary algorithm (EDEA) solution model is designed in this paper. Considering that the original population generated by the stochastic strategy is not ideal in terms of optimal solution quality and iterative convergence speed, EDEA uses the reverse learning strategy to generate the original population, which can greatly improve the efficiency of the algorithm without increasing the original search space. And by calculating the fitness of the original population and the reverse learning population, the individuals with higher fitness are selected as the final original population. In addition, the algorithm also tests the crossover operator and variation operator in DEA, from which the operator with the best fitness is selected. The specific steps of the algorithm are as follows.

1. Determine the population size, scale factor and hybridization probability of DEA and determine the fitness function.
2. Combine with the reverse learning strategy to generate populations of the specified size as the final initial populations of the EDEA.
3. Evaluate the quality of the previously generated initial population by calculating the fitness.
4. Determine whether the termination condition is satisfied or the evolutionary generation reaches the maximum. If the termination condition is reached, the evolution is terminated and the best individual is output as the optimal solution; otherwise, it continues to perform subsequent operations.
5. Based on the variation and crossover of the improved adaptive crossover variation operator, the intermediate population is obtained.

6. Select individuals from the original population and the intermediate population to get the new generation population.
7. Evolutionary algebra $g = g + 1$, go to step (4) to continue the operation.

The control of parameters in EDEA is very important, because each control parameter directly determines the convergence speed of EDEA. Due to the limitation of space, this section only introduces the crossover operation. If the crossover probability is set too low, the algorithm will converge too slowly; if the crossover probability is set too high, the search will be too random. The process of individual crossover between the gth generation population $\{x_i(g)\}$ and its mutation-generated intermediate population $\{v_i(g + 1)\}$ to generate the next generation $\{x_i(g + 1)\}$ is shown in Eq. (24).

$$x_{j,i}(g+1) = \begin{cases} v_{j,i}(g+1), \, rand(0,1) \le CR; j = j_{rand} \\ x_{j,i}(g), \, otherwise \end{cases} \tag{24}$$

where CR is the crossover probability, $CR \in [0,1]$; j_{rand} is a random integer in [1,2,...,D], where D is the dimension of the solution. The values of the parameter CR are given in Eq. (25).

$$cr_i = \begin{cases} cr_i(cr_a - cr_m)\frac{f_i - f_{min}}{f_{max} - f_{min}}, \, if \, f_i > f_a \\ cr_i, \, if \, f_i \le f_a \end{cases} \tag{25}$$

where f_i is the fitness, f_{min}, f_{max} are the minimum and maximum fitness values, respectively, and f_a is the average fitness value in the population in which it is located. Generally, $cr_a = 0.6$ and $cr_m = 0.1$.

The flow chart of the EDEA algorithm designed in this paper is shown in Fig. 3.

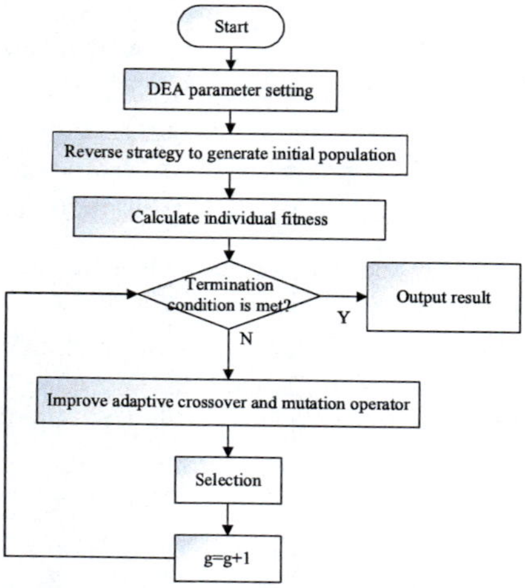

Fig. 3. EDEA flow chart.

4 Case Analysis

4.1 Raw Data

In this section, the optimization model proposed in this paper is validated with EDEA by an example of the self-driving vehicle delivery problem in a courier delivery center in Beilin District, Xi'an, Shaanxi Province, China. This example uses a soft time window constraint, if the CAV fails to deliver the goods to the customer within the time window $[e_i, l_i]$, the customer will not reject the goods, but the delivery service provider will be punished according to the length of the violation, so as to compensate the customer loss and improve the customer experience. Figure 4 shows the coordinate map, the line segment represents the road, point Q (3.5, 1.5) is the location of this distribution center in the map, points A~P are customers, and the location coordinates of customers are shown in Table 1. The main road intersections in the region are marked in Fig. 5, and the classification of the intersections is shown in Table 2.

In this case, it is assumed that the distribution center uses multiple CAVs with a maximum loading capacity of 100 kg and the distribution starts at 08:00 each day. Based on the research data and the actual situation, the speed of CAV in regular operation is set to 20 km/h and the carbon emission factor is 2.72 kg/L. The fuel consumption cost is calculated at the current fuel price of 8.19 yuan/liter. A half-day delivery plan is developed for the distribution center.

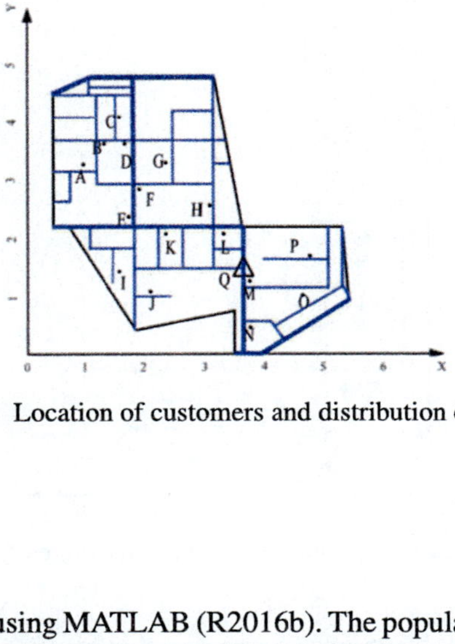

Fig. 4. Location of customers and distribution center.

4.2 Model Solving

The example was solved using MATLAB (R2016b). The population size is set to 300 and the maximum number of evolution is set to 400. In solving the model, the transformation rate is set to 0.5, the crossover probability is set to 0.6, and the scaling factor is set to 0.5. When the maximum number of evolutionary generations is not reached, the mutation, crossover, and selection operations are repeated. The final operation results show that in order to meet the maximum load limit, the CAV needs to make 3 deliveries, and finally 3 delivery routes are calculated. In the route expression, the capital letter represents the

Table 1. Customer information.

Customer Number	Coordinates	Weight (kg)	Service Time Window	Length of Service (min)	Shortest Distance from the Distribution Point (km)	Number of Separated Intersections	Adjacent Customer Points
A	(0.9,3.25)	3	[09:30–13:30]	10	2.1	9	-
B	(1.25,3.75)	8	[10:30–11:30]	15	2.5	9	C
C	(1.6,4)	12	[07:30–11:00]	20	4.2	9	B, D
D	(1.75,3.75)	4	[08:30–09:30]	5	2.2	8	B
E	(1.8,2.3)	4	[10:30–13:00]	5	2.3	7	F
F	(1.9,2.8)	18	[11:00–13:00]	30	1.0	6	E
G	(2.3,3.2)	5	[07:00–13:00]	10	1.5	5	-
H	(3.05,2.5)	33	[07:00–09:30]	70	0.7	3	L
I	(1.5,1.5)	24	[10:00–13:00]	30	1.7	6	J
J	(2.05,1.05)	20	[09:00–11:00]	30	2.2	5	I
K	(2.2,2)	23	[08:00–13:00]	35	1.3	4	-
L	(3.2,2)	6	[08:00–09:00]	10	0.7	2	H
M	(3.7,1.25)	12	[09:00–13:30]	20	0.7	0	-
N	(3.75,0.5)	27	[09:30–10:30] [12:00–13:00]	30	1.3	2	-
O	(4.6,1)	8	[10:00–11:30]	20	1.4	1	-

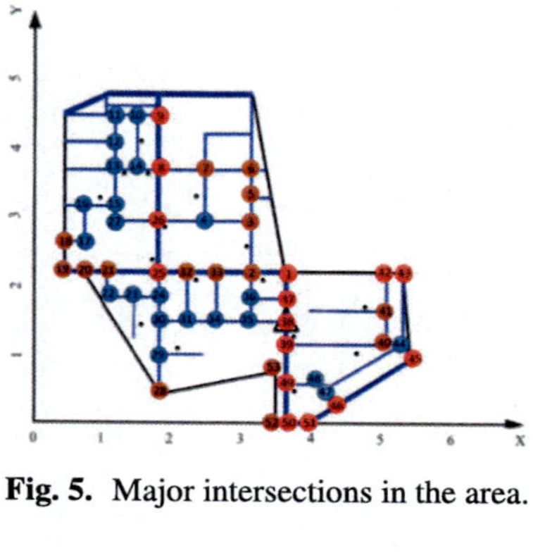

Fig. 5. Major intersections in the area.

customer point, the number represents the intersection junction through which the CAV passes, and the s in parentheses represents straight ahead, l represents left turn, and r represents right turn.

Table 2. Basic information of intersections.

Intersection color marking	Intersection red light time	Intersection green light time	Intersection yellow light time	Intersection lane number
●	70	55	5	6
●	40	30	5	4
●	20	20	5	2

Route 1: Q → 37(l) → 36(r) → 2(s) → H → 3(l) → 4(r) → G → 7(l) → 8(s) → D → 14(r) → C → 14(r) → B → 13(l) → 15(r) → A → 15(r) → 27(l) → 26(r) → F → E → 25(l) → 32(s) → 33(r) → 34(l) → 35(s) → 38(r) → Q;
Route 2: Q → M → 39(l) → 44(l) → 41(l) → P → 41(l) → 42(l) → 1(s) → 2(s) → 33(s) → 32(s) → 25(l) → 24(s) → 30(s) → 29(l) → J → 29(r) → 30(s) → 24(l) → 23(l) → I → 23(r) → 24(r) → 30(l) → 31(s) → 34(s) → 35(s) → 38(l) → Q;
Route 3: Q → 37(s) → 1(l) → L → 2(s) → 33(s) → K → 32(l) → 31(l) → 34(s) → 35(s) → 38(r) → 39(s) → 49(s) → N → 49(s) → 39(r) → O → 39(r) → Q.

The distribution cost of the 3 routes is shown in Table 3. The distribution center delivers 232 kg of goods to 16 customers along the 3 routes starting from 08:00 a.m., with a total distance of 16.8 km, passing through 58 intersections, fuel consumption of 0.982 L, and cost of 8.9 yuan. The solution results show that if CAV delivers according to the optimized route, it will generate 2 defaults, one for customer D and one for customer M. If CAV delivers to customer D and customer M strictly according to their time windows, it needs to deliver to them separately. At this time, the total cost $C_{trans} = 3.3$, the total mileage 5.7 km, and the unit mileage cost 0.58 yuan/km. Under the soft time window constraint, the penalty function is used to calculate the unit distance cost, and we get $f(t) = \frac{l}{1+e^{k}(AT-t)} = 0.58$, and the critical point AT = 2.17, that is, within 2.17 h, the default delivery strategy should be adopted, while in the case of delay time more than 2.17 h, then delivery should be made for them separately to avoid high default cost. The distance cost of default delivery in this example is 0.52 yuan/km, and the length of time beyond the time window is 1.7 h. Therefore, the default delivery strategy is adopted.

Table 3. Total cost of three routes.

	Route 1	Route 2	Route 3	Total
Distribution weight (kg)	87	81	64	232
driving distance (km)	5.1	7.5	4.2	16.8
Full load rate	87%	81%	64%	-
Intersection driving distance (m)	374	453	262	1,089

(continued)

Table 3. (*continued*)

	Route 1	Route 2	Route 3	Total
Number of straight crossings	4	10	8	22
Number of left turns at intersection	7	11	3	21
Number of right turns at intersection	9	3	3	15
Number of accelerated passes	5	7	2	14
Number of decelerated passes	8	9	4	21
Passes at a constant speed	7	8	8	23
Number of idle waiting times before acceleration	3	3	2	8
Fuel consumption (L)	0.312	0.431	0.239	0.982
Distribution costs (yuan)	1.9	2.6	1.4	5.9
Penalty fee (yuan)	0.3	0.4	0	0.7
Carbon dioxide Emissions (kg)	0.728	0.968	0.551	2.247
Carbon dioxide Emissions tax (yuan)	0.7	1.1	0.5	2.3

4.3 Algorithm Validation

To verify the effectiveness of the algorithm, it was compared with the standard genetic algorithm (SGA) and DEA, and the results are shown in Table 4 and Fig. 6. EDEA yielded a delivery mileage of 16.8 km, a delivery cost of only 5.9 yuan, and a fuel consumption of 0.982 L. The optimized target value was 18.3% lower than the DEA value and 35.9% lower than the SGA value. Therefore, EDEA has a clear advantage over other algorithms.

Table 4. Comparison of the results of three algorithms.

		Route 1	Route 2	Route 3	Total
SGA	Distribution mileage (km)	8.2	3.5	6.7	18.4
	Intersection mileage (m)	-	-	-	-
	Distribution costs (yuan)	4.4	2.1	2.7	9.2
	Fuel consumption (L)	0.768	0.367	0.473	1.608
	Total cost (yuan)	13.9			
DEA	Distribution mileage (km)	8.6	4.3	4.4	17.3
	Intersection mileage (m)	781	346.5	536.25	1663.75
	Distribution costs (yuan)	3.3	1.4	2.7	7.4
	Fuel consumption (L)	0.574	0.245	0.469	1.228
	Total cost (yuan)	10.9			

(*continued*)

Table 4. (*continued*)

		Route 1	Route 2	Route 3	Total
EDEA	Distribution mileage (km)	5.1	7.5	4.2	16.8
	Intersection mileage (m)	374	453	262	1089
	Distribution costs (yuan)	1.9	2.6	1.4	5.9
	Fuel consumption (L)	0.312	0.431	0.239	0.982
	Total cost (yuan)	8.9			

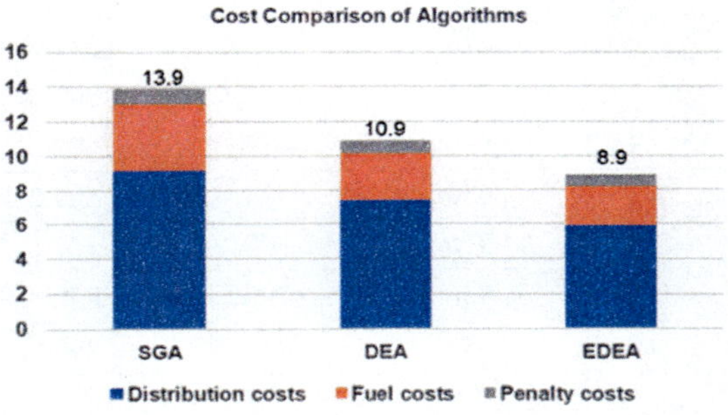

Fig. 6. Cost comparison of three algorithms.

The iteration plots of the three algorithms are given in Fig. 7. The results show that the target values calculated by all three algorithms shrink significantly with the in-crease of the number of iterations, among which EDEA converges the fastest, and when it reaches 45 iterations, the convergence speed of EDEA starts to lead the other two algorithms. The feasibility of EDEA is higher than the other methods for the same number of iterations.

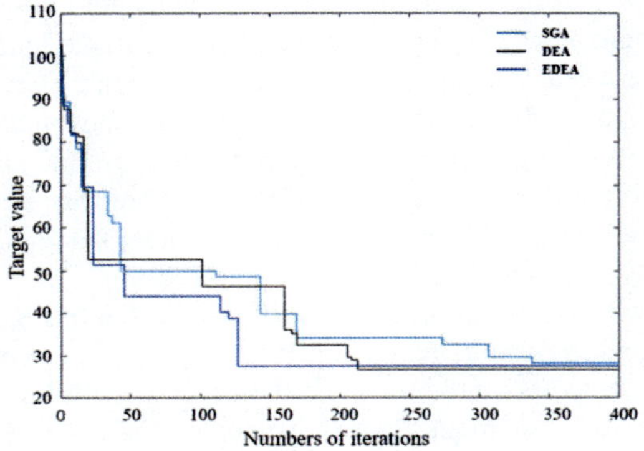

Fig. 7. Operation iteration diagram of the three algorithms.

Table 5 gives the optimal solution, shortest operation time, longest operation time and average operation time for 10 iterations using DEA and EDEA, respectively, for different population sizes. As shown in Table 5: EDEA always finds the best solution faster.

Table 5. Computing time for optimal solutions with population sizes.

Population size	Optimal solution(km)		DEA (the optimal solution of 10 times)	EDEA (10 times)		
	DEA	EDEA		Minimum time (s)	Maximum time (s)	Average time (s)
50	23.7	21.2	148	131	133	132
100	20.9	20.2	276	269.6	271.6	270.6
150	19.8	18.7	473.6	365.8	390.8	378.3
200	18.3	18.2	530.5	479	482	480.5
250	18.2	18.1	636	584	589.2	586.6
300	17.9	17.7	816	705	777	741
350	17.6	17.4	896	799	838	835.3
400	17.3	16.8	981	894	898	896

5 Conclusions

In this paper, we redefine the speed strategy of CAV at intersections and achieve the optimization of CAV fuel consumption cost by analyzing the relationship between CAV speed and fuel consumption. The fuel consumption optimization model is combined with the VRP optimization model with time window to achieve the optimization of the total cost of CAV distribution. Then the EDEA algorithm is designed to solve the model. EDEA uses a backward learning strategy to generate the original population, which can greatly improve the efficiency of the algorithm without increasing the original search space. And the model and algorithm proposed in this paper are experimented and verified by example study. The results show that EDEA has advantages in convergence time, solution size and solution accuracy compared with SGA and DEA, and has better performance in finding the optimal solution. The results of this study can effectively reduce the cost of CAV distribution and provide guidance for the low-carbon application of this technique.

However, the distribution optimization model proposed in this paper only considers the idealized optimization problem for a single vehicle at the intersection, but in reality, there are often many vehicles on the road and the traffic conditions are more complex. Multi-vehicle cooperative techniques can be considered to solve this problem in the future.

References

1. Dantzig, B., Ramser, H.: The truck dispatching problem. Manage. Sci. **6**(1), 80–91 (1959)
2. Sartori, S., Buriol, S.: A study on the pickup and delivery problem with time windows: matheuristics and new instances. Comput. Oper. Res. **124**(2), 105065 (2020)
3. Zhang, K.: Research on multi-objective vehicle routing optimization with soft time window constraints. Master thesis, Xi'an University of Architecture and Technology, Xi'an, China (2021)
4. Ye, R.: Research on optimization of vehicle routing problem with time window based on improved particle swarm optimization. Master thesis, Chang'an University, Xi'an, China (2021)
5. Liu, G.B., Hu, J., Yang, Y., Xia, S., Lim, M.: Vehicle routing problem in cold Chain logistics: a joint distribution model with carbon trading mechanisms. Resour. Conserv. Recycl. **156**, 104715 (2020)
6. Zhang, W., Li, H., Yang, W., Zhang, G., Gen, M.: Hybrid multiobjective evolutionary algorithm considering combination timing for multi-type vehicle routing problem with time windows. Comput. Ind. Eng. 108435 (2022)
7. Zhang, Q.: Research on speed guidance method at signalized intersection with energy consumption priority. Master thesis, Chang'an University, Xi'an, China (2020)
8. Zhao, H.: Simulation and optimization of vehicle energy consumption and emissions at urban road signalized intersection. Doctor thesis, Lanzhou Jiaotong University, Lanzhou, China (2019)
9. Meng, Z., Qiu, Z.: Research on eco-driving strategy of energy saving oriented signalized intersection. Traffic Inf. Saf. **36**(02), 76–84+92 (2018)
10. Lakouari, N., Oubram, O., Bassam, A., Pomares, H., Marzoug, R., Ez-zahraouy, H.: Modeling and simulation of CO2 emissions in roundabout intersection. J. Comput. Sci. **40**, 101072 (2020)
11. Wei, X., Leng, J., Sun, C., Huo, W., Ren, Q., Sun, F.: Co-optimization method of speed planning and energy management for fuel cell vehicles through signalized intersections. J. Power Sour. **518**, 230598 (2022)
12. Ma, C., Yu, C., Yang, X.: Trajectory planning for connected and automated vehicles at isolated signalized intersections under mixed traffic environment. Transp. Res. Part C Emerg. Technol. **130**, 103309 (2021)
13. Liu, C., Ma, T.: Green logistics management and supply chain system construction based on internet of things technology. Sustain. Comput. Inform. Syst. **35**, 100773 (2022)
14. Guo, Y., Chen, X., Yang, Y.: Multimodal transport distribution model for autonomous driving vehicles based on improved ALNS. Alex. Eng. J. **61**(4), 2939–2958 (2022)
15. Wen, X.: Driven by the epidemic, the value of unmanned delivery is increasingly prominent. Intell. Internet-Connected Veh. **01**, 26–29 (2021)
16. Wang, Z., Wu, G., Barth, M.: Cooperative eco-driving at signalized intersections in a partially connected and automated vehicle environment. IEEE Trans. Intell. Transp. Syst. **21**(5), 2029–2038 (2020)
17. Chen, K.: Research on distribution route optimization of driverless vehicle under intelligent network environment. Master thesis, Chang'an University, Xi'an, China (2020)
18. Lu, T., Wan, L., Li, Y., Zhang, X., Wang, R.: Design of intelligent intersection signal control system based on V2X. Comput. Technol. Dev. **31**(10), 161–167 (2021)
19. Yang, D.: Coordinated control of urban main channel signal in the environment of internet of vehicles. Master thesis, Guangdong University of Technology, Guangzhou, China (2022)
20. Shen, X., Zhang, X., Ouyang, T., Li, Y., Raksincharoensak, P.: Cooperative comfortable-driving at signalized intersections for connected and automated vehicles. IEEE Robot. Autom. Lett. **5**(4), 6247–6254 (2020)

21. Jin, Q., Song, G., Ye, M.: Optimization of eco-driving trajectories at intersections for energy saving and emission reduction. Saf. Environ. Eng. **22**(3), 75–82 (2015)
22. Han, M., Liao, S., Chang, J.: Dynamic group-based differential evolution using a self-adaptive strategy for global optimization problems. Appl. Intell. **39**(1), 41–56 (2013)

Sustainable Digital Marketing and the Digital Supply Chain Management Theoretical Aspects

Margarita Išoraitė(✉)

Vilniaus Kolegija/Higher Education Institution, Didlaukio Str. 49, 08303 Vilnius, Lithuania
m.isoraite@vvf.viko.lt

Abstract. The article examines how digital marketing and digital supply chain management can enhance the sustainability and competitiveness of businesses in the current environment. Digitization has become a necessity for businesses to cope with changing customer behavior, market dynamics, environmental challenges, and social expectations. Digitization of the supply chain has significant financial and sustainability benefits. This helps to increase productivity, reduce waste, improve traceability, and enhance welfare. Digital marketing is a type of marketing that uses online technologies to achieve business goals, while digital supply chain management is the application of digital technologies to optimize the flow of goods and information across the supply chain. Digitization helps bring products to market faster and with better quality while reducing waste. The article reviews the literature on digital marketing and digital supply chain management and identifies the key factors, benefits, challenges, and best practices for their successful implementation.

The article provides insights and guidelines for managers and policymakers to adopt and promote sustainable digital marketing and digital supply chain management practices.

Keywords: Digital marketing · Digital supply chain · Management

1 Introduction

Nowadays, there is not a single company or organization that does not think about the digitalization of business processes, sustainable digital marketing, and supply chain. Digitization has become a daily task, a challenge, and an opportunity for managers. Companies are affected by digital transformation. Digitization leads to higher productivity of industrial companies; the efficiency of operational processes helps to discover new sources of business growth and ensures real-time information acquisition and its targeted use. In Lithuania, digitization is perceived as the installation of software, production automation, or the use of analytical tools. A systematic approach to digitization allows for the growth of companies, the rational use of resources, ensures synergy between the implementation of technologies, the ability of employees to fully use them, the establishment of synergistic links with users and partners, and the meaningful selection and implementation of strategic alternatives. Managers associate a high growth

O. Prentkovskis et al. (Eds.): TRANSBALTICA 2023, LNITI, pp. 223–228, 2024.
https://doi.org/10.1007/978-3-031-52652-7_22

potential with companies' strategic digitization decisions and related organizational culture changes. There is a greater focus on the search for value chain management and digitization solutions. The ability to digitize business is inevitable. The aim of the article is to evaluate the theoretical aspects of sustainable digital marketing and digital supply chain management. The research subject is the sustainability of digital marketing and digital supply chains. The article employs the methods of analysis and comparison of scientific literature.

2 Digital Marketing

Nowadays, the number of Internet users around the world is still growing. In fact, it has been established that the number of regular Internet users increases steadily by about 4–5% every year. Significant changes are also evident in the field of retail trade. Therefore, it is not surprising that traditional marketing is becoming less efficient and effective. In view of this, marketers have turned to a new branch of marketing: digital marketing. Digital marketing is a type of marketing that uses online technologies to achieve business goals [9]. Over the past year, this type of advertising has grown significantly for businesses. Digital marketing aims to attract potential customers. Digital marketing comprises the following elements: search engine optimization (SEO), content marketing, social network marketing, pay-per-click (PPC) advertising, affiliate marketing, and email marketing.

Csordás et al. [5] explored dynamic innovation in digital marketing solutions, which attempts to gain a competitive advantage by combining scientific, technological, organizational trends and in order to make progress in the field of developing digital products, and services and ways of working of the organization in the aspect of modeling and knowledge management.

Bermeo-Giraldo et al. [2] stated that digital marketing is an excellent example of applying various technological changes. Altamira et al. [1] developed a digital strategy for a brand. Content marketing involves all content-related decisions that aim to attract consumer attention, promote brand awareness, and engagement, and drive sales or specific actions.

Masrianto et al. [15] analyzed the digital marketing process that a company uses to attract an audience and interest them in the company's products or services. The purpose of digital marketing is to provide distinctive value to potential customers and consumers through content, demonstrate product value, build brand loyalty, and increase sales. With the help of digital marketing, consumers are constantly studied and analyzed, and online shopping habits are studied.

Rodrigues et al. [19] and Altamira et al. [1] defined digital marketing as a social process in which individual organizations use digital technologies to attract new customers and retain existing ones. Digital marketing is a continuous process, so both new and existing customers are especially important. Good reviews about products, services, and even the communication of employees are a great way to attract new customers. People who are satisfied with the services or goods are likely to give good ratings and recommendations to others and thus attract potential users. It is very important to assess the needs of customers, pay attention to complaints and effectively solve the problems that arise, which cannot be avoided in any business area.

Bruce et al. [3] mentioned that effective digital marketing enables small and medium-sized businesses (SMEs) to compete with larger companies. Digital marketing provides an opportunity to reach not only a local but also a global audience. Social networks provide an opportunity to communicate directly with users and find out their opinion about the offered service or product. The target audience can be offered to visit the company's website, familiarize themselves with the offered products, evaluate them or leave a review. This gives you the opportunity to develop faster and create what exactly the person who will use your goods or services wants.

López García et al. [14] mentioned that digital technologies make it possible to run internal processes faster and more flexibly and contribute to their efficiency. Digital technologies also make it possible to diversify the income of companies by supplementing products with services. Digital technologies are equipment that provide prerequisites for changing objects, processes, and information related to them from an analog form to a binary form that can be understood by computerized systems.

Sakas et al. [21] mentioned that online channels create new opportunities for companies and create a personal image as the first step in improving business accessibility and visibility.

Matos-López [16] argued that the development of information and communication technology (ICT) in general, while Brzakovic et al. [4] mentioned that digital marketing shifts the focus from organizations to individuals.

Diez-Martin et al. [8] stated that digital marketing has changed how a company operates manages and communicates with its customers and the public worldwide. This becomes the key and a necessary tool to deal with the challenges that still exist in the field of marketing.

Desai [7] and Lavanya et al. [13] analyze that digital marketing includes any marketing effort that uses an electronic device or the Internet. The Internet medium allows consumers to search for and purchase goods or services at their own convenience, and it is a great opportunity for companies to reach those searching target customers very, very quickly.

3 Sustainable Digital Marketing and Digital Supply Chain Management

Joshi and Sharma [11] mentioned that the digital supply chain provides excellent connectivity in areas that were previously unavailable.

Pyun et al. [18] and Chauhan et al. [6] mentioned that the digital supply chain help to improve the stability, agility, and efficiency of supply chains.

Yerpude et al. [23] mentioned that the digital supply chain (DSC) delivers electronically from origin to destination. Customer behavior is constantly changing with the arrival of global market players. Key processes in the supply chain include relationship and collaboration management, customer service management, demand management, order fulfillment, product supply management, procurement, product development, and commercialization.

According to JSC Cargo Stream [22], the digital platform "Cargo Stream" created by Lithuanians already serves leading global shipping lines in Poland and the Baltic States.

This is how Cargo Stream has become a success story that shows that everyone in the supply chain can use universal digital solutions and benefit from them. The company provides an advanced automated logistics management platform to leading shipping lines to facilitate the processing of container redistribution and reuse requests, saving customers time and other resources. Cargo Stream is a digital logistics management workspace that enables you to create and manage container orders. The company's solution also allows all participants in the supply chain – shippers, forwarders, carriers, and shipping lines – to communicate easily in the digital space. Using a more efficient empty container redistribution system for all parties not only saves time but also contributes to reducing pollution and delays.

Joshi et al. [11], argued that understanding the sustainability perspective of systems requires organizations to explore new technologies that can help them succeed.

Romagnoli et al. [20], mentioned that the role of digital technologies is very important to obtain information about real value returns.

Hussain et al. [10], argued that, based on current information technology research, scholars are increasingly considering digital technology capabilities as an organizational resource to create greater ordering capabilities.

Kashem et al. [12], analyzed that artificial intelligence can make supply chains robust and sustainable.

Pyun et al. [21], examined the digital supply chain from the perspective of sustainable management.

Minculete et al. [17], mentioned that efforts have created aspects of digital SCM relationship consolidation based on agility, sustainability, and resilience as decisive factors that achieve and enhance your high performance, as well as the validity and resilience needed for its viability.

4 Conclusions

As society changes and develops, consumers increasingly choose to shop online, use digital marketing to find the necessary information about a product or service. This was influenced by the developing information technologies and the changing social and economic environment. Over the past three decades, the Internet has developed exponentially in both developed and developing countries. The presence of the Internet has affected all aspects of life and business. Recent advances in information technology (IT) have enabled exporters to improve their efficiency and overall performance. It is essential that sustainable digital marketing and digital supply chain management are competitive. Recently, the business environment in the world is changing, with unstable politics, non-compliance with environmental requirements, and unfavorable situations for business, which fall on supply chains. This unpreparedness of supply chain shortages, especially during the Covid-19 pandemic, has shown that traditional risk management strategies can be ineffective and can cause disruptions in the global supply chain. Therefore, proper management of sustainable social media marketing and a sustainable digital supply chain is essential.

References

1. Altamira, M.B., Putri, K.D.A.P., Samudra, R.M.R.T.: The role of creative content in digital marketing strategies in educational institution social media (case study: Instagram of Vocational Education Program, Universitas Indonesia). Proceedings **83**, 62 (2022). https://doi.org/10.3390/proceedings2022083062
2. Bermeo-Giraldo, M.C., Valencia-Arias, A., Ramos de Rosas, J.D., Benjumea-Arias, M., Villanueva Calderón, J.A.: Factors influencing the use of digital marketing by small and medium-sized enterprises during COVID-19. Informatics **9**, 86 (2022). https://doi.org/10.3390/informatics9040086
3. Bruce, E., Shurong, Z., Ying, D., Yaqi, M., Amoah, J., Egala, S.B.: The effect of digital marketing adoption on SMEs sustainable growth: empirical evidence from Ghana. Sustainability **15**, 4760 (2023). https://doi.org/10.3390/su15064760
4. Brzakovic, A., Brzakovic, T., Karabasevic, D., Popovic, G.: Empirical analysis of the influence of digital marketing elements on service quality variables in the small- and medium-sized enterprises sector in the Republic of Serbia. Sustainability **13**, 10264 (2021). https://doi.org/10.3390/su131810264
5. Csordás, A., Pancsira, J., Lengyel, P., Füzesi, I., Felföldi, J.: The potential of digital marketing tools to develop the innovative SFSC players' business models. J. Open Innov. Technol. Market Complex. **8**, 122 (2022). https://doi.org/10.3390/joitmc8030122
6. Chauhan, S., Singh, R., Gehlot, A., Akram, S.V., Twala, B., Priyadarshi, N.: Digitalization of supply chain management with industry 4.0 enabling technologies: a sustainable perspective. Processes **11**, 96 (2023). https://doi.org/10.3390/pr11010096
7. Desai, V.: Digital marketing: a review. Int. J. Trend Sci. Res. Dev. (IJTSRD) **5**(5), 196–200 (2019)
8. Diez-Martin, F., Blanco-Gonzalez, A., Prado-Roman, C.: Research challenges in digital marketing: sustainability. Sustainability **11**, 2839 (2019). https://doi.org/10.3390/su11102839
9. Drucker, P.F.: Management's new paradigms. Forbes Mag. **10**, 98–99 (1998)
10. Shahadat, M.H., Chowdhury, A.H.M.Y., Nathan, R.J., Fekete-Farkas, M.: Digital technologies for firms' competitive advantage and improved supply chain performance. J. Risk Financ. Manag. **16**, 94 (2023). https://doi.org/10.3390/jrfm16020094
11. Joshi, S., Sharma, M.: Sustainable performance through digital supply chains in industry 4.0 era: amidst the pandemic experience. Sustainability **14**, 16726 (2022). https://doi.org/10.3390/su142416726
12. Kashem, M.A., Shamsuddoha, M., Nasir, T., Chowdhury, A.A.: Supply chain disruption versus optimization: a review on artificial intelligence and blockchain. Knowledge **3**(1), 80–96 (2023). https://doi.org/10.3390/knowledge3010007
13. Lavanya, A., Radhikamani, M.: A study on digital marketing and it's impacts. Int. J. Creat. Res. Thoughts (IJCRT), 677–684 (2021)
14. López García, J.J., Lizcano, D., Ramos, C.M., Matos, N.: Digital marketing actions that achieve a better attraction and loyalty of users: an analytical study. Future Internet **11**, 130 (2019). https://doi.org/10.3390/fi11060130
15. Masrianto, A., Hartoyo, H., Hubeis, A.V.S., Hasanah, N.: Digital marketing utilization index for evaluating and improving company digital marketing capability. J. Open Innov. Technol. Market Complex. **8**, 153 (2022). https://doi.org/10.3390/joitmc8030153
16. Matosas-López, L.: The management of digital marketing strategies in social network services: a comparison between American and European organizations. J. Open Innov. Technol. Market Complex **7**, 65 (2021). https://doi.org/10.3390/joitmc7010065
17. Minculete, G., et al.: Relational approaches related to digital supply chain management consolidation. Sustainability **14**(17), 10727 (2022). https://doi.org/10.3390/su141710727

18. Pyun, J., Rha, J.S.: Review of research on digital supply chain management using network text analysis. Sustainability **13**, 9929 (2021). https://doi.org/10.3390/su13179929
19. Rodrigues, S., Correia, R., Gonçalves, R., Branco, F., Martins, J.: Digital marketing's impact on rural destinations' image, intention to visit, and destination sustainability. Sustainability **15**, 2683 (2023). https://doi.org/10.3390/su15032683
20. Romagnoli, S., Tarabu', C., Maleki Vishkaei, B., De Giovanni, P.: The impact of digital technologies and sustainable practices on circular supply chain management. Logistics **7**, 1 (2023). https://doi.org/10.3390/logistics7010001
21. Sakas, D.P., Reklitis, D.P., Terzi, M.C., Vassilakis, C.: Multichannel digital marketing optimizations through big data analytics in the tourism and hospitality industry. J. Theor. Appl. Electron. Commer. Res. **2022**(17), 1383–1408 (2022). https://doi.org/10.3390/jtaer17040070
22. JSC Cargo stream. Digitization is changing supply chain management: solutions are being used by the world's largest shipping lines and their customers. https://www.vz.lt/transportas-logistika/2023/02/20/skaitmenizacija-keicia-tiekimo-grandines-valdyma-sprendimus-pasitelkia-didziausios-pasaulio-jurines-linijos-ir-ju-klientai#ixzz7yqMqRdp5. Accessed 14 Apr 2023
23. Yerpude, S., Sood, K., Grima, S.: Blockchain-augmented digital supply chain management: a way to sustainable business. J. Risk Financ. Manag. **16**, 7 (2023). https://doi.org/10.3390/jrfm16010007

Green Logistics: From Theory to Practice

Aldona Jarašūnienė[1] (✉) and Margarita Išoraitė[2]

[1] Vilnius Gediminas Technical University, Plytinės Str. 27, 10105 Vilnius, Lithuania
aldona.jarasuniene@vilniustech.lt

[2] Vilniaus Kolegija/Higher Education Institution, Didlaukio Str. 49, 08303 Vilnius, Lithuania
m.isoraite@vvf.viko.lt

Abstract. Climate change has a profound impact on our way of life, prompting both individuals and companies to contribute to a greener future. The logistics sector is no exception, as green logistics plays an increasingly significant role within the supply chain. To practice sustainable and ecological logistics, it is crucial to implement green practices right from the initial movement of goods. This article delves into the concept of green logistics, supply chain trends, green supply chain practices, and tools for green logistics management. The adoption of green logistics by transportation companies is seen as a means to enhance their competitive advantage in today's economic, social, and environmental landscape. However, effectively implementing the principles of sustainable development requires the integration of different areas that are often treated separately, necessitating new methods and tools. Nevertheless, there exists a knowledge gap regarding the practical application of green logistics measures, highlighting the need for further understanding and insights in this area.

Keywords: Green logistics concept · Supply chain · Supply chain tools

1 Introduction

Currently, there is an increased need for sustainable development, the main element of which is the so-called green supply chain or, more broadly, green logistics. Green logistics are solutions that include everything related to production transportation activities, and supply chain. With the help of Green Logistics, the customers' needs are met with minimal costs. Despite current trends, managers responsible for supply chains are still skeptical of the changes proposed by the European Commission. The objective of this paper is to delve into the concept of green logistics, analyze the latest trends in supply chain management, and investigate the diverse array of tools available for implementing green logistics and fostering sustainable supply chains. Embracing green logistics within transportation companies is perceived as a strategic approach to enhance their competitive edge in the prevailing economic, social, and environmental landscape.

O. Prentkovskis et al. (Eds.): TRANSBALTICA 2023, LNITI, pp. 229–238, 2024.
https://doi.org/10.1007/978-3-031-52652-7_23

2 Green Logistics Theory

2.1 Green Logistics Concept

Green logistics forms the development prospects of the transport company linked to environmental issues. It can be said that transport is the activity in which the principles of green logistics are implemented, therefore it is important to evaluate the business activities in this sector progress and potential in the field of implementation of green logistics principles. Therefore, it is important to answer the question, how are the principles of green logistics implemented in the transport sector?

According to Cheng et al. [4], Green Logistics (GL) activities revolve around the environmentally conscious and efficient management of goods and information flows. These activities aim to generate added value for consumers while simultaneously preserving the environment and promoting the adoption of a circular business economy. By prioritizing sustainable practices, Green Logistics seeks to meet the evolving needs of consumers while minimizing negative environmental impacts. The principles of the circular economy, secondary use, sorting, and other areas of the green economy are of increasing concern to society and businesses. To be a strong market player, the business pays particular attention to sustainability and follows the path of the green course. Nowadays, the consumer chooses very responsibly. He wants to have a sustainable product that is made from recycled materials and is fully recyclable.

Based on the findings of Vienažindiene et al. [28], the integration of green logistics is of paramount importance in implementing the principles of the circular economy. By adopting green logistics practices, organizations are able to effectively address environmental, economic, and social concerns. These practices enable a more sustainable and holistic approach to resource management, waste reduction, and promoting a regenerative economy.

In contrast, Mak et al. [14], conducted an analysis and found that logistics, being largely invisible to most customers, poses difficulties in implementing green logistics policies effectively.

Wang et al. [29], expressed the belief that green logistics plays a key role in reducing the environmental impact associated with logistics activities. They argue that green logistics aims for balanced economic, social, and environmental development, distinguishing itself from traditional logistics, which focuses on cost reduction.

Discussing the broader impact of green logistics, Kwak et al. [12], noted that it considers the influence of transportation and environmental sectors on the entire logistics process. Furthermore, Kwak et al. [12], argue that green logistics, in a narrower sense, pertains to pollution reduction and management activities, including measures to decrease air pollution and gas emissions during the transportation of goods.

Richnák et al. [21], examined the directions of sustainable development from the point of view of logistics, ensuring a better quality of life for both current and future generations. Transport companies should realize and take responsibility not only for the success of their operations, but also for their contribution to the development and environmental impact of the community, region, or country.

Werner-Lewandowska et al. [31] posit that companies can achieve sustainable outcomes by prioritizing the implementation of logistics solutions. Factors such as advancing technological capabilities and the growing demand for innovation can contribute to making transport infrastructure more sustainable, adaptable to user needs, environmentally friendly, and resilient to climate change. However, it is essential to have a well-defined sustainable development policy to effectively pursue these goals.

In a similar vein, Trivellas et al. [26] emphasize the increasing interest in green logistics and sustainable supply chains. They note that green solutions are viable across various domains, including renewable energy, waste management, green transport, and logistics, and the implementation of the sharing economy. This highlights the wide-ranging possibilities for incorporating sustainable practices throughout different aspects of the logistics industry.

Tan et al. (2020) focus on the use of optimization techniques and mathematical programming models in green supply chain research. It is essential to better protect the interests of consumers, ensure fair competition between producers and facilitate the free movement of organic products in the EU. The organic product control system aims to give consumers certainty that when they buy EU or equivalent organic products, the rules have been applied at every stage of the supply chain. This should apply where these products are produced in the EU or imported.

Neto et al. [8] proposed a comprehensive framework for designing and assessing sustainable logistics networks. Their study highlights the growing significance of environmental factors in the design process of logistics networks. Recognizing the evolving environmental landscape, the framework takes into account the environmental influences that impact logistics operations, ensuring that sustainability considerations are effectively integrated into network design decisions.

Paužuolienė et al. [18] stated that the implementation of green logistics in the organization highlights a few advantages. According to Paužuolienė et al. [18] this practice reveals an increase in the volume of goods delivered on time, better use of capacity, promotes the quality of products or services, increases product variety, and reduces waste. Based on Paužuolienė et al. [18] research data, it can be concluded that price and delivery are still important to consumers consistency of time and quality of services provided and socially responsible activities of the company are practically irrelevant or efforts to contribute to the reduction of environmental problems.

Zowada [34] mentioned that the concept of green logistics is still poorly recognized by many companies. In most cases, it is due to the lack of knowledge of the managers about the assumptions of the concept itself and the methods of its implementation. It is widely recognized that logistics affects the environment, so green logistics becomes a key component for sustainable management.

Mala et al. [16] investigated that the advantage of green logistics is that its application makes assumptions to reduce general operating costs, reduce pollution, reduce fuel and energy consumption, optimize logistic flows, reduce emissions of toxic substances.

Vasiliauskas et al. [27] state that the concept of green logistics is related to sustainable development and is important on three equal levels: economic, ecological and social. This is also confirmed by the definition of green logistics. The interaction between the ecological, economic, and social levels is shown in Fig. 1.

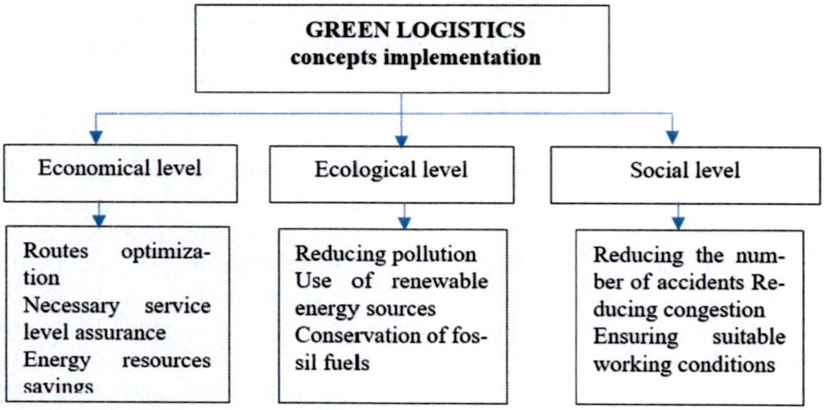

Fig. 1. The main levels of implementation of the green concept and the goals. Source: Vasiliauskas et al.

According to Vasiliauskas et al. [27] the interaction of economic, ecological, and social levels means that green logistics and the implementation of the concept in any organization should be carried out in accordance with the principles of economic, ecological, and social responsibility. The organization should prepare a green logistics implementation strategy specifically adapted to it, taking into account the external and internal factors affecting the organization's activities [27].

2.2 Supply Chain Trends

Jančiauskas et al. [19] argue that improving the supply chain is essential for a company to be competitive. The factors that determine that the company needs to be concerned and start improving its supply chain are as follows [19]:

- Rapid growth of transportation costs. Increased transportation costs mean that supply is a big loss for the company. The entire supply needs to be re-planned and transport optimized as much as possible.
- Peak production efficiency reached. This means that in order to continue to improve, it is necessary to strive for perfection not only in production but also in supply.
- Fundamental change in inventory management. When inventory management changes, the supply chain must also change.
- An increased range of products can also be a factor that forces supply chain improvement, as in this case supply needs to be re-planned and improved.
- The revolution in computer and communication technologies. To stay competitive, you need to improve your supply chain.
- Improve the supply chain to ensure supply optimization.
- Global competition. It is imperative to improve your supply chain and seek competitive advantages over your competitors.
- Foreign sources of supply. To look for cheaper and more competitive solutions, it is necessary to look abroad and at the same time to re-plan the supply chain more efficiently.

According to Rejeb et al. [20], the global impact of the pandemic has brought about visible social and economic effects worldwide. One of the sectors significantly affected by this crisis is the food supply chain (FSC), which has faced severe disruptions due to humanitarian threats and concerns about an unstable business environment.

Throughout the COVID-19 pandemic, serious challenges to the food supply have emerged, as highlighted by Rejeb et al. [20]. These challenges have had a notable influence on the four key principles of food security: accessibility, affordability, stability, and usability. The pandemic has significantly impacted these principles, underscoring the critical importance of ensuring a secure and reliable food supply chain in times of crisis.

Yuan et al. [32] mentioned that Industry 4.0 has revolutionized supply chain management. Thanks to Industry 4.0, logistics activities become more organized and simpler. The transportation company has better visibility into every process in the supply chain, making it more flexible and errors are easier to catch, and the ability to respond to last-minute changes is improved.

In terms of the importance of IT trends in supply chains, Kajba et al. [10] shed light on several trends categorized based on their temporal impact. Their research revealed that one trend holds short-term significance, namely big data analytics. Additionally, Kajba et al. [10] underscored the long-term importance of five trends, namely bionic enhancement, quantum computing, drones, 3D printing, and advanced vehicles.

Addressing the cost aspect of supply chain management (SCM), Wen et al. [30] noted that numerous practitioners have demonstrated the ability of effective SCM to reduce overall project costs. This highlights the financial benefits associated with efficient supply chain practices.

Discussing the risks faced by companies, Alkhatib et al. [1] emphasized that various risks can lead to disruptions in information and material flow. The logistics and transport sectors experienced significant disruptions during the pandemic, and stakeholders in the construction industry have raised concerns about the continued supply of raw materials, which has already had consequences for consumers. Alkhatib et al. [1] highlighted the vulnerability of the supply chain to internal and external disruptions, including economic downturns, challenges related to the loss of valuable customers, and the need for adaptation to new technologies and infrastructure quality.

2.3 Green Logistics and Supply Chain Management Tools

Gong et al. [6] argued that green supply chain management (GSCM) needs to be addressed systematically. Green supply chain management is not only limited to reducing the harmful impact on the environment and limiting carbon dioxide emissions, but also to achieving business goals. The most important goal is to reduce the costs of logistics processes, increase profits and at the same time adhere to the principles of sustainable development.

Hejazi et al. [7] mentioned that many supply chain and operations management practitioners and researchers have become interested in green supply chain methods. Green logistics includes all activities related to the transportation of products. Green logistics meet the needs of consumers at minimum costs, causing the lowest possible CO_2 impact on the environment. Green logistics is sustainable supply chain management.

Zhou et al. [33] put forth the argument that existing research on green logistics management (GLM) has primarily focused on assessing its impact on ecological sustainability. In contrast, Qu et al. [17] asserted that a secure, stable, and sustainable supply chain is not only vital for enterprise development but is increasingly becoming an integral component of sustainable industrial chain development. Their viewpoints underline the multifaceted nature of green logistics management and its broader implications for both environmental sustainability and the overall sustainability of industrial chains.

According to Shekarian et al. [23] and Qu et al. [17] integrating sustainable business practices not only aligns with the triple framework of sustainability, encompassing economic, environmental, and social aspects, but it also leads to increased profitability. By adopting sustainable practices, businesses can effectively balance their economic goals with environmental and social responsibilities, ultimately driving higher profits.

According to the findings of García Alcaraz et al. [5], the integration of green supply chains positively impacts cost reduction due to the increased involvement of green suppliers in environmental innovation. Similarly, Cerqueira-Streit et al. [2] highlighted the concerns of supply chain managers regarding sustainability issues. Jalali et al. [9] emphasized the significance of identifying barriers to the adoption and implementation of sustainable practices as a crucial step in their successful implementation.

Silva et al. [22] argued that supply chain management (SCM) is an increasingly important concept that companies must strategically integrate with suppliers and customers to achieve profitability. In line with this, Lazar et al. [13] emphasized the importance of sustainable development in companies, as it is expected to become a critical factor for international competitiveness, even within the interconnected domain of supply chains and logistics.

Chalmeta et al. [3] considered that one of the main challenges of SCM is the integration of sustainability principles in the supply chain, taking into account economic, environmental and social impacts and a multidimensional approach (institutional, geographical and temporal).

According to Tseng et al. [25], sustainable supply chain management can be defined as the implementation of strategies, transparent engagement, and the pursuit of organizational goals through an inter-organizational professional process.

Kumar et al. [11] analyzed that sustainable supply chain management can facilitate the adoption and implementation of supply chain integration using digital technologies.

Makarova et al. [15] mentioned that sustainable development is possible only through the implementation of complex projects in which transport infrastructure is an important component.

3 Research Methodology and Data Analysis

3.1 Research Methodology

The main instrument of the interview is a semi-structured questionnaire, which consisted of questions divided into the following question areas: the importance of green logistics for the organization, the implementation of green logistics solutions in the organization, experience and knowledge in the field of logistics sufficiency, green logistics implementation problems, improvement of green logistics. These areas have been identified based on the analysis of the scientific literature.

Purposive sampling was used to select the interview participants, only certain ones were included in the study employees of transport companies. The following criteria were used for the selection of informants: 1) position (top or middle manager); 2) awareness of the logistics situation in the company; 3) at least 3 years of work experience in the field of logistics. Sample volume using half the structured interview method consisted of 4 informants (Table 1).

Table 1. Rationale for the interview questionnaire.

Interview question areas	Questions	The authors who examined aspect
Green logistics concept and its implementation	What is green logistics and how it is implemented?	Cheng et al. (2023), Mak et al. (2022), Wang et al. (2018), Kwak et al. (2020)
Public-private cooperation	Whether the implementation of "Green logistics" measures would be far-reaching more effective through public-private cooperation?	Trivellas et al. (2020), Vasiliauskas et al. (2013)
Green logistics measures effectiveness	What green logistics measures that can be most effective?	Silva et al. (2019)
Green logistics risk	What are green logistics risk?	Mala et al. (2017)
Encourage the application of "Green logistics" measures	What would encourage the application of "Green logistics" measures	Mala et al. (2017)

3.2 Research Data Analysis

The experts were asked several questions about how they understand the term "Green logistics" and whether is there enough information about its implementation measures? Most experts understand the concept of "Green Logistics" correctly and say that there is enough information on the implementation of its measures.

The question was also asked whether the implementation of "Green logistics" measures would be far-reaching more effective through public-private cooperation. The majority of experts agree with the statement that the implementation of "Green logistics" measures would be much more efficient thanks to public-private cooperation.

From the question regarding "Green logistics" measures that can be most effective, experts singled out the promotion of intermodal transports, the application of information technologies, using vehicles that are less harmful to the environment, training employees, promotion of social responsibility, assessment of risks and extreme situations.

In response to the question, the experts distinguished the benefits of applying "Green logistics" measures improving the quality of work, increasing efficiency, better quality of services; investors attractiveness.

When asking the experts about the risks, the measures of "Green logistics" applied due to financial risk, psychological risk, risk of failure. When asked about the reasons why "Green logistics" measures are not applied, the answer was, that they are not applied due to lack of cooperation between the public and private sectors, lack of training, lack of implementation of advanced technologies.

When asking the question about what would encourage the application of "Green logistics" measures, experts answered that State support for companies or benefits, greater availability of information, technological advances.

Respondents highlighted that green logistics measures increase a company's competitive advantage, productivity, and profits and allow it to offer lower transport rates to customers due to cost savings. Experts agreed that the political and legal environment is important in encouraging companies to use green logistics measures. Another important point underlined was technological progress. Respondents identified that without technological progress, the implementation of green logistics measures in the logistics sector is not possible. Drivers, managers, and other responsible staff need to be responsible, educated, and understand the importance of going green. Educating employees is an important factor for a company to become greener. For this reason, it is important to integrate economic driving courses.

4 Conclusions

The theoretical issue of green logistics covers various research areas and challenges related to the integration of sustainability principles into logistics operations. The main aspects of the theoretical problem of green logistics are distinguished: environmental impact assessment; ecological transport and choice of transport modes; circular economy and waste management; green technology application, policy, and regulatory framework.

These aspects highlight the theoretical problem in green logistics, which encompasses multidisciplinary research involving environmental science, engineering, operations management, and policy analysis. An in-depth exploration of these issues can be found in academic journals and research publications dedicated to sustainable logistics and supply chain management.

Green logistics means methods that aim to improve the state of the environment. Since no single transport policy measure can solve all environmental problems, these measures the spectrum and scope are wider. By means of legislation, transport governments directly influence the transport sector, specifying the use of certain technologies or achieving a certain result. Market-based approaches and incentives use economic levers to balance, reduce or even eliminate the negative effects of transport on the environment. Planned green logistics measures mainly involve actions aimed at increasing the attractiveness of alternatives.

The interviews also showed that companies can more easily adopt green logistics measures due to their greater financial capacity. The interviews with the experts revealed that the implementation of green logistics measures depends not only on technological progress, the political and legal environment and society but also on the company's own internal strategy and the available human and financial resources. It is also important to pay attention to risk assessment and extreme situations.

References

1. Alkhatib, S.F., Momani, R.A.: Supply chain resilience and operational performance: the role of digital technologies in Jordanian manufacturing firms. Adm. Sci. **13**(2), 40 (2023). https://doi.org/10.3390/admsci13020040
2. Cerqueira-Streit, J.A., Endo, G.Y., Guarnieri, P., Batista, L.: Sustainable supply chain management in the route for a circular economy: an integrative literature review. Logistics **5**(4), 81 (2021). https://doi.org/10.3390/logistics5040081
3. Chalmeta, R., Santos-deLeón, N.J.: Sustainable supply chain in the era of industry 4.0 and big data: a systematic analysis of literature and research. Sustainability **12**(10), 4108 (2020). https://doi.org/10.3390/su12104108
4. Cheng, Y., Masukujjaman, M., Sobhani, F.A., Hamayun, M., Alam, S.S.: Green logistics, green human capital, and circular economy: the mediating role of sustainable production. Sustainability **15**(2), 1045 (2023). https://doi.org/10.3390/su15021045
5. García Alcaraz, J.L., et al.: Effect of green supply chain management practices on environmental performance: case of Mexican manufacturing companies. Mathematics **10**(11), 1877 (2022)
6. Gong, R., Xue, J., Zhao, L., Zolotova, O., Ji, X., Xu, Y.: A bibliometric analysis of green supply chain management based on the Web of Science (WOS) platform. Sustainability **11**(12), 3459 (2019). https://doi.org/10.3390/su11123459
7. Hejazi, M.T., Al Batati, B., Bahurmuz, A.: The influence of green supply chain management practices on corporate sustainability performance. Sustainability **15**(6), 5459 (2023). https://doi.org/10.3390/su15065459
8. Frota Neto, J.Q., Bloemhof-Ruwaard, J.M., van Nunen, J.A.E.E., van Heck, E.: Designing and evaluating sustainable logistics networks. Int. J. Prod. Econ. **111**, 195–208 (2008)
9. Jalali, M., Feng, B., Feng, J.: An analysis of barriers to sustainable supply chain management implementation: the fuzzy DEMATEL approach. Sustainability **14**(20), 13622 (2022). https://doi.org/10.3390/su142013622
10. Kajba, M., Jereb, B., Obrecht, M.: Considering IT trends for modelling investments in supply chains by prioritising digital twins. Processes **11**, 262 (2023). https://doi.org/10.3390/pr11010262
11. Kumar, A., Shrivastav, S.K., Shrivastava, A.K., Panigrahi, R.R., Mardani, A., Cavallaro, F.: Sustainable supply chain management, performance measurement, and management: a review. Sustainability **15**(6), 5290 (2023). https://doi.org/10.3390/su15065290
12. Kwak, S.-Y., Cho, W.-S., Seok, G.-A., Yoo, S.-G.: Intention to use sustainable green logistics platforms. Sustainability **12**(8), 3502 (2020). https://doi.org/10.3390/su12083502
13. Lazar, S., Klimecka-Tatar, D., Obrecht, M.: Sustainability orientation and focus in logistics and supply chains. Sustainability **13**(6), 3280 (2021). https://doi.org/10.3390/su13063280
14. Mak, S.-L., Wong, Y.-M., Ho, K.-C., Lee, C.-C.: Contemporary green solutions for the logistics and transportation industry—with case illustration of a leading global 3PL based in Hong Kong. Sustainability **14**, 8777 (2022). https://doi.org/10.3390/su14148777
15. Makarova, I., et al.: The role of multimodal transportation in ensuring sustainable territorial development: review of risks and prospects. Sustainability **15**(7), 6309 (2023). https://doi.org/10.3390/su15076309

16. Mala, D., Sedliačikova, M., Kaščakova, A., Benčikova, D., Vavrova, K., Bikar, M.: Green logistics in Slovak small and medium wood-processing enterprises. BioResources **12**(3), 5155–5173 (2017)
17. Qu, S., Ji, Y.: Sustainable supply chain management and optimization. Sustainability **15**(4), 3844 (2023). https://doi.org/10.3390/su15043844
18. Paužuolienė, J., Kavecké, I.: Žaliosios logistikos taikymo svarba: vartotojų nuomonės tyrimas. Reg. Formation Dev. Stud. **1**(36), 61–68 (2022). https://doi.org/10.15181/rfds.v36i1.2388
19. Jančiauskas, B., Treigienė, D.: Supply Chain Management. Technika, Vilnius (2003)
20. Rejeb, A., Rejeb, K., Appolloni, A., Iranmanesh, M., Treiblmaier, H., Jagtap, S.: Exploring food supply chain trends in the COVID-19 era: a bibliometric review. Sustainability **14**, 12437 (2022). https://doi.org/10.3390/su141912437
21. Richnák, P., Gubová, K.: Green and reverse logistics in conditions of sustainable development in enterprises in Slovakia. Sustainability **13**(2), 581 (2021). https://doi.org/10.3390/su13020581
22. Silva, W.H., Guarnieri, P., Carvalho, J.M., Farias, J.S., dos Reis, S.A.: Sustainable supply chain management: analyzing the past to determine a research agenda. Logistics **3**(2), 14 (2019). https://doi.org/10.3390/logistics3020014
23. Shekarian, E., Ijadi, B., Zare, A., Majava, J.: Sustainable supply chain management: a comprehensive systematic review of industrial practices. Sustainability **14**(13), 7892 (2022). https://doi.org/10.3390/su14137892
24. Tan, B.Q., Wang, F., Liu, J., Kang, K., Costa, F.: A blockchain-based framework for green logistics in supply chains. Sustainability **12**, 4656 (2020). https://doi.org/10.3390/su12114656
25. Tseng, M.-L., Bui, T.-D., Lim, M.K., Lewi, S.: A cause and effect model for digital sustainable supply chain competitiveness under uncertainties: enhancing digital platform. Sustainability **13**(18), 10150 (2021). https://doi.org/10.3390/su131810150
26. Trivellas, P., Malindretos, G., Rekliti, P.: Implications of green logistics management on sustainable business and supply chain performance: evidence from a survey in the greek agri-food sector. Sustainability **12**, 10515 (2020). https://doi.org/10.3390/su122410515
27. Vasiliauskas, A.V., Zinkevičiūtė, V., Šimonytė, E.: Implementation of the concept of green logistics refering to IT applications for road freight transport enterprises. Verslas: teorija ir praktika **14**(1), 43–50 (2013)
28. Vienažindienė, M., Tamulienė, V., Zaleckienė, J.: Green logistics practices seeking development of sustainability: evidence from lithuanian transportation and logistics companies. Energies **14**(22), 7500 (2021). https://doi.org/10.3390/en14227500
29. Wang, D.-F., Dong, Q.-L., Peng, Z.-M., Khan, S.A.R., Tarasov, A.: The green logistics impact on international trade: evidence from developed and developing countries. Sustainability **10**(7), 2235 (2018). https://doi.org/10.3390/su10072235
30. Wen, S., Tang, H., Ying, F., Wu, G.: Exploring the global research trends of supply chain management of construction projects based on a bibliometric analysis: current status and future prospects. Buildings **13**(2), 373 (2023). https://doi.org/10.3390/buildings13020373
31. Werner-Lewandowska, K., Golinska-Dawson, P.: Sustainable logistics management maturity—the theoretical assessment framework and empirical results from Poland. Sustainability **13**, 5102 (2021). https://doi.org/10.3390/su13095102
32. Yuan, X.-M., Xue, A.: Supply chain 4.0: new generation of supply chain management. Logistics **7**, 9 (2023). https://doi.org/10.3390/logistics7010009
33. Zhou, B., Siddik, A.B., Zheng, G.-W., Masukujjaman, M.: Unveiling the role of green logistics management in improving SMEs' sustainability performance: do circular economy practices and supply chain traceability matter? Systems **11**, 198 (2023). https://doi.org/10.3390/systems11040198
34. Zowada, K.: Green logistics: the way to environmental sustainability of logistics. Empirical evidence from Polish SMEs. Eur. J. Sustain. Dev. **9**(4), 231–240 (2020)

Challenges for Enhanced Military Mobility on the Eastern Flank of NATO

Jaroslav Kompan(✉) and Michal Hrnčiar

Armed Forces Academy of General M.R. Štefánik, Demänová 393, 03101 Liptovský Mikuláš, Slovakia

{jaroslav.kompan,michal.hrnciar}@aos.sk

Abstract. Russia's aggressive policy against Ukraine and growing instability in the Europe have renewed the need to fulfil NATO's three basic missions, as set out in the Alliance's Strategic Concept i.e., collective defence, crisis management and ensuring the security of all Allies. Military mobility in Europe has been identified as one of the key steps towards achieving credible deterrence. NATO has a priority interest in achieving the ability to move military troops across Europe as quickly as possible. In 2019, NATO members have already committed to increase their defence budgets to the 2% of their GDP by 2024, in line with previous commitments, a significant part of defence budgets should also be allocated to enhancing military mobility, especially on the NATO's eastern flank. The aim of the article is therefore to identify the specific challenges to enhanced military mobility on NATO's eastern flank and possible directions for improving infrastructure on NATO's eastern flank. At the same time, the findings could be used to propose areas that could positively influence the military mobility and forward presence of Alliance forces on NATO's eastern flank. The findings provide a framework for managing military mobility in Europe as a realistic endeavour. This endeavour may encounter some political contradictions, but these could be resolved by bilateral or multilateral agreements on technical and administrative issues of military mobility.

Keywords: NATO's eastern flank · Collective defence · Military mobility challenges · Mobility planning framework · Infrastructure development

1 Introduction

The principle of collective defence is regarded as one of the foundations of NATO to which individual members have committed themselves. Its validity and relevance are regularly proclaimed and reaffirmed at various political levels, and NATO's military planning processes are managed to fully reflect the political direction and statements of its Heads of State and Governments [1]. One of the most important factors in successfully supporting the Alliance's political commitment to collective defence is the way in which Allies shape the physical environment in which Alliance forces would operate if required, and to which comprehensive military planning must adequately respond. The movement of Alliance forces in times of crisis from their permanent locations to their potential areas

O. Prentkovskis et al. (Eds.): TRANSBALTICA 2023, LNITI, pp. 239–248, 2024.
https://doi.org/10.1007/978-3-031-52652-7_24

of operations is a challenging task requiring detailed and precise planning at all levels of NATO.

Despite the fact that enhanced military mobility is a priority for NATO, especially on its eastern flank, and that lessons learned from NATO and EU exercises are being analysed in detail and translated into measures to support the preparation of the operational environment, such as the reduction of border controls or the improvement of infrastructure (airports, ports, bridges and railways) [2], specific research was needed to identify a set of challenges that could provide a framework for the creation of effective control measures for enhanced military mobility.

However, an adequate response to the identified security threats [3] and the implementation of the above measures will require significant financial investments by individual member states, especially on NATO's eastern flank in the context of the war in Ukraine. Cooperation with Alliance and European Union partners, while actively engaging in joint projects, is proving to be one of the most effective ways of development [4], including significant cooperation in the areas of technological innovation of military systems and the introduction of standards for their widest possible compatibility [5]. Therefore, the article defines the basic requirements for planning the enhancement of military mobility imposed on the nations on NATO's eastern flank as equal partners and Allies in support of the collective defence of the Alliance's joint area of operations and, in addition, presents specific challenges for enhanced military mobility on NATO's eastern flank that have been identified as the critical framework for infrastructure development and subsequent force capability development.

The article is based on the results of the scientific research project "Optimisation of the possibilities for the execution of the military engineer support in land operating environment" and the European Union international scientific research project "Interdisciplinary Education and Training on Hybrid Warfare". Specifically, the article aims to identify the specific challenges to increased military mobility on NATO's eastern flank and possible directions for improving infrastructure on NATO's eastern flank. At the same time, the findings could be used to proposed the areas that could positively influence the military mobility and forward presence of Alliance forces on NATO's eastern flank. The individual findings and subsequent recommendations reflect the capacities and the potential ambitions of the Allies.

2 Military Mobility Planning Framework

Russia's aggressive policy towards Ukraine [6] and growing instability in the Balkans, the Middle East and Africa [7, 8] have renewed the need to fulfil NATO's three basic missions, as set out in the Alliance's Strategic Concept i.e., collective defence, crisis management and ensuring the security of all Allies [9]. In 2019, NATO members have already pledged to increase their defence budgets to the level of 2% of their GDP by 2024, in line with previous commitments [10]. However, the increased financial resources should be used as efficiently as possible, including to ensure not only the development and increase in the number of military forces and assets, but also their deployability in accordance with NATO requirements. All military activities are becoming increasingly dynamic and fast-paced, and military units must be prepared to respond more proactively to changes and to new threats [11]. Successfully meeting this requirement requires comprehensive management, organisation, synchronisation and execution of the improvement of NATO's military mobility. By improving military mobility, it is possible to achieve a high rate of advance, including decision-making, which enhances deterrence.

Maintaining and increasing the level of military mobility achieved in military operations is particularly important for land forces, since their deployment is directly affected by the traversability of the main lines of communication [12]. In the event of a threat to NATO's eastern flank, however, traversability is linked not only to operations under normal conditions and routine maintenance of infrastructure, but also to the ability to reach staging areas despite degraded infrastructure, whether caused by excessive traffic itself or by enemy activities. Therefore, military forces should strive to maximise the effectiveness of operations in a given operating environment, even in the face of limited infrastructure [13]. Therefore, military mobility planning should also be closely intertwined with planning for the use of infrastructure, considering limited resources [14], minimising costs [15, 16], reducing the time to reach designated areas, and maintaining NATO standards, while of course respecting the principles of environmental protection [17] and the safety of the civilian population [18]. NATO's military mobility planning process should therefore consist of the following steps:

- Political Guidance - Establishment of guidance from the political level as part of the political guidance for defence planning so that the required capabilities and necessary infrastructure can be specified.
- Identification of requirements - Determination of military mobility requirements, including required capabilities and infrastructure and their shortfalls.
- Distribution of requirements and development of goals - Initial distribution of overall requirements for minimum capabilities and infrastructure to ensure military mobility for individual countries and NATO components, as part of the "goals package", respecting the principle of fair sharing and delegation of reasonable challenges.
- Implementation - Support of national and international efforts and collective efforts to meet the stated goals for military mobility (including priorities) in an efficient and time-acceptable manner. The challenge is to focus on the most important identified capabilities and infrastructure gaps.
- Evaluation - Examining which policy objectives, ambitions and associated tasks have been met. Provides feedback and input into political guidance for the next cycle in the defence planning process.
- Revision - Examines and evaluates the defense and financial planning framework for military mobility. This review provides a key mechanism for generating feedback and input into the next planning cycle and is to be conducted periodically, at least on a biennial basis.

Comprehensive planning of military mobility as part of NATO's defence planning is necessary, especially as NATO's military capabilities have been greatly reduced in the changed security situation since 1989 [19]. For more than 30 years, the Allies have maintained only a limited presence on the Alliance's eastern border, and minimal attention has been paid to improving the infrastructure necessary to deploy larger formations of Allied forces. It was only after Russia's annexation of Ukrainian territory in 2014 and the subsequent outbreak of war in Ukraine in 2022 that NATO undertook the largest reinforcement of collective defence since the Cold War [20]. As a result, the readiness of the common operating environment, including the ability to rapidly reinforce forces throughout the area of responsibility and especially on NATO's eastern border, has become a sub-essential part of the deterrence and collective defence system.

3 Military Mobility Challenges

Military mobility in Europe has been identified as one of the key steps towards achieving credible deterrence and has become one of the most important issues in NATO-EU cooperation [21]. Both organisations, which include all the countries on NATO's eastern flank, have an interest in achieving the ability to move troops (including their equipment and subsequent full logistical support) across Europe as quickly as possible. Moreover, the common interest focuses on areas where the EU and NATO are highly interdependent. For example, the EU has greater authority in legal and regulatory matters, while also having financial resources and programmes for cross-border transport [22]. NATO, on the other hand, is fully capable of planning and calculating military needs for transport across Europe [23] to ensure the credible deterrent effect mentioned above. However, the two organisations remain distinct entities operating on different political, legal, governmental and membership bases.

Another dimension enters into the above parameters, namely that societies in the countries on NATO's eastern flank have changed fundamentally since 1989. There have been changes in ownership, some critical infrastructure has been sold or not maintained for long periods for economic reasons, and almost all critical infrastructure management systems are connected to virtual space. Connecting the various elements of critical infrastructure to the virtual space allows for efficiencies in transport operations, such as improved and faster logistical coordination, eliminating the need to immediately store unneeded stocks, thus saving a significant amount of money. Cyberspace is a modern and attractive platform for planning and controlling military mobility, but also for limiting it. Vulnerabilities in systems are precisely the target of adversary activities in hybrid warfare, especially to affect the overall functionality of individual systems and to influence public support for any activities [24]. Public opinion has now become an important factor influencing overall political decisions, including the direction of investment and support required to improve the infrastructure for enhanced military mobility in the Alliance's potential area of operations.

A likely scenario for a potential crisis situation on NATO's eastern flank would be insufficient Host Nation (HN) reception, staging, and onward movement (RSOM) capability or Troop Contributing Nation (TCN) capacity to provide military mobility, either in the form of strategic transport or subsequent bridging or breaching capability. Each country is responsible for transporting its own troops, equipment and materiel to the area of operations. This complex activity is directed and managed by bilateral agreements between the TCNs, the transporting country (company) and the HN. In a collective defence scenario where individual TCNs transport troops and materiel to the area of operations, the transit system would be under enormous pressure. This is one of the reasons why, in recent years, NATO member states and the EU have sought to reduce the bureaucratic burden in order to support the planning process by speeding up the border crossing of military units and their equipment [23].

If military mobility on NATO's eastern flank is to be feasible and secure, the participants in the cooperation need to be aware of the gaps and the ways in which they can be overcome together. These challenges, which can also be defined as the basic requirements for enhanced military mobility on NATO's eastern flank, fall into two main categories:

- Infrastructure supporting military mobility – identified routes and transport supporting infrastructure (e.g. bridges) with proper Military Load Classification (MLC) and technical status.
- Administrative support for military mobility - legal and regulatory constraints on the ease and speed of border crossings through established border crossing points, including the use of airspace and airports as an aerial port of debarkation.

3.1 Infrastructure Supporting Military Mobility

The Alliance's current focus on the ability to respond rapidly to possible threats to the territorial integrity and sovereignty of its eastern flank members can be seen as the logical culmination of changes in the security environment. It is therefore necessary to prepare plans for the deployment of declared units in case of need, including a plan for their logistical support, which will enable NATO to implement as soon as possible measures to fill the gap in defence planning and the improvement of military mobility from previous periods.

At the end of the twentieth century, until Russia's annexation of Ukrainian territory in 2014, NATO's main focus was on out-of-area operations such as those in Europe, which required a different kind of planning for military mobility and logistical support. These were left to the discretion of individual Allied nations to implement the redeployments and ensure the military mobility of their own contributions to international crisis management operations as they saw fit. Planning was adapted to the specific situation and area of operations, and military forces and equipment were transported relatively efficiently and easily by sea or air by commercial companies, often with ownership links in the Russian Federation or Ukraine. Therefore, there is a need for an increasingly massive deepening of the conclusions and outcomes of the NATO Brussels Summit [25] to increase airlift capacities and ensure rapid air mobility, including the building of infrastructure to ensure the aforementioned mobility and cross-border movement to NATO's eastern flank.

The need to identify the main and alternative supply routes (MSR/ASR) comes to the fore, but in some cases, those only partially meet the conditions required to support the transport of military units, equipment and material. This challenge is due to the fact that in Eastern Europe, new transport infrastructure, including roads, railways and structures on them (e.g. bridges), has been built without being designed for potential military use, e.g. without considering the requirement for sufficient MLC. Checking the load capacity of these defined supply routes is also necessary to identify weak points in the overall transport network in order to prevent a collapse of transport flows during the deployment of large groups of Alliance units on NATO's eastern flank.

One of the limiting factors in the use of road infrastructure for military transport or peacetime movement is the maximum permissible dimensions of vehicles and vehicles, the maximum permissible weights of vehicles and vehicles [26]. In response to the crisis, it is necessary to consider the above-mentioned facts, because the road network is also used by other entities and not only by military forces. In peacetime, in case of exceeding the set values, it will be necessary to ask the competent administrators of individual roads for permission or recommendation to carry out the transfer and, in case of permission, also to determine the route. This factor is clearly limiting for all military units, as a standard military convoy is likely to exceed the limits, requiring more extensive administration. In the case of oversized transport, it is therefore more advisable to use rail transport for planning and executing the deployment and transit of Alliance units, but military practice has shown that this method is still not preferred due to the limited availability, limited network, difficult recoverability and time-consuming response and flexibility of rail transport to respond to crisis situations [27].

Another important factor of mobility from a military point of view is the ability of units to move in a pre-combat dispersed formation, where the use of only the MSR/ASR can be considered a vulnerability, especially close to the line of contact. The development of weapon systems and innovative approaches to decision-making and force protection have also increased the need for speed of movement and, in particular, surprise in activies [28]. If we consider the recent trend in the development of armoured combat vehicles, in response to the above requirements, along with the improvement of combat parameters, especially the combination of passive and active protection of the crew, there is also an increase in the weight of combat vehicles. For the movement in combat formation, routes other than MSR/ASR are widely used, while their load capacity, including the load capacity of objects on them, may not be dimensioned for the weight parameters of the modern military vehicles.

On NATO's eastern flank, this challenge is even more pronounced because the transport infrastructure was largely built between World War II and 1989 and was designed, from a military point of view, to transport Warsaw Pact combat equipment (e.g. T-72 tanks), i.e. MLC 60. Therefore, it is expected that there will be an increased need to use military engineering assets for its reinforcement. We have identified a capability gap in this particular area due to the slow response of infrastructure improvement to the development of new weapon systems. Especially for the deployment of Alliance partners' main battle tanks and their requirements for possible support when crossing rivers or large dry obstacles. While the combat weight of the Leopard II (MLC 80), Challenger 3 (MLC 95) or Abrams M1A2 SEPv3 (MLC 91) is in the MLC 80–95 range, most military bridge systems (e.g. AAAB, EFA, PFM F2) can provide a load capacity of up to MLC 70. The M3 and IRB systems are capable of providing the required load capacity, but their numbers and representation in Member States' armed forces are insufficient for mass deployment.

The weight category and dimensional requirements of transport are not the only limiting factor in supporting the mobility of military units during combat operations. Taking into account the challenges already identified, as well as the lessons learned from previous military exercises and the way combat operations are conducted in Ukraine, we consider that the weakest point affecting military mobility will be bridges. The physical environment of the countries of NATO's eastern flank is characterised by a large number of rivers crossed by bridges of different load-bearing capacity, length and technical condition. For example, in the supply corridor through Germany, Poland to the Baltic countries, there are 6 main rivers (Oder, Wisla, Bug, Nemunas, Daugava, Parnu) with an average width of over 300 m. If only 1% of the bridges are damaged, regardless of the cause of the damage, the effort required to bridge with military bridges will be enormous [12, 29]. Therefore, improving and strengthening the compatibility of individual military bridge systems within Alliance countries could also make a fundamental contribution to collective defence efforts and, at the same time, significantly improve the readiness of the countries of NATO's eastern flank to a level that supports the use of effective weapon systems of partners in the defence of territorial integrity.

3.2 Administrative Support for Military Mobility

In the context of NATO's collective defence concept, Allies should be willing and able to accept and subsequently support further deployments of the Alliance's forces to destinations with a minimum of battalion-size land forces composed of military contributions from a number of Partners. Such deployments are currently being implemented in all countries on NATO's eastern flank in the form of "forward presence" [20]. Such efforts require the support of the HN armed forces or ministry of defence, as well as other government ministries and agencies. In addition to the required troop transport capabilities, the ability to adequately receive and provide for the enormous amount of military equipment, materiel, weapons and personnel (storages of equipment and materiel, accommodation of personnel and adequate provision for their normal lives) comes to the fore.

In the area of providing military mobility in the form of military transport, NATO can build on the EU's efforts to create connectivity in the form of a strong European transport network linking all member states, which is vital to the European economy and its global competitiveness. Identifying and agreeing on military requirements that reflect the needs of the EU and its member states is also an appropriate starting point for effective access throughout the EU [30] and for NATO forces. The Trans-European transport network [31] would also provide a space for more efficient troop transport and resupply.

Another area is cross-border movement permits, including diplomatic permits, where Member States can cooperate to increase the consistency and efficiency of their own procedures. The agreement on cross-border movement authorisations, focusing on land transport, is an opportunity to improve military mobility by road, rail and inland waterway. Administrative procedures could be partially integrated into the agreement to ensure common rules and procedures, in particular for the implementation of the Common Security and Defence Policy, including activities in other frameworks, exercises and routine movements. To support the simplification and standardisation of procedures for cross-border transport, in line with the conclusions of the EU Council of June 2018, PESCO (Permanent Structured EU Cooperation) projects have been established [23]. The projects aim, among other things, to streamline cross-border force transfers without lengthy bureaucratic procedures for EU operations and missions.

It is evident from the above that managing military mobility within Europe can be seen as a realistic endeavour, which may encounter some political contradictions, but which can ultimately be resolved through bilateral agreements. The efforts of the countries on NATO's eastern flank should therefore focus primarily on the exchange of experience and the implementation of joint European projects and agreements. Through such active participation in common projects, it is possible to deepen integration within NATO and the EU, which will also clearly strengthen common defence [32].

4 Conclusions

NATO members Heads of State and Governments have committed to improve the Alliance's ability to deploy and sustain forces and to enhance military mobility in all domains (especially land, sea, and air) by 2024 [33]. These efforts require the involvement of member governments, including national defence planning and intergovernmental cooperation of both civilian and military actors, as well as extensive efforts to coordinate all interests. By applying the basic framework for military mobility planning defined in the article, the planning and execution of these efforts could be synchronised and based on mutual considerations and interrelated infrastructure development projects rather than on conflicting plans.

The challenges defined above imply that the infrastructure already in place in the countries on NATO's eastern flank will be used for the purposes of NATO troop deployments. Therefore, any improvement of this infrastructure should also be closely linked to its possible use by potential partners engaged in NATO collective defence. Improvements to transport infrastructure should also be closely linked to projects to modernise armaments and introduce new weapon systems and equipment.

Individual challenges to military mobility on NATO's eastern flank should not be considered in isolation. This means that only a balanced response to all challenges can provide a decisive response to adversary actions, including hybrid warfare. The results presented in this article can serve as a basis for further research. This research should be directed, for example, at assessing the effectiveness and performance of military mobility promotion and at eliminating the quantitative and qualitative mobility limits of available military capabilities and weapon systems.

References

1. Spilý, P., Hanko, M.: NATO standardization of the operations planning. Vojenské reflexie **14**(1), 70–80 (2019)
2. Boeke, S.: Creating a secure and functional rear area: NATO's new JSEC Headquarters. https://www.nato.int/docu/review/articles/2020/01/13/creating-a-secure-and-functional-rear-area-natos-new-jsec-headquarters/index.html. Accessed 28 June 2023
3. Mušinka, M.: Možnosti hodnotenia bezpečnostných hrozieb. Vojenské reflexie **15**(1), 82–98 (2020)
4. Varecha, J.: Revolúcia vo vojenských záležitostiach. In: Národná a medzinárodná bezpečnosť 2018: 9. medzinárodná vedecká konferencia, pp. 500–506. Akadémia ozbrojených síl generála Milana Rastislava Štefánika, Liptovský Mikuláš (2018)
5. Turaj, M.: Zhodnotenie aktuálneho stavu bojového použitia predsunutých leteckých navádzačov v podmienkach Ozbrojených síl Slovenskej republiky. Vojenské reflexie **13**(2), 164–174 (2018)
6. Hrnčiar, M., Kompan, J.: Factors shaping the employment of military force from the perspective of the war in Ukraine. Vojenské rozhledy **32**(1), 69–82 (2023)

7. Jurčák, V., Marek, J.: The Russian geopolitical interests in Africa. Politické vedy **23**(4), 196–212 (2020)
8. Majchút, I.: Central African Republic: the West Contra Russia in the conflict solution. Politické vedy **25**(4), 213–230 (2022)
9. Strategic Concepts. https://www.nato.int/cps/en/natohq/topics_56626.htm. Accessed 28 June 2023
10. NATO: Why is spending 2% of GDP on defence so controversial? https://www.euronews.com/2023/04/07/nato-why-is-spending-2-of-gdp-on-defence-so-controversial. Accessed 28 June 2023
11. Nohel, J., Stodola, P., Flasar, Z.: Combat UGV support of company task force operations. In: Mazal, J., Fagiolini, A., Vasik, P., Turi, M. (eds.) MESAS 2020. LNCS, vol. 12619, pp. 29–42. Springer, Cham (2021). https://doi.org/10.1007/978-3-030-70740-8_3
12. Jančo, J., Kompan, J.: Influence of the bridge's status on the military mobility in the Slovak Republic. In: Prentkovskis, O., Yatskiv (Jackiva), I., Skačkauskas, P., Maruschak, P., Karpenko, M. (eds.) TRANSBALTICA 2022, pp. 470–478. Springer, Cham (2023). https://doi.org/10.1007/978-3-031-25863-3_44
13. Dologa, J., Rolenec, O., Hanáková, N.: Engineer mobility support on the territory of the Czech Republic as one of the host nation support tasks. In: Prentkovskis, O., Yatskiv (Jackiva), I., Skačkauskas, P., Maruschak, P., Karpenko, M. (eds.) TRANSBALTICA 2022, pp. 491–501. Springer, Cham (2023). https://doi.org/10.1007/978-3-031-25863-3_46
14. Varecha, J.: Nové výzvy pre Ozbrojené sily Slovenskej republiky. In: Národná a medzinárodná bezpečnosť 2017: 8. medzinárodná vedecká konferencia, pp. 576–582. Akadémia ozbrojených síl generála Milana Rastislava Štefánika, Liptovský Mikuláš (2017)
15. Frianová, V.: Uplatňovanie princípu ekonomickej efektívnosti vo verejnom sektore a osobitne v obrane. In: Národná a medzinárodná bezpečnosť 2022: medzinárodná vedecká konferencia, pp. 83–93. Akadémia ozbrojených síl generála Milana Rastislava Štefánika, Liptovský Mikuláš (2022)
16. Jirásková, S.: Ekonomika obrany: vybrané kapitoly: učebnica pre vysoké školy, 1st edn. Akadémia ozbrojených síl generála Milana Rastislava Štefánika, Liptovský Mikuláš (2022)
17. Hanakova, N., Manas, P., Rolenec, O., Palasiewicz, T.: National defense and environmental protection: on the Czech Armed Forces' approach to the development of environmental legislation in the Czech, EU and NATO context. Vojenské rozhledy **31**, 172–202 (2022)
18. Kazanský, R., Andrassy, V.: Conflict resolution approaches towards smart sustainability of internal relations. Entrepreneurship Sustain. Issues **6**(3), 1468–1484 (2019)
19. Rolenec, O., Kyjovský, J., Zelený, J., Palasiewicz, T.: Results of research to the issue of engineer support in combat operations. Vojenské rozhledy **26**, 51–67 (2017)
20. NATO's response to Russia's invasion of Ukraine. https://www.nato.int/cps/en/natohq/topics_192648.htm. Accessed 28 June 2023
21. Koster, T.S.: Reinforcement of NATO forces and military mobility. Atlantisch Perspectief **42**(4), 15–18 (2018)
22. Military Mobility: EU proposes actions to allow armed forces to move faster and better across borders. https://transport.ec.europa.eu/news-events/news/military-mobility-eu-proposes-actions-allow-armed-forces-move-faster-and-better-across-borders-2022-11-10_en. Accessed 28 June 2023
23. Action Plans on Military Mobility. https://defence-industry-space.ec.europa.eu/eu-defence-industry/military-mobility_en. Accessed 28 June 2023
24. Majchút, I., Vajda, M.: Hybrid warfare: military response options. Ante Portas **18**(1), 129–143 (2022)
25. Rapid Air Mobility (RAM). https://shape.nato.int/about/aco-capabilities2/rapid-air-mobility-ram. Accessed 28 June 2023

26. Directive (EU) 2015/719 of the European Parliament and of the Council of 29 April 2015 amending Council Directive 96/53/EC laying down for certain road vehicles circulating within the Community the maximum authorised dimensions in national and international traffic and the maximum authorised weights in international traffic. https://eur-lex.europa.eu/legal-content/EN/TXT/?uri=celex%3A32015L0719. Accessed 28 June 2023
27. Polach, M., Vlkovský, M.: Resilience of the railway infrastructure in an urban agglomeration with regard to the transport of specific material. In: Růžička, J. (ed.) 2023 Smart City Symposium Prague, p. 7. Institute of Electrical and Electronics Engineers, Danvers (2023)
28. Drozd, J., Rak, L., Zahradníček, P., Stodola, P., Hodický, J.: Effectiveness evaluation of aerial reconnaissance in battalion force protection operation using the constructive simulation. J. Defense Model. Simul. **20**(2), 181–196 (2023)
29. Jančo, J.: Mosty a ich vplyv na vojenskú mobilitu. Vojenské reflexie **17**(1), 89–107 (2022)
30. Europe needs a military Schengen. https://eda.europa.eu/webzine/issue12/cover-story/europe-needs-a-military-schengen. Accessed 28 June 2023
31. Trans-European Transport Network (TEN-T). https://transport.ec.europa.eu/transport-themes/infrastructure-and-investment/trans-european-transport-network-ten-t_en. Accessed 28 June 2023
32. Bystriansky, I.: Budovanie partnerstva EÚ a NATO. In: Národná a medzinárodná bezpečnosť 2022: medzinárodná vedecká konferencia, pp. 46–62. Akadémia ozbrojených síl generála Milana Rastislava Štefánika, Liptovský Mikuláš (2022)
33. Bessems, R.: Moving outside the box: military mobility as the key to enabling European security & defense. Atlantisch Perspectief **45**(4), 28–32 (2021)

Comprehensive Service of Refrigerated Containers in Intermodal Transport Chains

Ludmiła Filina-Dawidowicz[1](✉), Mykhaylo Postan[2], Paweł Grzelak[1], and Julia Sienkiewicz[1]

[1] West Pomeranian University of Technology, Al. Piastów 17, 70-310 Szczecin, Poland
lufilina@zut.edu.pl

[2] Odessa National Maritime University, Odesa, Ukraine

Abstract. Nowadays, growing demand for perishable food products delivery is observed. These products may require specific transport conditions and on longer distances could be carried out in refrigerated containers. In intermodal transport chains, a comprehensive refrigerated containers service needs to be provided to maintain the quality of the products transported and to ensure customer satisfaction with the service, which can be a challenge for forwarders or logistics operators. The purpose of the article is to analyse the services that constitute the comprehensive service of refrigerated containers within intermodal transport chains and to identify the criteria used to group these services. The grouping of these services according to selected criteria has been proposed. It was stated that refrigerated containers comprehensive service offer is influenced by market environment and customers' needs, as well as possibilities to perform the services by provider. The research results may be of interest to transport, forwarding and logistics companies involved in planning and implementing supply of perishable food products within intermodal transport chains.

Keywords: Perishable food product · Refrigerated container · Comprehensive service · Intermodal transport chain

1 Introduction

Perishable food products represent a specific cargo that requires special conditions during transport and storage (e.g., appropriate temperature, air humidity, ventilation) [1]. This cargo may lose the quality during its transportation; therefore, it should be delivered to the customers quickly and on time [2]. Intermodal transport of these goods means transporting cargo in specially designated units (e.g., refrigerated containers) using several modes of transport (e.g., road, sea, rail), without cargo reloading from the unit during the whole transport route [3, 4].

Within intermodal refrigerated container transport chains, the activities necessary for the smooth delivery of cargo can be defined. These activities include the services that can be carried out for the containers and the cargo contained therein within the individual links of the transport chains [5]. It should be emphasized that besides transportation,

O. Prentkovskis et al. (Eds.): TRANSBALTICA 2023, LNITI, pp. 249–260, 2024.
https://doi.org/10.1007/978-3-031-52652-7_25

handling and storage of containers, several additional operations may be required while delivering the cargo to destination (e.g., cargo veterinary and customs inspections, container weighing, providing information to the customer, etc.) [6, 7]. All these activities constitute a comprehensive service that a forwarder or logistics operator offers to the customers. The operator shapes the service offering, taking into account both the customer needs and the possibility to provide the services (e.g., availability, costs, profits, etc.). That fact deals with the need to assess the services by their provider to select those that will be launched to the comprehensive services offer.

Conducted literature review revealed that available research studies focus mainly on the separate groups of services performed for refrigerated containers in intermodal transport chains. Therefore, the aim of this article is to analyse the services that constitute the comprehensive service of refrigerated containers in intermodal transport chains and to identify the criteria used to group these services.

2 Literature Review

Services provided in intermodal transport chains include a complete range of services enabling the movement of cargo from the sender to the recipient [8]. These services usually include activities ordered by individual customers needed to deliver cargo efficiently to the destination. Comprehensive service of refrigerated containers is a set of organizational, technical, technological, financial, administrative and commercial processes, covering all transport and logistics services necessary during the shipment of cargo from the sender to the recipient, carried out by one operator, aimed at meeting customer expectations and maintaining the qualitative characteristics of perishable cargo during its transport by means of various transport modes (Fig. 1) [7].

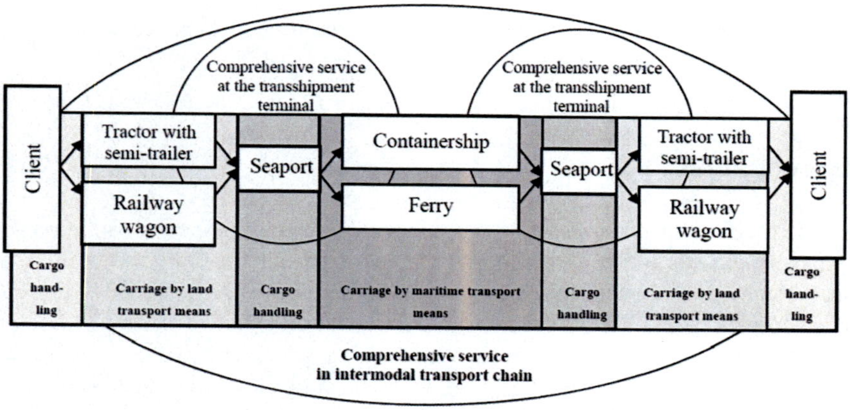

Fig. 1. An example of intermodal transport chain for refrigerated containers (own elaboration based on [9]).

Refrigerated containers transportation is usually performed using maritime and road transport means, railway transport is used less often in Europe. Refrigeration unit of the container should be plugged to the electrical power source during the whole transport process [10]. Dedicated infrastructure and equipment is used for this purpose [4, 11].

In the available literature the services provided at transshipment terminals, located i.e., in seaports (container and ferry terminals), are analysed in detail [10, 12]. These terminals offer a wide range of services for transport means, as well as transported cargo. Examples of services provided in seaports for maritime, road and rail transport means are presented in Table 1.

Table 1. Selected services provided for transport means in seaports [9].

Transport mean	Examples of services
Maritime	• mooring, piloting, towing, • minor repairs and maintenance, • delivery of utilities, food, fuel, spare parts, • waste collecting, • telecommunication and IT services, • fire protection and rescue, etc.
Road	• vehicle parking, • fuel supply, • cleaning and maintenance services, etc.
Rail	• wagons moving to/from the quay, • wagons moving along the quay, • cleaning services, • repair and maintenance, • train forming, etc.

In the available literature different criteria were applied to group the services [13, 14]. The services division considering the serviced object has been distinguished, which include the services provided for a perishable cargo, refrigerated container and transport vehicle. Both basic and specific services implemented by enterprises operating in the field of transport, forwarding and warehousing were mentioned [8]. Considering the type of performed process, technical and executive services (e.g., transport, warehousing, weighting, etc.), as well as organizational services (e.g., planning, managing, controlling) were specified [9]. It was noted that services may be provided directly by the enterprise or may be outsourced [15]. New services may be designed and launched to the provider's offer, considering the customers' needs [3]. To do this the managers of transport and logistics companies should possess knowledge related to new services implementation [16]. Each service requires the involvement of dedicated technical and technological, organizational, and financial facilities [17, 18].

It was highlighted that the services should be customer-oriented [19] and competitive [20]. The provision of services is influenced by local and international policies [21]. Services should be introduced in stages and sequentially, considering specific technology of their provision [22]. Moreover, their implementation to the services offer is associated with quick decision-making.

The way to perform the service should assure the efficiency of transport process [8]. Akkerman et al. [23] paid attention to the aspects of quality, safety and sustainability in food distribution and conducted a review of quantitative operations management approaches and challenges. In turn, Manzini and Accorsi noted that within design and management of supply chain the integrated approach should be used and simultaneous control of safety, quality, sustainability, and logistics efficiency of transported food products should be carried out [13]. Attention was also drawn to the need of traceability in supply chains of food to assure its quality and safety [24]. The issues of transport capacity planning and routes optimization were also investigated [25, 26].

Decisions to assess services are often based on their quality characteristics. A wide range of aspects related to port services evaluation was considered by Lu et al. [14], where aspects affecting the services quality were listed and involved, i.e., compliance with agreed time of ship service, customs work efficiency, efficiency of transshipment operations, availability of berths, accuracy of information, quality of port infrastructure, condition of hinterland infrastructure, professionalism of employees, etc. Evaluation criteria for refrigerated containers service complexity in transshipment terminals were also formulated and included: costs/profits, demands, reliability of a service/risk of failure to undertake a service, risk of cargo quality decrease, duration of a service launch (provision), importance of a service to a customer, feasibility of innovative solutions, compatibility of a service, flexibility of a service, possibility to undertake a service by means of outsourcing, energy efficiency and sustainable energy application [7]. However, these criteria were not applied to the entire intermodal transport chains.

Shaping a comprehensive service for refrigerated containers is a complex decision-making process that requires an appropriate approach to select the services, assess possibility to implement them, determine the link of transport chain where these services may be performed, and, consequently, define a plan for obtaining needed resources (including human, infrastructural, financial, etc.) to provide the service.

Based on the conducted literature review it was stated that comprehensive service of refrigerated containers in intermodal transport chains should be investigated in more details. Furthermore, it is important to organize and systemize knowledge related to services and criteria used to group them that may be useful for providers of refrigerated containers comprehensive service and allow to facilitate their decision-making process related to services selection.

3 Methodology

Presented research aimed to answer the following questions:

1. What services are performed for refrigerated containers within the intermodal transport chains and constitute comprehensive service offering?
2. What criteria may be applied to group these services?

Research methodology covered the following steps:

1. Conducting literature analysis and observations of perishable products deliveries in refrigerated containers within intermodal transport chains.

2. Identification of services performed for perishable food products in refrigerated containers within particular links of intermodal transport chains.
3. Identification of selected criteria for services grouping. During this step the possessed information was analysed and synthesised. The grouping of services according to selected criteria was proposed.
4. Drawing the conclusions.

4 Results

Based on conducted literature review and own observations the selected services performed for refrigerated container with cargo within the individual links of intermodal transport chains were identified (Table 2).

Table 2. Selected services performed for refrigerated container with cargo within individual links of intermodal transport chains (own elaboration).

Services	Maritime transport	Road transport	Railway transport	Transshipment terminal
Transportation	+*	+	+	–
Handling operations	–	–	–	+
Storage	–	–	–	+
Customs control	–	+	+	+
Weighting	–	+	+	+
X-ray (scanning)	–	+	+/–	+
Connection of a refrigerated container to the electrical supply source	+	+	+	+
Regular monitoring of cargo storage temperature	+	+	+	+
Changing the temperature setting in the container	+	+	+	+
Veterinary and phytosanitary inspection of cargo	–	+	+/–	+

(continued)

Table 2. (*continued*)

Services	Maritime transport	Road transport	Railway transport	Transshipment terminal
Chilling the container before loading or additional cooling to determine the desired temperature in the container with cargo	–	–	–	+
Repair and service of refrigerated container	+/–	+/–	+/–	+
Washing the container	–	–	–	+
Change/supplement of information labels	–	–	–	+
Information sharing	+	+	+	+
Documentation preparation	+	+	+	+
Cargo insurance	+	+	+	+
Organization and control of transport process	+	+	+	+
Cargo advertising	–	–	–	+
Consulting	+	+	+	+

where: * "+" - the service could appear in the offer, "–" - the service does not appear in the offer.

It should be noted that most of listed services are performed at transshipment terminals that play a key role in intermodal transport chains. However, the service offerings may change, new services may be launched and old one may be removed, considering customers' preferences.

Moreover, selected criteria for services grouping have been identified (Table 3).

Special attention may be paid to the selected criteria. Considering the criterion of service provision way, the services may be:

- provided by the company within its own resources (e.g., by a forwarder, transshipment terminal, etc.),
- commissioned to subcontractors (in outsourcing) when the service is carried out by a specialized company (e.g., cargo insurance, repair and service of refrigerated container, other specific logistics services, etc.).

Outsourcing of services is frequently used in transport and logistics activity. It has several advantages, including the lack of the need to purchase and maintain expensive

Table 3. Grouping the services provided for refrigerated containers within the intermodal transport chains according to selected criteria (own elaboration based on [7, 9]).

Lp.	Criterion	Services groups
1	Range of service	• international, • national, • regional, • intercompany
2	Company providing the service	• provided by the forwarder / multimodal transport operator, • provided by transport company (e.g., rail, road, maritime), • provided by seaport / transshipment terminal, • provided by specialized external company and other entities
3	Way of providing the service	• implemented within the company's resources, • provided by other entities (in outsourcing)
4	Profitability of service	• having an acceptable level of profitability, • having a conditionally acceptable level of profitability, • having an unacceptable level of profitability
5	Importance of service	• not important for an individual customer, • having current importance for the client, • having short-term importance (up to 1 year), • having a long-term importance (over 1 year)
6	Safety of service	• not affecting the quality of perishable cargo, • exerting a small impact on the quality of perishable cargo, • threatening the perishable cargo quality
7	Reliability of service	• long-term service performance without failing, • short-term service performance without failing, • service performance with failing
8	Availability of service	• unlimited availability of service, • limited availability of service, • unavailable service

(*continued*)

Table 3. (*continued*)

Lp.	Criterion	Services groups
9	Quality of service	• having an acceptable quality, • having a conditionally acceptable quality, • having an unacceptable quality
10	Sustainability of service	• having neutral impact on the environment, • having insignificant impact on the environment, • having significant impact on the environment
11	Service automation and/or digitalization level	• non-automated and/or digitalized service • having low level of automation and/or digitalization, • having medium level of automation and/or digitalization, • having high level of automation and/or digitalization
12	Feasibility of innovations implementation within the service	• feasible for innovations, • neutral for innovations, • unfeasible for innovations
13	Contact with container and load	• having direct contact with the container and cargo, • not having direct contact with the container and cargo
14	Number of entities involved in the service performance	• implemented by one entity, • implemented in cooperation with a larger number of entities
15	The number of entities offering the same service	• offered by one entity, • offered by several entities
16	Presence in the service offer	• existing, • new, • planned, • obsolete
17	The degree of necessity to carry out the transport task	• basic, • additional
18	The degree of necessity to perform the service	• mandatory, • on request

(*continued*)

Table 3. *(continued)*

Lp.	Criterion	Services groups
19	Integrity of service	• may be performed during selected link of transport chain, • may be performed during different links of transport chain
20	Mode of transport chain	• provided within road transport mode, • provided within railway transport mode, • provided within maritime transport mode, • provided within transshipment terminals

infrastructure or equipment in a case the demand for services is low. Among the disadvantages a high service price or high risk of failure to implement the service may be mentioned [15].

Applying the criterion of contact with the container and the cargo, the following services may be recognized:

- services having direct contact with the container and the cargo - these are usually technical and executive services (e.g., containers handling, transport, storage, weighing, repairing, loading container with cargo, etc.),
- services not having direct contact with the cargo but performed to meet the customer's expectations (e.g., sharing information about the stage of transport process, advertising of goods, consulting activity, etc.).

The quality of services that shape comprehensive service of refrigerated containers may be defined as a set of detailed features that were given to these services by the provider and are assessed by customers individually [10]. The services constituting the comprehensive service of analysed loading units can possess acceptable, conditionally acceptable, and unacceptable quality. Moreover, this quality may be divided into [9]:

- expected – the level of service quality is explicitly or implicitly expected by the customer,
- target – the level of service quality that the provider intends to achieve,
- delivered – the level of quality obtained and measured by the customer,
- perceived – the customer's viewpoint on the quality level of the service after its implementation.

In the available literature the division of services into new and existing could often be found [19]. However, services may be also grouped according to the criterion of their presence in the offer. The following services may be distinguished [7]:

- existing – which have been implemented in intermodal transport chain, they require decisions regarding their leaving in the offer, or providing possible improvements,
- new – which can be attached to an existing offer in short term and their detailed analysis is required, including the determination of the profitability and provision way,

- planned – their implementation is planned in the long-term perspective, they require a detailed analysis of implementation possibilities; however, it should be emphasized that the benefits of introducing these services can be both immediate and long-term,
- obsolete – services withdrawn from the comprehensive offer.

In addition, due to the degree of necessity to carry out a transport task, the services provided for refrigerated containers can be divided into:

- basic – necessary to carrying out the transport task (e.g., transportation, handling, storage, power supply for container refrigeration unit, etc.),
- additional – desired and recommended for carrying out a transport task, including a wide range of logistics services and activities performed at the customer's request.

Not all services constituting a comprehensive service can bring direct benefits and profits to the service provider. Some services provision may be enforced by law or market environment, while other may help to build the company positive image. Nevertheless, offered services should be significant for the provider's clients and allow to keep long-term cooperation with them. It should be mentioned that the provision of individual services, including handling, storage, informing the customer about the stage of transport process, may require, among others, the access to special infrastructure (e.g., cold stores), equipment (e.g., gantries, reachstackers), and IT systems (e.g., computers with specific software enabling data processing and storage). Moreover, the qualified staff is necessary to perform the services within the individual links of transport chain (e.g., repair and service of refrigerated container, customs control, cargo advertising, insurance, etc.). Therefore, services providers should carefully analyse the technical, organizational and economical possibilities to provide the services.

The knowledge about possible criteria used to group the services may be helpful for decision-makers while selecting and assessing services shaping comprehensive service offerings and company service strategy development.

5 Conclusions

Shaping of refrigerated containers comprehensive service within intermodal transport chains needs specific approach. Perishable products specificity should be considered while selecting the services and the way to perform them.

Based on conducted available literature review and own observations, the services constituting refrigerated containers comprehensive service were analysed and 20 selected criteria used to group these services were identified. The research results may be used in practice, giving the services providers an overview of possible services and criteria used to group them, while making decision about their launching to the refrigerated containers comprehensive service offer. It may be highlighted that this offer is influenced by market environment and customers' needs, as well as possibilities to perform the services by service provider.

Our future research will focus on the modelling of comprehensive services offer for perishable cargo carried out in intermodal transport chains. The research results may be of interest to transport, forwarding and logistics companies involved in planning and implementing supply of perishable food products in refrigerated containers.

References

1. Rusdiansyah, A., Adetio, I.R., Dewi, R.S.: The development of the refrigerated-container loading problem model for perishable fishery products considering internal temperature distribution. Int. J. Syst. Sci. Oper. Logist. **10**(1), 2051092 (2023)
2. Lin, X., Negenborn, R.R., Lodewijks, G.: Towards quality-aware control of perishable goods in synchromodal transport networks. IFAC-PapersOnLine **49**(16), 132–137 (2016)
3. Solak Fişkin, C., Akgül, E.F., Deveci, D.A.: New service development process in intermodal transport: the case of Turkey. In: Proceedings of XIII International Logistics & Supply Chain Congress Proceedings "Maritime Logistics: The New Port Projects of Turkey", Izmir, Turkey, 22–23 October 2015, pp. 759–769 (2015)
4. Szczepański, E., Jacyna, M., Jachimowski, R., Vašek, R., Nehring, K.: Decision support for the intermodal terminal layout designing. Arch. Civ. Eng. **67**(2), 611–630 (2021)
5. James, S.J., James, C., Evans, J.A.: Modelling of food transportation systems – a review. Int. J. Refrig. **29**(6), 947–957 (2006)
6. Postan, M., Filina-Dawidowicz, L.: Dynamic optimization model for planning of supply, production, and transportation of perishable product. In: Suchanek, M. (ed.) Sustainable Transport Development, Innovation and Technology. SPBE, pp. 235–244. Springer, Cham (2017). https://doi.org/10.1007/978-3-319-51427-7_20
7. Filina-Dawidowicz, L., Kostrzewski, M.: The complexity of logistics services at transshipment terminals. Energies **15**(4), 1435 (2022)
8. Fan, Y., Behdani, B., Bloemhof-Ruwaard, J.M.: Refer logistics and cold chain transport: a systematic review and multi-actor system analysis of an un-explored domain. Eur. J. Transp. Infrastruct. Res. **20**(2), 1–35 (2020)
9. Filina-Dawidowicz, L.: Wspomaganie podejmowania decyzji w zakresie kompleksowej obsługi kontenerów chłodniczych w zintegrowanych łańcuchach transportowych. Oficyna Wydawnicza Politechniki Warszawskiej, Warszawa (2018). (in Polish)
10. Filina-Dawidowicz, L., Gajewska, T.: Examination of importance and range of comprehensive service for refrigerated containers in seaports. Int. J. Appl. Manag. Sci. **10**(1), 26–43 (2018)
11. Haass, R., Dittmer, P., Veigt, M., Lütjen, M.: Reducing food losses and carbon emission by using autonomous control - a simulation study of the intelligent container. Int. J. Prod. Econ. **164**, 400–408 (2015)
12. Renko, S., Petljak, K., Naletina, D.: Food integrity throughout the chain: the case of good distribution practice. LogForum **15**(1), 53–69 (2019)
13. Manzini, R., Accorsi, R.: The new conceptual framework for food supply chain assessment. J. Food Eng. **115**(2), 251–263 (2013)
14. Lu, J., Gong, X., Wang, L.: An empirical study of container terminal's service attributes. J. Serv. Sci. Manag. **4**, 97–109 (2011)
15. Cai, X., Chen, J., Xiao, Y., Xu, X., Yu, G.: Fresh-product supply chain management with logistics outsourcing. Omega **41**(4), 752–765 (2013)
16. Shekar, A.: An Innovative Model of Service Development: a process guide for service managers. Innov. J. Public Sect. Innov. J. **12**(1), 4 (2007)
17. Abad, E., et al.: RFID smart tag for traceability and cold chain monitoring of foods: demonstration in an intercontinental fresh fish logistic chain. J. Food Eng. **93**(4), 394–399 (2009)
18. Manzoor, S., et al.: Carbon dots applications for development of sustainable technologies for food safety: a comprehensive review. Appl. Food Res. **3**(1), 100263 (2023)
19. Alam, I., Perry, C.: A customer-oriented new service development process. J. Serv. Mark. **16**(6), 515–534 (2002)

20. Yu, M., Nagurney, A.: Competitive food supply chain networks with application to fresh produce. Eur. J. Oper. Res. **224**(2), 273–282 (2013)
21. Rais, M., Kaul, A., Jain, A.: A comparison of co-regulation policies in the perishable food supply chain of India and China: an empirical approach. Int. J. Agric. Innov. Technol. Globalisation **1**(3), 285–303 (2020)
22. Leleń, P., Wasiak, M.: The model of selecting multimodal technologies for the transport of perishable products. Arch. Transp. **50**(2), 17–33 (2019)
23. Akkerman, R., Farahani, P., Grunow, M.: Quality, safety and sustainability in food distribution: a review of quantitative operations management approaches and challenges. OR Spectrum **32**(4), 863–904 (2010)
24. Aung, M.M., Chang, Y.S.: Traceability in a food supply chain: safety and quality perspectives. Food Control **39**(1), 172–184 (2014)
25. Zhao, Z., Li, X., Zhou, X.: Optimization of transportation routing problem for fresh food in time-varying road network: considering both food safety reliability and temperature control. ONE Open **15**(7), e0235950 (2020)
26. Freile, A.J., Mula, J., Campuzano-Bolarin, F.: Integrating inventory and transport capacity planning in a food supply chain. Int. J. Simul. Model. **19**(3), 434–445 (2020)

Digitalization of the Logistics Sector: The Case of Lithuania

Ieva Meidutė-Kavaliauskienė(✉), Urtė Antanaitytė, and Renata Činčikaitė

Business Management Faculty, Vilnius Gediminas Technical University (VILNIUS TECH), Vilnius, Lithuania

{ieva.meidute-kavaliauskiene,renata.cincikaite}@vilniustech.lt

Abstract. The logistics industry is an integral part of any country's economy and plays a major role in the movement of goods and services. In recent years, the transport and logistics sector has faced new challenges and technologies designed to facilitate and accelerate the movement of goods and products through supply chains. The "Industry 4.0" solutions in transport and logistics are aimed at simplifying logistics processes and managing information in real-time, providing timely data availability and accessibility to process managers throughout the supply chain. Implementing innovative technologies has become the main activity trend of current companies when many companies seek to make their operations more efficient and maintain competitiveness. Data from the Lithuanian Statistics Department show that in 2020 only 41.6% were interested in innovative solutions and their implementation in all companies in this sector. Taking into account this, the article analyzes the trends in the implementation of "Industry 4.0" technologies in the Lithuanian logistics sector.

Keywords: Logistics · Industry 4.0 · Transport sector · innovation

1 Digitalization 9n Logistics: The Concept of Industry 4.0

"Industry 4.0" is inseparable from logistics, as shown by the trends in logistics in recent years. According to Delfmann et al. (2018), logistics is a crucial driver of the fourth industrial revolution and will be a key industrial competitive factor in digitization [1].

Digitization is one of the most critical factors that help businesses succeed and help them achieve a competitive advantage [2, 3]. Innovative logistics technologies help solve efficiency and time reduction problems, thus increasing profit margins [4–6].

Logistics processes have been constantly changing and improving, driven by the need for fast and secure movement of goods. According to Jucys et al. (2022), logistics has changed from local to global, tactical to strategic, and invisible to decisive. Innovations in logistics help process a large amount of information, improve the integration of logistics processes, and promote the automation of processes in logistics and the entire supply chain [7]. Kahn (2018) defines supply chain innovation as a change (incremental or radical) in the supply chain network, supply chain technology, or supply chain processes (or combinations thereof) [8]. According to Griškauskas (2022), the

O. Prentkovskis et al. (Eds.): TRANSBALTICA 2023, LNITI, pp. 261–271, 2024.
https://doi.org/10.1007/978-3-031-52652-7_26

innovations implemented in logistics help to more efficiently manage the submission of orders, movement of goods, storage, invoicing, and other areas. The author singles out logistics activities in which innovations are created: transportation, company processes, storage, and service provision. Companies use various logistics 4.0 digitization tools and products offered on the market to remain competitive [9].

The term "Logistics 4.0" was first mentioned in 2011, which emerged from "Industry 4.0" because many new technologies apply to logistics activities, from processing market requirements and production planning to delivering intelligent products to end users. Radivojevič & Milosavljevič (2019) state that "Logistics 4.0" uses the latest innovations of the fourth industrial revolution [10]. According to Glistau and Machado (2018), Gökalp et al. (2017), the emergence of "Logistics 4.0" was motivated by the need to apply new technologies in this field, and each information and communication technology creates new solutions in specific parts of logistics [11, 12]. Szymanska (2017) points out that "Logistics 4.0" combines process and technical tools and technologies that support internal processes in supply chains [13]. Similarly, Bag et al. (2020) describe "Logistics 4.0", indicating that the three main characteristics of "Logistics 4.0" are vertical integration (integration of information technologies at various levels within the factory), horizontal integration (cooperation between corporations), end-to-end engineering integration (linking stakeholders, products, and machines together) [14]. Therefore, "Logistics 4.0" includes systematic planning and control of the movement of physical goods and the related information flow throughout the product life cycle.

Kayikci (2018) describes the features of digitization of the logistics system that help better understand the concept of "Logistics 4.0" [15]:

- Cooperation – digitization allows the creation of virtual logistics associations (clusters) where companies exchange data and information.
- Connectivity – Digitization enables horizontal and vertical integration in supply chains and the availability of information across individual chains.
- Adaptability – the flexibility of the system for connecting digital resources that allows you to respond to market changes.
- Integration – In the digital world, the integration of logistics systems is physically or functionally connecting different computer systems and software to ensure the coordination of logistics flows.
- Automation – Innovative technologies used in logistics systems can communicate with each other and make independent decisions based on their own environmental data processing.
- Cognition – The application of devices and systems to automate tasks that require human skills, knowledge, perception, and cognition (planning, reasoning, and learning).

Digitization of logistics processes will optimize processes, simplify workflow, and shorten delivery times. According to Torbacki and Kijewska (2019) and Parhi et al. (2022), the main features of the new approach to logistics are the extensive use of the Internet of Things, continuous transparency and visibility of the entire supply chain for all participants in the process, the ability to verify the consistency of the supply chain using blockchain technology and the ability to optimize the supply chain and suppliers dynamically [16, 17].

According to Barreto et al. (2017), an effective approach to “Logistics 4.0” should include decisions related to planning logistics processes, supporting warehouse operation and placement of goods in warehouses, managing a smart transport system and managing secure information flows [18].

2 Research Methodology

The secondary data analysis includes statistical data on the official websites of the Lithuanian Statistics Department, the European Parliament, and Eurostat. A descriptive and comparative analysis of statistical indicators is carried out, which are selected according to the relevance of implementing innovations and using “Industry 4.0” technologies in the Lithuanian logistics sector (see Table 1).

Table 1. Statistical indicators and applied analysis method.

Index	Methods of analysis
Cumulative innovation index	Descriptive analysis of the indicator
Total public funding for research and experimental development	Descriptive analysis of the indicator
Companies in the transport and storage sector have introduced innovations	Comparative analysis of indicators of the Lithuanian transport and storage sector and all sectors of economic activity in Lithuania
Transport and storage companies have implemented innovations, according to the number of employees	Comparative analysis of indicators of the transport and storage sector by company size
Costs of Lithuanian economic activity sectors for innovative activities	Comparative analysis of indicators of the Lithuanian transport and storage sector and all sectors of economic activity in Lithuania
Companies in the transport and storage sector that received financial support for innovative activities	Comparative analysis of indicators of Lithuanian companies in the transport and storage sector that received financial support, according to the source of funding
Use of Industry 4.0 technologies in the transport and storage sector	Comparative analysis of indicators of the Lithuanian transport and storage sector, companies that have implemented Industry 4.0 technology, according to different technologies, compared to the EU transport and storage sector

Source: compiled by the authors.

Statistical indicators will help review the innovative environment in Lithuania and the transport and storage sector, interpret the results of the survey and compare them, and evaluate the possibilities of implementing technologies “Industry 4.0”.

3 The Results of the Statistical Data Analysis of the Digitalization of Lithuanian Logistics Companies and the Environment of Industry 4.0 Technology and Innovation Implementation

First, the cumulative innovation index is evaluated, which shows the results of innovation activity in the countries of the European Union. Figure 1 shows Lithuania's relative EU innovation value score, that is, a composite indicator obtained by taking a weighted average of the EU-27 countries.

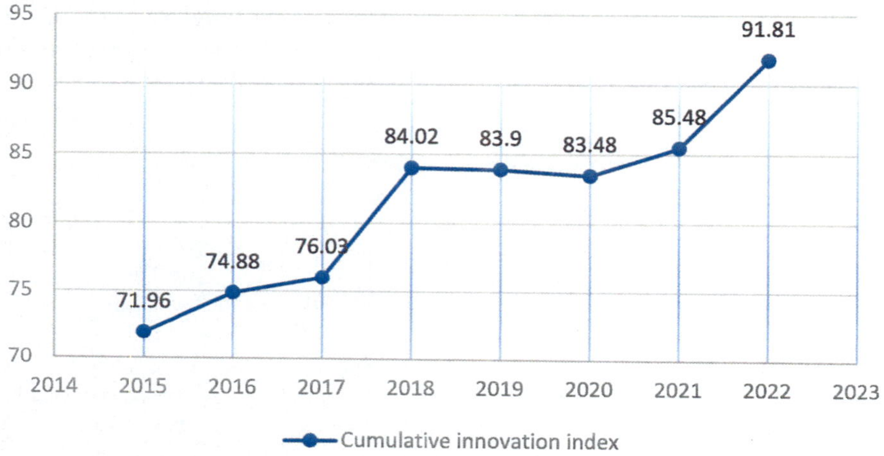

Fig. 1. Lithuanian cumulative innovation index (2015–2022). Source: compiled by the authors, based on the European Commission (2022) [19].

According to the estimates established by the European Commission, Lithuania is classified as an average innovator, the countries whose results fall between 70% and 100% of the EU average. Looking at the dynamics of innovation in Lithuania, it can be observed that from 2015 to 2022, Lithuania's total innovation index rose by 19.85% and was not lower than 70%, which means that Lithuania holds the position of an average innovator. The smallest change was observed in 2019–2020. The indicator almost did not change and decreased by 0.42%. From 2021 to 2022, the percentages rose the most, even by 6.33; it can be expected that Lithuania will reach 100% in a few years and rise to the position of a decisive innovator. According to data from the European Commission (2022), in 2022, compared to 2015, the productivity of research systems, innovators, interfaces, and intellectual property increased the most. The digitization and environmental sustainability indicators remained stable, while the human resources indicator decreased.

Another important indicator for developing innovation at the national level is state investment in research and experimental development (R&D).

Analyzing the diagram (see Fig. 2), there is a noticeable growth dynamic, which means Lithuania is steadily increasing R&D investments by about 50 million euros per year. Thus, the state supports and encourages interest in innovation at the national

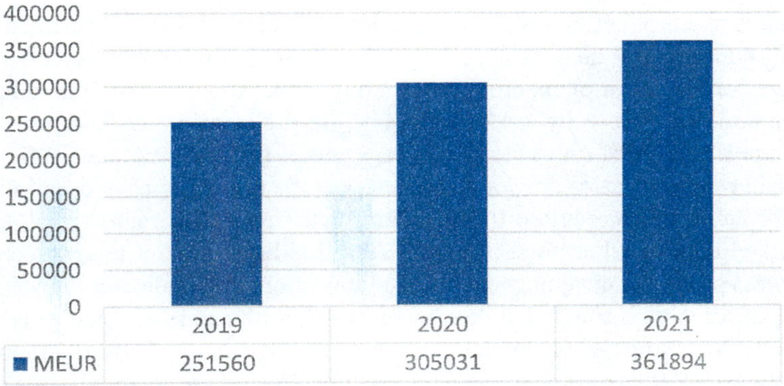

Fig. 2. Total state funding for R&D (2019–2021). Source: compiled by the authors, based on the Lithuanian Statistics Department (2023) [20].

level. According to data from the Lithuanian Statistics Department (2023), Lithuania 2021 invested 6.119 million euros in R&D for transport, communications, and other infrastructure activities, which is only 1.7% of all investments. It can be assumed that the scientific activity of the transport sector is not one of the areas the state supports the most. Thus, the state invests a lot in research and experimental development, which contributes to introducing innovations at the enterprise level. Therefore, it is important to analyze trends in innovation implementation at the level of Lithuanian industrial sectors. The transport and storage sector is most relevant to this study.

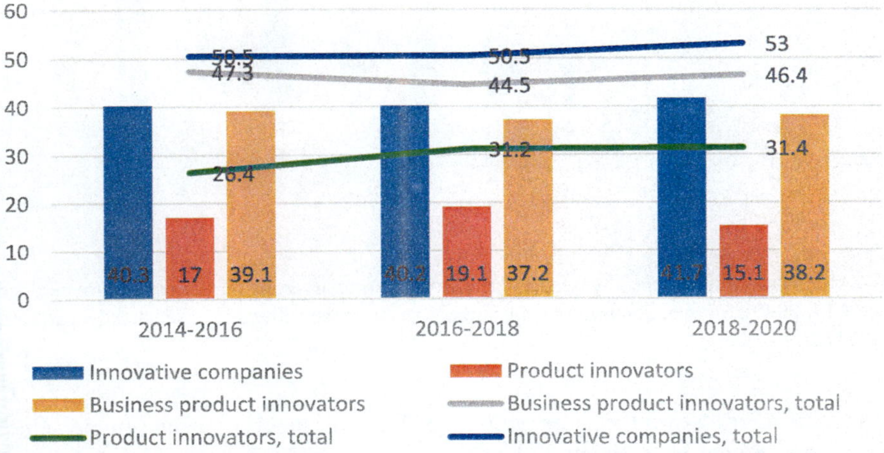

Fig. 3. Transport and storage companies implemented innovations (2014–2020). Source: compiled by the authors, based on the Lithuanian Statistics Department (2023) [20].

Figure 3 shows that the most innovative companies in the transport and storage sector are the most innovative, and all Lithuanian companies are the most innovative. The fewest companies are product innovators, presumably because inventing and bringing a new, innovative product to market is complex and requires a lot of cost and human resources. Additionally, there is a noticeable increase in the percentage of transport and storage companies implementing innovations. From 2014 to 2020, the indicator of innovative companies rose by 1.4%. However, the indicator of business process and product innovators in the transport and storage sector fell by 0.9% and 1.9%, respectively. Changes in indicators are not significant, but the drop may have been influenced by the substantial funds required for innovative activities and human capital. Looking at the general data of all sectors, it can be seen that in all areas - innovative companies, business process innovators, and product innovators - the indicator increased from 2018 to 2020. It can be concluded that, increasingly, Lithuanian companies are taking innovative actions, improving existing products and services, and offering innovations to the market. However, companies in the transport and storage sector are less and less enhancing existing products and business processes.

Figure 4 shows the innovative performance of companies in the transport and storage sector by number of employees.

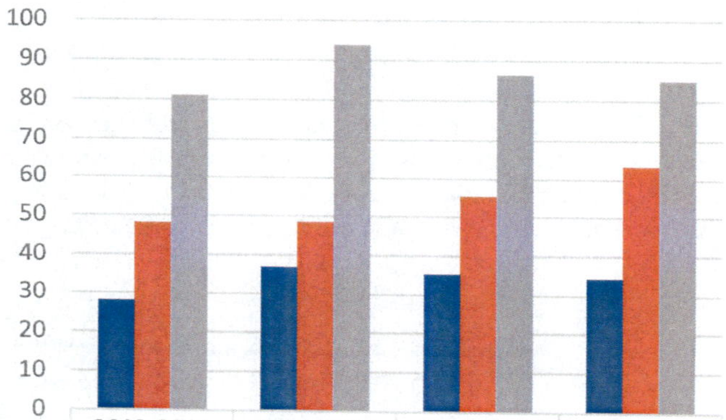

	2012-2014	2014-2016	2016-2018	2018-2020
10-49 employees	28	37	35.4	34.4
50-249 employees	48.1	48.5	55.3	63.5
250 and more emploees	81	93.8	86.4	85

Fig. 4. Innovations introduced by transport and storage companies, by number of employees, percent (2012–2020). Source: compiled by the author, based on the Lithuanian Statistics Department (2023) [20].

As seen in Fig. 4, over the entire period (2012–2020), small companies implemented the least innovations, from 28% to 37% of all companies of that size. Medium-sized companies implemented innovations dynamically and mostly implemented innovations in 2018–2020, 63.5% of all medium-sized companies. The diagram clearly shows that large organizations implemented the most significant number of innovations. The percentage of large companies that implemented innovations ranged from 81% to 93.8 percent; almost all large companies use innovations. Thus, it can be said that small transport and storage companies implement the least innovations, medium-sized companies on average, and large companies implement the most innovations.

Next, it is important to analyze the costs of companies allocated to innovative activities (see Fig. 5).

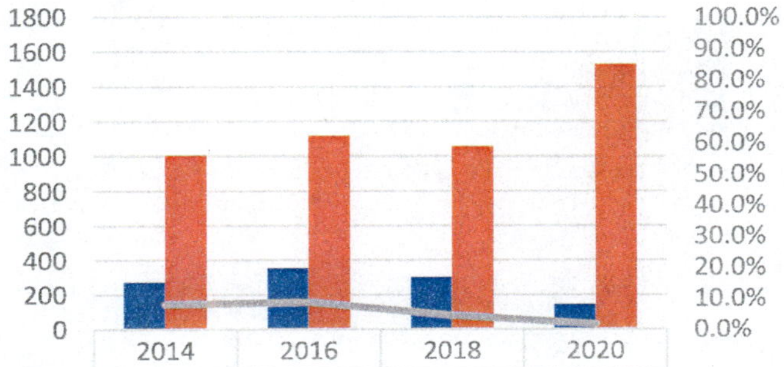

	2014	2016	2018	2020
Transport and Storage	272.9	352.3	300.6	144.5
Companies of all sectors, total	1005	1117.1	1055.7	1528.7
The share of transport and storage costs for innovative activities in circulation	8.0%	8.7%	4.5%	1.7%

Fig. 5. Companies' expenses for innovative activities, mln. EUR (2014–2020). Source: compiled by the authors, based on the Lithuanian Statistics Department (2023) [20].

The diagram shows that Lithuanian companies are investing increasingly in innovative activities; from 2014 to 2020, spending on innovative activities increased by EUR 523.7 million also reflected in the dynamics of companies engaged in innovative activities. As the indicator of inventive activity grows, the costs allocated to it also increase. However, there is a visible decline in the innovation rate of spending in the transport and storage sector. From 2014 to 2020, costs allocated to innovative activities decreased by up to EUR 128.4 million; from 2018 to 2020, they increased more than halved to EUR 156.1 million. A significant decrease in the indicator can also be seen in the share of turnover for innovative activities: from 2014 to 2020, it fell by up to 6.3%, and in 2020, the transport and storage sector devoted only 1.7% of turnover to innovative activities. The COVID-19 pandemic probably influenced the drop in these indicators.

Since the indicators of the investment of own funds in innovative activities by companies in the transport and storage sector are falling, it is important to analyze the activities of companies that used financial support for innovative activities.

Figure 6 shows a significant decline from 2016–2018 to 2018–2020; Support from all sources of funds decreased: municipalities, the state of Lithuania, EU support, and indicators of all financial support in the European Union. The visible decline is half or more. The only rising indicator in this period is the funds of the EU support program for R&D - the visible increase is 0.4%. Well, more and more transport and storage companies

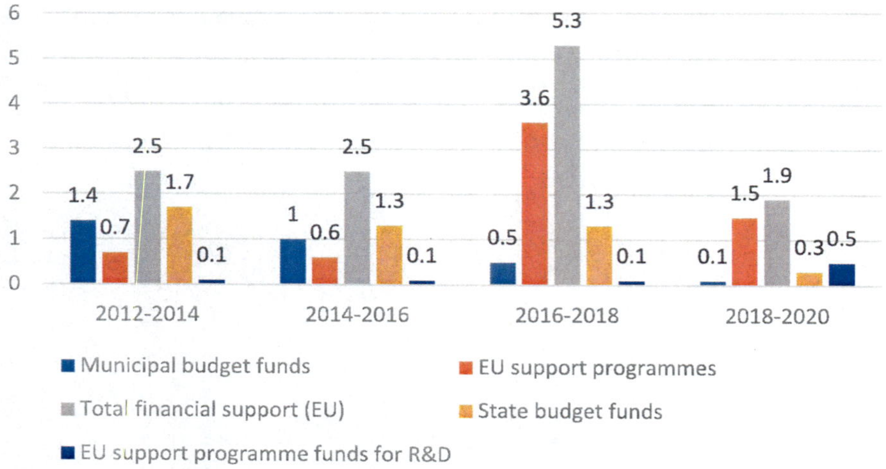

Fig. 6. Transport and storage companies that received financial support for innovative activities, percent (2012–2020). Source: compiled by the authors, based on the Lithuanian Statistics Department (2023) [20].

are receiving financial support. Possible reasons for this could be the lack of knowledge of companies about possible forms of financing, EU support programs, budget cuts at the state or municipal level, greater allocation of support funds to companies in other sectors, as well as the effects of COVID-19 when high costs were incurred for healthcare.

According to data from the Lithuanian Department of Statistics (2020), as the main factors hindering innovation implementation, companies in the transport and storage sector indicated a lack of funds for innovation implementation (19.8%), which is reflected in the turnover indicator for innovative activities. Equally significant obstacles are too much competition in the company's market (19.1%), lack of qualified personnel in the company (17.5%), and too high an innovation price (16.5%). All these reasons that hinder innovation implementation are also reflected in the literature analysis, where the problems of implementing "Industry 4.0" technologies raised by the authors coincide with the answers provided by the Lithuanian transport and storage sector.

According to data from the Lithuanian Department of Statistics (2020), companies in the transport and storage sector implemented innovations to improve existing products (18%), maintain high-quality products or services (17.6%), and satisfy the needs of existing customers (16.2%). The least relevant reasons for introducing innovations are price reduction (4.9%) and product range (5.3%). Thus, when introducing innovations, companies are more likely to improve their current operations, products, or services to retain existing customers rather than to introduce new products or services or reduce costs by reducing prices.

After analyzing the indicators of innovation activity in the transport and storage sector, it can be stated that the number of companies engaged in innovation in the transport and storage sector is decreasing, companies are devoting less and fewer turnover funds to innovative activities, and the number of companies using financial support is also decreasing.

Next, analyzing the use of Industry 4.0 technologies highlighted in the literature analysis in the transport and storage sector is essential (Fig. 7).

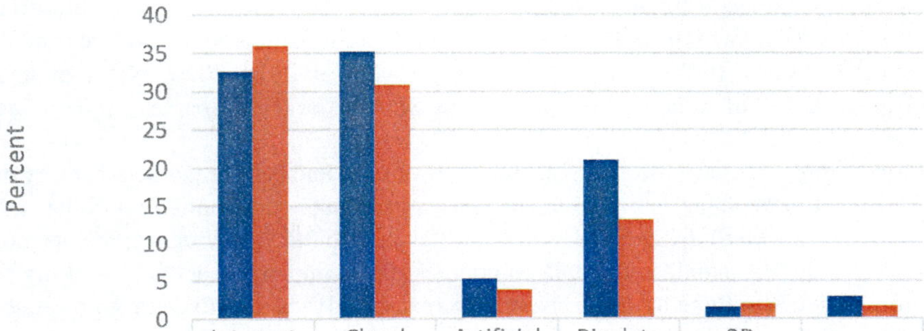

	Internet of Things (2021)	Cloud Computi ng (2021)	Artificial intelligen ce (2021)	Big data analysis (2020)	3D printing (2020)	Robotics (2020)
EU (27 states)	32.6	35.2	5.3	21	1.5	2.9
Lithuania	35.9	30.8	3.8	13.04	1.9	1.6

Fig. 7. Use of "Industry 4.0" technologies in the transport and storage sector, percent. Source: Compiled by authors based on European Commission (2023) [21].

As can be seen in the diagram, Lithuania exceeds the EU average in the number of companies using the Internet of Things and 3D printing in the transport and storage sector. Big data analytics has the most significant gap; Lithuania lags behind the EU average by 7.96%. Among Lithuanian companies in this sector, the Internet of Things is the most widespread; about 35.9% of companies use this technology. The use of cloud computing is not far behind - 30.8% of companies purchase these services. The least used technologies in Lithuanian transport and storage companies are robots - only 1.6%, 3D printing - 1.9%, and artificial intelligence - 3.8%.

The Internet of Things can be perfectly applied in transportation and storage, so it is not surprising that companies use this technology the most. Since the Internet of Things allows the collection and analysis of various data, cloud computing is required for their accumulation. Hence, the rate of this technology is not significantly behind that of the Internet of Things. One of the most cost-intensive "Industry 4.0" technologies is robots powered by artificial intelligence. It can be assumed that companies use it, at least due to this technology's high implementation and maintenance costs.

4 Conclusions

The statistical data analysis revealed that Lithuania invests more and more in R&D annually and provides companies with various financial support for innovative activities. Still, only 2.4% of companies in the transport and storage sector receive financial support. Furthermore, transport and storage companies that implemented innovations in 2020 were the least of all industries, 41.6%, and the share of company funds allocated to innovative activities from turnover in 2020 was only 1.7%. Thus, transport and storage sector companies implement innovations slowly, do not invest enough in creative activities, and do not take advantage of financial support opportunities. In addition, a significant difference was found between the number of companies that implemented innovations and the size of the company. In the transport and storage sector in 2018–2020, 34.4% of small companies, 63.5% of medium companies, and 85% of large companies implemented innovations.

Large companies have more funds, human resources, and better readiness to carry out innovative activities. This was reflected in the comparison of questionnaire survey results by company size. Medium and large companies evaluated the following statements more positively than very small and small companies: the company is actively looking for opportunities to introduce innovations; the company analyzes data to determine suitable areas; the company analyzes the impact of implemented innovations; the company invests in R&D; the company provides digital resources to employees; the company has a particular team or department responsible for innovation; the company uses data analysis, artificial intelligence, the Internet of Things, and manages cyber security related to these technologies.

Thus, very small and small enterprises are not as strongly prepared for technological changes; they lack resources and knowledge about existing financing support and innovation opportunities. Meanwhile, it is easier for large companies to embrace and take advantage of the changes in "Industry 4.0".

References

1. Delfmann, W., Ten Hompel, M., Kersten, W., Schmidt, T., Stölzle, W.: Logistics as a science: central research questions in the era of the fourth industrial revolution. Logist. Res. **11**(9), 1–13 (2018). https://doi.org/10.23773/2018_9
2. Philipp, R.: Digital readiness index assessment towards smart port development. Sustain. Manag. Forum: Nachhaltigkeits Manag. Forum **28**, 49–60 (2020). https://doi.org/10.1007/s00550-020-00501-5
3. Liu, S., Zhang, Y., Liu, Y., Wang, L., Wang, X.V.: An 'Internet of Things' enabled dynamic optimization method for smart vehicles and logistics tasks. J. Clean. Prod. **2019**(215), 806–820 (2019)
4. Raji, I.O., Shevtshenko, E., Rossi, T., Strozzi, F.: Modelling the relationship of digital technologies with lean and agile strategies. Supply Chain Forum Int. J. **22**(4), 323–346 (2021). https://doi.org/10.1080/16258312.2021.1925583
5. Sun, X., Yu, H., Solvang, W.D., Wang, Y., Wang, K.: The application of Industry 4.0 technologies in sustainable logistics: a systematic literature review (2012–2020) to explore future research opportunities. Environ. Sci. Pollut. Res. **29**, 9560–9591 (2022). https://doi.org/10.1007/s11356-021-17693-y
6. De Souza, E.D., Kerber, J.C., Bouzon, M., Rodriguez, C.M.T.: Performance evaluation of green logistics: Paving the way towards circular economy. Clean. Log. Supply Chain **3** (2022). https://doi.org/10.1016/j.clscn.2021.100019
7. Jucys, A., Budrys, T., Kreišmonienė, J.: Inovacijos logistikos versle. Verslas, Technologijos, Biomedicina: Inovacijų Įžvalgos 2022: Straipsnių Rinkinys **1**, 174–181 (2022)
8. Kahn, K.B.: Understanding innovation. J. Bus. Horiz. **61**(3), 453–460 (2018). https://doi.org/10.1016/j.bushor.2018.01.011

9. Griškauskas, R.: The impact of innovation on transport processes. In: Conference "Young Scientist 2022", pp. 91–94 (2022)
10. Radivojević, G., Milosavljević, L.: The concept of logistics 4.0. Logic.Sf.Bg.Ac.Rs. (2019). https://logic.sf.bg.ac.rs/wp-content/uploads/LOGIC_2019_ID_32.pdf
11. Glistau, E., Machado, N.I.C.: Logistics 4.0 and the revalidation of logistics concepts and strategies (2018). https://doi.org/10.26649/MUSCI.2018.023
12. Gökalp, E., Şener, U., Eren, P.E.: Development of an assessment model for industry 4.0: industry 4.0-MM. In: Mas, A., Mesquida, A., O'Connor, R.V., Rout, T., Dorling, A. (eds.) SPICE 2017. CCIS, vol. 770, pp. 128–142. Springer, Cham (2017). https://doi.org/10.1007/978-3-319-67383-7_10
13. Szymańska, O., Adamczak, M., Cyplik, P., Szymańska, O., Adamczak, M., Cyplik, P.: Logistics 4.0 - a new paradigm or set of known solutions? Res. Logist. Prod. **7**(4), 299–310 (2017). https://doi.org/10.21008/J.2083-4950.2017.7.4.2
14. Bag, S., Gupta, S., Luo, Z.: Examining the role of logistics 4.0 enabled dynamic capabilities on firm performance. Int. J. Logist. Manag. **31**(3), 607–628 (2020). https://doi.org/10.1108/IJLM-11-2019-0311/FULL/XML
15. Kayikci, Y.: Sustainability impact of digitization in logistics. Procedia Manuf. **21**, 782–789 (2018). https://doi.org/10.1016/J.PROMFG.2018.02.184
16. Torbacki, W., Kijewska, K.: Identifying Key Performance Indicators to be used in Logistics 4.0 and Industry 4.0 for the needs of sustainable municipal logistics by means of the DEMATEL method. Transp. Res. Procedia **39**, 534–543 (2019). https://doi.org/10.1016/J.TRPRO.2019.06.055
17. Parhi, S., Joshi, K., Gunasekaran, A., Sethuraman, K.: Reflecting on an empirical study of the digitalization initiatives for sustainability on logistics: the concept of sustainable logistics 4.0. Clean. Logist. Supply Chain **4** (2022). https://doi.org/10.1016/j.clscn.2022.100058
18. Barreto, L., Amaral, A., Pereira, T.: Industry 4.0 implications in logistics: an overview. Procedia Manuf. **13**, 1245–1252 (2017)
19. EC Homepage (2022). https://research-and-innovation.ec.europa.eu/statistics/performance-indicators/european-innovation-scoreboard_en
20. Lithuanian Statistics Department (2023). https://osp.stat.gov.lt
21. EC Homepage (2023). https://ec.europa.eu/futurium/en/system/files/ged/lithuania_211117.pdf

The Use of Fault Tree Analysis to Create Adverse Event Scenarios for the Purposes of Cargo Zone Crew Training at the Airport

Agnieszka A. Tubis(✉), Honorata Poturaj, Ewa Mardeusz, and Tomasz Kisiel

Department of Technical Systems Operation and Maintenance, Wroclaw University of Science and Technology, 50-370 Wroclaw, Poland
{agnieszka.tubis,honorata.poturaj,ewa.mardeusz, tomasz.kisiel}@pwr.edu.pl

Abstract. Virtual reality technology is now being used in many industries. It also plays an increasingly important role in training systems, which are also related to transportation in the broadest sense. The effectiveness of the use of Virtual reality technology in training processes is possible due to several factors relating to exposure to realistic scenarios, the ability to hone practical skills in a more engaging and immersive way than traditional methods, and the individualization of the training process. One of the elements of the training process is the trainee's exposure to adverse events. This article aims to present the results of a risk analysis to develop training scenarios for cargo handlers on the airside. The study used The Swiss Cheese Model and Fault Tree Analysis. Based on a 5-step procedure process that included analysis of the current process, identification of adverse events, identification of causes of adverse events, logical cause investigation, and brainstorming, results were obtained to develop training scenarios that use an immersive virtual reality system. Conclusions from the implemented research provide a theoretical basis for developing training scenarios for personnel responsible for cargo handling in the warehouse and apron area.

Keywords: Risk assessment · Virtual reality · Training scenario · FTA

1 Introduction

Virtual reality (VR) technology has been known and used for decades. In the initial development phase, VR technology was used primarily to improve those competencies for which training processes were associated with the risk of losing the health or life of training participants or very high implementation costs. The first applications of VR refer to the training of pilots, paratroopers, and military personnel [1]. However, the constant development of technology and related tools has made more and more scientists and practitioners look for opportunities to use them in their research and business areas. However, the training market is still the leading recipient of the effects of developing tools based on virtual reality. The scope of training conducted with its use is also becoming

O. Prentkovskis et al. (Eds.): TRANSBALTICA 2023, LNITI, pp. 272–286, 2024.
https://doi.org/10.1007/978-3-031-52652-7_27

richer as more and more sectors are starting to develop programs based on virtual training scenarios.

In most publications, VR technology is divided into two major types based on the level of interaction and immersive environment [2, 3]: (1) non-immersive virtual environment and (2) immersive virtual environment. In a non-immersive virtual environment, the training simulation is presented on a conventional personal computer and is typically explored using a keyboard, mouse, wand, joystick, or touchscreen. In an immersive virtual environment, the training is presented on multiple roomie screens or through a stereoscopic, head-mounted display unit [2]. For this reason, immersive VR (iVR) is defined as a highly interactive, 3-D computer-generated program in a multimedia environment that provides the effect of immersion to the users [4]. Conventional VR simulators are expensive and cumbersome, which raises concerns about availability [5, 6]. They also have limited ability to immerse users in a realistic operating environment [3]. On the other hand, immersive virtual reality (iVR) provides the advantages of conventional virtual reality and, at the same time, can be implemented on low-cost, portable, commercially available hardware [7]. iVR immersion is provided as an uninterrupted, scalable environment capable of simulating the full range of sensory stimuli and allows for 360° visual immersion and real-time manipulation of virtual objects [3, 8].

The article presents the results of research obtained as part of the project "Use of the virtual reality environment in the system for training handling employees in the handling of goods transported by air", funded by the National Center for Research and Development. One of the project's tasks is to develop training scenarios that consider the errors made by the crew handling air cargo loads. The scenarios developed in the described project will concern immersive virtual reality.

For the purposes of developing these scenarios, it is necessary to analyze the risk of adverse events in the process of handling cargo in the airport zone. Two research methods were used: a modified version of the HFACS (Human Factor Analysis and Classification) analysis and a qualitative fault tree analysis. The results of the analysis developed on this basis will make it possible to develop detailed scenarios of events that will allow the acquisition or consolidation of the appropriate behavior of the trained crew.

The article aims to present the risk analysis results to develop training scenarios for cargo handling employees in the airport zone. This goal will be achieved through the presentation and discussion of the results in the following sections of this article. The presented results' main contribution is identifying risks and the causes of their occurrence in air cargo handling operations, which should be considered in the prepared training scenarios for operators handling transport cargo. The structure of the article, along with the substantive scope, is presented in Fig. 1.

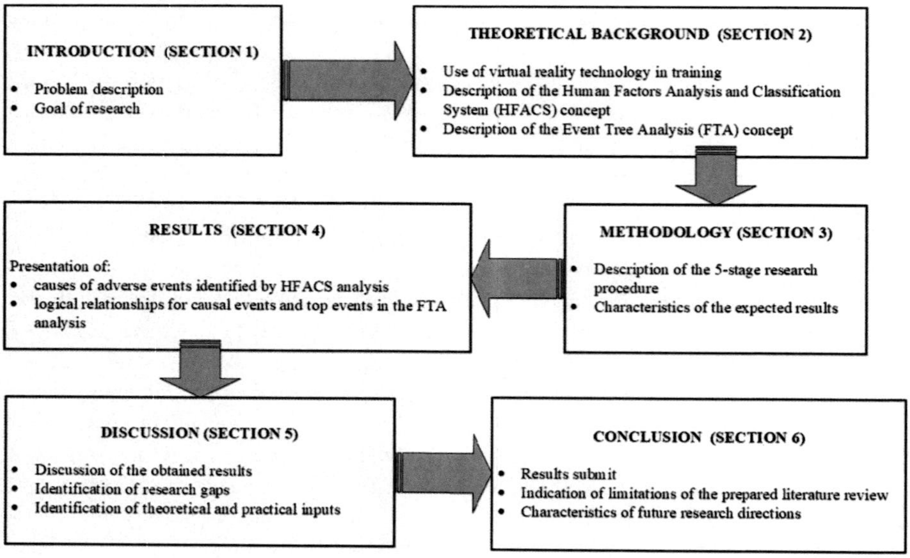

Fig. 1. The structure of the article and the objectives of individual sections.

2 Theoretical Background

2.1 Human Factors Analysis and Classification System

The Swiss cheese model is a known model for studying systemic causality of adverse events. This model explains how an accident could be regarded as the result of inter-relations between "unsafe acts' of front operators and latent conditions from a systems perspective [9]. It uses the concept of cheese slices, which symbolize four levels of protective barriers for the system. However, holes in these slices relate to unintended flaws in the mapped system in each slice. However, many researchers have emphasized that the lack of comprehensive explanations of the meaning of holes in cheese slices makes it challenging to apply the model to real systems [10]. For this reason, this model was constantly developed, and one of the concepts created on its basis is the Human Factors Analysis and Classification System (HFACS) [11].

The HFACS analysis refined the concept of "holes" in successive cheese slices, assigning them four areas of required analysis regarding the system/process under study. They are [11]: (1) Unsafe Acts, (2) Preconditions of Unsafe Acts, (3) Unsafe Supervision, and (4) Organization Influence. These categories consider the causes of human errors due to the individual behavior of employees, as well as constraints at the organizational level. Thanks to this, it is possible to research the causes of accidents at various levels of complexity [10]. The procedure assumes a detailed analysis of the occurring adverse events and assigning them to one or more categories of causes. This procedure, performed systematically, allows for identifying frequent causes of adverse events and their mutual correlation [12].

HFACS analysis is widely used in civil aviation and other domains to study human errors in accidents because of its high reliability [13]. Examples of its application are presented in [14]. The current wide application of HFACS analysis in many industries is due to its features, such as extensiveness, diagnostic functions, reliability, and usability. [12]. The characteristic distinguishing HFACS from the other accident causation model is the inclusion in the analytical procedure of the importance of management and organization in building safe systems [10]. Thanks to this, based on the results obtained, it is possible to identify various areas that can be improved by changing procedures or process organization and those that require employee training.

2.2 Fault Tree Analysis

Fault tree analysis (FTA) is a top-down deductive analysis, according to which the analyst, by way of deduction, identifies the reasons for the occurrence of a specific adverse event [15]. The analysis results are visualized as a tree that maps cause-and-effect relationships. The model uses logical gates and small events to present the path of an accident through different steps, whereby [15]:

- AND gate means that both the initial events are needed to occur in for the intermediate event to occur;
- OR gate means only one of the initial events may become the cause of the intimate event.

Ayyub, in his book [16], presented the procedure for FTA, which he described in the form of 8 steps: (1) Define the system of interest; (2) Define the top event of the system; (3) Define tree top Structure; (4) Explore each branch in successive levels of detail; (5) Solve the fault tree for the combination of events contributing to the top event; (6) Identify significant dependent failure potentials and adjust the model appropriately; (7) Perform quantitative analysis; (8) Use the results in decision-making.

FTA is one of the most popular probabilistic risk assessment methods used to assess system security and reliability [17]. For this reason, in the literature, one can find its application in various industries, e.g., marine [18], oil and gas [19], aerospace [20], and nuclear power plants [21]. Typically, FTA is applied on a quantified basis. It estimates the probability of any undesired events (top events), given the probability of the primary events, which are the lowest-level event causes. Therefore, it is necessary to quantify each primary event failure probability to measure the top event probability [17]. However, some authors use the tree model qualitatively without reliable quantitative data to map only logical cause-and-effect relationships between the causal events and the top event. Yazdi [22] recommend using the concept of fuzzy fault tree analysis without data or having unreliable information. Due to the lack of measurements.

3 Methodology

One of the tasks in the project "Use of the virtual reality environment in the system for training handling employees in the handling of goods transported by air" concerns the identification of undesirable events that may occur in the process of handling cargo loads in the area of the warehouse and airport apron. Detailed characteristics of these events are important from the point of view of prepared training scenarios. Their goal is not only to develop training preparing the operator for the performance of his future duties but also to prevent possible errors that he may make during the service. -consequence, which will allow to determine the relationships between the occurring factors and human error.

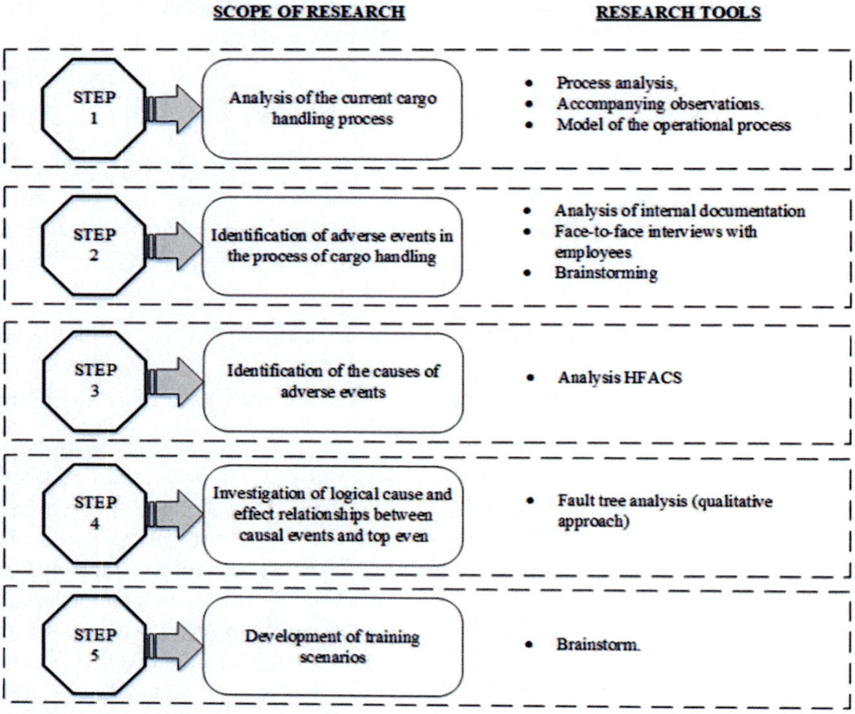

Fig. 2. Stages of the research procedure.

The research procedure includes five steps, which are presented in Fig. 2. The identification of adverse events was based on three research paths. First, the project team conducted a process analysis regarding handling various types of cargo transported by air. This analysis was prepared based on internal documents (including applicable procedures) and accompanying observations. On this basis, a process model was developed, which was used for further research. Then, the internal documentation regarding registered adverse events in the last 5 years, which took place at selected cargo airports, was analyzed. Then, direct interviews were conducted with the managerial staff and experienced employees (minimum 10 years of work experience) employed at the surveyed airports. Based on both of these studies, a catalog of adverse events that should be subject to risk analysis was developed. The last element of this stage was a discussion conducted following the brainstorming concept among team members. Its purpose was to identify potential adverse events that have not been recorded at the cargo airport so far and should be analyzed despite the low probability of occurrence.

In step 3, an HFACS analysis was performed for the identified adverse events. However, the scope of the analysis was limited to 2 slices: (1) Unsafe Acts and (2) Preconditions of Unsafe Acts. Such limitation is justified from the point of view of prepared training scenarios. Considering the high level of safety culture in force at airport service points and the limited influence of the operator on aspects related to the organization of the service system and the applied supervision mechanisms, both of these areas of analysis were excluded from the research procedure. The study assumed that the analyzed service system meets the required standards for organization and supervision. Face-to-face interviews with staff described in step 2 were used to identify the causes of adverse events, accompanying observations and a process audit. The identified causes were then classified according to the groups distinguished for this research stage.

The results of the conducted HFACS analysis were used to reflect the cause-and-effect relationships for the analyzed adverse event risks in the form of the fault tree technique. The identified root causes became the basis for conducting a qualitative risk analysis using the fault tree technique. The analysis was carried out only in the qualitative dimension due to the lack of comprehensive quantitative data on the frequency and effects of individual errors. For this reason, the event tree model was mainly used to reflect the logistic relationships between the causal events and the identified top events. For the needs of the developed scenarios, this dimension of the analysis turned out to be satisfactory.

4 Results

It is important to emphasize the validity of including both analytical tools indicated above. The HFACS analysis allowed the researchers to identify and classify the most common causes of errors made by cargo operations personnel. On the other hand, FTA analysis, made it possible to visualize the cause-and-effect relationships attributed to each of the identified adverse events. As a result of the research investigation, results were obtained to develop detailed scenarios for cargo employee training based on VR tools.

4.1 Human Factors Analysis and Classification System

The Human Factors Analysis and Classification System is presented in Table 1 (Unsafe Acts) and Table 2 (Preconditions of Unsafe Acts). In this article only part of the HFACS analysis is presented, for three adverse events:

- collision with another ground handling vehicle (1.1.1),
- forcing the right of way on another vehicle (1.3.2),
- incorrect connection of the transport set (1.6.3).

The entire analysis was carried out during the project research and consists of 54 adverse events, 17 Unsafe Acts and 24 Preconditions of Unsafe Acts.

In the group of Unsafe Acts the most common causes was:

1. operator fatigue, operator distraction, operator tiredness and stress (45 occurrences for each).

Table 1. HFACS analysis for Unsafe Acts.

Adverse Events	Symbols	Unsafe Acts
collision with another ground handling vehicle (1.1.1)	ZN_16	routine
	ZN_15	operator rush
	ZN_10	ignoring regulations
	ZN_11	ignoring signals reported by other traffic participants
	ZN_12	sudden change of direction
	ZN_3	incorrect assessment of the behavior of other road users
	ZN_6	mistimation of the distance to other objects;
	ZN_8	incorrect route selection
	ZN_2	misjudgment of speed
	ZN_4	erroneous perception of the displacement area
	ZN_1	terrorist attack
forcing the right of way on another vehicle (1.3.2)	ZN_16	routine
	ZN_15	operator rush
	ZN_10	ignoring regulations
	ZN_12	sudden change of direction
	ZN_6	mistimation of the distance to other objects;
	ZN_3	incorrect assessment of the behavior of other road users
	ZN_4	erroneous perception of the displacement area
	ZN_8	incorrect route selection
incorrect connection of the transport set (1.6.3)	ZN_16	routine
	ZN_15	operator rush
	ZN_17	operator negligence
	ZN_4	erroneous perception of the displacement area
	ZN_10	ignoring regulations

2. lack of knowledge about applicable regulations, loss of consciousness (38 occurrences for each).
3. physical indisposition of the operator (34 occurrences).

Table 2. HFACS analysis for Preconditions of Unsafe Acts.

Adverse Events	Symbols	Preconditions of Unsafe Acts
collision with another ground handling vehicle (1.1.1)	WW_22	stress
	WW_24	operator tiredness
	WW_19	operator fatigue
	WW_15	physical indisposition of the operator
	WW_20	operator distraction
	WW_23	loss of consciousness
	WW_16	operator micro-nap
	WW_18	incorrect operator vision
	WW_13	lack of knowledge about existing limitations
	WW_12	lack of knowledge about applicable regulations
	WW_14	lack of knowledge of procedures
	WW_3	no information about other operations carried out in the traffic zone
	WW_2	no driving experience
	WW_7	vehicle inoperability
	WW_21	poor visibility
	WW_1	very bad visibility
forcing the right of way on another vehicle (1.3.2)	WW_22	stress
	WW_24	operator tiredness
	WW_19	operator fatigue
	WW_20	operator distraction
	WW_23	loss of consciousness
	WW_16	operator micro-nap
	WW_18	incorrect operator vision
	WW_2	no driving experience
	WW_21	poor visibility
	WW_1	very bad visibility
	WW_12	lack of knowledge of regulations;

(*continued*)

Table 2. (*continued*)

Adverse Events	Symbols	Preconditions of Unsafe Acts
	WW_13	lack of knowledge about existing limitations
	WW_3	no information about other operations carried out in the traffic zone
	WW_7	vehicle inoperability
incorrect connection of the transport set (1.6.3)	WW_22	stress
	WW_24	operator tiredness
	WW_19	operator fatigue
	WW_12	lack of knowledge about applicable regulations
	WW_13	lack of knowledge about existing limitations
	WW_20	operator distraction
	WW_15	physical indisposition of the operator
	WW_9	inability to properly connect the transport set
	WW_8	lack of efficiency of individual transport elements
	ZN_17	operator negligence
	ZN_4	erroneous perception of the displacement area
	ZN_10	ignoring regulations

In the group of Preconditions for Unsafe Acts, the most common causes was:

1. ignoring regulations, operator rush, routine (45 occurrences for each).
2. erroneous perception of the displacement area, operator negligence (26 occurrences for each).
3. incorrect route selection (16 occurrences).

4.2 Fault Tree Analysis

As part of the work on adverse events analyzed in the project, 45 fault trees were developed. Individual trees present possible causes of occurrence of one or more adverse events. Three trees of variable complexity will be presented in the article. To prepare diagrams for fault tree analysis, elements described in Table 3 were used.

Table 3. Elements of Fault Tree diagrams.

fault tree element	description
	"OR" gate
ZN_8 WW_20	event/indirect effect (cause)
1.2.2 1.3.2	adverse event (fault)

For collision with another ground handling vehicle (1.1.1) event is presented (Fig. 3) tree of low complexity.

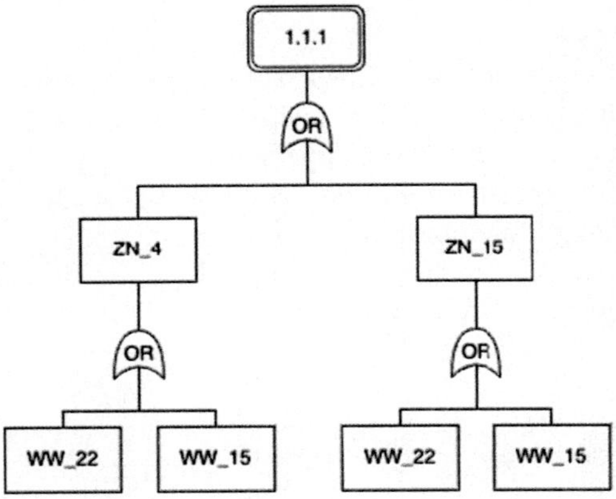

Fig. 3. Fault tree for collision with another ground handling vehicle.

For example, one cause and effect path of a fault tree should be read as follows: collision with another ground handling vehicle (1.1.1) could have occurred as a result of the mistaken perception erroneous perception of the displacement area (ZN_4), which was due to the stress (WW_22). Another possible scenario is that collision with another ground handling vehicle (1.1.1) have occurred as a result of the operator rush (ZN_15), which was result of the operator's physical indisposition (WW_15). This fault tree can also be used for events: forcing the right of way on another vehicle (1.3.2) and incorrect connection of the transport set (1.6.3), due to HFACS.

The second fault tree is of medium complexity. It was prepared for incorrectly connection of the transport set (1.6.3) event and is in Fig. 4.

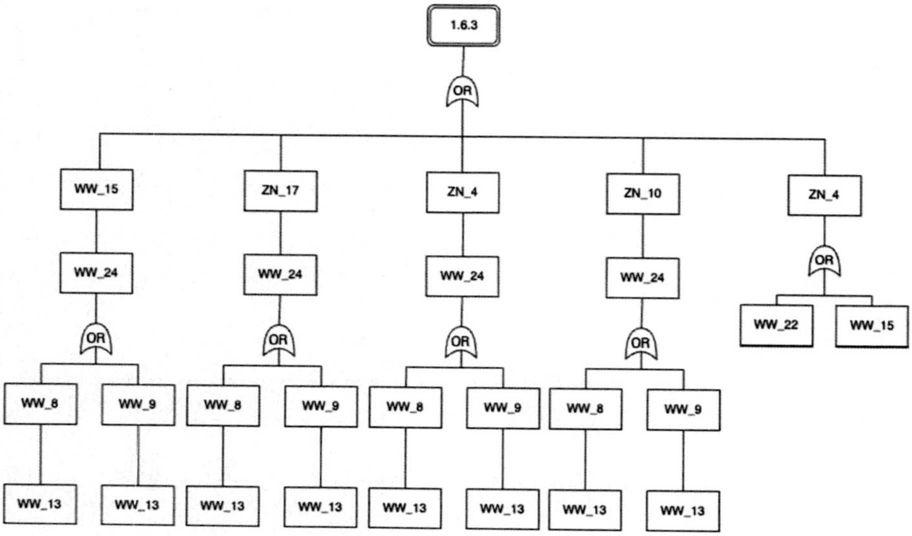

Fig. 4. Fault tree for incorrect connection of the transport set.

The fault tree presented in Fig. 4 shows that incorrect connection of the transport set may have occurred as a result of the following circumstances: physical indisposition of the operator (WW_15), resulting from operator tiredness (WW_24), resulting from inability to properly connect the transport set (WW_9) and this because of lack of knowledge about existing limitations (WW_13).

The last fault tree was prepared for forcing the right of way on another vehicle (1.3.2) and has the most complex structure, it is in Fig. 5.

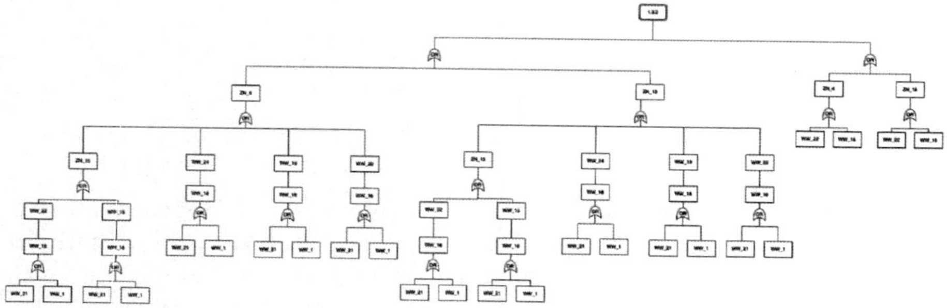

Fig. 5. Fault tree for forcing the right of way on another vehicle.

Situation when one vehicle forces the right way on another vehicle can occur when there is a lack of efficiency of individual transport elements (ZN_8). However, in a more complicated situation, the course of events may be as follows: operator ignore regulations (ZN_10), because he is tired (WW_24), fatigue (WW_19) or distracted (WW_20), the conditions of this operator are due to incorrect vision (WW_18), and this occurs due to poor (WW_21) or very poor visibility (WW_1).

At the fault trees from Fig. 4 and Fig. 5 is the part of fault tree presented in Fig. 3. This shows that all causes of each adverse event can be shown at one complex fault tree. However, the project that made it possible to carry out the described research required that the prepared fault trees be legible and applicable to the training scenarios. Therefore, the researchers decided to divide them into smaller and more readable elements.

5 Discussion

Issues related to the use of virtual reality technology are widely used in air transport. So far, scientific research is mainly concerned with air traffic control operators [23, 24]. The area of the cargo zone and the airport apron presented by the authors in the paper is an element of logistics service. The literature review showed that there need to be more training tools based on immersive virtual reality dedicated to the mentioned area of airport operations.

An essential element in the cargo zone and apron area is time. Minimizing the duration of operations means there is little room for error in logistics handling. High competence, backed by practical experience, determines the high efficiency of the implemented operations. Mistakes made by cargo zone and apron operators can cause significant financial losses. Irregularities in the area in question also generate disruptions in the operation of the entire airport. The development of training based on virtual reality technology that considers human error is crucial to the safety and organizational efficiency of the airport.

The fundamental problem of building training scenarios is that the subject of the study is events that do not always provoke the recording of negative consequences (reflecting the situation in the report) but only the occurrence of an unsafe situation. For this reason, the critical element of the research investigation conducted was face-to-face interviews with the airport's management and experienced airport personnel. The identification of most of the adverse events that were described in the training scenarios came from their knowledge and do-experience, which they collected over the years. This generates the problem of preparing a quantitative risk analysis due to the need for databases of recorded adverse events. A proposed solution to the problem presented in this article is the use of the fuzzy logic method in further stages of the research procedure.

The reasons for the events in question are situations that are not monitored by managers and may relate to fatigue, stress, and malaise. For this reason, the authors used HFACS analysis in their study. As noted in Chapter Three, due to the high safety standards of aviation, the scope of the analysis was limited to two elements Unsafe Acts and Preconditions of Unsafe Acts. The two levels of analysis indicated proved sufficient for the implementation of cargo operations. However, it is worth noting that this is due to the research objective of preparing operational scenarios. In a situation where the purpose of risk analysis would be improvement activities, the current system of HFACS analysis scope should include all levels.

The trees presented by the authors confirm the practical and theoretical added value in developing training scenarios. From a scientific point of view, the obtained results expand the knowledge regarding the causes of adverse events identified in the cargo area at the airport. On this basis, it will be possible to build simulation models to study the above phenomena. From the point of view of practice, the article provides knowledge about the causes of the adverse events that occur, which should inspire airport managers to monitor the psychophysical state of operational employees and the errors they make in cargo handling.

6 Conclusion

The use of immersive virtual reality in training systems is of interest in many industries. Its effectiveness is contingent on the development of training scenarios. A limited set of process information can reduce the effectiveness of using this technology. The detailed characterization of the process and training development to prepare the operator to perform activities requires a cause-and-effect analysis to identify human errors.

The research conducted, the conclusions formulated, and the discussion undertaken allow us to conclude that the growing interest in using VR in training systems will require the development of reliable scenarios. In identifying adverse events, there is still a lack of work enabling the application of the results obtained to broader immersive virtual reality technology applications. It is worth noting that despite this, the results developed in this work in the form of fault trees are sufficient for the preparation of training models.

An important limitation of the presented results is the lack of quantitative data to estimate the frequency of the identified causes of adverse events. This makes it impossible to identify a set of causes that, in particular, should be eliminated by properly prepared training scenarios. The analysis tools used in the described studies fulfilled their role as they allowed to formulate two sets of significant causes (two groups of HFACS) of adverse events occurring. However, for further investigation, analytical tools are needed to estimate the risk parameter of incidence. For this reason, it is planned to use fuzzy set theory in further proceedings to carry out risk assessment of occurring events. Due to the lack of historical data on the process of cargo handling in the area of the warehouse and apron, the next stage of realization of the research will be the preparation of fuzzy sets by the authors. The main goal of the effort is a set of data that allows the estimation of a quantitative measure of probability.

References

1. Howard, M.C., Gutworth, M.B., Jacobs, R.R.: A meta-analysis of virtual reality training programs. Comput. Hum. Behav. **121**, 106808 (2021). https://doi.org/10.1016/j.chb.2021.106808
2. Lee, E.-L., Wong, K.W.: A review of using virtual reality for learning. In: Pan, Z., Cheok, A.D., Müller, W., El Rhalibi, A. (eds.) Transactions on Edutainment I. LNCS, vol. 5080, pp. 231–241. Springer, Heidelberg (2008). https://doi.org/10.1007/978-3-540-69744-2_18
3. Mao, R.Q., Lan, L., Kay, J., Lohre, R., Ayeni, O.R., Goel, D.P., et al.: Immersive virtual reality for surgical training: a systematic review. J. Surg. Res. **268**, 40–58 (2021). https://doi.org/10.1016/J.JSS.2021.06.045. Accessed 05 May 2023

4. Pan, Z., Cheok, A.D., Yang, H., Zhu, J., Shi, J.: Virtual reality and mixed reality for virtual learning environments. Comput. Graph. **30**(1), 20–28 (2006). https://doi.org/10.1016/j.cag.2005.10.004
5. Bernier, G.V., Sanchez, J.E.: Surgical simulation: the value of individualization. Surg. Endosc. **30**(8), 3191–3197 (2016). https://doi.org/10.1007/s00464-016-5021-8
6. Parham, G., Bing, E.G., Cuevas, A., Fisher, B., Skinner, J., Mwanahamuntu, M., et al.: Creating a low-cost virtual reality surgical simulation to increase surgical oncology capacity and capability. Ecancermedicalscience **13** (2019). https://doi.org/10.3332/ecancer.2019.910
7. Radianti, J., Majchrzak, T.A., Fromm, J., Wohlgenannt, I.: A systematic review of immersive virtual reality applications for higher education: design elements, lessons learned, and research agenda. Comput. Educ. **147**, 103778 (2020). https://doi.org/10.1016/j.compedu.2019.103778
8. Moorthy, K., Munz, Y., Adams, S., Pandey, V., Darzi, A.: A human factors analysis of technical and team skills among surgical trainees during procedural simulations in a simulated operating theatre. Ann. Surg. **242**(5), 631–639 (2005). https://doi.org/10.1097/01.sla.0000186298.79308.a8
9. Kim, D.S., Yoon, W.C.: An accident causation model for the railway industry: application of the model to 80 rail accident investigation reports from the UK. Saf. Sci. **60**, 57–68 (2013). https://doi.org/10.1016/J.SSCI.2013.06.010. Accessed 29 June 2023
10. Zhan, Q., Zheng, W., Zhao, B.: A hybrid human and organizational analysis method for railway accidents based on HFACS-Railway Accidents (HFACS-RAs). Saf. Sci. **91**, 232–250 (2017). https://doi.org/10.1016/J.SSCI.2016.08.017. Accessed 29 June 2023
11. Shappell, S.A., Wiegmann, D.A.: Applying reason: the human factors analysis and classification system (HFACS). Hum. Fact. Aerosp. Saf. **1**, 59–86 (2001)
12. Diller, T., Helmrich, G., Dunning, S., Cox, S., Buchanan, A., Shappell, S.: The Human Factors Analysis Classification System (HFACS) Applied to Health Care. Am. J. Med. Qual. **29**(3), 181–190 (2014). https://doi.org/10.1177/1062860613491623
13. Olsen, N.S.: Coding ATC incident data using HFACS: Inter-coder consensus. Saf. Sci. **49**(10), 1365–1370 (2011). https://doi.org/10.1016/J.SSCI.2011.05.007. Accessed 29 June 2023
14. Tubis, A.A., Haładyn, S., Kierzkowski, A., Kisiel, T., Restel, F., Wolniewicz, Ł.: Analysis of human errors in the cargo logistics process in the airport zone. In: The Probabilistic Safety Assessment and Management, PSAM 2022. Honolulu (2022)
15. Baig, A.A., Ruzli, R., Buang, A.B.: Reliability analysis using fault tree analysis: a review. Int. J. Chem. Eng. Appl. **4**, 169–173 (2013). https://doi.org/10.7763/IJCEA.2013.V4.287
16. Ayyub, B.M.: Risk Analysis in Engineering and Economics. Chapman & Hall (2014)
17. Yazdi, M., Kabir, S., Walker, M.: Uncertainty handling in fault tree based risk assessment: state of the art and future perspectives. Process. Saf. Environ. Prot. **131**, 89–104 (2019). https://doi.org/10.1016/j.psep.2019.09.003
18. Acosta, H., Forrest, B.M.: The spread of marine non-indigenous species via recreational boating: a conceptual model for risk assessment based on fault tree analysis. Ecol. Modell. **220**(13–14), 1586–1598 (2009). https://doi.org/10.1016/J.ECOLMODEL.2009.03.026. Accessed 25 June 2023
19. Alkhaledi, K., Alrushaid, S., Almansouri, J., Alrashed, A.: Using fault tree analysis in the Al-Ahmadi town gas leak incidents. Saf. Sci. **79**, 184–192 (2015). https://doi.org/10.1016/J.SSCI.2015.05.015. Accessed 25 June 2023
20. Cheng, C.-Y., Li, S.-F., Chu, S.-J., Yeh, C.-Y., Simmons, R.J.: Application of fault tree analysis to assess inventory risk: a practical case from aerospace manufacturing. Int. J. Prod. Res. **51**(21), 6499–6514 (2013). https://doi.org/10.1080/00207543.2013.825744
21. Wang, J., Wang, F., Chen, S., Wang, J., Hu, L., Yin, Y., et al.: Fault-tree-based instantaneous risk computing core in nuclear power plant risk monitor. Ann. Nucl. Energy **95**, 35–41 (2016). https://doi.org/10.1016/j.anucene.2016.02.024

22. Yazdi, M.: Hybrid Probabilistic risk assessment using fuzzy FTA and fuzzy AHP in a process industry. J. Fail. Anal. Prevent. **17**(4), 756–764 (2017). https://doi.org/10.1007/s11668-017-0305-4
23. Ainakulov, Z., Pirmanov, I., Koshekov, K., Astapenko, N., Fedorov, I., Zuev, D., et al.: Risk assessment of the operation of aviation maintenance personnel trained on virtual reality simulators. Transp. Telecommun. J. **23**(4), 320–333 (2022). https://doi.org/10.2478/ttj-2022-0026
24. Updegrove, J., Jafer, S.: Optimization of air traffic control training at the federal aviation administration academy. Aerospace **4**(4), 50 (2017). https://doi.org/10.3390/aerospace4040050

Assessment of the Factors that Influence the Transport Sector Turnover in Lithuania

Algimantas Danilevičius[1] (✉) and Irena Danilevičienė[2]

[1] Faculty of Transport Engineering, Department of Mobile Machinery and Railway Transport, Vilnius Tech, Vilnius, Lithuania
algimantas.danilevicius@vilniustech.lt
[2] Faculty of Business Management, Department of Financial Engineering, Vilnius Tech, Saulėtekio Av. 11, Vilnius, Lithuania
irena.danileviciene@vilniustech.lt

Abstract. In today's ever-evolving but unpredictable world, the transportation of goods is important for every consumer if they want to have the right product at the right time and place. Transport companies also want to meet customers' needs as quickly and qualitatively as possible and continue to run their activities profitably. However, contingencies: pandemics and war have a significant impact on the speed, quality and revenue of the transport company. The purpose of the article is to evaluate the influence of factors on the turnover of the transport sector in the context of Lithuania. The aim is to analyze in detail both theoretical and practical aspects of transportation activities and to assess the influence of essential factors on the turnover and income of transport companies. After conducting the research, it was concluded that next to other internal and external environmental factors, export volume has the greatest impact on the income and turnover of transport companies.

Keywords: Export · Import · Logistic · Transportation · Turnover

1 Introduction

In today's world, the transportation of goods is important, as it aims to ensure that the goods reach their customers even in the farthest corner of the world. In the age of accelerating technological development, those companies are considered more competitive, which can meet customers' needs at the right time and with the lowest time costs and, at the same time, ensure the required level of income and operational efficiency. The modern customer is busy, constantly in a hurry and demands quality, timely and tailored services from service providers. The operating principles of traditional business are already outdated and require new, more efficient work methods.

The pandemic and the war's start also significantly impacted transport sector companies' performance. The transportation system is also affected by the restrictions created during the pandemic. So, the companies in the transportation sector decided to cancel flights to the affected countries, which led to the unemployment of many transportation workers [26]. Transport used during the pandemic has decreased due to a decrease

O. Prentkovskis et al. (Eds.): TRANSBALTICA 2023, LNITI, pp. 287–296, 2024.
https://doi.org/10.1007/978-3-031-52652-7_28

in total vehicle hours, which has led to a decrease in the amount of machinery used for transport. Global restrictions (border restrictions, travel bans, quarantine, stay-at-home, service closures) have reduced the transportation sector's demand and freight transportation efficiency [2].

When the amount of transportation turnover changes, it is important to correctly identify the factors that significantly influence the efficiency of the activity and evaluate their influence on the performance results. Transportation is defined by many factors, including the method, duration and size of the cargo, which must be carefully planned [11]. The problem of the article is identifying and evaluating the influence of factors on the turnover of the transportation sector.

The article aims to assess the influence of factors on the turnover of the Lithuanian transport sector.

The tasks of the article: to analyse the theoretical aspects of transportation activities; to perform an analysis of the Lithuanian transport sector; to assess the impact of the identified factors on the turnover of the transportation sector in Lithuania.

The methods of the research: analysis of scientific literature, comparing theoretical and practical statements, comparative analysis, correlation-regression analysis, and forecasting.

2 Theoretical Aspects of Transportation Activities

Special attention is paid to the efficiency of transportation activities in modern research and practice. In a world affected by globalization, the logistics industry has developed rapidly, has become particularly relevant in fostering the productivity of the production and trade sectors and has become an important source of economic life. This chapter examines theoretical aspects of transportation activities, and types of transportation and identifies factors that influence transportation efficiency.

2.1 Analysis and Features of Transportation Concepts

In a constantly changing world, companies that want to be competitive and attract as many customers as possible must meet their needs in a timely and quality manner. For the desired goods to reach the customers on time, it is important to transport them properly from one point to another. The importance of transportation has increased significantly during the global pandemic, when to do the job as well as possible, the activities of transport companies should be maximally adapted to changing needs. It follows that transport activities should be improved.

When analyzing the concept of transportation and its application, it is very important to discuss the significance of logistics. Logistics describes a dynamic and constantly evolving field of activity that involves the transportation, storage and handling of goods or the delivery of goods to the customer to obtain maximum profit [3, 17]. Logistics is the strategic process of acquiring, moving and storing materials, parts and finished inventory (and related information flows) through an organization and its marketing channels to maximize current and future profitability cost-efficiently and quality order fulfillment [5]. In this case, logistics is understood as a driver of business results, a profession and an object of scientific research, which daily has to face new challenges, apply requirements and restrictions.

Logistics can be equated with business because it is a very important business area [14]. The logistics activity chain ensures the smooth flow of the production process and the utilization of costs. Logistics activities occur in every supply channel, including customer service, warehousing, transportation, inventory management and information flow. The costs of logistics activities (such as premises rent, energy consumption, inventory storage, transport costs, etc.) must be monitored and optimized, as they are one of the most important parts of transportation costs. This ensures the fostering of the logistics economy.

Logistics economics studies [3] how logistics companies maximize their revenue with the lowest possible cost to meet the endless demands of customers. Naturally, good management is necessary to make this efficiency possible. So, logistics is concerned with how corporations achieve maximum returns with minimum costs at the micro and macro levels. This means productivity, employment, growth and contribution to the national economy could be improved. On the other hand, logistics is an inclusive term with quite a few branches of activity, so there is no single precise definition.

Logistics is a vast business field that includes many complex processes. Still, generally, it is described as an activity that ensures the movement of cargo and its control at the lowest cost. The modern logistics system consists of three main components: supply and procurement logistics, production logistics and sales logistics. How these logistics system components are managed determines which product will be used to meet the consumer's needs and when. It is the mutual interaction of two or more areas of activity, which aims to ensure the planning, production, control and efficient movement of raw materials, material resources and production from production to consumption.

We can't equate logistics and transportation because transportation is only one of the logistics functions [4]. To ensure the most efficient flow of material values, there must be a movement of material objects and the flow of information needed to make any management decision and estimate logistics costs. The efficiency of transportation ensures this.

Transportation is one of the most important logistics activities that help to ensure the movement of products from production to places of consumption [15, 27]. Transportation costs are often the highest compared to others, so this activity must be given much attention. Transportation gives the product a place and time value.

Transportation is the flow of material, information and money between consumers and suppliers [9]. A transportation system is a system that defines the movement of goods from one point to another, ensuring that the process will contribute to improving the economy and increasing the competitiveness of the company [25]. It is often difficult to plan how to deliver a product to the consumer quickly with quality and at minimal cost, so it must be a well-planned process. The war also significantly impacts transport sector companies' performance the end user. When planning a transportation route, it is important to consider the main criteria: price, assortment, time and possibilities [10].

Transportation is associated with many negative externalities, such as noise, air pollution, events [8]. Transportation has become more important in industrialized countries, where it has become an important activity in economic and social development that increases gross domestic product.

This activity includes selecting the type of transport and means of transport, determining the route, and selecting a specific carrier. The most important characteristics of transportation that influence the level of customer service – are reliability, delivery time, service area, the flexibility of product delivery, losses and damage to cargo, the ability of the transport company to provide not only transportation but also other services [20]. Regarding cargo transportation, we can choose from many types of transport – railway, road (car), air, water, and tube transport, but each has its advantages and disadvantages. Here the advantages and disadvantages of vehicles for transporting goods are distinguished [24].

The transportation system performing logistics activities benefits the quality of services and the company's competitiveness. A well-managed transportation system aims to meet customers' needs as well as possible, so goods should be sent to the right place at the right time and quality to create a close relationship between producers and consumers. Therefore, transportation is an efficient and economical basis for business logistics, expanding other logistics functions.

Globalization and economic development are important processes that promote well-organized logistics and supply chain management to improve performance [12]. The role of transportation in the logistics system is more complex than the usual delivery of goods to customers and its complexity can only be revealed through high-quality supply chain management [21].

It is also very important to respond flexibly to customers' changing needs and thus offer exclusive, specific services or differentiating prices. Thus, it is necessary to increase the value of the transport company in the context of increasing competitiveness. Planning, organization, management and control of all these processes can make up to 90% costs of the transport company. If the value chain works efficiently, there is an opportunity to reduce the amount of waste, promote energy conservation, protect the environment and experience lower costs (thus generating more profits). Ensuring such principles creates prerequisites for the transport company to be competitive and socially responsible.

2.2 Factors Affecting Transportation Activities

Timely and high-quality planning of the final result and its determining factors is important in transportation activities. Transportation activities are affected by many internal and external factors, which, if correctly identified and evaluated, can be managed and promote their positive impact on performance.

Internal factors influence the company's internal processes, while external factors influence the company's decision-making, the development of the supply chain, and the efficiency of relations with customers, suppliers, government or other institutions. The most important internal factors are human resource, institutional, microeconomic and external factors like political-legal, social, ecological, scientific-technological [7, 13].

As can be seen in Fig. 1, internal factors include human, institutional, technical, and microeconomic factors. At the enterprise level, the efficiency of operations is influenced by the size and location of the enterprise, working conditions, the employer's financial capabilities and work productivity, as well as the skills, professional competence and qualifications of the employees.

In more developed markets, working conditions are a very important factor for operational efficiency, because with more significant job opportunities, an employee can choose from more jobs, knows his value and is motivated to work efficiently for the benefit of the company, therefore one of the most important factors when choosing a job is the working conditions in which the employer, seeking to attract new employees must take care [11]. The better conditions the employer offers to the employee, the more highly qualified employees they can attract.

Another very important factor is work productivity [1, 18]. This factor shows the efficiency of the use of labour force because, with the growth of labour productivity, the amount of transported production increases simultaneously. Increased transportation amounts generate higher profits because costs do not increase, so companies can increase wages without raising prices or reducing profits. Due to the profit earned, companies expand their activities and increase the number of jobs and as the demand for labour increases, wages and operational efficiency also increases.

An unforeseen event that affected all countries and their inhabitants is the Covid-19 pandemic. The restrictions caused by the COVID-19 pandemic are increasing poverty levels in all countries, reducing the incomes of low-income earners and less affecting higher-income transportation firms, but harm firm performance [23].

After analyzing the internal factors, it is possible to move on to external factors, which in the scientific literature are usually divided into political-legal and social-economic, but other external factors affect wages (e. g., technological innovations or unforeseen events) [6, 13].

Political-legal factors include the influence of trade unions, the tax system, the setting of minimum wages and working time restrictions. Trade unions can become very important in securing better conditions for workers. Trade unions are favourable to employees when concluding collective agreements, which set specific hourly rate wages, monthly wages, other terms of work payment and negotiated with employers by trade unions that have a more significant advantage in negotiations.

Economic and social factors influence operational efficiency [16]. A group of economic factors is the basis of a country's competitiveness. In contrast to the previously discussed factors, most of which do not have an index expression and whose evaluation may have a considerable degree of subjectivity, indicators clearly express economic factors. Labour productivity is the gross domestic product produced in one hour of work (Parkin, 2010). Also, the gross domestic product per capita size can be used to assess the level of operational efficiency in the case of a specific activity. The inflation indicator is often used to describe the price level, which shows how much the general price level in the country has increased. The unemployment rate, often used in the scientific literature, is an indicator expressed through the unemployed to the labour force ratio. Operational efficiency is also influenced by the ability to plan routes to achieve maximum output (more cargo transported) with minimum costs (time, fuel, labour) [7].

After analyzing the factors discussed in the scientific literature, it can be stated that many factors influence the activity's efficiency, the influence of which and its importance and nature are quite different. Economic factors are primarily used in research in the scientific literature – the unemployment rate, inflation, an internal factor from the employee's side – labour productivity and a political factor – the minimum wage.

3 Methodology of Research

It is known from the analyzed scientific literature that mostly internal and external factors are used to study operational efficiency. This study analyses four main factors: export amount, import volume, gross domestic product, employment. Quarterly data for 2018–2022 were used to obtain more accurate results.

Correlation-regression analysis is used for the research. Using statistical methods for these studies, statistical hypotheses are put forward, which can be confirmed or refuted. A significance level of 0.05 is usually used if the data parameters differ, with this significance level they will not be more than 5% [19]. In practical calculations, the *p-value* is considered, a probability that shows how likely it is to make a mistake in saying that the parameters are significant. The decision whether to accept or reject the hypothesis is made by comparing the *p-value* with 0.05, if the *p-value* is lower than the possible rejection of the hypothesis the data are insignificant and it can be concluded that there are statistically significant areas. Thus, the following hypotheses will be followed when performing statistical research methods in this work: h0 – the data under study equals zero, h1 – the data under study is not equal to zero.

So, the first statistical method is correlation analysis. Correlation analysis makes it possible to determine whether there is a relationship between the considered factors expressed in quantitative indicators [22].

Pairwise correlation analysis aims to determine whether there is a relationship between factors Y and X. When performing pairwise correlation analysis it is necessary to: calculate the correlation coefficient r (Pearson's correlation coefficient) and assess its significance. The values of the correlation coefficient can be from −1 to 1, the closer the value is to one, the more significant it is. If the value is minus, it means that there is an inverse correlation and as one-factor increases, the other decreases and vice versa with a positive indicator.

The Pearson correlation coefficient is calculated using the Eq. 1 [22].

$$R = \frac{n\Sigma x_i y_i - (\Sigma x_i)(\Sigma y_i)}{\sqrt{n\Sigma x_i^2 - (\Sigma x_i)^2}\sqrt{n\Sigma y_i^2 - (\Sigma y_i)^2}}, \tag{1}$$

where n – the number of observations in the sample; R – correlations (Pearson coefficient); x_i – variable; y_i – another variable.

Whether the correlation coefficient is significant is decided by the significance level *p-value*. If the *p-value* < 0.01, the correlation coefficient R is significant, if the *p-value* > 0.01, the correlation coefficient R is insignificant.

After conducting a correlation analysis and finding out whether a relationship connects the studied data, it is possible to perform a regression analysis and determine what kind of relationship connects the indicators. The purpose of pairwise regression analysis is to determine the analytical expression of the relationship between the random variables X and Y. This is done by selecting the curve that best describes the set of statistical points and evaluating the adequacy of this curve [22]. Here is looking for the relationship between the variables. Regression curve $\hat{y} = f(x)$ shows whether there is a relationship between X and Y, if we are looking for a linear relationship.

To make sure whether the obtained coefficients are significant, the coefficient of determination is calculated, which shows what part of the dispersion of the dependent variable is explained by the independent variable. For pairwise regression, the coefficient of determination equals the square of the correlation coefficient.

After performing the pairwise regression analysis and finding out whether there is a linear relationship between the pairs of variables, it is possible to proceed to the multiple regression analysis. The aim of multiple regression analysis, as well as the pairwise regression analysis, is to study the influence between variables, only in the method of multiple analysis, when studying the influence, it is not studied with one variable, but with all their groups. The Eq. 2 presents this model [22].

$$\hat{y} = a_0 + a_1x_1 + a_2x_2 + \cdots + a_mx_m \tag{2}$$

where a_1, a_2, a_m – coefficients; a_0 – free coefficient; x_1, x_2, x_m – variables.

To check the significance of this equation, the coefficient of determination is also calculated and the presence of significant variables is checked according to the *p-value*. After performing these tests and verifying the assumptions, the equation can be considered valid and significant and conclusions can be drawn from it.

The research will aim to substantiate or disprove the proposed hypotheses, which have been drawn up based on the reviewed scientific studies:

Hypothesis 1: Growing exports increase transportation sector turnover and transport companies' income.
Hypothesis 2: Growing imports increase transportation sector turnover and transport companies' income.
Hypothesis 3: Increasing labour productivity increases the transportation sector's turnover and transport companies' income.
Hypothesis 4: Growing employment in the transport field increases goods turnover and transport companies' income.

Thus, the research carried out in accordance with these hypotheses will reveal whether the factors identified in the scientific literature influence the efficiency of the operations of transportation companies and what is the influence of these factors.

4 Evaluation of the Activity of Transport Companies

Based on the results of the correlation-regression analysis, the factors affecting the efficiency of the transport companies were determined and the regression equation was drawn up. Using the quarterly, seasonally adjusted data of 2018–2022 for the study, reliable results and a strong relationship between factors were obtained. It has been established that turnover of the transportation sector has the greatest impact (interrelationship) with exports, imports and gross domestic product (Table 1).

Table 1. Summary of analysis results.

Variable	Coefficient	*p-value*
Export	12.12	<0.05
Import	−7.58	<0.05
Gross domestic product	138.75	<0.05
Employment	***4089.4***	***0.13***

Table 1 shows that employment has no mutual relationship with turnover of the transportation sector, therefore, when creating the general regression Eq. (3), it is necessary to eliminate the employment component.

$$\hat{y} = 13.87x_1 - 8.83x_2 + 130.45x_3, \tag{3}$$

where x_1 – export; x_2 – import; x_3 – gross domestic product.

This equation shows that variable factors affect the turnover of the transportation sector. Exports increase turnover of the transportation sector, while imports and gross domestic product decrease turnover of the transportation sector. Based on the obtained results, it is possible to:

- confirm hypothesis 1 – growing exports increase turnover of the transportation sector and the income of transport companies;
- deny hypotheses 2 and 3 – growing imports and gross domestic product reduce turnover of the transportation sector and the income of transport companies;
- reject hypothesis 4 – employment does not affect transportation sector turnover and transport companies' income because the *p-value* is insufficient.

The next step is to evaluate the statistical reliability of the created equation. The results of the first tested hypothesis are presented in Table 2.

Table 2. The adjusted coefficient of determination and *p-value*.

Adjusted R2	*p-value*
0.9381	<0.001

Based on Table 2 results, it can be concluded that the constructed equation contains important data as the *p-value* of the equation is less than 0.001. The equation explains as much as 94% of the data in the equation, so judging by these primary criteria, it can be said that the equation has statistically significant data.

Summing up, several conclusions can be drawn from the constructed regression equation. First, only the first hypothesis is confirmed when analyzing the effects of the factors. Second, employment does not significantly impact the transportation sector's turnover and transport companies' income. Thirdly, exports have the greatest influence on the turnover of the transportation sector and the income of transport companies, therefore, to improve operations, ways to increase the volume of exports should be sought.

5 Conclusions

Transportation is managing the acquisition, movement and storage of goods to maximize profitability and efficiently fulfill orders. After analyzing the scientific literature, it was found that there are various internal and external factors for the transportation sector's turnover and the transport company's income. The main internal factors are human resource factors, institutional, technical, microeconomic and the external factors are political-legal, economic, social, scientific-technological.

Based on the analyzed scientific literature, quarterly data for 2018–2022 was chosen to be studied. The main influencing factors are the transportation sector's turnover and transport companies' income, export, import, gross domestic product, employment level. Also, 4 hypotheses about each of the variable factors are proposed.

Based on the results of the correlation-regression analysis, it can be concluded that the data are reliable and only the first hypothesis was confirmed that exports increase transportation sector turnover and transport companies' income. The other hypotheses are rejected because there is an inverse relationship. It is necessary to emphasize that employment does not affect the income of transport companies and turnover of the transportation sector, as the calculated *p-value* confirms this.

References

1. Akbari, M.: Logistics outsourcing: a structured literature review. Benchmarking: Int. J. **25**(5), 1548–1580 (2018)
2. Abu-Rayash, A., Dincer, I.: Analysis of mobility trends during the COVID-19 coronavirus pandemic: exploring the impacts on global aviation and travel in selected cities. Energy Res. Soc. Sci. **68**(101693), 1 (2020). https://doi.org/10.1016/j.erss.2020.101693
3. Bashir, R., Vijayalakshmi, H., Bashir, N.: A comparative study on the prices of products in logistics companies and their impact on economic growth. In: 8th International Conference on Reliability, Infocom Technologies and Optimization (Trends and Future Directions), 4–5 June 2020, pp. 934–939 (2020)
4. Budrys, N.: Vertės grandinės strategijos tobulinimo modelis tiekimo grandinės kontekste. In: 22-osios Lietuvos jaunųjų mokslininkų konferencija "Mokslas – Lietuvos ateitis", 2019 m. vasario 13 d. Vilnius, pp. 1–10 (2019)
5. Christopher, M.: Logistics and supply chain management. Financial Times, United Kingdom (2016)
6. Diao, M., Kong, H., Zhao, J.: Impacts of transportation network companies on urban mobility. Nat. Sustain. **4**, 494–500 (2021). https://doi.org/10.1038/s41893-020-00678-z
7. Erhardt, G.D., Roy, S., Cooper, D., Sana, B., Chen, M., Castiglione, J.: Do transportation network companies decrease or increase congestion? Sci. Adv. **5**(5), 1–11 (2019). https://doi.org/10.1126/sciadv.aau2670
8. Faulin, J., Grasman, S.E., Juan, A.A., Hirsch, P.: Sustainable Transportation and Smart Logistics: Decision-Making Models and Solutions. Elsevier, United Kingdom

9. Frazelle, E.: Supply Chain Strategy: The Logistics of Supply Chain Management. McGraw-Hill Companies, New York (2002)
10. Hidayatno, A., Destyanto, A.R., Fadhil, M.: Model Conceptualization on E-Commerce growth impact to emissions generated from urban logistics transportation: a case study of Jakarta. Energy Procedia **156**, 144–148 (2019). https://doi.org/10.1016/j.egypro.2018.11.119
11. Karami, Z., Kashef, R.: Smart transportation planning: Data, models, and algorithms. Transp. Eng. **2**(100013), 1–8 (2020). https://doi.org/10.1016/j.treng.2020.100013
12. Kherbach, O., Mocan, M. L.: The importance of logistics and supply chain management in the enhancement of Romanian SMEs. Procedia – Soc. Behav. Sci. **221**, 405–413 (2016). https://doi.org/10.1016/j.sbspro.2016.05.130
13. Krembsler, J., Spiegelberg, S., Kampf, N.L., Winter, T., Winter, N., Knappe, R.: Fare revenue forecast in public transport: a comparative case study. Elsevier **22**, 1–24 (2023)
14. Kucera, T.: Application of the activity-based costing to the logistics cost calculation for warehousing in the automotive industry. Communications **4**, 35–42 (2019). https://doi.org/10.26552/com.C.2019.4.35-42
15. Lemtaoui, M., Rochdi, M.H., Eloueldrhiri, S.: Measuring the supply chain performance in Morocco: application of the Edward Frazelle's model. In: International Colloquium on Logistics and Supply Chain Management, 27–28 April, 2017, Rabat, Morocco, pp. 188–192 (2017)
16. Mack, E.A., Agrawal, S., Wang, S.: The impacts of the COVID-19 pandemic on transportation employment: a comparative analysis. Transp. Res. Interdisc. Perspect. **12**(100470), 1–13 (2021). https://doi.org/10.1016/j.trip.2021.100470
17. McKinnon, A.: Decarbonizing Logistics: Distributing Goods In a Low-Carbon World. Kogan Page Limited, United States (2018)
18. Munawar, H.S., Khan, S.I., Qadir, Z., Kouzani, A.Z., Parvez Mahmud, M.A.: Insight into the Impact of COVID-19 on Australian transportation sector: an economic and community-based perspective. Sustainability **13**(1276), 1–24 (2021). https://doi.org/10.3390/su13031276
19. Murauskas, G., Čekanavičius, V.: Taikomoji regresinė analizė socialiniuose tyrimuose. Vilniaus universiteto leidykla (2014)
20. Nagulevičius, E., Jakubavičius, A.: Gamybos proceso modernizavimas, pramonės skaitmenizavimo iššūkių kontekste. In: 22-osios Lietuvos jaunųjų mokslininkų konferencija "Mokslas – Lietuvos ateitis", 2019 m. vasario 13 d. Vilnius, pp. 1–9 (2019)
21. Ojhaa, D., Struckellb, E., Acharyac, C., Pateld, P.C.: Supply chain organizational learning, exploitation, and firm performance: a creation-dispersion perspective. Int. J. Prod. Econ. **204**, 70–82 (2018). https://doi.org/10.1016/j.ijpe.2018.07.025
22. Pabedinskaitė, A., Činčikaitė, R.: Kiekybiniai modeliavimo metodai. Technika (2016)
23. Palomino, J. C., Rodríguez, J. G., Sebastian, R.: Wage inequality and poverty effects of lockdown and social distancing in Europe. Eur. Econ. Rev. **129** (2020). https://doi.org/10.1016/j.euroecorev.2020.103564
24. Palšaitis, R.: Tarptautinio verslo transportinis logistinis aptarnavimas. Vilnius: Technika (2011)
25. Putra, A.A., et al.: Model of logistic transport distribution in the urban area. In: IOP Conference Series: Earth and Environmental Science, vol. 419, pp. 1–12 (2020).https://doi.org/10.1088/1755-1315/419/1/012098
26. Siwiec, A.N.: Globalna pandemia a przewóz towarów drogą lotniczą w Europie. Acta Iuridica Resoviensia **3**(38), 356–370 (2022). https://doi.org/10.15584/actaires.2022.3.25
27. Zinkevičiūtė, V., Vasilis Vasiliauskas, A.: Gamybos logistika. Gamybos vadyba. Klaipėda: Socialinių mokslų kolegija (2013)

Reliability of Fuel Supply Chains During Military Operations – Case Study

Jacek Ryczyński(✉) and Artur Kierzkowski

Wroclaw University of Science and Technology, 50-370 Wroclaw, Poland
{jacek.ryczynski,artur.kierzkowski}@pwr.edu.pl

Abstract. The logistics structure of military factions engaged in combat operations should be organized following the guidelines in the NATO Directive MC 319/2. The "mixed variant" of material logistics is the most frequently applied approach, which blends several strategies, such as outsourcing, establishing a Multinational Integrated Logistic Unit (MILU), appointing a Lead Nation (L.N.) and a Role Specialist Nation (RSN), formalizing Mutual Support Agreements (MSA), or engaging Third Party Logistic Support Services (TPLSS). This piece details the findings from original research on the fuel logistics system during the Afghan conflict. The authors discuss the risk zones significantly influencing the logistics chain's successful functioning. The investigation concludes with assessing the fuel supply chain's reliability using the FMECA technique (Failure Mode Effects and Criticality Analysis).

Keywords: Army fuel operations · Supply chain · Failure Mode Effects and Criticality Analysis (FMECA)

1 Routes of Supply in Afghanistan

During the ISAF operation, the entire territory of Afghanistan was treated by NATO as a war zone. As a result of the activities, vast amounts of ammunition and material resources were used. All ammo types for small arms and artillery were transported by air. The goods were typically conveyed via maritime routes to Gulf ports and then ferried to Afghanistan as a cost-saving measure. There are two passages from Pakistan to Afghanistan (which were temporarily shut down after the Salalah incident in November 2011 but reactivated in July 2012). These paths commence from Karachi, in Pakistan's southern Sindh province, by the Arabian Sea. One route traverses the Khyber Pass, enters Afghan terrain at Torkham, and proceeds to Kabul, catering to northern Afghanistan. This route spans roughly 1,000 miles. The alternate route navigates through Baluchistan province, crosses the border at Chaman, and culminates in Kandahar in the southern part of Afghanistan.

NATO utilized these routes to convey fuel and various provisions, albeit not for weapons. Before their shutdown, the Pakistani ways catered to the majority of NATO's fuel requirements in Afghanistan. In 2007, the military consumed over 2 million litres of fuel daily, with 80% originating from refineries in Pakistan. The fuel storage facilities

O. Prentkovskis et al. (Eds.): TRANSBALTICA 2023, LNITI, pp. 297–306, 2024.
https://doi.org/10.1007/978-3-031-52652-7_29

at Bagram and Kabul air bases were meagre, holding less than 6 million litres, making NATO operations heavily reliant on the Pakistani supply channels. In 2010, owing to strained US-Pakistan relations, there was an upsurge in efforts to bolster fuel bases, substantially augmenting their capacity. Afghanistan shares its borders with Turkmenistan, Uzbekistan, and Tajikistan, where alternative supply routes, known as the Northern Distribution Network, were established. Nonetheless, these routes are more extensive, and the transportation costs are heftier than those via Pakistan. The Northern Distribution Network encompasses several diverse paths, with the most frequented and lengthiest, initiating at the port of Riga in Latvia on the Baltic Sea, spanning 5,169 km to the south via Russia using railways erected by Russia in the 1980s.

The supplies then transit through Kazakhstan and Uzbekistan, arriving at Afghanistan's northern frontier at Termes. To reach the country's southern part, cargo must be loaded onto trucks and transported across the rugged Hindu Kush through the Salang Tunnel. The Salang Tunnel, a key conduit connecting northern and southern Afghanistan, measures 2.5 km in length and is at an elevation of 3,350 m. The route is susceptible to avalanches and is relatively complex. Another southern way originates in Poti, Georgia, on the Black Sea, then heads to Baku, Azerbaijan, where goods are transferred onto barges and conveyed across the Caspian Sea. The subsequent phase involves a terrestrial journey through Turkmenistan and a railway through Uzbekistan until the Afghan border is reached. A third alternative, conceived to bypass the often turbulent Uzbekistan, extends from Kazakhstan to Kyrgyzstan and through Tajikistan before arriving in Termes.

2 Defense Logistics Agency Energy (DLA Energy) as the Governing Body for Fuel Supply in Afghanistan

The Defense Logistics Agency (DLA) is an agency of the US Department of Defense (DoD). It supervises the supply chain of liquid fuels right from the source to the point where the customer picks up. As a comprehensive fuel manager, DLA Energy takes on the responsibility of consolidating and reviewing fuel requirements, overseeing financing and budgeting, managing storage, and coordinating fuel distribution to meet the operational demands of the commander.

When DLA Energy purchases fuel, its contractors transport the fuel from the refinery to strategic storage points in all operational areas worldwide. Once the fuel reaches its destination, it can be stored or transferred directly to the customer. DLA Energy owns and tracks the entire supply chain until the fuel comes to its endpoint, e.g. it can be placed directly in military equipment or a storage unit on the battlefield. Typical means of moving fuel in an operational environment are pipelines, inland water tankers/barges, railways and fuel tanks. The level of control in the fuel supply chain varies depending on where the point of sale is located in the chain. Generally, DLA Energy supplies fuel to a centralized storage facility from which land customers collect and distribute fuel to subordinate units throughout the operation theatre. In this case, DLA Energy's supply chain control ends upon delivery to the central warehouse. The recipient can be a battalion or intermediary collecting the fuel at the landfill for distribution to its subordinate units in the field. As it gained experience in Afghanistan, DLA Energy changed its supply chain organization requirements to transport fuel directly from the refinery and distribute it to customers at remote forward operating bases. Since it is the final recipient who decides on the correctness of the delivery, DLA Energy relinquished control over such supply chains. The domain of military fuel distribution services started to be penetrated. According to NATO doctrine, the Operations Theater Maintenance Command (OTMC) is entrusted with the responsibility of orchestrating the transportation and distribution of fuel from the theater's main or intermediary bulk storage facilities to tactical ground clients. OTMC plans the distribution by considering factors such as customer demand, available distribution resources, operational capabilities, environmental conditions, and storage capacity. The fuel is transported by OTMC from the base by pipeline or truck and distributed to airports or operating units (brigades). Distribution tankers transport the fuel to battalion-sized units and independent combat teams. Figure 1 shows the organization of a fuel supply chain as part of DLA Energy's operations.

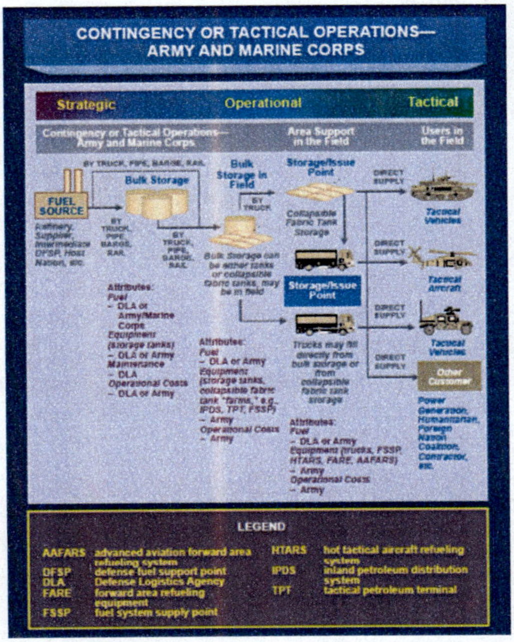

Fig. 1. Organization of the fuel supply chain in the Afghan war [2].

2.1 Fuel Outsourcing and Contracts at the Strategic Level

Since Afghanistan lacks refineries, all of the fuel it uses is imported. To fulfil service necessities, DLA Energy has entered into contracts with several Pakistani and Central Asian firms, including Red Star Enterprises Limited and Supreme Fuels Trading FZE. These contractors must deliver fuel to the primary fuel hubs in Afghanistan, from where it's further disseminated to military units via fuel tanks. Until 2009, most fuels were transported from Pakistan via the perilous Khyber and Salang Mountain ranges along the southern supply route. Since then, fuel transportation has shifted to the Northern Distribution Network, where over 70% of the fuel is conveyed by rail and truck from the Turkmen refinery over 1,000 miles. Most of Afghanistan's fuel is stored at two principal distribution centres: Bagram Airport in the north and Kandahar Airport in the south. In 2006, DLA Energy signed a contract to construct an additional fuel base in Kabul with a capacity of 50,000 barrels to avert delivery setbacks. The fuel transportation from the source to the distribution hub can take up to two weeks, and reaching the most remote parts of the area of responsibility can take up to 18 days [2].

A unique aspect of NATO ISAF's fuel operations was that local contractors assumed control of fuel supply chains, delivering fuel directly to the theatre of operations and most operational bases in Afghanistan. American fuel operations were also conducted in parallel with the contractors' management. Their mission was to provide adequate support to the forces of coalition partners, although on a larger scale and employing different contracting strategies.

2.2 Specificity of the Fuel Supply Chain in Afghanistan - Basic Ordering Agreement (BOA)

At the onset of the Afghan conflict, a prominent feature was the absence of national logistic support units. This circumstance compelled NATO to assume the duty of satisfying fuel requirements with restricted financial aid from shared NATO budgets. The ISAF Operations Command Europe and the Joint Forces Command Brunssum (JFC HQ) designed the idea of commercial fuel support arrangements for NATO ISAF members. These agreements are recognized as NATO Fuel Basic Ordering Agreements or simply Basic Ordering Agreements (BOAs). BOA was an innovative solution allowing NATO to offer fuel support to ISAF member states incapable of independently providing fuel support to their combat forces without financial aid from a common NATO budget.

BOA is a unique contract style - it doesn't align with the traditional contract concept. Instead, it comprised a set of commercial fuel support contracts provided to firms from NATO member countries under sourcing and contracting regulations. BOA was purely sales-oriented, a no-cost upfront contract without obligations for NATO member nations or ISAF partner countries. The contractors delivered the fuel and the required logistics infrastructure to the armies of the member nations.

An extraordinarily revolutionary aspect of the BOA was NATO's setting of a price per litre (PPL) for each contractor, subsequently standardizing the PPL to ensure a consistent cost per fuel type for all ISAF operations. This stipulation engendered unprecedented transparency, unmatched by other agreements, and facilitated open and fair negotiations regarding PPL value adjustment. Due to this transparency, PPL rates were subject to monthly review and adjustment. The offloading cost incorporated the expenses borne by the contractor related to transport operations and the actual fuel price.

3 FMECA as a Control Tool Over the Correct Functioning of the Supply Chain

As previously highlighted, the fuel supply chain faced a complex and high-risk operational environment in Afghanistan due to the country's unique supply circumstances and the ongoing conflict. Ensuring the proper functioning of the fuel supply chain was a daunting task. Risk measurement is usually subjective, as qualitative techniques assess risk. It is, therefore, essential to ensure systematic monitoring of existing risk and its continuous analysis. According to the tradition of risk management [7], probability and its impact are often used to assess it [4, 10, 11].

The abovementioned possibilities are met by the FMECA method, which can be a tool for the systematic organization and assessment of risk levels by experts. The FMECA method is a concept of exercising control over services or facilities and relates to techniques for avoiding or limiting the risk of those services or facilities. When using the FMECA method, it is possible to analyze risk in three dimensions: probability of occurrence, impact and control. To assess these three dimensions, the classic qualitative version of the FMECA method uses a 10–point scale (higher points mean worse cases in evaluating a given risk factor) – see Table 1.

Table 1. Relationship between qualitative and quantitative characteristics used to calculate Risk Priority Number [own study].

Value	Description
1	no significant impact on reliability or security
2	slight impact on reliability or safety
3	less than a very small effect, no damage, no injury, causes only maintenance-requiring operation (noticeable only with very careful examination)
4	very small effect, no damage, no injuries, causes only maintenance-requiring operation (noticeable only by very careful examination)
5	small effect, minor damage, only causes maintenance-requiring operation (it has a low impact on the system)
6	less than a critical effect (causes the loss of basic system functions, exceeding all safety margins, describes the event "one step" from the disaster)
7	critical effect (causes the loss of basic system functions, exceeding all safety margins)
8	more than a critical effect (causes the loss of basic system functions, exceeding all safety margins, describes the event "one step" from the disaster)
9	catastrophic effect (system stops working)
10	catastrophic effect (failure may cause completely dangerous system operation)

The Risk Priority Number (RPN), an indicator of the priority for corrective action, is obtained by multiplying the three dimensions of risk related to probability, impact and control. The higher its value, the faster the risk factor assigned to the risk priority number should be reduced. For this reason, the FMECA method has become an essential tool used in risk assessment in the supply chain and its control. The method analyses the cause-effect relationships of potential defects in the process or object and considers the analysis's criticality (risk) factor. The FMECA method allows subjecting the process or things to further research and then, based on the results obtained, introducing corrective actions and new solutions, effectively eliminating sources of defects, which aligns with the principle of continuous improvement [5].

It is not possible to eliminate the risk. However, you can limit them to an acceptable or tolerable, harmless level by eliminating their causes and effects, and at the same time, you can ensure complete supervision. Risk assessment in logistics, production facilities or any other links in the supply chain is critical because it can be associated with reducing broadly understood losses in terms of serviced units and infrastructure or human life.

3.1 FMECA – Fuel Supply Chain in Afghanistan

There are numerous ways in the literature to deal with uncertainty and risk. In the case of risk, we talk about risk management and conducting actions aimed at reducing the degree of risk impact on the functioning of the supply chain and securing the system's operational efficiency. Other risk mitigation methods in the supply chain and organizations include:

1. mitigation of supply risk by implementing behavioural inventory management techniques [11];
2. building strategic relationships with suppliers and managing these relationships [3];
3. early involvement of suppliers in the scope of implemented processes [4];
4. reduction of the complexity of the underlying supplies [1];
5. natural hedging of currency fluctuations and commodity prices [6, 7];
6. confidence building and the introduction of knowledge management into the supply chain [8], as well as the introduction of identification;
7. estimating supply chain gaps and identifying so-called "bottlenecks" [6].

The best way to identify and manage risk in the fuel supply chain in Afghanistan and to identify the factors determining the level of this risk was the "bottleneck" method [6]. The adopted assumption corresponds to the beliefs and objectives of the FMECA method [5]:

1. consistent and permanent elimination of defects ("weak" places) in the process by recognizing the natural causes of their occurrence and applying appropriate preventive measures;
2. avoiding the occurrence of recognized and still unknown flaws by using knowledge and experience from previously conducted analyses.

Table 2 presents the FMECA analysis for the fuel supply chain in Afghanistan. The first step was to identify the main risk areas and the most significant risk factors in each area. In the next step, expert knowledge and statistical data were used to calculate the values of the R, P and N coefficients. The final effect is calculated Risk Priority Number for each indicated risk factor.

Table 2. FMECA – fuel supply chain in Afghanistan [own study].

Area of risk	Risk factor	R*	P**	N***	RPN
Supply planning process	Accuracy of demand planning methods	5	5	8	200
	Information disturbances in the reporting of needs by fighting units	6	6	4	144
Storage of current fuel stocks	Quality changes – ageing processes of fuels	2	5	9	90
	Infrastructure failure	4	7	5	140
	Fuel evaporation processes	3	5	8	120
	Disasters and natural disasters	9	3	5	135
	Mixing good fuel with low-quality fuel	9	7	5	315
	Microbial contamination	9	3	5	135
	Destruction of inventory due to the shelling of warehouse bases	8	9	7	**504**
Operation of warehouse infrastructure	Tank leakage	3	7	4	84
	Overflow above tank level	4	6	4	96
	Evaporation	4	3	7	84
	Fuel contamination and distribution installations by microorganisms	3	7	7	147
	External pollution of fuels - a diversion	3	9	8	216
	Checking the tightness of liquid fuel storage tanks	3	8	1	24
	Adding incorrect quality fuels to the fresh fuel batch	9	7	4	252
	Failure of fire extinguishing systems	5	5	5	125
	Failure of Faraday mesh elements	5	5	5	125
Influence of the human factor	Mistakes in measuring the fuel level (fuel tank)	9	6	3	162
	Overfilling or underfilling of storage tank – unnoticed	8	6	5	240

(*continued*)

Table 2. (*continued*)

Area of risk	Risk factor	R*	P**	N***	RPN
	Inexact dosing of chemical additives to fuel – relating to the Single Fuel Policy concept	8	6	6	288
	Inaccurate mixing of additives with fuel – the concept of Single Fuel Policy	8	6	6	288
	Imprecision in evaluating the extent of valve opening and regulation of fuel transfer rate	7	5	3	105
	Accidental mixing of good fuel with low-quality fuel	9	7	6	378
	Targeted actions - a diversion				
Operation of the supply subsystem	**Problems with the delivery schedule**	7	9	8	**504**
	Supply disruptions resulting from supply routes being cut off	9	9	7	**567**
	Problems when accepting tankers at fuel bases, tank weighing, quality control of supplied fuels	7	7	6	294
	Explosion hazard when unloading tanks	7	8	8	448
	Non-compliance with OHS (Occupational Health and Safety) regulations by the organization (no OHS training for staff)	9	9	2	162
	Non-compliance with OHS regulations by employees	9	9	3	243
	External environmental factors – firing at convoys, bombings	9	9	7	**567**
IT support for logistics processes (digitization of supply chain processes)	Delays in the flow of information	9	4	4	144
	System failure	9	8	5	360
	IT systems compatibility	8	5	3	120
	Interruptions in the operation of IT infrastructure components due to shelling or attacks	8	9	5	360

* R – the frequency of occurrence of risk factor, ** P – the significance of risk factor defect, *** N – risk factor detection level.

As part of the analysis, 6 risk areas were identified, and as many as 36 risk factors within these areas. The riskiest factors requiring special control and monitoring were:

1. external environmental factors - firing at convoys, bombings - RPN value = 567;
2. supply disruptions resulting from supply routes being cut off - RPN value = 567;
3. Destruction of inventory due to the shelling of warehouse bases - RPN value = 504;
4. Problems with the delivery schedule - RPN value = 504.

According to the RPN values, the first two risk factors are the most important and require specific control and monitoring. All these factors are closely related and result directly from the specific nature of logistics operations in war zones. This means that a failure in one area can lead to problems in other areas: bombing can lead to supply disruptions by disrupting supply routes, leading to delivery scheduling problems.

4 Summary

During the International Security Assistance Force operation in Afghanistan, fuel supply chains played a crucial role in maintaining operational continuity. Fuel logistics covered complex operations, from purchasing and transporting fuel to its distribution to the most remote bases and active points. Each stage required careful planning and management to minimize the risk of operational downtime due to lack of fuel. One of the biggest challenges was that fuel transport was often the target of attacks, which created additional challenges for the security of convoys and storage infrastructure. Additionally, local conditions in Afghanistan, such as difficult terrain and often unpredictable weather conditions, further complicate the delivery process. In response to these challenges, NATO logistics deployed various strategies, such as using multiple supply routes, stockpiling fuel, and building secure storage facilities. Diversification of suppliers and fuel supply routes was one of the key strategies that helped maintain the continuity of supply despite difficult operating conditions. Another essential element was the close monitoring of fuel consumption, which allowed it to adjust its operations in response to supply fluctuations. Logistics operations, including fuel supply chains, have been a critical element in the success of ISAF operations, providing vital support to military operations in Afghanistan. Despite numerous challenges, managing fuel supply chains during ISAF operations in Afghanistan has taught important lessons for future operations in challenging environments. This shows how crucial it is to adapt logistics strategies to the specifics of the environment where operations occur.

References

1. Choi, T.Y., Krause, D.R.: The supply base and its complexity: implications for transaction costs, risks, responsiveness, and innovation. J. Oper. Manag. **24**, 637–652 (2006)
2. Evans, M., Masternak, S.: The silent revolution within NATO logistics: a study in Afghanistan fuel and future applications. NAVAL Postgraduate School (2012)
3. Giunipero, L.C., Eltantawy, R.A.: Securing the upstream supply chain: a risk management approach. Int. J. Phys. Distrib. Logist. Manag. **34**, 698–713 (2004)
4. Hallikas, J., Puumalainen, K., Vesterinen, T., Virolainen, V.M.: Risk-based classification of supplier relationships. J. Purch. Supply Manag. **11**, 72–82 (2008)
5. Hamrol, A., Mantura, W.: Quality management. Theory and practice. (original in Polish: Zarządzanie jakością. Teoria i praktyka). Wydawnictwo PWN, Warszawa (2005)

6. Ho, W., Zheng, T., Yildiz, H., Srinivas, T.: Supply chain risk management: a literature review. Int. J. Prod. Res. **53**, 5031–5069 (2015)
7. Hopkin, P.: Risk Management. Kogan Page Ltd., London (2003)
8. Pérez-Salazar, M., Lasserre, A.A., Cedillo-Campos, M.G., Hernández González, J.C.: The role of knowledge management in supply chain management: a literature review. J. Ind. Eng. Manage. **10**, 711–788 (2017)
9. Ryczyński, J., Biernikowicz, W.: The concept of uniform fuel of battlefield - selected problems. In: Transport Means - Proceedings of the 23rd International Conference, part 2, pp. 712–717, Kaunas University of Technology, Palanga (Lithuania) (2019)
10. Wald, Ch. F., Captain, T.: Energy security American's best defense: a study of increasing dependence on fossil fuels in wartime, and its contribution to ever higher casualty rates. Deloitte Development LCC (2009)
11. Zsidisin, G.A., Ellram, L.M.: Agency theory investigation of supply risk management. J. Supply Chain Manage. **39**, 15–27 (2003)

Operational Risks When Transporting Gas and Gas-Hydrogen Mixtures Through Existing Gas Pipelines

Liubomyr Poberezhnyi[2](✉), Liubov Poberezhna[1], and Pavlo Popovych[3]

[1] Ivano-Frankivsk National Technical University of Oil and Gas, Ivano-Frankivsk 76019, Ukraine

[2] Helmut Schmidt University/University of the Federal Armed Forces Hamburg, 22043 Hamburg, Germany
poberezl@hsu-hh.de

[3] West Ukrainian National University, Ternopil 46009, Ukraine
PPopovich@ukr.net

Abstract. Ensuring the safe transportation of energy sources is one of Europe's current problems. Many oil and gas pipelines, especially in Eastern Europe, have operated for over 40–50 years. The paper examines the effect of a corrosive medium (soil electrolyte) on the ability of pipe steels to resist long-time deformation. In acidic environments, the jump-like deformation was detected, which can cause corrosion cracking and subsequent depressurization of the pipe. One of the reasons for this behavior is hydrogen embrittlement, which can become a problem when transporting hydrogen and hydrogen-methane mixtures through existing gas pipelines. A method of predicting changes in the deformation of pipe steels based on the results of corrosion-mechanical tests is proposed.

Keywords: Exploitation degradation · Corrosion · Hydrogen pipeline transport · Deformation behavior

1 Introduction

During the long-term operation of main pipelines, under the influence of external forces, the working environment, and a number of other factors, changes in the mechanical properties of the metal occur, the performance of pipelines changes, and there is a tendency to develop processes of their destruction [1–4]. In this regard, to calculate the residual life of main pipelines, it is not enough to assess the technical condition of the pipeline only by non-destructive testing methods, since they do not consider the conditions and types of loading of pipelines, the state of the surface and geometry of pipes, etc. [5, 6].

A significant number of scientists were engaged in researching the problem of operational degradation of pipeline steel, in particular oil and gas pipelines. The main mass of them falls on Eastern Europe, in particular the countries of the former USSR, because they have a very developed network of gas pipelines, which were built in the 60s and 70s

O. Prentkovskis et al. (Eds.): TRANSBALTICA 2023, LNITI, pp. 307–316, 2024.
https://doi.org/10.1007/978-3-031-52652-7_30

of the last century. Researchers in recent years have established that it is most appropriate to determine the change in mechanical (strength, impact toughness, plasticity) and electrochemical (OCP, corrosion potential, corrosion current, polarization resistance) indicators to assess the level of material degradation. It was also established that the main role in degradation processes is played by a combination of influencing factors, primarily modes, duration of operation, corrosion activity of the transported product and soil electrolyte, etc. [7–9].

2 Materials and Methods

To study corrosion processes under stress, the previously developed KN-1 computerized installation [3]. Testing of gas pipeline pipe material samples in air and liquid working environments was carried out in static and repeated static loading by pure bending with automatic registration of the deflection of the sample and changes in the electrode potential using a computer, using 24-bit analog-digital conversion.

The study of long-term deformation of pipe steels was carried out in the zone of plastic (510 MPa), elastoplastic deformations (420 MPa), and near the yield limit (330 MPa).

To establish the level of influence of the corrosive environment on the long-term resistance to deformations of pipeline steels, they were first tested in air, and then in corrosive environments [10] (Table 1).

Table 1. Chemical composition of model soil electrolytes.

Model environment (ME)	NaCl, mol/l	HCl, mol/l
1	0.01	0.00001
2	0.05	0.0001
3	0.1	0.001

3 Results and Discussion

In acidic soils, corrosion of pipelines the oxygen and hydrogen depolarization occur. As the acidity of the soil electrolyte increases, the share of hydrogen depolarization increases [11]. This trend is dangerous for long-exploited pipe steels since the release of hydrogen on the surface of the metal leads to its flooding and deterioration of operational characteristics (hydrogen embrittlement). Steels that have been used for a long time are more sensitive to the penetration of hydrogen into their structure. As a result of operational degradation, microcavities are formed in the metal structure, which become traps for hydrogen atoms. With long-term contact with an acidic corrosive environment, degraded tubular steel undergoes hydrogen embrittlement, which significantly reduces its plasticity margin. This can lead to the initiation and rapid propagation of stress-corrosion and corrosion-fatigue cracks and cause the destruction of the pipe wall. Studying and monitoring the processes of deformation of the material of pipelines is one of the ways to timely detection of dangerous areas and implementation of maintenance measures [1, 8, 12, 13].

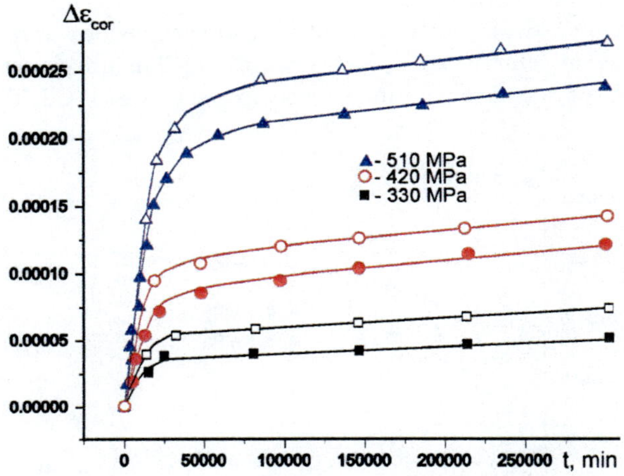

a)

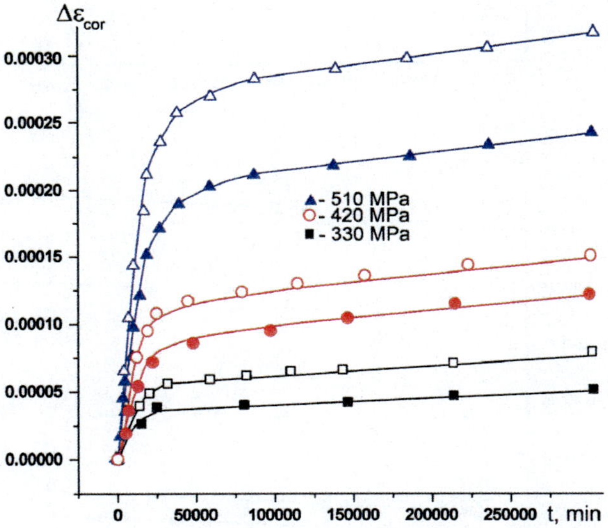

b)

Fig. 1. Kinetics of deformation of pipeline material in ME1: a) 17GS steel, b) 19G steel ■ – steel in the condition of delivery, □ – 41 years of operation.

At the lowest concentration of chlorides (Fig. 1), the deformation increased by 5…15 percent, proportional to the level of applied mechanical stresses [12]. Noticeable deformation jumps are not recorded, and the risk of emergency depressurization is insignificant, which is associated with the prevailing mechanism of oxygen depolarization, and insignificant hydrogen embrittlement.

For steel 19G in ME1, a characteristic deformation behavior was recorded with an increase in the rate of deformation at the first stage, which in turn causes an increase in absolute deformation and its relative increase compared to steel 17GS. The nature of the

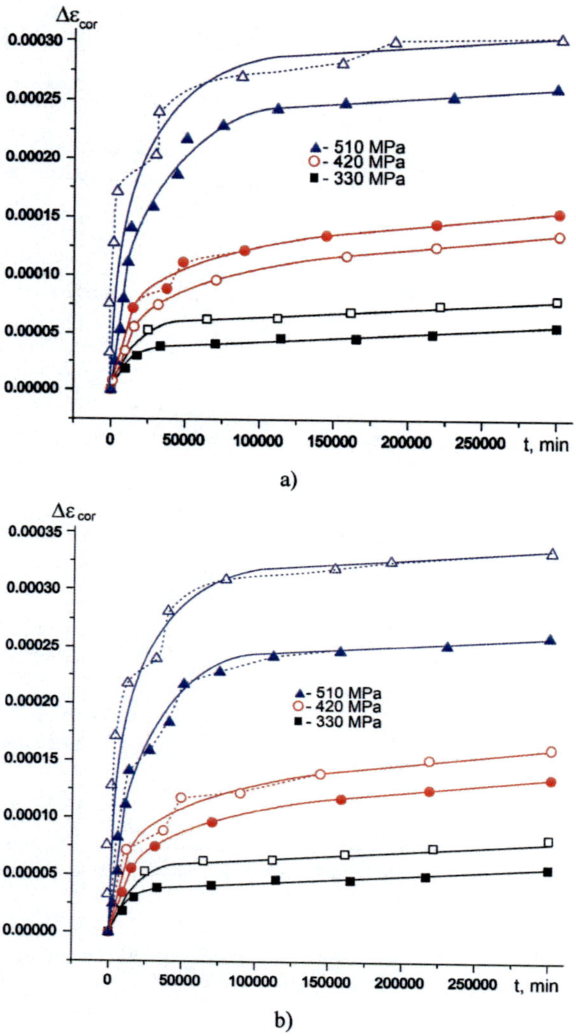

Fig. 2. Kinetics of pipeline material deformation in ME2: a) 17GS steel, b) 19G steel; ■ – steel in the condition of delivery, □ – 41 years of operation.

kinetic dependences is monotonous, there are no jump-like changes in the deformation. The margin of the strength of the pipelines allows to successfully continue the operation of pipes made of 19G steel for more than 40 years, subject to regular monitoring of electrochemical protection parameters and minimization of the unevenness of gas consumption. It is also necessary to pay more attention to areas near compressor stations, where the risk of damage to bitumen and bitumen-polymer insulation increases due to

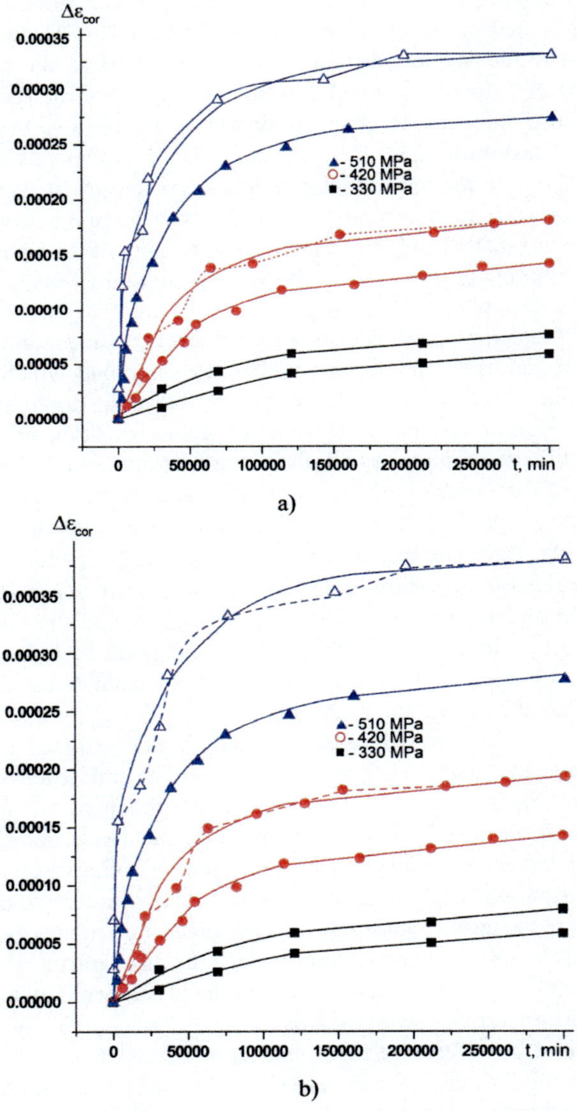

Fig. 3. Kinetics of pipeline material deformation in ME9: a) 17GS steel, b) 19G steel; ■ – steel in the condition of delivery, □ – 41 years of operation.

increased gas temperature. Deformation fluctuations are recorded when the pH is further reduced to 4 (ME2) (Fig. 2).

They are especially noticeable in the area of high stresses, which is explained by the influence of the mechanical factor against the background of hydrogen embrittlement. Cycles of deformation acceleration-deceleration are clearly observed, which correspond to crack growth cycles. They are most intense at the first exposure stage, then the deformation increase rate decreases. The increase in deformation is 7–18%, but there is a risk of sudden depressurization due to cracking caused by hydrogen embrittlement [14–17].

When moving to environments with a lower pH (ME2 and ME3), deformation jumps begin to occur due to the development and propagation of crack-like corrosion defects from the pipe surface. Moreover, in ME2 (pH4), the nature of the recorded deformation changes is most conducive to crack formation, as we observe the highest rate of development of corrosion-mechanical processes, which corresponds to the spread of surface defects. The deformation behavior of long-term service steel in ME3 at pH 3 is more predictable. Deformation jumps are fixed, but they are much smoother than in ME2 (Fig. 3). This is due to the significant chemical activity of the environment, which leads to the rapid blunting of the crack tip. However, here we record the largest absolute increase in deformation, which is 9–24%.

In ME3, for 19G steel, we predicted higher absolute values and relative increments of deformation, but the deformation fluctuations are more smoothed, which, in our opinion, is due to the blunting of the tips of crack-like defects due to the dissolution of the metal at their tip. The analysis of the data on the angles of inclination of the final sections of the kinetic curves and their comparison with the results obtained for 17GS steel indicates an increased risk of developing corrosion processes and the occurrence of emergency situations on pipelines made of 19G steel, which have been in operation for more than 40 years. In particular, the increase in deformation compared to that for 17GS is 15–21%.

To ensure the reliable operation of gas pipelines, it is also necessary to take into account significant changes in the logistics of gas supply and the unevenness of its consumption. Pressure fluctuations in the pipe are the cause of fatigue cracking and corrosion-fatigue damage [12]. Risks for gas pipelines from terrorist acts or active military actions should be highlighted separately.

Explosions near the pipeline will cause overloads, which will lead to the destruction of the long-used pipe steel, which has undergone hydrogen embrittlement.

After accumulating an array of experimental results, the question of forecasting the development of the investigated processes arises. In classical mechanics, the concept of creep is mainly related to the change in metal deformation at elevated temperatures. For the analytical description of such inelastic processes, logarithmic dependences are often used, which is where the term "logarithmic creep" came from. In the case studied by us, the term creep must be used with reservations, since the mechanisms of deformation of long-term used metal are associated not only with the plastic yielding but also with the opening of internal structural defects that arose as a result of degradation processes [18–23]. However, despite the differences in the mechanisms, the kinetics of the deformation process is similar to that of classical creep, so it was decided to use similar dependencies for the mathematical description of the obtained experimental results. It is proposed to use equations of the next form to describe deformation processes

$$y = a - b \cdot \ln(x),$$

or by substituting time and deformation parameters

$$\varepsilon = a - b \cdot \ln(t).$$

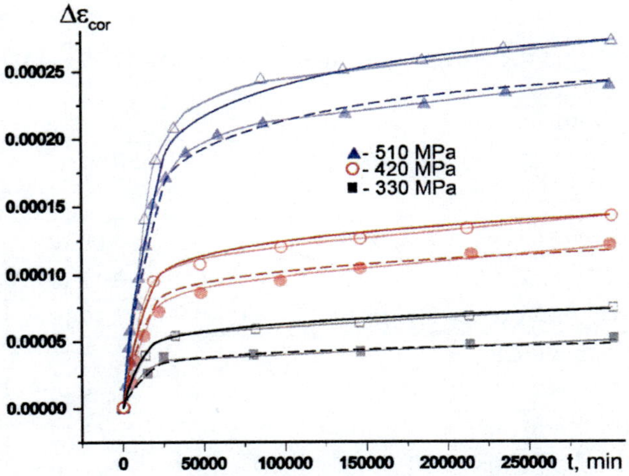

a)

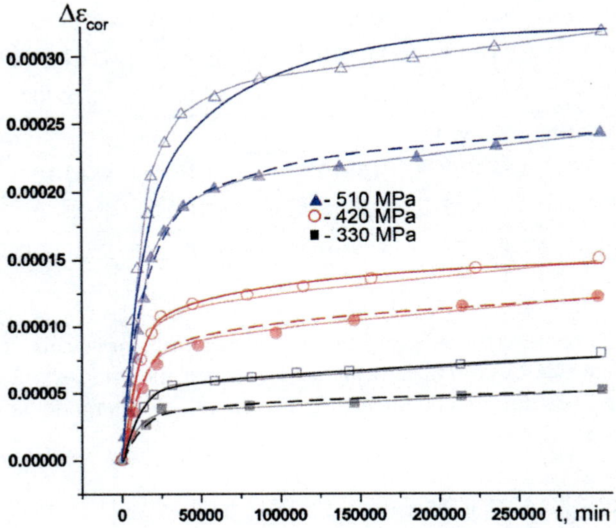

b)

Fig. 4. General view of experimental data with superimposed graphs of the obtained functional dependencies (ME1) for steel 17GS (a) and 19G (b).

The general view of the correlation of the obtained dependencies with experimental data is shown in Fig. 4.

Parametric identification was carried out based on the obtained laboratory experimental data and their comparison with data of tensometric measurements of deformation of operating pipelines by service organizations. Statistical processing of the results, which was performed in the Origin 2015 program, showed a reliable probability of mathematical description in the range of 0.96–0.98, which confirms the correct choice of the method of analytical description of the experiment (Table 2).

Table 2. Values of coefficients in analytical expressions for 17GS and 19G steel.

Non exploited 17GS steel						
ME	*330 MPa*		*420 MPa*		*510 MPa*	
	a	*b*	*a*	*b*	*a*	*b*
1	−2.66·10–5	$-6.03\cdot10^{-6}$	−9.56·10–5	$-1.70\cdot10^{-5}$	$-2.08\cdot10^{-4}$	$-3.62\cdot10^{-5}$
2	−5.47·10–5	$-8.62\cdot10^{-6}$	−2.03·10–4	$-2.68\cdot10^{-5}$	$-3.11\cdot10^{-4}$	$-4.65\cdot10^{-5}$
3	−2.24·10–4	$-2.28\cdot10^{-5}$	−3.60·10–4	$-4.10\cdot10^{-5}$	$-4.10\cdot10^{-4}$	$-5.73\cdot10^{-5}$
41 years of exploitation 17GS						
1	−3.83·10–5	$-8.73\cdot10^{-6}$	−6.86·10–5	$-1.66\cdot10^{-5}$	$-2.25\cdot10^{-4}$	$-4.02\cdot10^{-5}$
2	−5.10·10–5	$-1.01\cdot10^{-5}$	−1.85·10–4	$-2.72\cdot10^{-5}$	$-2.65\cdot10^{-4}$	$-4.62\cdot10^{-5}$
3	−2.13·10–4	$-2.34\cdot10^{-5}$	−3.63·10–4	$-4.46\cdot10^{-5}$	$-3.21\cdot10^{-4}$	$-5.47\cdot10^{-5}$
Non exploited 19G steel						
1	$-2.99\cdot10^{-5}$	$-6.31\cdot10^{-6}$	$-9.15\cdot10^{-5}$	$-1.67\cdot10^{-6}$	$-2.56\cdot10^{-4}$	$-4.07\cdot10^{-5}$
2	$-4.72\cdot10^{-5}$	$-8.03\cdot10^{-6}$	$-2.02\cdot10^{-4}$	$-2.68\cdot10^{-5}$	$-3.04\cdot10^{-4}$	$-4.63\cdot10^{-5}$
3	$-1.99\cdot10^{-4}$	$-2.04\cdot10^{-5}$	$-3.35\cdot10^{-4}$	$-3.81\cdot10^{-5}$	$-3.98\cdot10^{-4}$	$-5.48\cdot10^{-5}$
41 year of exploitation 19G steel						
1	$-4.73\cdot10^{-5}$	$-9.66\cdot10^{-6}$	$-1.20\cdot10^{-5}$	$-2.14\cdot10^{-5}$	$-2.83\cdot10^{-4}$	$-4.91\cdot10^{-5}$
2	$-6.70\cdot10^{-5}$	$-1.14\cdot10^{-5}$	$-2.23\cdot10^{-4}$	$-3.05\cdot10^{-5}$	$-3.08\cdot10^{-4}$	$-5.29\cdot10^{-5}$
3	$-2.22\cdot10^{-4}$	$-2.38\cdot10^{-5}$	$-3.79\cdot10^{-4}$	$-4.62\cdot10^{-5}$	$-3.21\cdot10^{-4}$	$-5.68\cdot10^{-5}$

Based on the results of predicting the deformation of pipe steels, it is possible to calculate the time until the critical deformation is reached, the value of which for pipelines is established by regulatory documents. This will allow emergency situations to be prevented in time.

4 Conclusions

Currently, the pipeline transport of hydrogen and methane-hydrogen mixtures through the network of operating gas pipelines is a quick and cost-effective solution for the development of the hydrogen energy sector. Research in recent years, including those conducted by the authors, has shown that as a result of the action of corrosive environments (acidic soil electrolytes, water condensate containing acidic gases (CO_2, H_2S, etc.), gaseous hydrogen) on the steel of a pipeline with a service life of more than 30 years, corrosion-hydrogen processes occur. Degradation They cause the pipe material to lose plasticity and accumulate scattered damage, which can result in stress-corrosion failure of the pipe followed by the formation of an explosive gas-air mixture.

The conducted studies recorded deformation jumps during long-term static loading, which indicate the exhaustion of the plasticity resource and increased operational risks due to hydrogen embrittlement.

Based on the results of corrosion-mechanical tests of pipe steels 17GS and 19G, mathematical dependencies were obtained for predicting deformations of pipelines in imitations of acidic soil electrolytes during their long-term influence.

In the future research is planned to expand the range of studied steels and modeling of the deformation behavior of pipe steels during the transportation of hydrogen and methane-hydrogen mixtures, as well as the use of the ionic strength of the soil electrolyte in combination with or instead of the concentration of corrosive components. This will make it possible to increase the level of reliability of main gas pipelines and identify areas that require priority modernization.

References

1. Nykyforchyn, H., et al.: Assessment of operational degradation of pipeline steels. Materials **14**(12), 3247 (2021)
2. Lesiuk, G., Correia, J.A., Krechkovska, H.V., Pękalski, G., de Jesus, A.M., Student, O.: Degradation Theory of Long Term Operated Materials and Structures. Springer, Cham (2021)
3. Poberezhny, L., Hrytsanchuk, A., Okipnyi, I., Poberezhna, L., Stanetsky, A., Fedchyshyn, N.: Minimizing losses during natural gas transportation. Strojnícky časopis-J. Mech. Eng. **69**(1), 97–108 (2019)
4. Marushchak, P.O., Salo, U.V., Bishchak, R.T., Poberezhnyi, L.Y.: Study of main gas pipeline steel strain hardening after prolonged operation. Chem. Pet. Eng. **50**(1–2), 58–61 (2014)
5. Sroka, M., Zieliński, A., Dziuba-Kałuża, M., Kremzer, M., Macek, M., Jasiński, A.: Assessment of the residual life of steam pipeline material beyond the computational working time. Metals **7**(3), 82 (2017)
6. Li, H., Huang, K., Zeng, Q., Sun, C.: Residual strength assessment and residual life prediction of corroded pipelines: a decade review. Energies **15**(3), 726 (2022)
7. Pobereznyi, L.Y., Poberezhna, L.Y., Maruschak, P.O., Panin, S.V.: Assessment of potential environmental risks from saline soils subsidence. In: IOP Conference Series: Earth and Environmental Science, vol. 50, no. 1, p. 012046. IOP Publishing (2017)
8. Poberezhnyi, L.Y., Marushchak, P.O., Sorochak, A.P., Draganovska, D., Hrytsanchuk, A.V., Mishchuk, B.V.: Corrosive and mechanical degradation of pipelines in acid soils. Strength Mater. **49**, 539–549 (2017)
9. Popovych, P., et al.: Corrosion-fatigue failure of tractor trailers metal materials in aggressive environments. KOM–Corros. Mater. Prot. J. **64**(2), 45–51 (2020)
10. Poberezhny, L., Stanetsky, A., Grytsuliak, G., Poberezhna, L., Kosmii, M., Hrytsanchuk, A.: Corrosion-mechanical behavior of gas main steel in saline soils. KOM–Corros. Mater. Prot. J. **63**(3), 105–112 (2019)

11. Wang, S., Yin, X., Zhang, H., Liu, D., Du, N.: Coupling effects of pH and dissolved oxygen on the corrosion behavior and mechanism of X80 steel in acidic soil simulated solution. Materials **12**(19), 3175 (2019)
12. Poberezhny, L., Poberezhna, L., Markhalevych, V.: Kinetics of deformation of pipe steels in acid soils. Published on 1998 (2020)
13. Roccisano, A., Nafisi, S., Ghomashchi, R.: Stress corrosion cracking observed in ex-service gas pipelines: a comprehensive study. Metall. Mater. Trans. A **51**, 167–188 (2020)
14. Zvirko, O.I., Savula, S.F., Tsependa, V.M., Gabetta, G., Nykyforchyn, H.M.: Stress corrosion cracking of gas pipeline steels of different strength. Procedia Struct. Integrity **2**, 509–516 (2016)
15. Zvirko, O.I., Nykyforchyn, H.M., Tsyrulnyk, O.T., Voloshyn, V.A., Venhrynyuk, O.I.: In-Service Degradation of Structural Steels under Cyclic Loading. Materials Science, pp. 1–7 (2022)
16. Nykyforchyn, H., Tsyrulnyk, O., Zvirko, O., Hredil, M.: Role of hydrogen in operational degradation of pipeline steel. Procedia Struct. Integrity **28**, 896–902 (2020)
17. Capelle, J., Gilgert, J., Dmytrakh, I., Pluvinage, G.: Sensitivity of pipelines with steel API X52 to hydrogen embrittlement. Int. J. Hydrogen Energy **33**(24), 7630–7641 (2008)
18. Kryzhanivskyy, Y., Poberezhny, L., Maruschak, P., Lyakh, M., Slobodyan, V., Zapukhliak, V.: Influence of test temperature on impact toughness of X70 pipe steel welds. Procedia Struct. Integrity **16**, 237–244 (2019)
19. Ohaeri, E., Eduok, U., Szpunar, J.: Hydrogen related degradation in pipeline steel: a review. Int. J. Hydrogen Energy **43**(31), 14584–14617 (2018)
20. Krechkovska, H., Student, O., Lesiuk, G., Correia, J.: Features of the microstructural and mechanical degradation of long term operated mild steel. Int. J. Struct. Integrity **9**(3), 296–306 (2018)
21. Masuyama, F.: Creep degradation in welds of mod. 9Cr-1Mo steel. Int. J. Press. Vessels and Pip. **83**(11–12), 819–825 (2006)
22. Zhuravel', I.M., Svirs' ka, L.M., Student, O.Z., Vorobel', R.A., Nykyforchyn, H.M.: Automated determination of grain geometry in an exploited steam-pipeline steel. Mater. Sci. **45**, 350–357 (2009)
23. Gabetta, G., et al.: In-service degradation of gas trunk pipeline X52 steel. Mater. Sci. **44**, 104–119 (2008)

Possibilities of Applying Biometric Data Scanning Tools in Vehicles

Margarita Prokopovič(✉) and Kristina Čižiūnienė

Faculty of Transport Engineering, Department of Logistics and Transport Management, Vilnius Gediminas Technical University, Plytinės Street 27, 10105 Vilnius, Lithuania
{margarita.prokopovic,kristina.ciziuniene}@vilniustech.lt

Abstract. Biometrics has become an increasingly popular technology in the transportation industry, as it can enhance road safety. Biometric data involves the use of physiological or behavioral characteristics to determine a person's condition. This article discusses the use of biometric technology in vehicles to ensure the safety of drivers and passengers. The article emphasizes that biometric technologies, such as sensory cameras and sensors, can be an effective tool for monitoring the state and actions of drivers, thereby reducing potential causes of accidents. In addition, it explores the legal aspects related to the collection of biometric data, aiming to ensure data security and privacy. The article provides several recommendations for using biometric data in vehicles, which include data security, access control, transparency, and compliance with relevant laws. Moreover, it is mentioned that drivers' different health and condition factors need to be considered to ensure reliable collection of biometric data. In the practical section, an expert evaluation is discussed that analyzes the five most commonly used biometric data reading technologies. The study found that the most effective technologies in use are a belt with sensors and a bracelet, a sensory camera, and a bracelet, using them together or replacing one with the other.

Keywords: Biometrics · Data protection · Biometric technologies · GDPR

1 Introduction

Biometric technologies are used in various areas that have access to and security of material objects and perform unique identification of a person. Analyzing scientific literature sources, biometric data scanning equipment, such as sensory cameras and various sensors, vehicles, which are used to ensure the safety of drivers, passengers, and road users, are becoming more and more noticeable. It has been established that one of the causes of car accidents is inappropriate driving when the driver's physical and/or psychological state is disturbed (e.g. fatigue, intoxication, stress) [1], as well as unsafe driving behavior (e.g. speeding, diverting attention to mobile) and violation of traffic rules [2]. Therefore, it is important to look for possible tools and actions to reduce such consequences. According to Arumugam S. (2019), the optimal way to prevent accidents is to monitor the driver, their driving style, changes in their physical, and psychological state and to warn the driver if any unexpected event occurs so that they can react and stop the transport decision [3].

O. Prentkovskis et al. (Eds.): TRANSBALTICA 2023, LNITI, pp. 317–326, 2024.
https://doi.org/10.1007/978-3-031-52652-7_31

In order to detect the disturbance of the driver's state in real-time data, biometric monitoring is one of the most effective tools that can detect the state change instantly. Some researchers are convinced that biometric scanning is an effective tool for driver impairment. Therefore, this article will explore the tools, methods, and possibilities for collecting biometric data, and after an expert evaluation, the most effective tools will be determined.

The subject of this article is biometric data scanning.

This article aims to determine the possibilities of scanning biometric data in vehicles.

To achieve this goal, the following tasks have been set:

1. At the theoretical level, analyze the aspects related to the screening of biometric data, such as data protection and the acts regulating it, and the application of biometric data systems.
2. Perform an expert assessment. To find out which biometric data scanning equipment experts think is more efficient and convenient for transportation.

2 Protection of Biometric Data

Despite the benefits, the use of biometrics in transport also raises privacy and data protection concerns. To protect against these problems, transport providers must ensure that appropriate measures are in place. Therefore, it is necessary to first discuss aspects of legal regulation in reading and using biometric data [4, 5].

The legal assessment of the driver's biometric data collection may differ from the legislation of the member states of the European Union [6]. Each member state has its own national law governing the collection of biometric data. Therefore, it is recommended to check your country's legislation regarding biometric data before starting to collect driver biometric data [7]. This ensures compliance with legal requirements for data collection and processing. It has been described that as a general principle, all data related to a person's health is considered personal information, and specific rules must be followed when handling personal health to ensure personal privacy and security [8]. In addition, the European General Data Protection Regulation (GDPR) also regulates biometric collection and processing. However, it is necessary to remember that according to GDPR, (2016), biometric data are sensitive and can only be processed under certain exceptions [9]. Their processing requires more protection, as this type of data may cause a greater human risk to fundamental rights and freedoms, so their use must be regulated by the relevant legislation, in this case, the GDPR. In addition, passengers must be given information about how their biometric data is used [10], to ensure that it is used only for the purpose, for which a consent has been issued. As mentioned by R. Jiang et al., it is important that biometric measures should be linked to an overall security program that includes other important initiatives such as training, the use of an insurance program, and protocols for handling situations related to misuse [11]. In addition, the use of biometric tools should comply with the law and be regulated by relevant authorities to ensure the proper processing and use of biometric data.

According to B. Woodward (2021), assessing the reliability of driver biometric data in the context of traffic safety must be done by taking into account many factors and context, including data collection methodology, sensor quality, and data analysis accuracy. In addition, it should be considered that individuals may have different health problems that may affect the reliability of the data, so it is important to ensure that the data is accurate [12].

In order to ensure security and reliability, several recommendations regarding the use of biometric data in transport vehicles are distinguished by analyzing the research and insights of other authors:

1. Data security: The security of biometric data must be of the highest level. It is recommended that data be stored encrypted and transmitted using secure communication systems [13].
2. Access control: It is necessary to ensure that only authorized and competent institutions and personnel have access to biometric data. It is also important to ensure that a reliable data storage and backup process is used [14].
3. Transparency: Public organizations must provide their customers with transparent and understandable information about the use of biometric data in vehicles, and it should also be clear how this data will be handled and used [15].
4. Systems level: A reliable and appropriately certified biometric data storage and verification system that complies with relevant legislation and standards must be used [16].
5. Compliance with legal and regulatory requirements: Drivers of a vehicle must comply with legal and regulatory requirements, and persons to whom biometric data belongs must be adequately protected.

It is crucial to ensure compliance with data protection laws, reliable scanning of biometric data, and the guarantee of personal privacy and security. This would increase road safety and reduce the number of traffic accidents.

3 Biometric Data Scanning Tools

Biometrics is becoming an increasingly popular field of technology in the transportation industry because of its potential to improve safety, efficiency, and passenger experience [17]. Biometric data refers to a physiological or behavioral characteristic for identifying individuals [18]. In transport, biometric data can be used for various purposes, such as passenger identification, access control, and payment verification. But one of the main advantages of collecting biometric data in transport is the possibility of increasing security. By using biometrics such as fingerprints, facial recognition, and iris scanning, transport providers can identify unauthorized access and reduce the risk of fraud or identity theft [19]. Biometrics can also improve efficiency by reducing boarding times, eliminating the need for physical tickets, and enabling automatic check-in on public transport [20]. In addition, biometrics can also help control driving and clearly ensure driving safety. For example, installing an eye tracking system can monitor the driver's activity and prevent reckless driving that can endanger passengers and other people on the road. Also, such data is useful in the context of safety security, as it can help determine the driver's state of fatigue, recognize his facial or voice features, as well as measure blood pressure and pulse quality [21]. There are sensors that can measure body temperature and perspiration, and this can also be used as biometric data. Thus, biometric data can still determine the physical condition of the driver and his ability to operate the vehicle. Biometric data can be used as a tool for employee health data. In this case, as emphasized by M.V. Natta et al. (2020), a thermographic camera is usually used for biometric identification, which measures a person's body camera. This may be important to ensure the safety of workers or to reduce the possibility of infection [22].

One biometric technology that is often used for security purposes is the facial recognition system. According to J. Kortly et al. (2020), facial recognition systems can be integrated into the vehicle management system as well as security systems to verify the identity of the driver when he logs into the vehicle management system [23].

Based on the above information, this paper will evaluate which of the proposed biometric data scanning tools are the most convenient and could be applied:

1. **Bracelet.** There are special bracelets that can record a person's pulse, blood pressure, and other stress indicators, which are unique and can be used as biometric data. In addition, bracelets can be equipped with special sensor technologies that can capture the shape of a person's fingers, fingerprints, etc. [24]. It can also be used as biometric data which can be used as a method of identity verification.
2. **Smart watch**. It can measure certain human body parameters including heart rate, blood pressure, body temperature, etc. This data is unique to each individual and can therefore be used as biometric data. In addition, some smart watch models have built-in sensors that can recognize unique features of the human body, such as fingerprints, which can be used as a biometric authentication method [25]. Various health data collected from the smartwatch can be used with the help of artificial intelligence algorithms to obtain personalized information about the health status of a person. Therefore, a smart watch is a very valuable tool for collecting biometric data.
3. **Sensors located on the steering wheel**. Some car manufacturers install light sensors on the steering wheel to measure the driver's body temperature and perspiration. In addition, some steering wheel knobs have built-in sensors that can capture the driver's fingerprints, as well as recognize the shape and size of the driver's hands [26]. It can also be used as a biometric data collection tool. After all, the steering wheel is an object that is frequently touched and used, so it can be used as a biometric authentication method.
4. **Sensor belt.** It can be used as a means to collect biometric data. Such a belt can measure certain parameters of the body, including lung activity, heart rate, blood pressure, computer, steps, etc. On-the-belt sensors, such as EKG sensors, can record the efficiency of failing heartbeats and increase the ability to recognize visual disturbances, or EEG Wave sensors, which can measure brain waves and help determine mood, stress, and more. In addition, belt sensors can be used in various fields, including sports and medicine [27]. For example, the belts used by athletes allow recognition of their movement level, pace, etc., and help prepare better for other sporting events. Belts with sensors can also be used as biometric methods of identity verification, as the data they can measure is unique to each individual.
5. **Sensory cameras.** They can be used as a tool to collect biometric data. Cameras can measure certain human body parameters or body activity that can be used as biometric data. For example, a light wave sensor measuring camera can capture the shape of a person's face, as well as eye movement patterns that are unique to each

individual. There are also special cameras that can measure a person's height and body shape [28]. It can also be used as biometric data which can be used as a method of identity verification. In addition, there are special cameras that can measure body temperature, etc. Various health data collected using the camera can also be used with the help of artificial intelligence algorithms to obtain personalized information about a person's health status [29]. Thus, the sensor camera can be used as a biometric data collection tool.

Various tools can be used to collect biometric data, but the most convenient and commonly used are bracelets, smart watches, steering wheel sensors, sensor belts, and sensor cameras. Therefore, these biometric data collection tools and their application possibilities in the transportation process will be further evaluated.

4 Methodology

The purpose of this study is to define a set of criteria that show which of the 5 devices mentioned above for collecting biometric data is the most applicable in the transportation process. First of all, the research methods that will be used in the study itself should be defined. An expert evaluation will be performed, the results of which will be ranked and the correlation coefficient will be calculated.

The essence of the expert assessment method is the rational organization of the quantitative assessment of the problem and the analysis carried out by experts in the processing of conclusions.

Expert assessment can be preliminarily ranked, if the experts' quality indices were not provided. Ranking is a procedure where the most important index is given a rank equal to 1, the next one according to importance - second, etc. The last index according to importance is given a rank m; where m is the number of indices to compare.

Relevance-based ranking models are used to generate a ranking list. To achieve this, objects are assigned weights that determine their rank [30].

Correlation analysis will be used to discover the relationship between variables. Pearson's linear correlation coefficient will be used.

The correlation coefficient is a criterion for quantifying the linear dependence between variables or a measure of the strength of the relationship. For variables measured on an interval scale, the Pearson correlation coefficient is calculated. When the observed categorical variables are measured on a rank or nominal scale, other measures of the strength of the relationship are used [31, 32].

Measures of association between rank variables are used to assess the strength of the association between variables measured on a rank scale. These are usually non-parametric measures of the strength of the relationship, independent of the nature of the distribution of the variables. These coefficients measure the linear dependence of variables: a positive correlation coefficient indicates a direct dependence of variables (higher values of one variable correspond to higher values of another variable), a negative one - an inverse correlation (higher values of one variable correspond to lower values of another variable) [33].

When calculating the dependence of variables on factors, weights are assigned. According to H. Sivilevičius (2017), the weights of the evaluation criteria mainly determine the evaluation result.

The average rating obtained during the expert evaluation is considered to be the solution to the problem (result of the evaluation) if the decisions of all experts are consistent. When a decision is made on the basis of the average assessment of experts or respondents, the level of consistency of their assessments should be determined. It is calculated using Kendall's coefficient of concordance W, which can only be used if series of estimates assigned by respondents or experts are provided. If these estimates are expressed in other units (e.g. points), they must be ranked [34].

5 Research Results and Discussion

The literature review identified 5 commonly used biometric scanners. In order to assess their applicability in the transport process, an expert evaluation was carried out with the participation of 8 competent experts with many years of experience in transport. The resulting rankings and their calculation are given in Table 1.

Table 1. Rating expert evaluation of biometric data scanning devices.

Calculations	Criteria				
	X_1	X_2	X_3	X_4	X_5
Sum of ratings	25	40	12	21	22
Mean of ratings	3.125	5	1.500	2.625	2.75
Difference between the sum of ratings and the fixed value	1	16	−12	−3	−2
Square of the difference	1	256	144	9	4

Criteria: X_1 – Bracelet, X_2 – Smart watch, X_3 – Sensors on the steering wheel, X_4 – Belt with sensors, X_5 – Sensory camera.

Since there are no correlated ranks, Kendall's coefficient of concordance (W = 0.6469) was calculated. This coefficient is greater than 0.5, which allows us to say that the opinions of all interviewed experts agree.

The fit coefficient weight is calculated using the chi-square Pearson test. $\chi^2 = 20.70$ or more than χ^{2kr} which is equal to 9.48773, so the expert opinions are still considered to be in agreement.

After calculating the smallest matching coefficient $W_{min} = 0.1186 << 0.6469$, it can be said that the opinions of the interviewed experts also agree. These obtained results allow us to conclude that the opinions of experts regarding 5 biometric data scanning devices are compatible and generalized.

Later, the indicators of the importance of the criteria were calculated and presented in Table 2.

According to the obtained results, biometric data scanning devices, according to their application in the transportation process, are arranged in the following order (from the most applicable to the least applicable):

Table 2. Indicators of biometric data scanning devices.

Indicator value	Criteria encryption symbol					Reference number
	X_1	X_2	X_3	X_4	X_5	
qj	0.2083	0.3333	0.1000	0.1750	0.1833	1
dj	0.7917	0.6667	0.9000	0.8250	0.8167	4
Qj	0.1979	0.1667	0.2250	0.2063	0.2042	1
Qj'	0.1917	0.0667	0.3000	0.2250	0.2167	1
Arrangement of factors by importance	4	5	1	2	3	

1. Sensors located on the steering wheel.
2. Belt with sensors.
3. Sensory camera.
4. Bracelets.
5. Smart watch.

Considering the obtained results, it was important to evaluate how these biometric data scanning tools interact with each other (Fig. 1).

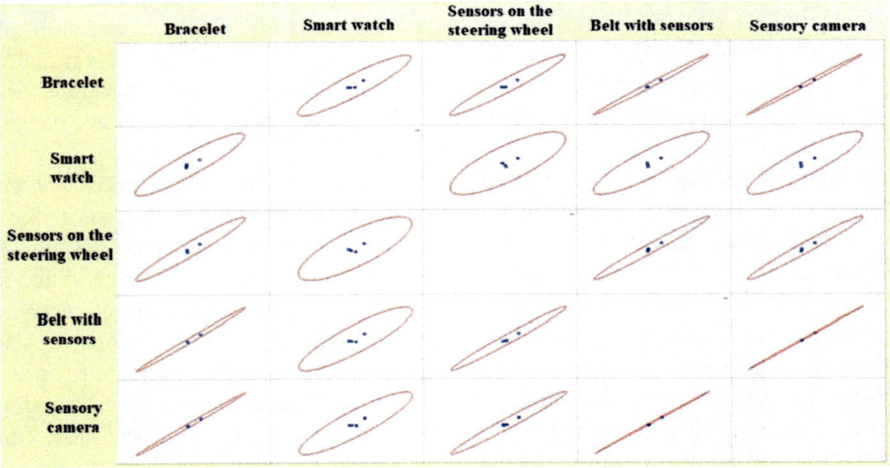

Fig. 1. Interaction of biometric data scanning tools.

Examining the correlations between the selected parameters, it can be stated that all measures ensuring biometrics were correlated with each other. Belt with sensors and bracelets, sensor camera and bracelets showed the highest correlation effect. This was also confirmed by the Pearson correlation coefficient (Table 3).

Table 3. Table of Pearson's correlation coefficient.

	Bracelet	Smart watch	Sensors on the steering wheel	Belt with sensors	Sensory camera
Bracelet	–	0.91781	0.96708	0.99602	0.9977
Smart watch	0.91781	–	0.78658	0.87877	0.88877
Sensors on the steering wheel	0.96708	0.78658	–	0.98592	0.98212
Belt with sensors	0.99602	0.87877	0.98592	–	0.99977
Sensory camera	0.9977	0.88877	0.98212	0.99977	–

These correlations (between sensor belt and bracelet and sensor camera and bracelet) were particularly strong and approached unity according to Pearson's correlation coefficient. Therefore, it can be concluded that these biometric monitoring tools are fully interchangeable and can be used both as a tool for duplicating biometric systems (if necessary) and as separate systems, the use of which ensures reliable data reading without the additional use of duplicating systems. The sensors on the steering wheel also showed a very strong correlation with other biometric reading devices (bracelet, sensor belt, and sensor camera), but the correlation was not close to unity, so their direct interchangeability is not recommended. Such systems can be used as support or backup, but not fully interchangeability. The smart watch showed the worst correlation of all biometric systems. Although the correlations were strong, for reliability, the correlation levels of Pearson coefficient values must approach unity if the elimination of duplication of biometric systems is desired. This biometric scanning system can be used as a backup or backup.

In any case, it can be said that biometric data scanning systems are necessary and effective. After the research, it can be seen that the most effective systems in use are a belt with sensors and a bracelet, a sensor camera and a bracelet, which do not require any additional, duplicating systems. It is only important to remember that in addition to efficiency, in order for the data to be accurate, it is also important to take into account such facts as the state of human health, which can be different for each person. It is also important to ensure compliance with the General Data Protection Regulation to ensure personal privacy and security. Overall, using the most efficient biometric data scanning systems, while complying with system quality, analysis accuracy, and data protection legislation, will allow monitoring of the state of a person's health at the moment and allow increasing road safety and reducing the number of traffic accidents.

6 Conclusion

The conducted analysis of scientific literature indicates that data security is a critical aspect and it is necessary to comply with all data protection laws and regulations to ensure that the scanned biometric data is reliable, its usage is transparent and trustworthy, to ensure the privacy and security of individuals.

The expert evaluation study showed that the most effective biometric data reading devices are a belt with sensors and a bracelet, a sensory camera, and a bracelet, using them together or replacing one with the other. This was also confirmed by the Pearson correlation coefficient, which also demonstrated that these biometric monitoring tools are fully interchangeable and can be used as a duplication tool for biometric systems (if necessary), as well as separate systems that ensure reliable data reading without the use of additional duplicating systems.

In conclusion, it can be asserted that biometric data can transform the organization of the transportation process by improving safety, efficiency, and the transportation process itself. However, it is crucial to balance the benefits with privacy and data protection aspects, to ensure that the use of biometric data in transportation is ethical and sustainable.

References

1. Chen, C.-L., Subramanian, R., Choi, E.-H., Liu, C.: Statistical methodology to make early estimates of motor vehicle traffic fatalities (Traffic Safety Facts Research Note). Transport. Res. Part A Policy Pract. **44**(5), 291–305 (2010)
2. Ivers, R., et al.: Novice drivers' risky driving behavior, risk perception, and crash risk findings from the drive study. Am. J. Public Health **99**(9), 1638–1644 (2009)
3. Arumugam, S., Bhargav, R.: A survey on driving behavior analysis in usage-based insurance using big data. J. Big Data **6**(86), 1–21 (2019)
4. Čižiūnienė, K., Matijošius, J., Čereška, A., Petraška, A.: Algorithm for reducing truck noise on via baltica transport corridors in Lithuania. Energies **13**, 6475 (2020)
5. Zaranka, J., Pečeliūnas, R., Matijošius, J.: Analysis of the influence of fatigue on passenger transport drivers' performance capacity. Transport **27**, 351–356 (2012). https://doi.org/10.3846/16484142.2012.750625
6. Sallavaci, O.: Strengthening cross-border law enforcement cooperation in the EU: the Prüm network of data exchange. Eur. J. Crim. Policy Res. **24**(3), 219–235 (2017). https://doi.org/10.1007/s10610-017-9355-0
7. Kilikevičius, A., Kilikevičienė, K., Fursenko, A., Matijošius, J.: The analysis of vibration signals of critical points of the bus body frame. Period. Polytech. Transp. Eng. **48**, 296–304 (2020). https://doi.org/10.3311/PPtr.12784
8. Di Minin, E., Fink, C., Hausmann, A., Kremer, J., Kulkarni, R.: How to address data privacy concerns when usingsocial media data in conservation science. Conserv. Biol. **35**(2), 437–446 (2021)
9. European parliament and council: general data protection regulation. J. Ref. **L119**, 1–88 (2016)
10. Jedliński, Ł, Caban, J., Krzywonos, L., Wierzbicki, S., Brumerčík, F.: Application of vibration signal in the diagnosis of IC engine valve clearance. J. Vibroengineering **17**, 175–187 (2015)
11. Jiang, R., Al-maadeed, S., Bouridane, A., Crookes, D., Beghdadi, A. (eds.): Biometric Security and Privacy. SPST, Springer, Cham (2017). https://doi.org/10.1007/978-3-319-47301-7
12. Woodward, B., Kliestik, T.: Driving car control algorithms in smart sustainable urban mobility systems. Contemp. Read. Law Soc. Justice **13**(2), 51–64 (2021)
13. Wilson, A.A.T., Bhattacharjee, N., Srinivasan, B.: Privacy-preserving biometrics authentication systems using fully homomorphic encryption. Int. J. Perv. Comput. Commun. **11**(2), 1742–7371 (2015)
14. Karlova, T.V., Bekmeshov, Y.A., Kuznetsova, N.M.: Protection the data banks in state critical information infrastructure organizations. In: International Conference "Quality Management, Transport and Information Security, Information Technologies" (2019)

15. Braun, T., Fung, B.C., Iqbal, F., Shah, B.: Security and privacy challenges in smart cities. Sustain. Cities Soc. **39**, 499–507 (2018)
16. Slusky, L.: Cybersecurity of online proctoring systems. J. Int. Technol. Inf. Manag. **29**(1), 56–83 (2020)
17. Caldwell, T.: Market report: border biometrics. Biometric Technology Today (15)5, 5–11 (2015)
18. Tripathi, K.P.: A comparative study of biometric technologies with reference to human interface. Int. J. Comput. Appl. **14**(5), 0975–8887 (2011)
19. Wilson, D.: Biometrics, borders and the ideal suspect. In: Borders, Mobility and Technologies of Control, pp. 87–109 (2006)
20. Pritchard, G., Vines, J., Oliver, P.: Your money's no good here: the elimination of cash payment on london buses. In: Proceedings of the 33rd Annual ACM Conference on Human Factors in Computing Systems, UK, pp. 907–916 (2015)
21. Magaña, V.C., Scherz, W.D., Seepold, R., Madrid, N.M., Gracia, R.: The effects of the driver's mental state and passenger compartment conditions on driving performance and driving stress. Sensors **20**(18), 5274 (2020)
22. Natta, M.V., et al.: The rise and regulation of thermal facial recognition technology during the COVID-19 pandemic. J. Law Biosci. **7**(1), 1–17 (2020)
23. Kortli, Y., Jridi, M., Falou, A.A., Atri, M.: Face recognition systems: a survey. Sensors. **20**(2), 342 (2020)
24. Liu, Y., Pharr, M., Salvatore, G.A.: Lab-on-skin: a review of flexible and stretchable electronics for wearable health monitoring. Am. Chem. Soc. **11**(10), 9614–9635 (2017)
25. Karatas, C., et al.: Leveraging wearables for steering and driver tracking. In: IEEE INFOCOM 2016 - The 35th Annual IEEE International Conference on Computer Communications. San Francisco, CA, USA (2016)
26. O. Abu-Faraj, Z., Chamaa, W., Hadchiti, A., Sraj, Y., Tannous, J.: Design and development of a heart-attack detection steering wheel. In: 11th International Congress on Image and Signal Processing, BioMedical Engineering and Informatics (CISP-BMEI). Beijing, China (2018)
27. Sanfilippo, F., Pettersen, K.Y.: A sensor fusion wearable health-monitoring system with haptic feedback. In: 2015 11th International Conference on Innovations in Information Technology (IIT). Dubai, United Arab Emirates (2015)
28. Neethirajan, S., Reimert, I., Kemp, B.: Measuring farm animal emotions—sensor-based approaches. Sensors. **21**(2), 553 (2021)
29. Matijošius, J., Vasiliauskas, A.V., Vasilienė-Vasiliauskienė, V., Krasodomskis, Ž: The assessment of importance of the factors that predetermine the quality of a service of transportation by road vehicles. Procedia Eng. **134**, 422–429 (2016)
30. Liu, T.Y.: Learning to Rank for Infromation Retrieval. Springer, Heidelberg (2011)
31. Garson, D.: Ordinal Association: Gamma, Kendall's tau-b and tau-c, Somers' d (2008)
32. Sakalauskas, V.: Duomenų analizė su STATISTICA. Margi raštai, Vilnius (2003)
33. Pukėnas, K.: Kokybinių duomenų analizė SPSS programa (2009)
34. Sivilevičius, H., Maskeliūnaitė, L.: Multiple criteria evaluation and the inverse hierarchy model for justifying the choice of rail transport mode. Promet Traffic Transport. **30**(1), 57–69 (2018)

Modelling the Potential Impact of the Application of Environmentally Friendly Transport Applied in Last-Mile Delivery on the National Economy: The Case of Latvia

Astra Auzina-Emsina(✉)

Riga Technical University, Kalnciema Iela 6-216, Riga, Latvia
astra.auzina-emsina@rtu.lv

Abstract. Green logistics or greener technologies in logistics compared to the current are widely discussed and promoted in the European Union (EU) level and national level energy, environmental, transport, economic recovery, economic transformation policies and initiatives. However, a limited amount of quantitative studies and report covers both impacts on specific sectors or industries and national level impact. Two development scenarios on technology application effects are modelled with input-output model for Latvia's economy obtaining results for 64 products of economic activity (NACE Rev2), and additional regional development examined (NUTS 3 disaggregation). The modelling results claim that if more land transport services (+10%), postal and courier services (+10%), and hence electricity (+10%), are demanded, but less warehousing and support services for transportation (−5%) due to digital solutions and more efficient technologies then the national economic output increase by 0.8%, amid aggregated sectors the largest impact is on wholesale (+1.7%) and industry (+1.3%) and regionally impact is between +0.85% (in Pieriga region) and 0.72% (in Vidzeme). Results on relatively noticeable demand decrease for petroleum products (−3%) and less air transport services due to emissions and replacement with land transport and glocalization (−2%), modelled impact is +0.5%. Results are highly sensitive on scenario assumptions. Companies and policies are tackling different areas; only identified are modelled.

Keywords: Last-mile delivery · Input-output · Intersectoral linkages

1 Introduction

Transport and storage sector together with hotels, restaurants and catering services has experienced the most severe negative impact in 2020–2022 due to pandemic policy measures, restrictions, changes in consumption pattern, lower income level. At the same time, some sub-industries as courier services and package stations had exploded. However, at the present, high and unstable energy prices together with severe emission reduction targets have resulted that green logistics or greener technologies in logistics compared to

O. Prentkovskis et al. (Eds.): TRANSBALTICA 2023, LNITI, pp. 327–336, 2024.
https://doi.org/10.1007/978-3-031-52652-7_32

the currently applied are widely discussed and promoted in the European Union (EU) and national governments concerning energy, environmental, transport, economic recovery, economic transformation, digitalization policies and initiatives.

Last-mile delivery services include activities from various fields that are also reflected by research activity. Numerous studies have investigated last-mile delivery and using drones in past decade, focusing on technological and economical, compering drone and land-based delivery services [1], general routing issues, drone routing issues [2] and [3], technologies of self-driving robots [4], even multi-vehicle truck-and-robot application in last-mile delivery [5].

Also other specific technologies used or potentially used in last-mile delivery are examined, as parcel stations and parcel lockers ([6] in Poland), home deliveries replaced with parcel lockers [7], customer preferences selecting last-mile delivery in e-commerce in general [8] and regarding sustainability of delivery [9], environmental sustainability of delivery considered by gender and/or age (as Norwegian female preferences in [10], for specific industry or products [11], in urban areas, as, sustainable urban last-mile [12], e-groceries [13], integrations of last-mile deliveries with passenger transport [14], passengers last-mile transport [15], sustainable urban last-mile for various stakeholders [16], using cargo bikes in urban deliveries [17]), mix of green and traditional delivery business models in urban areas [18].

However, there is limited research investigating the overall macroeconomic impact of last-mile delivery induced processes in the economy, as well on individual industries and regional perspective, that apply bottom-up approach rather than top-down approach.

The aim of this study is to model the potential impact of the application of environmentally friendly transport applied in last-mile delivery on the national economy in Latvia. The principal objectives are the following: 1) examine data availability, comparability, truthfulness, applicability in order to set the research time framework; 2) reveal tendencies and linkages that are applicable for assumptions elaboration; 3) build and test the model; ensuring that model can be used for the assumptions pre-defined; 4) check obtained modelling results and identify possible applied application options of research findings in various fields, including state policy-making. Statistical research methods, input-output is applied. The research involves analysing and modelling for Latvia's economy, however, there are no assertive limitations to apply the modelling instrument any EU member state with some modifications. Based on the findings of the literature review, comprehensive systematic and modelling methods produce truthful and useful results.

2 Materials and Methods

Pre-defined development scenarios and multisectoral macroeconomic model that is a comparative-static input-output model for one EU member state (Latvia) on latest (2015) data basis, in 64*64 economic activities level or disaggregation according to NACE Rev. 2 form the core of applied modelling instrument. Integration of input-output data in macroeconomic modelling has a long-standing history, and these data have been used in several studies also recently, as [19] and [20] for Latvia.

Postal and courier activities (H53) amid NACE 64 industries most closely represents last-mile delivery. Scenarios also focus on Land transport and transport via pipelines (H49) and Warehousing and support activities for transportation (H52). Regarding statistics, Latvia alike Lithuania, Poland, Estonia, Sweden, etc. applies product-by-product approach [21], however some (as the Netherlands, Denmark, Finland) publish input-output table set only in industry-by-industry level [22].

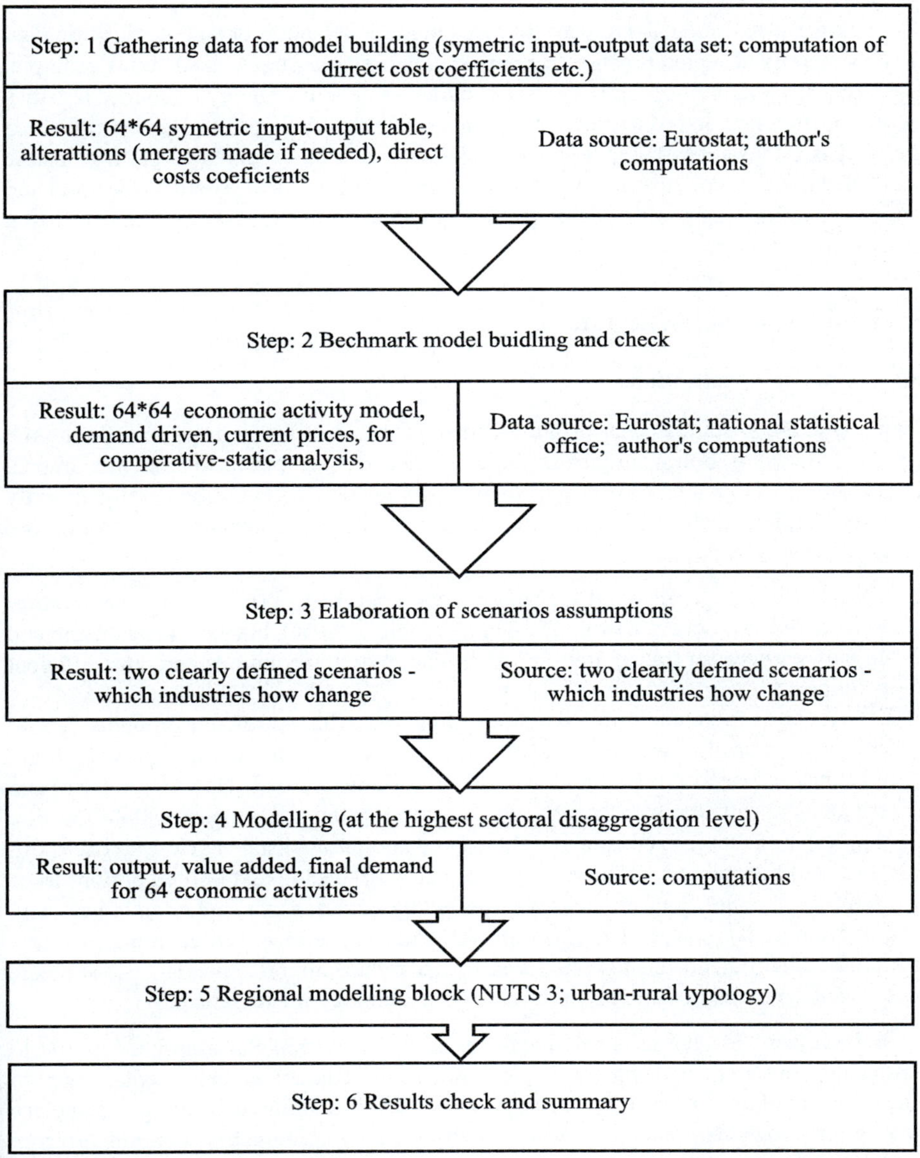

Fig. 1. Flowchart of multisectoral macroeconomic modelling process with regional module.

The database is Eurostat [21]. Latest officially published symmetric product-by-product input-output table set in current prices of 2015 for Latvia is used. Symmetric input-output (total, domestic, import) are published at minimum every five years with no more than 3 years after the reference year. The next symmetric tables' set for 2020 will be available in November 2023 according to national statistical office of Latvia [23]. MS Excel and EViews software are used.

Figure 1 is a graphical summary of steps in model elaboration and application, and it is a generally accepted practice to return to any previous step if needed, for example, if the scenario assumptions are revised by invited sustainable energy or courier logistics experts or for specific policy measure estimation, return to Step 2 and execute the next steps. It is impossible to skip a step, as all steps must be executed to get a plausible result that fits the economy, specific scenario under discussion and realistic sectoral and regional results. Regional modelling block (module) relies on NUTS 3 and Eurostat urban-rural typology.

3 Results and Discussion

3.1 Scenario Assumptions

Two comparative-static scenarios are estimated for Latvia's economy, considering the general tendencies, policy initiatives, target levels and assumptions to model the impact of sectoral development caused by potential application of environmentally friendly transport applied in last-mile delivery, evaluating the macroeconomic, sectoral, and regional perspectives:

1. 1st scenario – larger demand for land transport services (+10%), postal and courier services (+10%), and in a result also for electricity (+10%), but less warehousing and support services for transportation (−5%) due to digital solutions and more efficient technologies.
2. 2nd scenario - noticeable relative demand decrease for petroleum products (−3%) and less air transport services (−2%) due to emissions limitations and growing flight ticket prices, hard-hitting target levels due to environment (EU New Green Deal) and replacement with land transport and glocalization (in this study meaning the allocation of production closer to consumers; shorter transportation distances (less air transport, more land transport as water transport is relatively specific and limited options to be used for the majority of products transported by land transport inlands), larger share of local materials and products in produced product or service; gradual replacement of imported ingredients due to policy measures to reduce import dependence in strategic importance industries and products as high-technology manufacturing).

In both scenarios, it is assumed that the basic transportation consumption pattern and routine are consistent, and changes in technologies are gradual, no quick replacement or abandonment of existing patterns. In general, the society assumes that changes take place in long term towards greener, environmentally friendly, sustainable economy, greener logistics including. The other parameters (such as regional economic activity allocation by NUTS 3 level and economic activity by urban-rural typology classification, etc.) remain unaltered.

3.2 1st Scenario

Modelling results justify that demand increase results in domestic output increase and import increase. The output of land transport is forecasted to grow by 7.0% as considerable share of land transport services are imported (operated by foreign companies). Increase of land transport demand maintenance and repair, even reconstruction of roads and highways, and hence a notable impact is on mining of sand and alike road materials used (5.2%). Table 1 illustrates the largest changes in economic activity amid modelled 64 economic activities; amid economic activities that have impact larger than 2%, the majority are manufacturing activities, however, major service activities are evident as well. High-technology industries according to Eurostat [24] (as pharmaceutical products) are no amid the most affected economic activities.

Table 1. The largest modelled impact on sectoral performance of the 1st scenario (change, %).

Code	Economic activity	Change, %
H49	Land transport services and transport services via pipelines	7.0
B	Mining and quarrying	5.2
D	Electricity, gas, steam and air conditioning	3.7
C22	Rubber and plastic products	2.6
H53	Postal and courier services	2.5
N77	Rental and leasing services	2.3
C20	Chemicals and chemical products	2.2
C24	Basic metals	2.1
C29	Motor vehicles, trailers and semi-trailers	2.0

As modelling approach is bottom-up, it is possible to aggregate the detailed results into larger sectors and reveal that overall impact on economy. Table 2 illustrates the aggregated impact on sectors. Wholesale and retail trade, transport, accommodation and food service activities' output are modelled to increase by 1.7%, but industry, including, manufacturing, by 1.3%. The national economic output increase by 0.8%,

Even over-optimistic and possibly disputable assumptions (of 10% increase), actually leads to less than 2% impact on sectors as specific and narrow economic activities are shocked and due to import as well. For example, Coke and refined petroleum products (C19) is almost completely import dependent and hence any increase results is almost corresponding imports increase.

3.3 2nd Scenario

The 2nd scenario has a wider set of assumptions and hence the impact contains both increase and decline in demand effects. Table 3 illustrates the largest changes in economic activity amid modelled 64 economic activities that are above 1%. Mining (B), energy

Table 2. Modelled impact of the 1st scenario on aggregated sectors (change, %).

Code	Sector	Change, %
A	Agriculture, forestry and fishing	0.2
B–E	Industry (except construction)	1.3
F	Construction	0.3
G–I	Wholesale and retail trade, transport, accommodation and food service activities	1.7
J	Information and communication	0.5
K	Financial and insurance activities	0.3
L	Real estate activities	0.2
M_N	Professional, scientific and technical activities; administrative and support service activities	0.8
O–Q	Public administration, defence, education, human health and social work activities	0.0
R–U	Arts, entertainment and recreation; other service activities; activities of household and extra-territorial organizations and bodies	0.0
	Total	**0.8**

(D) and transport (H49, H53) economic activities are more affected, at the same time, manufacturing of chemicals (2.4%), rubber and plastic products (1.8%), basic metals and motor vehicles, mainly, trailers and semi-trailers (1.0%).

Table 3. The largest modelled impact on sectoral performance of the 2nd scenario (change, %).

Code	Economic activity	Change, %
B	Mining and quarrying	4.9
H49	Land transport services and transport services via pipelines	3.6
D	Electricity, gas, steam and air conditioning	3.5
H53	Postal and courier services	2.4
C20	Chemicals and chemical products	2.4
H50	Water transport services	2.3
C22	Rubber and plastic products	1.8
C24	Basic metals	1.2
C29	Motor vehicles, trailers and semi-trailers	1.0

Negative impact is modelled for three economic activities amid 64: for air transport (due to the direct assumptions concerning demand for) and modelled impact on, firstly, employment services (–0.1%) and, secondly, travel agency, tour operator and other reservation services and related services (−0.1%). Industry and wholesale and retail trade, transport, accommodation and food service activities are the most positively influenced sectors (see Table 4).

Table 4. Modelled impact of the 2nd scenario on aggregated sectors (change, %).

Code	Sector	Change, %
A	Agriculture, forestry and fishing	0.2
B-E	Industry (except construction)	1.2
F	Construction	0.2
G-I	Wholesale and retail trade, transport, accommodation and food service activities	0.9
J	Information and communication	0.3
K	Financial and insurance activities	0.2
L	Real estate activities	0.2
M_N	Professional, scientific and technical activities; administrative and support service activities	0.5
O-Q	Public administration, defence, education, human health and social work activities	0.0
R-U	Arts, entertainment and recreation; other service activities; activities of household and extra-territorial organizations and bodies	0.0
	Total	**0.5**

The results highlight that the average total impact on the economy (of 0.5%) is actually not relevant to individual economic activity performance as impact varies between 4.9% and −1.2% (for air transport) amid modelled 64 activities.

3.4 Regional Impact of Modelled Scenarios

Economic activity is distrusted unevenly, hence it is important to model how specific assumptions or shocks have an impact on regions, which regions are affected more (positively or negatively), which regions are less. Latvia according to NUTS 3 is subdivided into six regions, and Table 5 illustrates the computed bottom-up aggregated impact on these regions. Regional impacts (see Table 5 and Table 6) are represented with two signs after decimal separator to display the variance.

Table 6 illustrates the computed bottom-up aggregated impact on urban and rural regions (according to Eurostat typology [25]). Intermediate regions (representing approximately 33% of economic activity [23]) are influenced by largest scale compared to predominantly urban regions and predominantly rural regions.

Results are highly sensitive on scenario assumptions. Companies and policies are tackling different areas but only the identified ones are modelled. Future research will have to meet the challenge of growing and or unstable energy costs (petroleum products,

Table 5. Modelled regional impact of both scenarios (change, %).

Scenarious/NUTS3	TOTAL	Riga	Pieriga	Vidzeme	Zemgale	Kurzeme	Latgale
First scenario	**0.77**	0.76	0.85	0.72	0.75	0.79	0.75
Second scenario	**0.51**	0.48	0.58	0.53	0.55	0.55	0.52

Table 6. Modelled urban-rural impact of both scenarios (change, %).

Scenario/Territory	TOTAL	Predominantly urban regions	Intermediate regions	Predominantly rural regions
First scenario	**0.77**	0.76	0.81	0.74
Second scenario	**0.51**	0.48	0.56	0.54

electricity), how companies deal with these challenges and how short term and long-term decision are made and hence applied technologies altered or changed completely. Growing energy costs push towards use of greener energy (also double effect for companies as might be (more) sustainable for clients), cheaper energy or more subsidised energy (reduce costs directly).

4 Conclusions

Last-mile delivery is a complex and developing activity over time, that involves the application of present practices and also introduction of another new and innovative technical and digital solutions available today. In this study, the assumptions on last-mile delivery are stated on the basis of limitations of NACE 64 economic activity classification and interindustry linkages. That influences the sectoral and regional modelling results obtained. At the present, the scenario assumptions cover partly the application of environmentally friendly transport applied in last-mile delivery there are other aspects as emissions, available grants and even non-economic factors, including preferences, that also influence the selection and application of environmentally friendly technologies in practice. Future studies must consider to estimate energy prices elasticities and include price block in the model. The core elaborated input-output model can be easily updated with latest input-output data published by Eurostat or national statistical office. The modelling is performed for Latvia's economy, however, there are no assertive limitations to apply the modelling instrument (comparative-static input-output model with regional modelling block) for any EU member state with minor modifications.

References

1. Kirschstein, T.: Comparison of energy demands of drone-based and ground-based parcel delivery services. Transport. Res. Part D: Transp. Environ. **78**, 102209 (2020)

2. Dorling, K., Heinrichs, J., Messier, G.G., Magierowski, S.: Vehicle routing problems for drone delivery. IEEE Trans. Syst. Man Cybern. Syst. **47**(1), 70–85 (2017)
3. Schneider, M., Stenger, A., Goeke, D.: The electric vehicle-routing problem with time windows and recharging stations. Transp. Sci. **48**(4), 500–520 (2014)
4. Chen, C., Demir, E., Huang, Y., Qiu, R.: The adoption of self-driving delivery robots in last mile logistics. Transport. Res. Part E: Logist. Transport. Rev. **146**, 102214 (2021)
5. Ostermeier, M., Heimfarth, A., Hübner, A.: The multi-vehicle truck-and-robot routing problem for last-mile delivery. Eur. J. Oper. Res. **310**(2), 680–697 (2023)
6. Iwan, S., Kijewska, K., Lemke, J.: Analysis of parcel lockers' efficiency as the last mile delivery solution - the results of the research in Poland. Transport. Res. Procedia **12**, 644–655 (2016)
7. Van Duin, J.H.R., Wiegmans, B.W., Van Arem, B., Van Amstel, Y.: From home delivery to parcel lockers: a case study in Amsterdam. Transport. Res. Procedia **46**, 37–44 (2020)
8. Nguyen, D.H., de Leeuw, S., Dullaert, W., Foubert, B.P.J.: What is the right delivery option for you? consumer preferences for delivery attributes in online retailing. J. Bus. Logist. **40**(4), 299–321 (2019)
9. Ignat, B., Chankov, S.: Do e-commerce customers change their preferred last-mile delivery based on its sustainability impact? Int. J. Logist. Manag. **31**(3), 521–548 (2020)
10. Caspersen, E., Navrud, S.: The sharing economy and consumer preferences for environmentally sustainable last mile deliveries. Transport. Res. Part D: Transp. Environ. **95**, 102863 (2021)
11. Melkonyan, A., Gruchmann, T., Lohmar, F., Kamath, V., Spinler, S.: Sustainability assessment of last-mile logistics and distribution strategies: the case of local food networks. Int. J. Prod. Econ. **228**, 107746 (2020)
12. Bosona, T.: Urban freight last mile logistics—challenges and opportunities to improve sustainability: a literature review. Sustainability (Switzerland) **12**(21), 1–20 (2020)
13. Bjørgen, A., Bjerkan, K.Y., Hjelkrem, O.A.: E-groceries: Sustainable last mile distribution in city planning. Res. Transport. Econ. **87**, 100805 (2021)
14. Bruzzone, F., Cavallaro, F., Nocera, S.: The integration of passenger and freight transport for first-last mile operations. Transp. Policy **100**, 31–48 (2021)
15. Wang, H.: Routing and scheduling for a last-mile transportation system. Transp. Sci. **53**, 131–147 (2019)
16. Kiba-Janiak, M., Marcinkowski, J., Jagoda, A., Skowrońska, A.: Sustainable last mile delivery on e-commerce market in cities from the perspective of various stakeholders: literature review. Sustain. Cities Soc. **71**, 102984 (2021)
17. Nürnberg, M.: Analysis of using cargo bikes in urban logistics on the example of Stargard. Transport. Res. Procedia **39**, 360–369 (2019)
18. Perboli, G., Rosano, M.: Parcel delivery in urban areas: opportunities and threats for the mix of traditional and green business models. Transport. Res. Part C: Emerg. Technol. **99**, 19–36 (2019)
19. Ozolina, V., Auzina-Emsina, A.: Macroeconometric input-output model for transport sector analysis. Proc. Eur. Council Model. Simul. ECMS **35**(1), 82–87 (2021)
20. Auzina-Emsina, A., Ozolina, V.: Transportation, logistics and regional development in covid-19 era: Modelling sectoral shocks caused by policy and safety measures. Res. Rural Dev. **36**, 144–151 (2021)
21. Eurostat: Symmetric input-output table at basic prices (product by product), https://ec.europa.eu/eurostat/data/database. Accessed 01 July 2023
22. Eurostat: Symmetric input-output table at basic prices (industry by industry). https://ec.europa.eu/eurostat/data/database. Accessed 01 July 2023
23. Central Statistical Bureau of Latvia: Statistical database. https://data.stat.gov.lv/pxweb/en/OSP_PUB/. Accessed 12 July 2023

24. Eurostat: Aggregations of manufacturing based on NACE Rev. 2. https://ec.europa.eu/eurostat/cache/metadata/Annexes/htec_esms_an3.pdf. Accessed 13 July 2023
25. Eurostat: Urban-rural typology. https://ec.europa.eu/eurostat/statistics-explained/index.php?title=Glossary:Urban-rural_typology. Accessed 13 July 2023

Risk Assessment for the Preparation of Training Scenarios for Tram Drivers

Agnieszka A. Tubis(✉), Artur A. Kierzkowski, Łukasz Wolniewicz, Ewa Mardeusz, Franciszek J. Restel, Tomasz Kisiel, and Mateusz Zając

Department of Technical Systems Operation and Maintenance, Wroclaw University of Science and Technology, 50-370 Wroclaw, Poland
agnieszka.tubis@pwr.edu.pl

Abstract. Virtual reality (VR) is widely used in transportation systems. In railway, VR is used as an element supporting the personnel training process. The results presented in the article result from work under the project "Innovative training system for tram drivers based on a full-cabin simulator with the use of cognitive science," funded by the National Center for Research and Development. The first stage of this project's work was identifying all potential adverse events. Authors aim to present the results of the analysis and risk assessment of adverse events, which will form the basis for developing training scenarios for tram drivers. The activities undertaken in the article included the preparation of a risk matrix in two dimensions – the probability and consequences of an adverse event. The results showed that most of the identified events belong to the group of catastrophic and unacceptable risks. Therefore, they should be reflected as driver collocation scenarios.

Keywords: Risk assessment · Virtual reality · Tram driver · Training scenario · FTA

1 Introduction

Most training for tram drivers in Poland uses traditional training systems. These systems use driving simulators, mainly based on: (1) desktop structures, in which the image is projected onto a flat external screen, or (2) compact, in which the image is usually displayed on several independent monitors. In this case, the level of immersion (the degree of user involvement in the virtual reality experience) is low. The training participant sees progress in the implementation of tasks but does not feel full participation in the ongoing process. Therefore, the tasks performed are not accompanied by full emotions and reactions, especially in dangerous situations.

The article presents the results of research obtained as part of the project "Innovative training system for tram drivers based on a full-cabin simulator with the use of cognitive science" funded by the National Center for Research and Development. The main research problem is the development of a tram simulator that will meet the requirements for the assumed level of immersion. The goal is, therefore, to create a simulator in which

O. Prentkovskis et al. (Eds.): TRANSBALTICA 2023, LNITI, pp. 337–348, 2024.
https://doi.org/10.1007/978-3-031-52652-7_33

the trainee will be isolated from the outside world and immersed in the virtual world through the sense of sight (virtual projection of the environment) and the sense of hearing (sound projection). At the same time, the trainee will be in an overload state resulting from placing the simulator on the motion platform.

An equally important element of the created training system is the preparation of appropriate training scenarios. The training program aims not only to prepare the driver to fulfill the primary duties entrusted to him but also to enable him to develop correct reactions in difficult, dangerous, or unusual situations. For this reason, one of the stages of research work is to prepare a risk assessment related to a driver's job. Based on the identified undesirable events, training scenarios will be developed as a training ground for new and existing tram drivers participating in the training program.

Authors aim to present the analysis and evaluation results of the risk of dangerous events, which are the basis for developing school scenarios for tram drivers. This goal was achieved through the presentation and discussion of the results in the following sections of this article. The structure of the paper, along with the substantive scope, is present in Fig. 1.

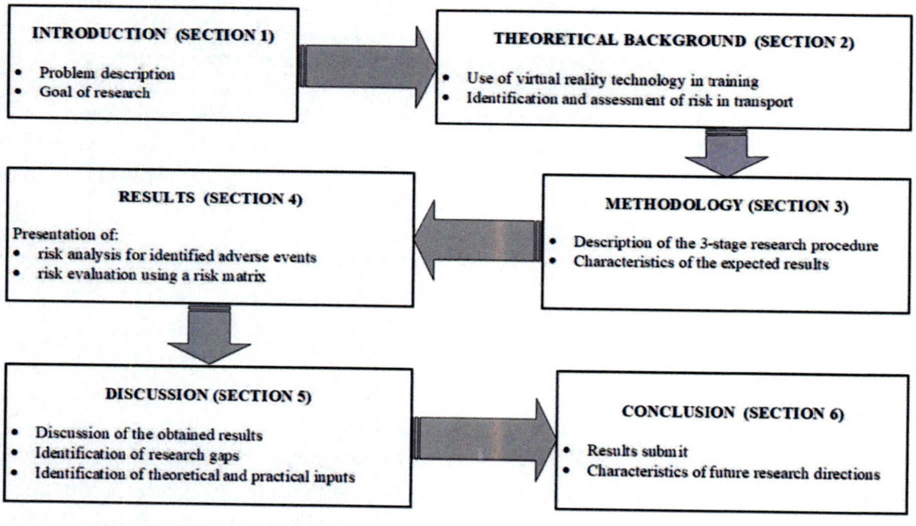

Fig. 1. The structure of the article and the objectives of individual sections.

The purpose of the article is to present the results of the analysis and risk assessment of adverse events that form the basis for the developing training scenarios for tram drivers, for this reason the structure of the article consists of six sections. Section 1 identifies a research problem. Section 2 reviews the literature. Section 3 describes the methodology of the conducted research, the results of which are presented in Sect. 4. Section 5 presents the discussion from the conducted research, while Sect. 6 presents the conclusions.

2 Theoretical Background

2.1 Virtual Reality in Training

Virtual reality technology is widely used in various sectors to achieve the assumed training and development goals effectively. Trainers use it to develop knowledge, skills, abilities, and other characteristics (KSAO) [1, 2]. In the initial stage of development, VR technology was used primarily to improve KSAOs, for which training processes were associated with the risk of loss of health or life of training participants or very high implementation costs [3]. Currently, however, training programs using VR are being developed to improve KSA-Os employees in such areas as rail transport [4], chemistry [5], marketing [6], nuclear power plants [7], and many others.

Bricken [8] gives several possible reasons for virtual reality's success in training. VR training tools enable you to offer trainees unique experiences consistent with instructional strategies, hands-on learning, and conceptual visualization while operating within the limits of system functionality. At the same time, they meet the requirements for an effective learning process today, which include [9]: focus, meaningful representation of information, multiple mappings of information, and reflective learning. At the same time, the currently created Virtual reality environments (VRE) have unique contributions to learning scientific visualization, instructions, sensory-motor performance, and for training [10]. It is also worth noting that VR performance metrics make it possible to identify mistakes made and, on this basis, to indicate the weaknesses of the trained person [11]. Thanks to this, it is possible to adjust the learning parameters to the needs of the training participants so that they can deliberately practice critical skills and increase exposure to a wide variety of both common and low-volume cases [12].

VR technology is classified in the literature according to the level of immersion in the environment used in a given training tool. According to this classification, we distinguish [11, 13]:

1. a non-immersive virtual environment means that the training simulation is presented on a conventional personal computer and is typically explored using a keyboard, mouse, wand, joystick, or touchscreen,
2. immersive virtual environment means the implementation of a training simulation on several screens the size of a training room or through a stereoscopic, head-mounted display unit. Thanks to this, a high level of user immersion in the virtual training environment is achieved.

It should also be emphasized that conventional VR simulators are expensive and cumbersome [14, 15]. This results in limited possibilities of moving them and making them available in convenient training locations. On the other hand, immersive virtual reality allows participants to achieve the assumed training effects as in traditional virtual reality while implementing low-cost, portable, commercially-available hardware [11].

The scenarios developed in the described project concerns immersive virtual reality. For this reason, there is a requirement to describe the detailed operating environment, which has to be reproduced in the form of created scenarios describing the recorded adverse events.

2.2 Risk Assessment in Transport

Risk is defined differently in the literature depending on the science that studies a given phenomenon. An interesting overview of concepts related to this issue can be found in Aven publications, e.g., [16–19]. Based on the literature review, Aven notes that in most cases, risk is defined in the context of the occurrence of specific adverse events and their effects [17]. For this reason, risk in mathematical notation is usually presented as the product of two variables: (1) a measure of the probability of a given event occurring and (2) a measure of the consequences of its occurrence.

When assessing the risk in transport, researchers focus their attention primarily on aspects related to driving safety [20]. This is understandable due to the overriding importance of protecting the health and life of participants in the transport process. This approach is so dominant that many authors define risk as the probability of an accident [21] and its consequences [22]. In transport systems, two basic types of risk are distinguished depending on the time horizon and management structures [23]:

- strategic risk – long-term risk, including social, group, and individual risk;
- operational risk – short-term risk related to current operations, including group and individual risk.

According to the ISO 31000:2018 standard, the risk assessment process should include three stages [24]: (1) identification of adverse events; (2) risk analysis; (3) risk evaluation. Identifying adverse events is a preliminary stage that allows determining the set of events that are the subject of further analysis. Therefore, it is necessary to create a possibly comprehensive list of potential events using various identification methods, both quantitative and qualitative. Among the most popular methods of identifying adverse events in the literature, the following are distinguished [25]: measurements, discussions, simulations, experience and expert assessment, laboratory tests, detection systems, modeling, scenarios, questionnaires, forecasts, hazard analyses, strengths and weaknesses analyses, historical data. The risk analysis stage includes the estimation of two primary measures of risk assessment, namely the probability, and consequences of the identified adverse events. Impact analysis can take many forms, from a simple description of the results expressed in terms of a linguistic variable to detailed quantitative modeling or vulnerability analysis [26]. However, the following approaches are used to estimate the probability [27]:

1. analysis of historical data;
2. probability forecasting using predictive techniques;
3. expert opinions.

The final stage of risk assessment is the evaluation. It includes a comparison of the level of risk for each adverse event identified in the analysis process with the adopted criteria. This stage significantly impacts the decision-making process and the planning of future activities. The risk matrix is a popular tool used at this stage [28].

3 Methodology

The first stage of research under the project "Innovative training system for tram drivers based on a full-cabin simulator with the use of cognitive science" is the identification of adverse events that may occur during the performance of duties by a tram driver. These events should be subject to a detailed analysis of the cause of their occurrence and the generated effects. Training scenarios for tram drivers are developed based on the obtained results. The risk assessment process was divided into two phases. In the first stage of the study, all potential adverse events were identified, and their likelihood and consequences were estimated. This stage aims to identify and evaluate adverse events for which training scenarios for new employees and improvement training for existing employees should be prepared. In the second stage of the research, detailed cause and effect analyses is carried out, allowing us to identify the sources of the analyzed adverse events to prepare training scenarios better. This article presents studies only from the first research stage, for which a research procedure consisting of 3 steps has been planned. These stages are shown in Fig. 2.

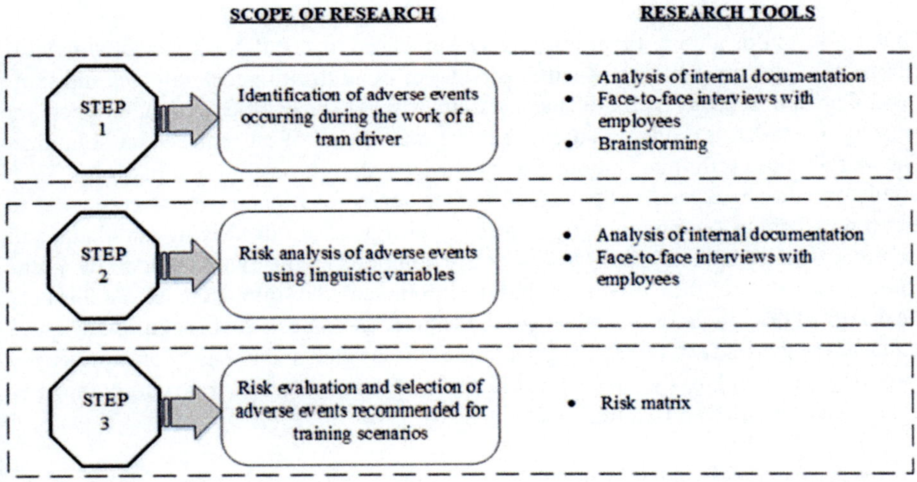

Fig. 2. Stages of the research procedure.

Three research methods were used to identify adverse events. First, the research team analyzed the documentation of selected urban carriers regarding registered accidents involving trams in the last five years. Then, direct interviews were conducted with the managerial staff and experienced tram drivers (minimum 10 years of experience), who, as part of a guided discussion, identified actual and potential adverse events that may occur during the implementation of tram transport. The last element of this stage was a discussion conducted following the brainstorming concept among team members. It aimed to identify potential adverse events that had not been identified in the previous two studies but were described, among other things, in the scientific literature. On this basis, catalogs of potential adverse events were developed by research team.

In the second stage of the study, the measure of the probability and consequences of individual adverse events was estimated. The analyses were carried out based on recorded events in the internal documentation of the selected urban transport operator and on the opinion of a team of experts chosen from people participating in the first stage of the research. Linguistic variables were used to estimate the risk measures because the analyzed events included situations not recorded in the internal documentation in the last five years. In their case, there was no quantitative data to carry out the analysis, and at the same time, experts confirmed that such situations may occur in the daily work of tram drivers.

Based on the results from stage 3, a risk evaluation was carried out, which became the basis for highlighting situations that should be mapped in training scenarios. A risk matrix was used for the evaluation. The obtained results were consulted with experts to confirm the priorities and scopes of the scenarios adopted for further research.

4 Results

4.1 Project Description

The project builds a tram driver training system based on a full-cab streetcar simulator on a motion platform, integrated with visual focus point tracking technologies and skin-galvanic response analysis for evaluating tram drivers. The explorers of the project aim to bring to market a product that exceeds the parameters of current market solutions. The market lacks training solutions for tram drivers that sit on a motion platform. Such simulators are characterized by a much higher degree of immersiveness compared to desktop or compact simulators. A proprietary design of a tram driving simulator will therefore be developed. There is also a lack of solutions on the market that allow multi-criteria evaluation of tram drivers during the implemented training courses. To this end, the design of the simulator will be integrated with technologies used in cognitive science, which allow the study of drivers' reaction time, stress level, and intensity of experienced emotions. The result of the project would aim at municipal public transport companies and streetcar motorman training centers.

4.2 Adverse Events

Identifying the causes of adverse events is the basis of the risk management process. As part of the implemented project, events were identified for which the probability of occurrence and their effects were determined. Probability and consequences were rated on a scale of 1 to 5. In the case of chance, a value of 5 means a very high probability of occurrence of the event, while 1 represents a very low probability. Regarding the consequences of the identified adverse events, 5 means fatalities and high financial losses, while 1 means negligible disruption to the transportation process. A detailed rating scale is presented in Table 1 for probability and Table 2 for consequences.

Table 1. Qualitative description of the scale of probability of adverse events.

1	infrequent	The probability of the event occurring is very low. The event occurs very rarely, with a frequency of once a month
2	unlikely	The probability of an event occurring is low. The event occurs infrequently, with a frequency of twice a month
3	possible	The probability of occurrence of the event is medium. The event occurs regularly, with a frequency of three times a month
4	likely	The probability of the event occurring is high. The incident occurs frequently, with a frequency of four times a month
5	frequent	The probability of the event occurring is very high. The event occurs very frequently, with a frequency of more than once a week (every day/every other day)

Table 2. Qualitative description of the magnitude of the consequences of adverse events.

1	very low	The consequences of the incident do not generate significant financial losses and have little impact on the ongoing transportation process. The effects of the event have minor consequences that can be easily mitigated or removed
2	low	The consequences of the incident generate minor financial losses and slightly affect the disruption of the transportation process. The consequences of the incident require corrective measures, but do not pose a serious threat to life and health
3	medium	The consequences of the incident generate minor financial losses and slightly affect the disruption of the transportation process. The consequences of the incident require corrective measures, but do not pose a serious threat to life and health
4	high	The consequences of the incident generate high financial losses and significantly affect the disruption of the transportation process. The consequences require corrective action and pose a threat to health and life
5	very high	The consequences of the incident lead to financial losses and fatalities. Cause significant risk to life and health, require immediate corrective action and long-term consequences

For further analysis, the probability and consequences of each identified adverse event in the analyzed process were determined. Linguistic variables were used to assess both measures. The results are presented in Table 3.

Table 3. Adverse events in the transport process with defined occurrence and consequences.

SYMBOL	ADVERSE EVENTS	OCCURRENCE OF EVENT	CONSEQUENCES OF EVENT
AE1	Derailment	infrequent	very high
AE2	Collision through the fault of a stranger	likely	high
AE3	A collision due to self-inflicted damage	unlikely	high
AE4	Vehicle damage	possible	medium
AE5	Pedestrian/bicyclist hit	unlikely	high
AE6	Fatal hit-and-run of a pedestrian	infrequent	very high
AE7	Overturning a passenger in a vehicle - sudden braking	possible	high
AE8	Slamming the passenger door	likely	high
AE9	Traffic stop - technical failure	possible	medium
AE10	Exceeding the speed limit	possible	medium
AE11	Route confusion	unlikely	low
AE12	Entering at a closed traffic light	unlikely	medium
AE13	Incorrect driving technique - sudden braking/acceleration	possible	medium

4.3 Risk Assessment Matrix

Adverse events were evaluated using a risk matrix. This method is widely used in transportation systems, as it allows prioritization for risk management. The advantages of using a risk matrix as a tool for presenting the result of the evaluation are discussed in [29]. A model risk matrix consisting of two dimensions – the probability of occurrence of an adverse event and its consequences is presented in Fig. 3. Due to the possibility of fatalities, all events with very high consequences were assumed to be critical risks.

The risk matrix presented describes the level of risk using the linguistic variables presented in Table 4.

The matrix identified adverse events according to the probability and consequence measures assigned to them. The route confusion event, the only one among the identified adverse events, was included in the acceptable risk category. Damage to the vehicle, traffic stop - technical failure, exceeding the speed limit, incorrect driving technique

IMPACT		VERY LOW	LOW	MEDIUM	HIGH	VERY HIGH
PROBABILITY	FREQUENT					
	LIKELY				AE2 AE8	
	POSSIBLE			AE4 AE9 AE10 AE13	AE7	
	UNLIKELY		AE11	AE12	AE3 AE5	
	INFREQUENT					AE1 AE6

Fig. 3. Risk matrix for the identified adverse events.

Table 4. Risk categories.

Risk level	Description
Low (L)	Level of risk acceptable. The impact on the implementation of the process is low. Available resources are sufficient to minimize the impact of the event.
Medium (M)	Level of risk unacceptable. The impact on process implementation is moderate. Risk mitigation measures implemented should have strong economic justification.
High (H)	Level of risk unacceptable. The impact on the implementation of the process is significant. It is necessary to implement risk mitigation measures within the established time horizon.
Critical (C)	Risk level critical. The impact on process implementation is catastrophic. It is necessary to immediately implement risk mitigation measures regardless of the cost of implementation.

- sudden braking/acceleration, entering at a closed traffic light is a medium level of risk. The occurrence of this category of events is an unacceptable risk, the impact of which on the implementation of the process is medium. Collision through the fault of a stranger, slamming the passenger door, overturning a passenger in a vehicle - sudden braking, collision due to self-inflicted damage, pedestrian/bicyclist hit – are events with a high level of risk. The occurrence of this category of events is an unacceptable risk, the impact of which on the implementation of the process is significant. The risk matrix identifies two events in the catastrophic risk category. These include a derailment and a fatal pedestrian hit-and-run.

5 Discussion

The training scenarios consider identified events and errors occurring during the process. This makes it possible to conduct training that considers specific errors and tailor the scenarios to tram drivers depending on the demand for a particular training area. The article presents undesirable events occurring during the implementation of the transportation process, which were given measures of probability and effects. Based on this, the authors developed a risk matrix, which allows the evaluation of identified adverse events. The use of the risk matrix to develop training scenarios for railway personnel is mainly due to the possibility of improving the safety and efficiency of system operation. During the implementation of tasks in the analyzed process, time is an essential element. Therefore, all undesirable events that disrupt the process should be eliminated.

It is believed that in developing training scenarios, events categorized as critical risks and unacceptable risks should be considered. The high level of competence and practical experience of personnel affect the effectiveness of the process. Including adverse events with catastrophic and unacceptable risks in training scenarios affects the safety of tram drivers and other traffic participants, the maintenance of operational continuity, and minimizing financial expenditures. Drivers' actions during an adverse event also affect comfort and passenger safety. It is also worth noting that the adverse events presented take into account regularity as well as unusual situations.

The literature still lacks studies that focus on examining the trainee's reactions to dangerous situations during training in such a way as to be able to influence the cause of the trainee's actions. The results obtained have both research and practical significance. Based on the conducted research, drivers' reactions to dangerous situations will be analyzed. This allows to increase knowledge of the factors affecting the work of tram drivers. At the same time, the obtained results make it possible to expand the knowledge of the cause and consequences of adverse events related to the work of tram drivers.

The practical application of the research is based primarily on the results obtained in the evaluation stage. The risk matrix provides knowledge regarding the prioritization of training to improve specific competencies. Events classified as catastrophic risk should be part of the training aimed at both those who start schooling and those who improve their competence (refresher training). Events classified as high-risk should be mandatory in the basic training program. Medium-level risk, on the other hand, can be part of additional training. Training personnel trained according to scenarios categorized as catastrophic and unacceptable risks will be better prepared to take action when an adverse event occurs. Not only will the personnel's response time be reduced, but efficiency will also be increased by minimizing the mistakes that can be made.

6 Conclusion

Training based on virtual reality technology is increasingly used in transportation systems, including railways. The effectiveness of realized training programs depends on the developed scenarios. Railway personnel participating in training that considers an undesirable event with catastrophic and unacceptable risks are prepared to take action in critical situations. Not only is the response time of the personnel minimized, but, above all, the efficiency of their actions is increased. The assumptions of presented paper provide a tool for identifying events that should be used in training programs.

In further stages of the work, a detailed cause-and-effect analysis should be carried out to identify the sources of adverse events. On this basis, it would be possible to obtain information on how to influence the attitudes and behavior of tram drivers through training.

References

1. Alaraj, A., Lemole, M., Finkle, J., Yudkowsky, R., Wallace, A., Luciano, C., et al.: Virtual reality training in neurosurgery: review of current status and future applications. Surg. Neurol. Int. **2**(1), 52 (2011). https://doi.org/10.4103/2152-7806.80117
2. Gigante, M.A.: Virtual reality: definitions, history and applications. In: Virtual Reality Systems, pp. 3–14. Elsevier, Amsterdam (1993). https://doi.org/10.1016/B978-0-12-227748-1.50009-3
3. Howard, M.C., Gutworth, M.B., Jacobs, R.R.: A meta-analysis of virtual reality training programs. Comput. Human Behav. **121**, 106808 (2021). https://doi.org/10.1016/j.chb.2021.106808
4. Nikitin, A., Reshetnikova, N., Sitnikov, I., Karelova, O.: VR training for railway wagons maintenance: architecture and implementation. Procedia Comput. Sci. **2020**(176), 622–631 (2020). https://doi.org/10.1016/j.procs.2020.08.064
5. Merchant, Z., Goetz, E.T., Keeney-Kennicutt, W., Cifuentes, L., Kwok, O., Davis, T.J.: Exploring 3-D virtual reality technology for spatial ability and chemistry achievement. J. Comput. Assist. Learn. **29**(6), 579–590 (2013). https://doi.org/10.1111/jcal.12018
6. Cheng, Y., Wang, S.-H.: Applying a 3D virtual learning environment to facilitate student's application ability – the case of marketing. Comput. Human Behav. **27**(1), 576–584 (2011). https://doi.org/10.1016/j.chb.2010.10.008
7. Masiello, I., Herault, R., Mansfeld, M., Skogqvist, M.: Simulation-based VR training for the nuclear sector—a pilot study. Sustainability **14**(13), 7984 (2022). https://doi.org/10.3390/su14137984
8. Bricken, M.: Virtual reality learning environments: potentials and challenges. ACM SIGGRAPH Comput. Graph. **25**(3), 178–184 (1991). https://doi.org/10.1145/126640.126657
9. Rupasinghe, T., Kurz, M.E., Washburn, C., Gramopadhye, A.K.: Virtual reality training integrated curriculum: an aircraft maintenance technology (AMT) education perspective. Int. J. Eng. Educ. **27**, 778–788 (2011)
10. Moreno, P., Mayer, R.E.: Virtual reality and learning: cognitive and motivational effects of students' sense of presence. In: Proceedings of Human-Computer Interaction, pp. 65–67. (2001)
11. Radianti, J., Majchrzak, T.A., Fromm, J., Wohlgenannt, I.: A systematic review of immersive virtual reality applications for higher education: design elements, lessons learned, and research agenda. Comput. Educ. **147**, 103778 (2020). https://doi.org/10.1016/j.compedu.2019.103778
12. Bowyer, M.W., Streete, K.A., Muniz, G.M., Liu, A.V.: Immersive virtual environments for medical training. Semin. Colon Rectal Surg. **19**(2), 90–97 (2008). https://doi.org/10.1053/j.scrs.2008.02.0052
13. Lee, E.A.-L., Wong, K.W.: A review of using virtual reality for learning. In: Pan, Z., Cheok, A.D., Müller, W., El Rhalibi, A. (eds.) Transactions on Edutainment I. LNCS, vol. 5080, pp. 231–241. Springer, Heidelberg (2008). https://doi.org/10.1007/978-3-540-69744-2_18

14. Parham, G., Bing, E.G., Cuevas, A., Fisher, B., Skinner, J., Mwanahamuntu, M., et al.: Creating a low-cost virtual reality surgical simulation to increase surgical oncology capacity and capability. Ecancermedicalscience **13** (2019). https://doi.org/10.3332/ecancer.2019.910
15. Bernier, G.V., Sanchez, J.E.: Surgical simulation: the value of individualization. Surg. Endosc. **30**(8), 3191–3197 (2016). https://doi.org/10.1007/s00464-016-5021-8
16. Aven, T.: Risk assessment and risk management: Review of recent advances on their foundation. Eur. J. Oper. Res. **253**(1), 1–13 (2016). https://doi.org/10.1016/J.EJOR.2015.12.023. Accessed 19 Sept 2022
17. Aven, T.: On how to define, understand and describe risk. Reliabil. Eng. Syst. Saf. **95**(6), 623–631 (2010). https://doi.org/10.1016/J.RESS.2010.01.011. Accessed 19 Sept 2022
18. Aven, T.: On the new ISO guide on risk management terminology. Reliabil. Eng. Syst. Saf. **96**(7), 719–726 (2011). https://doi.org/10.1016/J.RESS.2010.12.020. Accessed 19 Sept 2022
19. Aven, T., Renn, O.: On risk defined as an event where the outcome is uncertain. J. Risk Res. **12**(1), 1–11 (2009). https://doi.org/10.1080/13669870802488883
20. Tubis, A.: Risk assessment in road transport – strategic and business approach. J. KONBiN **45**(1), 305–324 (2018). https://doi.org/10.2478/jok-2018-0016
21. Hauer, E.: Traffic conflicts and exposure. Accid. Anal. Prevent. **14**(5), 359–364 (1982). https://doi.org/10.1016/0001-4575(82)90014-8. Accessed 19 Sept 2022
22. Haight, F.A.: Risk, especially risk of traffic accident. Accid. Anal. Prevent. **18**(5), 359–366 (1986). https://doi.org/10.1016/0001-4575(86)90009-6. Accessed 19 Sept 2022
23. Jamroz, K.: Miary ryzyka strategicznego w ruchu drogowym. J. KONBiN1 **13**, 197–208 (2010)
24. ISO 31000:2018 Risk management — Guidelines. International Organization for Standardization (2018)
25. Wróblewski, D.: Zarządzanie ryzykiem. Przegląd wybranych metod. Wydawnictwo CNBOP-PIB, Józefów (2015)
26. Tubis, A.: Metoda zarządzania ryzykiem operacyjnym w transporcie drogowym. Oficyna Wydawnicza Politechniki Wrocławskiej, Wrocław (2018)
27. Kaplan, S., Garrick, B.J.: On the quantitative definition of risk. Risk Anal. **1**(1), 11–27 (1981). https://doi.org/10.1111/j.1539-6924.1981.tb01350.x
28. Elmontsri, M.: Review of the strengths and weaknesses of risk matrices. J. Risk Anal. Crisis Response **4**(2), 49–57 (2014)
29. Szaciłło, L., Krześniak, M., Jasiński, D., Valis, D.: The use of the risk matrix method for assessing the risk of implementing rail freight services. Arch. Transp. **64**(4), 89–106 (2022). https://doi.org/10.5604/01.3001.0016.1185

Technology Advancement in Relation to Transport Poverty

Mariusz Kostrzewski[1](✉), Ahmed Eliwa[1], and Yahya Abdelatty[2]

[1] Warsaw University of Technology, 00-662 Warsaw, Poland
{mariusz.kostrzewski,01155770}@pw.edu.pl
[2] 31-586 Krakow, Poland

Abstract. Transport poverty (TP) is a widespread problem that affects individuals and communities all over the world, restricting access to services, education, job prospects, and social contacts. However, advancement of technology has emerged in easing TP by improving accessibility, affordability, and efficiency in transport systems. The advancements in electric and hybrid vehicles, ridesharing platforms, autonomous vehicles, and micro-mobility have contributed to addressing the challenges faced by communities experiencing TP. By promoting these technological innovations and inclusive transport policies, societies can potentially strive towards a more equitable and sustainable future. The paper studies the influence of technology on TP and its potential to ameliorate issues related. Moreover, it discusses various aspects of novel technologies on TP and presents carefully selected (dis)advantages of certain solutions' applications.

Keywords: Transport poverty · Accessibility · Affordability · Efficiency

1 Introduction

The advancement of technology has revolutionized various sides of human life, including transport related aspects. Technological innovations have brought about significant improvements in efficiency, accessibility, and convenience in the transport sector. However, amidst these advancements, a concerning issue has emerged – transport poverty (TP). TP refers to the inability of individuals or communities to access affordable and reliable transport options, leading to social exclusion and limited opportunities [1]. This paper examines the relationship between the advancement of technology and TP, exploring how technological developments both contribute to and address the issue.

The meaning of TP is not well defined [2]. Researchers concentrate on the fundamental social and distributive aspects of transport as well as their connection to more pronounced economic and social disparities [3]. There is a range of terms referring to the relation between (a lack of) transport and people's chances for a quality of life. According to [4], the terms "transport disadvantage", "transport poverty", "transport-related social exclusion", and "accessibility poverty" are often used. An individual is transport poor if at least one of the circumstances applies [3]:

Y. Abdelatty—Independent Researcher.

O. Prentkovskis et al. (Eds.): TRANSBALTICA 2023, LNITI, pp. 349–359, 2024.
https://doi.org/10.1007/978-3-031-52652-7_34

- Existing transport options do not get a person to place where they can satisfy their daily activity needs and maintain a decent quality of life.
- No form of transport exists that satisfies individual's condition and abilities.
- An excessive amount of time must be spent travelling.
- The travel conditions are hazardous, harmful, or unhealthy for individual.
- Household income falls below a poverty threshold after transport expenses [2].

The idea of social exclusion, on the other hand, emphasizes that a lack of material resources is just one of many potential causes of deprivation. Social exclusion is defined as the lack of or denial of resources, rights, products/services that prevents a person from participating in the typical interactions and activities that are available to most people in a society in the economic/social/cultural/political spheres [4].

Following this distinction, the phrase "transport poverty" would be suggested to describe a lack of resources for transport. In other words, lack of access to reliable transport reduces a person's potential mobility relative to the general population. Potential mobility is the term used to describe a person's capacity to move through space [5, 6]. A person may be unable to use a bus service since travel-related impairments due to financial limits (e.g., prohibiting the purchase of a car or even tickets), legal limitations (not having a driver's license), or limitations in mental or physical abilities [4].

Full electrical vehicles (EVs) and hybrid electric vehicles (HEVs) were introduced as a technology advancement of automotives. HEV technology, has been recognized as the ideal step between petroleum vehicles and EVs. According to Jorgensen [7], HEV technology combines an internal combustion engine (ICE) and an electric motor for propulsion. A rechargeable battery pack powers the motor in a HEV. This battery pack is often charged by the vehicle's onboard alternator and regenerative braking system. The plug-in hybrid electric vehicle (PHEV) is another type of HEV with the ability to recharge from outside sources, such as from grid electricity during off-peak hours. Unlike a typical HEV, a PHEV uses an electric motor as its primary source of power with ICE as a backup [8]. A PHEV transforms into a standard HEV and relies mostly on ICE for power once the storage battery reaches a certain state of charge [9]. Without giving up the advantages of an ICE-powered vehicle, a PHEV can offer environmental benefits comparable to those of EVs. The full range of an EV is often not exceeded by city dwellers, allowing them to completely forgo the need of petroleum-based fuels [10].

According to Touchstone Energy Cooperatives (TEC) in 2020, road vehicles with electric motors are not a recent invention. Although EVs may appear like the most recent and innovative creation, the technology has really been around for more than a century (1890s). TEC added that for a variety of reasons, EVs technology slackened with the development of gas-powered engines, but EVs are now once again surging to the forefront of the automotive industry. Significant developments in power electronics and energy storage technologies have sparked this revival [11]. Clean Vehicle Rebate Project (CVRP) [12] explained that EVs are powered by one or more electric motors.

In recent years, the use of ridesharing platforms (RPs) and other app-based transport has changed how people commute and travel contrary to conventional taxi services. RPs provide access to transport services (TSs) through apps by rides' requests [13].

One of the most well-known ridesharing services in the world is Uber. The Uber app offers safety features available to all users. In addition, every ride has a support group you may call on [14]. Uber has encountered legal issues and conflicts in different places, but its passenger transport sector's revolution is not denying. Lyft is another app-based TS that operates in US and Canada. It gained popularity owing to passenger-friendly experience. The app offers share feedback for functionality improvement [15].

Autonomous vehicles (AVs) have emerged as a transformative technology, offering immense potential for safety, efficiency, and sustainability. AVs employ a range of cutting-edge technologies to operate independently, without human intervention. Many transport, delivery, or manufacturing companies, especially those heavily reliant on logistics, can reap significant benefits from integrating AVs along with other advanced technologies into their operations [16]. Sensing and perception systems, they enable precise detection and tracking of the surrounding environment. Artificial intelligence and machine learning algorithms process the vast amount of sensor data, enabling object recognition and intelligent decision-making. Connectivity and communication systems facilitate real-time data exchange between vehicles and infrastructure, enhancing situational awareness and cooperative maneuvers. Mapping and localization techniques, such as simultaneous localization and mapping SLAM, ensure accurate positioning and effective navigation [17]. Redundant systems and fail-safe mechanisms ensure safety and reliability, while rigorous testing procedures address various scenarios. Continued research and collaboration are vital for realizing the full potential of AVs in revolutionizing transport for a safer and more sustainable future [18]. AVs can contribute to overcoming mobility and accessibility barriers based on tests focused on disadvantaged groups, showing interest in using AVs in their daily lives [19].

Micro-mobility (MM) options, including e-scooters (e-S) and bike-sharing systems (BS), have garnered significant attention in recent years as sustainable and flexible modes of transport [20]. Numerous studies have examined the impacts and challenges associated with these MM solutions. Research highlights their potential to enhance last-mile connectivity, reduce congestion and emissions, and improve accessibility and health outcomes. It also underscores concerns such as safety, equitable access, parking management, and regulatory frameworks. Overall, the literature emphasizes the importance of comprehensive planning, stakeholders' collaboration, and evidence-based policies to maximize the benefits of MM options while addressing potential drawbacks, ultimately contributing to more sustainable and inclusive urban transport systems [21].

2 Methodology

This paper employs a systematic research and study methodology to investigate the relation between technological advancement and TP. The primary objective is to critically analyze existing literature to examine how technological advancements impact transport accessibility and equity, as well as explore the potential consequences for individuals and communities experiencing TP. A rigorous search strategy was employed. Research databases were searched for publications in journals, conference proceedings, and reports, from 2000 to 2023 and were limited to those published in English.

It is important to acknowledge potential limitations of the study. This study relies on existing publications, which may introduce potential publication bias. Furthermore, the exclusion of non-English studies may introduce language bias.

3 Results and Discussion

This section focuses on 4 technology trends: EVs and HEVs, RPs and app-based TSs, AV technology, and MM options such as e-Ss and BS. The goal is to examine how these advancements have impacted TP and discuss their implications.

3.1 Development of Electric and Hybrid Vehicles

The development and widespread acceptance of EVs and HEVs (further: (H)EVs) can help to alleviate TP in a variety of ways. These cars provide advantages over regular gas or diesel-powered automobiles as lower fuel costs, greater accessibility, and sustainability. One of the major benefits of (H)EVs is their reduced dependency on fossil fuels. EVs are powered only by electricity, whereas HEVs combine an internal combustion engine and an electric motor. As a result, EVs and HEVs have much reduced fuel expenses compared to traditional vehicles [22]. Fuel cost reductions can have a significant impact on people and households facing TP by reducing transport expenses.

Studies have indicated that (H)EVs have the potential to lower gasoline expenses. In 2009, they have saved nearly 9.2 million barrels of fuel [23]. Fuel savings are predicted to rise as more (H)EVs enter the market. This prediction has been confirmed by 2 studies made in 2017 and 2020, which showed that (H)EVs can save from $11,000 to $15,000 fuel expenses during a 15-year ownership span [24, 25]. In addition to lower fuel costs, (H)EVs have the potential for long-term cost savings. Although the upfront purchase cost of these vehicles is often higher compared to traditional vehicles, the operational and maintenance costs can be significantly lower. EVs have fewer moving parts, resulting in reduced maintenance and repair requirements. Moreover, governments often offer incentives and subsidies for EV adoption, further reducing the financial burden for individuals experiencing TP [26]. TP is frequently caused by a lack of accessible and affordable transport options, especially in marginalised populations or rural areas. (H)EVs boost accessibility by broadening the number of transport options. Individuals who were unable to afford or access private cars can consider these alternatives, thanks to the rising acceptance of (H)EVs. This increased mobility has potential to improve prospects for work, education, healthcare, social integration, thereby mitigating the disadvantages of TP. (H)EVs have the potential to improve accessibility. International Council on Clean Transport (ICCT) states that EV-sharing programmes in low-income neighbourhoods can provide economical and dependable transport options [27]. Institute for Transport and Development Policy emphasized that e-bikes, e-Ss, and other forms of electric MM are gaining appeal in cities due to their ability to replace car trips, increase transport accessibility, and improve air quality [28].

Transport contributes significantly to air pollution and greenhouse gas emissions, both of which have a negative impact on the environment and human health. The use of

(H)EVs is an option for environmentally friendly transport [29]. TP is addressed while time lowering the environmental impact of mobility by the adoption of (H)EVs.

The environment advantages of EVs and HEVs have been proven in several studies. According to the European Environment Agency, EVs produce fewer emissions than conventional vehicles, which improves air quality and lowers noise pollution [30].

3.2 Emergence of Ridesharing Platforms (RP) and App-Based TSs

RPs have unique features that set them apart from conventional taxi services. They make use of technology to offer advantages including more accessibility, lower costs, and greater convenience. These systems use algorithms to connect drivers and passengers, optimise routes, and enable seamless transactions, making the experience more effective and user-friendly [31]. Passengers have benefited significantly from app-based TSs. They offer more convenience through on-demand services, shorter waiting times, and improved price and driver information transparency. Expanded service coverage has also benefited passengers, especially in places that are underserved by traditional taxis or public transport [13]. By providing on-demand rides and filling gaps in service coverage, RPs can enhance mobility options for individuals experiencing TP.

Compared to traditional taxis and owning a vehicle, app-based TSs may be less expensive [32]. They often provide competitive pricing and eliminate upfront costs associated with vehicle ownership, such as purchase, maintenance, and insurance expenses. Traditional taxi ridesharing practises may result in excessive diversion phenomena, increasing the cost and duration of the clients' journey. Intelligent scheduling and route selection for ridesharing taxis can offer more affordable prices and lower total operating costs, improving driver and passenger enthusiasm, making reasonable use of traffic resources, lowering tail-gas emissions, protecting the environment, and fostering the sustainable growth of urban traffic [33]. These cost reductions can benefit individuals facing financial constraints and help alleviate TP.

RPs and app-based TSs provide flexibility and convenience, catering to diverse transport needs. They offer on-demand services that can accommodate various trip purposes (commuting, grocery shopping, healthcare appointments, social activities). The flexibility and convenience of these services can help individuals overcome transport barriers and improve their access to services, education, employment, etc.

In 2018, the authors of [33] discussed the concept of flexible serving strategy which is primarily presented to categorise and group the passengers in the process of issues matching. They discovered that applying the flexible serving strategy, i.e., appropriately modifying the pickup and delivery locations, can save system costs and increase the passenger-to-vehicle matching rate without adding much to nuisance costs [33].

While RPs and app-based TSs have the potential to alleviate TP, they also face challenges and limitations. There is some evidence to support the idea that platforms for ridesharing have catered to the wealthy. Low-income people are less likely to use ridesharing because of barriers such as the requirement for a credit card and smartphone [34, 35]. According to [36], using RPs more frequently in the U.S. increases with income levels over USD 50,000. They asserted that although RPs are less expensive than taxis, the costs of excursions are typically far higher than for public transport. Similar claims were made about Uber and Grab drivers in Rangoon, Myanmar [37]. They noted that while RPs enabled obtaining employment for people with high levels of education and skill, individuals or drivers with less access to resources, such as less education or literacy or even a lack of a cell phone, will not be able to easily compete in this market.

To harness the full potential of RPs and app-based TSs in addressing TP, policy interventions and collaborations are needed. Governments and transport authorities can work towards creating a regulatory framework that promotes equitable access, affordability, and safety. Partnerships with public transit agencies and integration with existing transport systems can enhance the reach and effectiveness of services.

3.3 Advent of AVs Technology

AVs have the potential to revolutionize transport accessibility and address TP, but it is essential to consider both their benefits and potential challenges [38]. Here, we explore the impact of AVs on transport accessibility and analyze the potential implications they may have on disadvantaged communities. AVs can significantly improve mobility and accessibility, providing flexible and convenient transport options for individuals facing limited access to conventional transport systems. People with disabilities or limited mobility can benefit from specialized AVs features, offering increased independence and access to essential services. AVs integrated into public transport networks can enhance efficiency, optimize routes, and provide first- and last-mile connectivity, making public transit more accessible and attractive to a broader range of individuals [39].

AVs have the potential to bridge transport gaps in rural and underserved areas. AVs can reduce barriers faced by residents who rely on private vehicles or have limited public transport options. Shared AV services can enhance mobility, reduce congestion, and provide increased transport options for communities without personal vehicles [38].

The potential cost reduction associated with AVs technology can contribute to reducing TP [40]. Operational efficiency, shared mobility, and advances in EVs can lower transport costs. Shared AVs services can be more affordable than private car ownership, reducing expenses as fuel, maintenance, insurance, and parking fees. Affordable and accessible transport options can alleviate financial burdens for low-income individuals.

Equitable deployment strategies and supportive policies are vital to ensure that AVs do not exacerbate TP. Governments must prioritize accessibility, affordability, and inclusivity in AVs deployment plans. It is crucial to consider the needs of low-income individuals, older adults, and those with limited digital literacy to prevent a widening of the digital separation. Strategic placement of AVs infrastructure, along with targeted subsidies, incentives, and regulations, can help ensure that AVs services reach underserved communities and are priced in a manner that does not perpetuate TP [41].

While AVs hold promise, they can also present challenges. The initial cost of AVs technology and the development of required infrastructure may pose affordability barriers, limiting access for low-income individuals and exacerbating disparities. The digital divide could widen if AVs services primarily operate through digital platforms, excluding those without access or familiarity with technology. Moreover, disruptions to the labor market caused by AVs adoption may lead to job losses [42].

The environmental impact of AVs must be considered to ensure sustainability. While AVs can reduce congestion and optimize routes, empty or underutilized AVs could lead to higher energy consumption, congestion, and pollution. These adverse effects can disproportionately affect disadvantaged communities exposed to higher pollution [43].

AVs have the potential to enhance transport accessibility and reduce TP by improving mobility options, addressing gaps in underserved areas, and reducing costs. However, careful consideration must be given to equitable deployment strategies, affordability, and addressing potential challenges such as the digital divide and labor market disruptions. By prioritizing inclusivity, sustainability, and community engagement in policy development, AVs can contribute to a more accessible and equitable transport system while mitigating the risk of exacerbating existing disparities.

3.4 Emergence of MM Options: e-Ss and BS

MM transport, including e-Ss and BS, has gained popularity as alternative transport in urban areas. MM option offer convenient and efficient last-mile connectivity solutions. They fill the gap between public transport stations and final destinations, providing an accessible and affordable option for individuals. MM transport contributes to enhanced connectivity and accessibility for urban dwellers [44].

MM transport options are generally more affordable than private car ownership or traditional taxi services. E-Ss and BS offer pay-per-use or subscription-based models, allowing users to pay for transport as needed. This affordability can alleviate financial burdens for low-income individuals and contribute to reducing TP, as it is a cost-effective alternative to private car ownership or reliance on expensive transport options [45].

MM transport promotes active transport modes and physical activity. E-Ss and bicycles encourage individuals to engage in regular exercise while commuting, leading to health benefits such as increased fitness levels and improved mental well-being [46]. This aspect is particularly relevant in underserved communities with limited access to recreational facilities or affordable fitness options. By providing an affordable and healthy mode of transport, MM can contribute to addressing health-related inequalities and improving the overall quality of life for individuals in disadvantaged communities.

MM transport plays a vital role in reducing pollution and carbon emissions. As e-Ss and bicycles operate on human or electric power, they produce minimal or zero emissions during use. By encouraging individuals to shift from motorized vehicles to MM options, the overall carbon footprint can be significantly reduced. This environmental benefit is particularly relevant in densely populated urban areas with high traffic congestion and poor air quality; often areas where TP is prevalent [21]. MM transport contributes to reducing TP by providing accessible options for marginalized communities. E-Ss and BS can be strategically deployed in underserved neighborhoods, addressing gaps in public transport networks, and improving accessibility for non-motorized residents. It is crucial to consider the specific needs and challenges faced by e.g., individuals with disabilities or limited mobility to make MM services accessible.

MM transport includes challenges as well: safety concerns, accidents and injuries, must be mitigated through infrastructure planning, user education, and regulations.

4 Summary

The search for solutions to the TP problem continues. Selected ideas for its solution are articulated in this paper (briefly compared in Table 1). Some of them are applied for shorter distances (MM, RPs, and app-based TSs), others for longer ones ((H)EVs, AVs). The context of caring the environment expressed by reduction the negative impacts is important to the authors as well. On the other hand, all the presented solutions require ensuring a reduction in communication exclusion in terms of IT access both in terms of infrastructure (e.g., 5G is not available everywhere yet) and the apps UX.

Table 1. Selected (dis)advantages of described technologies.

	Shorter distances	Longer distances
Characteristics	MM: more affordable than private car or taxi services, promote active transport modes and physical activities (increase fitness levels and well-being), reduce environmental pollution and carbon emissions, exclude gaps in public transport networks and improving accessibility RPs and app-based TSs: connect drivers and passengers, optimize routes, enable seamless transactions, shorten waiting time, decrease operating costs, improving driver and passenger enthusiasm, lower tail-gas emissions, protect the environment; a credit card and smartphone are barrier for low-income people, more expensive than public transport, barrier for individuals of less education or literacy and not possessing cellphones	EVs, HEVs: reduce fuel expenses in comparison to traditional vehicles, boost accessibility by broadening the number of transport options, improve air quality and lower noise pollution AVs: shared services can be more affordable than private car ownership, reduce expenses as fuel, maintenance, insurance
References	MM: [21, 44–47] RPs and app-based TSs: [13, 31–37]	EVs, HEVs: [22–30] AVs: [38–43]

References

1. Eliwa, A., Kostrzewski, M.: Key indicators and measures of transport poverty. In: Lewczuk, K., Rosiński, A., Siergiejczyk, M. (eds.) Problemy transportu - wyzwania XXI wieku, pp. 57–68. Oficyna Wydawnicza Politechniki Warszawskiej, Warszawa (2022)
2. Lucas, K., Mattioli, G., Verlinghieri, E., Guzman, A.: Transport poverty and its adverse social consequences. Proc. Inst. Civil Eng. Transp. **169**(6), 353–365 (2016)

3. Lucas, K.: Editorial for special issue of European transport research review: transport poverty and inequalities. Eur. Transp. Res. Rev. **10**(1), 1–3 (2018). https://doi.org/10.1007/s12544-018-0288-6
4. Jeekel, J.F., Martens, C.J.C.M.: Equity in transport: learning from the policy domains of housing, health care and education. Eur. Transp. Res. Rev. **9**, 53 (2017)
5. Martens, K.: Accessibility and potential mobility as a guide for policy action. J. Transp. Res. Board **2499**(1), 18–24 (2015)
6. Sager, T.: Freedom as mobility: implications of the distinction between actual and potential travelling. Mobilities **1**(3), 465–488 (2006)
7. Jorgensen, K.: Technologies for electric, hybrid and hydrogen vehicles: electricity from renewable energy sources in transport. Utilities Policy **16**(2), 72–79 (2008)
8. Mustafi, Nirendra Nath: An overview of hybrid electric vehicle technology. In: Kalghatgi, G., Agarwal, A.K., Leach, F., Senecal, K. (eds.) Engines and Fuels for Future Transport. EES, pp. 73–102. Springer, Singapore (2022). https://doi.org/10.1007/978-981-16-8717-4_5
9. Singh, K.V., Bansal, H.O., Singh, D.: A comprehensive review on hybrid electric vehicles: architectures and components. J. Mod. Transport. **27**(2), 77–107 (2019). https://doi.org/10.1007/s40534-019-0184-3
10. Amjad, S., Subramanyan, N., Rangasamy, R.: Review of design considerations and technological challenges for successful development and deployment of plug-in hybrid electric vehicles. Renew. Sustain. Energy Rev. **14**(3), 1104–1110 (2010)
11. TouchStone Energy Cooperatives. https://shorturl.at/ciGX6. Accessed 25 June 2023
12. Clean Vehicle Rebate Project. https://shorturl.at/nzKV0. Accessed 25 June 2023
13. Rayle, L., Dai, D., Chan, N., Cervero, R., Shaheen, S.: Just a better taxi? a survey-based comparison of taxis, transit, and ridesourcing services in San Francisco. Transp. Policy **45**, 168–178 (2016)
14. Uber, About us. https://shorturl.at/inKWX. Accessed 27 June 2023
15. Lyft, Homepage. https://www.lyft.com/. Accessed 27 June 2023
16. Kostrzewski, M., Abdelatty, Y., Eliwa, A., Nader, M.: Analysis of modern vs. conventional development technologies in transportation—the case study of a last-mile delivery process. Sensors **22**(24), 9858 (2022)
17. Gupta, A., Fernando, X.: Simultaneous localization and mapping (SLAM) and data fusion in unmanned aerial vehicles: recent advances and challenges. Drones **6**(4), 85 (2022)
18. Choi, J.K., Ji, Y.G.: Investigating the importance of trust on adopting an autonomous vehicle. Int. J. Human-Comput. Interact. **31**(10), 692–702 (2015)
19. Faber, K., van Lierop, D., Wiegmans, B.: How will older adults use automated vehicles? assessing the role of AVs in overcoming perceived mobility barriers. Transport. Res. Part A: Policy Pract. **133**, 353–363 (2020)
20. Hamerska, M., Ziółko, M., Stawiarski, P.: A sustainable transport system—the MMQUAL model of shared micromobility service quality assessment. Sustainability **14**(7), 4168 (2022)
21. Abduljabbar, R.L., Liyanage, S., Dia, H.: The role of micro-mobility in shaping sustainable cities: a systematic literature review. Transp. Res. Part D: Transp. Environ. **92**, 102734 (2021)
22. Kurz, J.: Hybrid Electric Vehicles: A History of Technological Innovation. USC Viterbi School of Engineering (University of Southern California), Illumin Magazine (2000)
23. Bennion, K., Thornton, M.: Fuel Savings from Hybrid Electric Vehicles. National Renewable Energy Laboratory, Technical Report NREL/TP-540–42681, 11–12 (2009). https://www.nrel.gov/docs/fy09osti/42681.pdf. Accessed 14 July 2023
24. Letmathe, P., Suares, M.: A consumer-oriented total cost of ownership model for different vehicle types in Germany. Transport. Res. Part D: Transport. Environ. **57**, 314–335 (2017)
25. Borlaug, B., Salisbury, S., Gerdes, M., Muratori, M.: Levelized cost of charging electric vehicles in the United States. Joule **4**(7), 1470–1485 (2020)

26. Lévay, P.Z., Drossinos, Y., Thiel, C.: The effect of fiscal incentives on market penetration of electric vehicles: a pairwise comparison of total cost of ownership. Energy Policy **105**, 524–533 (2017)
27. Nicholas, M., Bernard, M. R.: Success factors for electric carsharing. International Council on Clean Transportation (2021). https://shorturl.at/klqsG. Accessed 14 July 2023/7/14
28. Yanocha, D., Medina, S.: Plugging e-bikes and e-scooters into sustainable transport networks. Institute for transportation & development policy (New York, United States of America). https://t.ly/AZ-Oc. Accessed 26 June 2023
29. Dižo, J., et al.: Electric and plug-in hybrid vehicles and their infrastructure in a particular European region. Transport. Res. Procedia **55**, 629–636 (2021)
30. EEA. Towards zero pollution in Europe. Luxembourg: European Environment Agency (EEA) (2020). https://t.ly/0tlK1. 2023/7/14
31. Cohen, A., Shaheen, S.A.: Planning for shared mobility. The American Planning Association (Chicago) (2018). https://t.ly/Q1gQC. 2023/7/14
32. Khavarian-Garmsir, A.R., Sharifi, A., Abadi, M.H.: The social, economic, and environmental impacts of ridesourcing services: a literature review. Future Transport. **1**(2), 268–289 (2021)
33. Zhao, M., Yin, J., An, S., Wang, J., Feng, D.: Ridesharing problem with flexible pickup and delivery locations for app-based transportation service: mathematical modeling and decomposition methods. J. Adv. Transp. **2**, 1–21 (2018)
34. Kirk, D.S., Cavalli, N., Brazil, N.: The implications of ridehailing for risky driving and road accident injuries and fatalities. Soc. Sci. Med. **250**, 112793 (2020)
35. Amey, A., Attanucci, J., Mishalani, R.: Real-time ridesharing: opportunities and challenges in using mobile phone technology to improve rideshare services. Transp. Res. Rec. **2217**(1), 103–110 (2011)
36. Deka, D., Fei, D.: A comparison of the personal and neighborhood characteristics associated with ridesourcing, transit use, and driving with NHTS data. J. Transp. Geogr. **76**, 24–33 (2019)
37. Notar, B.E., Min, K.S., Gautam, R.: Echoes of colonial logic in re-ordering "Public" streets: from colonial rangoon to postcolonial yangon. Transfers **8**(3), 55–73 (2018)
38. Dianin, A., Ravazzoli, E., Hauger, G.: Implications of autonomous vehicles for accessibility and transport equity: a framework based on literature. Sustainability **13**(8), 4448 (2021)
39. Beirigo, B., Schulte, F., Negenborn, R.R.: Overcoming mobility poverty with shared autonomous vehicles: a learning-based optimization approach for Rotterdam zuid. In: Lalla-Ruiz, E., Mes, M., Voß, S. (eds.) ICCL 2020. LNCS, vol. 12433, pp. 492–506. Springer, Cham (2020). https://doi.org/10.1007/978-3-030-59747-4_32
40. Bagloee, S.A., Tavana, M., Asadi, M., Waller, S.T.: Autonomous vehicles: challenges, opportunities, and future implications for transportation policies. J. Mod. Transport. **24**(4), 284–303 (2016)
41. Epting, S.: Automated vehicles and transportation justice. Phil. Technol. **32**(3), 389–403 (2018). https://doi.org/10.1007/s13347-018-0307-5
42. Pettigrew, S., Fritschi, L., Norman, R.: The potential implications of autonomous vehicles in and around the workplace. Int. J. Environ. Res. Public Health **15**(9), 1876 (2018)
43. Silva, Ó., Cordera, R., González-González, E., Nogués, S.: Environmental impacts of autonomous vehicles: a review of the scientific literature. Sci. Total Environ. **830**, 154615 (2022)
44. Wanganoo, L., Shukla, V., Mohan, V.: Intelligent micro-mobility E-S: revolutionizing urban transport. In: Agrawal, P., Madaan, V., Sharma, A., Sharma, D., Agarwal, A., Kautish, S. (eds.) Trust-Based Communication Systems for Internet of Things Applications, pp. 267–290. Wiley, Hoboken (2022)
45. Sherriff, G., Lomas, M., Blazejewski, L., Larrington-Spencer, H.: A micromobility buffet: e-Ss in the context of multimodal spaces and practices in Greater Manchester. Active Travel Stud. Interdisc. J. **3**(1) (2023)

46. Bretones, A.: Marquet, ORiding to health: Investigating the relationship between micromobility use and objective physical activity in Barcelona adults. J. Transp. Health **29**, 101588 (2023)
47. Sanguinetti, A., Alston-Stepnitz, E.: Using emerging mobility data to advocate equitable micromobility infrastructure in underserved communities. Transport. Res. Part D Transp. Environ. **117**, 103613 (2023)

Manual Waste Sorting Study in Conveyor Transport System Based on Virtual Reality

Robert Giel(✉) and Alicja Dąbrowska

Wroclaw University of Science and Technology, Wroclaw, Poland
{robert.giel,alicja.dabrowska}@pwr.edu.pl

Abstract. Presented study explores the application of Virtual Reality (VR) technology in enhancing manual waste sorting in conveyor-based systems. The research addresses the inefficiencies in manual sorting processes and the limited scope of current studies. VR is proposed as a solution to replicate real sorting lines for experimental research, offering benefits like parameter variation and worker safety. The study outlines the components of the VR system, including virtual sorting cabin representation and haptic gloves. It emphasizes the possibilities and limitations of VR-based research, aiming to improve waste management and material recovery rates.

Keywords: Waste management · Virtual support · Internal transportation · Waste stream sorting

1 Introduction

Conveyor-based transport systems are widely used in for material handling in domains such as mining [1], warehousing [2] or manufacturing [3]. One of the application areas is also waste management. Collected waste are transported to waste treatment facilities to enhance recovery materials with the aim of minimizing landfill and thereby reducing the negative impact on the environment and human health. At these facilities, the waste is subjected to physical, thermal, chemical or biological processes and transported between processing points by a complex conveyor belt system. Despite the many technological solutions for separating the waste stream into usable and unusable fractions for recycling, manual sorting is still widely used [4, 5].

Research on manual waste sorting mainly focuses on ergonomics and human health, such as ergonomic assessment in waste sorting jobs [6], ergonomic evaluation of the work table for waste sorting [7], health risk assessment of sorters [8] or assessment of occupational exposure to dust, endotoxin and microorganisms [9]. In other works (e.g., [10]), manual waste sorting is considered one of many elements of a sorting plant based on the adopted simplifications. There is a lack of research addressing the impact of selected factors (in particular, the highly variable characteristics of the waste stream) on the efficiency of manual waste sorting. The challenges of studying manual sorting include the unavoidable occurrence of human factors, common limited access to the real system with a limited range of system-specific parameters and also time, space and cost limitations of research in laboratory conditions.

O. Prentkovskis et al. (Eds.): TRANSBALTICA 2023, LNITI, pp. 360–368, 2024.
https://doi.org/10.1007/978-3-031-52652-7_35

For example, studies of the effect of changes in the size and quality of the stream on sorting efficiency are conducted in the mining industry. However, due to the number of fractions processed and the nature of waste sorting, these studies are difficult to carry out under real conditions in this area.

The insufficient research in the field of manual sorting and the limitations of carrying it out in real or laboratory conditions justify the consideration of studies according to a new approach. New opportunities are provided by the technologies supporting the concept of Industry 4.0. Within these technologies, Internet of Things, Cloud and Big Data, Cyber Physical Systems, advanced robotics and blockchain are mainly indicated [11]. Among them, Extended Reality (XR) is also highlighted as a group of technologies focused on enhancing the user experience in terms of digital visualization and interaction [12]. The XR group includes Virtual Reality (VR), Augmented Reality (AR) and Mixed Reality (MR) (Fig. 1).

The development of a sorting system based on VR technology allows to eliminate difficulties in studying the process of manual sorting occurring in real conditions. This is because the station model developed in VR will allow arbitrary control of system parameters. These parameters include speed, length, width and height of the conveyor belt, on which waste is transported. Under real-world conditions, research is conducted with a limitation to fixed technical parameters of the conveyor. It is also possible to control the time of exposition of waste for humans, for example, the previous insertion of virtual walls.

In addition to the technical aspects of conveyors, any control of the waste stream is also possible. In real systems, the input stream depends on many factors. The morphology of waste from multi-family houses, single-family houses and those from services or industry is different. Another difficulty in studying the sorting process also concerns the data collection process. Currently, the process is evaluated post-factum based on data from an entire month of sorting. Because of this, the literature practically does not analyze the impact of stream size and quality change on the manual sorting process. Despite the fact that such studies are conducted in systems of lower complexity, such as: mining.

As a result, conducting research in VR will allow us to develop models of the manual sorting process, which in turn will allow to develop strategies to deploy workers at the belt and better manage the sorting process.

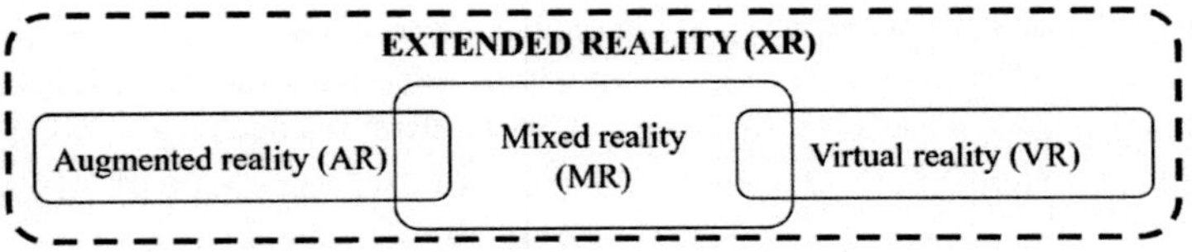

Fig. 1. Spectrum of Extended Reality [13].

VR allows users to completely immerse themselves in a simulated digital environment, detaching from the physical world. AR, on the other hand, supplements the physical world by adding virtual digital information, enriching the user's experience. MR involves merging real and virtual worlds, resulting in hybrid environments and visualizations. This technology enables the coexistence of virtual and real objects, permitting users to interact with both simultaneously [13].

From the characterized capabilities of ER, VR fits best into the identified research problem and will eliminate the limitations of current research options in the context of manual sorting.

VR has many application areas such as entertainment, manufacturing [14], education [15], psychology [16] or medicine [17]. Among these areas is also transport.

Virtual Reality Applications in Transport Systems

Numerous review articles confirm the widespread use of VR in the transportation field. In [18] a short review was provided in the context of VR for intelligent city (road planning and design in a 3D virtual city, drivers training), logistics (virtual logistics laboratory, container operation management system, e-commerce logistics distribution path) and test systems. In [19] immersive technologies for transportation fields were investigated and classified into theoretical construction and algorithm optimization, immersive visualization-enabled user studies, enhanced urban design and development, and digital training and education. In addition to articles treating the transport sector as a whole, there are also papers limited to a specific transport issue. Among them Virtual Reality for automated driving (summary of VR applications in: navigation and routing, driving behavior, safety, education and training, UI design and prototyping, passenger experiences, trust and acceptance, vulnerable road users, traffic simulation) [20], road safety research (summary of research oriented on drivers, pedestrians, passengers and cyclists) [21] and planning for sustainable smart cities (potentials of VR in: visualizing pedestrian thermal comfort in cities, visualizing smart transportation, data management, visualizing the cognitive behavior of urban dwellers) [22] were considered.

The advantages of using VR in transportation based on selected examples are shown in Table 1.

Table 1. Advantages of selected VR applications in transportation.

VR application	Advantages
education (teaching students [23], vehicle operator training [24], traffic controllers training [25])	increased immersion, higher motivation, deeper learning, no financial or social costs of a potential mistake during training, a recreation of multiple training settings
analysis of pedestrian behavior [26, 27]	minimal ethical concerns, more diverse traffic scenarios, safe environment for studying pedestrian-driver relations
tourism [28]	low cost and environmentally friendly way of traveling, no accessibility barriers
transport developments testing (e.g. autonomous vehicles [29])	risk and cost reduction, accelerated development
infrastructure design [30]	user experience and interaction with future transport infrastructure

Given the outlined limitations of research in the area of manual sorting (limited access to the real system, time, space and cost constraints in laboratory conditions, harmful working environment for human health, limited possibility to change the parameters of the system) and VR possibilities, it was considered to perform the study with the use of a virtual sorting line. Therefore, the paper aims to present the concept of a manual waste sorting study using VR technology. The main components of the VR system for manual sorting, possibilities for its use in experimental research and limitations will be presented.

2 VR System for Manual Waste Sorting Study

When describing the components of the VR system, the following are indicated: users with their goals and tasks, virtual representation of a system/process/product, bidirectional connection between user and virtual model for user interaction with model and for sensing [31] (Fig. 2). Bidirectional connectivity is provided by input (e.g., controllers, gloves, tracking sensors) and output devices (e.g., visual, audio, haptic, smell and taste).

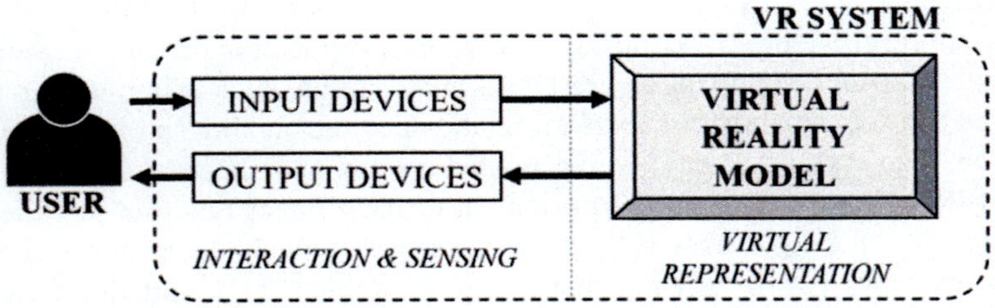

Fig. 2. VR system main components [31, 32].

Depending on the input and output devices applied that determine the level of system advancement, there are three types of VR in terms of immersion (sense of being present in a virtual environment) level: non-immersive, semi-immersive and fully-immersive [33]. The higher the level, the greater the immersion, but also the higher the cost due to more specialized devices.

Referring the VR system architecture to manual waste sorting, it is necessary to consider the following research stages:

Stage I. Virtual Representation of the Waste Sorting Cabin

The manual waste sorting process takes place in waste sorting cabins. Workers (sorters) are located on one or both sides of the conveyor belt on which the waste is transported. Each workstation is equipped with receivers of picked materials, i.e. chutes and/or waste bags. Typically, the farther away the workstation, the more fractions to be sorted due to the so-called " post-cleaning" of previous sorters. Components of the sorting cabin indicate the necessity of replicating the following in the virtual model: conveyor belt, waste stream, workstations with stream receivers. An example representation of a ten-station sorting cabin in a simulation model is shown in the Fig. 3.

Fig. 3. Virtual representation of a manual waste sorting cabin developed based on Flexsim software.

Stage II. Selection of Input and Output Devices and Their Connection to the Virtual Model

The basic set of dedicated VR devices used in entertainment consists of a computer, VR headset and controllers. In the context of the research under consideration, the indicated elements are insufficient. The controllers only allow replication of hand movement by relocating it in space and picking by pressing a single button. Such a simplification would disrupt the considered study of hand sorting, in which the picking action is crucial. For this reason, haptic gloves should be considered instead of standard controllers. In this way, it will be possible to map the movement of each finger (and not just the whole hand). In addition, the gloves will stimulate the sense of touch. An example of a haptic gloves is shown in Fig. 4. Whereas the Fig. 5 shows their representation in a virtual model.

Fig. 4. Hands in haptic gloves in real world.

Fig. 5. Hands represented in a simulation model by haptic gloves.

Stage III. Validation of the Developed VR System for the Manual Waste Sorting Study
The VR system will substitute for the actual sorting cabin in experimental research. Due to this, it is necessary to check the consistency degree between the results from the real system and the virtual system. Data required to be collected from both systems include:

- detection time of the waste after it appears in the field of view,
- average time to pick up the waste,
- amount of waste selected per unit of time under ideal conditions (sorter theoretical productivity),
- the amount of waste picked per unit of time (sorter productivity) under variable conveyor speed and variable waste stream morphology.

Results from both systems can be compared using statistical tests or measures such as RMSE. Successful validation of the developed VR system will allow it to be used for experimental research of manual waste sorting.

3 Possibilities and Limitations of VR-Based Manual Waste Sorting Study

The proposed VR system for manual waste sorting is intended for experimental research. With its use, it is possible to study the effect of variation of selected groups of input parameters on the efficiency of the system. Experiments can be oriented to study:

- stream-related parameters: morphology (amount of individual fractions in the waste stream), heterogeneity, conveyor belt occupancy,
- conveyor-related parameters: speed, width, height,
- workstations-related parameters: arrangement of workstations (on both sides or on one side of the conveyor), number of workstations, number of selected fractions at a given workstation, spacing between successive workstations.

The multiplicity of parameters that can have a significant impact on the efficiency of manual sorting translates into a high complexity of the layout of experiments under real or laboratory conditions. The proposed VR system overcomes this limitation and thus provides a wide range of tests in a short time resulting from the absence of test preparation (collection of waste samples, adjustment of parameters of real objects). In addition, the virtual environment eliminates the exposure of the tested sorters to harmful working conditions.

In addition to many advantages, the proposed solution also has several limitations:

- no representation of harsh working conditions (unpleasant odor, waste dirt, dust), which can translate into reduced productivity of workers in real conditions,
- the required amount of computer equipment and VR devices adequate to the number of sorters to be tested - the more people to be tested at the same time, the more computers and devices should be provided (one sorter = one computer + one VR set). However, it should be noted that it is possible to study only one sorter, and the others can be replaced by virtual objects,
- the cost and availability of VR devices.

Despite the indicated limitations, the proposed solution significantly outperforms the currently available options for manual waste sorting research.

4 Summary and Conclusions

The article is part of the area of waste management, as well as the conveyor-based transportation sector. More specifically, the process of manual waste sorting carried out by people based on the use of conveyor transport was considered. Limitations of studies both in real-world systems and in those replicated in laboratory conditions were reported. As a way to address the outlined limitations, the concept of research using VR technology was proposed.

With regard to the overall architecture of the VR system, the following were considered: virtual representation of waste sorting cabin, selection of input/output devices and system validation. Additionally, the possibilities and limitations of using the proposed VR system in the context of experimental research of manual waste sorting were identified.

The proposed VR-based study concept provides missing so far: unlimited modifications of system parameters, significant reduction of research time and no contact with actual waste. This may translate into the development of research on manual waste sorting. This in turn can contribute to the reduction of waste sent to landfill and significantly increase the achieved recovery rates.

References

1. Kawalec, W., Suchorab, N., Konieczna-Fuławka, M., Król, R.: Specific energy consumption of a belt conveyor system in a continuous surface mine. Energies **13**(19), 5214 (2020)
2. Boysen, N., Briskorn, D., Fedtke, S., Schmickerath, M.: Automated sortation conveyors: a survey from an operational research perspective. Eur. J. Oper. Res. **276**(3), 796–815 (2019)

3. Elahi, M., Afolaranmi, S.O., Mohammed, W.M., Lastra, J.L.M.: Energy-based prognostics for gradual loss of conveyor belt tension in discrete manufacturing systems. Energies **15**(13), 4705 (2022)
4. Bourtsalas, A.C., Themelis, N.J.: Materials and energy recovery at six European MBT plants. Waste Manag. **141**, 79–91 (2022)
5. Lubongo, C., Alexandridis, P.: Assessment of performance and challenges in use of commercial automated sorting technology for plastic waste. Recycling **7**(2), 1–26 (2022)
6. Cunha, J., Carneiro, P., Colim, A.: Ergonomic assessment in waste sorting jobs with different methods. In: Studies in Systems, Decision and Control, pp. 461–469 (2020)
7. Emmatty, F.J., Panicker, V.V., Baradwaj, K.C.: Ergonomic evaluation of work table for waste sorting tasks using digital human modelling. Int. J. Ind. Ergon. **84**, 103146 (2021)
8. Barkhordari, A., et al.: Characteristics and health effects of particulate matter emitted from a waste sorting plant. Waste Manag. **150**, 244–256 (2022)
9. Eriksen, E., Afanou, A.K., Madsen, A.M., Straumfors, A., Graff, P.: An assessment of occupational exposure to bioaerosols in automated versus manual waste sorting plants. Environ. Res. **218**, 115040 (2023)
10. Ip, K., Testa, M., Raymond, A., Graves, S.C., Gutowski, T.: Performance evaluation of material separation in a material recovery facility using a network flow model. Res. Conserv. Recycl. **131**, 192–205 (2018)
11. Efthymiou, O.K., Ponis, S.T.: Industry 4.0 technologies and their impact in contemporary logistics: a systematic literature review. Sustainability (Switzerland) **13**(21), 11643 (2021)
12. Cárdenas-Robledo, L.A., Hernández-Uribe, Ó., Reta, C., Antonio Cantoral-Ceballos, J.: Extended reality applications in industry 4.0 – a systematic literature review. Telemat. Inf. **73**, 10183 (2022)
13. Wong, K.C., Sun, Y.E., Kumta, S.M.: Review and future/potential application of mixed reality technology in orthopaedic oncology. Orthop. Res. Rev. **14**, 169–186 (2022)
14. Malik, A.A., Masood, T., Bilberg, A.: Virtual reality in manufacturing: immersive and collaborative artificial-reality in design of human-robot workspace. Int. J. Comput. Integr. Manuf. **33**(1), 22–37 (2020)
15. Rojas-Sánchez, M.A., Palos-Sánchez, P.R., Folgado-Fernández, J.A.: Systematic literature review and bibliometric analysis on virtual reality and education. Educ. Inf. Technol. **28**(1), 155–192 (2023). https://doi.org/10.1007/s10639-022-11167-5
16. Yaremych, H.E., Persky, S.: Tracing physical behavior in virtual reality: a narrative review of applications to social psychology. J. Exp. Soc. Psychol. **85**, 103845 (2019)
17. Javaid, M., Haleem, A.: Virtual reality applications toward medical field. Clin. Epidemiol. Glob. Health **8**(2), 600–605 (2020)
18. Vrbani, F., Ivanjko, E.: Virtual reality and transport : a short review. In: 3 Virtual Reality Based Simulators, pp. 457–462 (2022)
19. Li, F., Trappey, A.J., Lee, C.H., Li, L.: Immersive technology-enabled digital transformation in transportation fields: a literature overview. Expert Syst. Appl. **202**, 11759 (2022)
20. Riegler, A., Riener, A., Holzmann, C.: A systematic review of virtual reality applications for automated driving: 2009–2020. Front. Human Dyn. **3**, 1–29 (2021)
21. Vankov, D., Jankovszky, D.: Effects of using headset-delivered virtual reality in road safety research: a systematic review of empirical studies. Virtual Real. Intell. Hardware **3**(5), 351–368 (2021)
22. Jamei, E., Mortimer, M., Seyedmahmoudian, M., Horan, B., Stojcevski, A.: Investigating the role of virtual reality in planning for sustainable smart cities. Sustainability (Switzerland) **9**(11), 1–16 (2017)
23. Pense, C., Tektaş, M., Kanj, H., Ali, N.: The use of virtual reality technology in intelligent transportation systems education. Sustainability (Switzerland) **15**(1), 300 (2023)

24. Ćwil, M., Bartnik, W.: Physically Extended Virtual Reality (PEVR) as a new concept in railway driver training. In: Chen, J.Y.C., Fragomeni, G. (eds.) HCII 2019. LNCS, vol. 11575, pp. 230–242. Springer, Cham (2019). https://doi.org/10.1007/978-3-030-21565-1_15
25. Walters, W.T., Walton, J.: Efficacy of virtual reality training for pilots: a review of links between user presence, search task performance, and collaboration within virtual reality. Proc. Human Fact. Ergon. Soc. Ann. Meet. **65**(1), 919–922 (2021)
26. Deb, S., Carruth, D.W., Sween, R., Strawderman, L., Garrison, T.M.: Efficacy of virtual reality in pedestrian safety research. Appl. Ergon. **65**, 449–460 (2017)
27. Schneider, S., Bengler, K.: Virtually the same? analysing pedestrian behaviour by means of virtual reality. Transport. Res. Part F: Traffic Psychol. Behav. **68**, 231–256 (2020)
28. Beck, J., Rainoldi, M., Egger, R.: Virtual reality in tourism: a state-of-the-art review. Tour. Rev. **74**(3), 586–612 (2019)
29. Yao, S., Zhang, J., Hu, Z., Wang, Y., Xilin, Z.: Autonomous-driving vehicle test technology based on virtual reality. J. Eng. **16**, 1768–1771 (2018)
30. Birrell, S., Payre, W., Zdanowicz, K., Herriotts, P.: Urban air mobility infrastructure design: using virtual reality to capture user experience within the world's first urban airport. Appl. Ergon. **105**, 103843 (2022)
31. Nikolić, D., Whyte, J.: Visualizing a new sustainable world: toward the next generation of virtual reality in the built environment. Buildings **11**(11), 546 (2021)
32. Zhang, H.: Head-mounted display-based intuitive virtual reality training system for the mining industry. Int. J. Min. Sci. Technol. **27**(4), 717–722 (2017)
33. Albuha Al-Mussawi, R.M., Farid, F.: Computer-based technologies in dentistry: types and applications. J. Dentistry (Tehran, Iran) **13**(3), 215–222 (2016)

Development of Rail Freight Transport Considering the International Intermodal Transport and Logistics: Lithuanian Case

Aldona Jarašūnienė(✉), Dmitrij Ševaldin, and Stasys Steišūnas

Vilnius Gediminas Technical University, Plytinės Street 27, 10105 Vilnius, Lithuania
{aldona.jarasuniene,stasys.steisunas}@vilniustech.lt

Abstract. Rail transport allows transporting large and heavy loads at relatively low transport costs, in observance of precise departure and arrival schedules, offering safety, reliability, regularity of transport, low dependence on weather conditions, and environmental friendliness, which have a positive impact on logistics decisions. More extensive use and development of railways is essential to meet the demand for sustainable transport, reduce road occupancy, environmental pollution and energy consumption. The Authors examine the importance of the synergies of railway transport, logistics and intermodal transport, as well as the analysis of the indicators and trends of freight transport by rail in Lithuania and recommendations for the development of rail freight transport. The application of expert judgment in order to create a plan for increasing the infrastructure capacity of railway lines, as well as for properly selected measures to increase the capacity of the infrastructure are emphasized.

Keywords: Lithuanian railways · Intermodal transport · Infrastructure capacity · Multi-criteria decision making · Freight traffic flows

1 Introduction

Lithuania's transport sector determines the country's macroeconomic indicators and is one of the most important sectors in the country. The development of the transport system is aimed at ensuring safe, convenient, and efficient movement and transport of goods not only domestically but also abroad, while complying with all environmental standards and requirements. Over the past few years, freight transport volumes have been steadily increasing in Lithuania. However, freight transport itself is a rather volatile business. In the development of its freight transport, Lithuania has faced the COVID-19 coronavirus pandemic, as well as the complicated geopolitical situation and international tensions in the Baltic region. Given the importance of the Lithuania's transport sector and the existing circumstances, the efficiency of freight transport must be improved through quality planning and organisation of transport processes, ensuring efficient functioning of logistics chains. Rail transport allows transporting large and heavy loads at relatively low transport costs, in observance of precise departure and arrival schedules, offering

O. Prentkovskis et al. (Eds.): TRANSBALTICA 2023, LNITI, pp. 369–378, 2024.
https://doi.org/10.1007/978-3-031-52652-7_36

safety, reliability, regularity of transport, low dependence on weather conditions, and environmental friendliness, which have a positive impact on logistics decisions. More extensive use and development of railways is essential to meet the demand for sustainable transport, reduce road occupancy, environmental pollution and energy consumption. In order to strengthen Lithuania's competitive advantage in freight transport, it is necessary to continuously develop the rail transport sector in order to exploit its potential even more efficiently. The development of the rail transport sector definitely has a positive impact on logistics.

This article deals with the problem of reduced volume of freight transport by the State Company "Lietuvos Geležinkeliai" (Lithuanian Railways) after the im-position of international sanctions on Russia and Belarus in 2022.

The aim of the article is to evaluate freight transport indicators of Lithuanian Railways in 2022, to assess the opportunities for diversification of freight transport activities and to propose a solution for the development of a plan to increase the capacity of the congested sections of the railway lines of the current freight transport routes.

The development of intermodal transport and a plan for increasing the capacity of Lithuanian railway lines is analysed in this article.

2 The Importance of Synergies Between Rail Transport, Logistics and Intermodal Transport

International business environment is inevitably changing in the face of intensive globalisation processes, and logistics operators have new specific needs determined by the intensive development of economic relations in different business environments. However, there is a lack of application of new solutions and organisational forms, as well as of more detailed research and analyses in order to better coordinate logistics solutions in rail freight transport, not only in Lithuania, but also on a global scale [1]. Rail transport is usually associated with national rail networks and is poorly developed internationally, unable to operate in a coordinated manner due to the existing technical, organisational and technological barriers between different regions of the world. These factors hinder the efficiency of rail transport, effective functioning of logistic chains in freight transport and do not allow to become even more competitive in long-distance carriage.

Poorly developed international marketing does not allow to take advantage of the diversity of transport and logistics services in certain regions and countries, including Lithuania, and the services offered in foreign markets [2].

The Lithuanian rail transport sector needs to adopt international practices using a modern portfolio of rail transport and related services, including the arrangement of rail transport, logistics services, transhipment, processing, and distribution of freight between different modes of transport, storage of freight, as well as additional services related to the development of partial production and packaging services. This would allow creating a fully developed, appealing model of rail transport activity adapted to Lithuanian conditions and operating on a system basis, and drafting a strategy for the development of activities in foreign markets. The level of integration of rail transport into the transport chain is determined by the role of rail transport in the intermodal transport system. One of the most common modes of intermodal transport is rail-road intermodality, aiming to create a seamless transport chain by combining rail and road modes of transport [3].

Christopher, M. [4] analyses thematic areas related to freight transport and logistics, supply chain management, including intermodal transport, technology deployment and logistics innovation. McKinnon, A., Cullinane, S., & Browne, M. [5] examine strategic aspects of establishment of logistics hubs and intermodal transport, also discussing the technological and environmental sustainability peculiarities that can affect the organisation of freight transport. Rushton, A., Croucher, P., & Baker, P. [6] emphasize the importance of analysing problematic issues faced in the development of a network of logistics hubs, including the integration of modes of transport and the use of technology in logistics processes. Dablinskas, A., & Rutelionė, I. [7] argue that the most efficient rail transport process and the high quality of logistics processes is achieved when it is managed in a coordinated manner. The economic aspect of problem solutions includes activities such as proper routing, use of multimodal transport, proper pricing of services, ensuring quality, and competitiveness. The aspect of solving environmental problems is focused on the use of renewable Transportation energy sources, saving fossil fuel energy, minimising emissions, and other measures related to the concept of environmental friendliness. The social aspect of problem solving is focused on the formation of a "conscious" user and organisation, which includes social responsibility and employee competence (Vienažindiene & et al. [8]).

The social aspect of problem solving is focused on the formation of a "conscious" user and organisation, which includes social responsibility and employee competence. Wang, S., & Hu, Z-H. [9].

The integrated management of the transport process in a freight route allows achieving a high level of service quality in rail transport.

In order to ensure high-level efficiency in the rail transport value chain, railway companies must coordinate their actions and decisions, applying a methodology for managing the transport process and resources, and planning joint actions. In the freight transport value chain, the quality of service provision is determined by interaction and cooperation between railway companies in a single railway network, i.e. in a corridor as an integrated chain of the transport process.

In the case of Lithuania, foreign direct investment should be directed towards the development and integration of logistics services in the rail transport sector into international logistics chains, creating preconditions for establishing a supporting and geographically convenient freight distribution point in the global logistics system and a freight consolidation base abroad.

In order to exploit the potential of rail transport more efficiently, it is important to conduct an analysis of the indicators and trends of rail freight transport in Lithuania.

3 Analysis of Indicators and Trends of Rail Freight Transport by Lithuanian Railways

The main indicator of freight transport is the total amount of freight transported during the year in tonnes (thousands or millions of tonnes). This figure includes the amount of freight transported domestically and internationally [10].

In Lithuania, the average annual rail freight transport accounts for around 49 million tonnes. The indicators of rail freight transport of Lithuanian Railways are presented Table 1.

Table 1. Freight carriage and freight turnover of Lithuanian Railways [11].

Year	Freight carriage by rail (thousand t)	Rail freight turnover (thousand tkm)
2022	30,977.4	7,375,058
2021	51,079.6	14,565,850
2020	53,429.7	15,864,679
2019	55,209.2	16,180,599
2018	56,775.7	16,884,825
2017	52,638.2	15,413,521
2016	47,650.6	13,790,109
2015	48,053.2	14,036,238
2014	49,000.1	14,306,751
2013	48,028.1	13,343,682
2012	49,377.2	14,171,644

During the period from 2012 to 2021, annual rail freight transport in Lithuania increased by 4%. However, with the outbreak of the war in Ukraine and the international sanctions imposed on Russia and Belarus, annual rail freight transport in Lithuania fell to 31 million tonnes of freight, which is 39% less than the average freight transported in 2012–2021.

According to the Ministry of Transport and Communications of the Republic of Lithuania, oil and oil products, various construction materials, fertilisers and agricultural products accounted for the major share of freight carried by rail in 2022 (Fig. 1).

In terms of domestic freight flows, the majority of rail freight was transported to Klaipeda Seaport, accounting for 10.7 million tonnes. Volumes of domestic carriage were higher than planned, with around 6.4 million tonnes of freight transported in Lithuania. In comparison, domestic traffic accounted for 5.8 million tonnes in 2021. Counting all freight transported to and from Kaliningrad through the territory of Lithuania, regardless of its origin and destination, 8 million tonnes of freight was carried in Kaliningrad transit, and another 5.9 million tonnes headed in other directions [12].

In the current geopolitical context, LTG Cargo, a freight transport company of Lithuanian Railways Group, has been actively diversifying its activities to the West in order to expand the North–South freight route and develop it instead of the East–West route which has been pre-dominant for many years.

Efficient diversification and expansion of freight transport business requires optimising freight transport costs through quality planning and organisation of logistics processes. The first step in the freight transport is to select the right transport technology, carrier, mode of transport and vehicle type. Then, it is necessary to ensure the

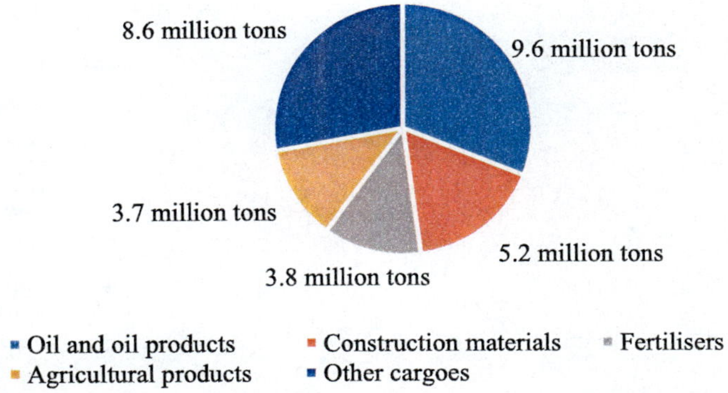

Fig. 1. Main products transported by Lithuanian Railways in 2022.

interoperability between the means of transport and the infrastructure structures that support it, followed by proper handling and securing of the goods and selection of the transport route. Finally, it is necessary to ensure transport control in order to preserve the characteristics of the goods during transport and handling operations. To meet these challenges, an increasing number of countries around the world are switching to container freight transport, which significantly reduces freight costs, shortens working time and simplifies transhipment [13].

This situation has led to the emergence of intermodality as a phenomenon. Intermodal transport is a mode of freight transport where freight is transported from the point of origin to the point of destination by several consecutive modes of transport, using transhipment infrastructure, i.e. an intermodal terminal, without unloading the freight itself. The goods transported this way are packed into intermodal load units such as containers or semi-trailers. The intermodal mode of transport has been developed not only as a way of simplifying, standardising and optimising the transport process, but also as a means of transporting goods in a more environmentally friendly way. Containerised freight is therefore a potential for rail transport.

Globally, the share of containerised freight has been increasing by around 10% each year [14]. Such growth is natural, as the carriage of goods in containers is much more advanced compared to traditional transport. Although containerised transport in Lithuania has not yet reached global extent, preparations are underway for the more advanced transport of goods. Lithuanian Railways has a well-developed infrastructure and the Kaunas and Vilnius intermodal terminals established in 2014. The Kaunas intermodal terminal is capable of handling up to 55,000 twenty-foot containers (TEUs) per year, with container capacity of 1,120 TEU. The Vilnius Intermodal Terminal has an annual capacity of up to 100,000 TEU, with container capacity of 1,465 TEU [15]. LTG Cargo already offers its existing and potential customers a sufficiently broad portfolio of container train services, with its transport and service areas covering the Lithuanian, Western European, Scandinavian and Asian markets (Fig. 2).

Current routes of intermodal freight transport by Lithuanian Railways and their lengths:

1. Vilnius (Lithuania)–Klaipeda (Lithuania) – 350 km.

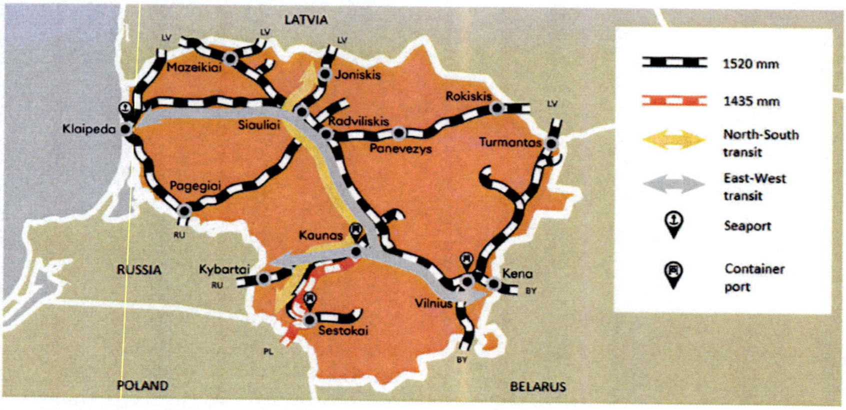

Fig. 2. Directions of freight transport by Lithuanian Railways [16].

2. Kaunas (Lithuania)–Gdansk (Poland) – 549 km.
3. Kaunas (Lithuania)–Malaszewicze (Poland) – 557 km.
4. Šeštokai (Lithuania)–Muuga (Estonia) – 671 km.
5. Kaunas (Lithuania)–Sławków (Poland) – 740 km.
6. Kaunas (Lithuania)–Duisburg (Germany) – 1,520 km.

In order to diversify freight transport, Lithuanian Railways Group actively cooperates with Polish, Latvian and Estonian railways.

To strengthen rail links between the Baltic States, the *Amber Train* project was implemented in 2018. The first international intermodal train was formed on the route Šeštokai–Riga–Tallinn, connecting the three Baltic States by rail and opening up new business opportunities for freight carriers. Today, goods can be transported across Lithuania, Latvia and Estonia in less than 24 h. Another unique feature of the project is the link between Western and Northern Europe. The intermodal terminals in Šeštokai and Kaunas allow transhipment of goods from the 1435 mm gauge to the 1520 mm gauge and vice versa [17], allowing the Baltic States to further develop freight transport with other European Union countries.

In order to strengthen the rail links between Lithuania and Poland, LTG Cargo has established a subsidiary in Poland LTG Cargo Polska. The company develops rail freight transport activities in Poland, strengthens logistics supply chains and expands freight mobility between Lithuania, Ukraine and Poland [13].

Since 2023, freight trains on the Kaunas–Sławków route have had a stopover at Warsaw, which allows freight to be carried to the Pruszków terminal near the capital of Poland. The Sławków Euroterminal near Katowice is an important hub of the Polish broad gauge rail network, linking rail lines between Lithuania and Poland and Ukraine and Poland. The terminal also offers regular freight trains to Schwarzheide (Germany), Maddaloni (Italy), Koper (Slovenia) and Gdansk (Poland). The Kaunas–Pruszków–Sławków route is a part of the European transport corridor linking the countries from Finland to Italy by rail. It allows to expand this network with daily connections to Germany (the port of Bremerhaven, Hamburg and Duisburg), Croatia (Rijeka) and Slovenia (Koper) [14].

Currently, freight trains with a capacity of 36 semi-trailers or containers depart from Kaunas intermodal terminal to Duisburg intermodal terminal and back through the territory of Poland three times a week [18].

In the development of intermodal freight transport to the Northern Poland region, a possibility has been created to transport freight from Kaunas intermodal terminal to Gdansk. Customers shipping freight to and from Gdansk have access to one of the most important deep-sea ports in the Baltic Sea with the largest container ships from North and South America and Asia arriving to the port. It is the only port in the region having regularly arriving freight from the Far East [19].

As part of the development of intermodal freight transport to the Eastern Poland region, transporting freight from Kaunas intermodal terminal to Malaszewicze has become possible. Located in the east of Poland, the Malaszewicze intermodal terminal has a distinctive capacity to receive freight carried by road from East Asia and to tranship it onward to European Union countries. Malaszewicze is home to a number of freight forwarding, logistics and transport companies important for the region [20].

Drivers of LTG Cargo transport intermodal freight in the territory of Lithuania, and drivers of LTG Cargo Polska – in the territories of Poland and Germany (LTG Cargo, 2023d). Last year, LTG Cargo transported 84,000 tonnes of agricultural products from Ukraine to the Klaipeda Seaport via Poland and 120,000 tonnes of oil to Ukraine via Poland. 66,000 tonnes of cargo were transported on the new Kaunas–Duisburg route in April–December of 2022 [12]. In 2022, LTG Cargo Polska, a subsidiary of LTG Cargo, carried 583 freight trains. In comparison, the company transported only 68 trains in 2021 [21].

The long-term strategy for freight transport of the Lithuanian Railways Group provides for the aim to increase the volume of freight transport to 64 million tonnes in 2030 and then – to 103 million tonnes in 2040 [22].

When increasing rail freight volumes, it is important to consider the capacity of the existing railway infrastructure and the available capacities, which is one of the most important tasks for the development of the logistics chain. To solve this problem, it is possible to use the Multi-Criteria Decision Making methods, based on the results of surveys of railway experts.

According to Simanavičienė, R. [23], Multi-Criteria Decision Making methods are used to find the optimal solution and are divided into two groups: multi-object and multi-objective methods. Multi-Criteria Decision-Making methods that use vector optimisation based on a decision-making process model are called Multi-Objective Decision Making methods.

According to Sivilevičius, H. [24], the initial information for processing expert survey results is the numerical data that shows expert priorities and the justification of these priorities (importance).

4 Recommendations for the Development of Rail Freight Transport

The capacity of railway lines is increased by organisational and technical measures or by reconstructing parts of the infrastructure. Organisational and technical actions are less pricy and faster to implement than reconstruction, but such capacity increases are limited. Moreover, organisational and technical actions cannot always be implemented without reconstructing the rail network or individual lines.

The most important organisational and technical actions include the optimisation of train timetables, the use of the double-stack method and several locomotives in one train, and shorter intervals in the train schedule. The construction of second and third tracks on interchanges, the lengthening of arrival and departure tracks in railway stations, the electrification of railway lines and improvement of the efficiency of the layout of traffic lights on interchanges are the main steps of a reconstruction.

The measures for increasing the capacity according to the period of their implementation can be divided into three groups: measures that require a short implementation period – measures are implemented in up to 3 years; measures that require a medium-term implementation period – measures are implemented in 3 to 7 years; measures that require a long-term implementation period – measures are implemented in 7 years and more.

All measures for increasing the capacity of railway infrastructure must ensure train traffic safety, the required level of infrastructure capacity, the reliability of the operation of the rail network, and the reduction of freight and passenger transport costs. A key condition for increasing the capacity of railway infrastructure is a timely completion of works and the observance of the principle of staging. The staging principle involves prioritising works on the most congested sections of railway lines. Thus, increasing the capacity of the railway infrastructure requires a plan with a sequential order of implementation of capacity improvement measures.

In order to draw up a plan to increase the capacity of railway infrastructure, experts must be consulted to properly assess certain measures to increase the capacity of infrastructure. Therefore, using expert survey methods for the preparation of a capacity improvement plan is recommended.

Having received expert survey results, data are processed in application of multi-criteria evaluation methods. These methods are used to solve problems involving many functions of an objective that are optimised simultaneously.

The aim of the processing is thus to obtain aggregated data and new information in the form of implicit expert evaluations. The results of the processing are used to formulate a solution to the problem, or, in the case of the problem under consideration, a plan for increasing the capacity of the railway infrastructure.

5 Conclusions

The conducted analysis of the literature sources showed that there is a lack of application of new organizational forms, more detailed research, and analysis in order to better coordinate logistic solutions between different countries developing of freight transport by rail.

The conducted analysis of the rail transport of freight by Lithuanian Railways in 2022 revealed that the freight transport rates fell by 39%.

A conducted analysis of the long-term strategy for freight transport of the Lithuanian Railways Group shows that the aim is to increase freight transport volumes from 31 to 64 million tonnes by 2030.

As part of the diversification and development of international freight transport, Lithuanian Railways has been developing access to Poland, Germany, Italy, Slovenia and Croatia. In the context of the development of intermodal transport in the North–South direction, assessing the levels of infrastructure capacity on individual railway lines of the current routes and identifying congested sections is substantiated.

By applying multi-criteria decision-making methods, it is possible to identify problematic areas of railway infrastructure more accurately, their influence on cargo transportation and the functioning of logistic chains.

Multi-criteria decision-making expert survey methods should be used to draw up a plan for increasing the capacity of the congested sections of individual lines, offering measures to increase the capacity of the railway infrastructure in the order of priority.

The development of the railway transport sector, the promotion of intermodal freight transport, and the aim of its more efficient use, emphasize the positive impact on logistics.

References

1. Batarlienė, N., Župerkienė, E.: Analysis of factors influencing the competitiveness of rail freight transport in the context of sustainable development. Sustainability **13**(3), 1231 (2021)
2. Jatkuvienė, A., Rekis, E.: Assessment of freight transportation management efficiency in Lithuanian railway network. Transport **36**(2), 195–207 (2021)
3. Vasiliauskas, A., Paulauskas, V.: Optimization of transport flow in a transportation system of geographically distributed objects. Transport **36**(4), 576–589 (2021)
4. Christopher, M.: Logistics and Supply Chain Management. Financial Times/Prentice Hall, London (2016)
5. Mckinnon, A., Cullinane, S., Browne, M.: Green Logistics: Improving the Environmental Sustainability of Logistics. Kogan Page, Londres (2010)
6. Rushton, A., Croucher, P., Baker, P.: The handbook of logistics and distribution management. Online resources: The Chartered Institute of Logistics and Transport, Kogan Page. (2022)
7. Dablinskas, A., Rūtelionė, I.: Sustainable development of logistics centers and their impact on regional economy. Inzinerine Ekonomika-Eng. Econ. **29**(2), 187–194 (2018)
8. Vienažindienė, M., Tamulienė, V., Zaleckienė, J.: Green logistics practices seeking development of sustainability: evidence from Lithuanian transportation and logistics companies. Energies **14**(22), 7500 (2021). https://doi.org/10.3390/en14227500
9. Wang, S., Hu, Z.-H.: Green logistics service supply chain games considering risk preference in fuzzy environments. Sustainability **13**(14), 8024 (2021). https://doi.org/10.3390/su13148024
10. Vaičiūnas, G., et al.: Railways. General course. Technika, Vilnius (2018)
11. Freight transport by rail. Retrieved from Statistics Lithuania (2023). https://osp.stat.gov.lt/statistiniu-rodikliu-analize?indicator=S5R048/
12. Annual results of LTG Cargo engaged in active diversification to the West: reports 31 million tonnes of transported freight. Retrieved from Ministry of Transport and Communications of the Republic of Lithuania (2023). https://sumin.lrv.lt/lt/naujienos/ak

13. Ševaldin, D.: Investigation of the interaction between the capacity of the Lithuanian railway infrastructure network and increased length and weight freight train traffic. Master's Final Degree Project [Lietuvos geležinkelių infrastruktūros tinklo pralaidumo bei padidinto ilgio ir masės prekinių traukinių eismo sąveikos tyrimas]. Vilnius Gediminas Technical University, Vilnius (2022). In Lithuanian
14. Services. Freight transport. Intermodal transport. Containers. Kaunas-Slawkow. Retrieved from LTG Cargo (2023a). https://cargo.litrail.lt/kaunas-slawkow
15. Jokubauskienė, B.: Opportunities for the development of railway intermodal terminals. Master's Final Degree Project [Geležinkelių intermodalinių terminalų plėtros galimybės]. Kaunas: School of Economics and Business, Kaunas University of Technology (2023). In Lithuanian
16. Directions of freight carriage. Retrieved from LTG Cargo (2022a). https://cargo.litrail.lt/
17. Services. Freight transport. Intermodal transport. Containers. Amber Train. Retrieved from LTG Cargo (2022b). https://cargo.litrail.lt/amber-train/
18. Services. Freight transport. Intermodal transport. Kaunas–Duisburgas–Kaunas. Retrieved from LTG Cargo (2023d). https://cargo.litrail.lt/puspriekabiu-pervezimai
19. Services. Freight transport. Intermodal transport. Containers. KIT–Gdansk. Retrieved from LTG Cargo (2023b). https://cargo.litrail.lt/kit-gdanskas
20. Services. Freight transport. Intermodal transport. Containers. KIT– Malaszewicze. Retrieved from LTG Cargo (2023c). https://cargo.litrail.lt/235
21. ELTA. Lithuanian Railways Group announces 2022 results: beating the forecasts. Retrieved from delfi.lt (2023). https://www.delfi.lt/verslas/transportas/lietuvos-gelezinkeliu-grupes-skelbia-2022-m-rezultatus-pranoko-prognozes-93845473
22. Long-term strategy of AB LTG Cargo for 2021–2040. Retrieved from LTG Cargo (2021). https://cargo.litrail.lt/kroviniu-vezimo-direkcija
23. Simanavičienė, R.: Sensitivity analysis of quantitative multi-objective decision-making methods: doctoral dissertation. Technika, Vilnius (2011). http://dspace1.vgtu.lt/handle/1/926
24. Sivilevičius, H.: Application of expert evaluation method to determine the importance of operating asphalt mixing plant quality criteria and rank correlation. Baltic J. Road Bridge Eng. **6**(1), 48–58 (2011). https://doi.org/10.3846/bjrbe.2011.07

New Sustainable and Economical Tank Shapes for the Oil and Gas Transport Infrastructure of Ukraine

Anton Makhinko[1(✉)], Nataliia Makhinko[1], Viktor Karpov[1], Katerina Krayushkina[1], and Oleksandr Kordun[2]

[1] National Aviation University, Liubomyra Huzara 1, Kyiv 03058, Ukraine
{makhinko.anton,nataliia.makhinko,karpov.viktor, kateryna.kraiushkina}@npp.nau.edu.ua

[2] V. Shimanovsky Ukrainian Institute of Steel Construction, Shimanovsky Str 2/1, Kyiv 02125, Ukraine

Abstract. New constructive forms of drop-shaped tanks for oil and oil products have been invented. The stability and economy of the invented forms of tanks consists in saving materials and resources for their manufacture, installation and operation. Comparison of new forms of reservoirs with well-known analogues is made. Emphasis is placed on the advantages of new design solutions for tanks and the shortcomings of existing solutions are described. The use of new sustainable and economic forms of tanks for oil and oil products makes it possible to increase the reliability and efficiency of the energy complex of Ukraine and sustainable development of the transport industry under martial law.

Keywords: Constructive form · Drop-shaped tanks · New design · Oil storage · Pressure · Meridional and ring stress

1 Introduction

Transport infrastructure is one of the basic sectors of the economy. The functioning of the transport sector of Ukraine in the conditions of martial law is impossible without an effective system of supply and storage of fuel. This is especially true for road, aviation, maritime and inland water transport. The gas and oil supply system is one of the main elements of the fuel and energy complex of Ukraine, which includes 34,000 km of main gas pipelines, 8,000 km of oil pipelines, 13 underground gas storage facilities with a capacity of 43 billion m^3, 360 oil depots and 12,000 tanks with a total capacity of 5 million m^3. The provision of a reservoir park is an urgent problem for fuel producers, logistics companies, and gas station networks. It is also an important factor characterizing the country's energy supply and energy independence. It was established that one ton of oil production or processing per year should correspond to 0.3–0.4 m^3 of tank capacity. Every year, the volume of construction in the oil and gas transport infrastructure of Ukraine increases. New materials, new technologies are emerging, highly efficient and

O. Prentkovskis et al. (Eds.): TRANSBALTICA 2023, LNITI, pp. 379–385, 2024.
https://doi.org/10.1007/978-3-031-52652-7_37

technological construction methods are being introduced. All this contributes to the construction of large-volume and economical steel tanks of a new constructive form [1–10].

The nature itself creates forms that are rational from the point of view of structural strength. Examples are the forms of shells of mollusks, the skeletons of sea urchins, etc. One of the perfect natural forms is the egg shell. As a mathematical model of drop-shaped shells, the form of a liquid drop lying quietly on a plane is taken (Fig. 1). Median surfaces of this group of shells belong to the class of surfaces of revolution.

The main purpose of drop-shaped tanks is the storage of products with high saturated vapor pressure under excess pressure, which can significantly reduce evaporation losses compared to "atmospheric" tanks. However, the cost of a cylindrical "atmospheric" tank is much less than a drop-shaped one of the same volume. This is due to the complexity of the construction of the drop-shaped shell. Therefore, an indispensable condition for the widespread introduction of drop-shaped tanks is its efficiency, which is determined by comparing the additional cost and savings from reducing losses over the depreciation period. Since the cost of metal structures is determined to a large extent by its own weight, as little metal as possible should be spent on drop-shaped tanks. This condition can be met by using the ability of momentless shells of double curvature, as well as by balancing the loads distributed over their surface by tension or compression, simultaneously acting in the direction of the main curvature. The basis for determining the rational form of the tank body is the requirement of full use of the bearing capacity of the metal of a thin shell of revolution.

Fig. 1. The drop-shaped tank "Smiling Jack" – decorated 80000-barrel storage tank.

Now several constructive solutions of drop-shaped tanks have been proposed. For example, a drop-shaped tank is known, consisting of a shell and a device for ensuring rigidity, made outside the tank [11]. However, the disadvantage of this tank is that the supporting structures in the form of columns, stiffening rings, and cables are bulky and have a significant metal capacity [1, 12–17]. These structures do not participate in the operation of the tank with excessive internal pressure, and their installation requires more labor and careful inspection. In such a system, it is difficult to create the same stiffness over the entire surface of the tank. Reservoirs consisting of a wall, a flat bottom, and roof elements are also known [18]. The disadvantages of these tanks include the fact that the load-bearing structures are inside the tank, which reduces its reliability, makes it impossible to control the condition of the load-bearing structures, and reduces the useful volume. In addition, with a significant volume and height of the tank, the condition of uniform strength of the torus walls is not fulfilled, because the hydraulic pressure of the liquid varies with height, and the radius of rounding of the torus shell remains constant [1, 15].

The purpose of the work is to simplify the designs of drop-shaped tanks, increase their reliability and reduce the metal capacity. For this purpose, three new sustainable and economic structural forms of drop-shaped tanks are proposed.

2 The Drop-Shaped Tank with a Membrane Top

In this tank, the reliable operation of the shell is ensured by a stiffening ring and support struts. The shell is made of two parts fixed in a ring of rigidity, while the upper part has the appearance of a membrane (Fig. 2).

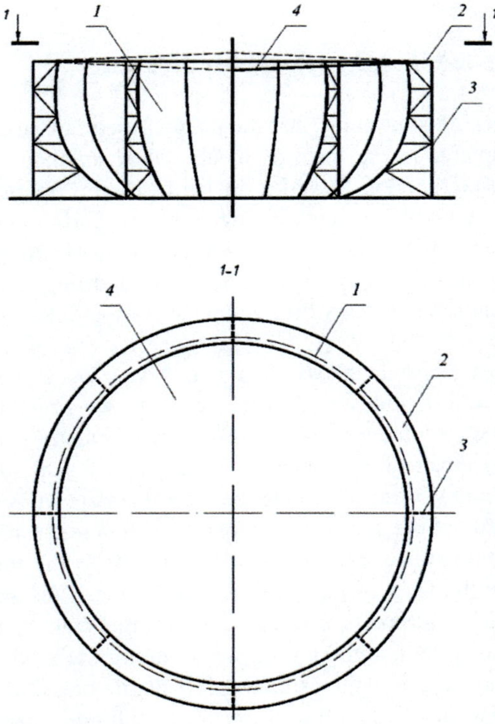

Fig. 2. The drop-shaped tank with an upper part in the form of a membrane: 1 – shell; 2 – stiffness ring; 3 – lattice struts; 4 – membrane.

The construction of the tank from two parts, one of which has the form of a membrane, ensures its optimal operation: under vacuum and excess pressure, the membrane works in tension - this increases the reliability of the tank and allows you to get rid of other supporting structures, which in general simplifies its construction. In Fig. 1 schematically shows the main view of the proposed tank. The drop-shaped tank includes a shell 1 attached to a stiffness ring 2, lattice struts 3 that support a stiffness ring 2, and a membrane 4 attached to a stiffness ring 2 [4].

The tank works as follows: when the pressure inside the tank body decreases, which is caused by draining the liquid and lowering the temperature, the membrane 4 bends inside the tank. With excess pressure inside the tank, which occurs when the temperature rises, the membrane bends outward. The dotted line shows the extreme positions of the membrane. At the same time, the volume of the tank partially increases.

The tensile forces that arise in the membrane in the first and second cases are received by the stiffness ring 2, which through the struts 3 transmits the force to the foundation. Racks 3 have a thin-walled tubular construction. A partial increase in the volume of the tank compensates for the excess pressure. This allows you to reduce the loss of liquid vapors, such as oil products, through the tank's "breathing" valve. In addition, the proposed design makes it possible to make the most of the industrial approach of rolling and high-strength steel, which reduces labor costs for manufacturing and installation.

3 The Drop-Shaped Tank with a Guy System

The essence of this design solution is that the drop-shaped tank includes a shell made of interconnected elements in the form of sheets, which in shape resemble mutually convex lenses ("watermelon skins") and a device to ensure operational rigidity. The liquid storage tank is assembled using welding from two parts - upper and lower. The lower part is assembled in the design position with the help of lifting mechanisms. The upper part is mounted on a pre-assembled rack, cables and rings. After that, the upper part is lowered onto the lower part by lifting mechanisms and they are welded together [5].

In Fig. 3 shows the proposed tank arranged in this way. Inside the shell 1 is a system that provides rigidity and includes a telescopic rack 2, two rings 3 and 4, a movable sleeve 8 and steel ropes - cables 6 and 7. Rack 2 is two elements - outer 5 and inner 9. The outer element 5 is attached to the bottom of the tank. At the same time, the inner element rests freely on the spring 10 inside the outer element. A system of cables 6 is attached to the upper end of the internal element 9, which pass through the upper ring 4, the movable sleeve 8 and are attached to the lower ring 3. At the same time, the lower ring 3 is placed above the level of the maximum diameter of the shell. In the process of filling the tank, the liquid pressure is perceived by the shell 1, which balances the previous tension of the system, and the shell works in normal mode. When the tank is emptied, a vacuum is created, and the external atmospheric pressure is perceived by the stiffening system and the shell. The forces from the shell are transmitted to the rings, and from them through the cables to the movable element of the telescopic rack. Due to the fact that the movable rack is located on a spring, the operation of the entire structure ensures the uniform and optimal operation of the strength of all its elements.

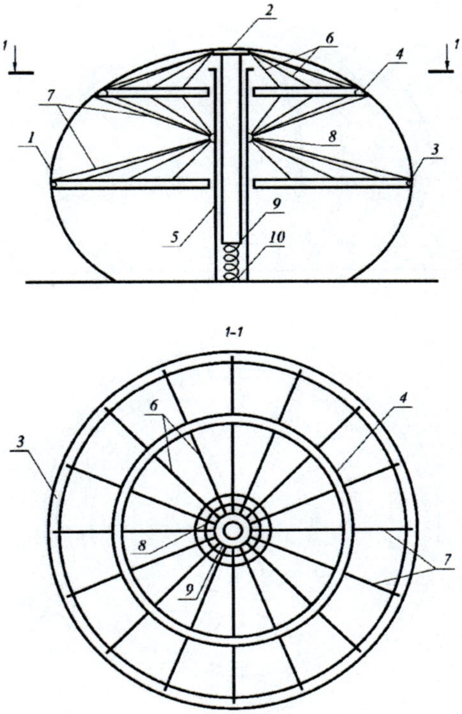

Fig. 3. The drop-shaped tank with a guy system: 1 – shell; 2 – telescopic struts; 3, 4 – stiffness rings; 5 – external element; 6, 7 – guys; 8 – movable sleeve; 9 – internal element; 10 – spring.

4 The Drop-Shaped Tank with External Ring and Meridional Trusses

The purpose of this new structural form is to reduce steel consumption and increase the reliability of the tank by providing a rigidity that is evenly distributed over the entire surface of the tank. The goal is achieved by fact the tank is made in form of a lattice structure, including circumferential and meridional trusses with connections that cover the shell and are rigidly connected to it. (see Fig. 4). The liquid storage tank consists of a drop-shaped shell 1, covered by outer ring 3 and meridional trusses 2, connected to each other and to the shell. To increase the stiffness of the trusses, the trusses have bracing 4 in the form of steel bars [6]. The tank works like this. When a vacuum is formed inside the container, which occurs when the liquid drains and the temperature drops, the external atmospheric pressure tries to compress the shell. The force of external pressure is perceived by the shell 1 and transmitted through the ring 3 and meridional 2 trusses to the foundations. In case of excess pressure inside the tank, the ring 3 and meridional 2 trusses perceive tensile stresses that occur in the shell, which increases the reliability of the tank and allows reducing the thickness of the shell wall.

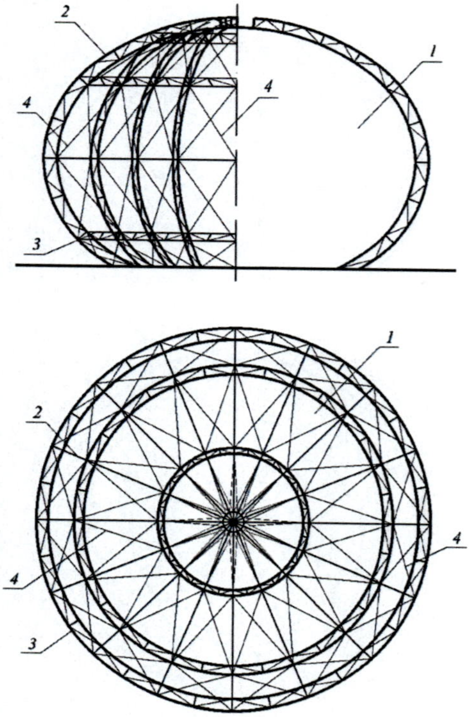

Fig. 4. The drop-shaped tank with external ring and meridional trusses 1 – shell; 2 – meridional trusses; 3 – circular trusses; 4 – bracings.

5 Conclusions

The proposed design solutions of drop-shaped tanks are more economical in terms of metal consumption compared to the known ones, contribute to the reduction of labor costs during manufacture and installation due to the possibility of industrialization and factory production of structures.

Note that all the proposed tank designs have an important advantage – the ease and clarity of creating design models in programs for finite element analysis of structures. It should always keep in mind that the design model is just an abstraction created for certain purposes, and therefore it should always be as simple and understandable as possible. The rather complex and excessively detailed design models are becoming increasingly popular for designing modern structures. It is assumed that in this way the most accurate result is guaranteed. However, it should be taken into account that this result may still be missed or incorrect due to the difficulty of analyzing and comprehending redundant information. Therefore, modern buildings should not only have a beautiful shape. They should be of simple design to minimize errors when they are calculated in finite element analysis programs [19, 20]. All proposed new designs satisfy these principles.

References

1. Storozhenko, L., Makhinko, A.: Steel cylindrical tanks. PoltNTU, Poltava (2012)
2. Permyakov, V., Perelmuterr, A., Yurchenko, V.: Optimal design of steel core structures. Steel, Kyiv (2008)
3. Bu, F., Qian, C.: On the rational design of the top wind girder of large storage tanks. Thin-Walled Struct. **99**, 91–96 (2016)
4. Agboola, O.O., Akinnuli, B.O., Kareem, B., Akintunde, M.A.: Optimum detailed design of 13,000 m^3 oil storage tanks using 0.8 height-diameter ratio. Mater. Today: Proc. **44**, 2837–2842 (2021)
5. Pan, J., Liang, S.: A study on the buckling behavior of in-service large open-topped oil-storage tanks. Structures **29**, 211–224 (2020)
6. Koshevyi, O.: Optimal design of cylindrical tanks with rigid coating shells. Strength Mater. Theory Struct. **103**, 253–265 (2019)
7. Burgos, C.A., Batista-Abreu, J.C., Calabro, H.D., Jaca, R.C., Godoy, L.A.: Buckling estimates for oil storage tanks: effect of simplified modeling of the roof and wind girder. Thin-Walled Struct. **91**, 29–37 (2015)
8. Makhinko, A., Makhinko, N.: To the calculation of the optimal level of reliability by using economic indicators. In: Blikharskyy, Z., Koszelnik, P., Mesaros, P. (eds.) Proceedings of CEE 2019, CEE 2019, Lecture Notes in Civil Engineering, vol. 47, pp. 251–259. Springer, Cham (2020). https://doi.org/10.1007/978-3-030-27011-7_32
9. Krayushkina, K., Oliynyk, O.: Stabilized soil - new material for road construction. In: IOP Conference Series: Materials Science and Engineering, vol. 918, no. 1, p. 012117 (2020)
10. Krayushkina, K., Khymeryk, T., Dubik, O., Fedorenko, K.: Behavior of the soil layer of rigid airport structure under load. In: AIP Conference Proceedings, vol. 2684, p. 030022 (2023)
11. Sgorozhenko, L.I., Badov, A.V., Kharchenko S.A.: Drop-shaped tank. Copyright certificate SU 1679021 (1991)
12. Volmyr, A.: Flexible plates and shells. Gostehteorlit, Moscow (1956)
13. Lessig, E., Lileev, A., Sokolov, A.: Sheet-metal structures. Stroiizdat, Moscow (1970)
14. Vedernikov, G., et al.: Metal structures. General course. Stroiizdat, Moscow (1998)
15. Globa, V.: Standard calculations for the construction of oil and gas storage facilities. IFDTUNH, Ivano-Frankivsk (2000)
16. Godoy, L.A.: Buckling of vertical oil storage steel tanks: review of static buckling studies. Thin-Walled Struct. **103**, 1–21 (2016)
17. Sun, B., et al.: Wind buckling analysis of a large-scale open-topped steel tank with harmonic settlement-induced imperfection. Buildings **12**(11), 1973 (2022)
18. Artanov, A.I.: Droplet-shaped tank for storing liquids. Copyright certificate SU 950880 (1982)
19. Barabash, M.S., Kostyra, N.O., Pysarevskiy, B.Y.: Strength-strain state of the structures with consideration of the technical condition and changes in intensity of seismic loads. In: IOP Conference Series: Materials Science and Engineering, vol. 708, no. 1, p. 012044 (2019)
20. Gorodetsky, A.S., Barabash, M.S., Romashkina, M.A., Tomashevsky, A.V.: "Characteristic load" principle. Int. J. Comput. Civ. Struct. Eng.Struct. Eng. **16**(2), 50–62 (2020)

How the War in Ukraine Impacts Global Air Transportation Ecosystem: Assessment and Forecasting of Consequences

Viktoriia Ivannikova[1](✉), Olena Sokolova[2], and Kostiantyn Cherednichenko[2]

[1] Dublin City University, Glasnevin, Dublin 9, Ireland
viktoriia.ivannikova@dcu.ie

[2] National Aviation University, 1 Liubomyra Huzara Ave, Kyiv, Ukraine
{olena.sokolova,kostiantyn.cherednichenko}@npp.nau.edu.ua

Abstract. Russian invasion of Ukraine on 24th February 2022 leds to the restriction of air traffic in Ukrainian and Russian air spaces due to the military activity and war-related sanctions. Airspace closures have forced airlines to take detours and seek alternative routes. All these causes considerable increasing of flight time on routes connecting Europe and Asia, North America and Asia and North America and Middle East. In turns longer travel times lead to the increasing of aircrafts' fuel burn, which causes higher airlines operating costs and CO_2 emissions. This paper assesses the impact of the Russian invasion of Ukraine on the global air transportation ecosystem on the example of three European airlines, i.e. Finnair, LOT Poilish and Lufthansa Airlines, analysing their most popular routes between Europe and Asia before and after flight restrictions. Performed analysis shows increasing of the airlines operations costs and CO_2 emissions from 18% to 40% on the certain routes after closure of air spaces. This research also contains a forecast of global aviation CO_2 emissions for the next five years under the two different scenarios of military conflict development on the territory of Ukraine, constructed with the help SARIMA Model.

Keywords: War in Ukraine · Global air transportation ecosystem · CO_2 emissions · Airline operating costs · Forecast

1 Introduction

On 24th February 2022 Russia invaded Ukraine just after aviation industry started to recover from COVID-19 pandemic. Immediately after Russia invasion Ukrainian air space was closed and all flights of commercial aviation were suspended for ensuring aviation safety and avoiding human losses. Being in solidarity with Ukraine, Great Britain, U.S., Canada and EU countries banned any transportation of Russian air carriers in their air space. In response to these sanctions, Russia limited the opportunities for airlines from 36 countries, including all 27 members of the European Union to cross its borders [1].

O. Prentkovskis et al. (Eds.): TRANSBALTICA 2023, LNITI, pp. 386–401, 2024.
https://doi.org/10.1007/978-3-031-52652-7_38

Ukraine comprises rather small part of the global air transportation market [2]. Its share of total air passenger traffic in Europe is 3.3%, and 0.8% in the world, as per 2021 [3] (Table 1).

Table 1. Traffic shares for selected markets impacted by the conflict.

% share of passengers numbers in 2021	Total European Traffic (excl. Russia domestic)	Global traffic
Russia International	5.7%	1.3%
Ukraine	3.3%	0.8%
Moldova	0.4%	0.1%
Belarus	0.3%	0.1%

Source: IATA Economics using DDG

However, the impacts of the war extend much beyond air travel to and from Ukraine. Closure of airspaces have forced airlines to seek alternative routes and take detours. This causes considerable increasing of flight time on routes connecting Europe and Asia, North America and Asia and North America and Middle East. In turns longer travel times lead to the increasing of aircrafts' fuel burn, which cause higher airlines' operating costs and CO_2 emissions. The effect of flights restrictions on European airlines have been different for various companies [4]. The impacts vary between airlines, depending on the location of hub airports, routes network and country of registration. The most affected European airline by Russian-Ukrainian war is Finnair, having a hub in Helsinki for connecting flights between Asia and Europe. Its proximity to the Russian Federation meant that for the most of its direct air routes to Asia, operated through Russian airspace, Finnair had to find alternative routes trajectories. For example, flights between Helsinki and Tokyo offered by Finnair, now take up to 3.5 h longer (see Fig. 1).

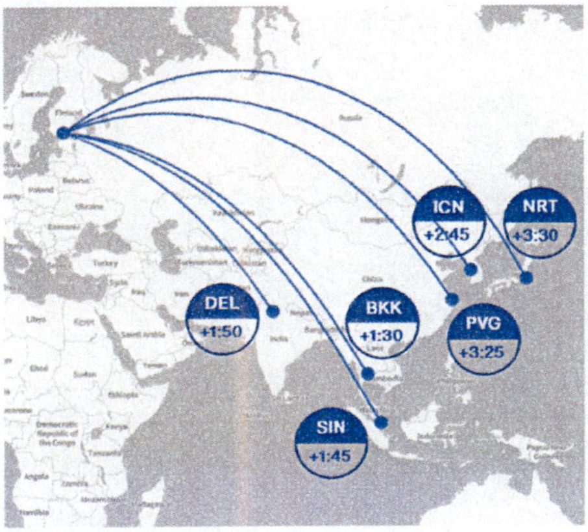

Fig. 1. Duration of the Finnair alternative routes from Helsinki to Asia after flights restrictions [4].

The airlines based in Asia have been less impacted by the conflict. They have not been barred from Russian airspace, but some of them (e.g. Japan Airlines, Korean Air, etc.) though overfly Russia.

The paper addresses the following research questions:

- **How much do airlines operating costs change** on the selected direct routes between Europe and Asia after flights restrictions due to the Russian invasion of Ukraine?
- **How much do airlines CO_2 emissions change** on the selected direct routes between Europe and Asia after flights restrictions due to the Russian invasion of Ukraine?
- **How global aviation CO_2 emissions will be changed in five years perspective** under two different scenarios of military conflict development of the territory of Ukraine (the war will end within one year or within three years)?

The paper's findings help to better inform scholars and practitioners of extremely negative consequences of the Russian-Ukrainian conflict for aviation, which can prevent achievement of sustainable development goals in air transport within the established time frame [5, 6]. Moreover, the study makes a theoretical contribution to the research regarding the war impact on the aviation industry.

The paper is organized as follows. Section "Closed Ukrainian air space and global air traffic" analyses pertinent literature. Section "Research methodology" describes the paper's research approach, developed to answer on the research questions. The analysis and discussion of findings regarding the impact of the war in Ukraine on the global air transportation ecosystem is presented in the section "Findings and discussions", followed by the research conclusions in the section "Conclusion".

2 Closed Air Spaces and Global Air Traffic

Ukrainian and Russian airways network comprise a huge part of global air transportation system. Closure of their airspaces causes serious problems for aviation (Fig. 2).

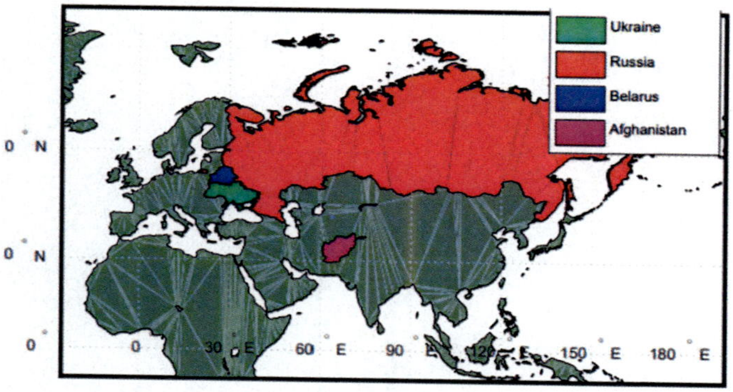

Fig. 2. Closed air space after Russian invasion of Ukraine [7].

The huge territory of Russia means large detours for many airlines overflying Russia during operation of their flights. This comprises more than 10% of pre-COVID-19 international traffic volumes, measured in RPKs, including travel from Europe to Asia (4.5%), North America to Asia (3.0%) and North America to the Middle East (4.0%) [8].

There are just a few research papers providing a preliminary estimation of the war impact in Ukraine on the global air transport. Ostroumov, I. Ivashchuk, O. and Kuzmenko, N highlighted in their papers [7, 9] that the geometry of closed airspace due to the military conflict in Ukraine seriously affects direct connections from Asia to North America and European countries as well as polar flights from North to South directions. Many flights have been cancelled due to closed airspace and special restrictions and trajectories of transit flights have been re-planned to overfly the territories of Russia and Ukraine [7] (Fig. 3).

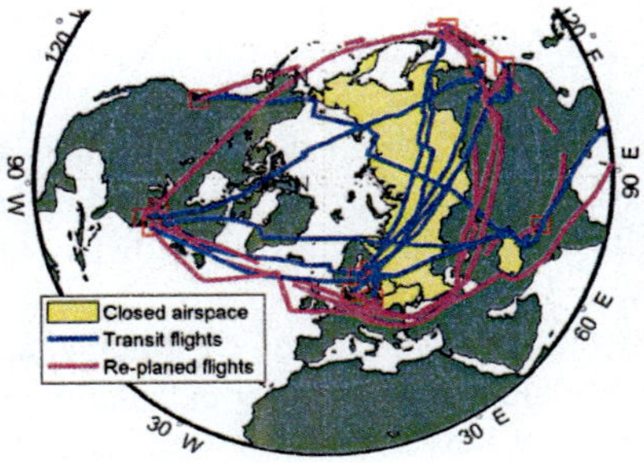

Fig. 3. Transit and replanned flights trajectories after Russia invasion of Ukraine [7].

Ostroumov, Ivashchuk and Kuzmenko [7, 9] provided a statistical analysis of scheduled air traffic losses inside Ukraine for different airlines due to the Russia-Ukrainian war as well as predicted for seven weeks the losses in total trajectories length, flight time, available seats, and available seat-kilometer for flights of different airlines, connected with closed Ukrainian airports.

The extent of literature featuring impact of the war in Ukraine on aviation is relatively constrained to the analysis of the increase in trajectory length and flight time. The changes of airlines operating costs and CO_2 emissions on the routes between Europe and Asia after flights restrictions due to the Russian invasion of Ukraine and also prediction of the changes in global aviation CO_2 emissions under the different scenarios of military conflict development on the territory of Ukraine have not been researched yet. This paper addresses found research gap.

3 Research Methodology

To assess impact of the war in Ukraine on the global air transportation ecosystem three questions have been researched:

1. **airlines operating costs** on the selected routes between Europe and Asia before and after flights restrictions due to the Russian invasion of Ukraine;
2. **CO_2 emissions** on the selected routes between Europe and Asia before and after flights restrictions due to the Russian invasion of Ukraine;

3. **global aviation CO_2 emissions in the next five years** under two different scenarios of the military conflict development on the territory of Ukraine.

Three European airlines, i.e. Finnair, LOT Poilish and Lufthansa Airlines, operating direct flights from Europe to Asia have been selected as case studies. Fourteen different routes, operated by Chinese and European airlines, have been chosen for the investigation (see Table 2 and Fig. 4, correspondently). The initial data, used during routes analysis were obtained with the help of Flightradar 24 resource [10]. It is an internet-based service that shows real-time aircraft flight tracking information on a map, including origins and destinations, flight numbers, aircraft types, positions, altitudes, headings and speeds. The data presented in Table 2 show that at the moment there is a huge competition between European and Chinese airlines, specializing in the shortest routes between Europe and Asia. The competition is unequal as Chinese airlines have permission to fly within Russian air space, which give then advantages in flights time and routes distances (see Table 2).

The flights routes, having the most increased flight times now is the Finnair flights between Helsinki and Tokyo Narita Airport in Japan and Helsinki and Seoul Incheon Airport in South Korea, with the increase of the routes distances after flights restrictions on 39.9 and 49.3%, correspondently.

$ACMIF_i$ (Aircraft depreciation AD_i, Crew CC_i, Maintenance M_i, Insurance I_i, Jet Fuel consumption F_i,) methodology was used for calculation of the airlines' operating costs of the selected single flights between Europe and Asia.

$$ACMIF_i = AD_i + CC_i + M_i + I_i + F_i, \tag{1}$$

Aircraft depreciation costs (AD_i) were calculated on the bases of a new aircraft price, depreciation rate and annual aircraft utilization. Crew costs (CC_i) consist of the costs of cabin and flight crew salary and allowances. Aircraft maintenance costs (M_i) consist of technicians' salary and spare parts costs (usually it is 10% of a new aircraft price per a year). Insurance costs (I_i) are a combination of aircraft insurance costs, crew members insurance and air carrier liability insurance. During calculations annual insurance rate was taken as 1.5%. Aviation fuel costs (F_i) were determined by applying the average hourly fuel consumption for each type of aircraft per the cost of 1 ton of fuel. The aviation fuel price in March 2023 was $123 per barrel, which corresponds to 852.96 euros per ton.

Airline operating costs for a single flight (OC_{ij}) were calculated according to the following formula:

$$OC_{ij} = ACMIF_i \cdot t_j + AN_{ij}, \tag{2}$$

where t_j – flight time; AN_{ij}– airport and navigation fees.

CO_2 emissions on the selected routes before and after flights restrictions were determined using the ICAO Carbon Emissions Calculator Methodology in the specialized software [11, 12].

The final stage of the research was a forecast of the global aviation CO_2 emissions for the next 5 years under the two different scenarios of military conflict development on the territory of Ukraine, i.e.: First Scenario - Optimistic: war ends, airspace opens

Table 2. Characteristics of the investigated flights, between Europe and Asia.

No	Route	Airline	Flight number	Aircraft type	Flights frequen-cy per a week	Route distance			
						Before restrictions, km	After restrictions, km	Difference, km	Difference, %
1	Helsinki (HEL) - Shanghai (PVG)	Finnair	AY87	A-350–900	2	7,636	10,666	3,030	39.7
2	Helsinki (HEL) - Shanghai (PVG)	Juneyao Airlines	HO1608	B-787–9	2	7,636	7,636	0	0.0
3	Helsinki (HEL) - Токyo, Narita Airpot (NRT)	Finnair	AY73	A-350–900	4	7,838	10,969	3,131	**39.9**
4	Helsinki (HEL) - Seoul (ICN)	Finnair	AY41	A-350–900	4	7,039	10,506	3,467	**49.3**
5	Warsaw (WAW) - Beijing (PEK)	LOT Polish	LO91	B-787–8	7	6,952	8,678	1,726	24.8
6	Warsaw (WAW) - Beijing (PEK)	Air China	CA738	A-330	2	6,952	7,534	582	8.4
7	Warsaw (WAW) - Токyo, Narita Airpot (NRT)	LOT Polish	LO79	B-787–9	6	8,623	11,062	2,439	28.3
8	Warsaw (WAW) - Seoul (ICN)	LOT Polish	LO97	B-787–8, B-787–9	4	7,734	9,606	1,872	24.2
9	Frankfurt (FRA) - Beijing (PEK)	Lufthansa	LH720	A-340–300	7	7,858	9,236	1,378	17.5
10	Frankfurt (FRA) - Beijing (PEK)	Air China	CA932	B-777–300	7	7,858	7,858	0	0.0
11	Frankfurt (FRA) - Токyo, Haneda Airpot (HND)	Lufthansa	LH716	B-747–8	6	9,379	11,469	2,090	22.3

(continued)

Table 2. (*continued*)

No	Route	Airline	Flight number	Aircraft type	Flights frequen-cy per a week	Route distance			
						Before restrictions, km	After restrictions, km	Difference, km	Difference, %
12	Frankfurt (FRA) - Seoul (ICN)	Lufthansa	LH712	B-747–400	7	8,554	10,101	1,547	18.1
13	Frankfurt (FRA) - Seoul (ICN)	Asiana Airlines	OZ542	A-350–900	7	8,554	10,110	1,556	18.2
14	Frankfurt (FRA) - Seoul (ICN)	Korean Air	KE946	B-777-300ER	5	8,554	10,112	1,558	18.2

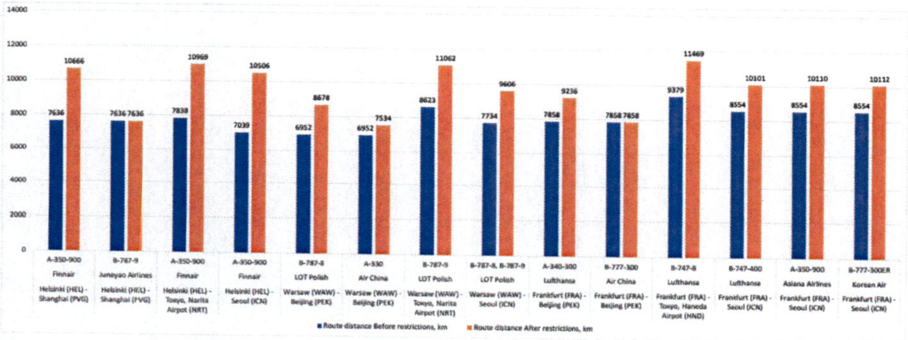

Fig. 4. Comparison analysis of the investigated flights routes distances before and after flights restrictions.

and air traffic restores within 1 year; First Scenario - Realistic: war ends, airspace opens and air traffic restores within 3 years.

Statistical data for determination of the forecasted volumes of the global aviation CO_2 emissions were taken from the www.statista.com (see Fig. 5) [13].

Time series forecasting is a difficult problem with not an easy answer. There are countless number of statistical models, used for forecasting, however it is never clear which model is best. One of the most frequently used time series model (due to its accuracy) is the autoregressive integrated moving average (ARIMA) model. The ARIMA model is described by the formula 3 below.

$$Y_t = \beta_2 + \omega_1 \cdot \varepsilon_{t-1} + \omega_2 \cdot \varepsilon_{t-2} + \cdots + \omega_q \cdot \varepsilon_{t-q} + \varepsilon_t, \tag{3}$$

The main assumption made for this model implementation is that the considered time series is linear and follows a normal statistical distribution.

The ARIMA model consist of other subclasses, such as autoregressive (AR), moving average (MA) and autoregressive moving average (ARMA) models.

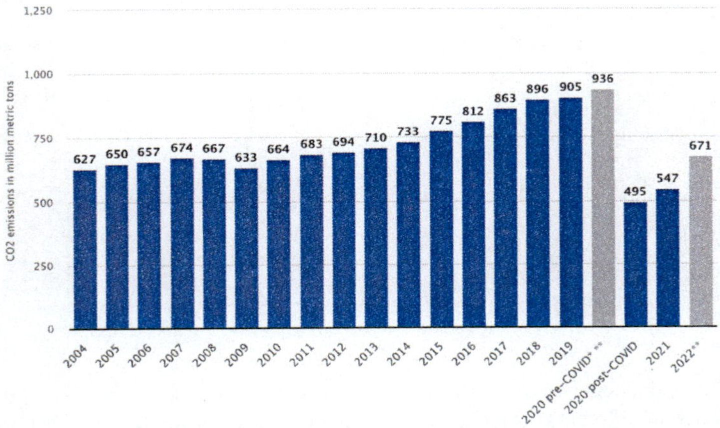

Fig. 5. Carbon dioxide emissions from commercial aviation worldwide from 2004 to 2022 [13].

For seasonal time series forecasting, which is more suitable for the aviation industry, characterised by cyclicity, there is a very successful variant of the ARIMA model – SARIMA (seasonal autoregressive integrated moving average) model. SARIMA model is described by the formula 4 below.

$$y_t = c + \sum\nolimits_{n=1}^{p} a_n y_{t-n} + \sum\nolimits_{n=1}^{q} \theta_n \epsilon_{t-n} + \sum\nolimits_{n=1}^{P} \varnothing_n y_{t-sn} + \sum\nolimits_{n=1}^{Q} \mu_n \epsilon_{t-sn} + \epsilon_t, \tag{4}$$

This model is a quite similar to ARIMA one, except that there is an additional set of autoregressive and moving average components. Additional lags are offset by the frequency of seasonality. SARIMA models allow differencing data by seasonal and non-seasonal frequencies. The RStudio software was used for the development of different predictive scenarios for global CO_2 emissions using SARIMA model.

4 Findings and Discussions

This section presents results of calculation of Finnair, LOT Polish and Lufthansa Airlines operating costs and CO_2 emissions before and after flights restrictions on the selected routes between Europe and Asia. Figures 6, 7, 8, 9, 10, 11, 12, 13 and 14 provide the summary of the performed calculations.

Average Finnair Airline operating costs on the selected routes between Europe and Asia have been increased by 38.60% after flights restrictions, while average CO_2 emissions on the investigated routes have been increased by 40.15% after the restrictions of flights (see Fig. 6–8).

Average LOT Polish Airline operating costs on the selected routes between Europe and Asia have been increased by 23.29% after flights restrictions, while average CO_2 emissions on the investigated routes have been increased by 23.57% after the restrictions of flights (see Fig. 9, 10 and 11).

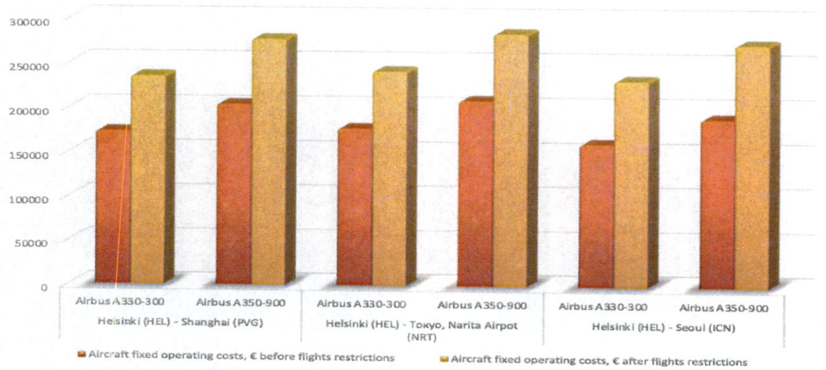

Fig. 6. Airline's operating costs on the investigated routes before and after flights restrictions during operation of different types of aircraft: A Case of Finnair Airlines.

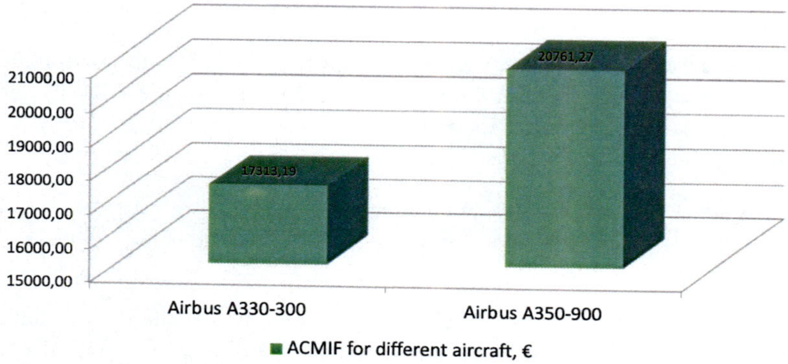

Fig. 7. ACMIF of the Finnair Airline long-haul fleet.

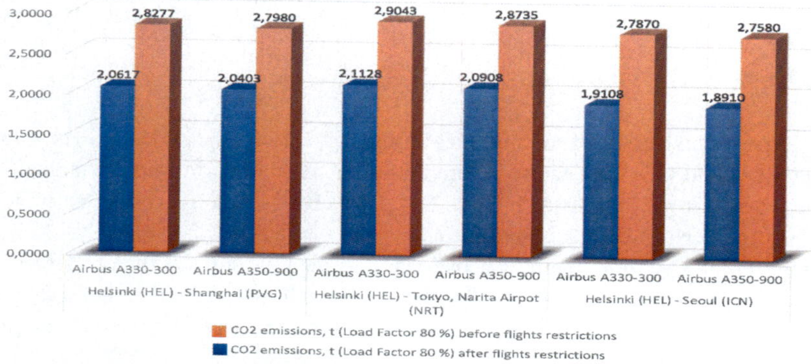

Fig. 8. CO2 emissions on the investigated routes during operation of different types of aircraft before and after flights restrictions: A Case of Finnair Airlines.

Average Lufthansa Airline operating costs on the selected routes between Europe and Asia have been increased by 18.62% after flights restrictions, while average CO_2 emissions on the investigated routes have been increased by 18.21% after the restrictions of flights (see Fig. 12, 13 and 14).

The conducted analysis showed that overflight of Ukrainian and Russian airspaces due to the military conflict contributes to a significant increase of airlines operating costs on the selected routes between Europe and Asia, varying from 38.6% to 18.62% and increase of CO_2 emissions varying from 40.15% to 18.21% depending on airlines. The performed research proved that Finnair Airlines is the most impacted airline by the Ukrainian conflict.

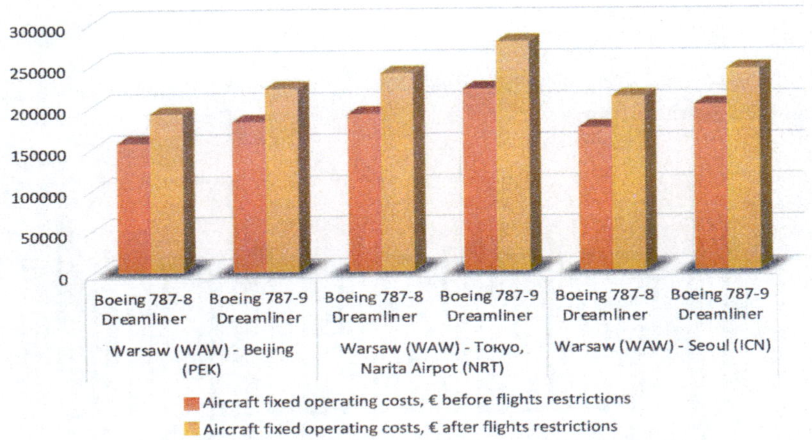

Fig. 9. Airline's operating costs on the investigated routes before and after flights restrictions during operation of different types of aircraft: A Case of LOT Polish Airlines.

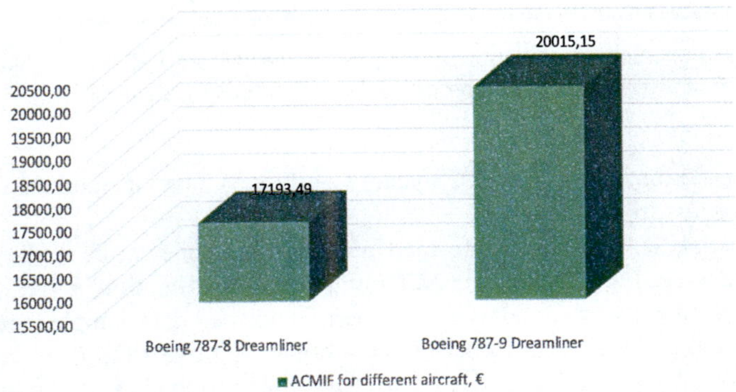

Fig. 10. ACMIF of the LOT Polish Airlines long-haul fleet.

At the final stage of the research, we developed a forecast of global aviation CO_2 emissions for the next 5 years under the two possible scenarios of military conflict development on the territory of Ukraine. Global aviation CO_2 emissions monthly statistics

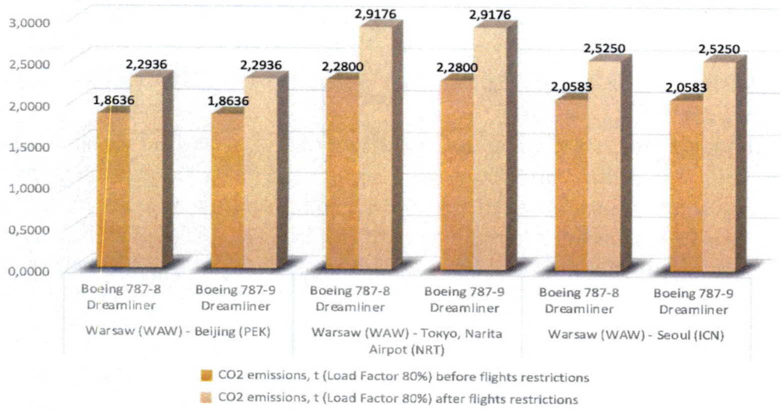

Fig. 11. CO2 emissions on the investigated routes during operation of different types of aircraft: A Case of LOT Polish Airlines.

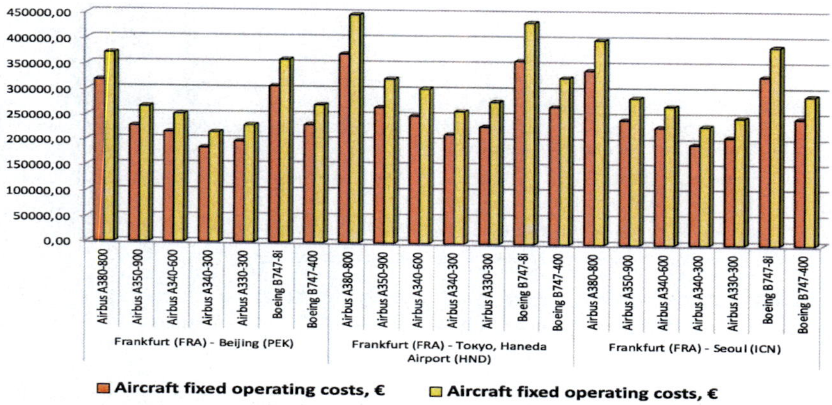

Fig. 12. Airline's operating costs on the investigated routes before and after flights restrictions during operation of different types of aircraft: A Case of Lufthansa Airlines.

(in million metric tons) for 2020–2022 years (see Fig. 5) have been taken as an initial data for the forecasting model.

This initial data sample was transformed into logarithmic view to be applicable for the SARIMA model (see Fig. 12), then the ACF (autocorrelation function) and PACF (partial autocorrelation function) were analysed for the transformed data sample (see Fig. 13).

After this the logarithmic forecast with confidence intervals has been constructed, with the help of SARIMA model (see Fig. 14), which was coded in RStudio (see Fig. 15) (Figs. 17 and 18).

Main forecasting finding after the modelling were as follows:

- p-value of the model is 0.91 ($>=0.85$), which confirms the sufficiently high accuracy of the forecast;

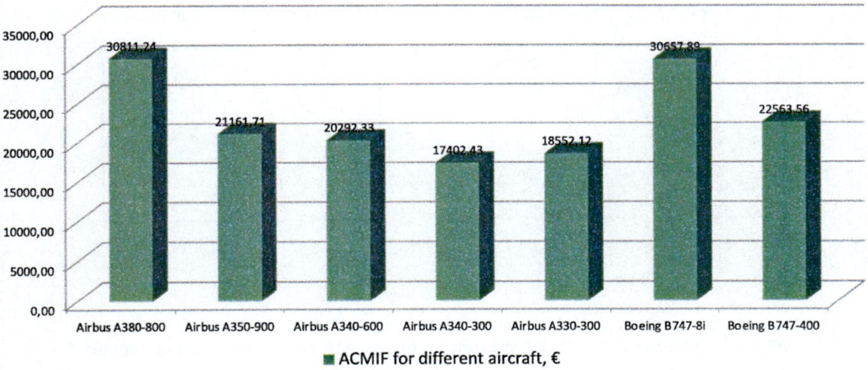

Fig. 13. ACMIF of the Lufthansa Airlines long-haul fleet.

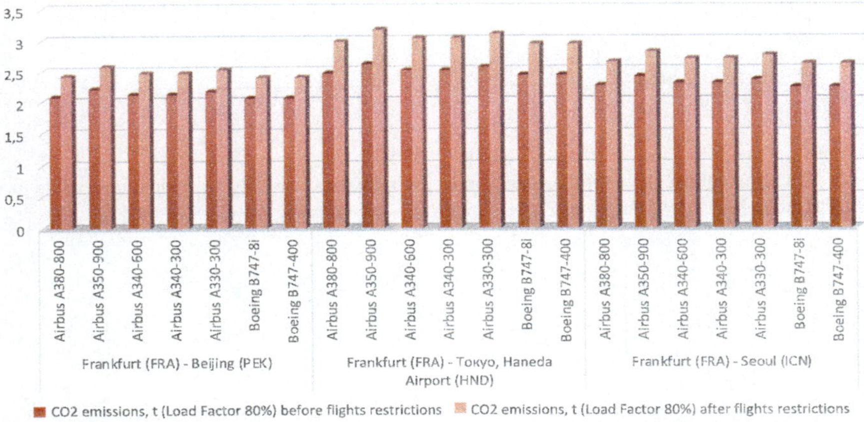

Fig. 14. CO2 emissions on the investigated routes during operation of different types of aircraft: A Case of Lufthansa Airline.

- global aviation CO_2 emissions in «realistic 2025 year» will reach ≈ 72 million metric tons per month (increased by 28.8% comparing with 2022), in «optimistic 2025 year» global aviation CO_2 emissions will reach ≈ 68 million metric tons per month (increased by 22.6%);
- by the end of 2023 the difference between two forecasts (optimistic and realistic) will be approximately 3% and by the end of 2025 it will be approximately 6 %.

The forecasted values of the global aviation CO_2 emissions for the next 5 years under two different scenarios of military conflict development of the territory of Ukraine determined with the help of SARIMA model in RStudio are presented on the Fig. 16 (Fig. 19).

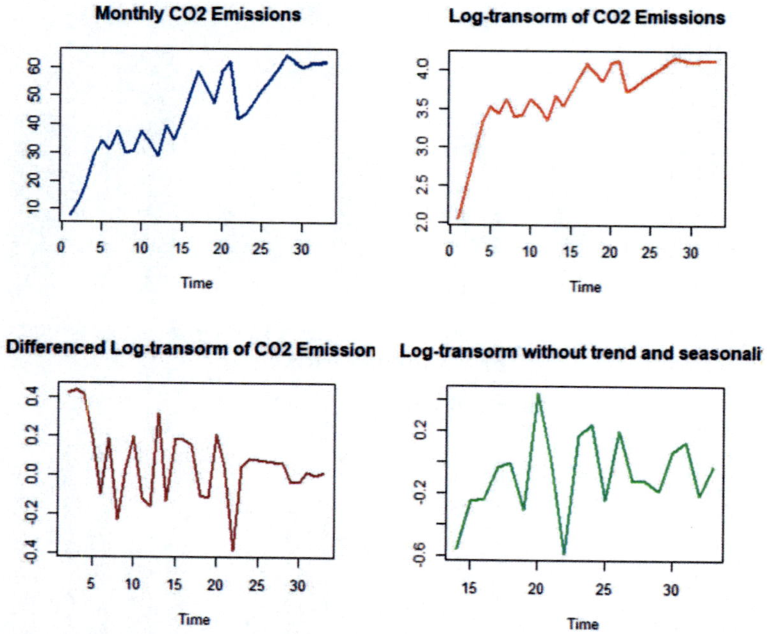

Fig. 15. Transformation of the global aviation CO_2 emissions monthly statistical data to be used in SARIMA model.

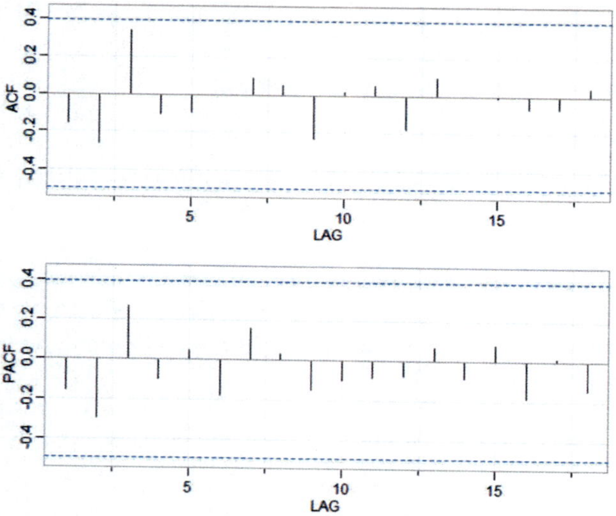

Fig. 16. Autocorrelation function (ACF) and partial autocorrelation function (PACF) of the transformed data sample.

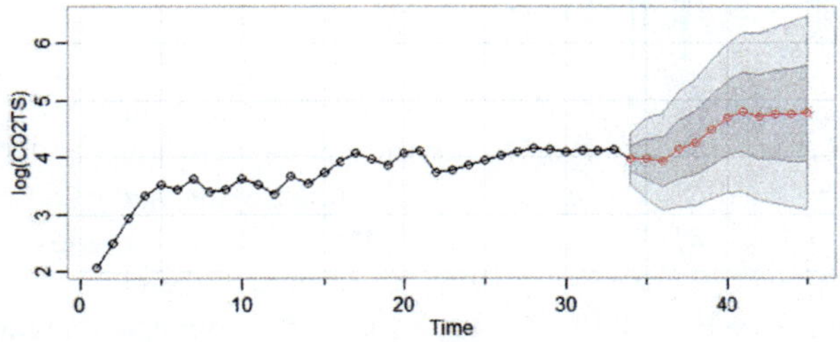

Fig. 17. SARIMA model and logarithmic forecast with confidence intervals.

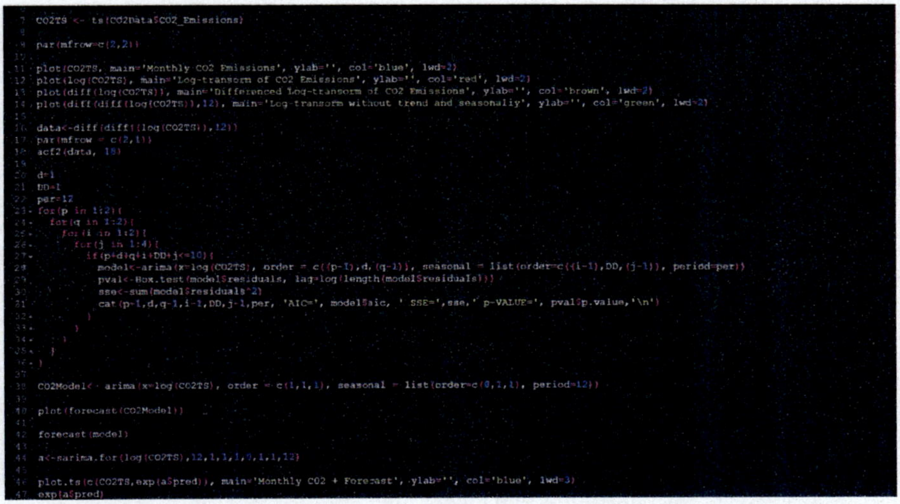

```
CO2TS <- ts(CO2Data$CO2_Emissions)

par(mfrow=c(2,2))

plot(CO2TS, main='Monthly CO2 Emissions', ylab='', col='blue', lwd=2)
plot(log(CO2TS), main='Log-transorm of CO2 Emissions', ylab='', col='red', lwd=2)
plot(diff(log(CO2TS)), main='Differenced Log-transorm of CO2 Emissions', ylab='', col='brown', lwd=2)
plot(diff(diff(log(CO2TS)),12), main='Log-transorm without trend and seasonaliy', ylab='', col='green', lwd=2)

data<-diff(diff((log(CO2TS)),12))
par(mfrow = c(2,1))
acf2(data, 18)

d=1
DD=1
per=12
for(p in 1:2){
  for(q in 1:2){
    for(i in 1:2){
      for(j in 1:4){
        if(p+d+q+i+DD+j<=10){
          model<-arima(x=log(CO2TS), order = c((p-1),d,(q-1)), seasonal = list(order=c((i-1),DD,(j-1)), period=per))
          pval<-Box.test(model$residuals, lag=log(length(model$residuals)))
          sse<-sum(model$residuals^2)
          cat(p-1,d,q-1,i-1,DD,j-1,per, 'AIC=', model$aic, ' SSE=',sse,' p-VALUE=', pval$p.value,'\n')
        }
      }
    }
  }
}

CO2Model<- arima(x=log(CO2TS), order = c(1,1,1), seasonal = list(order=c(0,1,1), period=12))

plot(forecast(CO2Model))

forecast(model)

a<-sarima.for(log(CO2TS),12,1,1,1,0,1,1,12)

plot.ts(c(CO2TS,exp(a$pred)), main='Monthly CO2 + Forecast', ylab='', col='blue', lwd=3)
exp(a$pred)
```

Fig. 18. Coded in RStudio SARIMA model.

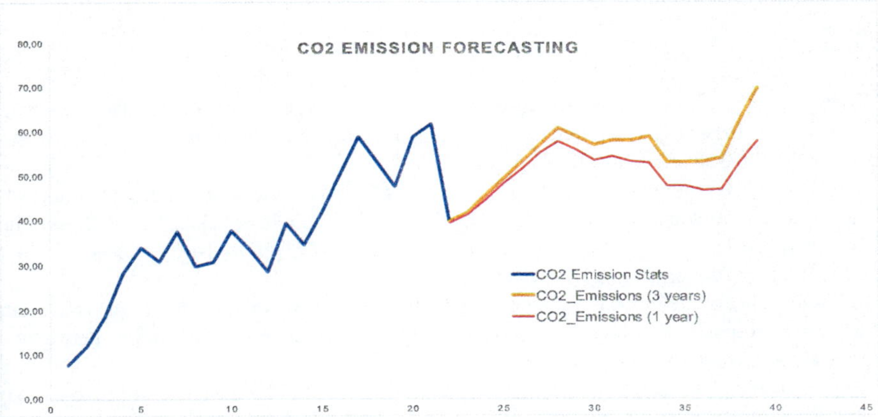

Fig. 19. Forecast of the global aviation CO_2 emissions for the next 5 years under two different scenarios of military conflict development of the territory of Ukraine.

5 Conclusions

This paper has investigated how war in Ukraine impacted global air transportation ecosystem on the example of three European airlines, i.e. Finnair, LOT Polish and Lufthansa Airlines, analysing their most popular direct routes between Europe and Asia before and after flight restrictions. The study considerably adds to the aviation industry research by investigating of the current and future state of the global air transportation ecosystem, impacted by the war in Ukraine. Result of the conducted research proves that Russia invasion of Ukraine led not only to the changes in air routes network between Europe and Asia, North America and Asia and North America and Middle East, but also significantly worsen the financial state of certain airlines and negatively affects environment. Performed analysis shows increasing of the airlines operating costs on the certain routes between Europe and Asia after flights restrictions from 38.6% to 18.62% and rising of CO_2 emissions from 40.15% to 18.21%, depending on airlines. Finnair Airlines seems to be the most impacted airline by the Ukrainian conflict.

This research also significantly contributes to the scientific approach for forecasting of the global aviation CO_2 emissions under the influence of war with the help of SARIMA model. Received data show that the overflight of Russian and Ukrainian air spaces due to the continuation of military conflict in Ukraine lead to the increasing of CO_2 emissions in "realistic 2025 year" (war will end within 3 years) on 28.8% and in "optimistic 2025 year" (war will end within one year) on 22.6% comparing with 2022, which can prevent the achievement of sustainable development goals in air transport within the established time frame.

Together with the airspace closures and flight restrictions, leading to the changing of air routes trajectories, increasing of airlines operating cost and CO_2 emissions, Russian invasion of Ukraine also affect slots at key airports worldwide. Many went unused, and others were affected by rescheduling and longer flight times. This is another topic, which needs further research.

References

1. Russian flights bans hit airlines from 36 countries - aviation authority. https://www.reuters.com/business/aerospace-defense/russia-imposes-sweeping-flight-bans-airlines-36-countries-2022-02-28. Accessed 03 Aug 2023
2. Sokolova, O., Grygorak, M., Ivannikova, V.: "Green" sector of the air transport of Ukraine sustainable development. In: Prentkovskis, O., Yatskiv (Jackiva), I., Skačkauskas, P., Junevičius, R., Maruschak, P. (eds.) TRANSBALTICA 2021. LNITI, pp. 448–455. Springer, Cham (2022). https://doi.org/10.1007/978-3-030-94774-3_44
3. IATA Factsheet: The impact of the war in Ukraine on the aviation industry. https://www.iata.org/en/iata-repository/publications/economic-reports/the-impact-of-the-conflict-between-russia-and-ukraine-on-aviation. Accessed 04 Aug 2023
4. Impact of Russian Airspace Sanctions on Flight Routes and Flight Times. https://www.oag.com/blog/impact-russian-sanctions-flight-times. Accessed 04 Aug 2023
5. Ivannikova, V., Boldyrieva, M., Konovalyuk, V.: Modern ecological approach to air transportation management. In: Prentkovskis, O., Yatskiv (Jackiva), I., Skačkauskas, P., Junevičius, R., Maruschak, P. (eds.) TRANSBALTICA 2021. LNITI, pp. 658–670. Springer, Cham (2022). https://doi.org/10.1007/978-3-030-94774-3_64

6. Ivannikova, V., Kochura, O.: Research tools used to formulate sustainable development strategy for airlines. In: Transport engineering and management: 19-th Conference for Junior Researchers "SCIENCE – FUTURE OF LITHUANIA", 6 May 2016, Vilnius, pp. 160–167 (2016)
7. Ostroumov, O. Ivashchuk and Kuzmenko, N.: Preliminary estimation of war impact in Ukraine on the global air transportation. In: 2022 12th International Conference on Advanced Computer Information Technologies (ACIT), Ruzomberok, Slovakia, pp. 281–284 (2022). https://doi.org/10.1109/ACIT54803.2022.9913092
8. War in Ukraine and air transport. https://airlines.iata.org/2022/03/25/war-ukraine-and-air-transport. Accessed 04 Aug 2023
9. Ivashchuk, O., Ostroumov, I.: Impact of closed Ukrainian airspace on global air transport system. In: Faure, E., Danchenko, O., Bondarenko, M., Tryus, Y., Bazilo, C., Zaspa, G. (eds.) Information Technology for Education, Science, and Technics, ITEST 2022, vol. 178, pp. 51–64. Springer, Cham. (2023). https://doi.org/10.1007/978-3-031-35467-0_4
10. Flightradar24 Homepage. https://www.flightradar24.com. Accessed 04 Aug 2023
11. Clarke, D., Flachenecker, F., Guidetti, E., Pionnier, P.-A.: CO_2 emissions from air transport. A near-real-time global database for policy analysis. OECD. No. 2022/04, p. 72. https://doi.org/10.1787/ecc9f16b-en. Accessed 20 June 2023
12. ICAO Carbon Emissions Calculator Methodology. https://applications.icao.int/icec/Methodology%20ICAO%20Carbon%20Calculator_v11-2018.pdf. Accessed 20 June 2023
13. Carbon dioxide emissions from commercial aviation worldwide from 2004 to 2022. https://www.statista.com/statistics/1186820/co2-emissions-commercial-aviation-worldwide. Accessed 04 Aug 2023

Analysis of Compliance of the Lithuanian Railway Infrastructure as a Unified North Atlantic Treaty Organization Supply System with the Requirements

Ieva Meidutė-Kavaliauskienė(✉), Renata Činčikaitė, and Almantas Liepuonis

General Jonas Žemaitis Military Academy of Lithuania, Šilo Street 5A, Vilnius, Lithuania
{ieva.meidute,renata.cincikaite}@lka.lt

Abstract. Better military mobility in Europe is considered one of the most important aspects in ensuring better deterrence. The member states of the North Atlantic Treaty Organization (from now on referred to as NATO) are interested in the rapid transfer of defense forces, supplies, and equipment across Europe. When transporting military cargo by railways from point A to point B, various challenges are encountered; therefore, the article aims to analyze the compliance of Lithuanian railways with the requirements set by NATO, identifying infrastructure deficiencies and threats in the transport of military equipment. In implementing the objective of the work, an analysis of scientific literature on the topic of the railway transport sector is carried out, the capacity of the Lithuanian railway transport sector to receive military cargo from allied countries, and the compliance of the Lithuanian railway infrastructure with NATO requirements are analyzed. The results found that Lithuanian railways do not meet the criteria set by NATO but can carry out military cargo transportation following the legal act SMGS (Agreement on International Carriage of Goods by Rail).

Keywords: North Atlantic Treaty Organization · Lithuanian railway transport · Transport of military equipment · NATO standards · Length of railways

1 Theoretical Aspects of the Railway Transport Sector

Every year, the transportation of foreign military cargo by railways is increasing. The movement of military cargo by rails currently faces various legal, physical and regulatory obstacles [1]. The North Atlantic Treaty Organization member states are interested in reassessing their infrastructure to quickly move equipment, defense forces, and supplies. Developed railways can be deployed effectively in crises or conflicts [2]. Lithuania is also interested in improving the railway infrastructure network; therefore, the main activity directions are to review the Lithuanian railway infrastructure and assess its compliance with North Atlantic Treaty Organization (NATO) requirements.

Well-organized military mobility in Europe is considered an essential step to ensure better deterrence. Every year, NATO military units are deployed in Lithuania, which is

O. Prentkovskis et al. (Eds.): TRANSBALTICA 2023, LNITI, pp. 402–410, 2024.
https://doi.org/10.1007/978-3-031-52652-7_39

particularly important for Lithuania's security [3]. When the armies of NATO countries go to Lithuania, they also transport their military equipment and resources, usually transported by Lithuanian railways. Scientific articles do not assess the condition of the railway infrastructure in Lithuania according to NATO requirements. There is a lack of information on the problems of military cargo transportation. Taking this into account, the article analyzes the Lithuanian railway infrastructure and its compliance with NATO requirements.

Some authors emphasize the strategic role of the railway in national security and defence [4, 5]. Others emphasize that it is an economical and time-saving mode of transportation [6], it is a vehicle that can quickly transport large amounts of materials and equipment that would be impossible to move with other vehicles [3]. It is one of the safest modes of transport [1]; this vehicle ensures a high level of transport safety, low vulnerability to adverse weather conditions and total dependence on infrastructure availability [7], it is a faster means of transport than sea transport and cheaper than air transport [8].

In recent years, significant development trends have been observed in rails. They are determined by the need to increase military logistics operations' speed, efficiency, and safety. One of the most prominent trends in the development of military railways is the application of advanced technologies, the integration of railway systems with other modes of transport, the attention to sustainability and the development of specialized military railways [9].

One of the main trends is the application of advanced technology in railways to improve efficiency and safety. This includes the adoption of automation and control systems and the use of sensors and data analytics. These factors enable real-time tracking and analysis of train movements, road conditions, track conditions, and loads [10]. According to [11], accidents or other accidents on the railways are mainly caused by old, worn rails and climatic conditions. Modernization by introducing advanced technologies would make it possible to notice and eliminate defects faster and to plan transportation based on weather conditions. Automation and control systems can improve cargo security and reduce human errors [7].

Another trend is the integration of railways with other modes of transport. For example, the use of air and sea transport to expand the logistics network. This integration can help military forces respond more quickly to changing situations and increase overall efficiency of logistics operations [12].

The rail transport sector increasingly focuses on sustainability by developing green locomotives and using renewable energy sources to power trains. This positively impacts the environment and reduces the dependence on fossil fuels, which can be a strategic vulnerability [13]. In recent years, there has been a growing awareness of the environmental impact of military activities, and many military forces are now working to reduce their CO_2 footprint. Electrification is essential to reduce dependence on fossil fuels and achieve a more favorable effect on the environment. The electrified railway is characterized by high traction quality, high running speed, low transportation costs, reduction of environmental pollution, and other technical and economic characteristics [14]. By 2025 by the end, many military forces will be investigating the use of electric locomotives, which may be superior to traditional diesel locomotives. Electric locomotives are quieter, emit less emissions, and require less maintenance, resulting in savings in the long run. In addition, electric locomotives can be more powerful and efficient than diesel locomotives, which can help military forces transport heavy equipment and supplies over long distances faster and more reliably. Adopting electric locomotives is also compatible with broader trends in sustainable transport and energy use. Although initial electrification costs may be high, the long-term benefits are clear [15].

Improvements in equipment can also significantly improve the development of railways. Specialized military trains, such as wagons and armored trains, are being developed to quickly and safely transport equipment and soldiers to different battle locations [16].

In summary, the development trends of military railways are driven by the need to increase the speed, efficiency, and safety of military logistics operations. Through continued investment in advanced technology, sustainment, and specialized military trains, the military can improve its ability to respond to changing situations and achieve strategic objectives. Military mobility and its development are primarily a deterrent, but its benefits, a well-developed national transport infrastructure, also serve the public good.

When transporting military cargo by rail, it is essential to ensure that it is adequately secured and protected from damage. This may require the use of specialized rail cars or containers for the safe storage of military equipment and supplies. In addition to securing the load, it is also important to ensure that the transport process is secure and that the load is protected from theft or sabotage. This may include using armed guards or other security measures to prevent unauthorized access to the cargo. Generally, transporting military cargo by rail requires careful planning and coordination to ensure that it reaches its destination safely and on time [17]. Armed forces' rail transport depends on operational, training, and logistics needs. Depending on its nature and purpose, military transport is divided into three types: operational, supply, and evacuation [6].

Railway transport is characterized by high capacity, so it can be used for mass transport of equipment over medium and long distances. Railways are also gaining popularity because of their minimal sensitivity to weather conditions. Railways are also valued for their ability to transport heavy and bulky cargo over long distances and relatively quickly. The capacity of the railway depends on the loading gauges, the passing facilities, the available motive power, and the condition of the rolling stock [18]. Specialized vehicles are required for transportation, which must be identified before movement. According to Pecina and Dufek (2016), the logistics capabilities of operations must be designed to reduce national resources and be used appropriately per established requirements [19]. There cannot be any shortcomings in the implementation of logistic operations to receive the support of North Atlantic Treaty Organization (NATO) allies. Properly developed road and railway infrastructure networks must be used according to standards requirements.

According to Dailydka (2010), improving the railway infrastructure, reducing rolling stock shortages and reducing the irrational use of rolling stock in local transport sections is very important [20]. To ensure their defensive and deterrence posture, NATO countries must demonstrate the ability to rapidly deploy, reinforce, and maintain their forces in all countries. Allied forces seek to remove legal and administrative barriers and improve the condition of the rail infrastructure to deploy troops. Both NATO and the European Union have recognized the shortcomings of military mobility as a strategic vulnerability [4]. Therefore, it can be said that the rational use of rolling stock depends on the development of the railway transport system. By allocating more resources to railway development, discrepancies will be reduced faster and the country will develop more quickly.

The NATO armed forces use rail transport because it has the highest mass transport capacity compared to other forms of land transport. Rail transport quickly transports personnel, heavy weapons, and military equipment over long distances. In peacetime, the military uses rail transport mainly for range exercises and training. In addition to supplying supplies, weapons, ammunition, food, rockets, lubricants, or other cargo belonging to the army and transporting military equipment to repair factories [6].

The movement of heavy equipment (tanks, howitzers, infantry fighting vehicles, armored vehicles) by road would be more expensive and slower. It would threaten civilian road users and damage road infrastructure, requiring more complex coordination actions. Rail transport is important in Allied military logistics. The use of railways reinforces their importance in the field of defense, especially when it comes to implementing the decisions taken at the NATO summits in Newport and Warsaw, where member states decided to strengthen the Alliance's eastern flank for non-rotating Allied forces [6].

In conclusion, railways are an unrivalled means of transport for military equipment due to their ability to move heavy and bulky cargo over long distances and at relatively high speeds. Without the ability to quickly transport equipment, it can be difficult to defend yourself during a military conflict and is therefore a critical element of defense. In peacetime, the army uses railway transport for exercises, training, and transportation of military supplies.

2 The Framework of the Study

The study includes North Atlantic Treaty Organization (NATO) documents (STANAG 2468 AMOVP-4 (A) "Technical aspects of the transport of military equipment by rail" [21], STANAG 4062 "Lifting and anchoring devices for military equipment intended for transport by land and sea" [22], AJP-4.4 "Allies movement and transport doctrine" and the analysis of the requirements for the transportation of military cargo by rail.

The Lithuanian railway infrastructure is also analyzed according to the NATO standard AMOVP-4 - "Technical aspects of the transportation of military equipment by railways". Basic requirements for railways (which were included in this study):

1. Uniform track width connecting with other states;
2. Track construction and maintenance.

In order to fulfill the purpose of the study, a secondary data analysis was performed. Statistical data from the Ministry of Transport of the Republic of Lithuania and AB "LTG Cargo" were used for the analysis of the Lithuanian railway infrastructure in carrying out military cargo transportation, and NATO standards were used for the evaluation.

3 Results of the Analysis of Lithuanian Railway Infrastructure that Meets NATO Requirements

NATO rail standards - ensuring the safe and efficient transport of military cargo. One of the main standards for transporting military cargo by rail is (STANAG 2468) "Technical aspects of the transport of military equipment by rail". This standard provides guidelines for the safe and efficient transportation of military cargo by rail, including rail gauge compatibility, loading and unloading procedures, load securing, tagging, and emergency response procedures. Furthermore, STANAG 2468 provides the rules and procedures to be followed to transport military cargo by rail in various NATO countries.

In summary, STANAG 2468 provides detailed technical requirements for rails used in military operations to ensure their safe and efficient operation in multiple situations. Another important standard is AJP–4.4, Allied Movement and Transportation Doctrine. This standard states that in every country where military cargo is transported by rail, problems should be identified well before shipment. In the future, trains will be required to switch to national and regional networks with new regulations and rules. For example, track widths, bridge and platform heights, and speed limits must all comply. Therefore, it is important to analyze the situation of the Lithuanian railway track network and identify problems. According to data from the Ministry of Transport of the Republic of Lithuania (2021), many non-standard 1520 mm wide tracks dominate in Lithuania.

AJP-4.4 "Doctrine of Movement and Transportation of the Allies" states that in the future, rolling stock operated with non-standard width must be renovated into new (standard) ones, which are increasingly used by the countries of the European Union. Therefore, the rails in Lithuania will have to be renovated when switching to the standard width of 1435 mm. The Ministry of Transport of the Republic of Lithuania (2021) also states that it is important to modernize the railway infrastructure as soon as possible, as this would contribute to the successful operation of the European railway system [23]. Most of the attention is paid to expanding the technical interaction of railways with other European countries, increasing the speed of trains, reducing the load on roads and promoting combined vehicles. During the cooperation, modern safety and environmental protection requirements must be implemented.

In the Republic of Lithuania, a nonstandard 1520 mm wide railway track is used; therefore, all military equipment transportation operations are carried out under the Railway Cooperation Organization (OSJD), the International Agreement on the International Carriage of Goods by Rail (SMGS) [24]. This agreement stipulates that if the width of the railway track is different, depending on the technical possibilities, the cargo can be transported by transshipment from the platforms of one width to the platforms of the other width. This agreement allows the transportation of military equipment by rail, even though it does not meet the requirements of NATO standards.

Railways can transport heavy and bulky goods over long distances and at relatively high speeds. According to the Allied Joint Movement and Transport Doctrine (2013), capabilities depend on loading facilities, roadway construction, propulsion, and rolling stock. It is important to ensure maintenance of railway tracks and, if necessary, quick repairs. Often in remote and uneven terrain, the tracks are not properly maintained, and installing a new road takes a lot of money and time. Therefore, according to the Ministry of Transport of the Republic of Lithuania (2021), technical interaction is one of the ways to develop safe and uninterrupted railway traffic, depending on the regulatory documents that must be followed.

In conclusion, NATO rail standards are vital in ensuring military cargo safe and efficient transport. All NATO countries must ensure that their infrastructure is safe, efficient, and meets all requirements following the standards. Currently, standard gauge tracks in Lithuania make up only 6.4% of the entire railway network.

Bureika and Žabelovič (2013) claims that when the Baltic countries joined the European Union, the integration of railway transport into the European railway system began [25]. The railway infrastructure was updated and the most modern and modern rolling stock was put in operation. However, the railway track network in the Baltic countries is not sufficiently developed to meet NATO standards. The standard states that Europe must have a unified rail network to transport military cargo quickly in times of peace or conflict. Almost all of Europe uses the standard 1435 mm wide tracks; apart from the Baltic countries, only Spain, Ireland, Finland, and Portugal do not have them. The European standard gauge (1435 mm) is narrower, while the railway network in the Baltic countries consists of a more comprehensive gauge (1520 mm), which does not directly allow Estonia, Latvia, and Lithuania to connect with the rest of Europe by rail [26]. According to data from the Ministry of Transport of the Republic of Lithuania (2021), the total length of the railway network in Lithuania is currently 1868.8 km. However, it was observed that old rolling stock with a width of 1520 mm is dominant. When soldiers from NATO countries arrive in Lithuania with equipment by rail, due to the difference in gauge at the Lithuanian border in Šeštokai and Mockava, they have to reload all their equipment into wide-gauge rolling stock [26–28].

The units of the allies that arrived on the territory of Lithuania were transported on broad-gauge rails and platforms specially adapted for transporting military cargo. From a strategic point of view, the deployment of allies may take time at this stage, as there is no possibility of arriving directly at the assigned locations. Replacing rolling stock with standard ones that connect to the west would allow for faster and more economical transportation of goods. Another important problem is that the transportation of military cargo requires a military escort, which must be provided with appropriate service and rest conditions, i.e., typically the train must have a wagon adapted for protection with minimum domestic and hygienic conditions.

Reloading of platforms from broad gauge to standard gauge is carried out by Lithuanian Railways employees who have the competence to perform these works. The stations are equipped with ramps and cranes that allow reloading of all military equipment. Lithuanian Railways also has 20 modernized platforms to transport military cargo on the territory of Lithuania on broad gauge tracks. At the Mockava station, the loading time of heavy equipment, such as a tank, is estimated to be about 35 min on average, so in favorable weather conditions it would be possible to reload about 190 pieces of military equipment.

Currently, the standard gauge in Lithuania is built from the borders of Poland and Lithuania through Marijampole to the Palemonas station. However, projects have already been planned to facilitate the transportation process. One of the proposed methods is to set up four military logistics sites where it would be possible to conveniently unload/reload the transported equipment. The plans envisage two sites at the existing European track on the railway section at the Polish-Lithuanian border and at the Kaunas Palemonas terminal. Two more should appear at the Lithuanian-Latvian border (near Jonava and Panevežys). With the installation of such military equipment logistics sites, it would be possible to receive as many as three formations up to 750 m long, transporting 300 allied heavy equipment. It will be possible to conveniently and quickly reload all the delivered equipment, even in cars, and vice versa, as all parking lots will be connected to the Lithuanian car network to improve interaction. Cargo handling at the terminal will also be facilitated by newly installed security systems, fences, and other necessary equipment.

In summary, it can be said that the railway infrastructure in Lithuania does not fully meet the AMOVP-4 NATO standard because, during the analysis, it was found that the old 1520 mm wide rails dominate, which poses a significant risk in the transportation of goods and will require financial resources for renovation to meet the requirements set by NATO. One of the problems is that loading works are being carried out at the Mockava and Šeštokai stations, so there is a risk of quickly transporting military cargo to units located on the territory of Lithuania or the Baltic countries. There is also a lack of unique platforms for transporting heavy equipment, as Lithuanian Railways has a small number of them.

4 Conclusions

In summary, according to the analysis of the scientific literature, railway transport is positively evaluated due to speed, environmental reasons, low transportation costs, and resistance to weather conditions. This type of transport is convenient for transporting large, heavy, or bulky loads. Development trends are observed in the railway infrastructure, which increases the speed, efficiency, and safety of military logistics operations. The main tools used are applying advanced technologies, integrating the railway system with other modes of transport, attention to sustainability, and developing specialized military railways. Railway transport is important in the military logistics system because the rapid transport of equipment is essential for the defense of the country. In times of war, rail transport can be a key tool for deploying troops, conducting evacuations, military mobilization, and delivering supplies, weapons, ammunition, and food.

Secondary data analysis showed that, according to the AJP-4.4 doctrine, Lithuanian railways could transport military cargo by transshipping equipment in a broader gauge (1520 mm) but could not ensure fast military cargo transportation from North Atlantic Treaty Organization (NATO) allies.

Based on the AMOVP-4 NATO standard, the Lithuanian railway infrastructure does not meet their requirements. In the current situation, a small part of the entire railway track network consists of standard gauges that meet NATO requirements; therefore, great attention must be focused on the further modernization of railway lines.

References

1. Jarašūnienė, A., Čižiūnienė, K.: Krovinių vežimo geležinkelių ir jūrų transportu sąveikos probleminių aspektų vertinimas Lietuvoje. Darnios aplinkos vystymas **16**(1), 16–22 (2019)
2. Maisel, A.: Baltic trainspotting: railways and NATO'S logistics problem in northeastern Europe (2018). https://mwi.usma.edu/baltic-trainspotting-railways-natos-logistics-problem-northeastern-europe/
3. Heinrich, J., Hodges, B., Lindley-French, B.: The CEPA mobility project. Moving Mountains for Europe's Defences (2021)
4. Păunescu, D.-M.: Military mobility in the european theatre". Strateg. XXI - Secur. Def. Fac. **17**(1), 208–212 (2021). https://doi.org/10.53477/2668-2001-21-25
5. Bieger, J., Ferrer, J., Riedlinger, D., Xu, W., Demarest, J.: Simulating army rail yard operations at the port of bremerhaven. Ind. Syst. Eng. Rev. **6**(2), 95–100 (2019). https://doi.org/10.37266/iser.2018v6i2.pp95-100
6. Stanczyk, K., Pietrzyk-Wiszowaty, K.: Possibilities to use rail transport potential for the needs of the army. Eur. Res. Stud. J. **XXIV**(1), 1173–1188 (2021). https://doi.org/10.35808/ersj/2016,
7. Pietrzak, O., Pietrzak, K.: The role of railway in handling transport services of cities and agglomerations. Transp. Res. Procedia **39**, 405–416 (2019). https://doi.org/10.1016/j.trpro.2019.06.043
8. Bureika, G., Gaidamauskas, E., Kupinas, J., Bogdevičius, M., Steišūnas, S.: Modelling the assessment of traffic risk at level crossings of lithuanian railways. Transport **32**(3), 282–290 (2016). https://doi.org/10.3846/16484142.2016.1244114
9. Acero, R., Torralba, M., Pérez-Moya, R., Pozo, J.A.: Value stream analysis in military logistics: the improvement in order processing procedure. Appl. Sci. **10**(1), 106 (2019). https://doi.org/10.3390/app10010106
10. Indraratna, B., Sun, Q., Heitor, A., Grant, J.: Performance of rubber tire-confined capping layer under cyclic loading for railroad conditions. J. Mater. Civ. Eng. **30**(3), 06017021 (2018). https://doi.org/10.1061/(ASCE)MT.1943-5533.0002199
11. Orlovienė, J., Žygienė, R.: Railway tracks forecasting with special rdm – 22 program, no. January 2010 (2014)
12. Zitrický, V., Černá, L., Abramovič, B.: The proposal for the allocation of capacity for international railway transport. Procedia Eng. **192**, 994–999 (2017). https://doi.org/10.1016/j.proeng.2017.06.171
13. Maskeliūnaitė, L., Sivilevičius, H.: Expert evaluation of criteria describing the quality of travelling by international passenger train: technological, economic and safety perspectives. Technol. Econ. Dev. Econ. **18**(3), 544–566 (2012). https://doi.org/10.3846/20294913.2012.710178
14. Chen, C.-L., Hall, P.: The impacts of high-speed trains on British economic geography: a study of the UK's InterCity 125/225 and its effects. J. Transp. Geogr. **19**(4), 689–704 (2011). https://doi.org/10.1016/j.jtrangeo.2010.08.010
15. Zhou, X., Lin, X., Ji, X., Liang, J.: Effects of high-speed railway construction and operation on related industries in China. Sustainability **13**(11), 6119 (2021). https://doi.org/10.3390/su13116119
16. Krykavskyy, Y., Stasiuk, K.: Railway freight transportation liberalization as a solution for improving client services quality. SHS Web Conf. **67**, 03007 (2019). https://doi.org/10.1051/shsconf/20196703007
17. Otzulis, V., Ozoliņa, Ž.: Baltijos šalių gynybos strategijos formavimas: priimančiosios šalies parama, Liet. metinė Strateg. apžvalga, pp. 75–97 (2017). https://doi.org/10.47459/lmsa.2017.15.4

18. Biernikowicz, W.: Rail transport in NATO's logistics system: the case of Poland. Eur. Res. Stud. J. **XXIV**(1), 748–761 (2021). https://doi.org/10.35808/ersj/2071
19. Pecina, M., Dufek, R.: Use of logfas tools in logistics planning in Nato. L. Forces Acad. Rev. **21**(2), 120 (2016). https://www.armyacademy.ro/reviste/rev2_2016/Pecina.pdf
20. Dailydka, S.: Choosing railway vehicles for carrying passengers. Transport **25**(1), 11–16 (2010). https://doi.org/10.3846/transport.2010.02
21. STANAG-2468 AMOVP-4. Technical Aspects of the Transport of Military Materials by Railroad (2011)
22. STANAG 4062. Slinging and tie-down Facilities for lifting and tying down military equipment for movement by land and sea (2016)
23. Lietuvos Respublikos susisiekimo ministerija. Apie sektorių. retrieved from https://sumin.lrv.lt/lt/veiklos-sritys/gelezinkeliu-transportas-1/apie-sektoriu-2, (2021)
24. Geležinkelių bendradarbiavimo organizacija (OSJD). Tarptautinio krovinių vežimo geležinkeliais susitarimas (SMGS) (2019). https://www.litrail.lt
25. Bureika, G., Žabelovič, A.: Eismo saugos ekonominis įvertinimas AB „Lietuvos geležinkeliai, pp. 247–252 (2013)
26. Antrasis operatyvinių tarnybų departamentas prie krašto apsaugos ministerijos, Grėsmių nacionaliniam saugumui vertinimas (2022). https://www.vsd.lt/wp-content/uploads/2022/04/LT-el-_.pdf
27. Rail Baltica pritaikys ir karinės technikos gabenimui (2020). http://www.rail-baltica.lt/news/268/75/Rail-Baltica-pritaikys-ir-karines-technikos-gabenimui/
28. Linking people, nations and places (2023). https://www.railbaltica.org/wp-content/uploads/2022/11/RB-TEN_T-broschure_210x210mm_EN_Preview.pdf

Assessment of the Correct Distribution of the Selected Type of Goods on the Loading Area of the Semi-Trailer: Case Study

Arnold Jančár(✉) and Ján Ondruš

Faculty of Operation and Economics of Transport and Communications, Department of Road and Urban Transport, University of Žilina, Univerzitná 8215/1, 010 26 Žilina, Slovakia
arnold.jancar@stud.uniza.sk, jan.ondrus@uniza.sk

Abstract. In Europe, the most common method of transporting goods is via road transport. Lorries account for a large portion of this, since is the most effective and capable type of vehicle. It has a big loading area, so it can carry lot of goods. Based on statistics from road accidents, specifically accidents where the transported goods were damaged, half of them were not correctly secured or positioned on loading area by driver. The purpose of this article is to propose the correct position and design a layout for safely securing specific cargo in any given lorry. First, one vehicle was selected, in order to be weighted with portable measuring weights. In addition, the dimensions of the vehicle were also measured using a measuring tape. The collected data was analysed and calculated. Based on calculations, no problems have been discovered when positioning the cargo on the loading area. Provided that in the future more attention should be paid to correctly secure and position the cargo in the loading area, in order to improve transport safety.

Keywords: Freight · Semi-trailer · Weight distribution

1 Introduction

When loading and storing any goods in the vehicle, the maximum permitted dimensions and weights for vehicles must be observed in accordance with the Decree of the Ministry of Transport and Construction of the Slovak Republic No. 134/2018 Z. z.

According to this Decree [1], the maximum permissible dimensions of vehicles of categories M2, M3, N and O are set at 2.55 m for the maximum permissible width and 16.50 m for the maximum permissible length of the tractor-trailer combination. In addition, the maximum permissible height of the tractor-trailer combination (vehicles of categories N3 and O4) is 4.00 m plus 2% of the height. The maximum distance between the centre line of the semi-trailer kingpin and the rear outline of the semi-trailer is set at 12.00 m.

Also, according to this decree, the maximum permissible weights for combinations of vehicles of all categories are set at 40.00 tonnes. The maximum permissible weights per individual drive axle, if the drive axle is equipped with dual tyre mounting and air

O. Prentkovskis et al. (Eds.): TRANSBALTICA 2023, LNITI, pp. 411–420, 2024.
https://doi.org/10.1007/978-3-031-52652-7_40

suspension or suspension recognised as equivalent within the European Union, are set at 11.50 tonnes for all categories of vehicles.

Minimum axle loads must also be observed to ensure that the vehicle is stable and can be steered and braked. Manufacturers provide charts showing the maximum payload of the vehicle, depending on the position of the centre of gravity of the load. These graphs can also be determined by calculation. Adherence to the load distribution diagram on the vehicle will help to avoid exceeding the maximum permissible axle loads on the vehicle. For correct load distribution on the vehicle, it is necessary to have load distribution diagrams from the body or vehicle manufacturer. If these are not available, they can be calculated using software based on vehicle geometry and maximum axle loads. Correct load distribution according to the diagram will help to comply with the maximum permissible axle loads of the vehicle [2].

The even distribution of container loads in terminals and on roads, including analyses and diagrams of different containers and chassis, is described in the article [3–5].

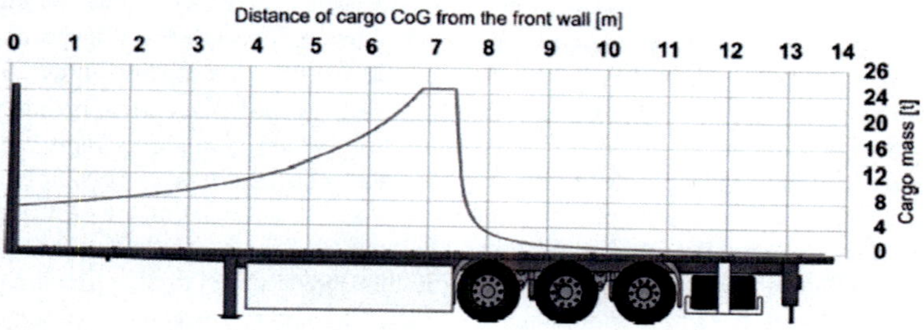

Fig. 1. Load distribution diagram for a tri-axle semi-trailer [6].

The article [7] deals with the quantification of the influence of intensive braking on axle load changes during the transport of cement and selected types of gravel. The effect of vehicle load also causes changes in the centre of gravity position. The centre of gravity displacement consequently affects the driving characteristics of the vehicle [8]. The issue of traffic accidents in Slovakia was also dealt with in the articles [9, 10], which assessed road safety on the basis of statistical data.

When operating a vehicle on the road, it is important to follow the rules regarding the permissible weight of the vehicle and the combination to ensure safety on the road and on the vehicles themselves. The driver should also keep in mind the rules regarding the maximum speed limit in relation to the weight of the vehicle and load [11]. It is also required that the weight on the drive axle or axles of the vehicle should not fall below 25% of the total weight of the loaded vehicle and the total weight of the loaded vehicle or tractor should not be less than 35% of the total weight of the power unit.

The problems associated with improper load securing and lateral accelerations in transport have been addressed in various articles [12–15]. One study found, that the "European best practice guidelines on cargo lashing for road transport" meet the security requirement of cargo's road transport [16, 17].

2 Analysis of the Semi-Trailer and Measurements in Empty State

This chapter analysed the transport company's semi-trailer, which includes several important elements. The combination of vehicles consisted of a double-axle truck of category N3 and brand name Volvo FH 500 EURO VI, which was equipped with air suspension on both axles, while the rear drive axle had dual tyre mounting. In addition, the air suspension and on-board software on the vehicle allowed the weight to be measured using a built-in axle load indicator that recalculates the air pressure in the system and evaluates the load on each axle. This meant that the driver had real-time information on how the vehicle was loaded. Also attached to the vehicle was a three-axle category O4 semi-trailer, which had a three-sided roll-up canvas structure and was a Schwarzmüller brand. This semi-trailer was an important part of the combination as it allowed the transport of larger quantities of goods and was also equipped with a weight measurement system.

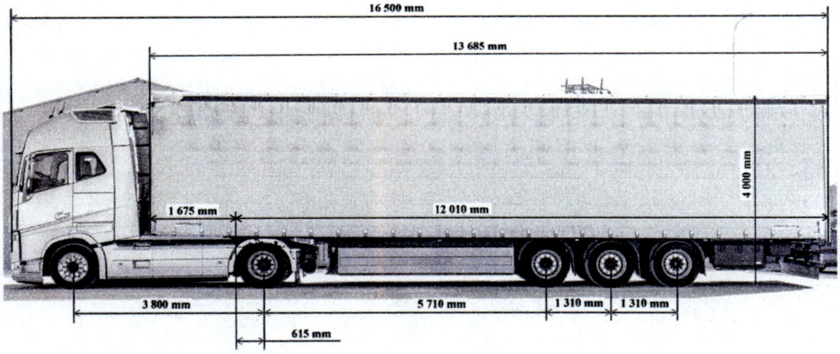

Fig. 2. Measured dimensions of the semi-trailer [author].

2.1 Dimensions of Semi-Trailer

At the beginning, it was necessary to measure the external dimensions of the semi-trailer in order to continue with the calculation process and to assess the correct distribution of goods in Truck Stow. This process involved the use of two types of measurements. Firstly, we used a YATO YT-71580 20-m measuring tape and then a BOSCH DLE 40 Professional laser rangefinder to measure the dimensions. In the work we decided to use the values measured with the measuring tape as we considered them to be more accurate. The measured dimensions are shown in Fig. 2. And were used for further calculations and assessments of the correct layout of the goods in Truck Stow. This procedure ensured the accuracy and reliability of the calculations (Fig. 3).

Fig. 3. Measuring band and laser rangefinder [author].

2.2 Weights Measurements in Empty State

Two different methods were used to measure the mass of the semi-trailer. The first way was to measure using the built-in axle load indicator, which allows the weight to be measured directly on the vehicle. The second method was measurement using the PW-10 portable tensile scales, which allow accurate weight measurement even outside the vehicle. These scales have an accuracy class IIII for static weighing, which means they are very accurate [18]. These measurement methods have been used to ensure that accurate and reliable weight data is obtained for the combination vehicle (Fig. 4).

Fig. 4. Portable weighing scales PW-10 [author].

Measurement Using the Built-in Axle Load Indicator. When using these scales, deviations in weighing can occur because their principle is to measure the load based on the air pressure in the suspension system. This measurement can be affected by various factors. The measured weights are then displayed on the on-board computer, where the values are displayed in tonnes and remain rounded to one decimal place. This accuracy may be insufficient for some users and it is therefore recommended to use other scales that offer higher accuracy. The values in the table have already been converted from tonnes to kilograms to make them easier for users to read and understand (Table 1).

Table 1. Measurement of semi-trailer in empty state using the built-in scales and portable scales [author].

Type	1. Axle	2. Axle	3. Axle	4. Axle	5. Axle	Total
Built-in scales	6,500 kg	8,300 kg	1,900 kg	1,900 kg	1,900 kg	**16,000 kg**
Portable scales	6,300 kg	3,920 kg	1,800 kg	1,980 kg	2,020 kg	**16,020 kg**

After checking with the built-in axle load indicator, we found that the front steering axle of the tractor (1st axle) was loaded with a mass of 6,500 kg. The load on the rear drive axle of the tractor (2nd axle) was 3,800 kg. We have also revealed that the tri-axle of the semi-trailer (3rd, 4th, 5th axle) was loaded with a mass of 5,700 kg.

Measurement With Portable Scales PW-10 Tensile Balances. The measurement was carried out on a flat concrete surface on the company's premises. It was ensured that there was no external influence that could affect the accuracy of the measurement. The measurement results are shown in the table below, where you can see detailed information about the measured values.

Using PW-10 portable scales, we found that the front steering axle of the tractor (1st axle) had a mass of 6,300 kg. On the other hand, the rear drive axle of the tractor (2nd axle) was loaded with a mass of 3,920 kg. The tri-axle of the semi-trailer (3rd, 4th, 5th axle) had a load of 5,800 kg.

3 Loading of the Semi-Trailer and Measurement in Loaded State

When goods were loaded, important data was recorded for each pallet unit that was loaded onto the lorry. To ensure accuracy, the dimensions of the pallet units were measured with a tape measure and the weights of the pallet units were compared with the markings on the pallet using forklift scales. In addition, information on the distribution of the pallets on the loading area was recorded, as well as the number of layers that were stacked on top of each other. This process ensured that the goods were loaded thoroughly and accurately onto the skid, ensuring the goods were transported safely to their destination.

A total of 63 pallet units with a total weight of 21,702 kg were loaded. All pallet units had the same dimensions, length 1,230 mm, width 830 mm and height 980 mm. The only difference between the pallet units was their weight, which is shown in Fig. 5 below.

front wall of the trailer

Row	1. layer		2. layer	
1	329 kg	325 kg	329 kg	325 kg
2	325 kg	325 kg	325 kg	325 kg
3	329 kg	325 kg	329 kg	325 kg
4	325 kg	325 kg	329 kg	325 kg
5	329 kg	325 kg	329 kg	325 kg
6	346 kg	350 kg	342 kg	350 kg
7	346 kg	350 kg	346 kg	350 kg
8	346 kg	346 kg	346 kg	346 kg
9	359 kg	359 kg	359 kg	359 kg
10	365 kg	356 kg	350 kg	356 kg
11	359 kg	359 kg	359 kg	359 kg
12	359 kg	356 kg	359 kg	362 kg
13	359 kg	359 kg	359 kg	359 kg
14	359 kg	359 kg	359 kg	359 kg
15	350 kg	350 kg	350 kg	350 kg
16	303 kg	350 kg	350 kg	

Fig. 5. Individual weights and placement of pallet units [author].

The table shows that the lightest pallet unit weighed 303 kg and the heaviest 365 kg. The heaviest pallets were staggered over the tri-axle of the semi-trailer to avoid overloading, i.e. exceeding the maximum permissible weight of the tractor itself and its individual axles. For this reason, the lightest pallets were distributed in the first half of the trailer.

3.1 Weights of Loaded Semi-Trailer

When the combination was loaded, 63 pallet units were loaded with a total weight of 21,702 kg. The total length of the loaded load was 13,280 mm and the total width of the load was 2,460 mm. After loading, the semi-trailer was measured using the built-in axle load indicator and also using the PW-10 portable weighing scales. The measured values are shown in the table below (Table 2).

Table 2. Weight measurement of the semi-trailer in loaded state via built-in and portable [author].

Type	1. Axle	2. Axle	3. Axle	4. Axle	5. Axle	**Total**
Built-in scales	7,400 kg	10,500 kg	6,700 kg	6,700 kg	6,700 kg	**38,000 kg**
Portable scales	7,420 kg	10,400 kg	6,540 kg	6,680 kg	6,760 kg	**37,800 kg**

For a better overview, the measured weights of the loaded combination are compiled in the following summary table, which shows the data measured using the built-in axle load indicator, the PW-10 portable weighing scales and the maximum permissible weights according to the legislation (Table 3).

Table 3. Comparison of measured values of the loaded semi-trailer [author].

Axle	Built-in scales	Portable scales	Maximum permissible weights
Tractor front steering axle	7,400 kg	7,420 kg	7,500 kg
Rear tractor drive axle	10,500 kg	10,400 kg	11,500 kg
Trailer tri-axle	20,100 kg	19,980 kg	24,000 kg
Total weight of semi-trailer	**38,000 kg**	**37,800 kg**	**40,000 kg**

The measured values of the loaded combination show that the maximum permissible axle loads have not been exceeded. Also the maximum gross combination mass has not been exceeded. The proportion of the total weight of the combination on the tractor driving axle, when considering the built-in load indicator, is 10,500/38,000 = 28%, which is 3% more than the required 25%. When considering the PW-10 portable scales, the proportion is 10,400/37,800 = 28%, so that the condition is also met.

From the table it is possible to notice the differences between the measured values. The biggest difference was at the load of the tri-axle semi-trailer, approximately 120 kg. This difference confirms the fact that the scales in the vehicle are only informative for the driver because of the deviations.

In order to compare the changes in the loads on each axle, a graph was produced using the above tables for the combination vehicle, showing the maximum permissible axle loads, operating weights and the weights of each axle for the loaded vehicle. These weights were taken from measured data obtained using the PW-10 portable scales (Fig. 6).

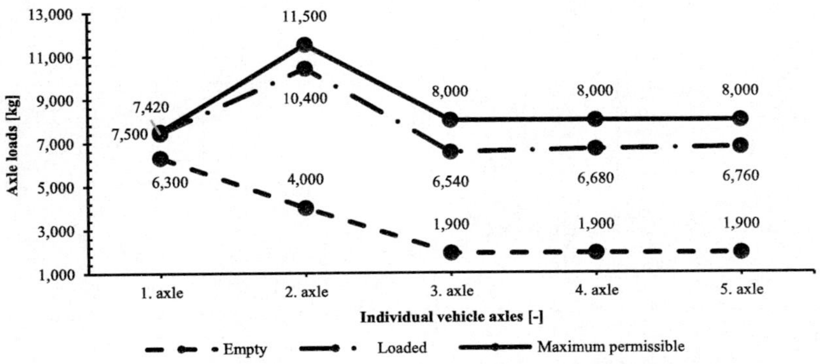

Fig. 6. Comparison of load changes on individual axles of the semi-trailer [author].

4 Assessing the Distribution of Goods in Truck Stow

Truck Stow is a software that is used to optimise the distribution of goods in vehicles [19]. The software takes into account the weight and dimensions of the goods and determines the best place where each load unit should be placed. Thanks to this functionality, the risk of overloading individual axles can be minimised and the safe and efficient transport of goods can be ensured. In addition, Truck Stow allows you to create plans for the placement of goods in the three-dimensional cargo space. These plans are created based on the weights and dimensions of the individual load units and allow for more accurate and efficient placement of goods in the cargo area of a vehicle, trailer or semi-trailer. Truck Stow also allows you to create load units and create vehicle load diagrams according to defined parameters. In this way, it is possible to plan the transport of goods more accurately and minimise the risk of overloading vehicles. With the help of Truck Stow, the handling of load units in the load compartment of a vehicle, trailer or semi-trailer can be managed efficiently.

4.1 Checking the Load Distribution

When checking the distribution of the load on the loading area of the vehicle, the load units were successively stacked according to how they were actually loaded on the vehicle. A total of 63 pallets were loaded with a total weight of 21,702 kg. In Fig. 7 you can see the load diagram and the views of the load distribution control for the vehicle. In addition, the centre of gravity of the load is clearly indicated in the views, which is important in the correct distribution of the individual loads on the vehicle. The check ensures that the centre of gravity is evenly distributed to prevent an unbalanced vehicle that could compromise road safety. For example, the study found that the distribution of the load weight on the loading surface of the vehicle affects the braking process, with higher rear axle loads leading to better deceleration and shorter braking distances [20].

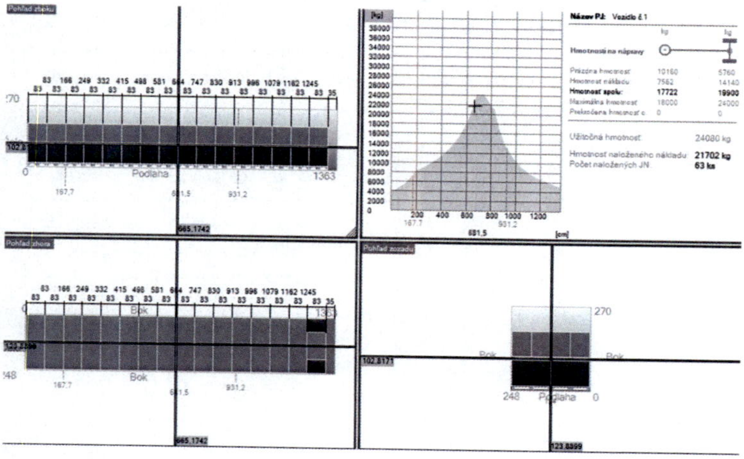

Fig. 7. Checking the load distribution in Truck Stow [author].

The analysis of the load control of the semi-trailer shows that the centre of gravity of the load is located at a distance of 6,651,742 mm from the front wall of the semi-trailer and at a height of 102,8171 mm from the floor of the semi-trailer. This finding is important because we could ensure that the load is evenly distributed over the whole surface of the semi-trailer and that the risks of unbalanced loading are minimised. Looking at the load distribution, it can be concluded that with this load arrangement, the load mass of the load on the trailer kingpin is 7,562 kg and the load mass of the load on the tri-axle of the trailer is 14,140 kg. The total weight of the load on the semi-trailer kingpin is 17,722 kg and on the tri-axle of the semi-trailer 19,900 kg. This information could be useful when planning the transport of heavier loads, which would have to be distributed exactly evenly over the whole surface of the trailer. The value of the maximum load per pin is 18,000 kg because of the fact that the maximum permissible total weight of the tractor is considered. This information could be important for drivers who would like to transport large loads and would need to know what the maximum permissible weights are to avoid breaking road regulations and to ensure the safe transport of goods.

5 Conclusion

In order to assess the correctness of the distribution of the selected type of goods, it was first necessary to analyse the selected semi-trailer of the transport company. The individual dimensions of the semi-trailer, the total weight and the weights of the individual axles of the semi-trailer were determined in the framework of this analysis. All dimensions were within the limits set by law. Using the PW-10 Portable Tensile Scales and the built-in axle load indicators, it was determined that the loaded semi-trailer did not exceed even the maximum allowable weight of 40 tons. At the same time, the permissible weights per axle were not exceeded either. Going forward, it is recommended that palletized goods be loaded at the same height to avoid uneven weight distribution. It is also recommended to test the use of new technologies for the distribution of goods that can help to increase the efficiency of the loading and unloading process.

Acknowledgement. This work was supported by the Slovak Research and Development Agency under contract no. APVV-20-0626. This work was created within the project APVV-20-0626: Biofidelic human body surrogate to increase the objectivity within the forensic analysis of road traffic accidents.

References

1. Slov-lex. 134/2018 Z.z. - Vyhláška Ministerstva dopravy a výs.... Slov-lex (2018). https://www.slov-lex.sk/pravne-predpisy/SK/ZZ/2018/134/20230201
2. Jagelčák, J.: Loading and Securing of Cargo in Road Transport. EDIS. Publishing center of the University of Zilina, Zilina (2015)
3. Jagelčák, J, Kubasáková, I.: Load distribution in general purpose maritime container and the analysis of load distribution on extendable semitrailer container chassis carrying different types of containers. NAŠE MORE : znanstveni časopis za more i pomorstvo **61**(5–6), 106–116 (2014). https://hrcak.srce.hr/clanak/192576

4. Dižo, J., Blatnický, M., Droździel, P., Melnik, R., Caban, J., Kafrik, A.: Investigation of driving stability of a vehicle-trailer combination depending on the load's position within the trailer. Acta Mechanica et Automatica **17**, 60–67 (2023). https://doi.org/10.2478/ama-2023-0007
5. Ramos, A.G., Silva, E., Oliveira, J.F.: A new load balance methodology for container loading problem in road transportation. Eur. J. Oper. Res. **266**(3), 1140–1152 (2018). https://doi.org/10.1016/j.ejor.2017.10.050
6. Directorate-General for Mobility and Transport (European Commission), Cargo securing for road transport :2014 European best practices guidelines. LU: Publications Office of the European Union (2014). https://data.europa.eu/doi/10.2832/80373
7. Vrabel, J., Jagelcak, J., Zamecnik, J., Caban, J.: Influence of emergency braking on changes of the axle load of vehicles transporting solid bulk substrates. Procedia Eng. **187**, 89–99 (2017). https://doi.org/10.1016/j.proeng.2017.04.354
8. Rievaj, V., Vrábel, J., Synák, F., Bartuška, L.: The effects of vehicle load on driving characteristics. Adv. Sci. Technol. Res. J. **12**(1), 142–149 (2018). https://doi.org/10.12913/22998624/80896
9. Ballay, M., Macurová, Ľ.: Development of road safety status and the evaluation criterion causes of specific traffic accidents. In: Transport Means – Proceedings of the International Conference. PTS II. 3.-5.October, Kaunas, Litva, pp. 765–770. (2018)
10. Macurová, Ľ, Kohút, P., Čopiak, M., Imrich, L., Rédl, M.: Determinig the energy equivalent speed by using software based on the finite element method. Transport. Res. Procedia **44**, 219–225 (2020). https://doi.org/10.1016/j.trpro.2020.02.050
11. Marienka, P., Frančák, M., Jagelčák, J., Synák, F.: Comparison of braking characteristics of solo vehicle and selected types of vehicle combinations. Transport. Res, Procedia **44**, 40–46 (2020). https://doi.org/10.1016/j.trpro.2020.02.007
12. Jagelčák, J., Gnap, J., Kostrzewski, M., Kuba, O., Frnda, J.: Calculation of an average vehicle's sideways acceleration on small roundabouts. Sensors **22**(13), 13 (2022). https://doi.org/10.3390/s22134978
13. Jagelčák, J., Gnap, J., Kuba, O., Frnda, J., Kostrzewski, M.: Determination of turning radius and lateral acceleration of vehicle by GNSS/INS sensor. Sensors **22**(6), 6 (2022). https://doi.org/10.3390/s22062298
14. Jagelcak, J., Vrabel, J., Nieuwesteeg, M.: Draft for revision of the standards EN 12640 and EN 12641 regarding the securing of cargo on road means of transport. Logi Sci. J. Transp. Logist. **8**(2), 2 (2017). https://doi.org/10.1515/logi-2017-0015
15. Gnap, J., Jagelčák, J., Marienka, P., Frančák, M., Kostrzewski, M.: Application of MEMS sensors for evaluation of the dynamics for cargo securing on road vehicles. Sensors **21**(8), 8 (2021). https://doi.org/10.3390/s21082881
16. Cheng-qiang, Z., Hong-wei, Z., Chao-zhi, H., Jin-song, D.: Research on the influence of cargo securing force with typical road alignments and vehicle working conditions. In: 2017 4th International Conference on Transportation Information and Safety (ICTIS), pp. 27–32 (2017). https://doi.org/10.1109/ICTIS.2017.8047737
17. Çalık, H., Juwet, M., Yaman, H., Vanden Berghe, G.: Cargo securing under multi-drop and axle weight constraints'. Eur. J. Oper. Res. **307**(1), 157–176 (2023). https://doi.org/10.1016/j.ejor.2022.08.031
18. Přenosné váhy. https://www.tenzovahy.cz/prenosne-vahy
19. Truck Stow. https://fpedas.uniza.sk/~truckstow/obsah.html
20. Skrúcaný, T., Vrábel, J., Kendra, M., Kažimír, P.: Impact of cargo distribution on the vehicle flatback on braking distance in road freight transport. MATEC Web Conf. **134**, 00054 (2017). https://doi.org/10.1051/matecconf/201713400054

Peculiarities Traffic Accidents with the Participation of Motorcyclists

Ľudmila Macurová[1](✉), Pavol Kohút[1], Ján Ondruš[2], and Michal Ballay[3]

[1] Institute of Forensic Research and Education, University of Žilina, Ul. 1. Mája 32, 010 26 Žilina, Slovak Republic
{ludmila.macurova,pavol.kohut}@uzvv.uniza.sk

[2] Faculty of Operation and Economics of Transport and Communications, Department of Road and Urban Transport, University of Žilina, Univerzitná 8215/1, 010 26 Žilina, Slovak Republic
jan.ondrus@fpedas.uniza.sk

[3] Faculty of Security Engineering, Department of Fire Engineering, University of Žilina, Ul. 1. Mája 32, 010 26 Žilina, Slovak Republic
michal.ballay@fbi.uniza.sk

Abstract. As road users, motorcyclists are the most vulnerable participants on high-speed roads. The presence of motorcyclists on roads is also influenced by the nature of the seasons, when, especially during the summer months, an increased number of motorcycles is recorded. Driving motorcyclists presents a wide range of risks affecting the occurrence of various traffic collisions (they often drive at a higher speed than the permitted speed, overtake road vehicles from the right and left, do not maintain a sufficient distance). The consequences of traffic collisions involving motorcyclists are generally much more serious compared to traffic collisions involving road vehicles. The article deals with the peculiarities of traffic accidents involving motorcyclists. In addition to the safety of motorcyclists, the causes of the occurrence and distribution of traffic accidents are also described, as well as the most common types of motorcycle collisions. The paper contains an analysis of a specific accident event between a passenger vehicle and a motorcycle that occurred on a Class I road, in the outskirts of the village. This traffic accident was caught on the onboard camera. During the analysis of the accident event, the accident scene, the driving technique of the drivers, the cause of the accident, as well as the possibilities of averting the accident by road users.

Keywords: Analysis · Accident event · Motorcycle · Road traffic · Vehicle

1 Introduction

In several developing countries, motorcycles are the main mode of transport because they are cheaper, more accessible and have greater mobility on congested roads. Motorcycles are two-wheeled, trackless means of transport, which are designed to transport, as a rule, one or two people sitting behind each other. According to content and weight, motorcycles are divided into very light (up to 100 cm^3 content), light (up to 200 cm^3 content), heavy (up to 500 cm^3 content) and others [7].

O. Prentkovskis et al. (Eds.): TRANSBALTICA 2023, LNITI, pp. 421–430, 2024.
https://doi.org/10.1007/978-3-031-52652-7_41

Various traffic collisions between road vehicle drivers and motorcycle drivers occur very often on roads (especially expressways). Some traffic collisions are minor, while others can cause a serious traffic accident. The time of year also has a significant impact on the occurrence of various traffic collisions, when, especially during the summer months, in spring or autumn, the number of motorcyclists increases, who are the bearers of many risks and specific accident events. The most common causes of traffic collisions involving motorcyclists include errors due to inattention, overestimating one's abilities, fast driving, dangerous maneuvers and inconsiderateness towards other road users. Important factors affecting the number of motorcycle accidents can also be various technical problems of motorcycles, the condition of the road or any surprising fact (for example, an unpredictable turn, sudden obstacle, unclear section), which has a significant impact on the driver's ability to anticipate and adapt driving. [6, 13].

Traffic accidents involving motorcyclists usually occur when a road vehicle and a motorcycle collide, a pedestrian and a motorcycle, a motorcycle and a motorcycle, a motorcycle and a bicycle, or when they collide with an obstacle (fixed, unstable, moving) [3]. The diversity and development of these traffic accidents are evaluated through detailed special parameters that are important in assessing the causes and consequences of traffic accidents. The mentioned contribution deals with the analysis of a specific traffic accident between a passenger motor vehicle (Peugeot brand) and a motorcycle (Kawasaki brand), which occurred during the day, in a section outside the built-up area of the municipality. During the accident, the motorcycle was moving in the section in question at a speed (approx. 142 km/h), which was significantly higher than the maximum speed allowed for the given Sect. (90 km/h). The mentioned speed of the motorcycle significantly played a negative role in the occurrence of the accident in question.

2 Safety of Motorcyclists as Road Users

The issue of the safety of road users is highly topical and important due to the increase in traffic intensity and the development of other alternative modes of transport. These important attributes of road traffic are ensured by each participant on the road by respecting the rules, regulations and recommendations on road traffic, by using protective elements, by their responsibility and consideration towards other road users or by sufficient foresight and timely reaction to traffic situations.

Traffic accident rate indicators are an important parameter for assessing the safety of road users, which evaluate the seriousness of traffic accidents as well as the relative degree of safety. The consequences of various traffic collisions are usually much more serious for motorcycle drivers compared to road vehicle drivers. Compared to drivers of road vehicles, motorcyclists are more vulnerable, they are not sufficiently protected (especially when falling on the road, sliding on the road or after hitting a fixed obstacle) and they are at risk of very serious, even fatal injuries [7].

The probability of dying in an accident on a motorcycle is about 30 times higher than when riding a road vehicle. Up to 80% of traffic accidents result in injuries or deaths. The most common consequences of motorcyclists' injuries are serious head injuries, spine damage, complicated fractures of the upper and lower limbs, pelvic fractures, internal bleeding, abrasions of various nature and severity depending on the speed of the motorcycle, road surface and protective clothing, etc. [1].

Riding motorcycles has completely different parameters than driving road vehicles. Motorcycles and road vehicles are equipped with different elements of active and passive safety, the primary function of which is to prevent and minimize injuries to road users. Active elements of motorcycle safety are operability (easy brake operation, automatic or semi-automatic transmission, clutch booster), visibility and distinguishability (for example, reflective clothing and driver's helmet, reflective elements on the motorcycle, automatic switching on of dipped headlights), comfort (for example, protection of the driver from the wind, rain and dirt using clothing, reduction of vibrations acting on the hands and feet of the driver) and driving characteristics (for example, the introduction of the ABS system). As for the passive elements of motorcycle safety, protective bags, airbags, spine protectors, helmets, protective clothing, etc. are important [3, 4].

In congested road traffic conditions, there is a fundamental difference in the movement of a road vehicle and a motorcycle in the lane, as well as in the movement outside the lane. A motorcyclist often changes speed and direction of travel, may ride in the same lane alongside other road vehicles, may swerve, or may follow a leading vehicle diagonally. These movements of motorcyclists riding outside the traffic lanes significantly affect the probability of the occurrence of various traffic conflicts in conditions of road traffic congestion, which represents the main safety problem of motorcycle traffic. [8].

3 Traffic Accidents with the Participation of Motorcyclists

Traffic accidents involving motorcyclists can occur in various configurations, such as a motorcycle falling to the ground, a motorcycle hitting a fixed obstacle, a motorcycle colliding with a road vehicle. After the motorcycle falls to the ground, it can be the movement of the motorcycle and the crew without hitting an obstacle, as well as the movement of the motorcycle and the crew with the impact of the motorcycle, the passenger, or at the same time. Among the most common causes of a motorcycle crash are the motorcyclist's driving technique and speed (related to his riding experience), weather conditions, road condition, road pollution (mud, sand, dust, oil, diesel), technical malfunction of the motorcycle. A motorcyclist can also fall from horizontal road markings (skid on a wet surface), manhole covers and tram tracks (especially when approaching the tracks at an angle) [4, 5].

If the motorcycle hits a fixed obstacle, the motorcycle and the crew continue to move, or the motorcycle does not move further, but the crew continues to fly. In some cases, after hitting an obstacle, the motorcycle and the crew stop moving. In the case of a collision between a motorcycle and a road vehicle, it can be an impact of the motorcycle into the front or rear of the vehicle with the impact of the crew on the vehicle, an impact of the motorcycle into the side of the vehicle with movement of the crew outside the vehicle (with or without contact with the vehicle), as well as without further movement of the crew. It can also be the impact of a road vehicle into the rear of the motorcycle with the impact of the crew on the vehicle, or the impact of a road vehicle into the side of the motorcycle with partial or complete overlap [4, 5].

Various traffic collisions involving motorists are also influenced by the fact that road vehicle drivers automatically assume that motorcyclists will stay in the right lane as much as possible. In reality, however, a motorcyclist requires the same space as a road vehicle, as he needs to avoid possible obstacles on the road, in sharp turns he must get as far as possible from the shoulder and also he must use the space of the right lane sensibly. The rule of keeping a safe distance applies not only between two passenger vehicles, but also between a road vehicle and a motorcycle. Although the motorcycle is fast, its braking effect is limited, as the contact surface of the tires with the road is small. Due to its dimensions, the motorcycle can remain in the so-called blind spot of the vehicle, which can be the cause of a collision situation [7, 9].

The analysis of various traffic collisions involving motorcyclists is handled by traffic accident analysts who use various forensic engineering methods with software equipment in the form of various simulation programs. For the computational analysis of a traffic accident, it is important to remember that the crew and the motorcycle do not form one unit after the collision. In traffic accidents of road vehicles, in most cases the crew remains in the cabin, while in traffic accidents of motorcycles the crew is usually separated from the motorcycle. Due to the lower coefficient of friction between the motorcycle and the road compared to the coefficient of friction between the crew and the road, the deceleration of the motorcycle is smaller [2, 5, 11] (Fig. 1).

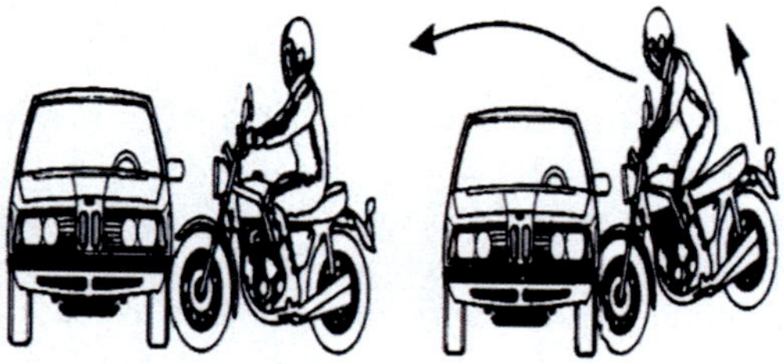

Fig. 1. Correct movement (left) and incorrect movement (right) of the motorcyclist's body when hitting the side of the vehicle [4].

In expert practice, in motorcycle accidents with passenger vehicles, it turns out that the size of the injuries is significantly smaller if the crew flies over the vehicle compared to if they hit some part of the vehicle. When braking with the rear wheel locked, the motorcyclist can maintain directional stability, but when the front wheel is locked, the motorcycle falls. In the case of skidding on a wet bituminous road, the front wheel of the motorcycle usually skids, and on a dry bituminous road, the rear wheel of the motorcycle usually skids. When the front wheel of a motorcycle skids, the motorcyclist falls over the handlebars and his further movement takes place in front of the motorcycle, after the fall the motorcycle moves in the tangent direction to the original trajectory of movement (a motorcycle sliding on the ground may start rotating around some point of contact with the road, or the wheel may rotate on on a dirty road and subsequently the wheel slips). The main factor that affects the parameters of the fall is the initial speed of the motorcycle, each fall is original and it is impossible to achieve identical conditions for its repetition. The fall of the motorcycle and the motorcyclist, as well as their movement after the fall, is very complicated due to the large number of input parameters. When a motorcycle falls, an experienced driver will behave differently than an inexperienced driver, and therefore even the consequences of a seemingly identical fall can be different [2, 5, 9, 13].

4 Analysis of the Accident Eent – Vehicle and Motorcycle

The traffic accident between a passenger motor vehicle and a motorcycle occurred during the day, on a Class I road, in a section outside the built-up area of the municipality. Two motorcycles were moving behind each other on the main road (a camera was placed on the second motorcycle, which recorded the course of the accident scene). The driver of the vehicle entered the main road from the side road, who did not see the movement of the motorcyclist on the main road before entering. The motorcyclist reacted to the entry of the vehicle onto the main road by braking intensively, leaving a brake mark on the road. During intensive braking, a loss of directional stability began to occur, to which the motorcyclist responded by reducing the braking effect (as a result of which there was no brake mark just before the collision). The collision between a vehicle and a motorcycle occurred in the area of the central dividing line. After the collision, the motorcyclist's body rotated over the front of the vehicle and landed on the road to the right of the rear of the vehicle. The course of this accident scene is shown in Fig. 2, which contains 6 frames from the video recording.

Fig. 2. Course of the accident - vehicle and motorcycle [12].

1 – Position of the vehicle and motorcycle at the time approx. 1.14 s before the collision, 2 – Position of the vehicle and motorcycle at the time approx. 0.47 s before the collision, 3 – Position of the vehicle and motorcycle at the time of the collision, 4 – Position of the vehicle, motorcycle and motorcyclist at the time approx. 0.3 s after the collision, 5 – Position of the vehicle, motorcycle and motorcyclist at the time of approx. 0.5 s after the collision, 6 – Position of the vehicle, motorcycle and motorcyclist at the time of approx. 1 s after the collision.

From a forensic point of view, the motorcyclist suffered severe injuries (rupture of the bladder and pelvic pelvis, a comminuted fracture of the lower part of the spindle bone with displacement of fragments, contusion of the lungs and peritoneum with a healing time of 6–7 months. These injuries were caused with great force and violence by blunt objects of a narrower and wider contact surface when a motorcyclist hits than a motorcycle driver in a collision with a road passenger vehicle [12].

The mutual movement of the vehicle and the motorcycle in the time just before the collision could be determined with relatively high accuracy by analyzing the video recording, using a spatial model of the traffic accident site created using the Agisoft Metashape software program. The scene of the traffic accident was documented using the method of sequential photography from an unmanned vehicle (drone) and subsequently a three-dimensional (3D) model of the scene of the traffic accident was created from this photo documentation (Fig. 3).

Fig. 3. The place of the traffic accident - drone photo (left) and 3D model (right) [12].

The analysis of the video recording was carried out in such a way that the identifiable points on the corresponding image of the video recording correspond with the points in the spatial model of the traffic accident site [11]. The mutual movement of the vehicle and the motorcycle during the accident (values of time (s) and speed (km/h)) was also detected in the mentioned way.

Fig. 4. View from the direction of travel of the vehicle driver (left) and from the direction of travel of the motorcyclist (right) [10].

According to Fig. 4. it follows that the vehicle was moving at a speed of approx. 30 km/h and the motorcyclist was moving at a speed of approx. 140 km/h. For the first time, the driver of the vehicle was able to recognize the motorcyclist, who was in the area of the obscured view, approximately 2.9 s before the collision. This is a decisive moment in terms of assessing the issue of giving priority in driving.

4.1 Evaluation of the Driving Technique of the Motorcycle Driver

The motorcycle driver was traveling at a speed of approximately 142 km/h at the beginning of the accident. During the movement of the motorcycle, the driver of the vehicle entered the main road in a transverse direction (from the point of view of the direction of the motorcyclist from right to left), while the vehicle entered the movement corridor of the motorcyclist. After recognizing the collision situation, the motorcyclist reacted to this situation in about 1.87 s before the moment of collision, while the longitudinal distance between the front wheel of the motorcycle and the left part of the vehicle was about 68m. After the reaction time of 0.6 s, the motorcyclist started to brake, and at the moment of the reaction, he was moving at a speed of about 133 km/h. The motorcyclist braked intensively on the approx. 14.2 m track, but later partially released the braking effect due to loss of directional stability. Subsequently, there was a collision with a vehicle. At the time of the collision, the speed of the motorcycle was approximately 100 km/h. According to the performed analysis of the accident scene, it follows that the motorcyclist reacted to the movement of the vehicle in time, and therefore the motorcyclist's driving technique was technically correct in terms of the timeliness of the reaction. It also follows that the driver was moving in the section in question at a speed of approximately 142 km/h, which was a speed higher than the maximum permitted speed, which is 90 km/h. The motorcyclist's riding technique in terms of speed was technically incorrect.

4.2 Evaluation of the Driving Technique of the Driver of the Vehicle

The driver of the vehicle was entering the main road, turning left (oblique to the longitudinal axis of the main road). At that moment, a motorcyclist was moving in the lane towards the village. The driver entered the main road without stopping, while continuously reducing the driving speed. Just after entering the main road, the vehicle was moving at a speed of approximately 14 km/h.

For the first time, the driver could see the motorcyclist at a mutual distance between the vehicle and the motorcycle of approx. 109 m, while approx. 2.9 s remained to elapse after the moment of the collision. The driver of the vehicle could only start reacting at the moment of recognizing that the motorcyclist was driving at a speed significantly exceeding the speed of 90 km/h. The mere recognition of the presence of a motorcyclist on the main road 2.9 s before the collision did not mean that the driver of the vehicle had to react. In view of the above, it can be concluded that if the motorcyclist was driving at a speed of approximately 90 km/h, the driver of the vehicle could have continued entering the main road and, under these circumstances, the right-of-way would have been given.

The driver of the vehicle (according to her description of the accident scene) did not recognize the motorcyclist before entering the main road. In the time 2.9 s before the collision, the driver of the vehicle could have spotted the motorcyclist, provided she correctly evaluated the traffic situation. She could recognize the collision situation approximately 1.9 s before the collision. The fact that the driver did not recognize the motorcyclist can be considered an incorrect element of the driver's driving technique. Her reaction to the motorcyclist was delayed.

In order to recognize the fact that the motorcyclist is driving at a speed of about 142 km/h, it was necessary to follow the motorcyclist in a time interval of about 1 s. After this time, the driver of the vehicle could recognize the occurrence of a collision situation and subsequently react (Fig. 5).

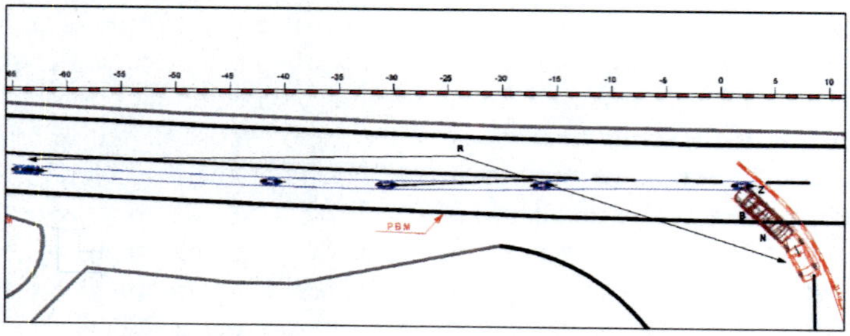

Fig. 5. Calculation of the course of the accident scene with the timely reaction of the driver of the vehicle [10].

R – the relative position of the vehicle and the motorcycle at the moment of the beginning of the reaction of the driver of the vehicle, N – the position of the vehicle at the moment of the beginning of the braking effect, B – the position of the vehicle at the moment of the beginning of full braking of the vehicle, Z – the position of the vehicle at the moment of stopping this vehicle - in case timely reaction of the driver of the vehicle.

By applying the calculation, it was found that at the given time (1.9 s before the collision), the front part of the vehicle was located approx. 2 m before the entrance to the main road, while the vehicle was moving at a speed of approx. 23 km/h. If the driver of the vehicle reacted in time (at the moment of recognition of the collision situation) and after the reaction time she braked intensively, she would still not be able to stop the movement of the motorcyclist in front of the corridor and thus prevent a traffic accident.

In the sense of the above, it should be noted that the incorrect driving technique of the motorcyclist (driving at a speed higher than the maximum allowed for the given section) was an element of the accident, which made it difficult for the driver of the vehicle to evaluate the traffic situation at the moment of deciding to enter the main road. The final assessment of the fact whether the driver's entry onto the main road was correct is subject to a legal question, namely a question of foresight.

4.3 Discussion and Results in the Analysis of the Accident Event

The incorrect driving technique of the motorcyclist was an element of the accident scene, which caused a collision situation and at the same time made it impossible for the motorcyclist to prevent the traffic accident. In general, it can be concluded that the higher the rate of exceeding the maximum permitted speed of the motorcycle, the more difficult the conditions for the driver of the vehicle to evaluate the situation were. She had less time available and had to evaluate the speed of the motorcycle at a greater distance, which was a smaller object for the driver of the vehicle.

The driver of the vehicle entered the main road in such a way that she did not give way to the motorcyclist in relation to his actual driving speed, namely 142 km/h (in relation to the incorrect driving technique of the motorcyclist). However, this female driver entered the main road by giving priority to the motorcyclist in relation to his correct driving technique (at a speed of 90 km/h). If the motorcyclist was moving at a driving speed of 90 km/h, he would have reacted to the situation in time and braked by smooth deceleration, then there would have been no collision between the vehicle and the motorcycle. As part of the assessment of other elements of the technical cause of the traffic accident, there was also the way the road vehicle entered the main road, which was conditioned by a legal question, namely to what extent the driver of the vehicle (before entering the main road) should have foreseen the speed limit of vehicles on the main road.

In the event that the female driver of the vehicle has the obligation to anticipate the driving of vehicles moving along the main road at a speed of over 90 km/h (oncoming vehicles cannot yet be seen), then the way the female driver entered the main road was performed in a collision and this element can be considered as another element of the technical cause of the traffic accident. If the driver of the vehicle does not have to anticipate exceeding the speed of vehicles moving on the main road at a speed of over 90 km/h when entering the main road (oncoming vehicles cannot yet be seen), then the way the driver entered the main road was performed correctly and the technique the driving of the female driver has no causal connection with the occurrence of the traffic accident in question.

5 Conclusion

Traffic accidents involving motorcyclists usually represent events that occur when the movement of a motorcycle meets a vehicle, a pedestrian, a motorcycle, a bicycle or an obstacle (fixed, unstable, moving). The analysis of traffic accidents as a field of forensic engineering deals with the origin and course of various traffic collisions. When processing a complex analysis of the accident scene, it is necessary to solve tasks of a technical and legal nature. The diversity of the course of the accident scene requires complex evaluations of a number of special parameters, the results of which have an important impact on the evaluation of the degree of fault in a traffic accident.

The paper analyzed a specific traffic accident in which a motorcycle (Kawasaki brand) and a passenger motor vehicle (Peugeot brand) collided. The cause of the traffic accident was the incorrect driving technique of the motorcyclist, who was moving along the main road at a speed of 142 km/h, which was much higher than the maximum permitted speed for the given Sect. (90 km/h). This element of the accident scene made it difficult for the driver of the vehicle to evaluate the traffic situation at the moment of deciding to enter the main road. The fact that the driver did not recognize the motorcyclist can be considered an incorrect element of the driver's driving technique. Her reaction to the motorcyclist was delayed. The final assessment of the fact whether the driver's entry onto the main road was correct is subject to a legal question, namely a question of foresight. According to the results of the analysis of the traffic accident, it follows that if the motorcyclist (at the moment of the beginning of his reaction) was moving at the prescribed speed of 90 km/h, there would not have been a collision between the vehicle and the motorcycle during continuous deceleration. The driver of the vehicle could continue to enter the main road and in these circumstances the right of way would occur.

Considering the analysis of a specific traffic accident between a passenger motor vehicle and a motorcycle, it is important in terms of reducing the number of traffic accidents and their negative consequences to pay attention to the important attributes of road traffic, especially road users and means of transport moving on roads.

Acknowledgements. This work was supported by the Slovak Research and Development Agency under contract no. APVV-20-0626. This work was created within the project APVV-20-0626: Biofidelic human body surrogate to increase the objectivity within the forensic analysis of road traffic accidents.

References

1. Hirt, M., et al.: Traffic Accidents in Forensic Medicine and Forensic Engineering, p. 160. Grada Publishing (2012). ISBN 978-80-247-4308-0
2. Hugemann, W., et al.: Accident Reconstruction. Wolfgang Hugemann (Hrsg.) Muenster. p. 1254 (2007). ISBN 3-00-0194-19
3. Jurina, R., Kasanický, G.: Selected Chapters of Forensic Engineering. College in Sládkovičovo, Fakulty of Law J. Jesenský, p. 380 (2013). ISBN 978-80-89267-91-0 (2013)
4. Kasanický, G., Kohút, P., Priester, J.: Analysis of Single-Track Vehicle Accidents, p. 488. University of Žilina, EDIS, Žilina (2003). ISBN 80-7100-599-1
5. Kohút, P., Kasanický, G.: Evalution of crash tests of the institute of forensic engineering. University of Žilina, aimed at ess. Commun. Sci. Lett. Univ. Žilina **15**(2), 124–130 (2013)
6. Lin, M.R., Kraus, J.F.: A review of risk factors and patterns of motorcycle injuries. Accid. Anal. Prev. **41**(4), 710–722 (2009)
7. Minh, C., Sano, K., Matsumoto, S.: The speed, flow and headway analyses of motorcycle traffic. J. Asia Soc. Transport. Stud. **6**(2005), 1496–1508 (2005)
8. Nguyen, X., Hanaoka, S., Kawasaki, T.: Traffic conflict assessment for non-lane-based movements of motorcycles under congested conditions. IATSS Res. **37**(2), 137–147 (2014)
9. Ondruš, J., Skrúcaný, T., Kolla, E.: Comparative braking tests of the motorcycles. In: Transport Means-Proceedings of the International Conference, pp 291–296 (2017)
10. SW PC Crash (14.0), SW XL Vision ™
11. Tarkowski, S., Nieoczym, A., Caban, J., Gardyński, L., Vrábel, J.: Reconstruction of road accident using video recording. In: MATEC Web of Conferences, vol. 252, p. 05023. EDP Sciences. (2019)
12. The case study – Analysis of the accident event vehicle and motorcycle (Archive ÚSI ŽU)
13. Varin, B.: Factors related to motorcycle fatalities. In: Proceedings EVU – European Association for Accident Research and Analysis, Strasbourg, France, 26–28 October 2022, pp. 103–109 (2022). ISBN 978-80-554-1962-6

Simulation of Truck Customs Terminal Work

Georgii Prokudin, Oleksii Chupaylenko, Tanya Khobotnia, Kateryna Hilevska(✉), and Irina Prokudina

National Transport University, Kyiv 01010, Ukraine
katerinagui@ukr.net

Abstract. The article studies the operation of a cargo customs terminal, which is an element of the system of international multimodal transportation. The object of the study is the operation of a cargo customs terminal as a queuing system network. The main problem is to manage the operation of a cargo customs terminal to ensure an adequate response to production situations arising in the course of its operation. A methodology for analyzing a queuing network that has been developed provides an adequate response to situations arising in the course of its operation and allows adjusting the values of some parameters and characteristics of the network (number of channels, service intensity, etc.) in order to improve the network operation and ensure an optimal process of servicing requirements. To implement the analytical model of the customs terminal operation in the management of variable cargo modules, the system for analyzing the customs terminal mass service network (SACTMSN) was developed in the Excel environment, which consists of two parts – the part for entering initial data and the calculation part. Calculations of the operation mode of a cargo customs terminal as a queuing system using real-life examples showed ways to optimize terminal operations.

Keywords: Customs terminal · Mass service network · Variable · Cargo modules · Excel environment · Transport operation

1 Introduction

The volume of international transportation is growing, and the role of transport terminals, which have also taken on the functions of customs clearance and customs clearance, is growing. The volume of functions and operations performed by customs terminals is increasing. The work of customs terminals during international transportation should provide for an adequate response to situations arising in the course of their operation. A description of the logistics approach is needed, which, with the help of customs technologies, solves the problem of optimal organization of the work of customs terminals in the service of international cargo transportation. The effective functioning of cargo customs complexes and terminals should ensure the optimization of the rolling stock of transport enterprises engaged in cargo transportation. The technology of cargo delivery with the use of cargo customs complexes as intermediate temporary points of its preservation during regular transportation will allow increase the number of motor vehicle trains on the routes and increase the efficiency of their use.

C. Prentkovskis et al. (Eds.): TRANSBALTICA 2023, LNITI, pp. 431–439, 2024.
https://doi.org/10.1007/978-3-031-52652-7_42

2 Peculiarities of Customs Cargo Terminals

In order to analyze and develop a new approach to improving the logistics of the customs terminal, its functioning in the process of crossing the border with the change of cargo modules was analyzed.

The model of the operation of the customs terminal was presented within the framework of the theory of mass service systems, the structure of the object is a mass service network [1].

Demand flows are characterized by appropriate intensities, which form flows of tractors, the service of which consists in the crossing of cargo modules that have reached the border from the other side, passed customs control and are ready for transportation to their destinations [2].

Systems of mass service of passing customs care, which include the performance of customs operations. Failures may occur at this stage. If the defects are insignificant, this can be corrected on the spot. From the point of view of mass service theory, it is necessary to formalize such a system, presenting it in the form of a generalized structure. This structure can represent either a multiphase mass service system or a network of mass service systems. Orders are accepted for service from an unlimited number of "customers", for which there is a correspondence of exogenous and endogenous characteristics [3–5].

For the specified class of transport problems, quite perfect methods of their solution have been developed (E.V. Dijkstra, L.V. Kantorovich, A.P. Kozhyn, M.N. Bidniak, A.I. Vorkut, N.T. Kunda, G.S. Prokudin and others) [1–7]. But a rather significant drawback of matrix optimization methods for solving transport problems is that they are able to provide an optimal transportation plan only on a transport network (TN) with oriented graphs [8].

Effective functioning of cargo customs terminals (CCT) and terminals ensures optimization of the rolling stock of transport enterprises engaged in cargo transportation [9]. The technology of cargo delivery with the use of CCT as intermediate temporary points of its preservation during regular transportation will allow increasing the number of motor vehicle trains of motor transport enterprises (MTE) on the routes and increasing the efficiency of their use [10].

In international cargo transportation on the Ukraine-EU and EU-Ukraine routes, the transshipment point can be CCT, which is located on the Ukrainian side of the border [11]. The western regions of Ukraine bordering the EU member states, under the condition of effective cross-border cooperation, make it possible to use the beneficial geopolitical potential of the state [12]. Long-term projects regarding the use of CCT along the western border are one of the options for the rationalization of freight transportation, therefore, the rational organization of work and cooperation of MTE with terminal and warehouse complexes is very important and relevant.

Studies have shown that the main problem is the management of the cargo customs terminal to ensure an adequate response to production situations arising in the course of its operation. To solve the problem of the cargo customs terminal operation, it is necessary to represent it as a queuing system network and develop a methodology for analyzing the queuing network The analysis methodology should ensure an adequate response to situations arising during its operation and allow adjusting the values of some parameters and characteristics of the network (number of channels, service intensity, etc.)

in order to improve the network operation and ensure an optimal process of servicing requirements.

3 Development of a Technological Scheme for Crossing the Border with a Change of Cargo Modules at the Cargo Customs Terminal

To develop a new approach to the analysis of the work of the customs terminal, we will analyze the functioning of the technological scheme of crossing the border (from state A to state B and vice versa) with a change of cargo modules (CM). Technological scheme of crossing the border with change BM shown in Fig. 1, where: road train – RT; cargo module – CM; truck tractor – TT; state A – sA; state B – sB (Fig. 1).

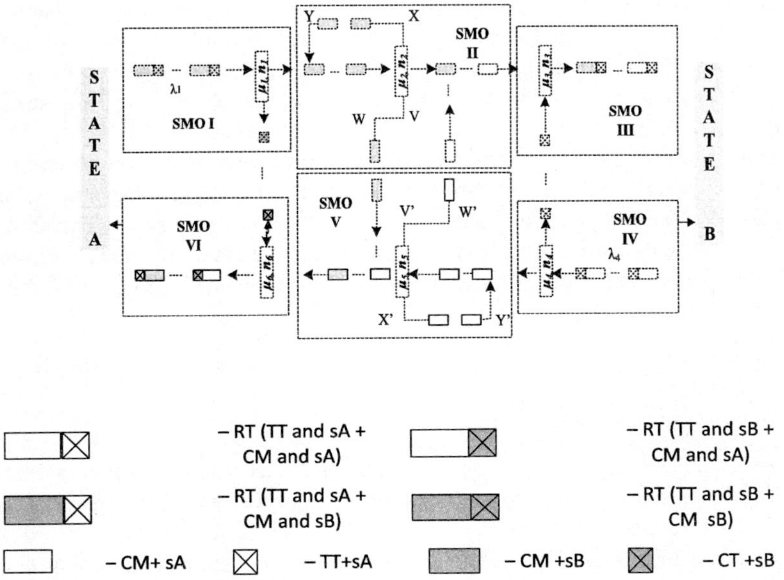

Fig. 1. Technological scheme of crossing the border with a change of cargo modules.

MSS I, MSS IV - mass service systems (MSS), to which requirements are received, which are tractor vehicles with variable CMs. Demand flows are characterized by corresponding intensities λ1 and λ4. In MSS I and MSS IV, CMs are unhooked from their own tractors, which must be attached to CMs that have arrived from the other side of the border and are ready for transportation to their destination. Other requirements streams formed in MSS I and MSS IV are flows of variable CMs aimed at customs control, designated as MSS II and MSS V.

MSS II, MSS V - MSS of passing customs control, which include the implementation of customs operations. Crashes may occur at this stage. If there are minor defects that can be eliminated on the spot, the cargo module is sent for re-inspection (respectively sections X-Y and X'-Y'). If the cargo can be passed through the border, it forms a flow of losses (respectively sections V-W and V'-W').

MSS III, MSS VI - MSS on which CMs that have already passed customs control are joined to free tractor vehicles. Each MSS is characterized by its service intensity (μi) and number of service channels (ni).

When processing incoming flows for service, customs terminals are complex systems in which the request is sequentially serviced at several stages. The complexity of formalizing the processes of servicing such a border point is explained by the content of the service itself, as it consists of several stages.

From the point of view of mass service theory, it is necessary to formalize such a system, presenting it in the form of a generalized structure. This structure can be either a multiphase MSS or a MSS network.

To determine the direction in which the structure will be studied, we note that both multiphase MSS and mass service networks consist of several typical nodes, that is, they represent a collection of several systems [13]. However, multiphase systems are MSSs consisting of several typical nodes located in series and all requests served in one node are sent to the next. In other words, the output stream of one node is the input stream of the next one.

If we talk about mass service networks, they also consist of several typical nodes, but the order of traversing the nodes can be different for different requirements. Some of the applications require service in all nodes, and for others - only in some. The presented scheme for crossing the border is very effective, because the procedure itself is simplified in many ways (drivers and tractor-trailers on both sides of the border do not pass customs control), but it has some significant drawbacks, namely:

- the presence of oncoming cargo flows of the same power is assumed, i.e. $\lambda 1$ should be approximately equal to $\lambda 4$;
- cargo modules must also be unified.

We can conclude that formalizing the work of the customs terminal, within the framework of the theory of mass service systems, the structure of the object represents a mass service network.

An important means of improving the cargo delivery system is the creation of an effective transport process, which includes the rational organization and coordination of all its elements [13]. Solving this problem is possible on the basis of a comprehensive study of cargo delivery. The methodological basis of the complex approach to research is system analysis, which has become widely used as a tool for solving complex problems in various spheres of activity.

The goal of system analysis is to find ways to improve the efficiency of transport [14]. Such a statement of the problem requires the determination of the problems of transport functioning, the place of their detection, their content and influencing factors.

4 Research of the Cargo Customs Terminals as a Mass Service System

The main task of the description and analysis of the research object is its presentation in a systematic form, convenient for further research. As a result of the implementation of this stage, the following should be determined: the purpose of functioning of the systems; organizational and functional structure; performance indicators and criteria; problems and assessment of their impact on final results.

The first step of decomposition is the specification of the research object as a single system. It consists in defining the cargo delivery system and its elements; allocation of the external environment, which represents a set of all systems with which the system under consideration enters into certain relations in the process of functioning; formulation of the requirements of the external environment for the object of research, definition of its purpose and main tasks.

The second step of decomposition of the system into "production" and "management" is performed by presenting it in the form of a management object and a management system. The description of the control object consists in highlighting the elements of the transport process and their interaction. When describing the management system, the methods and factors are defined, with the help of which it is possible to manage the transport process. The main results of this stage are the creation of a model of the transportation process and the determination of system performance evaluation indicators for further research into ways to improve transportation efficiency.

To develop the mathematical apparatus of the mass service network, we will present its possible states (Fig. 2).

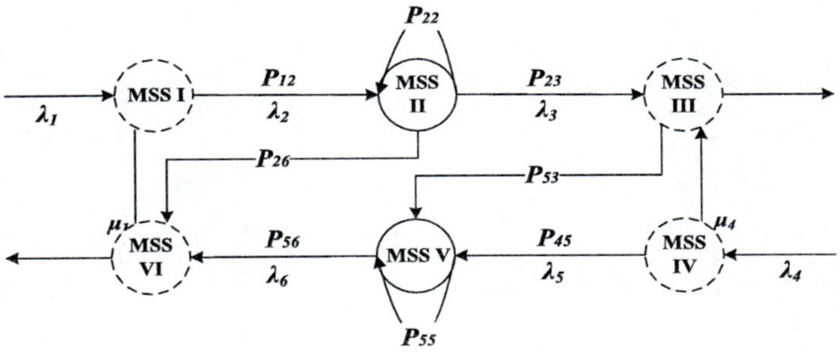

Fig. 2. Original mass service model.

In order to simplify the model [8], the incoming flows of variable cargo modules arriving from both sides of the border at the customs terminal are combined into one external source (0), which is allowed according to (Fig. 2). On the basis of Fig. 3, the transfer matrix (1) is presented, which allows determining the intensities of internal flows.

$$T = \begin{vmatrix} \lambda_0 & \lambda_1 & \lambda_2 & \lambda_3 & \lambda_4 & \lambda_5 & \lambda_6 & \\ 0 & P_{01} & 0 & 0 & P_{04} & 0 & 0 & \lambda_0 \\ 0 & 0 & P_{12} & 0 & 0 & 0 & 0 & \lambda_1 \\ 0 & 0 & P_{22} & P_{23} & 0 & 0 & P_{26} & \lambda_2 \\ P_{30} & 0 & 0 & 0 & 0 & 0 & 0 & \lambda_3 \\ 0 & 0 & 0 & 0 & 0 & P_{45} & 0 & \lambda_4 \\ 0 & 0 & 0 & P_{53} & 0 & P_{55} & P_{56} & \lambda_5 \\ P_{60} & 0 & 0 & 0 & 0 & 0 & 0 & \lambda_6 \end{vmatrix} \quad (1)$$

where P_{ij} is the probability of the demand leaving the i-th system and its arrival in the j-th system.

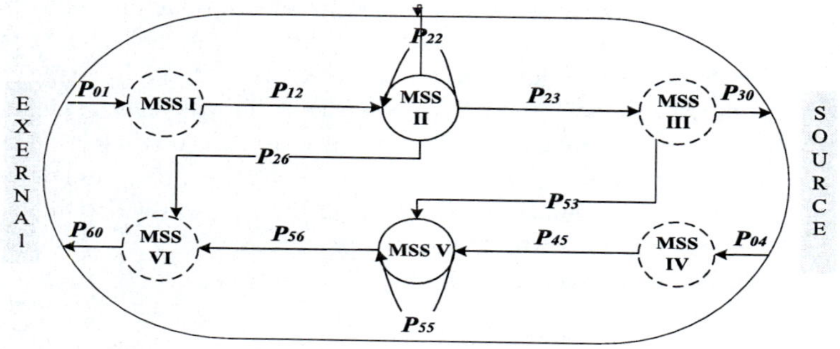

Fig. 3. The original mass service model.

Obviously, there is:

$$\sum_{j=0}^{M} P_{ij} = 1, \quad i = 0, 1, 2, \ldots, M. \quad (2)$$

The intensity of the incoming flow is

$$\lambda_0 = \lambda_1 + \lambda_4. \quad (3)$$

And $\lambda_1 \approx \lambda_4$, , that will ensure a stable regime. The limits of fluctuation of probability and probability values are 15–20%. These limits of the intensity of incoming flows are provided by the information flow that accompanies the cargo during the delivery process. On the basis of (2, 3), we obtain:

$$\begin{cases} P_{01} = 0.5 \\ P_{04} = 0.5 \\ P_{12} = 1 \\ P_{30} = 1 \\ P_{45} = 1 \\ P_{60} = 1 \end{cases} . \quad (4)$$

Let the total intensity of the flow of demands in the system be j, i.e. the average number of demands arriving in this system per unit of time in a stable mode. The average number of requests leaving the same system is equal to the average number of requests entering it. On the other hand, it is assumed that the probability that a request leaving system i will go to system j does not depend on the previous path of this request and the state of the network as a whole. Therefore, if there is a stable mode, then

$$\lambda_j = \sum_{i=0}^{M} \lambda_i P_{ij}, \quad j = 0, 1, 2, \ldots, M. \tag{5}$$

In this expression, denotes the total intensity of the source. Where do we get:

$$\begin{cases} \lambda_1 = \lambda_0 P_{01} \\ \lambda_2 = \lambda_1 P_{12} + \lambda_2 P_{22} \mapsto \lambda_2 - \lambda_2 P_{22} = \lambda_1 P_{12} \mapsto \lambda_2(1 - P_{22}) = \lambda_1 P_{12} \mapsto \lambda_2 = \lambda_1 P_{12}/(1 - P_{22}) \; [1] \\ \lambda_3 = \lambda_2 P_{23} + \lambda_5 P_{53} \quad [2] \\ \lambda_4 = \lambda_0 P_{04} \\ \lambda_5 = \lambda_4 P_{45} + \lambda_5 P_{55} \mapsto \lambda_5 - \lambda_5 P_{55} = \lambda_4 P_{45} \mapsto \lambda_5(1 - P_{55}) = \lambda_4 P_{45} \mapsto \lambda_5 = \lambda_4 P_{45}(1 - P_{55}) \; [3] \\ \lambda_6 = \lambda_2 P_{26} + \lambda_5 P_{56} \quad [4] \end{cases} . \tag{6}$$

Since λ_1 and λ_4 are known, we can use the system of equations to find the values of other intensities (the sequence of transformations is placed in square brackets). Having calculated the corresponding values of λ_i, it is possible, at the corresponding intensities μ_1, μ_2, μ_3,..., μ_6 (throughput capacity of service mechanisms) to obtain the main characteristics of the mass service network, such as the total time the application is in the system, the waiting time of the application in the i-th system, the average number of applications in the network, etc.

When receiving these characteristics, it is possible to increase or decrease the number of channels or the number of places in the queues of various mass service systems in order to improve the functioning of the network.

In the light of the above-mentioned structure of the mass service network on the example of the customs terminal, it is clear that the input intensity of the i-th system is equal to the absolute throughput of MSSi-1, i.e. λi = Ai-1.

The developed method of analysis of the mass service network provides an adequate response to situations arising during its operation and allows to adjust the value of some parameters and characteristics of the network (number of channels, service intensity, etc.) in order to improve the operation of the network and ensure the optimal process of servicing requirements [14].

In order to implement an analytical model of the customs terminal's operation when managing variable cargo modules, a customs terminal mass service network analysis system (SAMTMO) was developed in the Excel environment, which consists of two parts - a part for inputting output data and a calculation part.

After starting to execute the "Search for a solution" option of the Excel table processor, for some time (about a minute), the system will automatically select the optimal number of channels and cities in the waiting queues of all MSSs in the complex, and to the system's "Continue" or "Stop" message, you must respond " Stop" and press "OK". The result will appear on the screen.

As a result of the analysis of the obtained data, it is possible to state the fact that the average number of requirements in the network is determined by Little's formula and is 19 cargo modules, and the optimal structure of the customs cargo terminal will include:

- number of channels in MSS1 n1 = **1**; y MSS2 n2 = **1**; y MSS3 n3 = **1**;
- the number of places in the queue MSS2 l2 = **4**; MSS3 l3 = **4**;
- number of channels in MSS4 n4 = **2**; y MSS5 n5 = **1**; y MSS6 n6 = **2**;
- the number of places in the queue MSS5 l5 = **4**; MSS6 l6 = **4.**

5 Conclusions

The work of the cargo customs terminals, which is an element of the system of international multimodal transportation, has been studied. We can conclude that the structure of the research object is a mass service network. The goal of system analysis is to find ways to improve the efficiency of transport. A method of mass service network analysis has been developed, which provides an adequate response to situations that arise during its operation, and allows adjusting the value of some network parameters and characteristics (number of channels, service intensity, etc.) in order to improve the network and ensure an optimal service process requirements In order to implement an analytical model of the customs terminal's operation in the management of variable cargo modules, a system of analysis of the customs terminal mass service network (SACTMSN) was developed in the Excel environment, which consists of two parts - a part for inputting output data and a calculation part. Calculations of the mode of operation of the cargo customs terminal as a mass service system based on real examples showed ways to optimize the operation of the terminals.

The co-authors took part in all the research and writing of the article. All co-authors have equal contributions to the article's materials and conclusions.

References

1. Prokudin, G., Chupaylenko, O., Dudnik O., Oliskevych, O.: Development of vehicle speed forecasting method for intelligent highway transport system. East.-Eur. J. Enterp. Technol. **4/3**(100), 6–14 (2019)
2. Sharai, S., Oliskevych, M., Roi, M.: Development of the procedure for simulation modeling of interrelated transport processes on the main road network. East.-Eur. J. Enterp. Technol. **5/3**(101), 70–83 (2019)
3. Kunda, N., Lebid, V.: Interconnection of transport services while cargo transportation. J. Adv. Res. Law Econ. **10.8**(46), 2394–2406 (2019)
4. Prokudin, G., Chupaylenko, O., Dudnik, O., Prokudin, O., Dudnik, A., Svatko, V.: Application of information technologies for the optimization of itinerary when delivering cargo by automobile transport. East.-Eur. J. Enterp. Technol. **2**(3(92)), 51–59 (2018)
5. Apfelstädt, A., Dashkovskiy, S., Nieberding, B.: Modeling, optimization and solving strategies for matching problems in cooperative full truckload networks. IFAC-PapersOnLine **49**(2), 18–23 (2016)
6. Harris, I., Wang, Y., Wang, H.: ICT in multimodal transport and technological trends: unleashing potential for the future. Int. J. Prod. Econ. **159**, 88–103 (2015)

7. Shin, S., Roh, H.-S., Hur, S.: Characteristics analysis of freight mode choice model according to the introduction of a new freight transport system. Sustainability **11**(4), 1209 (2019)
8. Ritzinger, U., Puchinger, J., Hartl, R.F.: A survey on dynamic and stochastic vehicle routing problems. Int. J. Prod. Res. **54**(1), 215–231 (2015)
9. Danchuk, V., Bakulich, O., Taraban, S., Bieliatynskyi, A.: Simulation of traffic flows optimization in road networks using electrical analogue model. In: Murgul, V., Pukhkal, V. (eds.) EMMFT 2019. AISC, vol. 1258, pp. 238–254. Springer, Cham (2021). https://doi.org/10.1007/978-3-030-57450-5_22
10. Lebid, I., Medvediev, I., Eliseyev, P., Sakno, O.: A modelling approach to the transport support for the harvesting and transportation complex under uncertain conditions. In: CONFERENCE 2020, 26th Technical and scientific conference «Transport, ecology – sustainable development» Volume 977, EKO. IOP Publishing (2020)
11. Prokudin, G., Chupaylenko, O., Lebid, I., Luzhanska.: Logistics approach to the organization of unbalanced freight transportation in transport networks. In: 24th International Scientific Conference. Transport Means. Sustainability: Research and Solutions. PART I, pp. 22–26. Kaunas, Lithuania (2020)
12. Gryshchuk, O., Petryk, A., Yerko, Y.: Development of methods for formation of infrastructure of transport units for maintenance of transit and export freight flows. Technol. Audit Prod. Reserv. **1**(2(63)), 26–30 (2022)
13. Prokudin, G., Chupaylenko, O., Khobotnia, T., Remekh, I., Lyamzin, A., Kovalenko, O.: Optimizing unbalanced freight deliveries in transportation networks. East.-Eur. J. Enterp. Technol. **2/3**(116), 22–32 (2022)
14. Silantieva, I., Katrushenko, N., Kushym, B.: Ensuring effectiveness in handling the movement of goods and passengers by enhancing information and communication technologies. In: Proceedings of the 1st International Scientific Conference, Current Problems of Transport, Ternopil, pp.75–83 (2019)

Sustainable Supply Chain Management Tools

Margarita Išoraitė(✉)

Vilniaus Kolegija/Higher Education Institution, Didlaukio Str. 49, 08303 Vilnius, Lithuania
m.isoraite@vvf.viko.lt

Abstract. Suppliers not only ensure the smooth operation of the company but also influence the sustainability of the company's activities, products, and services. The company's sustainability is also evaluated in terms of its entire supply chain, so it is very important to choose responsible and sustainable suppliers. The article analyzes sustainable supply chain and supply chain management concepts, sustainable supply chain tools, and sustainable supply chain trends. To achieve the goal, the analysis of scientific literature and the method of comparative analysis are applied. Analysis of scientific literature that sustainable supply chain management reduces negative environmental impacts, and increases consumer satisfaction and competitiveness. Sustainable supply chain management and sustainable supply chain management indicators can improve sustainable supply chain management processes.

Keywords: Supply chain · Sustainable supply chain · Supply chain management

1 Introduction

Sustainable supply chain management is the oversight of the entire delivery of goods from one location to another. A supply chain is thus a business link. This includes the contribution of each business to the entire chain from production to production and final consumption. The main objective of supply chain management is to ensure that it is efficient and effective in order to reduce costs and ensure customer satisfaction. Adequate management skills and appropriate training related to supply chain management would be required for any organization to be successful.

Investigation Problem. One of the most important sectors in European countries is the transport sector. Transport and logistics play an important role in Lithuania. The main means of human communication is transport, which is why more and more attention is being paid to it. The supply chain plays an important role nowadays, especially in a sustainable supply chain. Therefore, the application of various practices such as supply chain management, and sustainable supply chain management plays an important role. The concept of supply chain can be heard in various countries. A sustainable supply chain is examined by authors such as Tundys [21], Sánchez-Flores et al. [17], Saeed et al. [16], Panigrahi [13], Bernal et al. [1], Osei et al. [12].

O. Prentkovskis et al. (Eds.): TRANSBALTICA 2023, LNITI, pp. 440–446, 2024.
https://doi.org/10.1007/978-3-031-52652-7_43

Scientific Novelty. When examining the field of sustainable supply chain, sustainable supply chain management measures, this is not yet a widely studied subject, very little is discussed in the scientific literature. Sustainable supply chain management tools are not widely explored in the concept of sustainable supply chain. Sustainable supply chain management is still considered a novelty in Lithuania.

The purpose of the article is to evaluate sustainable supply chain management tools.

Research Contribution. Nowadays, there are trends to reduce the negative impact of transport on the environment, so more attention is paid to a sustainable supply chain, its management and measures. This concept is explored through a sustainable supply chain and its benefits. As the business environment is constantly changing, it is important for the supply chain to adapt to disruptions seamlessly. It is important for the supply chain to have the ability to adapt quickly to disruptions. Therefore, it is essential that a coherent supply chain ensures a smooth process.

1.1 Sustainable Supply Chain and Supply Chain Management Concept

Kashem et al. [7] mentioned that the supply chain connects customers and suppliers. A sustainable supply chain management includes many business processes of the company, which include the management of orders to suppliers, the supply of necessary raw materials, supply management, planning the supply of manufactured products according to orders to the customer, transportation, and supply requests. A sustainable supply chain is an interconnected or interconnected network, processes, and activities, the path of which starts from the submission of an order or supply request, and continues through production, services that reach the end user, to analysis and reporting. A sustainable supply chain also includes logistics management, delivery, and removal of goods. A sustainable supply chain includes suppliers who take delivery and transportation services.

Sustainable supply chain and sustainable supply chain management concept comparison is presented in Table 1.

Yuan et al. [19] considered that a sustainable supply chain 4 also facilitates the transition to sustainable innovation within the organization, increasing the contribution of employees to the creation of sustainable innovation. Sustainable supply chain 4 provides knowledge about product and process innovations in the market. Sustainable Supply Chain 4.0 facilitates partnership processes and helps align the needs of different stakeholder groups. A sustainable supply chain 4 is becoming a priority for both internal and external business entities.

Lengyel et al. [8] mentioned that a sustainable supply chain includes processes from order receipt to order completion. A sustainable supply chain consists of manufacturers, logistics companies, wholesalers, and retailers. In a sustainable supply chain, they operate using information, human and other resources. In a sustainable supply chain, the flow of goods and services is managed to ensure maximum value for customers and gain a competitive advantage in the market, therefore it is important to avoid unnecessary costs and maintain optimal speed of processes.

Joshi et al. [15] opined that changing market dynamics and consumer needs are causing organizations to rethink and adapt existing processes.

Table 1. Sustainable supply chain and sustainable supply chain management concept comparison.

Author	Definition	Highlights
Tundys [21]	A sustainable supply chain involves the integration of social, economic, and environmental indicators	Effective management of logistics processes to improve flow profitability
Sánchez-Flores et al. [17]	Sustainable supply chain management covers social responsibility, environmental care, and economic feasibility in all functions of the supply chain	The balance that can exist between social responsibility, environmental care, and economic feasibility in all functions of the supply chain
Saeed et al. [16]	Organizations aiming for sustainability must transform their current supply chains so that sustainability goals are incorporated into their operations from purchase to distribution	The aim is to provide maximum value to all stakeholders and meet customer requirements in the area of sustainability
Panigrahi [13]	The concept of Sustainable Supply Chain Management (SSCM) is important for environmental, social, and corporate responsibility through economic performance	Environmental, social, and corporate responsibility
Bernal et al. [1]	SSCM indicators are usually presented in the literature	Environmentally friendly indicators have been set
Osei et al. [12]	Improvements in a sustainable supply chain can be seen as improvements in environmental, social and economic outcomes	Improvements in a sustainable supply chain

According to Luo et al., the long-term sustainability of SC is still uncertain and requires further research.

Mandičák et al. [10] mentioned that supply chain technologies will reduce costs and shorten material delivery times.

Brandenburg et al. [2] indicated that process orientation, customer focus, and collaboration are critical elements of a sustainable supply chain.

According to Chauhan et al. [4] SCM and Industry 4.0 apply technologies such as the Internet of Things (IoT), cloud computing, big data, artificial intelligence (AI), and blockchain.

Silva et al. [18] argue that a sustainable supply chain can provide strategic and transparent social, environmental, and economic dimensions to systemic inter-organizational coordination projects and business processes.

Gružauskas et al. [6] argue that a supply chain management system evolve into sustainable development – capacity, collaboration, and complexity management.

1.2 Sustainable Supply Chain Tools

A sustainable supply chain includes functions such as planning, sourcing, manufacturing, shipping, and returns. A sustainable supply chain is run by a team of supply chain managers. Optimum management of a sustainable supply chain depends on all process participants communicating well and sharing real-time information promptly to identify and resolve issues as effectively as possible.

Correia et al. [5] state that there are several tools for sustainable supply chain management: (i) provide a comprehensive approach to sustainability; consider three aspects of sustainability; ii) considers various governance-oriented functional/activity areas and specific actions (action area structuring) aimed at improving sustainability; (iii) is not limited to practices that need to be developed for supply chains and other stakeholders.

Reyes-Soriano et al. [14] mentioned that nowadays companies must promote and contribute to environmental sustainability in their production activities.

Tseng et al. [20] believed that sustainable supply chain management (SSCM) helps companies control costs and maximize economic performance.

1.3 Sustainable Supply Chain Trends

Much of the sustainability impact of products occurs in the supply chain, which cannot be directly controlled. Developing long-term relationships with suppliers and improving their sustainability, including health and safety, environment, and social and business ethics, is essential. All suppliers are expected to adhere to the standards of the code of conduct, and all those defined as "at-risk suppliers" are subject to regular sustainability audits and corrective action plans.

Budrys [3] analyzed that supply chain participants must be familiar with each other's business vision when cooperating and strategic goals that should be acceptable to each other. A collaborative strategy is a win-win for all parties – there are no losers, and there are partners who create a common value chain and deliver higher value to the end user. The researchers say that the supply chain is limited and focuses more on efficiency than on the end user. Therefore, in this context, it would be very wrong to focus only on your functions in product development. It is important to understand the continuity of the process and how the value chain changes at the next stage of the supply chain. It is necessary to constantly exchange information and cooperate. Authors yes also notice that more and more organizations are changing their orientation and approach to supply chain management from a broader point of view – value chain management.

It is important to emphasize that more and more companies are starting to switch to a sustainable business model. In such a large sector as logistics and trade, the principle of sharing significantly reduces the ecological footprint of companies, which is also in line with EU expectations and legislation. Constantly working with leading retail chains and thousands of customers, companies are looking for solutions to reduce the impact of their business on the environment. Therefore, the goal of every company is to help increase the sustainability of the supply chains of customers and business partners, saving time, the environment, and money.

Mikuckienė et al. [11] stated that production raw materials or products must be delivered by transport companies if the business is to operate. The supply chain starts from this process. Properly managing a sustainable supply chain and forming optimal production plans in the company help to avoid high operating costs and disruptions in production processes that may occur due to improper management of logistics processes taking place inside and outside the company.

Budrys [3] mentioned that from a supply chain management perspective, it can be argued that organizations focus on their immediate customer fulfilling orders on time. Given the rapid dynamics of the market, it is often observed that the flexibility of organizations' responses is rather limited due to the lack of knowledge about the end user.

Budrys [3] stated that a value chain management approach strengthens sustainability and collaboration, allows an understanding of the dynamics of demand, and understands who creates value for the end user. Continuous cooperation between suppliers allows to increase productivity and reduces demand forecast deviation and planning costs.

2 Conclusions

The approach to sustainability combines values, ongoing activities, strategic directions, and principles of sustainable (responsible) activities that contribute to the goals of sustainable development. Sustainable supply chain management principles are an inseparable and integral part of the company's long-term strategy, realized through the areas of economic impact, social responsibility, and environmental protection. The transport company is based on the following principles of responsible activity: Transparency - base activities on responsibility and zero tolerance for corruption; Respect - attention to the human and labor rights of employees; Social responsibility – application of measures that create long-term value within the company and for society; Reducing the impact on the environment - implementing preventive measures for environmental protection and ensuring their impact; Honest reporting - voluntary disclosure of the company's performance results to the public and interested groups. The transport company bases its activities on the goals of sustainable development, good practices of sustainable development and sustainability, and social responsibility.

By engaging with suppliers who are guided by values and meet sustainability standards, companies ensure that all supply chains have the same goals: a sustainable supply chain. In cooperation with suppliers, companies constantly improve the quality and safety level of their products and strive to create a supply chain that meets the requirements of suppliers. Suppliers who value sustainability also value quality.

References

1. Bernal, J.P.G., Castro, J.A.O., Mantilla, C.E.M.: The sustainable supply chain: concepts, optimization and simulation models, and trends. Revista Ingernieria (2020)
2. Brandenburg, M., Gruchmann, T., Oelze, N.: Sustainable supply chain management—a conceptual framework and future research perspectives. Sustainability **11**(24), 7239 (2019). https://doi.org/10.3390/su11247239
3. Budrys, N.: Vertės grandinės strategijos tobulinimo modelis tiekimo grandinės kontekste, 22-osios Lietuvos jaunųjų mokslininkų konferencijos "Mokslas – Lietuvos ateitis" teminė konferencija Proceedings of the 22nd Conference for Junior Researchers "Science – Future of Lithuania" EKONOMIKA IR VADYBA/ECONOMICS AND MANAGEMENT (2019)
4. Chauhan, S., Singh, R., Gehlot, A., Akram, S.V., Twala, B., Priyadarshi, N.: Digitalization of supply chain management with industry 4.0 enabling technologies: a sustainable perspective. Processes **11**(1), 96 (2023). https://doi.org/10.3390/pr11010096
5. Correia, E., Garrido-Azevedo, S., Carvalho, H.: Supply chain sustainability: a model to assess the maturity level. Systems **11**, 98 (2023). https://doi.org/10.3390/systems11020098
6. Gružauskas, V., Burinskienė, A.: Managing supply chain complexity and sustainability: the case of the food industry. Processes **10**(5), 852 (2022). https://doi.org/10.3390/pr10050852
7. Kashem, M.A., Shamsuddoha, M., Nasir, T., Chowdhury, A.A.: Supply chain disruption versus optimization: a review on artificial intelligence and blockchain. Knowledge **3**, 80–96 (2023). https://doi.org/10.3390/knowledge3010007
8. Lengyel, P., et al.: Development of the concept of circular supply chain management—a systematic review. Processes **9**(10), 1740 (2021). https://doi.org/10.3390/pr9101740
9. Luo, J., Huang, M., Bai, Y., Li, J.: Supply chain management during a public health emergency of international concern: a bibliometric and content analysis. Processes **11**, 713 (2023). https://doi.org/10.3390/pr11030713
10. Mandičák, T., Mésároš, P., Kanáliková, A., Špak, M.: Supply chain management and big data concept effects on economic sustainability of building design and project planning. Appl. Sci. **11**(23), 11512 (2021). https://doi.org/10.3390/app112311512
11. Mikuckienė, D., Šateikienė, D.: Tiekimo grandinės valdymo tobulinimas UAB "Home Group"- Improvement of supply chain management in UAB "Home Group", Mokslas ir edukaciniai procesai science and processes of education, pp. 73–86 (2022)
12. Osei, M.B., Papadopoulos, T., Acquaye, A., Stamati, T.: Improving sustainable supply chain performance through organisational culture: a competing values framework approach. J. Purchasing Supply Manag. (2023). https://doi.org/10.1016/j.pursup.2023.100821
13. Panigrahi, S.S., Bahinipati, B., Jain, V.: Sustainable supply chain management: a review of literature and implications for future research. Manag. Environ. Qual. **30**(5), 1001–1049 (2019). https://doi.org/10.1108/MEQ-01-2018-0003
14. Reyes-Soriano, F.E., Muyulema-Allaica, J.C., Menéndez-Zaruma, C.M., Lucin-Borbor, J.M., Balón-Ramos, I.D.R., Herrera-Brunett, G.A.: Bibliometric analysis on sustainable supply chains. Sustainability **14**, 13039 (2022). https://doi.org/10.3390/su142013039
15. Joshi, S., Sharma, M.: Sustainable performance through digital supply chains in industry 4.0 era: amidst the pandemic experience. Sustainability **14**, 16726. (2022). https://doi.org/10.3390/su142416726
16. Saeed, M.A., Ishfaque, W., Wolfgang, K.: Literature review of drivers of sustainable supply chain management, In: Jahn, C.K., Ringle, W., Christian, M. (ed.) Digitalization in Maritime and Sustainable Logistics: City Logistics, Port Logistics and Sustainable Supply Chain Management in the Digital Age. Proceedings of the Hamburg International Conference of Logistics (HICL), vol. 24, epubli GmbH, Berlin, pp. 159–184 (2017). ISBN 978-3-7450-4332-7. https://doi.org/10.15480/882.1488

17. Sánchez-Flores, R.B., Cruz-Sotelo, S.E., Ojeda-Benitez, S., Ramírez-Barreto, M.E.: Sustainable supply chain management—a literature review on emerging economies. Sustainability **12**(17), 6972 (2020). https://doi.org/10.3390/su12176972
18. Silva, W.H., Guarnieri, P., Carvalho, J.M., Farias, J.S., Reis, SAd.: Sustainable supply chain management: analyzing the past to determine a research agenda. Logistics **3**(2), 14 (2019). https://doi.org/10.3390/logistics3020014
19. Yuan, X.-M., Xue, A.: Supply chain 4.0: new generation of supply chain management. Logistics **7**, 9 (2023). 10.3390/ logistics7010009
20. Tseng, M.-L., Bui, T.-D., Lim, M.K., Lewi, S.: A cause and effect model for digital sustainable supply chain competitiveness under uncertainties: enhancing digital platform. Sustainability **13**(18), 10150 (2021). https://doi.org/10.3390/su131810150
21. Tundys, B.: Sustainable supply chain management – past, present and future. Research papers of Wroclaw University of Economics and Business 2020, vol. 64, nr 3. https://doi.org/10.15611/pn.2020.3.15

Research on the Relationship Between the Length of Superior Road Infrastructure and Foreign Direct Investment in the Slovak Republic

Martin Zuzaniak(✉) and Vladimír Konečný

Department of Road and Urban Transport, Faculty of Operation and Economics of Transport and Communications, University of Žilina, Univerzitná 8215/1, 010 26 Žilina, Slovakia
zuzaniak@stud.uniza.sk, vladimir.konecny@uniza.sk

Abstract. High-quality and functional transport infrastructure is one of the basic prerequisites for the rapid development of Slovak regions. Due to the previous problems with planned projects, Slovakia has major shortcomings in the construction of the highway networks, and its quality varies significantly between individual regions. The importance of road infrastructure for the development of regions and the attraction of foreign investments is currently only estimated. The aim of this research is to study this relationship and provide the answers. This study analyzes statistics related to the length of road transport infrastructure, especially highways and expressways, and foreign investments for individual regions in the horizon of the last 9 years before COVID-19 pandemic. The research considers the different size of the territory of the regions, so the density values of the superior road infrastructure in the regions are calculated and used. The results of the correlation analysis showed a strong dependence between the road infrastructure density and foreign direct investment across the study period in Slovakia. However, there was not only a difference in the strength of the relationship between regions, but also in the response time to the change in infrastructure. This finding demonstrates the importance and relevance of superior road infrastructure for the attraction of foreign investors and development of regions, which has its own specific progression for each region.

Keywords: Superior road infrastructure · Foreign direct investment · Correlation analysis · Slovak regions

1 Introduction

The economic growth and development of any country is greatly influenced by the quality of its transport infrastructure. The level of transport infrastructure is also reflected in other indicators of economic performance and plays a vital role in the decision of foreign investors to invest in a given region and country [1]. Therefore, questions regarding the measurement of transport infrastructure performance and its correlation with economic

O. Prentkovskis et al. (Eds.): TRANSBALTICA 2023, LNITI, pp. 447–456, 2024.
https://doi.org/10.1007/978-3-031-52652-7_44

growth are a topic of discussion in both academic and non-academic circles [2]. Transport infrastructure is seen as essential for economic development, but it´s impact on the region is much broader and therefore difficult to analyze using only quantitative indicators [3].

1.1 Previous Research in This Area

Transport improvements are likely to have a positive impact on a country's economic growth and this is achieved by improving productivity, reducing production costs and introducing technology [4]. Economic growth can on the contrary positively affect the transport network back through technological improvements and enhanced financial support [5]. There can be found several detailed related studies for different countries [6, 7]. Numerous empirical studies have found a long-run equilibrium relationship between economic development and the transport sector [8]. Using Granger causality, few studies identified a bidirectional relationship between the transport infrastructure and economic growth [8, 9]. Several studies suggest a positive impact of the road transport on economic growth [10, 11], but some studies have also identified the opposite impact of economic growth on the transport sector [12, 13]. The authors of another study detected cointegration between transport infrastructure and economic growth through autoregressive and vector model, which confirms the occurrence of long run equilibrium relationships between them [14]. As a condition for attracting foreign direct investments and generating additional economic growth, there is an urgent need to promote transport infrastructure development [14, 15].

However, especially in the developed countries of Western Europe, the level of superior road infrastructure is not increased by growth in the length, but by adding more lanes and introduction of intelligent transport systems [16]. Antonova et al. [17] investigated the impact of highway infrastructure density on economic development and GDP in the Czech Republic through correlation and regression analysis. Infrastructure is constantly expanding and upgrading due to the increasing number of users, which causes an increase in user utility in terms of reduction of travel times, increasing safety, meeting higher quality standards in informing infrastructure users and financing infrastructure development through increased state budget revenues via taxes and infrastructure charges [18].

2 Superior Road Infrastructure in Slovakia

In the context of road infrastructure, the research deals only with highways and expressways, i.e. superior infrastructure. This is because these networks are the main links and routes for the exchange of goods and labor mobility between regions, or in international road freight and passenger transport as part of international transport corridors for trucks and are therefore the most interesting for investors. High-quality and dense road infrastructure is a prerequisite for the development and progress of any region. However, the construction of highways has been a long-term problem in Slovakia. This is also why the 2 largest economic centers of the country, Bratislava and Košice, are still not permanently connected by highway. Figure 1 below shows the development of the superior road infrastructure in Slovakia between 2011 and 2019. The total length of motorways and expressways increased by 14.17% between 2011 and 2019, which is comparable to Hungary, but significantly less than in Poland and the Czech Republic, where the growth was 57% and 71% respectively over the period. Length alone is not the only important factor in terms of roads, quality is also a decisive element.

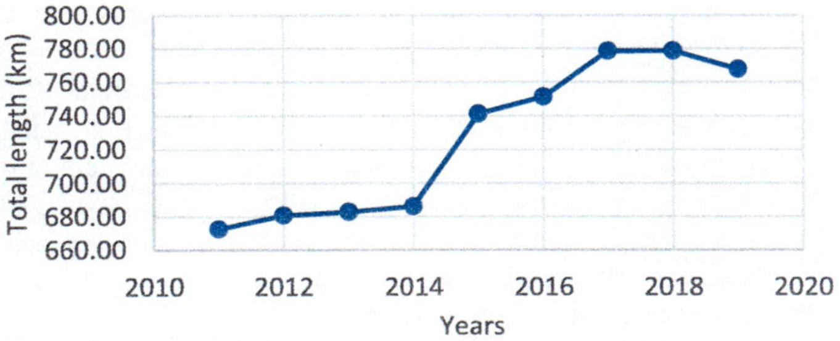

Fig. 1. Total length of highways and expressways in Slovakia between 2011–2019, source: authors according to [19].

The evolution of density of superior road infrastructure within individual regions in SR is shown in Fig. 2. As with the national trend, individual regions have seen small increases in the length and density of superior road infrastructure between 2011 and 2019. The most significant increase was recorded in central Slovakia in the Banská Bystrica (BB) and Žilina (ZA) regions and in the east in the Prešov (PO) region between 2014 and 2017. The Košice (KE), Trenčín (TN) and Trnava (TT) regions also recorded an increase. On the contrary, a decrease was registered in the Bratislava (BA) and Nitra (NR) regions, although it was not a big drop, it is still an undesirable situation. In the long term, the Bratislava region has the densest superior road infrastructure, which reaches incomparably higher values compared to other regions. Other parts of western Slovakia, namely the Trnava and Trenčín regions, have also maintained high density, but the Žilina region has moved closer to them. The Košice region has the lowest density and the Nitra region started to fall behind the others in the researched period.

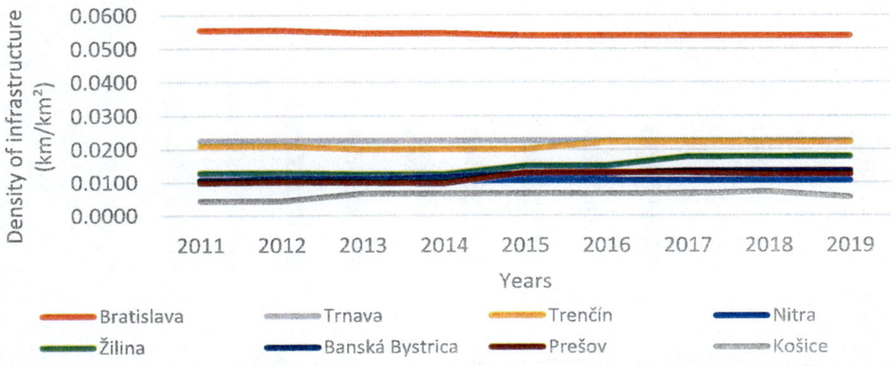

Fig. 2. Density of superior road infrastructure in Slovak regions between 2011–2019, source: authors according to [19].

3 Foreign Direct Investment

Foreign direct investment (FDI) is a major means of economic integration because it links economies by creating long-term stable international relations. FDI is the ownership interest of a company or government from a different country in a foreign project or company. FDI represents the decision of a business entity to buy or acquire a significant stake in a foreign enterprise with a view to expanding into a new territory [20]. Foreign direct investment also enhances the competitiveness and development of the national economy by transferring new technologies and managerial skills, increasing labor productivity and foreign trade [21]. The development of international economic activities and the inflow of foreign capital into economy through the liberalization of economic conditions allows foreign entities to be involved in the internal economy, which enables the recovery of the economy by increasing the share of foreign capital, but on the other hand underlines the need for the establishment of a fair legal framework on a contractual basis for the protection of investors and their investments [22].

The degree of development of the regions within the SR is determined, for example, by the quality of the superior road infrastructure in the form of highways and expressways for motor vehicles, the inflow of foreign direct investment and the related wage level and unemployment rate of the population. In connection with more developed infrastructure (highways), proximity to trans-European transport networks and goods outlets, several foreign investors prefer the west of Slovakia. In the Žilina, Trnava and Bratislava regions were formed automotive clusters around KIA, PSA Peugeot, Volkswagen. In the Trnava and Nitra regions, an electronics cluster was formed around Samsung and Sony, which was complemented by AU Optronics in Trenčín region [23].

The development of foreign direct investment in Slovakia over the period 2011–2019 is shown in Fig. 3. The amount of foreign investment within Slovakia between 2011 and 2019 increased by 34.28%, which is more than a third of the original value at the beginning of the period. Inflation may have played a partial role in this rise, but it is nonetheless a significant increase and a positive trend for the country. A more significant increase was recorded particularly in the second half of the period, which may portend a continuation of this trend in the years to come, although as we now know, this was followed by the COVID-19 pandemic and a slowdown of the whole economy.

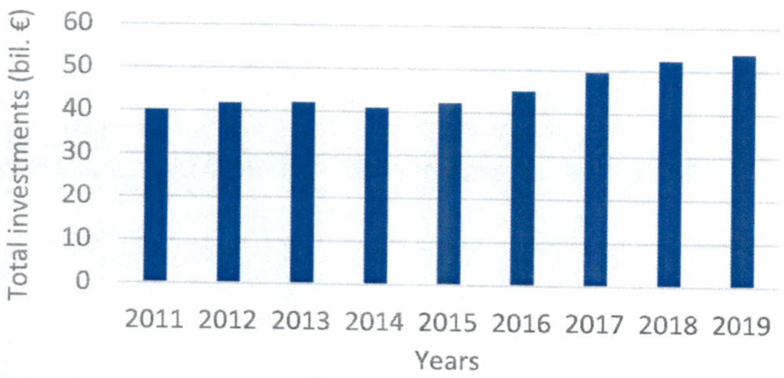

Fig. 3. Total foreign direct investment in Slovakia between 2011–2019, source: authors according to [24].

Figure 4 tracks the share of foreign direct investment by each region for the year 2019, when values in many cases reached their peak. This year, as always before, the BA region had the largest share with 68.79%, and the ZA region also reached its maximum values, ranking 2nd in share. The decrease in values compared to 2018 was recorded by Trenčín and Trnava region, and the smallest part of foreign investments includes Prešov region with only 1.42%. These results show big differences between the individual regions, mainly between the Bratislava region and other parts of Slovakia. Main reasons are the capital city of Bratislava, where many foreign companies have their headquarters and its position in the center of Europe, where many roads cross and nearness of other countries. Investors can also find a lot of educated workforce and dense traffic infrastructure there. One of the preconditions for fast building of highways is the simplicity of a terrain, which the western part of Slovakia is known for. In the regions with smallest share of foreign investment, these are also the reasons for its bad situation.

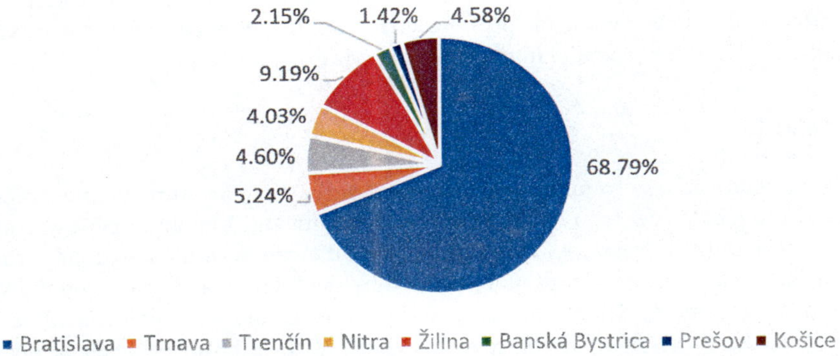

Fig. 4. Share of the foreign direct investment for each region of Slovakia in 2019, source: authors according to [24].

4 Methodology

This study analyzes statistics related to the length of road infrastructure, especially highways and expressways, and foreign direct investment for whole country of Slovakia and also individual regions in the horizon of the last 9 years before COVID-19 pandemic through correlation analysis. The data were taken from the Statistical Office of the Slovak Republic and the National Bank of Slovakia. The last year studied is 2019, because in 2020 came the COVID-19 pandemic, which significantly affected all sectors and so could bias the researched results.

Correlation analysis and Pearson correlation coefficient are used to find the strength of the relationship between road infrastructure and foreign direct investment. Using correlation analysis, we wanted to determine dependence between these two variables and find out if a change in one variable affects the other variable. Pearson Product-Moment Coefficient given in formula (1) does not depend on the dimension in which the variables have been measured and is therefore a dimensionless coefficient with no bounds. The correlation coefficient reaches values in the interval <−1,1>. Positive values of the correlation coefficient indicate a direct dependence, negative values an indirect dependence and the value 0 implies independence between variables [25].

$$r_{xy} = \frac{n * \sum xy - \sum x * \sum y}{\sqrt{(n * \sum x^2 - (\sum x)) * (n * \sum y^2 - (\sum y))}}. \tag{1}$$

When calculating the correlation coefficient, the length of the infrastructure was divided by the total area for each region. This has been done to allow for cross comparison as each region is different in size and therefore the length of road infrastructure required will vary. This will give the length of road infrastructure per unit area of the region and is therefore comparable for all regions. In this case, FDI was not divided by the number of inhabitants for each region. However, such a calculation was also carried out and the results were very similar in both cases, thus this does not have a major impact on the final conclusions. Infrastructure and its changes can have different impact and, in particular, time effects on other economic factors for each region. Therefore, a calculation was also carried out to shift the potential impact time of infrastructure on FDI.

5 Results

Using the methodology from the previous chapter, the Pearson correlation coefficient between the length (density) of the road infrastructure and foreign direct investment was calculated for Slovakia. As mentioned above, the values of the length of superior road infrastructure for each region are divided by the total area of that region, giving a comparable value of kilometers of superior infrastructure per kilometer of region territory. Based on this, it is possible to assess the actual relationship between the density of the overlying infrastructure and the interest of foreign investors in a given location. As a first step, the correlation coefficient was calculated considering all regions of SR for each year between 2011–2019. The results of the calculation are shown in Table 1.

The resulting coefficient values are very similar over the whole period ranging in a small interval between 0.93–0.95, which is indicative of a strong direct relationship between the variables under study. Therefore, throughout the observation period, the length and the density of the superior road infrastructure had a large connection with the inflow of foreign investment into the regions and the amount of FDI increases with the growing length of the infrastructure. Since all values are not equal to 1, there are still some differences between regions. In Žilina and Košice region, even though the road transport infrastructure is not as dense, the foreign direct investment is relatively large. While in Žilina the difference is not that big, Košice region has average results in FDI, but the lowest density of superior road infrastructure out of all regions. This could be caused by the second largest economic center in the SR municipality of Košice located in this region, which can distort the results. A denser superior road infrastructure therefore increases the attractiveness for both multinational and domestic companies and motivates them to invest more. Better connectivity reduces costs and promotes efficient supply chain management, while poor transport infrastructure is detrimental to regional competitiveness. A second calculation was also carried out examining the relationship between the change in the length of superior road infrastructure and foreign direct investment for the period 2011–2019 for each Slovak region separately. The results in the form of Pearson's correlation coefficient can be found in Table 2.

Table 1. Pearson correlation coefficient considering all Slovak regions for each year.

Year	Values of the correlation coefficient r between the density of superior road infrastructure and foreign direct investment
2011	0.93182286
2012	0.93279383
2013	0.94090037
2014	0.93858608
2015	0.93796135
2016	0.93236218
2017	0.93681981
2018	0.94222265
2019	0.93096181

Table 2. Pearson correlation coefficient between the change in the superior road infrastructure and foreign direct investment for each Slovak region between 2011–2019.

Region	SR	BA	TT	TR	NR	ZA	BB	PO	KE
r	0.87	−0.67	0.76	0.74	−0.47	0.86	0.24	0.898	−0.07

In this case, the results are not as clear-cut as they were before. This is understandable given the data in Fig. 2 and 3, as the length of infrastructure grew in bursts and not everywhere, while foreign investment grew more evenly across the whole country. For whole Slovakia, this coefficient reaches value 0.87, indicating a direct strong dependence. A similar result was observed in 4 regions, namely Trenčín, Trnava, Žilina and Prešov, where the growth of the superior road infrastructure was recorded over the researched period. In Bratislava and Nitra region the result was the opposite in the form of medium indirect dependence. This result was expected because of decrease in the length of highways and expressways in these regions. In the remaining 2 regions, values indicating weak, or no dependence were obtained. In general, FDI grows with the rise in the length of the superior road infrastructure, especially if the growth is more pronounced, but the situation and values are not the same in every region.

The impact of transport infrastructure on FDI may not show up in the same year and therefore it is necessary to examine this relationship even at a certain time lag, namely 1 and 2 years. There is not much change in the results for Table 1 when examining the possible shift in the impact of transport infrastructure on FDI in the case of the density of superior road infrastructure in the SR. However, with the impact of a change in the density of regional infrastructure on FDI when shifted by 2 years, the values of the correlation coefficient across regions in Table 3 reach comparable to slightly higher values than in Table 2 without the shift. While in some regions such as Košice, Banská Bystrica and Nitra the correlation values have become significantly stronger, in Trenčín and Prešov region, the correlation values have vastly decreased. This may suggest a different speed of the impact of the change in the length of the superior road infrastructure on FDI, which may occur faster in some regions than in others and may cause a different effect for different parts of the country.

Table 3. Pearson correlation coefficient between the change in the superior road infrastructure and foreign direct investment for Slovak regions between 2011–2019 with a lag of 1 and 2 years.

Region	SR	BA	TT	TR	NR	ZA	BB	PO	KE
$r + 1$	0.96	−0.66	0.34	0.31	−0.58	0.90	0.97	0.27	−0.17
$r + 2$	0.97	−0.88	−0.67	−0.04	−0.67	0.95	0.89	−0.10	−0.71

6 Conclusion

Transport infrastructure is one of the fundamental pillars of a country's economy. In the case of Slovakia, road transport is the dominant mode of transport, therefore a high quality and dense road infrastructure is essential for the development of the country. Potential investors are particularly interested in connections by highways and expressways, as these networks can meet their logistical requirements, so these types of roads were examined in this study. The researched period is 2011–2019 due to the arrival of the COVID-19 pandemic in 2020, which has changed the economic market to a great extent. During the studied period, there was an increase of 14.17% in the length of superior road infrastructure in Slovakia, which is lower compared to the neighboring countries and the government's intentions. Moreover, this growth was not recorded in all regions or across the entire study period. Foreign direct investment is an important part of the country's market and linkages with foreign economies, providing employment opportunities and bringing the possibility of economic growth. Between 2011–2019, the amount of FDI in the SR increased by more than a third, and this growth was recorded across all regions. However, even this growth has not been uniformed and there have been differences between regions.

Through the Pearson correlation coefficient, a strong direct relationship between the length (density) of the superior road infrastructure and foreign direct investment has been found. This means the strong connection between the density of superior road infrastructure and amount of foreign investment a given region will receive. As the length of highways and expressways increases, the willingness of foreign entities to invest finance in a given region increases at a similar rate. The calculations for each region, with exceptions, also confirmed an increase in FDI when the length of the superior road infrastructure increased over the period 2011–2019. The exceptions are regions where a decrease in the length of road infrastructure, but an increase in FDI occurred. In this case, the location of the 2nd largest city in the Košice region and the quality of the infrastructure, which was not considered, may play a role. Extension research has also confirmed the theory that a change in the density of motorways and expressways in individual regions can bring about an impact on FDI with a time lag, namely by 2 years. However, the actual magnitude of the impact and the time lag is different for each region.

Based on these results, especially in less developed regions such as the east of Slovakia (Prešov, Košice) or the south of central Slovakia (Banská Bystrica, Nitra), it is necessary to try to increase the connectivity and quality of the superior road infrastructure. Improved transport infrastructure increases regional accessibility, reduces transport costs, and makes the region more attractive to investors. More companies and jobs are created in locations with better connections to major agglomerations. Quality transport infrastructure also increases the competitiveness of local businesses and their productivity. Highways can be a decisive factor for making a large investment when choosing a destination country or region with a similar business environment. However, they may not be sufficient if the quality of the business environment is poor. The availability of skilled human capital is another aspect that enhances the competitiveness of a region, and this is also why regions with the largest cities and sufficient labor force may attract more FDI than the density of their superior road infrastructure would suggest. This study can serve as a basis for addressing the appropriateness and need for highways and expressways for each region and possible further follow-up studies that would expand on this issue. Future research will continue with the comprehensive assessment of the positive externalities from road transport.

Acknowledgement. This publication was realized with support of Operational Program Integrated Infrastructure 2014–2020 of the project: Innovative Solutions for Propulsion, Power and Safety Components of Transport Vehicles, code ITMS 313011V334, co-financed by the European Regional Development Fund.

References

1. Sdoukopoulos, A., Pitsiava-Latinopoulou, M., Basbas, S., Papaioannou, P.: Measuring progress towards transport sustainability through indicators: analysis and metrics of the main indicators initiatives. Transp. Res. Part D Transp. Environ. **67**, 316–333 (2019). ISSN 1361-9209. https://doi.org/10.1016/j.trd.2018.11.020
2. Skorobogatova, O., Kuzmina-Merlino, I.: Transport infrastructure development performance. Procedia Eng. **178**, 319–329 (2019). https://doi.org/10.1016/j.proeng.2017.01.056
3. Filčák, R., Rochovska, A., Hornak, M.: Evaluation of Slovakia's R1 expressway enhancement impacts on local socio-economic development: expert panel approach. Geografie **126**, 29–53 (2021). https://doi.org/10.37040/geografie2021126010029
4. Beyzatlar, M.A., Karacal, M., Yetkiner, I.H.: Granger causality between transportation and GDP: a panel data approach. Transp. Res. Part A Policy Pract. **63**, 43–55 (2014)
5. Yu, N., Jong, M.D., Storm, S., Mi, J.: Transport infrastructure, spatial clusters and regional economic growth in China. Transp. Rev. **32**, 3–28 (2012)
6. Deng, T.: Impacts of transport infrastructure on productivity and economic growth: recent advances and research challenges. Transp. Rev. **33**(6), 686–699 (2013). https://doi.org/10.1080/01441647.2013.851745

7. Nihayah, D., Kurniawan, G.: Impact of road infrastructure and foreign direct investment to asean economy. Econ. Dev. Anal. J. **10**(2), 233–242 (2021). https://journal.unnes.ac.id/sju/index.php/edaj/article/view/43193
8. Sahoo, P., Dash, R.: Economic growth in South Asia: role of infrastructure. J. Int. Trade Econ. Dev. **21**, 217–252 (2012). https://doi.org/10.1080/09638191003596994
9. Pradhan, R.P, Bagchi, T.P.: Effect of transportation infrastructure on economic growth in India: the VECM approach. Res. Transp. Econ. **38**(1), 139–148 (2013). ISSN 0739-8859. https://doi.org/10.1016/j.retrec.2012.05.008
10. Saidi, S.: Impact of road transport on foreign direct investment and economic growth: empirical evidence from simultaneous equations model. E3 J. Bus. Manag. Econ. **7**, 064–071 (2016). https://doi.org/10.18685/EJBME(7)2_EJBME-16-012
11. Poliak, M., Kubáňová, J., Zhuravleva, N.A., Jaśkiewicz, M., Lakhmetkina, N.: Identification of the impact of transport performance on the economy of particular area. LOGI Sci. J. Transp. Logist. **14**(1), 3923, 192–202 (2023). https://doi.org/10.2478/logi-2023-0018
12. Hakim, M., Merkert, R.: The causal relationship between air transport and economic growth: empirical evidence from South Asia. J. Transport Geogr. **56**, 120–127 (2016)
13. Maparu, T.S., Mazumder, T.N.: Transport infrastructure, economic development and urbanization in India (1990–2011): Is there any causal relationship? Transp. Res. Part A Policy Pract. **100**, 319–336 (2017)
14. Pradhan, R.P., Norman, N.R., Badir, Y., Samadhan, B.: Transport infrastructure, foreign direct investment and economic growth interactions in India: the ARDL Bounds testing approach. Procedia Soc. Behav. Sci. **104**, 914–921 (2013). ISSN 1877-0428. https://doi.org/10.1016/j.sbspro.2013.11.186
15. Muvawala, J., Sebukeera, H., Ssebulime, S.: Socio-economic impacts of transport infrastructure investment in Uganda: insight from frontloading expenditure on Uganda's urban roads and highways. Res. Transp. Econ. **88**, 100971 (2021). ISSN 0739-8859. https://doi.org/10.1016/j.retrec.2020.100971
16. Gnap, J., Varjan, P., Ďurana, P.: Research on the relationship between freight transport and transport infrastructure in selected European countries. Transp. Proc. **14**, 63–74 (2019)
17. Antonova, B., Hyrslova, J., Drahotsky, I.: Highway infrastructure in the Czech Republic and its impact on GDP growth. In: SGEM 2015: Political Science, Law, Finance, Economics and Tourism, vol. 3, pp. 357–364 (2015)
18. Jánošková, K., Král, P.: Network Externalities in Transport, Conference: 2nd International Conference on Management Innovation and Business Innovation (ICMIBI 2014), vol. 44, pp. 3–8 (2014). ISSN 2251-3051. https://doi.org/10.5729/lnms.vol44.3
19. Statistical Office of the Slovak Republic (SOSR). https://datacube.statistics.sk/. Accessed 28 July 2023
20. Hayes, A.: Direct Foreign Investment (FDI): What It Is, Types, and Examples (2023). https://www.investopedia.com/terms/f/fdi.asp. Accessed 28 July 2023
21. Gaspareniene, L., Kliestik, T., Sivickiene, R., Remeikiene, R., Endrijaitis, M.: Impact of foreign direct investment on tax revenue: the case of the European Union. J. Competitiveness **14**, 43–60 (2022). https://doi.org/10.7441/joc.2022.01.03
22. Ministry of Finance of the Slovak Republic (MFSR). https://www.mfsr.sk/sk/medzinarodne-vztahy/medzinarodna-ochrana-investicii/. Accessed 28 July 2023
23. Slovak Investment and Trade Development Agency (SARIO). https://www.sario.sk. Accessed 28 July 2023
24. National Bank of Slovakia (NBS). https://nbs.sk/statisticke-udaje/statistika-platobnej-bilancie/priame-zahranicne-investicie/. Accessed 28 July 2023
25. Konečný, V.: Economic analysis in road transport and freight forwarding - tutorials for exercises. EDIS, Žilina (2018). ISBN 978-80-554-1470-6

Visualization Creation of the Klaipeda Seaport for the Navigational Simulator

Arvydas Jankauskas[1(✉)], Nijolė Batarlienė[2], and Vytautas Dubra[1]

[1] Lithuanian Maritime Academy, I. Kanto str. 7, 94221 Klaipeda, Lithuania
{a.jankauskas,v.dubra}@lajm.lt

[2] Vilnius Gediminas Technical University (VILNIUS TECH), Plytines 27, 10105 Vilnius, Lithuania
nijole.batarliene@vilniustech.lt

Abstract. Ensuring safety in seaport navigation is of paramount importance for the protection of lives, accident prevention, and environmental preservation. This paper highlights the significance of regular training and qualification upgrades for port personnel, pilots, and mariners to enhance their knowledge and understanding of navigational safety procedures, local regulations, best ship handling practices, and situational awareness. It emphasizes the importance of realistic and up-to-date navigation simulations, as outdated data and inaccurate simulation results can lead to erroneous conclusions. The study focuses on the Klaipeda Seaport and explores its navigational challenges. It also delves into the development of a detailed simulation model using advanced tools and methodologies. The effectiveness of the simulator in training, qualification improvement, and research is showcased, underscoring its contribution to navigational safety and seaport development.

Keywords: Simulation · Maritime · Safety

1 Introduction

The safety of seaport navigation is crucial for accident prevention, safeguarding lives, and preserving the environment. Regular training and qualification upgrades are essential for port personnel, pilots, and mariners to enhance their expertise in navigational safety procedures, local regulations, best ship handling practices, and situational awareness. Accurate simulation and modeling of the navigational environment are crucial elements in achieving these goals. Maritime safety is influenced by several factors that can to some extent affect the safety of ships, people, property and the environment [1].

Safety is often defined as independence from unacceptable risk. Since its establishment, the International Maritime Organization (IMO) has been working to reduce risks at sea through the implementation of specific legal measures. The main IMO document focusing on this objective is the 1974 International Convention for the Safety of Life at Sea, as amended (SOLAS). SOLAS Chapters provide the necessary mitigation measures by identifying and addressing a number of hazards that must be addressed in

O. Prentkovskis et al. (Eds.): TRANSBALTICA 2023, LNITI, pp. 457–466, 2024.
https://doi.org/10.1007/978-3-031-52652-7_45

order to ensure the safe operation of ships at sea. This document further discusses three specific risks mitigated by SOLAS: (a) risks related to structural integrity and stability; (b) fire hazard and (c) navigational hazard. The purpose of this article is to analyze the options that ensure the reduction of signs of the third risk by implementing high accuracy data-based models.

Regular data updates, data accuracy and monitoring real environment changes are essential practices to maintain the integrity of any data. Using outdated data in any research project, including simulations carried out with a stimulator, can indeed lead to significant issues and potentially produce inaccurate or misleading conclusions.

Nothing stands unchanged. The activity of the port is constantly intensifying, the sizes of ships are increasing. Port infrastructure changes according to the changing situation. Generally, the manufacturer modeled navigation areas in simulators, not always accurately reflecting the latest local changes or providing a detailed visual representation. Simulator managers play a vital role in maintaining and updating the latest changes in port infrastructure. The goal of the simulator manager is to create a virtual environment that closely mirrors the real port infrastructure and conditions. The reliability of computer simulation depends on own ship models, virtual training area, etc. [2].

The best and relatively cheapest way to ensure the safety of navigation in the port is to have a properly modelled, realistic navigation environment, to achieve foreseen objectives. New technologies have made modeling very cost-effective and attractive not only for teaching but also for research purposes [3].

2 Navigational Conditions in the Klaipeda Seaport

Klaipeda seaport serves as a vital hub for shipping and cargo operations. As the northernmost ice-free port in the Baltic region, it has a strategic advantage in the field of international trade and maritime transportation. A multimodal, universal, deep-water seaport can accommodate vessels 400 m in length, up to 59 m in width and with a maximum draught of 13.8 m. [4, 5]. The port accepts around 6,000 ships each year, cargo turnover is up to 36 million tons every year [6].

However, the port faces unique navigational challenges owing to its geographical location. West winds and sea cross current dominate in the port of Klaipeda, the navigation situation for ships arriving the port is quite difficult. The most difficult places from a navigation point of view are the approach channel, port gates, inner navigation channel from buoy No. 5 to buoy No. 9 and Malku Bay (see Fig. 1) that require careful manoeuvring by vessels within the port.

The access channel begins 0.7 nautical miles from the Klaipeda seaport gate. It is 150 m in width and 15.5 m in depth. Ships entering the port are influenced by the wind, swell and current the direction of which depends on the direction of wind. It is very complicated place for ships leaving the port, as ships making the turn of 23 degrees at the port gates encounter extra external influences of swell and cross current (see Fig. 1).

Vessels passing the port gates get considerable drift depending on wind and current, as the shipping channel turns at 23 degrees and the current reaches 1.5 times higher speed due to narrowing of the channel. The distance between the heads of the port breakwaters is 235 m, but the navigation channel is only 150 m in width.

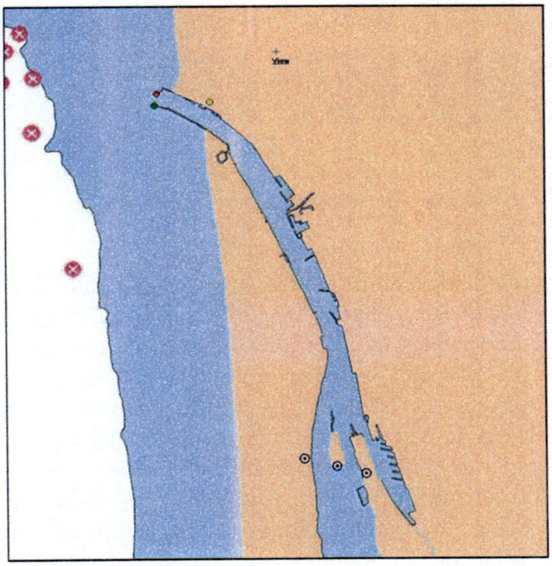

Fig. 1. The Klaipeda Seaport chart view.

The length of inner channel from buoy No. 5 to buoy No. 9 is one nautical mile with the traffic lane of 250 m in width and 15 m in depth. The biggest challenge for proceeding ships is the side wind, speed limit, 25 degrees bend at the buoy No. 9 and the vessels interaction on vessels staying along berths (see Fig. 1).

Inner channel from buoy No. 9 to buoy No. 19 is 2 nautical miles in length. In the narrowest place is 150 m in front of quay No. 62. The channel turns 15 degrees at buoy No. 19 (see Fig. 1).

The width of the channel from the western edge to the pier of the Central Klaipeda Passenger Cargo Terminal is 220 m, and to Klaipedos Smelte Container Terminal quays Nos. 82 – 86 are 330 m. The turning basin 800 m in length and 600 m in width is at Klaipedos Smelte.

The entrance into the Malku Bay is 90 m wide. The inner basin is dredged up to 14,5 and can safely accommodate Panamax and Post-Panamax vessels type (see Fig. 1). The Klaipeda State Seaport Authority now is implementing the project of the navigation channel reconstruction and deepening. During the reconstruction the port entrance direction will be changed 10 degrees to the North to minimize the turn in the entrance. The Klaipeda seaport channel will be dredged to 17.0 m to ensure the competitiveness of the port.

3 Tools and Methodology

3.1 Model Wizard - Scene Editor

Model Wizard - Scene Editor software tool and modelling sequence (see Fig. 2) were used to create Klaipeda seaport detail area.

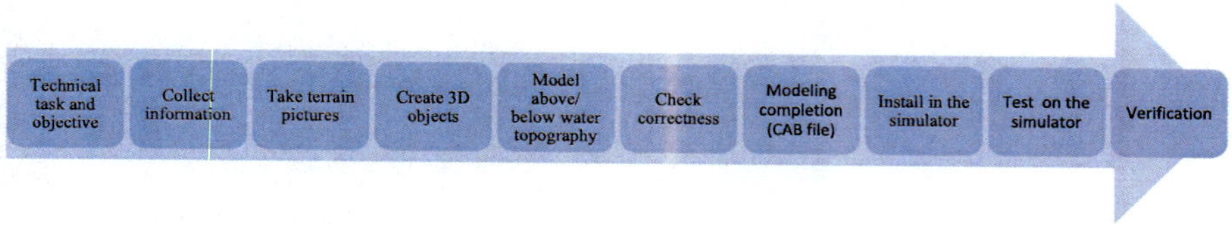

Fig. 2. Modelling procedure sequence [7].

The Model Wizard - Scene Editor software tool is a powerful solution for generating accurate spatial models of the Klaipeda seaport area for the Navi-Trainer simulators (NTPro). This software enables the creation of a realistic navigational environment, by utilizing electronic chart data, satellite imagery of the land relief earth's terrain, and accurate photos of the area. The resulting simulator provides training and research opportunities that closely imitate real-world conditions, facilitating effective learning and preparation. In order to run the simulator on a typical PC, the mathematical model used in the simulator must meet the requirements of both realism and computational speed [8]. The ship control navigation training should be conducted under conditions that are close to real [7].

At the beginning of an area modelling, it is necessary to determine the scene technical task assignment, it is necessary to have information containing boundary of the area to be modelled, boundaries of levels of detail, a list of required navigational charts and information on piers, which will be used for training in the ship mooring and the purpose of use. All required information should be collected and represented in the form of paper charts and diagrams or in the digital form.

The purpose of use Klaipeda port area model is to conduct pilots, VTS operators, seafarers training and upgrading of qualification, specific ship's type handling and research. These tasks require very detail modelling to feel a sense of reality. The four important elements ensure realism: very responsible modelling of visual, radar, electronic chart views and sensors interaction.

The most important indicator of modelling correctness is correct modelling of scene sensors components.

Sensor component – interconnected set of polygons designed for simulation of physical interaction of simulator scene elements, i. e. vessels, and other objects with terrain: mooring operations, bank/ shallow effect, ship-to-ship interaction [7].

To optimise software algorithms for the calculations of interactions, the sensor component should represent a set of polygons with tied apexes and positive values of normal. The accuracy of the solution improves as the number of elements (and nodes) increases (see Fig. 3).

Topographic surfaces can be represented as a vector triangular irregular mesh or as a raster (high-resolution bitmap). FEM is used as a numerical tool to perform abstraction to obtain topographical coordinates of terrain points [9].

From automobiles, aircraft, marine structures, bridges, highways, and high-rise buildings, FEM has revolutionized scientific modelling and engineering design [10]. Finite elements are usually formulated to establish sufficient conditions for convergence and good accuracy [11].

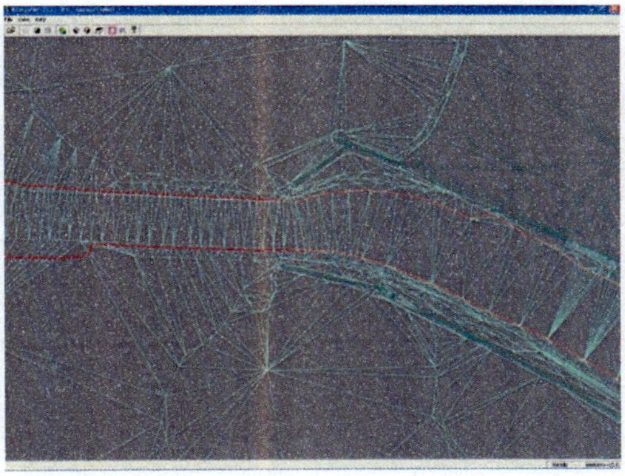

Fig. 3. Klaipeda seaport entrance sensor view.

FEM involves breaking down a system into smaller, simpler pieces called finite elements, connected by nodes, and obtain an approximate solution, which are then analysed using mathematical equations to simulate the behaviour of the entire system [12].

FEA was developed in the 1950 in the Aerospace Industry by Boeing and Bell Aerospace in the US and Rolls Royce in the UK. One of the earliest applications of FEM was in the aerospace industry. Engineers used the technique to analyse the behaviour of aircraft structures under different loading conditions.

3.2 Navigational Simulator NTPro 5000

The Transas NTPro 5000 simulator, equipped with Full Mission Bridge simulators, plays a crucial role in training, qualification improvement, and research. It offers various bridge configurations, visualizations, and ship control units, allowing for comprehensive and realistic simulations [13]. The simulator's high-quality visual representations and accurate ship controls provide an immersive training experience for pilots, VTS operators, and seafarers. This enables them to develop the necessary skills and familiarity with the specific navigational challenges encountered in the seaport.

The Full Mission Bridges simulator Transas NTPro 5000 at the Lithuanian Maritime Academy (LMA) consists of six separate bridges. Two bridges are equipped with Full Mission Bridge conning station with 150- and 200-degrees horizon visualisation and full set of real ship controls (see Fig. 4).

Four bridges are equipped with 120-degrees horizon visualization and with simplified ship control units.

Two Instructor places have full control and monitoring capability of each simulator navigation bridge.

Fig. 4. Bridges with 200 (on the left)- and 120-degrees (on the right) horizon visualization respectively.

4 Klaipeda Seaport Exercise Area Development

The development of a highly detailed and accurate simulation environment for the Klaipeda seaport exercise area involved collaboration between specialists from the Lithuanian Maritime Academy and Transas. Extensive data collection was conducted, including paper and electronic charts, satellite photos, and precise GPS measurements. The Model Wizard - Scene Editor tool and AutoCAD software were utilized to accurately position and model visual objects and navigational features within the exercise area.

The following should be carefully modelled:

1. Above the water part (terrain visualisation, geographical positions and boundaries of berths', terminals', jetties').
2. Underwater part (seabed, depths, depth contours, depth lines, depths along berths).

LMA and Transas researchers developed the Klaipeda seaport exercise area with 3D visualization in the end of 2012. The size of the area is 36.2 × 54.0 nautical miles with the following coordinates: NE corner 56°08,00'N 21°31,00'E and SW corner 55°14,00'N 20°26,00'E (see Fig. 5).

In the area were assigned parts which should satisfy the High Level of Detail to conduct the tasks requiring data and interaction precision, that includes manoeuvring close to the coastline and execution of mooring operations (see Fig. 5).

Klaipeda seaport shipping area (scene) was developed using the Model Wizard – Scene Editor tool based on electronic chart data, bathymetric data, satellite data on the earth terrain (see Fig. 6).

2D scene modelling was created using paper and electronic charts provided by Hydrographic Office of the Lithuanian Maritime Safety Administration.

3D modelling requires exact terrain pictures. Pictures step-by-step were made with precise GPS position and direction/ bearing (see Fig. 7). Sailing on the boat along all port area were made 1,500 photos.

The 3D visualization was created using high-quality and accurate positioning photos of the port terrain and using the AutoCAD software all 3D visual objects were placed in their exact position using Model Wizard-Scene Editor Ver. 5.0 program.

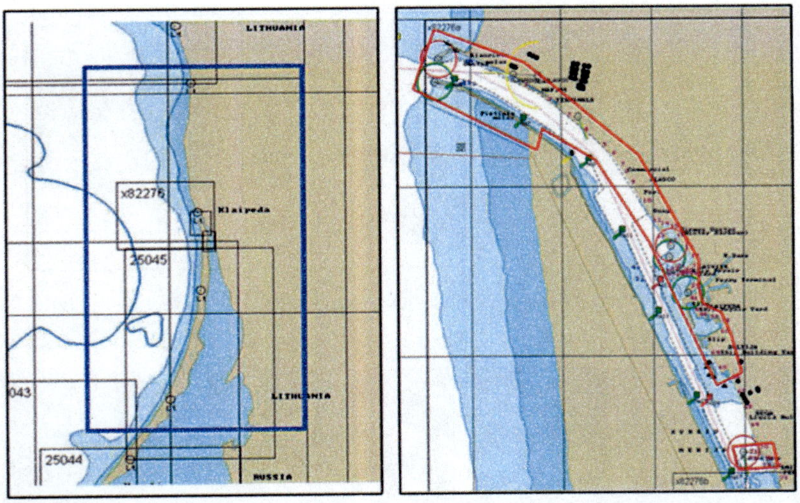

Fig. 5. Klaipeda port simulation full area (blue) and High Level of Detail area (red).

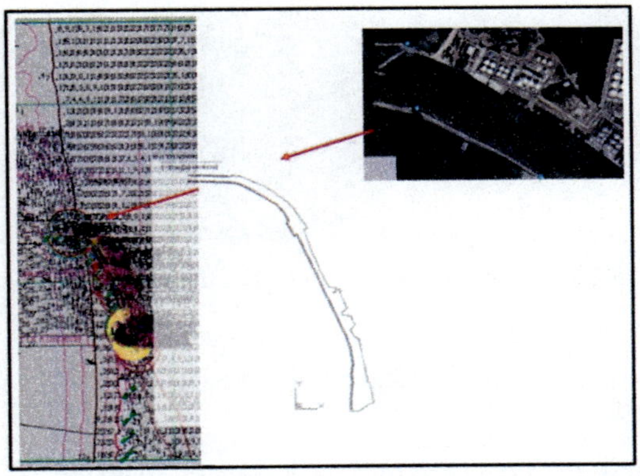

Fig. 6. Modelling the Klaipeda seaport area.

The coordinates of the simulated quays, piers and terminals have been adjusted according to topographical satellite photos. Aids to Navigation were modelled based on the publication of the Lithuanian Maritime Safety Administration "Lithuanian navigational signs and lights. The Baltic Sea". Klaipeda seaport inner channel, depths, berths and jetties are modelled according to the port plan and bathymetric data, provided by Klaipeda State Seaport Authority.

Seabed and depth modelling is very important to conduct precise navigational tasks.

Errors that may occur during simulation: uneven depth, depth variation anomalies, design depth not reached, ship-shore interaction not working, uneven depth transitions (see Fig. 8).

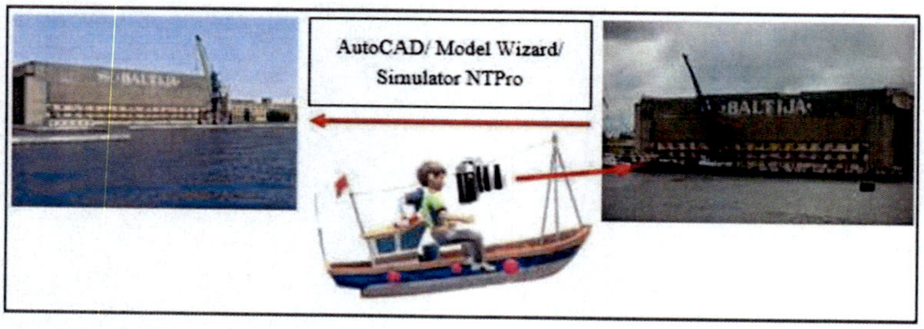

Fig. 7. Photographing the area: real view on right, simulator view on the left.

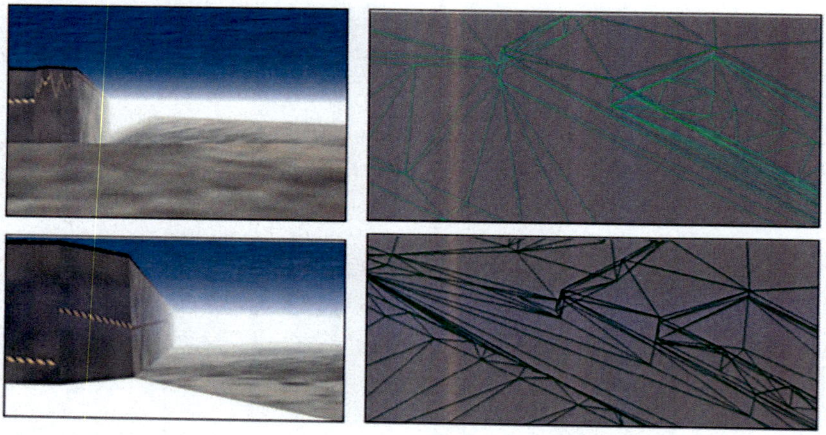

Fig. 8. The modelling errors (at the top) and modelling correction (at the bottom).

The main causes of errors are the neglect of modelling rules. To avoid mistakes, you need to know the basic notes:

1. The depth contour of the smaller value must be enclosed by the contour of the greater depth in sequence order.
2. Depth contours and depth lines cannot intersect.
3. Check whether the type of berth and the depth to them are correctly selected.
4. Coastlines should not intersect each other or themselves.

The check of modelling visual and sensor correctness using Model Wizard-Scene Editor should be made at any stage of modelling and of course at the final stage before creation of installation file. It is recommended to check correctness at an early stage to detect errors as early as possible and avoid remodelling work.

At the final stage of scene construction necessary to check if visual, sensor, radar, relief, seabed and electronic charts component are marked for generation. The scene construction result is CAB file to install in a simulator.

5 Conclusions

The creation of a realistic simulation and modelling environment in the Klaipeda seaport has proven to be a valuable investment in enhancing navigational safety. The detailed and accurate exercise area, incorporating up-to-date data and advanced tools, has facilitated effective training, qualification improvement, and research activities. The fidelity and accuracy of the simulator have been validated through successful replication of real-world scenarios, providing users with a sense of realism and preparing them to navigate the challenges of the seaport. The Klaipeda seaport exercise area serves as a critical resource for training pilots, VTS operators, and seafarers, significantly contributing to the overall safety and development of the seaport. Continuous updates and improvements to the simulation model based on the latest information will further enhance navigational safety and the preparedness of seaport personnel. The mentioned above is especially important in order to meet the actual and forthcoming legal statements and requirements of the IMO.

It is important to note that during the conducted research and modelling, a model of Klaipeda port visualization and adaptation for training purposes was created, which is based on the basic software solutions offered by the company Transas.

This resulted with the Klaipeda seaport exercise area model with 3D real visualization using accurate photos of the area, the latest topographic, bathymetric data, which fully reproduces the real view and conditions. During the creation of the detailed exercise area of the Klaipeda seaport, 1,500 high-quality photos were taken, based on which an accurate image was obtained in the simulator.

Full Mission Bridge simulator can capture and analyse data during simulation exercises. This data provides valuable insights into vessel behaviour, navigational patterns, human factors, and safety performance. By analysing the collected data, seaport developers can identify trends, evaluate safety measures, and make informed decisions to enhance navigational safety within the port.

LMA simulator with realistic Klaipeda seaport simulator exercise area offer a safe and cost-effective platform to train, assess, and improve navigational skills, make research, evaluate port infrastructure, and enhance safety measures during seaport development. Simulator plays a crucial role in ensuring navigational safety, mitigating risks, and promoting efficient and secure vessel operations within seaport.

Errors that can occur during the research and simulation are estimated. They may occur because of uneven depth, depth variation anomalies, design depth not reached, ship-shore interaction not working and uneven depth transitions.

Using the Klaipeda seaport exercise area many research projects related with new infrastructure in the port and entering of new type of vessels were successfully completed. When in 2014 October 27 the LNG "Independence" entered in the Port of Klaipeda, the captain was surprised: "Wow, just like in the simulator".

References

1. Bošnjak, R., Kezić, D., Vidan, P., Kavran, Z.: Collision prevention in Singapore strait by using timed petri net. Transport **35**(3), 273–282 (2019). ISSN 1648-4142/eISSN 1648-3480. https://doi.org/10.3846/transport.2019.11623

2. Zwolan, P., Czaplewski, K.: The virtual model of deep water container terminal T2 in Port Gdańsk. TransNav Int. J. Mar. Navig. Saf. Sea Transp. **12**(3), 551–555 (2018). https://doi.org/10.12716/1001.12.03.15
3. Malik, A.A., Zafar, N.: Applications of simulation technology - pitfalls and challenges. TransNav Int. J. Mar. Navig. Saf. Sea Transp. **9**(3), 391–396 (2015). https://doi.org/10.12716/1001.09.03.12
4. Port of Klaipeda. Orders of Harbour Master. https://portofklaipeda.lt/en/orders-of-harbour-master/. Accessed 15 July 2023
5. Gudelis, V., Benetyte, R.: Analysis of the competitiveness of the performance of Baltic ports in the context of economic sustainability MDPI. Sustainability **13**(6), 3267 (2021). https://doi.org/10.3390/su13063267
6. Port of Klaipeda 2022 results. https://portofklaipeda.lt/en/naujienos/port-of-klaipeda-2022-results-optimism-despite-a-challenging-year/. Accessed 15 July 2023
7. Transas MIP Ltd: Model Wizard. User Manual, version 5.00 (2012)
8. Tarasov, S.V., Lebedev, D.V., Nikolaev, I.A.: Component modelling of ship systems in TRANSAS liquid cargo handling simulators. IFAC Proc. Vol. **46**(9), 2197–2200 (2013). https://doi.org/10.3182/20130619-3-RU-3018.00286
9. Conde López, E., Salete Casino, E., Flores Escribano, J., Vargas Ureña, A.: Application of finite element method to create a digital elevation model. Mathematics **11**, 1522 (2023). https://doi.org/10.3390/math11061522
10. Liu, W.K., Li, S., Park, H.S.: Eighty years of the finite element method: birth, evolution, and future. Arch. Comput. Methods Eng. **29**, 4431–4453 (2022). https://doi.org/10.1007/s11831-022-09740-9
11. Perumal, L., Koh, W.H.: Techniques for element formulation and quadtree-based triangular mesh generation for strain-based finite elements. MethodsX **10**, 102027 (2023). ISSN 2215-0161. https://doi.org/10.1016/j.mex.2023.102027
12. Fish, J., Belytschko, T.: A First Course in Finite Elements. Wiley, Chichester (2007)
13. Sencila, V., Zažeckis, R., Jankauskas, A., Eitutis, R.: The use of a full mission bridge simulator ensuring navigational safety during the Klaipeda seaport development. TransNav Int. J. Mar. Navig. Saf. Sea Transp. **14**(2), 417–424 (2020). https://doi.org/10.12716/1001.14.02.20

Safety Behaviour of Heavy Truck Drivers in International Transport

Sebastjan Škerlič[1](✉), Robert Muha[1], and Vanja Erčulj[2]

[1] Faculty of Maritime Studies and Transportation, University of Ljubljana, Pot pomorščakov 4, 6320 Portorož, Slovenia
sebastjan.skerlic@fpp.uni-lj.si

[2] Faculty of Criminal Justice and Security, University of Maribor, Kotnikova Ulica 8, 1000 Ljubljana, Slovenia
vanja.erculj@fvv.uni-mb.si

Abstract. This study aims to provide insight into the safety behaviour of heavy truck drivers in the European Union. A survey of 249 truck drivers in international road transport was carried out with the goal of improving the state of safety in the transport industry in the EU. Individual truck driver behaviours that could potentially compromise the safe operation of a truck were examined. The results obtained highlight that heavy truck drivers are getting enough sleep and that any potential fatigue does not affect their regular work duties. Furthermore, drivers are consistent in logging their driving hours and the majority of them follow the speed limits. The results represent one of the important steps in the regulation of the transport industry from a safety, labour and social perspective and therefore build on the current area of research.

Keywords: Truck drivers · Safety · Safety behaviour · International transport · International study

1 Introduction

Heavy truck driving is an occupation that carries a high risk of fatigue. There is a high risk of accidents and safety becomes an important issue for all those involved in the road transport sector. Any work-related accident causes businesses to suffer losses such as workplace injuries, lower employee morale and motivation, damaged assets, losses in production, penalties and, last but not least, reputational damage [1].

Truck driving involves many occupational risks. According to Smith [2], 756 truck drivers died from work-related injuries in the USA in 2015, and more than 65,000 truck drivers in the private sector suffered work-related injuries. Smith also highlights that truck drivers are more than seven times more likely to suffer a fatal injury at work than the average worker. Most of these fatal work injuries are caused by road traffic incidents, typically road accidents. In many cases, accidents involving trucks can be directly linked to the working conditions of the drivers. Pressure from supply chain

O. Prentkovskis et al. (Eds.): TRANSBALTICA 2023, LNITI, pp. 467–475, 2024.
https://doi.org/10.1007/978-3-031-52652-7_46

participants, coercion, poor earnings and unbalanced power relations in general lead drivers to adopt riskier driving behaviour, which has an impact on road safety [3].

A number of studies on safety have therefore attempted to analyse the factors that influence the safety behaviour of truck drivers. The factors that are most frequently highlighted are: employer pressures [4], pay-for-productivity [5–10], working time violations [11], poor working conditions [5], and job dissatisfaction [9, 12, 13]. All of these factors, in turn, contribute to fatigue in heavy truck drivers, which not only affects quality of life, but also increases the risk of human errors and, consequently, the risk of road accidents.

In order to improve the state of safety in the transport industry in the EU, a survey was carried out on a sample of 249 truck drivers working in international road transport. The objective of the survey was to determine the rate at which truck drivers are involved in road accidents and assess their safety behaviour while driving a truck. The results will be compared with similar studies in the conclusion and represent one of the important steps in the regulation of the transport industry from a safety, labour and social perspective. The results obtained therefore provide a further contribution to the current field of research.

The paper is structured as follows: in the second part, a literature review is carried out on the factors that affect the safety behaviour of truck drivers. This is followed by a presentation of the research methods used. The paper then proceeds with a presentation and discussion of the results, highlighting the contribution to theory and practice in the field of safety behaviour of heavy truck drivers. The limitations of the study and recommendations for further research are outlined in the conclusion of the paper.

2 Literature Review

A number of studies on safety have attempted to analyse the factors that influence the safety behaviour of truck drivers. A fundamental problem in the transport industry is the pressure to maximise productivity. In practice, this means that financial compensation for truck drivers is productivity-dependent on a global scale [3, 4, 6–8, 14–17]. Higher pay for work performed is therefore associated with more kilometres travelled. Increased mileage is more likely to expose drivers to higher risks. The amount of kilometres driven has been shown to be systematically associated with road deaths and injuries [9]. Viscelli [18] points out that current economic incentives in the transport industry are counterproductive for long-term road safety and driver health.

This is confirmed by the leading authors who have studied the impact of financial compensation on the safety behaviour of truck drivers. Indeed, all leading authors specifically emphasise the negative impact of productivity-based pay on the health of truck drivers and on road safety [4–11]. However, they also point out that better-paid truck drivers are also safer drivers, resulting in fewer road accidents [5, 10, 11, 19–21].

Safety behaviour is also significantly influenced by the working conditions at individual companies and, in particular, by the truck drivers' level of satisfaction. Overall, data on subjective well-being (SWB) show that employees working in the transport sector have well below average life and job satisfaction [22]. The struggles of truck drivers are illustrated by an interview of Apostolopoulos et al. [23] with drivers who report being exposed to constant stress. Furthermore, drivers emphasise that this is not the kind of life they really want to live. The reason a truck driver gives for being in this profession is that it provides him with what he needs to support his family and maintain himself and his lifestyle. Low levels of well-being among truck drivers can have various adverse effects, including lower work productivity, poor health outcomes [24], and reduced personal and public safety [23]. In addition, due to the prospect of low truck driver happiness, logistics companies have difficulties attracting new talent. These adverse effects are particularly pertinent because many Western countries currently face a shortage of transportation workers because of a combination of high voluntary turnover [12, 25], a rapidly aging workforce [26], and difficulties finding young, capable drivers [27].

Indeed, truck drivers' safety behaviour is influenced by a variety of factors that have been studied thoroughly and, most importantly, systematically by researchers over the years. However, it is worth recognising that truck drivers are like any other individuals in the workplace, acting on the basis of their own sense of accountability or convictions. They have their own views on specific situations on the road, just as they have their own views on their own safety behaviour and, above all, on their own truck driving abilities. For this reason, individual truck driver behaviours that could potentially compromise the safe operation of a truck will be examined. The following individual behaviours are highlighted:

- driving at the legal speed limit,
- not being careful when getting out of the truck,
- driving without sufficient sleep,
- being in a difficult situation with the truck with no way out (exit),
- making sure that the hours of driving are logged consistently and
- occasionally failing to carry out a daily vehicle inspection due to fatigue and lack of time.

3 Methods

A total of 731 heavy truck drivers, mainly operating in the EU, were invited to take part in the survey. 249 drivers agreed to participate, a response rate of 34%. 220 responded to all the questions. An online questionnaire was set up for this purpose and sent to the drivers via text message, which they mainly filled in on their smartphones. This method ensured that the responses were kept anonymous. Drivers were given the option of completing the questionnaire in English, Italian, Slovenian, Croatian and Serbian.

The questionnaire was reviewed by five truck drivers and two CEOs of major international trucking companies before it was sent out. Most of the respondents were approached at the truck terminals of the Port of Trieste (Italy) and the Port of Koper (Slovenia), and some of them at various truck stop locations. This survey is part of a larger study that analysed truck maintenance in international transport companies [28] and the impact of financial and non-financial work incentives on truck drivers' behaviour in international transport [29]. The survey is divided into two parts. The first part of the questionnaire consists of questions related to demographic data, while the second part focuses on questions about the safety behaviour of heavy truck drivers. The questionnaire was developed by incorporating standard demographic data based on recent studies in the transport sector [6, 7, 10, 15]. The classification of road accidents is based on the Internal Code of Traffic Accidents [30].

The final sample comprises 220 heavy truck drivers, including one female driver, who participated in the study. The median age (SD) of the participants was 40.8 (10.5) years. Approximately one third (37.7%) had more than high school education. The average (SD) number of years working as a truck driver was 13.7 (10.1) years. On average (SD), they are employed at the current employer for 5 (5.1) years.

4 Results and Discussion

The findings of the survey on heavy truck drivers involved in road accidents over the last three years are shown in Table 1. On average, drivers have been involved in less than one or 0.8 ($n = 7$) accidents in the last three years where only material damage was sustained. The average rate of involvement in accidents in which one or more persons suffered minor physical injury is even lower, at 0.2 ($n = 5$). Similarly, for accidents in which one or more persons were seriously injured, as 82% of drivers were not involved in such a traffic accident in the last three years. The average is 0.1. Only one out of 220 respondents had been involved in an accident in which someone had lost their life.

Table 1. Involvement in road accidents in the last three years.

How many times have you been involved in a road accident in the last three years?	Valid	No. of units	Average	Standard deviation	Min	Max
Accidents resulting in material damage only						
	220	249	0.8	1.21	0	7
Accidents resulting in one or more persons suffering minor physical injury						
	219	249	0.2	0.59	0	5
Accidents resulting in one or more persons seriously injured						
	218	249	0.1	0.43	0	3
Accidents that resulted in someone's death						
	218	249	0	0.07	1	1

Figure 1 shows the rate of involvement of heavy truck drivers in road accidents over the last three years in European Union. As can be observed, the majority of accidents resulted in material damage only.

Table 2 presents the results of the part of the questionnaire where respondents rated statements related to their safety behaviour. The majority of drivers in the sample drive within the prescribed speed limits and consistently log their driving hours. Very rarely do drivers report not getting enough sleep. Even when tired or in a rush, truck drivers very rarely skip their daily vehicle inspection. Drivers rarely find themselves in a difficult situation without a way out or a suitable exit. The data obtained suggest that truck drivers display good safety behaviours.

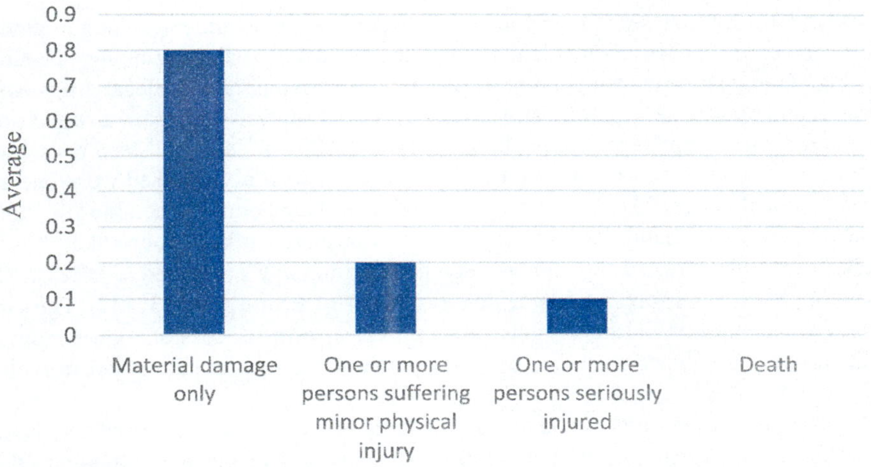

Fig. 1. Involvement in road accidents in the last three years (distribution of the number of accidents, according to the type of accident).

Table 2. Safety behaviour of heavy truck drivers.

Rate the following statements. From 1 (I completely disagree) to 5 (I completely agree)	n	Average	Standard deviation	Min	Max
I always respect speed limits while driving	217	3.9	0.99	1	5
I occasionally jump from the cab when exiting the truck to save time	216	2.8	1.32	1	5
I occasionally drive even though I haven't had enough sleep	216	2.2	1.11	1	5
Sometimes I find myself in a difficult situation with the truck without a way out	217	2.6	1.15	1	5
I make sure that the hours of driving are logged consistently	217	4.3	0.81	1	5
When I'm tired or in a hurry, I sometimes skip the daily check of the vehicle	216	2.7	1.27	1	5

Some conclusions can be drawn when comparing the individual results with other scientific studies, especially as regards the answers relating to sleep deprivation and increased fatigue. Yassierli et al. [1], like most experts, point out that fatigue is a normal biological state that every human being experiences. However, having fatigue while driving is one of the main causes of road accidents. It is estimated that fatigue is involved in around 10–20% of serious accidents [1]. Studies have documented fatigue among other dangerous behaviours that lead to a driving accidents, such as drinking, speeding, driving while distracted, using a mobile phone while driving, taking illegal drugs while driving, and being a new driver. Fatigue in drivers is strongly associated with conditions associated with sleep, such as circadian rhythm (i.e. time of day) and sleep restriction, as well as with conditions associated with work tasks, such as increased workload and the duration of driving [1]. A survey by the National Sleep Foundation in the US, which looked at the sleep habits of key categories of transport workers, found that 17% of truck drivers sleep less than 6 h on average per working day, compared to 16% of train operators, 10% of bus/taxi/limousine drivers and 9% of airline pilots [31, 32]. The same survey found that 44% of truckers get a full night's sleep on rare occasions or never during their workdays, outperformed only by train operators, who report getting a restful night more often.

The results of the study conducted on a sample of 220 heavy truck drivers in the EU show that drivers are getting enough sleep and that any fatigue they may experience does not affect their regular work duties. Other comparable studies highlight different results. The study by Yassierli et al. [1] conducted on fuel truck drivers shows moderate to high levels of fatigue. Sleep deprivation in heavy truck drivers was the subject of a study dating back to 1997. Mitler et al. [33] reported that Canadian drivers are chronically sleep deprived and the majority of drivers indicated that they had at least one 6-min interval of drowsiness while driving over a 5-day study. This suggests that driver fatigue contributes to reduced road safety.

A key factor in the prevention of the risk of fatigue-related accidents is the explicit observance of regulations on work and driving time in truck driver scheduling [34]. Since the study of heavy truck drivers indicates lower levels of fatigue, it can be concluded that EU initiatives to regulate working conditions have achieved their purpose. In 2006, the new EU rules on driving times, breaks and rest periods for drivers [35] who transport goods and passengers by road came into force. The new rules were aimed at reducing truck driver fatigue, improving working conditions in the transport industry and reducing road accidents. This is supported by the results obtained, as drivers are taking care to consistently log their driving hours and are driving at the speed limit most of the time. A comparison with a study carried out in the USA [15] shows that European drivers are more consistent in logging and observing driving hours in accordance with EC Regulation 561/2006. In the study by Hege et al. [15], only 39.2% of drivers always comply with the Hours of Service (HOS) regulations, which limit the duration of driving and impose rest periods for long-haul commercial vehicles in the US. Incidentally, in the same study, Hege et al. [15] report an improvement in fatigue levels of truck drivers, as the truck drivers in this study reported more hours of sleep at night than the researchers had expected. They tie their findings in with results showing that drivers may be getting more sleep due to changes in HOS regulations. They further point out that several crucial work-related factors have a significant impact on the levels of fatigue and sleep quality of long-haul truck drivers, and show that longer working hours and more kilometres driven result in less sleep.

The upside of EU legislation is that the rules on driving times, breaks and rest periods for drivers (Regulation (EC) No 561/2006) allow for a maximum of 9 h of driving, unlike in other countries where driving for more hours is legally allowed. This may be an important factor underscoring the proper safety behaviour of the truck drivers involved in the present study.

5 Conclusions

On a global scale, there have been many studies looking at the different factors influencing the safety behaviour of truck drivers. The most commonly highlighted factors are the pressures of the transport industry, the impact of pay on productivity, working time violations, poor working conditions and job dissatisfaction. All of these factors not only affect quality of life, but also increase the risk of human error and, consequently, of road accidents. For this reason, the present paper examined individual truck driver behaviours that could potentially compromise the safe operation of a truck. The results showed that, compared to similar studies, the safety behaviour of truck drivers in the EU is at a satisfactory level. Heavy truck drivers are getting enough sleep and any potential fatigue does not affect their regular work duties. Furthermore, drivers are consistent in logging their driving hours and the majority of them follow the speed limits.

The results therefore provide an insight into the current field of ensuring safety in the transport industry. Regulating this field can reduce road safety issues in the long term, as well as improve the job status of heavy truck drivers. The negative factors associated with the job are the reason why there is less and less interest in the profession, while the demand for transport services is increasing. Looking ahead, it would seem worthwhile to replicate this study in the future and compare the results. This would allow us to determine whether the initiatives adopted by the European Parliament on 8 July 2020, which include better working conditions for drivers, have achieved their objective.

References

1. Mahachandra, M., Sutalaksana, I.Z.: Fatigue evaluation of fuel truck drivers. Procedia Manufact. **4**, 352–358 (2015)
2. Smith, S.M.: Workplace hazards of truck drivers in 2012. Monthly Labor Rev. 1–7 (2015)
3. Cruz Ross, A.: Upping the quality of employment of Mexican truck drivers. J. Labor Soc. **22**(1), 103–128 (2019)
4. Thompson, J., Stevenson, M.: Associations between heavy-vehicle driver compensation methods, fatigue-related driving behavior, and sleepiness. Traffic Inj. Prev. **15**(Suppl. 1), 10–14 (2014)
5. Belzer, M.H.: Work-stress factors associated with truck crashes: an exploratory analysis. Econ. Labour Relat. Rev. **29**(3), 289–307 (2018)
6. Belzer, M.H., Sedo, S.A.: Why do long distance truck drivers work extremely long hours? Econ. Labour Relat. Rev. **29**(1), 59–79 (2018)
7. Mahajan, K., Velaga, N.R., Kumar, A., Choudhary, A., Choudhary, P.: Effects of driver work-rest patterns, lifestyle and payment incentives on long-haul truck driver sleepiness. Transp. Res. Part F **60**, 366–438 (2019)
8. Girotto, E., et al.: Working conditions and sleepiness while driving among truck drivers. Traffic Injury Prev. **20**(5), 504–509 (2019)
9. Lee, J., Huang, Y., Sinclair, R.R., Sinclair, R.R., Cheung, J.H.: Outcomes of safety climate in trucking: a longitudinal framework. J. Bus. Psychol. **34**, 865–878 (2019)

10. Faulkiner, M.R., Belzer, M.H.: Returns to compensation in trucking: does safety pay? Econ. Labour Relat. Rev. **30**(2), 262–284 (2019)
11. Williams, E., Monaco, K.: Accidents and hours-of-service violations among over-the-road drivers. J. Transp. Res. **40**(1), 105–115 (2001)
12. Prockl, G., Teller, C., Kotzab, H., Angell, R.: Antecedents of truck drivers' job satisfaction and retention proneness. J. Bus. Logist. **38**(3), 184–196 (2017)
13. Zaranka, J., Pečeliūnas, R., Žuraulis, V.: A road safety-based selection methodology for professional drivers: behaviour and accident rate analysis. Int. J. Environ. Res. Public Health **18**, 12487 (2021)
14. Williamson, A., Friswell, R.: The effect of external non-driving factors, payment type and waiting and queuing on fatigue in long distance trucking. Accid. Anal. Prev. **58**, 26–34 (2013)
15. Hege, A., Perko, M., Johnson, A., Yu, C.H., Sönmez, S., Apostolopoulos, Y.: Surveying the impact of work hours and schedules on commercial motor vehicle driver sleep. Saf. Health Work **6**(2), 104–113 (2015)
16. Chen, G.X., Sieber, W.K., Lincoln, J.E., et al.: NIOSH national survey of long-haul truck drivers: injury and safety. Accid. Anal. Prev. **85**, 66–72 (2015)
17. Kudo, T., Belzer, M.H.: The association between truck driver compensation and safety performance. Saf. Sci. **120**, 447–455 (2019)
18. Viscelli, S.: The Big Rig: Trucking and the Decline of the American Dream. University of California Press, Oakland (2016)
19. Golob, T.F., Hensher, D.A.: Driving behaviour of long distance truck drivers: the effects of schedule compliance on drug use and speeding citations. Int. J. Transp. Econ. **23**, 267–301 (1996)
20. Rodríguez, D.A., Rocha, M., Khattak, A.J., Belzer, M.H.: Effects of truck driver wages and working conditions on highway safety: case study. Transport. Rec. **1883**, 95–102 (2003)
21. Rodríguez, D.A., Targa, F., Belzer, M.H.: Pay incentives and truck driver safety: a case study. Ind. Labor Relat. Rev. **59**(2), 205–225 (2006)
22. De Neve, J.E., Ward, G.: Happiness at Work (No. CEP Discussion Paper No 1474). Centre for Economic Performance London School of Economics and Political Science, London (2017)
23. Apostolopoulos, Y., Sönmez, S., Hege, A., Lemke, M.: Work strain, social isolation and mental health of long-haul truckers. Occup. Therapy Mental Health **32**, 50–69 (2016)
24. Apostolopoulos, Y., Sönmez, S., Shattell, M.M., Gonzales, C., Fehrenbacher, C.: Health survey of US long-haul truck drivers: work environment, physical health, and healthcare access. Work **46**, 113–123 (2013)
25. Staats, U., Lohaus, D., Christmann, A., Woitschek, M.: Fighting against a shortage of truck drivers in logistics: measures that employers can take to promote drivers' work ability and health. Work **58**, 1–15 (2017)
26. American Trucking Association: Turnover Rate at Large Truckload Carriers Rises in First Quarter. Accelerating Churn Rate Indicates Continued Tightness in Driver Market (2018). http://www.trucking.org/article/Turnover-Rate-at-Large-Truckload-Carriers-Rises-in-First-Quarter
27. Schulz, S.A., Luthans, K.W., Messersmith, J.G.: Psychological capital: a new tool for driver retention. Int. J. Phys. Distrib. Logist. Manage. **44**, 621–634 (2014)
28. Škerlič, S., Sokolovskij, E., Erčulj, V.: Maintenance of heavy trucks: an international study on truck drivers = Obsługa techniczna samochodów ciężarowych: międzynarodowe badanie kierowców ciężarówek. Eksploatacja i Niezawodność **22**(3), 493–500 (2020)
29. Škerlič, S., Erčulj, V.: The impact of financial and non-financial work incentives on the safety behavior of heavy truck drivers. Int. J. Environ. Res. Public Health **18**(5), 1–14 (2021)
30. SURS: TR_PromNesr - TR - Internal code list of traffic accidents, V1 (2020). https://www.stat.si/Klasje/Klasje/Tabela/2226

31. Sullman, M.J.M., Stephens, A.N., Pajo, K.: Transport company safety climate: the impact on truck driver behavior and crash involvement. Traffic Inj. Prev. **18**(3), 306–311 (2017)
32. National Sleep Foundation. Sleepy pilots, train operators, and drivers. National Sleep Foundation, Arlington (VA) (2012)
33. Mitler, M.M., Miller, J.C., Lipsitz, J.J., Walsh, J.K., Wylie, C.D.: The sleep of long-haul truck drivers. N. Engl. J. Med. **337**(11), 755–762 (1997)
34. Goel, A.: The Canadian minimum duration truck driver scheduling problem. Comput. Oper. Res. **39**(10), 2359–2367 (2012)
35. Regulation (EC) No 561/2006 of the European Parliament and of the Council of 15 March 2006 on the harmonisation of certain social legislation relating to road transport and amending Council Regulations (EEC) No 3821/85 and (EC) No 2135/98 and repealing Council Regulation (EEC) No 3820/85 (Text with EEA relevance) (2023)

Supply Chain Digital Maturity Modeling – A Case Study of a Wood-Based Supply Chain

Natalia Gnacy(✉) and Sylwia Werbińska-Wojciechowska

Wroclaw University of Science and Technology, Wybrzeze Wyspianskiego 27, 50-370 Wroclaw, Poland
{natalia.gnacy,sylwia.werbinska}@pwr.edu.pl

Abstract. The proposed chapter focuses on the problem of digital maturity modeling in a supply chain from the wood-based industry. Digital transformation means changes in all parts of the supply chain by replacing traditional practices with new practices that are digitally based. As a result, it improves any organization's operational efficiency and decreases operational risk in the chain. Therefore, the paper addresses the problem of digital maturity modeling in a selected company operating in a wood-based supply chain. Indeed, the paper presents a short literature review on the defined research area. Later, the wood-based supply chain is characterized. Based on this, a digital maturity model for the case company is investigated. The conclusions and directions for further research are presented in the last section.

Keywords: Wood-based industry · Digital supply chain · Digital maturity · Digital maturity modeling

1 Introduction

The current economic environment, in which supply chains perform their functions and tasks is characterized by high dynamics of change as well as operational uncertainty. Globalization, unstable economy, shortages in supply, or complexity of the general business setting make today's supply chains unsustainable. Indeed, companies need to be adaptive to this fast-changing environment, e.g., by applying sustainability principles or turning to local suppliers to overcome staff and material shortages. The solution may lie in using new technologies connected with Industry 4.0 concept development. In turn, this demands supply chain operations to become more intelligent. As a result, the smart transformation process is based on such attributes as digitalization, transparency, and network collaboration [1].

Especially supply chain digitalization is highlighted in the literature as a solution to the fundamental operational problems occurring in the chain [2, 3]. The denotation 'digital' refers to *the use of electronic data sets, which stay in contrast to analog ones* [4]. Indeed, digital transformation is defined as *the process that is used to restructure*

O. Prentkovskis et al. (Eds.): TRANSBALTICA 2023, LNITI, pp. 476–485, 2024.
https://doi.org/10.1007/978-3-031-52652-7_47

economies, institutions, organizations, and society on a system level [5, 6]. Indeed, it can be perceived as system changes that result in technological improvements to increase supply chain efficiency [7].

The need for digital transformation is directly linked to the expectations of all supply chain participants and stakeholder groups. These expectations include both operational and strategic requirements for the supply chain, such as, for example, ensuring the availability of products or raw materials. This trend can also be observable in the forestry and wood-based industries [8, 9].

The traditional wood supply chain generally involves the various stages of the material and information flow process, starting with the planning of timber harvesting and ending with logistics and sale to the end customer [9]. In the case of the wood-based industry, the main products transformed and offered to the customer are so-called waste wood, i.e., the waste material generated during sawmilling and production (e.g., shavings, sawdust, chips), as well as processed round wood. The effective management of material and informational flows in such supply chains involves dealing with several problems related, on the one hand, to the product itself (e.g., weight of wood, size of material) and, on the other, to the effective and reliable information providing (e.g., digitalization of operational processes). As a result, this paper outlines the possibility of digital maturity modeling for supply chain performing in the wood-based materials sector.

Following this, in the next section, the authors present a literature review on digital maturity in supply chains is presented. Later, the wood-based supply chain is characterized. This gives the possibility to focus on digital maturity modeling for the case company. The chapter concludes with a summary and indication of potential directions for further research work in the analyzed area.

2 Review on Digital Maturity Modeling in Supply Chains

Recent reports on digital supply chains (e.g. [3, 10]) underline that digital transformation mainly focuses on reducing costs and increasing operational efficiency by implementing selected Industry 4.0 technologies. Therefore, the traditional supply chain changes at strategical, tactical, and operational levels due to technological improvements that are implemented in physical infrastructure and superstructure as well as in the intercommunication layers [3, 7].

Traditional supply chains typically operate along so-called linear pathways, which transfer information with raw materials and finished goods from one end of the supply chain through the production system to the point of distribution [4, 10]. Moreover, the information flow is based on physical documentation in the supply chain and traditional communication channels within the company and between the different links in the supply chain. In parallel, the digital supply chain structure is a flexible, interconnected matrix that enables a non-linear flow of data and information to meet consumer and market demands more agilely and flexibly and reduce the response time needed to meet changing business conditions [11]. In addition, the digital supply chain may be viewed as a virtual structure with all its processes and data collected through sensors and stored in digital databases. Such data can be used to reduce costs and increase efficiency, which is the goal of any supply chain, whether digital or traditional [7].

Digital technologies' role in supporting sustainability and efficiency in supply chains is well recognized [12]. The literature review on digital supply chain research trends is presented, e.g., in [13]. The conceptual model for a digital supply chain following a literature review from 1989 to 2019 is given in [14]. Later, the research work on investigating the links between supply chain resilience and digital transformation is presented in [15]. The aspect of building a supply chain 4.0 model in relation to, among others, resilience is also investigated in, e.g., [1]. The readiness assessment regarding digital transportation in supply chains is summarized in [16]. In addition, the impact of digitalization on a company's business model elements is defined in [4].

The digitalization level of a supply chain is defined in relation to its digital maturity level and the progress in adopting digital tools [15]. According to the Cambridge Dictionary, maturity is defined as *a very advanced or developed form or state.* Maturity model (MM) is *a technique to measure several aspects of processes in an organization* [17]. Typically, maturity models can illustrate linear and non-linear progression in implementing digital transformation. They generally consist of 4–6 linear steps representing a typical sequence-based evolutionary development pathway in discrete steps [18]. The digital maturity model is defined as *an effective tool to provide guidelines for a clear path throughout the digital transformation journey* [19]. In relation to business-related maturity, maturity models' general purpose is to *assess the state of maturity of a particular maturity factor regarding a specific domain of interest based on a set of criteria in a comprehensive manner* [5].

Recently, in the literature, there is no one uniform structure for supply chain digital maturity modeling [7]. Following [20], a maturity model for supply chains is usually based on five components: assessment procedure, assessment criteria, assessment areas, framework matrix, and maturity levels. The main dimensions of digitalization included in the known models are business, organizational, process & method, and technological [21]. A detailed review of the known digital maturity supply chain models is given, e.g., in [5], where 28 maturity models are compared and discussed.

The existing maturity models allow companies operating in supply chains to self-assess their digital transformation readiness. However, the digital maturity modeling for the wood-based industry sector still needs development.

3 Wood-Based Supply Chain Digital Maturity Modeling

3.1 Wood-Based Supply Chain Operation

In the known literature, we may define different types of supply chains related to the wood industry. The most commonly referred to are forest supply chain [22], wood supply chain [23], and wood-based supply chain [24].

The forest supply chain generally relates to the time-based sequence of activities and processes from standing trees to end-users that transform the wood raw material into final forest-based products[25]. In a given supply chain, raw wood can be transformed into products such as logs, roundwood, sawn timber, panel and construction timber, biomass, pulp, paper, or bioenergy (for electricity and heat). The literature emphasizes that this type of supply chain mainly focuses on meeting industrial needs in an environmentally sustainable manner [26].

In the literature, the term wood supply chain mainly refers to the supply chain management of roundwood, which is used, among other things, as biomass or for industrial production [27].

Wood-based supply chain refers to the networks and supply chains of wood-based products for which so-called waste wood is used, i.e., waste from sawn timber production (e.g., sawdust, shavings, woodchips) and round timber that has been processed. The main wood-based finished product types include laminated board, chipboard, and worktops [24]. The main performance of such a wood-based supply chain can be generally illustrated by the example of the production company (with its supply/distribution links) operating in the global market (Fig. 1).

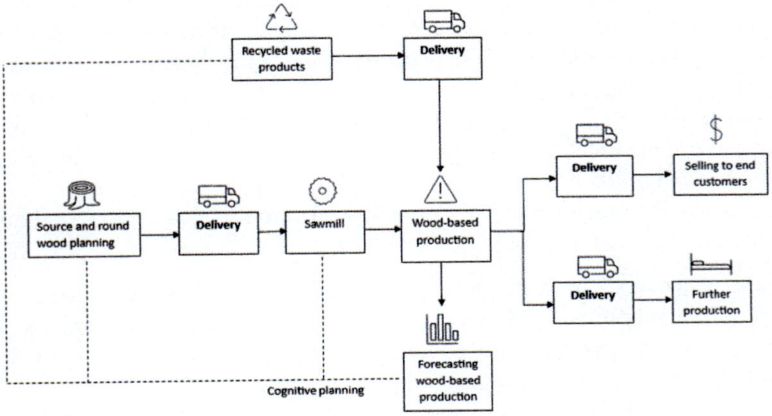

Fig. 1. A wood-based supply chain for the analyzed case company.

The material flow in such SC starts during the process of wood harvesting from forests and delivering timber to the sawmill. Harvesting timber from the forest requires proper planning, taking into account aspects such as the harvestable volume and the tree's age. The roundwood is then stored and transported for processing. The waste from wood processing is the raw material for the wood-based products industry. Products that result from wood processing, such as sawn timber, also undergo further processing. During further production, waste is also generated, which is used as raw material for the wood-based products industry. The municipal waste recycling industry is another supply source of raw materials for the case company. Municipal waste recycling companies buy municipal waste, like furniture, and transform it into pre-processed wood-based materials. The finished products offered by the case company (e.g., wood board) are mainly used for furniture production and in the construction industry. So, they are sold to these secondary wood industries or wholesalers. The retailers sell the products to the end customers.

The performed material and information flows are undergoing digital transformation. It needs to be supported by, e.g., the necessary business strategy, technologies, or infrastructure. Therefore, the level of digital maturity for such a supply chain should be analyzed to define the possible future directions for the organizational development of the case company.

3.2 Digital Maturity Modeling in the Forest/Wood-Based Industry

Forest and wood-based supply chains face with many challenges in Industry 4.0 technologies implementation. The main two are technical challenges and socio-economic challenges. The first ones are related to the application and device environment. The second ones are related to cooperation and trust between organizations and financial capacity [9]. Digitalization will generate problems in both areas.

Following [28], a maturity modeling approach can measure any supply chain's digitalization level. In addition, the authors state that along the maturity levels, the digital orientation of a company depends on enterprise interoperability. This is confirmed, e.g., by [12], where interoperability and integration are one of the dimensions of forest-based supply chain digitalization. The current status of digitalization in wood supply chains is investigated in [29], where the authors present a bibliometric literature review of the well-known literature dated from 2016 to 2021. Additionally, the digitalization in wood supply is reviewed in [9, 12].

When analyzing wood-based or forest-based supply chains, the digitalization dimensions should encompass at least four aspects:

- **business processes** – how the performance of internal and external processes in an organization is supported by digital transformation, i.e., technologies implementation, better coordination in the areas of product planning and development, dissemination of information to increase efficiency and effectiveness,
- **customer support** – how the collaboration between the supply chain and customer can be supported by, e.g., increasing in transparency, improved performance of distribution processes and customer service activities, technologies supporting an exchange of information and material flows,
- **sustainability goals** – how the sustainability requirements of the industry sector can be fulfilled by new technologies implementation (e.g., concerning circular economy in the SC, sustainable transport performance),
- **organization resilience** – how the digital transformation affects operational resilience and cybersecurity.

Taking one step further, the supply chain digital transformation can be analyzed by maturity assessment. To this aim, it is necessary to introduce the main SCs' digital maturity assessment areas and define maturity levels.

According to [7, 20, 28, 30], the maturity model should satisfy at least descriptive, prescriptive, and comparative purposes and support the digital transformation roadmap definition in a supply chain. Therefore, at least two assessment areas should be investigated: modern technology implementation to provide a necessary infrastructure for digital transformation. The second area is connected with interoperability and collaboration within a supply chain. In addition, the proposed model introduces six maturity levels (Table 1).

According to Table 1, the supply chain digital maturity is divided into six levels. Every level has different characteristics. In addition, achieving a higher level of maturity means

Table 1. Definition of maturity levels for the wood-based supply chains.

Maturity level	Digital maturity assessment areas	
	Technology	Interoperability & collaboration
0	Not conducted digitalization. A supply chain is traditionally organized, and processes are performed based on employee knowledge. There is no infrastructure for supporting management processes (the process is very poorly controlled or is not controlled at all). Lack of risk management rules and procedures	Communication is performed using traditional ways (e.g., e-mails, phones, and documents). Process management is reactive
1	Processes are partially planned and implemented, but still, there is a lack of enabling technologies in an organization. Sustainability goals are to be defined but not supported by modern technologies. Critical risks are managed in a supply chain but lack available IT (information technology) tools to gather knowledge and monitor operational process implementation	Single objects integration, management based on the experience of the planner, and communication supported by the traditional solutions between organizations, but organizations in the supply chain start implementing digital transformation individually
2	Process management is constrained by some limitations concerning the primary organizational responsibilities and enabling technologies, planning, and project integration based mostly on employee experience. Partial support in supportability achievement (e.g., in transportation area). Risk assessment implemented and supported by available IT infrastructure	There is still a lack of integration and interoperability in the applications, and the information exchange customer support is performed using traditional ways. Each organization uses new technologies to support its task and is prepared to create synergy within the supply chain
3	Processes management and performance are partially supported by digital infrastructure (e.g., technologies for goods tracking in transportation processes and condition monitoring of devices and equipment). The infrastructure supports sustainability plans. Systematic monitoring of the critical risks	Partial integration and interoperability of some applications for information exchange and material flow performance. Organizations start creating synergy within a supply chain, focusing on efficiency and effectiveness

(*continued*)

Table 1. (*continued*)

Maturity level	Digital maturity assessment areas	
	Technology	Interoperability & collaboration
4	Processes thoroughly planned and implemented, the company's primary functions, procedures, and operations are fully supported by digital technologies such as e.g. IoT, AI, cloud computing, and software solutions such as computer-aided programs for order prediction, simulation, etc. Infrastructure ready for circular economy principals implementation. Continuous monitoring of risks in a chain with the use of IT infrastructure	Integration and interoperability are based on well-known and shared standards within the company concerning the industry's best practices. Use of collaborative equipment to reduce the need for human labor. Integration of customer requirements. Business partners involved in risk management processes and sustainability goals achievement
5	Processes are digitally oriented and based on solid technology infrastructure and high potential growth of an organization. Technologies used to predict further developments in the market, optimization of supply chains, and to predict risks (operational and cyber-related ones)	A high level of integration and interoperability provides increased robustness and security in information exchange, full collaboration among the company functions, and decision-making. Human-centred operation

greater synergy between implemented technologies and sustainability goals achievement or supply chain resilience.

As the level of maturity increases, the risk of replacing humans from the system and the possibility that AI can manage the whole system also increases. It results in lower supply chain resilience in relation to cyber threats. On the other hand, operational resilience concerning system recovery possibilities is higher. Indeed, it is essential to emphasize that digital transformation has various advantages associated with reducing costs and increasing profits. However, there are also many different types of challenges and problems related to digital transformation in supply chains.

The digital maturity assessment in a supply chain is meant to be performed by site managers or accomplished alternatives within the business partners with insights into operations and digital initiatives. Researchers or permanent firm delegates can mediate the assessment procedure, allowing detailed questions and discussions within specific area-criteria-combinations.

3.3 Case Study

The proposed maturity model was used to assess the digital maturity level of selected production company from the wood-based sector. The company under investigation is a producer of finished products made from wood processing waste and from pre-processed wood materials obtained from municipal waste companies (according to Fig. 1). The company belongs to an international concern that is one of the leading European manufacturers of wood-based products, specializing in the production of materials for the furniture, interiors, and construction industries. The company continuously improves its production process and invests resources to ensure the highest level of environmental safety in its plant operations. Quality, cost, and product innovations are implemented as part of the company's developed innovation culture. The company has implemented standards from the ISO 14001, ISO 9001, ISO 5001, and ISO 45001 series. It provides customer-focused service and a consistent economic, ecological, and social sustainability orientation.

The company operates as a participant in global supply chains. Raw materials are supplied by local (domestic) suppliers and international companies from European countries. The direct recipients of the manufactured products are mainly Polish customers (e.g., the performance of individual projects and projects for business organizations). The company's management has developed partnership relations, particularly with domestic suppliers and customers. In the face of supply problems (closed markets from eastern countries, increased demand for wood-based products), the company continuously looks for new supply sources. It highly depends on suppliers from the municipal waste management sector.

The evaluation of the digital maturity of the analyzed company has been prepared based on the following:

- face-to-face interviews with senior and middle management, based on a structured interview aimed at identifying processes that are or can be supported by digital transformation,
- document workflow analysis with a particular focus on information flow management and performance,
- analysis of the scope of prepared reports and indicators in the area of business process efficiency, supportability goals achievement, and risk assessment.

According to the SC digital maturity dimensions, the company's digital maturity characteristics are presented in Table 2.

Table 2. Maturity levels assessed for the case company in relation to SC digital maturity dimensions and assessment areas.

SC digital maturity dimensions	Digital maturity assessment areas	
	Technology	Interoperability & collaboration
Business processes	3	3
Customer support	4	3
Sustainability	4	4
Resilience	3	4

According to the conducted reviews and analysis of the investigated company and its relations with its suppliers and customers, the necessary infrastructure for obtaining a high level of digital maturity is insufficient. In addition, the staff is not trained enough. Every participant operating in the analyzed wood-based supply chain receives different levels of digital maturity and implements selected advanced technologies as a separate decision focused on the internal optimization of performed processes. Basic logistic processes, such as supply processes planning, cost conversion, material requirements planning, or selected customer relationship management, are generally automated and digitalized. The overall maturity assessment places the investigated company at level 3 based on the introduced model.

4 Conclusions

The article presents the digital maturity modeling problem for the selected production company operating in the supply chain from the wood-based products sector. According to the conducted literature review, digital maturity for wood-based supply chains still needs development, especially at the technological level. The proposed digital maturity model has been implemented in the case company. However, its applicability will be investigated in future research works. On the one hand, the resilience of the investigated supply chain concerning digital maturity and Logistic 4.0 technologies will be investigated. On the other, a methodology for maturity assessment based on fuzzy logic will be developed.

References

1. García-Reyes, H., Avilés-González, J., Avilés-Sacoto, S.V.: A model to become a supply chain 4.0 based on a digital maturity perspective. Procedia Comput. Sci. **200**(2019), 1058–1067 (2022)
2. Dillmann, R.: Digital Transformation in Supply Chain Management (2016)
3. Gezgin, E., Huang, X., Samal, P., Silva, I.: Digital transformation : raising supply-chain performance to new levels. In: McKinsey & Company Operations, pp. 1–11 (2017)
4. Rachinger, M., Rauter, R., Müller, C., Vorraber, W., Schirgi, E.: Digitalization and its influence on business model innovation. J. Manuf. Technol. Manag. **30**(8), 1143–1160 (2019)
5. Hellweg, F., Janhofer, D., Hellingrath, B.: Towards a maturity model for digital supply chains. Logist. Res. **16**(1) (2023)
6. Unruh, G., Kiron, D.: Digital transformation on purpose. MIT Sloan Manag. Rev. (2017)
7. Buntak, K., Kovačić, M., Mutavdžija, M.: Measuring digital transformation maturity of supply chain. Tehnički glasnik **15**(2), 199–204 (2021)
8. Molinaro, M., Orzes, G.: From forest to finished products: the contribution of Industry 4.0 technologies to the wood sector. Comput. Ind. **138**, 103637 (2022)
9. Müller, F., Jaeger, D., Hanewinkel, M.: Digitization in wood supply – a review on how Industry 4.0 will change the forest value chain. Comput. Electron. Agric. **162**, 206–218 (2019)
10. Laaper, S., Yauch, G., Wellener, P., Robinson, R.: Embracing a digital future – how manufacturers can unlock the transformative benefits of digital supply networks. Deloitte Insights **28** (2018)
11. Mussomeli, A., Laaper, S., Gish, D.: The rise of the digital supply network. Deloitte University Press **45**(3), 1–21 (2017)

12. Scholz, J., et al.: Digital technologies for forest supply chain optimization: existing solutions and future trends. Environ. Manag. **62**(6), 1108–1133 (2018)
13. Weerabahu, W.M.S.K., Samaranayake, P., Nakandala, D., Hurriyet, H.: Digital supply chain research trends: a systematic review and a maturity model for adoption. Benchmark. Int. J. **30**, 3040–3066 (2022)
14. Garay-Rondero, C.L., Martinez-Flores, J.L., Smith, N.R., Caballero Morales, S.O., Aldrette-Malacara, A.: Digital supply chain model in Industry 4.0. J. Manuf. Technol. Manag. **31**(5), 887–933 (2020)
15. Yin, W.: Identifying the pathways through digital transformation to achieve supply chain resilience: an fsQCA approach. Environ. Sci. Pollut. Res. **30**(4), 10867–10879 (2023)
16. Lassnig, M., Müller, J.M., Klieber, K., Zeisler, A., Schirl, M.: A digital readiness check for the evaluation of supply chain aspects and company size for Industry 4.0. J. Manuf. Technol. Manag. **33**(9), 1–18 (2018)
17. Angreani, L.S., Vijaya, A., Wicaksono, H.: Systematic literature review of Industry 4.0 maturity model for manufacturing and logistics sectors. Procedia Manuf. **52**(2019), 337–343 (2020)
18. Grooss, O.F., Presser, M., Tambo, T.: Balancing digital maturity and operational performance - progressing in a low-digital sme manufacturing setting. Procedia Comput. Sci. **200**, 495–504 (2022)
19. Anderson, C., William, E.: Digital maturity model. Deloitte, 1–24 (2018)
20. Wernicke, B., Stehn, L., Sezer, A.A., Thunberg, M.: Introduction of a digital maturity assessment framework for construction site operations. Int. J. Constr. Manag. **23**(5), 898–908 (2021)
21. Hellweg, F., Lechtenberg, S., Hellingrath, B., Thomé, A.M.T.: Literature review on maturity models for digital supply chains. Braz. J. Oper. Prod. Manag. **18**(3), 1–12 (2021)
22. Santos, A., Carvalho, A., Barbosa-Póvoa, A.P., Marques, A., Amorim, P.: Assessment and optimization of sustainable forest wood supply chains – a systematic literature review. Forest Policy Econ. **105**, 112–135 (2019)
23. Palander, T.: Outsourcing issues of wood supply chain management in the forest industry. Forest Sci. **68**, 521–532 (2022)
24. Odlanicka-Poczobutt, M., Szmal, A.: Analiza potrzeb w obszarze dystrybucji jako etap projektowania sieci logistycznej na przykładzie wybranego producenta produktów drewnopochodnych. Zeszyty Naukowe Politechniki Śląskiej. Seria: Organizacja i Zarządzanie **78**(1928), 319–333 (2015)
25. He, Z., Turner, P.: A systematic review on technologies and Industry 4.0 in the forest supply chain: a framework identifying challenges and opportunities. Logistics **5**(4), 88 (2021)
26. She, J., Chung, W., Vergara, H.: Multiobjective record-to-record travel metaheuristic method for solving forest supply chain management problems with economic and environmental objectives. Nat. Res. Model. **34**(1), 12256 (2021)
27. Eriksson, P., Roos, A., Mark-Herbert, C.: The role of harvester measurement in the wood supply chain. Int. J. For. Eng. **00**(00), 1–13 (2022)
28. De Carolis, A., Macchi, M., Negri, E., Terzi, S.: A maturity model for assessing the digital readiness of manufacturing companies. IFIP Adv. Inf. Commun. Technol. **513**, 13–20 (2017)
29. Gharaibeh, L., Eriksson, K.M., Lantz, B., Matarneh, S., Elghaish, F.: Toward digital construction supply chain-based Industry 4.0 solutions: scientometric-thematic analysis. Smart Sustain. Built Environ. **13**, 42–62 (2022)
30. Werner-Lewandowska, K., Kosacka-Olejnik, M.: Logistics 4.0 maturity in service industry: empirical research results. Procedia Manuf. **38**(2019), 1058–1065 (2019)

Challenges of Implementing Reverse Logistics in Ensuring Circular Economy Goals

Aidas Vasilis Vasiliauskas(✉) and Olga Navickienė

General Jonas Žemaitis Military Academy of Lithuania, Šilo Str. 5A LT-10322, Vilnius, Lithuania
{aidas.vasilisvasiliauskas,olga.navickiene}@lka.lt

Abstract. Our planet is struggling to provide the necessary natural resources for about 8 billion people, and our Earth's important ecosystems are on the verge of collapse. In 2015, the Department of Economic and Social Affairs of the United Nations presented in its annual report that worldwide, as population increases, the need for land, food, and important natural resources in 2030 will be twice as high as the same needs in 2010. The developing world economy contributes to a large extent to the destruction of the Earth's resources. Most companies still tend to operate on the basis of a linear take-make-dispose economic model. Meanwhile, the principles of circular economy not only encourage the reduction or elimination of waste and pollution, maximizing the efficiency of the use of products and materials, but also promote the natural regeneration of systems [4]. Researchers believe that one of the most difficult parts of the circular economy is reverse logistics, where the hardest part is managing the collection of waste from consumers to capture value and convert materials back into resources. The focus of this article is on the role of reverse logistics in the circular economy and closed-loop supply chains. The purpose of this article is to reveal the importance of reverse logistics and the most important challenges in ensuring the implementation of the principles of circular economy.

Keywords: Circular economy · Closed cycle supply chain · Reverse logistics · Reverse logistics activities

1 Introduction

The developing global economy contributes to a large extent to the destruction of the earth's resources. Most companies still tend to operate based on a linear take-make-dispose economic model, where they turn the raw material, receive it into a finished product, and sell it on the market to the end user. In turn, the consumer discards the product (completely or partially consumed), so the manufactured product reaches the end of its life cycle [5]. With this linear approach, companies not only don't worry about what happens to the product when it is discarded after its final use, but they also don't worry about the resources used to produce the product. Therefore, it is not surprising that more and more attention is paid to those who are looking for more efficient use of

O. Prentkovskis et al. (Eds.): TRANSBALTICA 2023, LNITI, pp. 486–494, 2024.
https://doi.org/10.1007/978-3-031-52652-7_48

resources and more efficient processes in different areas of production and at different stages of consumption and who use the principles of circular economy. The circular economy is well known for boosting economic growth, generating new companies and employment opportunities, lowering prices, enhancing supply security, and minimizing environmental impact while saving on material costs. Esposito et al. believe that one of the most difficult parts of the circular economy is reverse logistics, where the most difficult part is managing the collection of waste from consumers to capture value and turn materials back into resources [3]. Such a circular model of reverse logistics allows us not only to protect the environment, save raw materials, but also to achieve maximum financial results.

The Circular Economy Action Plan of the European Union is not integrated in Lithuania's strategic documents or transposed into legislation. There is also no institution responsible for the implementation of circular economy principles. This, along with limited financial instruments, forms a barrier for Lithuania to move toward a circular economy and implement the goals of the European Green Course. Furthermore, it has been observed that the literature on reverse logistics enablers for firm-wide competitiveness is still developing [14]. To eliminate this gap, the article aims to reveal the importance of reverse logistics and the most important challenges in ensuring the implementation of the principles of circular economy. To this end, the first chapter of the article introduces the concept of circular economy. In the second, the features of closed-loop supply chains are revealed, and in the third, the essence of reverse logistics and the main activities that ensure the functioning of the closed-loop supply chain are revealed. In the fourth chapter, the problem of implementing reverse logistics activities is presented. The article ends with conclusions on the contribution of reverse logistics to ensuring the implementation of circular economy goals. The article is based on the analysis, systematization, and summarization of scientific literature examining the concept of the circular economy, the principles of closed-loop supply chain operation, and the challenges of implementing reverse logistics.

2 Concept of Circular Economy

A new European growth plan built on the European Green Deal and the idea of competitive sustainability was unveiled by the European Commission in 2020 in the Annual Plan. The European Union sets an ambitious goal for itself and for all of us, to transition to a fully circular economy by 2050.

The circular economy offers ways and solutions to achieve a sustainable lifestyle and an environment-friendly economy. The European Commission has been working on an action plan since 2015 to implement the circular economy, which consists of both industrial policy and regulatory measures. All levels of government, industry, inventors, investors, and consumers have a crucial role to play in advancing the circular economy's objectives.

The European Commission released a new circular economy action plan in March 2020 with the goal of making Europe greener and more competitive. To ensure that resources are used as long as possible in the EU economy, it encompasses activities that cover the entire life cycle of products, such as product design, promotion of circular economy processes, and the promotion of sustainable consumption. To ensure the sustainability of the European Union economy and to turn climate and environmental problems into economic opportunities, companies are encouraged to switch to a clean circular economy, whose cycle aims to preserve the value of raw materials, materials, and products as long as possible, and to reduce the amount of waste generated as much as possible. Companies are encouraged to pursue sustainable development through efforts to expand value through improved or new business models and services.

Using creative products and solutions that can be reintroduced into biological and technological cycles and processes, the circular economy seeks to avoid the depletion of resources and discover ways to protect and repair the environment [17].

To define the concept of circular economy, different authors present different definitions of the concept of circular economy. Korhonen et al. present the circular economy as a closed-loop economic system, where raw materials, products and their components lose their value the least, when renewable resources are continuously used, and systemic thinking forms the basis of the system [13]. Hofmann states that it is an economic model that effectively uses various resources to reduce waste, maintain the value of materials, assets, and reduce the use of primary resources and products [9]. Kirchherr et al. argue that it is an economic system in which manufactured products with an end-of-life cycle must be replaced by products that can be used in production more than once by recycling them [12].

In line with the idea of sustainable development, the circular economy converts the conventional one-way linear economic model of "take-produce-throw" into a feedback circular economy mode of "raw material - product - waste - renewable raw material," which uses resources and protects the environment more effectively to achieve the greatest economic and social benefits with the fewest resource and environmental costs. The circular economy includes five main areas: 1) make (product design and production processes); 2) use; 3) reuse; 4) remake using secondary raw materials; 5) recycle.

According to Ying and Zhou, the circular economy is basically an ecological economy that requires human economic activity according to the 3R principle (reduce, reuse, and recycle [10].

The circular economy pursues multifaceted goals - it encourages protecting the environment, saving materials, creating additional jobs, and saving money for producers and consumers. The circular economy not only reduces companies' dependence on limited natural resources, but at the same time helps companies create greater value. Regarding the benefits for companies following the principles of the circular economy, Jabbour states that "the most important benefits for companies are: 1) reduction of raw material waste by reducing the use of energy and materials in production, which helps companies reduce waste and carbon emissions, reduce energy-related costs, waste management and emissions control; and 2) increasing competitive advantage through process innovation [11].

The advantages provided by the circular economy are analysed by various scientists, but in practice, various challenges of the development and implementation of the circular economy are encountered. The main challenges faced by companies in implementing circular economy goals are the lack of knowledge and competence and resources, confusing and unstable market demand, and complicated cooperation between different participants.

2.1 Closed-Loop Supply Chain

Because it integrates sustainability with corporate growth, the circular economy is particularly appealing to businesses. Ritzen and Sandstrom claim that, due to the continued overconsumption of resources on a worldwide scale and the subsequent harm to the environment, such integration is essential for many businesses. However, the circular economy is seldom, and often only partially, implemented in practice [16]. This is presumably due to the significant changes, time and work needed to go from a conventional linear model to a circular economic one. From a logistic point of view, the circular economy can be seen as the integrated direct and reverse management of the movement of products or materials in the supply chain. Therefore, companies that have decided to apply/or are applying the principles of circular economy in their operating models must also apply these principles when adapting their supply chains [15].

Given the closed loop of the circular economy model, the same principle must apply to supply chains: supply chains must also become closed. The presented supply chain model (see Fig. 1) shows that in a circular economy, products and materials used in the supply chain that would already be considered waste in the direct chain are taken back into the chain for new processes and are reused.

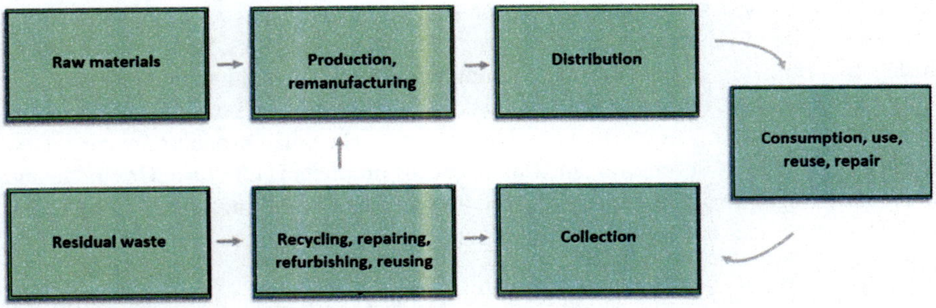

Fig. 1. Circular economy supply chain [8].

In this way, closed-loop supply chains can be considered economically and ecologically sustainable. The purpose of such chains is to close the resources in the supply chain ring to achieve the highest level of material efficiency, creating added value in the supply chain [7].

According to Defee et al., the closed-loop supply chain consists of five main processes: product acquisition, reverse logistics, inspection, disposition, and reordering. Reconditioning), distribution, and sales [2].

3 Concept and Activities of Reverse Logistics

When the shortage of raw materials began to appear in the global economy and environmental requirements appeared in legislation, companies were forced to respond to this by adapting their supply chain systems and processes. The circular economy emphasized the maximum circulation of end-of-life products and encouraged the reuse or recycling of products. As part of this aspiration, reverse logistics has become a key component of the circular economy [1].

Esposito et al. believe that one of the most difficult parts of the circular economy is reverse logistics [3]. Therefore, to understand how important reverse logistics is for the circular economy (and at the same time for the reverse supply chain), it is necessary to clearly understand how reverse logistics is defined, what activities are assigned to it, and what reverse logistics goals and processes are related to the circular economy.

The scope of the original definitions of reverse logistics is limited to the movement of materials and products in the opposite direction of the direct flow, such as from a customer to a retailer or from a supplier to a factory. Bernon et al., note that the new definitions of reverse logistics that emerged in the 1990s not only described reverse flow, but also emphasized the activities of this flow, such as disposal, remanufacturing, and remanufacturing, reuse [1].

Rogers and Tibben-Lembke stated that reverse logistics refers to the movement of materials and products from their normal destination to capture or gain value from the materials and products or dispose of them properly [18]. This means that end-use materials are treated as valuable industrial nutrients and not as waste.

Analyzing the activities carried out in the context of reverse logistics and closed-loop supply chain, activities mentioned by various authors can be observed, which are also attributed to closed-loop supply chain and separate reverse logistics. Bernon et al. distinguishes the following activities when defining reverse logistics: the management action of logistics functions, reuse and recovery actions, distribution, value recovery (recapturing values), reverse flow or return, where the return can be when products are cancelled, returned due to repair or renewal, and due to excess products [1]. In turn, Govindan and Bouzono distinguished reverse logistics activities aimed at extracting value from returned products - these are: collection, inspection and separation, reprocessing, disposal, and redistribution [6].

There is no unified agreed list of reverse logistics activities; activities can be broken down and supplemented by other processes that contribute to the goals of reverse logistics. The reverse supply chain and reverse logistics activities collected in Table 1 show the complexity and interdependence of these processes.

After evaluating the closed-loop supply chain and reverse logistics activities discussed in the scientific literature, we see that the activities are intertwined and there is no clear separation, which activities should be treated only as closed-loop supply chain activities and which are classified only as reverse logistics activities. It is therefore not surprising that some authors use the concepts of closed-loop supply chain and reverse logistics interchangeably. This is also because some activities may be performed at different points/stages of the closed-loop supply chain, and some activities may be performed multiple times, at different points in the closed-loop supply chain, and at different times, with different purposes (e.g., transportation/checking/sorting).

To ensure the implementation of circular economy principles in a closed-loop supply chain, a theoretical model of reverse logistics activity in a closed-loop supply chain can be created by performing selected basic functions of reverse logistics (see Fig. 2).

Table 1. Activities classified as reverse logistics and closed-loop supply chain.

Activity	Closed loop SC	Reverse logistics
Collection/acquisition	Defee et al., (2009)	Fleischmann et al., (2000)
Presorting/disposition	Defee et al., (2009)	Goltsos et al., (2019)
Inspection	Defee et al., (2009);	Fleischmann et al., (2000)
Sorting out	Guide Jr., Van Wassenhove, (2002)	Fleischmann et al., (2000)
Separation/extraction	Guide Jr., Van Wassenhove, (2002)	Fleischmann et al., (2000)
Transportation	Guide Jr., Van Wassenhove, (2002)	Fleischmann et al., (2000)
Storing/storaging	Guide Jr., Van Wassenhove, (2002)	Fleischmann et al., (2000)
Reprocessing	Guide Jr., Van Wassenhove, (2002)	Fleischmann et al., (2000)
Distribution	Guide Jr., Van Wassenhove, (2002)	Bernonas et al., (2018);
Return		Bernonas et al., (2018)
Returns management		Bernonas et al., (2018)
Recovery	Defee et al., (2009);	Bernonas et al., (2018)
Repairing/maintenance	Defee et al., (2009);	Fleischmann et al., (2000) Bernonas et al., (2018)
Refurbishing/reconditioning	Guide Jr., Van Wassenhove, (2002)	Fleischmann et al., (2000) Bernonas et al., (2018)
Recycling		Fleischmann et al., (2000) Bernonas et al., (2018)
Disassembly		Fleischmann et al., (2000)
Shredding		Fleischmann et al., (2000)
Remanufacturing	Guide Jr., Van Wassenhove, (2002) Defee et al., (2009)	
Reassembly		Fleischmann et al., (2000)
Product testing		Fleischmann et al., (2000)
Repacking/repackaging		Fleischmann et al., (2000)
Reuse	Guide Jr., Van Wassenhove, (2002)	Bernonas et al., (2018)
Cleaning		Fleischmann et al., (2000)
Waste management		Fleischmann et al., (2000)
Remarketing	Defee et al., (2009)	Remarketing
Resale	Defee et al., (2009)	
Disposal	Guide Jr., Van Wassenhove, (2002)	Bernonas et al., (2018)

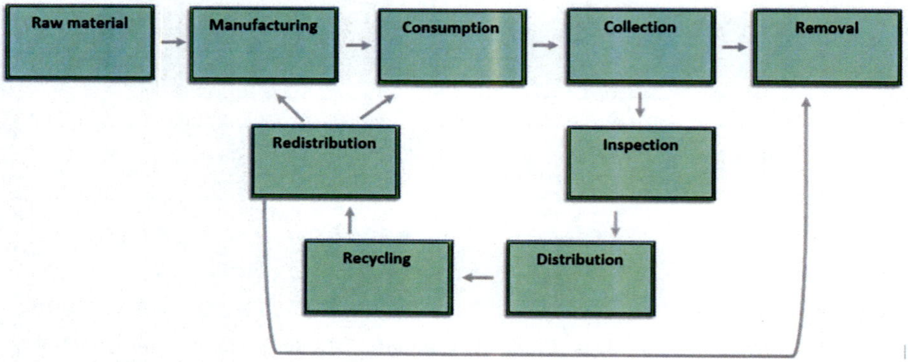

Fig. 2. Reverse logistics activity model in a closed-loop supply chain [2].

4 Challenges of Implementation of Reverse Logistics Activities

Reverse logistics is becoming an increasingly used practice to close the loop and become part of a closed-loop (annular) supply chain. According to Govindan and Bouzon, as in many fields, in reverse logistics, there are many factors that influence the successful implementation of reverse logistics. Some factors can determine/accelerate the implementation of reverse logistics, while others, on the contrary, can slow it down. The factors identified by them form a circle of success for the successful implementation of reverse logistics, where each factor has influence on the other factor.

Economic benefit and financial situation. Benefits can be direct or indirect. Direct benefits are achieved when materials are returned to the process, reducing the need for raw materials and returning value to the product. The indirect benefit of improving a company's social image is by attracting attention and attracting consumers. However, the company's limited financial resources can limit the implementation of reverse logistics, and too little understanding of the benefits of reverse processes (direct and indirect) can stop companies from obtaining economic benefits.

Legal framework or legislation. These factors are considered mandatory for companies, as they regulate and establish environmental requirements and other restrictions or instructions of activity. The absence of a legal base or unclear regulation are considered obstacles to achieving the goals of reverse logistics. Also, the introduction of the legal framework can result in an increased financial burden on the company to meet the requirements.

Corporate social responsibility. Improving social life and fostering environmental protection is one of the success factors of the company's activities, often in the long term. In this way, companies not only contribute to the development of sustainability, improve their image, but also attract users/customers who support it. Customer expectations for the company (to be socially responsible) are also one of the factors that force the company to choose the path of reverse logistics implementation, which in turn can cause the company's financial burden.

Leadership and management. The participation of the company's top management in the processes of implementation of the reverse logistics system can be a stimulating factor, and the absence of this leadership, in contrast, can stop these processes.

In turn, the authors mentioned in their article single out three areas in which logistics specialists are faced with the challenges of implementing reverse logistics:

Communication or communication challenges. For a company that implements the principles of a circular economy through the processes of a reverse logistics system, it is very important to properly communicate both with customers and with suppliers who provide the company with various raw materials and products. Even poor communication within the company can have a negative impact on processes.

Information challenges. Information is of great importance in the reverse logistics system. To prepare the process of reverse logistics, to manage it properly, it is necessary to have the right information, which allows you to see and evaluate demand and consumption needs, information on used and returned products, material and product movement times, their movement costs, restoration, and write-off costs. Information collection and processing becomes an essential factor in a successful reverse logistics system.

Process-driven challenges. In short, it is the absence of a standard process/system for reverse logistics. Since reverse logistics is a relatively new field, companies often tend to treat reverse logistics as direct logistics - only the other way around. This is a fundamental mistake. Reverse logistics must be built and designed from the very beginning of the operational process, considering all the contributions of the participants of the reverse logistics system. Undoubtedly, to evaluate the activity of such a created chain, criteria must also be created for evaluating the processes of the reverse logistics system for this. Therefore, from the beginning of the process, the company's employees, customers, and partners must have a clear picture of how the reverse logistics system works and how its processes are evaluated. Such a reverse logistics system must be standardized and approved within the company.

5 Conclusions

The circular economy is a social, environmental, and economic paradigm that pursues multifaceted goals - it encourages protecting the environment, saving materials, and creating additional jobs. To ensure the sustainability of the European Union economy, to turn climate and environmental problems into economic opportunities, companies are encouraged to switch to a clean circular economy.

The circular economy strongly depends on closed-loop supply chains that have the purpose of closing resources in the supply chain ring to achieve the highest level of material efficiency. To do so, the closed-loop supply chain must consist of several processes: product acquisition, reverse logistics, inspection and disposition, reordering, reconditioning, distribution and sales.

In order to maintain the value of assets in the supply chain as long as possible, it is necessary to ensure the operation of a closed-loop supply chain, where reverse logistics activities have a great influence. The success of these reverse activities is nothing more than a properly arranged model of the reverse logistics system, clear responsibilities, and resources of the participants of this model.

References

1. Bernon, M., et al.: Aligning retail reverse logistics practice with circular economy values: an exploratory framework. Prod. Plan. Control **29**, 483–497 (2018)
2. Defee, C.C., et al.: Leveraging closed-loop orientation and leadership for environmental sustainability. Supply Chain Manag. Int. J. **14**(2), 87–98 (2009)
3. Esposito, M., et al.: Reverse logistics for postal services within a circular economy. Thunderbird Int. Bus. Rev. **60**, 741–745 (2018)
4. Geissdoerfer, M., et al.: The circular economy – a new sustainability paradigm? J. Clean. Prod. **143**(4), 757–768 (2017)
5. Ghisellini, P., et al.: Exploring environmental and economic costs and benefits of a circular economy approach to the construction and demolition sector. J. Clean. Prod. **178**, 618–643 (2018)
6. Govindan, K., Bouzon, M.: From a literature review to a multi-perspective framework for reverse logistics barriers and drivers. J. Clean. Prod. **187**, 318–337 (2018)
7. Guide, V.D., Jr., Van Wassenhove, L.N.: The evolution of closed-loop supply chain research. Oper. Res. **57**(1), 10–18 (2009)
8. Habibi, M.K.K., et al.: Collection-disassembly problem in reverse supply chain. Int. J. Prod. Econ. **183**, 334–344 (2017)
9. Hofmann, F.: Circular business models: business approach as driver or obstructer of sustainability transitions? J. Clean. Prod. **224**, 361–374 (2019)
10. Ying, J., Zhou, L.: Study on green supply chain management based on circular economy. Phys. Procedia **25**, 1682–1688 (2012)
11. Jabbour, A.B.: Going in circles: new business models for efficiency and value. J. Bus. Strateg. **40**(4), 36–43 (2018)
12. Kirchherr, J., et al.: Conceptualizing the circular economy: an analysis of 114 definitions. Res. Conserv. Recycl. **127**, 221–232 (2017)
13. Korhonen, J., et al.: Circular economy as an essentially contested concept. J. Clean. Prod. **175**, 544–552 (2018)
14. Larsen, S.B., et al.: How the reverse supply chain impacts the firm's financial performance: a manufacturer's perspective. Int. J. Phys. Distrib. Logist. Manag. **48**(3), 284–307 (2018)
15. Lüdeke-Freund, F., et al.: A review and typology of circular economy business model patters. J. Ind. Ecol. **23**(1), 36–61 (2018)
16. Ritzén, S., Sandström, G.Ö.: Barriers to the circular economy – integration of perspectives and domains. Procedia CIRP **64**, 7–12 (2017)
17. Sandoval, V., et al.: Key elements in assessing circular economy implementation in small and medium-sized enterprises. Bus. Strateg. Environ. **27**(8), 1–10 (2018)
18. Tibben-Lembke, R.S., Rogers, D.S.: Differences between forward and reverse logistics in a retail environment. Supply Chain Manag. Int. J. **7**(5), 271–282 (2002)

Analysis of Theoretical Aspects of Supply Chain Resilience Determinants and Strategies

Aidas Vasilis Vasiliauskas(✉) and Olga Navickienė

General Jonas Žemaitis Military Academy of Lithuania, Šilo Street 5A, 10322 Vilnius, Lithuania
{aidas.vasilisvasiliauskas,olga.navickiene}@lka.lt

Abstract. Today, you can find quite many various paper on the topic of supply chain resilience. Although the topic of supply chain resilience is fully covered, the rapid development of humanity creates new opportunities and obstacles for supply systems and thereby affects their sustainability, and therefore there is a need for its further research from the point of view of today. The emergence of new technologies, such as AI, increasing popularity of certain concepts, such as the concept of sustainability, can shake the current understanding of the concept of supply chain resilience and become challenges for the resilience of existing systems and enterprises. Such novelties lead to changes in company activities, national legislation, and trade relations. These externalities and related issues may have relevance over the next few years and possibly the next decade, so it is worth investigating them and countermeasures to ensure sustainability now to minimize future negative risks and their impact. The main objective of this paper is supply chain disruptions and resilience strategies. The paper seeks to provide a theoretical overview of the concept of supply chain resilience, identify all possible challenges for supply chain resilience, and uncover the main strategies intended at preventing disruptions in the supply chains.

Keywords: Supply chain · Supply chain resilience · Supply chain disruptions · Supply chain uncertainty

1 Introduction

Today, the concept of globalization is raging in the world. Elements of the same product are produced and assembled into finished products in different countries and transported to the final consumer anywhere in the world. The final products contain foreign and domestic resources added at various stages of production and delivered through international supply chains. Supply chains enable global production and international trade. However, the further the concept of the global economy develops, the more complex and multifaceted the chains become.

Looking at statistics on the global supply chain system and some chains of international companies, we can see a trend that production capacity is more concentrated in Asia, but innovation and new technologies tend to originate in Europe and North America. This means that in most cases other countries have a close relationship with the

O. Prentkovskis et al. (Eds.): TRANSBALTICA 2023, LNITI, pp. 495–503, 2024.
https://doi.org/10.1007/978-3-031-52652-7_49

regions mentioned above. In addition, supply chain costs are one of the largest groups of elements in the total cost of the product. Supply chain logistics accounts for 5% to 50% of the total cost of a product. This category of costs may include costs for transportation, storage, packaging, and other services. The more complex the product, the more chains it needs, which accordingly increases the cost and complicates the implementation process.

Understanding the above, we conclude that the supply chain plays an important role in the world economy. Supply chain issues can cause production interruptions, temporary shortages of products or materials, an increase in the final price, a stop or slowdown in trade, social or environmental problems, and a slowdown in the pace of development of the world economy.

Disruptions in supply chains have a negative impact not only on the company but also on other stakeholders.

In the literature, there is a classification of uncertainties. There are two categories: low chance causing high impact and high chance causing low impact [6]. The second type occurs more often and can be a result of demand fluctuation, error in planning or minor issues during production or distribution processes, etc. The damage is small and can be resolved quickly. The first type (low chance but high impact) is quite a rare uncertainty, but when it appears, it results in negative output and requires a longer time to solve it and restore the company's activities to the previous or better level. The damage caused by such events is greater than from usual operational risks.

To deal with uncertainty, managers use different strategies and techniques of strategic risk management. Nevertheless, each uncertainty needs its own strategy to solve it, but traditional methods cannot deal with modern issues, so we need to find new solutions. Today in the literature and practice, the concept of supply chain resilience has become popular and appeared to be an effective tool to overcome challenges in the supply chain. This concept involves not only managing uncertainty but also improving the efficiency of supply chain networks and competitive advantage for companies.

2 Concept of Supply Chain Resilience

The first mentions of the concept of supply chain resilience can be found in the early 2000s. However, the concept of "resilience" was used earlier in other sciences, such as psychology, ecology, sociology, and others. The concept of resilience is of Latin origin and means "bounce back". In the Middle Ages it had a more negative meaning, in the Middle French (résiler) it meant "to retract" or "cellar" [1].

In social sciences, resilience refers to the ability of individuals, communities, or societies to cope with and recover from various challenges, such as natural disasters, economic crises, or social unrest. It includes the ability to mobilize necessary resources in optimal time, adapt, rebuild social systems for better effectiveness, and maintain social cohesion. Social resilience is more related to the resilience of society, rather than specific individuals or groups. This means the distinct values and traditions of societies: their customs, languages, religions, and spiritual heritage, ethnic distinctions, and much more.

Economic resilience includes the capability to lessen possible weaknesses, to encourage more robust and inclusive growth, to improve the macro-economy's ability to resist shocks, and to reallocate resources in order to take advantage of new growth possibilities.

The type, amount, and direction of these connections, as well as the degree of diversity in the scope and geography of its commerce and the scale of its external imbalances, all affect a country's ability to withstand economic shocks.

Scientists highlighted resilience of an organization in the management field. The ability of an organization to foresee potential dangers, to successfully deal with unfavorable occurrences, and to adapt to changing conditions should be understood as organizational resilience. S. Duchek argued that in the scientific literature organizational resilience is defined using 3 different approaches or a combination of 2 of them [3]. She mentioned such perspectives:

- defensive, from this point of view, resilience is the ability to resist adverse situations and/or the ability to recover after disturbances and return to a normal state;
- improving, focused on the need to adjust any changes to come out of the crisis stronger than before;
- anticipation (notion); resilience involves activities of identifying possible risks and taking prevention steps.

Supply chain resilience refers to the supply chain's ability to withstand disruptions and other issues and recover from them while maintaining its core functions and meeting customer needs. This involves the ability to anticipate, adapt, and effectively respond to various changes that may affect the flow of goods, services, and information in the supply chain.

The term "resilience" suggests that, despite the "outrage", the supply chain system keeps unchanged those properties and characteristics that make it a given system. The resilience of the supply chain system can also be understood as the ability of the system to maintain a dynamic balance with the environment, or as the ability of the system to compensate for the influence of external environmental factors.

We have collected several definitions of supply chain resilience, which scientists have given in their works, in Fig. 1. Placing the publications in chronological order, it is possible to trace how the scientists' opinions about the phenomenon changed. The range of years of the taken literature is 17 years (2005–2022). Examining the works of scientists, we observed that their opinion was influenced not only by the literature studied by them, but also by the specifics of their field of activity and their background.

In these definitions, it is possible to trace the changes in approaches that Stephanie Duchek noted in her works. Indeed, in the early definitions, the use of a defensive approach is observed, since 2013, we can observe that scientists began to think not only about restoring the system to the previous state, but also about the growth of indicators to the new desired state (improving perspective). Closer to 2020, the opinion shifts to anticipation approach and in definitions we can notice relation to risk management.

In the literature, there is a classification of uncertainties. There are two categories: low chance causing high impact and high chance causing low impact [6]. The second type occurs more often and can be a result of demand fluctuation, error in planning or minor issues during production or distribution processes, etc. The damage is small and can be resolved quickly. The first type (low chance but high impact) is quite a rare uncertainty, but when it appears, it results in negative output and requires a longer time to solve it and restore the company's activities to the previous or better level. The damage caused by such events is greater than from usual operational risks.

Authors	Definition of Supply chain resilience	Decade
Estrada, 2005	SCR implies the ability of a system to return to its original or desired state after being disturbed.	2001-2010
Shaffi, 2005	In the corporate world, resilience refers to the ability of a company to bounce back from a large disruption—this includes, for instance, the speed with which it returns to normal performance levels (production, services, fill rate, etc.).	
Datta, Christopher and Allen, 2007	SCR is the ability of the production–distribution system to meet each customer demand for each product on time and to quantity.	
Datta, 2007	SCR is defined as not only the ability to maintain control over performance variability in the face of disturbance, but also a property of being adaptive and capable of sustained response to sudden and significant shifts in the environment in the form of uncertain demands	
Barroso et al., 2010	SCR is the supply chain's ability to react to the negative effects caused by disturbances that occur at a given moment in order to maintain the supply chain's objectives	
Li and Zhao, 2010	SCR is self-adaption and self-recovery abilities.	
Ponis and Koronis, 2012	SRC is the ability to proactively plan and design the Supply Chain network for anticipating unexpected disruptive (negative) events, respond adaptively to disruptions while maintaining control over structure and function and transcending to a post-event robust state of operations, if possible, more favorable than the one prior to the event, thus gaining competitive advantage	2011-2020
Pettit et al., 2013	SCR is the ability to survive, adapt, and grow in the face of turbulent change	
Hohenstein et al, 2015	SCR is the supply chain's ability to be prepared for unexpected risk events, responding and recovering quickly to potential disruptions to return to its original situation or grow by moving to a new, more desirable state.	
Li et al., 2017	SCR refers to a supply chain's capability to cope with changes, which is formed through being prepared to endure future changes, being alert to changes and being agile in response to changes.	
Sangari and Dashtpevma. 2019	SCR is the capability of being responsive to unplanned changes and disruptions.	
Dormady et al., 2019	SCR is able to endure the crisis and able to adjust flexibility to retrieve back to its sustainable state as soon as possible.	
Dixit et al., 2020	SCR is an adaptive potential of a SC to prepare and get ready for unpredictable future events and correspondingly retaliate quickly towards disruptions and recuperate back by sustaining continuity of SC processes at the desired level of excellence.	
Um and Han, 2021	SCR is the ability to tolerate supply chain risks in sourcing, manufacturing and delivery.	2021-
Grzybowska and Tubis, 2022	SRC is the capability to anticipate crises, identify risks, increase their impacts, adapt and respond quickly to threatening disruptions or vulnerabilities in the supply chain, and return to normal	

Fig. 1. Definition of supply chain resilience.

To deal with uncertainty, managers use different strategies and techniques of strategic risk management. Nevertheless, each uncertainty needs its own strategy to solve it, but traditional methods cannot deal with modern issues, so we need to find new solutions. Today in the literature and practice, the concept of supply chain resilience has become popular and appeared to be an effective tool to overcome challenges in the supply chain. This concept involves not only managing uncertainty, but also improving the efficiency of supply chain networks and competitive advantage for companies.

3 Factors of Resilience

Before going to the factors of resilience, it is necessary to look closer to the resilience triangle (see Fig. 2). By looking at R-triangle, we can understand what resilience is and how it works. A visual concept of resilience is based on two criteria (time and performance) and illustrates the fluctuation of performance after disruption and the time needed to recover to the previous state. When a crash event occurs, you may see a dramatic drop in performance, and after that it takes time for SC to restore its profit or other performance indicators. The main aim of supply chain managers is to reduce the area of the R-triangle to minimum. The larger the area of this figure, the lower the resilience ability of the enterprise or system. Different resilience mechanisms can adjust the dimensions of the figure to the optimal and help maintain performance indicators.

In the literature, several approaches can be found to determine the factors (elements) of supply chain sustainability. First, we suggest considering the approach of division into elements in the work of Petit et al. [7]. They offer to look at such elements as vulnerabilities and capabilities; further authors develop concept of zone of resilience which occurs as a balance between those two factors. Both vulnerabilities and capabilities—attributes that provide a company with the ability to foresee and overcome disruptions—are important elements that make an enterprise sensitive to disruptions. The increase of one of them will lead to an increase of negative outcomes (see Fig. 3). Not having sufficient capabilities given the vulnerabilities a firm faces exposes the firm to risk, but investing in capabilities that are not needed can reduce profits. The company should find the optimal ratio between those two factors to obtain resilience.

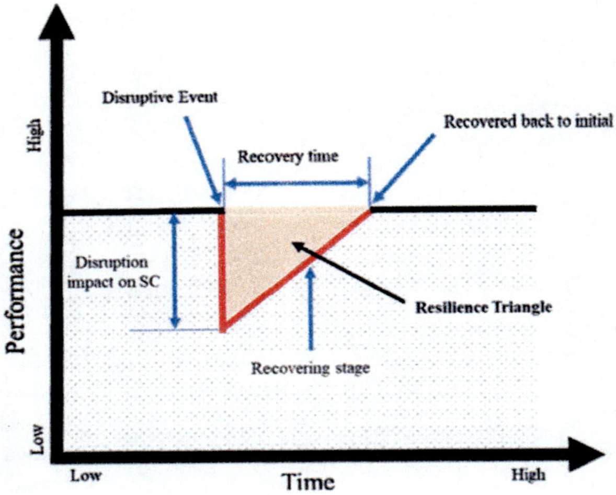

Fig. 2. Resilience triangle [4].

Petit et al. divide these two factors into 21 sub-factors, which in turn are divided into an even greater number of smaller sub-factors. We suggest that you familiarize yourself with the grouping in Fig. 4. The division into sub-factors was created to subjectively measure factor. Later, in their next work in 2013, they developed a measurement tool and conducted an experiment that revealed critical linkages between factors and suggests a correlation between increased resilience and improved supply chain performance.

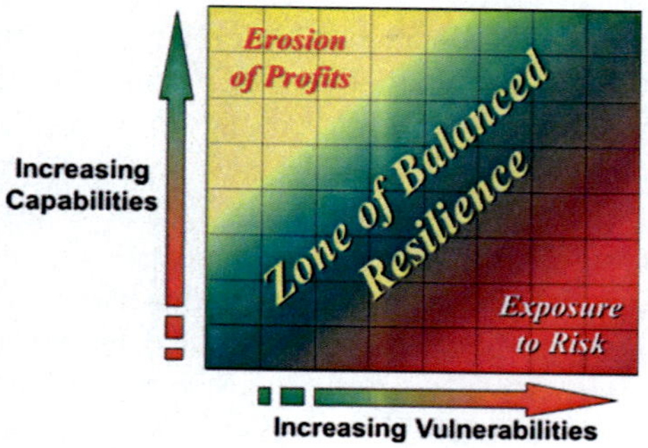

Fig. 3. Resilience fitness space [7].

Let us consider another approach to determining the factors of supply chain resilience. Juttner and Maklan offered their vision of factors which they called 'formative elements' [5]. According to the authors, formative factors are important from a supply chain management viewpoint because they explain how the preparation, reaction, and recovery of supply chain event can be ensured. They also mentioned that they used the capability approach to identify key elements. Due to their study, there are 4 formative elements of SCR: flexibility, velocity, visibility, and collaboration.

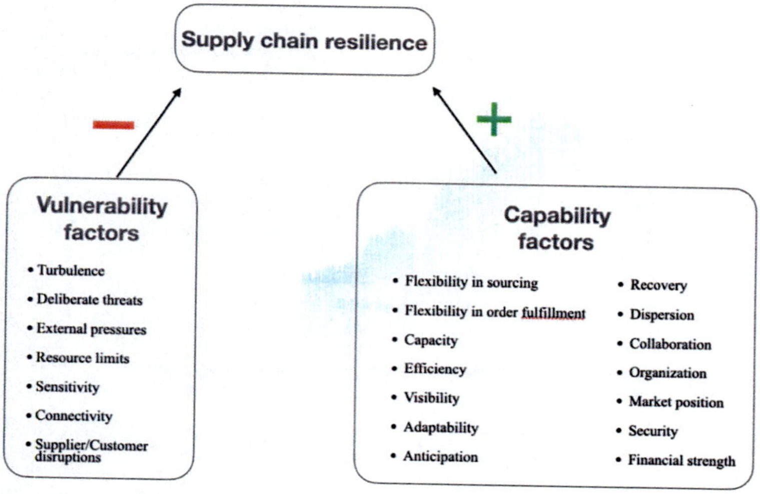

Fig. 4. Supply chain resilience factors [5].

Flexibility is defined as "being able to bend easily without breaking". Velocity is described as the "speed of motion, action, or operation, rapidity and swiftness" and is measured as a function of time and distance.

Visibility refers to the capability of "being perceived by the eye or mind". Collaboration means "working with another or others on a joint project".

4 Supply Chain Resilience Strategies

In the literature, we can find quite a wide range of supply chain resilience strategies. All of them are divided into proactive and reactive groups of strategies. Both types of product are developed to combat disruption. The difference between those two groups is the approaches and timing of the actions taken. Proactive strategies are applied before the disruption event occurs; usually they are used during the planning stage. In such strategies, optimum supply chain network structures are considered. Reactive strategies are applied when a collapse has happened. In such strategies, the optimum control policies are managed. Joshi and Huynh showed the previously mentioned division (see Fig. 5).

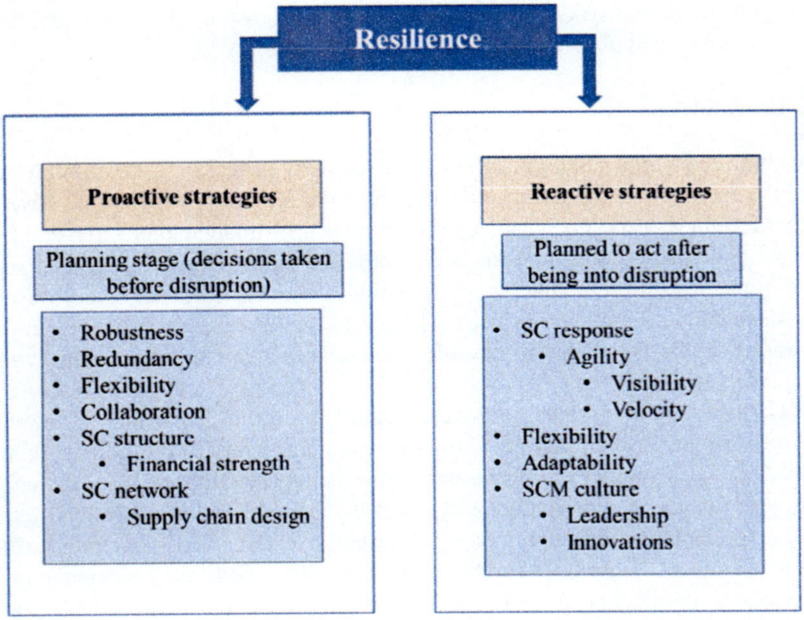

Fig. 5. Categories of supply chain resilience strategies [4].

Visibility refers to the capability of "being perceived by the eye or mind".

It is worth noting that the division is conditional, as some of the strategies can move from one category to another. Attribution of the strategy to a certain category in this case will depend on the purpose and time of its application.

However, both types of strategy are often used together. Scientists mentioned that there is an interrelation between some of the strategies and few of them support or/and reinforce each other.

Tukamuhabwa et al. made a table of 24 strategies [8]. They divided them into 2 categories mentioned above. In the work, Tukamuhabwa et al. highlights four key strategies: flexibility, redundancy, supply chain collaboration, and supply chain agility.

The flexibility strategy aimed to increase the ability to response effectively to changes caused by external and internal circumstances.

The redundancy strategy aims to ensure the usual activity of the company with the help of additional resources (inventories). Redundancy is the deliberate and selective use of inventory and spare capacity that can be used during a crisis to deal with, for example, supply shortages or demand spikes [2]. This strategy needs to be applied carefully, as extra capacity means extra expense.

The agility strategy aims to improve the reaction of the supply chain network to disruption. Supply chain agility, according to Christopher and Peck, is the capacity to react quickly to unexpected changes in demand or supply [2].

The collaboration strategy aims to use relationships with other stakeholders to combat disruption. Supply chain cooperation, according to Pettit et al., is the capacity to collaborate successfully with other organizations for mutual gain in areas like forecasting, postponement, and risk sharing. The collaboration strategy is an effective tool in complex structures supply chain networks [7].

5 Conclusions

Supply chain resilience can be described as a recovery process after the failure that is measured by performance and time indicators. The definition of resilience has developed through different spheres. Each sphere brought novelty to understanding and created the possibility to look at the term from other points of view. Moreover, we can see the positive impact of different spheres, such as the effect of risk management on supply chain management. Three core aspects of SCR (defense, recovery and improvement, anticipation) should be taken into consideration today when building or improving the supply chain network.

Today, there is no agreement between scientists on the list of factors (elements) that form resilience. Some authors said that resilience refers to the balance between capability and vulnerability of the company, others suggested that resilience consists of 'formative elements': flexibility, velocity, visibility, and collaboration. In the literature, 13 formative elements are mentioned by different scientists. From our point of view, the list of factors will change in further years as we are a new real knowledge and the development of civilization provokes us to develop our vision of various phenomena. Therefore, further research and discussions of this topic will be of relevance.

Supply chain resilience strategies are sets of ideas and actions that aim to improve the resilience of the supply chain network or company. Strategies are categorized into two groups: proactive and reactive. Although scientists pay more attention to reactive strategies, in practice managers use a combination of several different strategies. The ability of SCR strategies to enforce each other stimulates managers to use a mix of such measures.

References

1. Alexander, D.E.: Resilience and disaster risk reduction: an etymological journey. Nat. Hazards Earth Syst. Sci. **13**, 2707–2716 (2013)
2. Christopher, M., Peck, H.: Building the resilient supply chain. Int. J. Logist. Manag. **15**(2), 1–14 (2004)
3. Duchek, S.: Organizational resilience: a capability-based conceptualization. Bus. Res. **13**(1), 215–246 (2019). https://doi.org/10.1007/s40685-019-0085-7
4. Joshi, S., Huynh, L.T.: Review of supply chain resilience: phenomena, modelling techniques and framework of resilience building strategies with future research directions. Int. J. Supply Chain Oper. Resil. **5**(2), 150–184 (2022)
5. Juttner, U., Maklan, S.: Supply chain resilience in the global financial crisis: an empirical study. Supply Chain Manag. Int. J. **16**(4), 246–259 (2011)
6. Nezamoddini, N., et al.: A risk-based optimization framework for integrated supply chains using genetic algorithm and artificial neural networks. Int. J. Prod. Econ. **225**, 107569 (2020)
7. Pettit, T.J., et al.: Ensuring supply chain resilience: development of a conceptual framework. J. Bus. Logist. **31**(1), 1–21 (2010)
8. Tukamuhabwa, B.R., et al.: Supply chain resilience: definition, review, and theoretical foundations for further study. Int. J. Prod. Res. **53**(18), 1–32 (2015)

Choosing Optimal Maintenance Service Level Depending on Financial Model of an Airline

Roman Fedorov, Danila Larin(✉), Vladimir Perekrestov, and Timur Tyncherov

Transport and Telecommunication Institute, 1 Lomonosova Street, Riga 1019, Latvia
danila.larin1487@gmail.com

Abstract. Highly accurate spare part management is a prerequisite for airlines and aircraft maintenance and overhaul organizations to survive if they want to be profitable and ensure regular flights. Poor planning may result in financial resources being frozen in the form of spare parts that have been bought as well as render timely aircraft maintenance and overhaul impossible due to the lack of spare parts. Today, many airlines and aircraft maintenance and overhaul organizations have learnt how to accurately forecast the demand for spare parts in order to maximize spare parts stock service level. However, reaching the maximum of service level may require financial costs, which can eliminate all benefits from the operational efficiency of the airline's fleet. This work considers the issues of finding the optimal service level, depending on the conditions in which the airline operates.

Keywords: Aviation spare parts · Planning uncertainty · Inventory management

1 Introduction

Aviation business is extremely exposed to external factors. Political instability, fuel prices, currency exchange rates, etc. have a destabilizing effect on airlines. Even volcanic eruptions have a strong effect on profitability. In such circumstances, airline cost management is a necessary condition for the stable development of an airline.

A significant part of the airline's expenses is occupied by aircraft maintenance. According to the International Air Transport Association (IATA), the share of maintenance costs is 10–12% and plays a remarkable role. In turn, in the structure of maintenance costs, a significant part is spent on spares and inventory.

The global stock of spare parts reaches 80 billion US dollars, and the annual cost of used parts is only about 40 billion US dollars. The inefficiency of warehouse usage leads to the "freezing" of large sums of money. Added to this are the costs of organizing storage (personnel, storage facilities, equipment, electricity, and the cost of maintaining the specified parameters for temperature and humidity). In addition, revealed cases of overproduction in the aviation industry, being a significant part of the distribution, falls on the shelves of warehouses. And most importantly, this problem is manifested in the severity of the impact on the environment. From financial management point of view the

O. Prentkovskis et al. (Eds.): TRANSBALTICA 2023, LNITI, pp. 504–516, 2024.
https://doi.org/10.1007/978-3-031-52652-7_50

average carrying cost has been estimated at 29% [1]. This ensures additional expenses of an airline and reduces the profit of an airline.

The number of spare parts and inventory of an airline depends on different factors. The level of inventory can have a significant impact on an airline's operation. Availability of spare parts allows to fix defects quickly and prevent or reduce time delay. That is very important for an aircraft operator as long flight delays may be costly due to special fees that airlines should pay to passengers in such a case. However, the level of inventory varies from airline to airline.

According to IATA Maintenance Cost Technical Group (MCTG) [2] report, the average inventory value is $3.23M per aircraft. Important to note that this value varies between regions (see Fig. 1). It is seen from Fig. 1 that European and American operators have similar levels. The Middle East and North Africans (MENA) and the Asia Pacific operators' warehouse averages are around $4.5M per aircraft, which is more than twice bigger than the European and American levels. Such differences may be explained by airline location factors. MENA and the Asia Pacific regions are far from the location of main airplanes and airplane parts manufacturers (so-called original equipment manufacturer - OEM). The amount of shop for repair of aircraft spare parts in Europe and America regions are bigger than in other regions. Therefore, lead times (for purchased) and turnaround (for repaired) parts are longer for MENA and the Asia Pacific airlines. On the other side, the European Union laws in terms of flight delay fees are stricter than for example in the Middle East. It means that airlines to prevent additional costs must pay more attention to prevention measures, and home base stock is one of them.

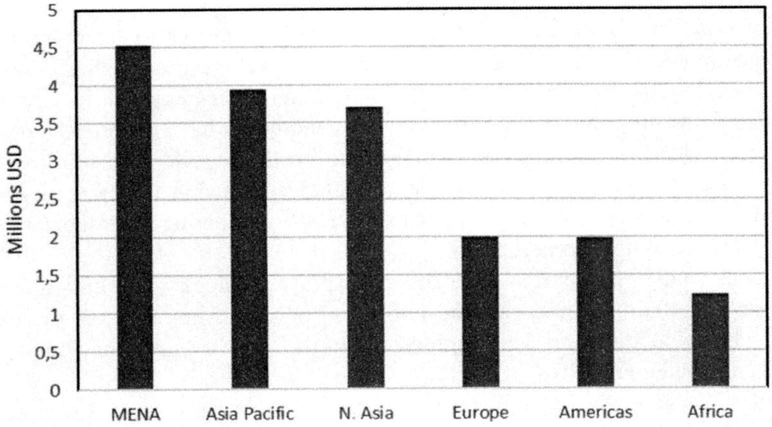

Fig. 1. Spare parts and inventory held per 1 aircraft in area (source: IATA MCTG report [2]).

The process of formation of the list of the spare parts, purchasing, storage and reordering are referred to area of inventory management. The goal of inventory management normally is to provide the highest possible level of service at the lowest total cost. The number of spare parts required for an operation of an aircraft depends on several factors, such as operational or reliability goals, financial factors, removal rates for scheduled and unscheduled removals, parts repair turnaround time, scrap rates for the removed asset, replenishment lead times for replacement assets, flight schedules, freight options, and a lot of other factors [1].

In practice, due to the weak interaction between the technical and commercial divisions of airlines, the level of warehouses is often beyond the point of optimum in relation to the level of service required at the moment.

As well there is another statement in industry that service level must be as high as possible to maximize the profit. But it is only consideration, and this statement is a subject to be verified and that is exactly the problem that this paper tackles.

The purpose of this work is to find the optimal service level model in terms of the ratio of income and expenses for the maintenance of the airline; determine which parameters have the greatest impact on service level model; and what are the limitations of using the optimal service level model in connection to airline industry.

2 Literature Review

Not so many articles could be found concerning the maintenance service level and its impact on airline company operation and its profitability. As for profitability the following works that identify the key factors for success could be revealed: in the study of US Regional Airlines operations [3] the authors have identified that internal factors like operation efficiency, cost control and aircraft utilization are of higher importance and influence on profitability, than external factors like oil prices or the cost of capital. In the study on effects of internal resources on airline competitiveness [4] the researchers pointed out that in order to perform well an airline have to ensure the best ration between adequate level of service and relatively low cost.

In scientific literature the term service level is mostly referred to inventory management and availability of stock when dealing with spare parts or any other types of the goods and under condition that stock is run on continuous review policy. Service level is not a fixed equation that is used in every case and could have different approaches. Each time service level is established depending on the requirement and goal of the company, analyst, or researcher. Service level could be defined in a number of ways like the probability that stockout will not occur or average number of stockouts per year or average delay in filling backorders [5].

In recent studies on service level the availability of stock was studied basically focused on calculation of the gap and allocation of the inventory level to customer demand. The calculations are well represented in the work on inventory management and production scheduling [6].

In today's studies the researchers are trying to find a way on how to improve the service level without buying extra stock [7] and how to implement lean inventory management in the strategy of the company to minimize loses and identify the demand [8]. For example, in oil and gas industry some researchers are trying to identify the best inventory model for company performance using ABC approach and different service level for every part depending on its price [9]. Other companies could use vital, essential and desirable (VED) or high price, medium price and low price (HML) to work out the best inventory strategy for company operation [10].

Summarizing above it could be stated that to work out a proper ratio of service level required and operation performance it is of vital importance to identify the key success parameters that are important for business. A good review on the breakdown of key parameters for airline companies across various articles are listed in the literature review of an article [11] that devited to the impact of operational performance on profitability in Turkish Airlines.

The aircraft maintenance aspect has been extensively studied in terms of its impact on the financial model. This research work was carried out in order to study the impact of the aircraft maintenance process and especially the impact of inventory management on the level of service and efficiency of the airline, as well as the search for the optimal level of service.

3 Research Methodology

The authors of this study operate under the premise of rational behavior by an airline company and the airline's aim to maximize profit. Therefore, it is reasonable to hypothesize that the optimal level of service for the airline would be the one that maximizes the airline's profit. To do this, it is necessary to ascertain the nature of the relationship between the airline's profit and the level of service. To define the optimal service level (SL) as a first step an influence of service level itself on the revenue and cost of airline was examined. As a core function a classical exponential dependency was used: an increase of service level causes the growth of investment in stocks of spare parts and expenses involved and at the same time this will cause the growth of available flight hours for operations. The dependency is represented on the Fig. 2 and Fig. 3.

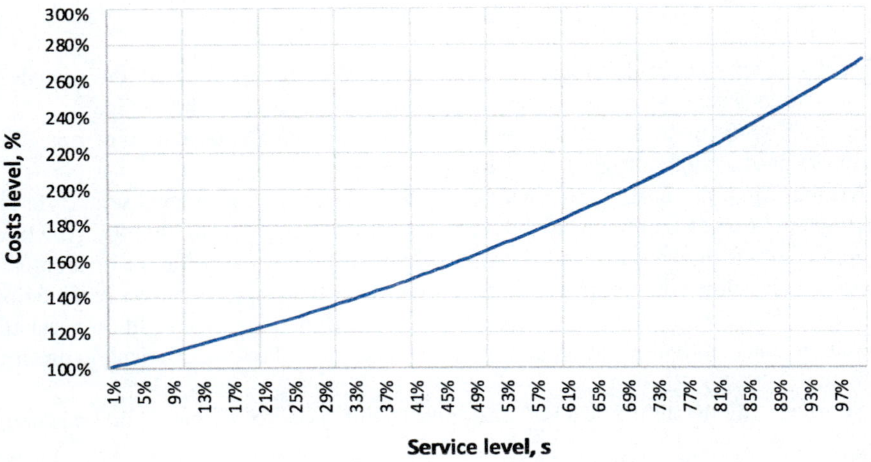

Fig. 2. Dependency of the airline costs on the service level.

Service level in our case shows the availability of flight hours depending on the availability of the spare parts (cost) and at the same time it shows the dependency of flight hours on the airline profit. Summarizing the above said service level in this article defines how available flight hours influence the profit of an airline.

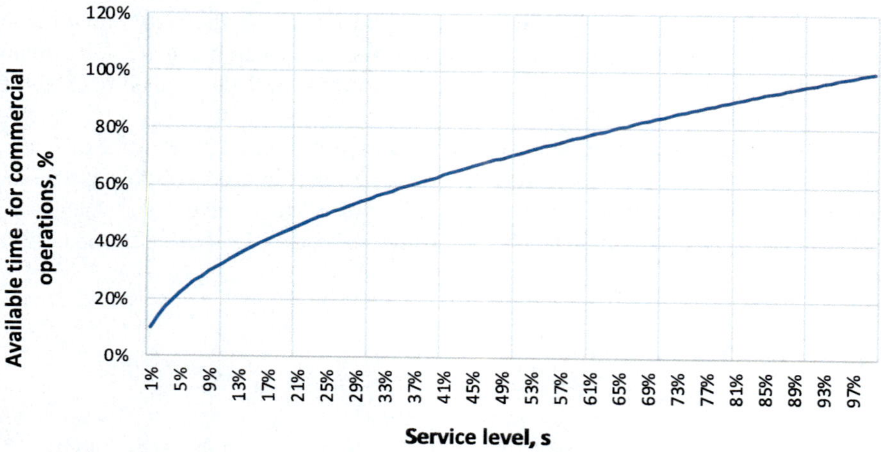

Fig. 3. Dependency of available flight hours on the service level.

Airline cost in these calculations stands for cost on spare parts required for aircraft operations. Available flight hours stand for the availability of aircraft in hours to be used in commercial operations carrying passengers.

In the most basic view, the profit function is simple and equals to a difference of revenue and cost. In our case the profit function was defined the following basing on the result of empirical analysis of airline operations:

$$A\sqrt{s} - Be^{s}, \quad (1)$$

where A – coefficient of commercial effectiveness of an airline showing the growth of revenue when service level of available flight hours changes, s – service level,

B – coefficient of cost effectiveness of an airline showing the level of expenses when service level of available flight hours changes.

On the Fig. 4 the change in dependency of profit from service level with different meanings of A and B is represented. Examples shown at Fig. 4 state that the growth of service level will make costs and revenue behave differently at different meanings of service level: below some points revenue grows faster than costs and over some points costs grow faster than revenue. These different influences of service level on costs and revenue means that there is an optimal point of service level that maximize the profit of an airline.

Summarizing the above said and examples in Fig. 4 the following could be stated:

- the function has the optimal point,
- the optimal point is different with different meanings of A and B,
- at some meanings for A and B optimal meanings of service level could be outside of the range for service level that is limited from (0) and (1) in our case,
- meaning of the function in optimal point depends on both parameters A and B, but the optimal location on SL axe will remain the same when with the change of parameters, A and B the relation in-between them stays stable.

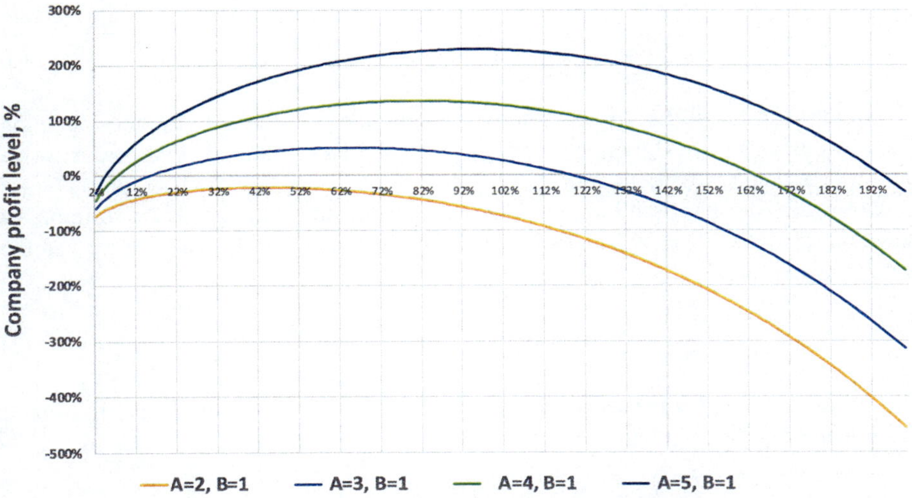

Fig. 4. Profit function at different meanings of A and B.

The general view of airline profit dependency is represented in the following equation:

$$A\sqrt[a]{s} - Be^{bs}, \tag{2}$$

where s – service level, $A\sqrt[a]{s}$ – airline revenue, Be^{bs} – expenses of an airline, a – operational efficiency coefficient showing the influence of service level change on the available flight hours, A – commercial efficiency coefficient of an airline showing growth of revenue with the growth of flight hours, b – inventory efficiency coefficient showing the level of service level change because of the growth of the spare part stock, B – cost efficiency coefficient of an airline showing the costs and price level of spare parts consumption.

The problem of optimal service level search basing on the goal to maximize profit was represented the following way: $A\sqrt[a]{s} - Be^{bs} \rightarrow max$ (s), where s = [0;1].

To solve this task Eq. (2) was differentiated and equaled to 0 to find the optimum:

$$\frac{\left(A\sqrt[a]{s} - Be^{bs}\right)d}{dS} = \frac{A}{a\sqrt[a]{s^{a-1}}} - Bbe^{bs} = 0 \text{ or } \frac{A}{a\sqrt[a]{s^{a-1}}} = Bbe^{bs}, \tag{3}$$

or modified:

$$\frac{A}{abB} = s^{\frac{a-1}{a}} e^{bs}. \tag{4}$$

Then left and right part was on exponentiation with a/(a-1) and afterwards multiplied with $\frac{ab}{a-1}$. Result is represented in the Eq. 5 and 6 below:

$$\left(\frac{A}{abB}\right)^{\frac{a}{a-1}} = se^{\frac{ab}{a-1}s}, \tag{5}$$

$$\frac{ab}{a-1}\left(\frac{A}{abB}\right)^{\frac{a}{a-1}} = \frac{ab}{a-1}se^{\frac{ab}{a-1}s}. \tag{6}$$

Multiplication $\frac{ab}{a-1}$ Eq. (6) was done to get it in line and similar to Lumbert function [12]. In our case it has graphical and equation similarity and suit our goal to solve multitask and complex problem.

For received Eq. (6) the Lumbert identity was applied meaning that the solution for equation $xe^x = q$ is $x = W(q)$, where $W()$ – Lumbert function. In our case we had the following solution:

$$\frac{ab}{a-1}s = W\left(\frac{ab}{a-1}\left(\frac{A}{abB}\right)^{\frac{a}{a-1}}\right). \tag{7}$$

Out of this equation the service level s^* that maximizes the profit of an airline could be defined the following way:

$$s^* = \frac{a-1}{ab}W(\frac{ab}{a-1}\left(\frac{A}{abB}\right)^{\frac{a}{a-1}}), \tag{8}$$

where $W()$ – Lumbert function.

For visual representation and for understanding of function behavior the following meanings of parameters have been used: a = 2 (revenue function becomes a square root from service) and b = 1 (revenue function follows exponential growth with growth of service level). As a result have received the following optimal meaning of service level:

$$s^* = \frac{1}{2}W\left(\frac{2}{1}\left(\frac{A}{2B}\right)^{\frac{2}{1}}\right) \text{ or } s^* = \frac{1}{2}W\left(\frac{A^2}{2B^2}\right). \tag{9}$$

Lumbert function is represented on the Fig. 5, and has the following meanings in the key points:

$$W(0) = 0,\ W(e) = 1,\ W(1) = 0.567(\text{constant omega}).$$

As meanings $\frac{A^2}{2B^2}$ couldn't be negative (<0), then only the right side, which is positive, of the function was examined $x = \frac{A^2}{2B^2}$.

Then $s = W(x)/2$ (see Fig. 6) and following:

Based on the identified dependency of the optimal service level on x ($\frac{A^2}{2B^2}$), and considering that the maximum service level is equal to 1, we can draw the following conclusions:

1. If $\frac{A^2}{2B^2}$ is equal to or greater than 14.778, then the optimal service level is 1. In other words, airline companies should increase their existing level of service to the maximum extent possible.
2. If $\frac{A^2}{2B^2}$ is less than 14.778, then the optimal service level is less than 1. The airline company, by calculating this value and knowing the current level of service, should strive to reduce the gap between the current and optimal service levels.

Note: This meanings of a and b were used only to represent visually the function. In case other a and b meanings are used, optimal relationship of service level and coefficients (A, B) could be different.

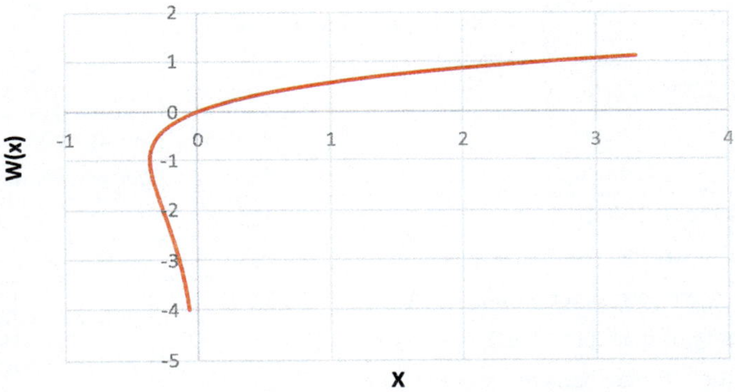

Fig. 5. General view of Lumbert function.

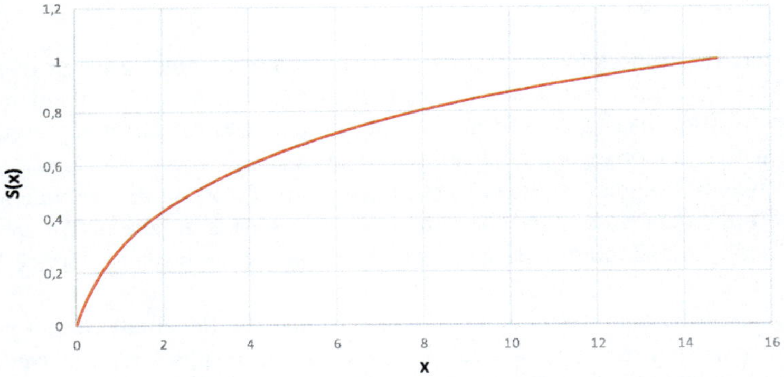

Fig. 6. Function $s^*(x)$, based on the Lumbert function.

4 Data Description

The calculation done in this research was based on the analysis of 425 work orders in a real airline company that caused AOG situation at line maintenance operations for aircraft type Airbus 320. 168 work orders out of 425 work orders have required 2,662 part numbers (p/n) to be installed on an aircraft (see Table 1). Work orders prepared for aircraft maintenance requiring materials contained the following type of materials: consumables, expendables and rotables. Some defects could contain only retables or only expendables and consumables or a mixture of different spare part types. As well for analytical purposes the downtime of an aircraft on the ground was taken and studied.

All the maintenance data was collected from airline maintenance information system (MIS) including the information on requests for materials, warehousing operations, purchase orders placed, pick slip issued, safety stock inventory, aircraft (AC) downtime, man hours required for maintenance and so on.

Table 1. Data description.

	Description of data	Meaning
1	AC type	Airbus 320 family only
2	Type of maintenance	AOG situation, line maintenance
3	Work orders (WO)	425
4	Work orders that required spares	168
5	Total spare parts items used in this WO	2,662
6	Quantity of spare parts purchased	351
7	Quantity of spare parts that were on stock	2,311

5 Results

As it was stated in methodology in order to do the complete practical research a researcher or analyst have to know and calculate A, B, a and b coefficients. To find them, nonlinear regression function has to be solved with the proper data on different service level meanings within one or several companies or datasets.

In practice the company tends to have statistics only for a concrete service level and doesn't pose the data from other companies. That is why as an alternative approach a company could use simulation of service level assuming the possible service level on the statistical data.

Having used the data from company A and having run the simulation to proof the theoretical part from methodology above the following results were received for the service level of available flight hour (FH) revenue and material cost engaged.

(Table 2).

Service level in the Table 2 stands for the available flight hours out of the whole flight hours available. Revenue was calculated as an average revenue from 1 flight hour multiplied by the available flight hours. Material cost indicate how much company A invested or should have been invested in stocks to minimize available flight hour loss because of non-availability of spare parts.

Table 2. Revenue and cost involved at a certain service level.

Service level	97.7%	98.3%	98.7%	99.3%	99.9%
Revenue from available FH, $	23 279 600	23 419 600	23 519 600	23 649 600	23 799 600
Material cost, $	3 100 000	3 120 000	3 150 000	3 450 000	3 900 000
Revenue Difference, $		140 000	100 000	130 000	150 000
Material cost difference, $		20 000	30 000	300 000	450 000

The data in Table 2 illustrates the variance in revenue and material costs, highlighting distinctions among various service level interpretations. Visually the information is represented on the Fig. 7.

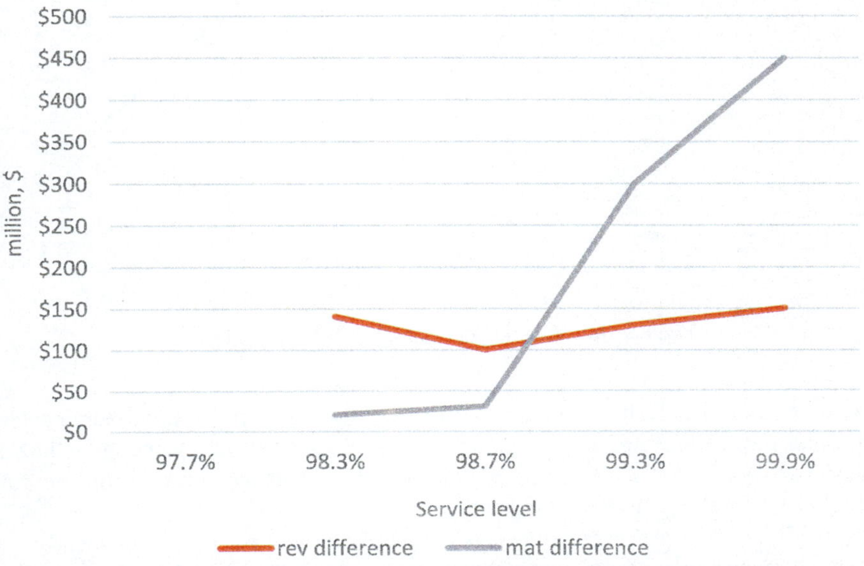

Fig. 7. Revenue possibility versus cost required to reach desired service level.

On the Fig. 7 the classical service level rule is presented. Higher the service level, higher the costs are. But in our case as well it could be seen that if company A spends more money in stock to reach the level of 98.7% then the higher revenue could be received from operations and flight hour in particular.

Another view of profit function and visual representation of optimum is represented on the Fig. 8.

The function in the Fig. 8 shows the existence of the optimum and confirms the theoretical and methodological approach.

On the basis of the results of the experiment above and methodology chosen the following results were received:

1. Having done the methodological search and description of the function, it could be stated that the main focus has to be done on the service level and its optimal meaning.
2. Theoretically and practically, it was proved that there is an optimum of the service level.
3. Parameters a and b influence significantly the optimum of service level and must be carefully calculated.

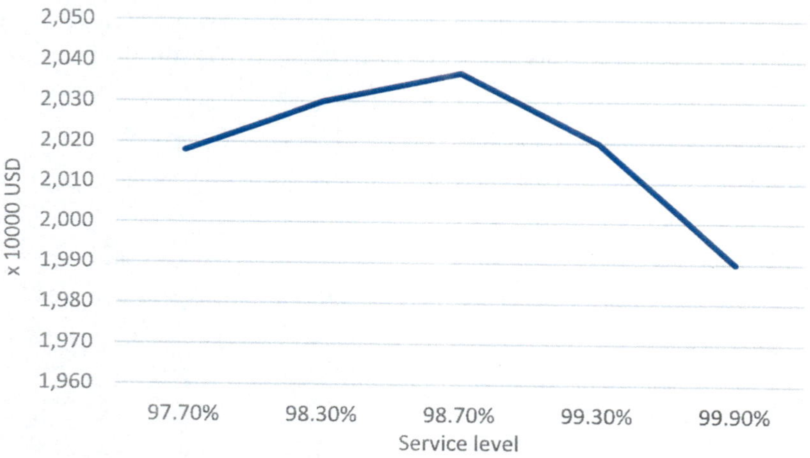

Fig. 8. Profit function depending on the service level.

4. Real business rarely has enough data to calculate properly optimal service level because the business basically stays with the same service level. At the same time the world-wide statistics based on other operational and financial criteria can't be fully applicable for the particular company.
5. In real life and on real parameters cost and revenue function behave differently.
6. Data of an airline used in the experiment have showed that the optimal service level to be 98.9%.
7. Starting from 98.9% expenses grow faster than the revenue.
8. The higher parameter A (growth of profit from service level growth) over B (growth of expenses to cover higher service level), the higher service level s is optimum. And vice versa.
9. Depending on the characteristics of an airline and its change during the time, the optimal meaning of service level will change too. Table 3 describes the dependence between such characteristics and optimal service level.

Table 3. Characteristics of an airline and it influence optimal service level.

№	Characteristic	Influence on the optimal service level s^*
1	High vs low flight season	In a high season parameter A growths rapidly. As a result, the optimal service level point moves right (increasing). For low and high season, the different service level is optimal
2	Growth of competition on the market of AC operation	Growth of competition reduces A and as a result reduces optimal service level
3	Growth of AC fleet	The growth of the fleet leads to significant growth of A with the lower growth of B. As a result, higher service level is optimal
4	Growth of fleet utilization	Leads to significant growth of A with a lower growth of B. As a result, higher service level is optimal
5	Fleet age: young vs old	The older fleet leads to growth of B with the same level of A. With the aging fleet the optimal service level is reducing
6	Fleet unification	Fleet unification leads to the reduction of B and growth of the optimal service level s. And vice-verse: the less unified fleet leads to a lower optimal service level
7	Hub chain or point to point (P2P) operational model	To keep the same service level for P2P model incurs higher costs (B) than it is required for Hub model. That is why for P2P airline optimal service level is lower than for a hub one
8	Currency devaluation (growth of spare parts cost is higher than the growth of fares)	Decreases the relation A to B and as a result reduces optimal service level
9	Supply chain disruption, growth of delivery time for spare parts	Leads to the growth of B and as a result to drop of optimal service level

6 Conclusions

In this article, the authors study the problem of determining the optimal level of service and the factors influencing it. The authors proposed to determine the optimal level of service based on the problem of maximizing the airline's profits and formulated this approach both at the model and practical levels. As a result of the study, it was found that the optimal level of service depends on the operational and financial characteristics of the airline, as well as the market in which it operates. These parameters are different for different airlines and markets and are not constant over time and change the optimal level of service. The authors showed that the widespread opinion that a higher level of service leads to increased airline profits is only sometimes valid. The study proves the need for regular assessment of the chosen level of service optimality.

As well the authors underline the importance of building interaction between the commercial services of the airline responsible for revenues and the technical and logistics departments responsible for ensuring the level of service. The lack of proper interaction and regular adjustment of the optimal service to changing conditions can lead to the fact that the airline will not be at the optimum point and may move further and further away from it. The authors noted the high complexity of calculating the optimal level of service. They proposed a tool for deciding to change the level of service based on the developed model.

In the following studies, the authors plan to study the proposed approach in more depth and the possibility of its application for MRO organizations.

References

1. IATA. Guidance Material and Best Practices for Inventory Management. Montreal, Geneva (2015)
2. Maintenance Cost Technical Group Home page. https://www.iata.org/en/programs/ops-infra/techops/mctg/
3. Genc, I., Miller, J.R., Gursoy, D.: The macroeconomic environment and airline profitability: a study of us regional airlines. Tour. Anal. **11**(6), 381–395 (2006)
4. Low, J., Lee, B.: Effects of internal resources on airline competitiveness. J. Air Transp. Manag. **36**, 23–32 (2014)
5. Hillier, F., Lieberman, G.: Introduction to Operations Research, 9th edn. McGraw-Hill, New York (2010)
6. Silver, E., Pyke, D., Peterson, R.: Inventory Management and Production Scheduling. John Wiley and Sons, Hoboken (1998)
7. Galmann, F., Belvedere, V.: Linking service level, inventory management and warehousing practices: a case-based managerial analysis. Oper. Manag. Res. **4**(1), 28–38 (2010)
8. Perez, I., Gómez, M., Lopes-Martinez, I., Hernández, J.: Strategies for the preservation of service levels in the inventory management during COVID-19: a case study in a company of biosafety products (2020). https://doi.org/10.21203/rs.3.rs-92185/v1
9. Ali, U., Salah, B., Naeem, K.: Improved MRO inventory management system in oil and gas company: increased service level and reduced average inventory investment. Sustainability **12**(19), 8027 (2020)
10. Tripaty, S., Gupta, A.: Inventory Management using FSN Analysis and Demand Forecasting. LAP LAMBERT Academic Publishing, Beau Bassin (2018)
11. Atioğlu, E.: The impact of operational performance on profitability: an ARDL approach on Turkish airlines. Afyon Kocatepe Üniversitesi Sosyal Bilimler Dergisi. (2021)
12. Corless, R.M., Gonnet, G.H., Hare, D.E., Jeffrey, D.J., Knuth, D.E.: On the Lambert W Function. Adv. Comput. Math. **5**, 329–359 (1996)

Railway Transport

Reduction of Energy Consumption by Electric Rolling Stock of Quarry Railways

Liliia Kondratieva[1(✉)], Liliia Overianova[1], Ievgen Riabov[1], Bagish Yeritsyan[1], and Sergey Goolak[2]

[1] National Technical University "Kharkiv Polytechnic Institute", 2 Kyrpychova Street, Kharkiv 61002, Ukraine
kondratieva.lilia@gmail.com, {overanova,bagish_ericjan}@ukr.net
[2] State University of Infrastructure and Technologies, 9 Kyrylivska Street, Kyiv 04071, Ukraine

Abstract. In this work, the reduction of traction energy consumption with an on-board energy storage system on an electric locomotive of quarry railway transport trains has been investigated. To carry out the calculations, a mathematical model of energy processes in the traction system of the electric locomotive has been developed following modes: traction, electrodynamic braking, and maneuvering. For the determination of energy savings, the method which is based on the analysis of the tangential power dependences of the electric locomotive has been proposed. It was established, that the use of an on-board energy storage system on an electric locomotive provides a reduction of energy consumption by approximately 10% and it is practically independent of the energy storage power. The maximum power consumed from the traction network is reduced during the exploitation the on-board energy storage system. The reduction of energy consumption has been also established by about 10% if applied recuperative electrodynamic braking of the electric locomotive without on-board energy storage system. Taking into account that reduction of energy consumption from the traction network contributes to the reduction of losses in it, as well as the possibility of providing autonomous movement, the use of an on-board energy storage system becomes a priority. The practical value is a development of methods for estimating energy consumption by electric rolling stock in the case of using on-board energy storage systems and modeling the systems with energy storage systems.

Keywords: Electric locomotive · Energy storage system · Energy efficiency

1 Introduction

Nowadays, electrified railway transport is widely used in the mining enterprises of Ukraine to transport iron ore from the quarry to the processing factory. The main traction rolling stock is traction units manufactured by the JSC "Dnipro Electric Locomotive Plant" (Dnipro, Ukraine). The traction units are characterized by long service lives and insufficient energy efficiency. Therefore, the renewal of such rolling stock with high energy efficiently tractions electric drives, implementation of recuperation or on-board storage devices, development of energy-saving modes of movement, etc. is relevant.

O. Prentkovskis et al. (Eds.): TRANSBALTICA 2023, LNITI, pp. 519–528, 2024.
https://doi.org/10.1007/978-3-031-52652-7_51

Today, the use of on-board energy storage systems (ESS) is widespread. This makes it possible to accumulate energy during electrodynamic braking and to provide power to rolling stock systems, reducing energy consumption from the traction network. This will contribute to achieving the goals of climate neutrality and increasing the competitiveness of products.

The aim of the work is to study the reduction of energy consumption using an on-board energy storage system on an electric locomotive for quarry railway transport.

2 Literature Review

Reducing energy consumption by rolling stock is a priority task for its improvement [1, 2]. One of the directions is the use of on-board ESS, which has become widespread in rail transport [3]. On-board ESSs are used to accumulate energy during electrodynamic braking, improve the operating conditions of the main energy source and ensure autonomous movement of rolling stock. In [4], a study was carried out to determine the parameters of the on-board ESS, taking into account the movement according to the last-mile principle, when the rolling stock must move without a contact network. According to the results of calculations, it is shown that the use of ESS reduces losses in the traction network by 43%, peak current - by 32%, extends the battery life by 16.3%, and stabilizes the voltage of the traction network. In [5], the application of kinetic and accumulator energy storage devices on a freight locomotive has been investigated. According to the results of the calculations, the locomotive with the energy storage consumed 12.40% and 20.65% less fuel, and also had lower gas emissions into the atmosphere. In work [6], the use of three types of storage devices - supercapacitor, lithium batteries, and inertial - was studied. According to the results of the simulation of the train movement, it was determined that the energy consumption from the traction network is reduced by 23.6%, 22.9%, and 23.7%, respectively, compared to the train movement without the on-board energy storage system. The paper [7] investigated the potential of energy saving using on-board energy storage devices on a quarry dump truck. It has been established that fuel savings can be 10…12%. The GE company created a battery locomotive [8], the use of which as part of a multi-section diesel locomotive allows to save 10…15% of fuel. In [9], energy processes in the traction system of a subway train equipped with supercapacitors were investigated. The reduction in energy consumption was 5% for the 1.73 kWh battery, 10.3% for the 3.46 kWh battery, and 15.7% for the 15.7 kWh battery. In [10], it is shown that the use of an on-board ESS as part of the power plant of a diesel locomotive will reduce fuel costs by 22% to 30% compared to a locomotive without an on-board ESS. In work [11] it is shown that the use of regenerative braking will allow achieving energy storage equal to 21…55% of the total energy consumption.

The works of many researchers, applying various criteria to optimize the parameters and structure of energy storage system are devoted to the solution of similar tasks. In this work, the impact of the on-board energy storage power on the reduction of energy consumption for traction is investigated.

3 Research Methodology

The specialized electric locomotive proposed by the authors for quarry railway transport [12] was used for the study. A 16-axle two-section electric locomotive with a mass of 2×200 tons is required to replace the old electric locomotive. The electric locomotive is equipped with an asynchronous traction electric drive with a power of 6,700 kW, maximum traction effort is 1,300 kN, maximum speed – 60 km/h. Figure 1 shows the tangential power during the passage, which was determined as a result of solving the traction task. The mathematical model given in [13] was used for the solution, taking into account the clauses of the theory of locomotive traction [14, 15]. For an electric locomotive without ESS, the only available technology to reduce traction energy consumption is recuperative braking, which was taken into account in the calculation.

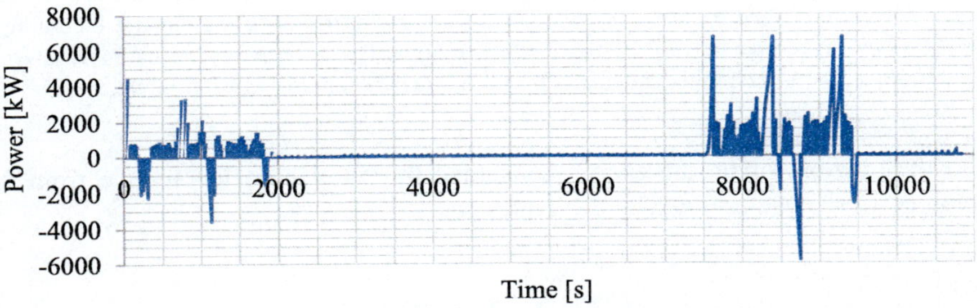

Fig. 1. Dependence of the tangential power of the electric locomotive.

Further, the model of energy processes in the traction system of an electric locomotive equipped with an on-board ESS will be considered. The power flow diagram is shown in Fig. 2. In the traction mode, powering of the traction electric drive and auxiliary systems of the electric locomotive can be carried out jointly from the catenary network and the on-board ESS. In the electrodynamic braking mode, energy is stored in the on-board ESS and can be recuperated to the traction network.

The energy consumed by an electric locomotive with an on-board ESS is determined by the expression

$$E_T = E_C - E_R + E_I, \tag{1}$$

where E_C – the energy consumed by the electric locomotive from the traction network, E_R – energy that is returned to the traction network during recuperation, E_I– the energy which it is necessary to charge ES additionally.

The subtraction of energy recuperated in the traction network corresponds to the domestic practice of calculating energy consumption for traction. Recharging with ES energy E_I is necessary to ensure the execution of the on-board ESS operation algorithm when the electric locomotive is moving. Therefore, we add energy E_I to the total consumption.

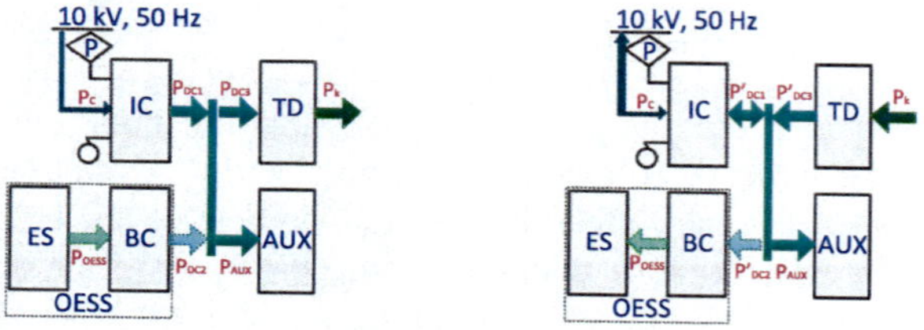

a) b)

Fig. 2. Flow diagrams in the traction system: a – traction mode, b – electrodynamic braking mode: P – pantograph, IC – input converter, TD – traction electric drive, OESS – on-board energy storage system, ES – energy storage, BC – matching converter, AUX – auxiliary systems; P_C – power that is consumed from the traction network, P_{DC1} – power that enters the intermediate circuit from the input converter, P_{DC2} – the power that comes from to the intermediate circuit from OESS, P_{OESS} – power that is consumed from ES, P_{DC3} – power that is transmitted to the traction electric drive, P_{AUX} – power consumed by auxiliary systems, P_k – tangent power of the electric locomotive, P'_{DC3} – power that comes from the traction electric drive, P'_{DC2} – the power that is transferred to the OESS, P'_{DC1} – the power consumed from the traction network in EDB mode.

In the traction mode, the energy consumed by the electric locomotive from the traction network is determined by the expression

$$E_{c1} = \sum_{p=1}^{q} \int_{0}^{\tau_p} p_{Lp}(t)dt, \tag{2}$$

where $p_{Lp}(t)$ – is the dependence of the power consumed by the electric locomotive from the catenary network at the p-th stage of consumption, τ_p – duration of the p-th stage of consumption, q – the number of stages of consumption.

The power consumed by the electric locomotive from the contact network is determined by the expression

$$p_{Lp}(t) = \frac{p_{DC1p}(t)}{\eta_1}, \tag{3}$$

where $p_{DC1p}(t)$ – the power that is consumed from the input converter, η_1 – input converter efficiency.

The power consumed from the input converter is determined as follows

$$p_{DC1p}(t) = \begin{cases} \frac{p_{kp}(t)}{\eta_{TD}} + p_{ap}(t), & \left(\frac{p_{kp}(t)}{\eta_{TD}} + p_{ap}(t)\right) \le P_{IN} \\ P_{IN}, & \left(\frac{p_{kp}(t)}{\eta_{TD}} + p_{ap}(t)\right) > P_{IN} \end{cases}, \tag{4}$$

where $p_{kp}(t)$ – dependence of the tangential power, $p_{ap}(t)$ – dependence of the power of auxiliary systems, which is consumed from the intermediate circuit, P_{IN} – input converter power.

At a given power, the power of the input converter is determined using expression (1).

The energy consumed from ES is determined by the expression

$$E_{ES} = \sum_{u=1}^{v} \int_{0}^{\tau_u} p_{ESu}(t)dt, \tag{5}$$

where $p_{ESu}(t)$ – the dependence of the power consumed by an electric locomotive with ES at the u-th stage of consumption, τ_u – the duration of the u-th stage of consumption, v – the number of stages of consumption.

The power consumed from ES

$$p_{ESu}(t) = \frac{p_{DC2u}(t)}{\eta_{BC}}, \tag{6}$$

where $p_{DC2u}(t)$ – the power consumed from the matching converter, η_{BC} – the efficiency of the matching converter. In the calculations, we establish a constant value equal to 0.98.

The power in the intermediate circuit consumed by the traction electric drive and auxiliary systems with ES

$$p_{DC2u}(t) = \begin{cases} 0, & \left(\frac{p_{ku}(t)}{\eta_{TD}} + p_{au}(t)\right) \leq P_{IN} \\ \frac{p_{ku}(t)}{\eta_{TD}} + p_{au}(t) - P_{IN}, & \left(\frac{p_{ku}(t)}{\eta_{TD}} + p_{au}(t)\right) > P_{IN} \end{cases}, \tag{7}$$

where $p_{ku}(t)$ – dependence of the tangential power, $p_{au}(t)$ – dependence of the power of auxiliary systems, which is consumed from the intermediate circuit.

In coasting mode, the energy consumed from the contact network goes to power the auxiliary systems of the electric locomotive and is determined by expression (5), in which the power is calculated as

$$p_{Lp}(t) = \frac{p_{ap}(t)}{\eta_1}, \tag{8}$$

ES does not take part in energy exchange, therefore power and energy are equal to zero.

In the electrodynamic braking mode, the auxiliary systems of the electric locomotive are powered by the energy supplied from the traction electric drive (we assume that the power from the generating traction electric drive is always sufficient to power the auxiliary systems). The rest of the energy is ES used for charging, and if the power is ES insufficient, it is sent to the traction network.

In this mode, energy from the traction network is not consumed, so the consumed energy is zero.

The energy stored in ES is determined by the expression

$$E'_{ES} = \sum_{h=1}^{g} \int_{0}^{\tau_h} p'_{ESh}(t)dt, \tag{9}$$

where $p'_{ESh}(t)$ – dependence of the power transmitted to the ES at the h-th stage, τ_h – duration of the h-th stage of consumption, g – the number of stages of consumption.

The ES power is

$$p'_{ESh}(t) = \eta_{BC} p'_{DC2h}(t), \tag{10}$$

where $p'_{DC2h}(t)$ – power available for charging the on-board ESS.

The power that is available for charging the on-board ESS

$$p'_{DC2h}(t) = \begin{cases} 0, & p'_{DC3h}(t) \le p_{ah}(t) \\ p'_{DC3h}(t) - p_{ah}(t), & 0 < \left(p'_{DC3h}(t) - p_{ah}(t)\right) \le \frac{P_{ESnom}}{\eta_{BC}} \\ P_{ESnom}, & \left(p'_{DC3h}(t) - p_{ah}(t)\right) > \frac{P_{ESnom}}{\eta_{BC}} \end{cases}, \tag{11}$$

where $p'_{DC3h}(t)$ – the power supplied to the intermediate circuit from the traction electric drive, $p_{ah}(t)$ – dependence of the power of auxiliary systems, which is consumed from the Intermediate circuit, P_{ESnom} – rated ES power.

It is accepted here that if the power supplied to the intermediate circuit is less than the power of the auxiliary systems, then the power of the auxiliary systems is reduced.

The power supplied to the intermediate circuit from the traction electric drive

$$p'_{DC3h} = p_{kh}(t)\eta_{TD}, \tag{12}$$

where $p_{kh}(t)$ – dependence of the tangential power.

The energy that is recovered to the traction network is determined similarly to (7). The power of recovery is determined as follows.

$$p'_{Rj}(t) = p'_{DC1j}(t)\eta_1, \tag{13}$$

where $p'_{DC1j}(t)$ – power at the terminals of the input converter, which is returned to the contact network.

The power at the terminals of the input converter, which is returned to the catenary network

$$p'_{DC1j}(t) = \begin{cases} 0, & p'_{DC3j}(t) \le p_{aj}(t) \\ 0, & \left(p'_{DC3j}(t) - p_{aj}(t)\right) < \frac{P_{ESnom}}{\eta_{BC}} \\ -\left(p'_{DC3j}(t) - p_{aj}(t) - \frac{P_{ESnom}}{\eta_{BC}}\right), & p'_{DC3j}(t) > p_{aj}(t) + \frac{P_{ESnom}}{\eta_{BC}} \end{cases}, \tag{14}$$

where $p_{aj}(t)$ – the power of auxiliary systems, $p'_{DC3j}(t)$ – the power supplied to the intermediate circuit from the traction electric drive.

The amount of energy that needs to be supplied in ES before starting the movement

$$E_I = \frac{E_F - E_S}{\eta_1 \eta_{BC} \eta_{ES}}, \tag{15}$$

where E_S – energy that must be stored in the ES before the start of the movement, E_F – the energy that remains in the ES after the passage, η_{ES} – energy storage efficiency. When calculating, we establish it as a constant value equal to 0.9.

In the case of maneuvering during loading and unloading, the power supply of the electric locomotive systems is provided by the on-board ESS.

The energy consumed from the on-board ESS is determined by expression (10). The power consumed from ES is determined for this case by the expression

$$p_{ESu}(t) = \frac{\frac{p_{ku}(t)}{\eta_{TD}} + p_{au}(t)}{\eta_{BC}}, \quad (16)$$

When calculating according to expression (16), the traction drive efficiency is assumed to be 0.5.

Thus, the model defined by expressions (1)–(16) allows you to calculate the energy consumption during the movement of an electric locomotive equipped with an on-board ESS.

4 Results and Discussion

The data of Table 1 are the results of calculations of energy consumption by an electric locomotive without on-board ESS, which is determined using the above-described mathematical model when the power of on-board ESS is equal zero. The data of Table 2 are the results of calculations during the movement of an electric locomotive with an on-board ESS (in Table 2, the largest power consumed by an electric locomotive from the traction network of increased volume P max, energy accumulated in the ES is E_A, the rest of the designations are according to model (4)–(16). As it is seen from Table 1, the use of recuperative braking provides a return to the traction network of 148 kWh, which is 10.3% of the energy consumed for traction.

Table 1. The results of the calculation of energy consumption during the movement of a train with an electric locomotive without an on-board ESS.

Energy consumed from the traction network, kWh	1,438.0
Energy that is returned to the traction network, kWh	148.0
Energy spent for traction, kWh	1,290.0
The largest power consumed by an electric locomotive from the traction network, kW	8160

It follows from Table 2 that the energy consumption from the traction network during the movement of a train with an electric locomotive with an on-board OESS decreases to 1,193...1,281 kWh, which is (82.9...89)% of the energy consumption during the movement of an electric locomotive without the use of recuperative braking. This is because during maneuvering, the traction drive and auxiliary systems are powered from the on-board OESS.

In comparison with the version of the movement in which recuperative braking is used, the energy consumption does not differ significantly: 1,292.2...1,277.9 kWh in the variant with on-board OESS – 1,290 kWh with recuperative braking.

Table 2. The results of the calculation of energy consumption during the movement of a train with an electric locomotive equipped with an on-board ESS.

	ES power, kW						
	500	1,000	1,500	2,000	2,500	3,000	3,500
E_C, kWh	1,281.0	1,276.0	1,267.0	1,254.0	1,238.0	1,218.0	1,193.0
E_R, kWh	97.7	60.8	37.0	22.6	14.6	9.0	5.0
E_I, kWh	98.9	67.2	51.8	49.0	56.3	70.0	89.9
E_T, kWh	1,292.2	1,282.4	1,281.8	1,280.4	1,279.7	1,280.0	1,277.9
$P_{\max}$, kW	7,615	7,078	6,541	6,003	5,467	4,930	4,393
E_A, kWh	50.2	87.1	110.6	125.4	133.4	139.0	143.0

The data presented in Table 2 are the results when using the on-board ESS, the maximum power consumed by the electric locomotive from the traction network is reduced. This, firstly, ensures reduction of losses in the traction network. Secondly, it contributes to the adherence to the traffic schedule, as it reduces the influence of the limitation of power available from the traction network on train movement. In the proposed algorithm, the traction drive and auxiliary systems of the electric locomotive are powered from the on-board ESS.

This eliminates the operation of the input converter in modes characterized by a decrease in the power factor and therefore reduces losses in the traction network. Thus, a decrease in losses in the traction network is expected, the calculation of which is one of the areas of further research.

5 Conclusions

The Author investigated the reduction of energy consumption by an electric locomotive for a quarry railway when applying energy recovery to the traction network and energy accumulation in the on-board storage device.

Both in the case of recuperation to the traction network and in the case of energy accumulation in the on-board accumulator, the reduction of energy consumption from the traction network is reduced by about 10% relative to the consumption for the case where electrodynamic braking of the electric locomotive is not applied. The maximum power consumed from the traction network by an electric locomotive without an on-board storage unit is estimated as 8,160 kW, and the maximum power consumed by an electric locomotive with an on-board storage unit is 4,393…7,615 kW, depending on the capacity of the on-board storage unit.

The obtained results are showing that the reduction of energy consumption from the traction network practically does not depend on whether recuperation is applied to the traction network or energy is accumulated in the on-board storage. However, taking into account the fact that when using an on-board storage device, the power consumed from the traction network can be significantly reduced, the use of energy storage in the on-board storage device is considered a priority. In this case, the losses in the traction power supply system are reduced, and the unevenness of energy consumption from the external power supply system is reduced, which contributes to increasing the efficiency of electric traction on quarry railways.

References

1. Corlu, C.G., de la Torre, R., Serrano-Hernandez, A., Juan, A.A., Faulin, J.: Optimizing energy consumption in transportation: literature review, insights, and research opportunities. Energies **13**, 1115–1130 (2020). https://doi.org/10.3390/en13051115
2. Sablin, O., Bosyi, D., Kuznetsov, V., Lewczuk, K., Kebal, I., Myamlin, S.S.: Efficiency of energy storage control in the electric transport systems. Arch. Transp. **62**(2), 105–122 (2022). https://doi.org/10.5604/01.3001.0015.9569
3. Fedele, E., Iannuzzi, D., del Pizzo, A.: Onboard energy storage in rail transport: Review of real applications and techno-economic assessments. IET Electr. Syst. Transp. **11**(4), 279–309 (2021). https://doi.org/10.1049/els2.12026
4. Graber, G., Galdi, V., Calderaro, V., Piccolo, A.: Sizing and energy management of on-board hybrid energy storage systems in urban rail transit. In: 2016 International Conference on Electrical Systems for Aircraft, Railway, Ship Propulsion and Road Vehicles & International Transportation Electrification Conference (ESARS-ITEC), pp. 1–6 (2016). https://doi.org/10.1109/ESARS-ITEC.2016.7841408
5. Spiryagin, M., Wu, Q., Wolfs, P., Sun, Y., Cole, C.: Comparison of locomotive energy storage systems for heavy-haul operation. Int. J. Rail Transport. **6**(1), 1–15 (2018). https://doi.org/10.1080/23248378.2017.1325719
6. Wu, C., Lu, S., Xue, F., Jiang, L., Chen, M.: Optimal sizing of onboard energy storage devices for electrified railway systems. IEEE Trans. Transport. Electrificat. **6**(3), 1301–1311 (2020). https://doi.org/10.1109/TTE.2020.2996362
7. Terblanche, P.J., Kearney, M.P., Hearn, C.S., Knights, P.F.: Technology selection and sizing of on-board energy recovery systems to reduce fuel consumption of diesel-electric mine haul trucks. In: Awuah-Offei, K. (ed.) Energy Efficiency in the Minerals Industry. GET, pp. 301–333. Springer, Cham (2018). https://doi.org/10.1007/978-3-319-54199-0_17
8. Leading The Charge: Battery-Electric Locomotives Will Be Pushing US Freight Trains Further. https://www.ge.com/news/reports/leading-charge-battery-electric-locomotives-pushing-us-freight-trains. Accessed 31 Jan 2023
9. Radu, P., Szelag, A., Steczek, M.: On-board energy storage devices with supercapacitors for metro trains—case study analysis of application effectiveness. Energies **12**, 1291–1300 (2019). https://doi.org/10.3390/en12071291
10. Cipek, M., Pavkovi, D., Krznar, M., Kljai, Z., Mlinari, T.: Comparative analysis of conventional diesel-electric and hypothetical battery-electric heavy haul locomotive operation in terms of fuel savings and emissions reduction potentials. Energy **232**, 121097 (2021). https://doi.org/10.1016/j.energy.2021.121097
11. Knibbe, R., et al.: Application and limitations of batteries and hydrogen in heavy haul rail using Australian case studies. J. Energy Storage **56**(Part A), 105813 (2022). https://doi.org/10.1016/j.est.2022.105813
12. Riabov, I., Mosin, S., Overianova, L., Kondratieva, L., Demydov, O., Goolak, S.: Evaluation of technical parameters locomotive for railway career transport. Transp. Syst. Technol. **39**, 83–100 (2022). https://doi.org/10.32703/2617-9040-2022-39-9

13. Goolak, S., Kondratieva, L., Riabov, I., Lukoševičius, V., Keršys, A., Makaras, R.: Research and optimization of hybrid on-board energy storage system of an electric locomotive for quarry rail transport. Energies **16**, 3293 (2023). https://doi.org/10.3390/en16073293
14. Yi, S.: Principles of Railway Location and Design. Academic Press, Cambridge (2017)
15. Riabov, I., Kondratieva, L., Overianova, L., Goolak, S.: Assessment of the on-board energy storage parameters of the locomotive for rail quarry transport. In: TRANSBALTICA XIII: Transportation Science and Technology. TRANSBALTICA 2022. Lecture Notes in Intelligent Transportation and Infrastructure. Springer, Cham. (2023). https://doi.org/10.1007/978-3-031-25863-3_65

Development Genesis of Functional Safety on the Example of an Element of the Railways Infrastructure Subsystem

Iryna Bondarenko[1](✉) and Larysa Neduzha[2]

[1] Lviv Polytechnic National University, Profesorska St. 1, Lviv 79000, Ukraine
iryna.o.bondarenko@lpnu.ua

[2] Ukrainian State University of Science and Technologies, Lazaryan St. 2, Dnipro 49010, Ukraine
nlorhen@i.ua

Abstract. The degree of protection against the danger arises as a result of incorrect (unforeseen by the developers) operation of a technical object is determined by the issues of functional safety, which have acquired the greatest relevance in this period of human development and have become the object of research. Authors consider the problems of functional safety of dynamic systems on the example of an element of the railway infrastructure subsystem – a railway track. It is shown that an attempt to create a single universal system for classifying tracks, considering the satisfaction of heterogeneous requirements to ensure the conditions for interoperability of various railway systems, is unrealistic in the current circumstances. The missing components for the classification system of the railway track structure operability are considered. An analysis of the historical understanding of various processes in relation to real-time is given. The main direction for improving the quality of the functional safety of a railway track is formulated – the need to combine the theory of elasticity with new theories that change the approaches to studying time-varying processes from quasi-dynamic to dynamic.

Keywords: Genesis · Functional safety · Railway infrastructure · Infrastructure subsystem · Railway track

1 Introduction

To ensure their own life, men have created many technical objects and systems, the totality of which is the Technosphere. All objects and systems of the Technosphere undergo significant changes in the process of operation. The degree of protection against the danger that may arise as a result of incorrect (unforeseen by the developers) operation of a technical object is determined by the issues of functional safety, which have acquired the greatest relevance in this period of human development and have become the object of study. In this paper, the authors consider the problems of functional safety of dynamic systems on the example of an element of the railway transport infrastructure subsystem – a railway track. Historically, there is a railway track with a track gauge of

O. Prentkovskis et al. (Eds.): TRANSBALTICA 2023, LNITI, pp. 529–538, 2024.
https://doi.org/10.1007/978-3-031-52652-7_52

1,435 mm and 1,520 mm, for which design, operation, and control systems have been developed in parallel in various countries. Modern integration processes provide for the interoperability of these structures and require compliance with the same technical standards. With all the apparent variety of regulatory documents developed for two railway track structures, they do not cover the uniqueness of each individual situation associated with changing national standards to common ones established by technical standards for interoperability.

Therefore, the objective of the study is to show the incapacity of the approach to manage the process of ensuring functional safety, which responds to changes occurring throughout the functionality during all stages of the life cycle, by imposing regulated dependencies obtained on the basis of a deterministic approach to solving local problems for each functionality.

2 Research Method

Since the genesis of development is the origin and subsequent development process, which led to a certain state that exists at the time in question, the paper considers the formation of concepts and the approaches developed on their basis that are used today to address issues of ensuring functional safety.

A functional attribute determines the purpose of the system, as well as its main goals, objectives and functions. Comprehensive and sustainable management of rail infrastructure requires a structured and methodical approach. Therefore, the requirements for ensuring the functional safety of the railway as an element of the transport infrastructure subsystem are considered bearing in mind the various areas of requirement given in Table 1 at all stages of the life cycle.

Solving problems related to ensuring the functional safety of any system can be compared with the creation of various objects using the Lego constructor, where each module is an accumulator of modern theoretical and practical knowledge in various fields. When considering issues of functional safety, in addition to the usual coordination of classical knowledge, methods and approaches, such areas as project management (organization of efficient use of resources), system diagnostics (the process of identifying problems in the system and solving them), reliability of systems (setting the boundaries of the values of all parameters characterizing the ability to perform the required functions in specified modes and operating conditions), system safety (taking measures against unwanted intrusions into the operation of the systems themselves, and from undesirable effects of the operation of systems on the surrounding systems (in the broad sense, both technology and people and ecology), combined possibilities are created. As an example, it is possible to cite the option of describing algorithms by mathematical methods built on a combination of principles for constructing algorithms in programming (describing a sequence of actions) and principles for the construction of structural diagrams in reliability (describing structural diagrams of the operation of objects in various loading systems and duplication of functions performed) with the binding of the technical states in the operation of elements and their systems, adopted in diagnostics, to the states classifying the ability to perform work, adopted in reliability. Thus, consideration of the safety issues in the functioning of systems makes it possible to combine knowledge of

Table 1. General requirements for the railway track.

The area of requirement	Description
Reliability	Design, construction or assembling, maintenance and control of all parameters adopted for both the elements and the structure of the track must guarantee safety at a level corresponding to the operating conditions for the network, including considering changes in parameters during operation up to the maximum allowable values
	The wheel-rail contact parameters must meet the stability requirements necessary to ensure safe movement at the maximum allowable speed. The parameters of the braking equipment must guarantee the possibility of stopping within the specified braking distance at the maximum permitted speed
	The elements used in the structure and the structure of the railway track itself must comply with the conditions of all types of impacts during the period of their operation. The safety implications of any random failures shall be limited by appropriate means
	The choice of materials used should be directed towards limiting the formation, spread, fire exposure and smoke in the event of a fire
	All devices intended for use by users must be designed so as not to compromise the safe operation of the devices or the health and safety of users
Availability	Monitoring and maintenance of a railway track should be organized in such a way as to carry out and quantify its performance under given conditions
Health	Materials that, by virtue of the way they are used, may pose a risk to the health of those who have access to them should not be used in railway infrastructure
	Materials must be selected, located and used in such a way as to limit the emission of harmful and dangerous vapors or gases, especially in the event of a fire
Environmental protection	At the design stage of the system, in accordance with the legislation, the impact of the railway track on the environment must be evaluated and considered
	The materials used in the infrastructure must prevent the emission of vapors or gases that are harmful and dangerous to the environment, especially in the event of a fire
	When operating the rail system, the prescribed noise levels must be observed. The operation of the rail system must not cause an unacceptable level of ground vibration for activities and areas close to the infrastructure and in normal condition

both humanitarian and technical fields. However, the main task of ensuring safety during the functioning of systems is reviewing and evaluating the probability of various hazards occurring during the life cycle of the railway track.

The classic track design consists of only five elements: rails, rail fastenings, under-rail base (sleepers, beams, plates), ballast layer and sub-ballast base (protective layers and roadbed). In the process of using these elements, the national standards of all countries have formed requirements both for the geometry of each element of the track structure and for the track structures themselves. Some national standards have come to the need to establish stiffness (elastic) and geometric relationships in the track design for various operating conditions.

Regulated states of these systems are considered as hazards of technical systems. The basis that served to determine the technical conditions of the railway track is the famous empirical law [1] on the proportionality between the acting force and the resulting deformations marked the beginning of the theory of elasticity. It is used as a foundation for engineering and architecture. The provisions of the theory of elasticity are the basis for modeling the behavior of physical bodies under the impact of various forces in modern software products [2–7]. Thus, two criteria appeared in engineering that evaluate the state of objects:

1. The first state when an object can lose its bearing capacity, this state is called the first limit state. The first limit state is for loss of load-bearing capacity or unsuitability for use (strength, stability, durability). This state evaluates the general ability of the object to perform its functions and answers the question whether or not the object can perform its functions under such operating conditions. This criterion is widely used in various fields of engineering [8–13].
2. The second limit state is for unsuitability for normal use (deflection, deformation, crack opening size). This state evaluates how well an object performs its functions under certain operating conditions. Basically, the second critical state in engineering requires evaluating the behavior of a structure or system [14–21] or the appropriateness of using a structure or system [22–26].

Exactly the development of the theory of reliability led to the ability for evaluating the quality of the object's performance of its functions. This was achieved by introducing the concept of "risk" as a definition of all possible hazards (risks) during the operation of an object, and "risk level" as a characteristic for evaluating the probability of occurrence of hazardous events that entail dangerous consequences [27–32]. Thus, there has been a leap in the development of event forecasting: from predicting the occurrence of individual specific events to predicting the occurrence of events during the life cycle, given the functions that are inherent in the stages of the system life cycle. For engineering objects, this is forecasting by the process of managing reliability, availability, maintainability, and safety (RAMS), based on the life cycle of the system, considering the tasks solved at various stages of this cycle [33–36]. However, there is a need to integrate the changes occurring at each stage of the life cycle into one process. To explore this possibility, the algorithm of the research methodology for the "Reliability" and "Availability" requirement areas is presented below:

$$L_F = \{\{a_1, a_2, a_i, \ldots\}, \{b_1, b_2, b_i, \ldots\}, \{\delta_1, \delta_2, \delta_i, \ldots\}\},$$

$$P_F = f(L_F, V, F),\ T = \{\{t_1, t_2, t_i, \ldots\}\},\ T^0 = \left\{\left\{t_1^0, t_2^0, t_i^0, \ldots\right\}\right\},$$

$$\omega_i = \left\{\left\{g_{11}, g_{i2}, g_{ij}, \ldots\right\}, \left\{\phi_{11}, \phi_{i2}, \phi_{ij}, \ldots\right\}\right\},$$

$$K = \{\{\omega_1, \omega_2, \omega_i, \ldots\}, \{s_1, s_2, s_i, \ldots\}\},$$

$$S_i = \left\{\left\{\sigma_{11}, \sigma_{12}, \sigma_{pi}, \ldots\right\}, \{y_{11}, y_{12}, y_{sm}, \ldots\}\right\}, \tag{1}$$

$$[S_i] = \left\{\left\{[\sigma_1], [\sigma_2], \left[\sigma_p\right], \ldots\right\}, \left\{[y_1], [y_2], [y_m], \ldots\right\}\right\},$$

$$S_i = f\left(K, P_F, T^0\right) \le [S_i],\ Y = \{S_1, S_2, \ldots, S_n\},$$

$$D = \{\{\Delta_1, \Delta_2, \ldots, \Delta_n\}, \{L_1, L_2, \ldots, L_n\}, \{R_1, R_2, \ldots, R_n\}\},$$

$$S_i = f(D, T),$$

$$\exists(S_i \wedge Y) : (S_i \wedge Y)$$
$$\Rightarrow (\textit{good state} \ \vee \textit{faulty state} \ \vee \textit{partial up state} \ \vee \textit{disablet state}),$$

$$\forall(\textit{good state} \ \vee \textit{faulty state} \ \vee \textit{partial up state} \ \vee \textit{disablet state}) \Rightarrow$$
$$\Rightarrow S_i = f(K, P_F) \wedge S_i = f(D, T), \forall\omega_i \in K \Rightarrow Y = f(K, P_F, D, T),$$

where L_F, a_i, b_i, δ_i – respectively: the trajectory of the wheel, the width of the rail/wheel contact, the gap between the wheel and the rail at the time of contact;

$V, F, P_F, T, T^0, S_i, Y$ – set, respectively: speed of movement, loads, load impulse parameters, time, track elements, temperature, reaction waves by direction, stress-strain state of elements, stress-strain state of the structure;

$K, \omega_i, s_i, g_{ij}, \phi_{ij}$ – respectively: assembly of track elements, connections between elements, j-th element of the set of geometric dimensions and physical and mechanical properties of the element ω_i;

$\sigma_{pi}, y_{sm}, [\sigma_n], \left[y_m\right], [S_i]$ – respectively: ℓ-th, m-th element of the set of stress and strain magnitude parameters and their limit values in the elements, the limit value of the element state,

D, Δ_i, L_m, R_k – respectively: the type of malfunction or defect, the ℓ-th, m-th element of the set of parameters of the magnitude of the malfunction and the location of the malfunction, the k-th element of the deformation states of the element ω_i.

3 Research Results

The analysis of existing methods for determining the technical conditions of the railway track carried out from the point of view of the development genesis of ensuring the functional safety of the railway revealed the following results.

Today there is a tendency of staying ahead in the development of intelligent technologies based on knowledge extraction and administering them. There is an actualization and intellectualization of the initial information used in the decision-making process in various subject areas with a further transition to automation of the decision-making process by developing correct mathematical models and modeling methods for information systems that allow solving optimization problems. Thus, mankind is good at ensuring the functional safety of simple systems at the stages of design, operation, and forecasting. But there are problems in solving the functional safety of complex systems, and especially big problems in the process of forecasting can be observed.

First of all, this is due to the fact that not a single existing national system for classifying the performance of a track structure has a system for evaluating the performance of a railway track structure depending on the state of each of its elements. In addition, although both the transportation process (in terms of losses due to speed limitation) and the nature of track oscillations (as an indicator of changes in its state) under the rolling stock are affected not by the maximum, but by the actual speed, the task of determining the actual the speed of trains and the actual level of force load on the track remains very relevant. And although there is enough data on the passage of measuring trains in the same direction for a long time, a methodology has not yet been created for evaluating how the indicators of the state of the track under trains circulating in the section with different axial loads and speeds will change. And there is still no methodology for how the indicators measured by a track-testing car change when driving at different speeds with different elasticity of the track.

The geometric and physical-mechanical characteristics of the elements of the track structure affect the performance of the track structure itself and its service life. This indicator cannot be chosen without considering the economic component. So, with short service life, there is a danger of loss of stability of the track due to the reduction of measures for routine maintenance. In addition, a larger budget will be needed to restore the track to working capacity. And with a long service life, there will be a need to increase the cost of current maintenance of the track, since there is also a risk of loss of track stability due to the accumulation of track breakdown. However, today there is no work evaluation of the railway track under the impact of rolling stock by developing modern methods for determining the rational, from the viewpoint of the technical use of the track, the ratio of the actual operating conditions of the track, the frequency of overhaul schemes and the characteristics of track construction elements.

Although applying the theory of elasticity in the development of modeling methods and experiments has led to the need to study functional safety, but today there is a restriction for the application of technical standards in the design of a common functional safety system and the exchange of existing methods and approaches between functional systems of railway track structures with a track gauge of 1435 mm and 1520 mm.

The main reason is the historical understanding of various processes in relation to real-time. So, there is a universal classification of all systems with respect to the changes occurring in them in relation to real-time:

1. Processes that are relatively stable over time, and are considered static.
2. Processes that change in a certain fixed time interval, and are considered quasi-dynamic.

3. Processes, where the change is fixed in conjunction with the information of the measuring equipment in real time, and they are considered as dynamic.

With the first type of processes, the parameters characterizing it are easily calculated and there are no problems in predicting its behavior. In this case, the researcher understands the physical nature (reasons that caused it, and regularities as reactions to incoming impacts) of the course of this process in the full volume for the research. Accounting for real-time does not affect the flow of the process under investigation within the framework of the study.

The second type of processes, as a rule, has clear patterns of change over a fixed time interval, so it is not difficult to set a function that describes the changes in the best way and find its parameters using differential or integral calculus. However, when forecasting its behavior, problems arise that can be solved using a measure of the probability of the system being in a given state. In this case, the researcher partially understands the physical nature of the process and adjusts the behavior of the process to that obtained in the experiment. The velocity of the process under investigation within the research partly depends on the velocity of the physical process in real time.

The third type of processes is interesting in itself since it is really created only with a continuous flow of information from the registered devices in a particular experiment. In this case, the experiment should give real values, but how can one assert that they are real if there is no understanding of the true value? Consequently, there are difficulties in forecasting its behavior on the basis of any data whatsoever, since there is no way to understand the pattern of change and accept a function for research over the entire time interval of the research. In this case, the researcher does not understand the physical nature of the process, makes assumptions, and plans experimental confirmation of his/her assumptions using the second type of processes. The course of the physical process under study within the framework of the study completely depends on real time accounting.

The above gaps, by determining the missing dependencies to improve the quality of the functional safety of the railway track, show the need to change the theoretical provisions adopted on the basis of the use of the theory of elasticity. One should take an example in the field of visualization, where raster graphics are replaced by vector ones. And on the basis of using such a method, to describe the propagation of impacts (impulses) using the propagation of waves (the theory of elastic waves and electromagnetic waves), as energy carriers for performing processes [15, 16, 18], and create new software systems. That will allow us not to assign which function will change, but to create these functions in time, taking into consideration the impact characteristics and the propagation time of this impact in the structure.

4 Conclusion

The essence of modern managing the functional safety of the railway track is not to process all the results with the help of artificial intelligence, but to formulate comprehensive criteria for evaluating the achievement of the main goal, and to define this goal in such a way that it would be achieved with available technical means. Given that modern technical means exceed the capabilities necessary to verify the results (determined on the basis of the provisions of the theory of elasticity, which is embedded in all methods and approaches, both theoretical and experimental research), it is necessary to change the system concept. The general aspiration of the new concept is the use of various fields of science with a focus not on ways to solve the problem, but on ways to describe the nature of the change in physical processes that take place both in the structural elements and in the structure of the railway track in real time for the processes under investigation.

The complexity of the situation lies in the fact that it is impossible to achieve effective management of functional safety by considering and solving only local problems, for example, such as the organization of the transportation process, the organization of the diagnostic process, the organization of repair work, etc. Although the whole range of actions necessary to ensure the functional safety of the railway track is carried out by various structures and organizations that do not contact each other in real life, only the formation of a common understanding of the process will allow developing the criteria for local tasks. Therefore, there is a need to create an information center, on the basis of the data of which experts from various fields in cooperation will create a new classification of track design performance. It will be based on criteria common to all types of railway tracks, for example, on the deformability work performed by the track under the rolling stock impact, and further, depending on the type of track, both geometric and mechanical requirements for these structures will be recalculated.

References

1. Hooke's law. http://www.femto.com.ua/articles/part_1/0889.html. Accessed 03 May 2023
2. Vieira, P.C.S, Silva, W.T.M.: Elastoplastic analysis of reinforced concrete space frames using the uniaxial and biaxial bending method. Acta Scientiarum Technol. **45**(1) (2023). https://doi.org/10.4025/actascitechnol.v45i1.60188
3. Besekar, S.S., Pimpalkar, P.R., Gorantiwar, V.S.: Design and analysis of leaf spring using Ansys. Int. J. Res. Appl. Sci. Eng. Technol. **11**(3) (2023). https://doi.org/10.22214/ijraset.2023.49583
4. Liu, Y., Zhang, S., Yan, P., Li, H.: Finite element modeling and test of piezo disk with local ring electrodes for micro displacement. Micromachines **13**, 951 (2022). https://doi.org/10.3390/mi13060951
5. Jadhav, S., Schoiswohl, M., Buchmayr, B.: Applications of finite element simulation in the development of advanced sheet metal forming processes. Berg Huettenmaenn Monatsh **163**, 109–118 (2018). https://doi.org/10.1007/s00501-018-0713-0
6. Wang, G., et al.: Recent advances in the mechanics of 2D materials. Int. J. Extreme Manuf. (2023). https://doi.org/10.1088/2631-7990/accda2
7. Cramer, N.B., et al.: Elastic shape morphing of ultralight structures by programmable assembly. Smart Mater. Struct. **28**(5), 055006 (2019). https://doi.org/10.1088/1361-665X/ab0ea2
8. Shi, S., Pang, M., Lou, T.: Numerical assessment on continuous reinforced normal-strength concrete and high-strength concrete beams. Buildings **13**, 1157 (2023). https://doi.org/10.3390/buildings13051157
9. Sagybekova, A., Kiyalbay, S., Belov, A., Kiyalbayev, A., Tursumbekova, K.: Comparison of test results to determine the parameters of soil strength to ensure the stability of earth slopes. EUREKA Phys. Eng. **6**, 3–11 (2022). https://doi.org/10.21303/2461-4262.2022.002691
10. Kumar, S.D.V., Karuppanan, S., Perumal, V., Ovinis, M.: Failure pressure prediction of corroded high-strength steel pipe elbow subjected to combined loadings using artificial neural network. Mathematics **11**, 1615 (2023). https://doi.org/10.3390/math11071615

11. Murawski, K.: Technical lateral buckling with stress and strain analysis of semi-slender thin-walled cylindrical pinned column made of steel St35 simplified with A=Ael, Jz=Jzel, E=Ec, Epl=Esc (2023). https://doi.org/10.54769/RDZMM6VGS1
12. Peng, P., Li, Z., Zhang, X., Zhang, W., Dong, W.: Efficient method for calculating slope failure risk based on element failure probability. Appl. Sci. **13**, 4806 (2023). https://doi.org/10.3390/app13084806
13. Franklin, V.A., et al.: Evaluation of critical buckling strength of columns: an experimental study. In: AIP Conference Proceedings 30, vol. 2520, no. 1, p. 030003 (2022). https://doi.org/10.1063/5.0103818
14. Yuan, P., Cai, C.S., Liu, K., Wang, X., Ke, L.: Dynamic performance-based assessment for tied-arch bridges subjected to heavy multi-axial tractor-trailers. Struct. Infrastruct. Eng. (2022). https://doi.org/10.1080/15732479.2022.2155975
15. Bondarenko, I., Severino, A., Olayode, I.O., Campisi, T., Neduzha, L.: Dynamic sustainable processes simulation to study transport object efficiency. Infrastructures **7**, 124 (2022). https://doi.org/10.3390/infrastructures7090124
16. Bondarenko, I., Campisi, T., Tesoriere, G., Neduzha, L.: Using detailing concept to assess railway functional safety. Sustainability **15**, 18 (2023). https://doi.org/10.3390/su15010018
17. Farhangi, V., Zadehmohamad, M., Monshizadegan, A., Izadifar, M., Moradi, M.J., Dabiri, H.: Effects of geogrid reinforcement on the backfill of integral bridge abutments. Buildings **13**, 853 (2023). https://doi.org/10.3390/buildings13040853
18. Bondarenko, I., Lukoševičius, V., Keršys, R., Neduzha, L.: Investigation of dynamic processes of rolling stock-track interaction: experimental realization. Sustainability **15**, 5356 (2023). https://doi.org/10.3390/su15065356
19. Le, T.-C., Luu, T.-H.-T., Nguyen, H.-P., Nguyen, T.-H., Ho, D.-D., Huynh, T.-C.: Piezoelectric impedance-based structural health monitoring of wind turbine structures: current status and future perspectives. Energies **15**, 5459 (2022). https://doi.org/10.3390/en15155459
20. Magalhães, K.M.M., Brasil, R.M.L.R.F., Wahrhaftig, A.M., Siqueira, G.H., Bondarenko, I., Neduzha, L.: Influence of atmospheric humidity on the critical buckling load of reinforced concrete columns. Int. J. Struct. Stabil. Dyn. **22**, 2250011 (2022). https://doi.org/10.1142/S0219455422500110
21. Abderezak, R., Daouadji, T.H., Rabia, B.: New solution for damaged porous RC cantilever beams strengthening by composite plate. Adv. Mater. Res. **10**(3), 169–194 (2021). https://doi.org/10.12989/amr.2021.10.3.169
22. Tan, J., Xiang, P., Zhao, H., Yu, J., Ye, B., Yang, D.: Stochastic analysis of train running safety on bridge with earthquake-induced irregularity under aftershock. Symmetry **14**, 1998 (2022). https://doi.org/10.3390/sym14101998
23. Al-Lami, A., Hilmer, P.: Implementing LEAN and six-sigma: a case study in developing the composites production process economically and ecologically. Engineering (2015)
24. Abebe1, S., Mohammed, T.A.: Performance assessment of reinforced concrete frame under close-in blast loading. Adv. Civ. Eng. 3979195 (2022). https://doi.org/10.1155/2022/3979195
25. Jin, Z., Liu, W., Pei, S.: Probabilistic evaluation of railway vehicle's safety on bridges under random earthquake and track irregularity excitations. Eng. Struct. **266**, 114527 (2022). https://doi.org/10.1016/j.engstruct.2022.114527
26. Contreras Lopez, J., Chiachío, J., Saleh, A., Chiachío, M., Kolios, A.: A cross-sectoral review of the current and potential maintenance strategies for composite structures. SN Appl. Sci. **4**, 180 (2022). https://doi.org/10.1007/s42452-022-05063-3
27. Li, R.Y.M., Chau, K.W., Zeng, F.F.: Ranking of risks for existing and new building works. Sustainability **11**, 2863 (2019). https://doi.org/10.3390/su11102863
28. Tang, Y., Jing, J., Zhang, Z., Yang, Y.: A Quantitative risk analysis method for the high hazard mechanical system in petroleum and petrochemical industry. Energies **11**, 14 (2018). https://doi.org/10.3390/en11010014

29. Bahamid, R.A., Doh, S.I., Khoiry, M.A., Kassem, M.A., Al-Sharafi, M.A.: The current risk management practices and knowledge in the construction industry. Buildings **12**, 1016 (2022). https://doi.org/10.3390/buildings12071016
30. Iqbal, S., Ehtisham, N., Bukhari, S.F.K., Mahmood, S.: The assessment of risk management & engineering management practices at project planning phase on performance of construction projects. J. Bus. Soc. Rev. Emerg. Econ. **6**(4), 1369–1378 (2020). https://doi.org/10.26710/jbsee.v6i4.1446
31. Greenberg, H.R. Cramer, J.J. (eds.): Risk Assessment and Risk Management for the Chemical Process Industry, 392 p. Stone & Webster Engineering Corporation (1991). ISBN 978-0-471-28882-4
32. Altavilla, A., Garbellini, L.: Risk assessment in the aerospace industry. Saf. Sci. **40**(1–4), 271–298 (2002). https://doi.org/10.1016/S0925-7535(01)00049-2
33. Pandit, R., Astolfi, D., Hong, J., Infield, D., Santos, M.: SCADA data for wind turbine data-driven condition/performance monitoring: a review on state-of-art, challenges and future trends. Wind Eng. **47**(2), 422–441 (2023). https://doi.org/10.1177/0309524X221124031
34. Zhicheng, G., Liang, R.Y., Patnaik, A.K.: Probabilistic lifetime performance and structural capacity analysis of continuous reinforced concrete slab bridges. Int. J. Adv. Struct. Eng. (IJASE) **9**(3), 1–15 (2017). https://doi.org/10.1007/s40091-017-0160-2
35. Wu, J.: Study on Performance monitoring system of civil engineering structure. Adv. Eng. Technol. Res. **1**, 482 (2023)
36. Ułanowicz, L., Jastrzębski, G., Jóźko, M., Szczepaniak, P.: Method for improving the aircraft hydraulic drives ability to work. J. KONBiN **50**(2), 127–144 (2020). https://doi.org/10.2478/jok-2020-0031

Study of Effect of Oil Replenishment on the Amount of Mechanical Impurities in Multi-unit Diesel Engine Oil

Gediminas Vaičiūnas(✉) and Stasys Steišūnas

Vilnius Gediminas Technical University, Saulėtekio al. 11, 10223 Vilnius, Lithuania
gediminas.vaiciunas@vilniustech.lt

Abstract. Practical tests were carried out to evaluate the efficiency of diesel engine oil additional filtration equipment. When tests were performed on non-new diesel engines, oil was periodically replenished. Due to the oil replenishment, it was difficult to evaluate the change in the characteristics of the oil. The authors of the article proposed a methodology for evaluating the effect of oil replenishment on the effectiveness of oil filtration equipment.

Keywords: Diesel engine · Oil filtration systems · Mechanical impurities

1 Introduction

Diesel engine oil filtration systems are being modernized for several reasons. On the one hand, when adapting engines to alternative fuels, it is necessary to make a correction in the oil filtration system of the diesel engine [10]. For example, using fuel made from biomass [2]. On the other hand, to improve the performance of the engine, corrections are also made to the oil system [5]. After making corrections to the technical parameters, corrections must also be made to the oil system, which requires appropriate research [3]. When examining not just one vehicle, but a fleet of them, it is necessary to evaluate the probable effect of oil system corrections [12]. As the safety requirements of the application increase, the requirements for engine oils change, and therefore their operating equipment changes as well [6]. After installing new filtering devices, the characteristics of the oil filtering systems change, and they need to be adjusted again [4]. Oil parameters such as viscosity determine the possibility of changes in the engine oil system [1]. Adjustments to the engine oil system produce results such as improved exhaust gas composition [8]. Used oil can be used as fuel for diesel engines if the oil system of its year of operation met the necessary conditions [9]. Used oil can also be used as fuel for furnaces, but it must meet the requirements for this [7].

This study aims to evaluate the efficiency of diesel engine oil additional filtration system.

O. Prentkovskis et al. (Eds.): TRANSBALTICA 2023, LNITI, pp. 539–546, 2024.
https://doi.org/10.1007/978-3-031-52652-7_53

2 Methodology of the Study

The studies were carried out on a diesel multi-unit (DMU) DR1A. The DR1A consists of two motor units (front and rear) and several passenger wagons between them. The design speed of the DMU is 120 km/h, all units are four-axle, and the axle formula is 2_0–2. Motor units are equipped with M-756 B engines. 736 kW diesel engine, four-stroke, 12-cylinder, crankshaft speed 1,500 rpm, fuel specific consumption 215 g/kWh. One of the engines is equipped with additional oil filtration equipment and the other is not. The scheme of the additional oil filtration system is shown in Fig. 1.

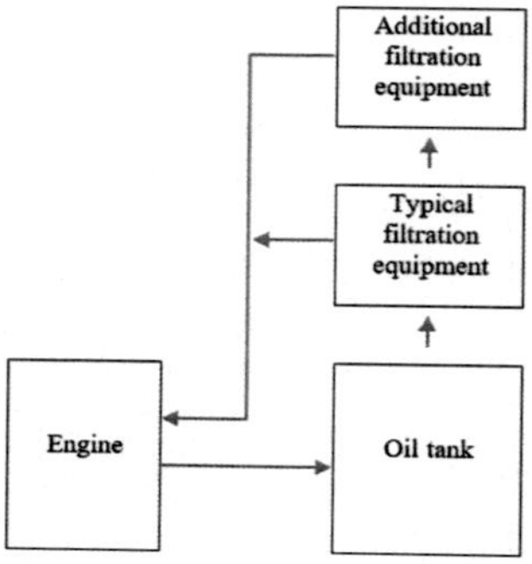

Fig. 1. The scheme of the additional oil filtration system.

Additional oil filtration equipment parameters are:

1. partial-flow filters (Kidney Loop);
2. designed to filter small stream of oil (about 7 l/min);
3. removes 1–40 micron particles;
4. the filter element of the equipment is made of cellulose and cotton.
5. the filter absorbs water, antifreeze and other foreign particles.

The course of the research is described in more detail in the previously published article by the authors, therefore, saving volume, this material can be consulted in the relevant literature [11].

The results of the tests of the samples of mechanical impurities taken from the engines with (Engine-1) and without additional filtration equipment (Engine-2) in oil are given in Fig. 2 [11].

The amount of mechanical impurities in diesel Engine-2 oil became significantly higher than in diesel Engine-1 oil (this pattern is very pronounced when mileage is from 5,000 to 20,000 km). This hypothesizes that the additional fine partial filtration equipment may have reduced the amount of mechanical impurities in the oil.

The authors do not address all possible contaminants in this article. The authors discuss only the pollutants of greatest importance. The percentage distribution of pollutants is shown in the figure below (it is the result of research done by the authors).

The data in Fig. 3 show that iron is the largest pollutant (more than 80%), copper - about 10% and chromium – about 5%. Therefore, the change in the amount of iron, copper

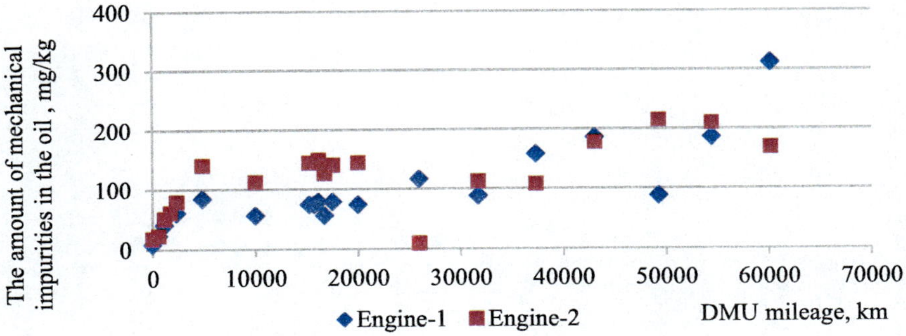

Fig. 2. Test results for mechanical impurities in the oil.

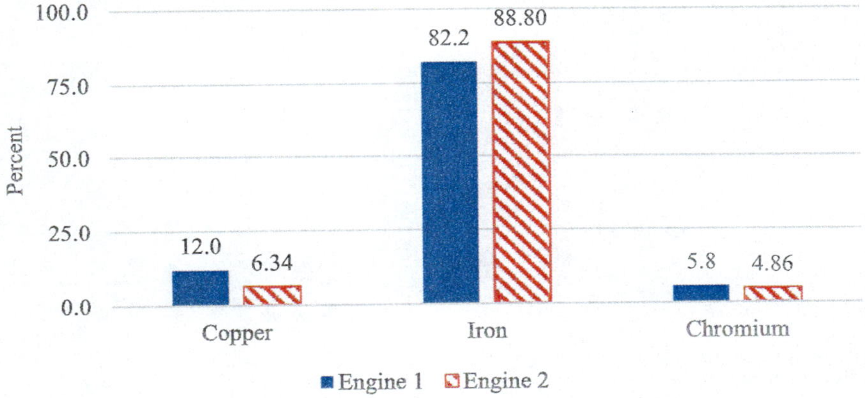

Fig. 3. The percentage distribution of pollutants.

and chromium are further examined. The variation of the amount of iron according to the DMU mileage is shown in Fig. 4.

The average iron content in the oil is 13 mg/kg in the Engine-1 and 30 mg/kg in the Engine-2.

The average copper content (Fig. 5) in the oil is 1.7 mg/kg in the Engine-1 and 2.7 mg/kg in the Engine-2.

The average chromium content in the oil is 0.7 mg/kg in the Engine-1 and 1.7 mg/kg in the Engine-2. Examining the change of iron, copper, and chromium content in the oil according to the DMU mileage shows that the content of these elements increases from 28,000 km mileage. Examining the change in the amount of copper, its increase is noticeable in the Engine-1 at a mileage of 40,000 km (this is shown in Fig. 6). This is the only "anomaly" in the analysis of changes in the amount of iron, copper, and chromium in the oil in the range of 20,000 to 60,000 km. On average, the iron content is about 2, copper 1.6, chromium 2.4 times lower in Engine-1 oil with additional oil filtration equipment.

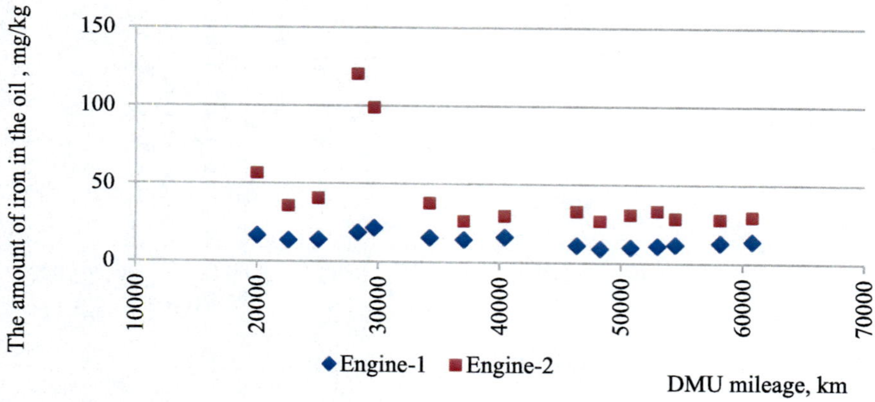

Fig. 4. Dependence of the amount of iron in the oil on the DMU mileage.

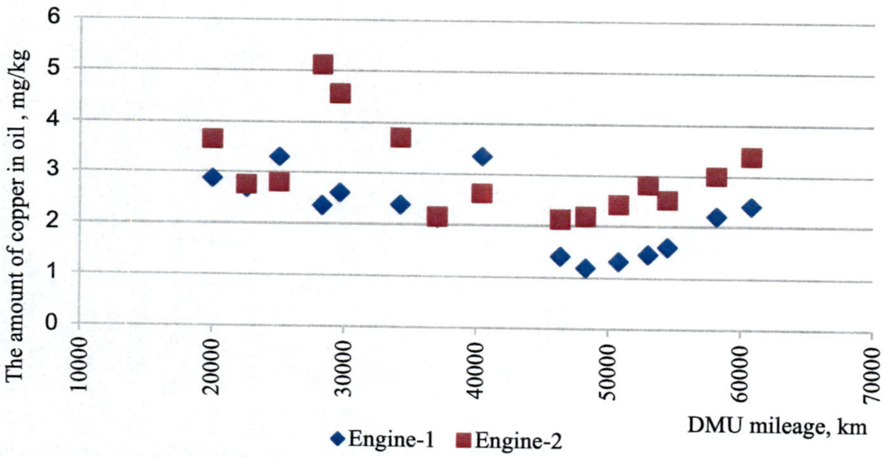

Fig. 5. Dependence amount of copper in oil on DMU mileage.

It is important to pay attention to the fact that oil is added periodically during operation. For this reason, Fig. 4 and 6 show a decrease in emissions in some mileage intervals (in addition, the DMU has oil filtration equipment). Amount of replenished oil according to the mileage of the DMU is shown in Fig. 7.

The authors propose a methodology for estimating the amount of added oil in order to more correctly analyze the change in the amount of pollutants.

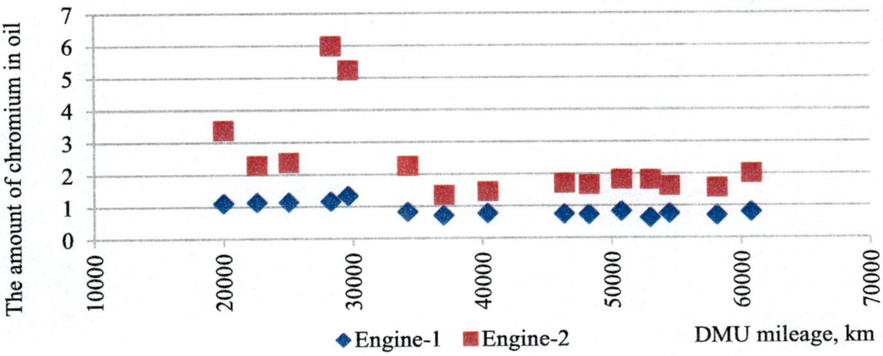

Fig. 6. Dependence amount of chromium in oil on DMU mileage.

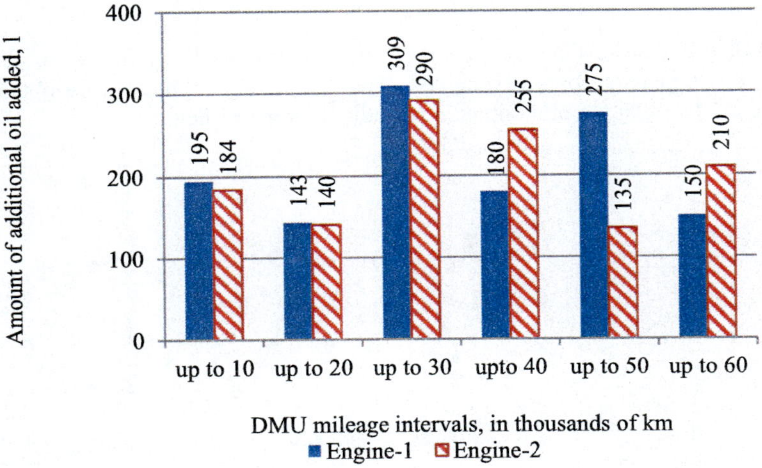

Fig. 7. Amount of replenished oil according to the DMU's mileage.

3 Evaluation of Effect of Oil Replenishment

When examining the amount of pollutants in the oil, the authors suggest using the formula (1) below to estimate the effect of additional oil filling. The authors assume that the resulting amount of pollutants can be calculated according to the following principle:

$$N = \frac{C_{2.1} \cdot M}{M - \Delta M} \cdot M - C_{1.2} \cdot M, \tag{1}$$

where N – Change in the amount of iron in the oil, mg/kg; M – oil mass in the engine, kg; ΔM – replenished oil mass, kg; $C_{1.2}$ – pollutant concentration at the beginning of the mileage interval after adding oil, mg/kg; $C_{2.1}$ – pollutant concentration at the end of the mileage interval before adding oil, mg/kg.

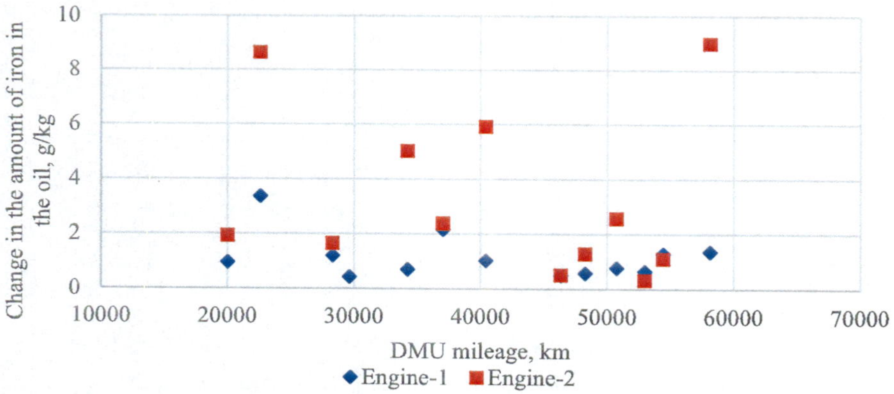

Fig. 8. Change in the amount of iron in the oil, considering oil replenishment.

Data of Fig. 8 show that the increase in iron content in the engine with additional filtration equipment is obviously higher. The data in Fig. 9 and 10 are obtained by assessing the fact that the oil content is periodically replenished.

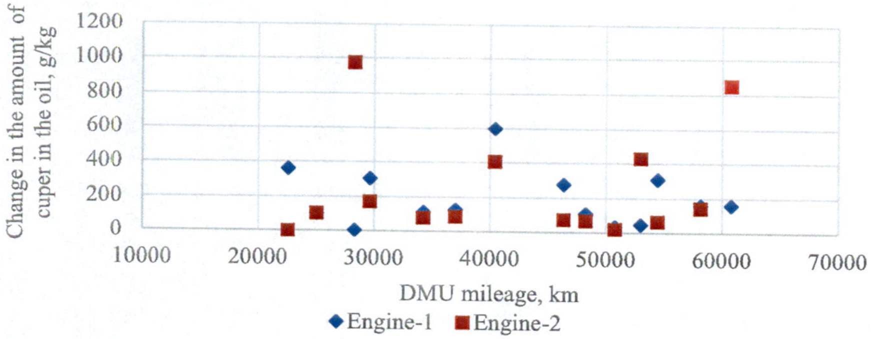

Fig. 9. Change in the amount of cupper in the oil, considering oil replenishment.

After analyzing the amount of pollutants according to the methodology proposed by the authors (their results can be seen in Fig. 8, 9 and 10), it was found that the additional oil filtration equipment has a particularly large effect on the amount of iron impurities.

For copper and chromium impurities, additional filtration equipment is less important. However, using this equipment neutralizes the peaks of copper and chromium impurities generation.

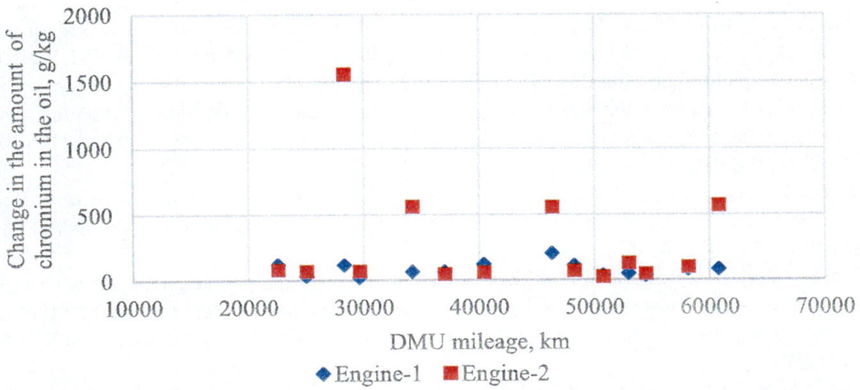

Fig. 10. Change in the amount of chromium in the oil, considering oil replenishment.

4 Conclusions

After installing additional oil filtration equipment in Engine-1, the amount of mechanical impurities in its oil became significantly higher than in Engine-2 (this pattern is very pronounced when the mileage is from 5,000 to 20,000 km). However, these results are affected by oil replenishment.

The authors proposed a methodology for estimating the amount of added oil in order to more correctly analyze the change in the amount of pollutants. After analyzing the amount of pollutants according to the methodology proposed by the authors, it was found that the additional oil filtration equipment has a particularly quite great effect on the amount of iron impurities.

After evaluating the oil replenishment, it can be seen that the intensity of impurity formation in engine oil without additional filtering equipment is higher than in engine oil with additional filtering equipment. For example, the maximum amount of iron in the engine with additional filtration equipment was up to 3 g/kg, and in the oil of the engine without additional filtration equipment up to 9 g/kg. Using this equipment neutralizes the peaks of copper and chromium impurities generation.

References

1. Alimova, Z., Makhamajanov, M.I., Magdiev, K.: The effect of changes in the viscosity parameters of engine oils on the operation of engine parts. Eur. J. Acad. Res. **2**(10), 151–154 (2022). https://doi.org/10.5281/ZENODO.5884973
2. Chen, X., Wang, Zh. Pan, Sh., Pan, H.: Improvement of engine performance and emissions by biomass oil filter in diesel engine. Fuel. **235**, 603–609 (2019). https://doi.org/10.1016/J.FUEL.2018.08.038
3. Gritsenko, A., Zadorozhnaya, E., Shepelev, V., Gimaltdinov, I., Shepelev, V.: Parameters of internal combustion engine efficiency while introducing additives in the oil. Tribol. Ind. **41**(4), 592–603 (2019). https://doi.org/10.24874/ti.2019.41.04.11

4. Hujo, Ľ., Michalides, M., Nosian, J., Simikić, M.: Measurement of flow and pressure characteristics of combustion engine oil filters. In: MATEC Web of Conferences, vol. 351, p. 01004 (2021). https://doi.org/10.1051/MATECCONF/202135101004
5. Kim, K., et al.: Characterization of engine oil additive packages on diesel particulate emissions. J. Mech. Sci. Technol. **34**(2), 931–939 (2020). https://doi.org/10.1007/S12206-020-0142-3/METRICS
6. Hujo, L., et al.: Design of laboratory test equipment for automotive oil filters to evaluate the technical life of engine oil. Appl. Sci. **11**(2), 483 (2021). https://doi.org/10.3390/APP11020483
7. Li, H., Zhang, F., Feng, Z., Li, W., Zou, X.: Study on waste engine oil and waste cooking oil on performance improvement of aged asphalt and application in reclaimed asphalt mixture. Constr. Build. Mater. **276**, 122138 (2021). https://doi.org/10.1016/J.CONBUILDMAT.2020.122138
8. Al Sheikh Omar, A. Motamen Salehi, F., Farooq, U., Morina, A., Neville, A.: Chemical and physical assessment of engine oils degradation and additive depletion by soot. Tribol. Int. **160**, 107054 (2021). https://doi.org/10.1016/J.TRIBOINT.2021.107054
9. Olimjonov, R., Gulmatov, A.: Cleaning the lubricant from mechanical impurities using a magnet-cotton filter. In: IOP Conference Series: Earth and Environmental Science, vol. 1076, no. 1, p. 012008 (2022). https://doi.org/10.1088/1755-1315/1076/1/012008
10. Vaičiūnas, G., Steišūnas, S., Bureika, G.: Specification of estimation of a passenger car ride smoothness under various exploitation conditions. Eksploatacja niezawodnosc. **23**(4), 719–725 (2021). https://doi.org/10.17531/EIN.2021.4.14
11. Vaičiūnas, G., Steišūnas, S.: Investigation of the efficiency of additional engine oil fine filtration equipment installation on diesel trains. In: Transport Problems 2022 : XIV International Scientific Conference. IX international Symposium of Young Researches, Katowice – Silesia: Proceedings, pp. 758–763 (2022)
12. Zhang, R., et al.: A study into the impact of engine oil on gasoline particulate filter performance through a real-world fleet test. SAE Technical Papers. 2019-April (2019). https://doi.org/10.4271/2019-01-0299

Comparison of the Development of Private Car and Railway Transport Systems in Europe

Gediminas Vaičiūnas(✉)

Vilnius Gediminas Technical University, Saulėtekio al. 11, 10223 Vilnius, Lithuania
gediminas.vaiciunas@vilniustech.lt

Abstract. With the motive of protecting the environment, many countries have put forward a vision to replace private car transport with railway transport as much as possible. The author of the article proposes a method to determine in which countries it would be most appropriate. The article presents an analysis of the number of private cars and the length of railway lines in countries, assessing the relationship of these indicators with the country's area, population and gross domestic product.

Keywords: Railway transport · Car · Development of systems · Area · Population

1 Introduction

One of the obstacles in the development of green transport is the relatively high number of polluting cars [7]. Therefore, it is proposed to develop railway transport more widely as a means of environmental friendliness of transport. Rail transport is considered to be a suitable mode of transport for the development of green transport [3]. Both railway and road networks are an important part of every country's economy and an important factor in economic development [6]. The road and rail network serves production and industry on the one hand, and creates tourism opportunities on the other [8]. It is very important to rationally predict train speeds when building new railway lines [13]. Train speeds that are too low will lead to insufficient line capacity, slow passenger train traffic, and unreasonably high speeds require costs that may not pay off [16]. When building new railway lines, it is necessary to ensure that they are sustainable and reliable, as well as ensuring that they are properly loaded [2]. Traffic safety on future railway lines must be taken into account when planning railway development. The reliability, availability and maintenance of the rail network should be taken into account in the safety analysis of the rail network [17]. The railway infrastructure project management system must not only focus on its composition, but also take into account the changes and effects of the internal and external environment [9]. Safe transportation of dangerous goods is also an important part of green transport [10]. Any such initiatives involve the creation of strict and robust regulatory and organizational structures that create and maintain a clear separation between the state and the railways, strongly supported by the ideas of rail

O. Prentkovskis et al. (Eds.): TRANSBALTICA 2023, LNITI, pp. 547–552, 2024.
https://doi.org/10.1007/978-3-031-52652-7_54

financing and rail finance. [5]. The right balance between different modes of transport is very important in developing a smart transport system [11]. Analyzing the latest literature shows that special attention is paid to public transport [1, 15]. The integrity of the topics covered is also important for summarizing the body of research [4]. In this article Author proposed a methodology to determine in which countries it would be most appropriate.

2 Research Methodology

The author proposes to evaluate both the development of railways and the development of car transport in countries by calculating the ratio of the relevant indicators to the area of the country, the population figure and gross domestic product [12]. For this purpose, it is necessary to calculate the ratio of the number of private cars in the countries to the area of the countries, the number of inhibitors and gross domestic product [14]. It is also necessary to calculate this ratio according to the length of the railways in the countries. Since all the results obtained are measured in different units of measurement, it is possible to compare them with each other only after their normalization. Only after normalization of data it is possible to calculate the arithmetic or geometric mean from normalized values (the ratio of the number of private cars to the above indicators and the ratio of the length of the railways to the above indicators).

3 Discussion on Results of the Study

Following the steps described in the methodology above, the results were obtained. This chapter reviews the intermediate and final results of this study. The relationship between the number of private cars and the country's area, population and gross domestic product is shown in Fig. 1.

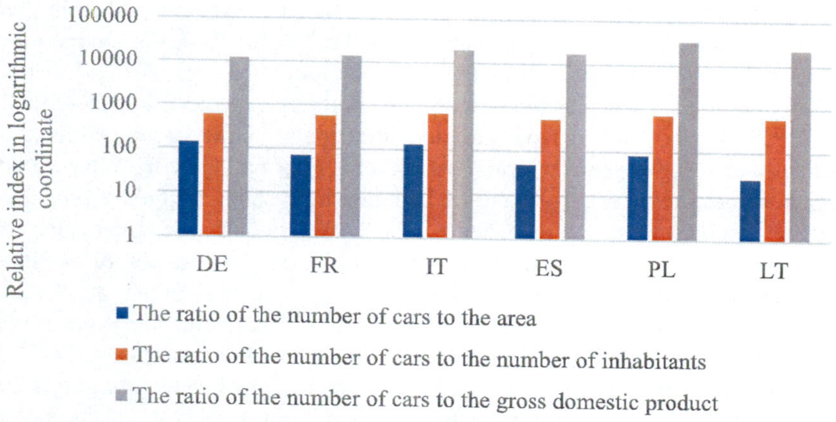

Fig. 1. The ratio of the number of private cars to the country's area, population and gross domestic product.

Since there is a very large difference between the values, the Fig. 1 and 2 are presented in logarithmic scales. When calculating relative indicators to compare the number of private cars in countries, the results of the comparison differ depending on what the number of private cars is analyzed in terms of (country area, population or gross domestic product). Data of Fig. 1 show that the number of private cars per country's area is highest in Germany and Italy, while the lowest in Spain (ES) and Lithuania (LT). The extremum number of private cars per population is not visible. In terms of the ratio of the number of private cars to gross domestic product, Poland (PL), Italy (IT), Spain and Lithuania are in the lead.

The relationship between the length of railways and the country's area, population and gross domestic product is shown in Fig. 2.

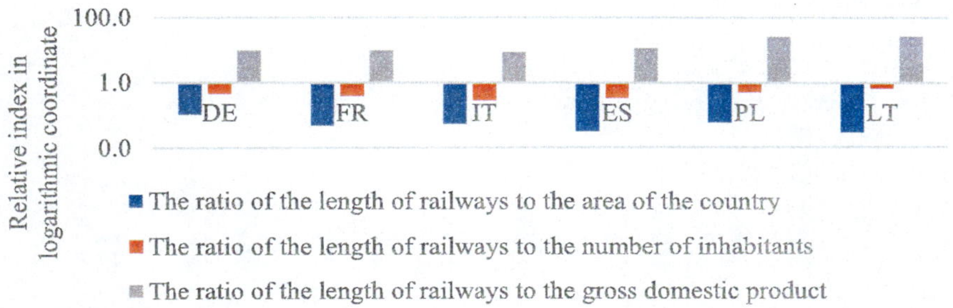

Fig. 2. The relationship between the length of railways and the country's area, population and gross domestic product.

The data of Fig. 2 clearly show that Poland and Lithuania are in the lead in terms of the ratio of the length of the railway to the domestic product. And in terms of the ratio of the length of the railway to the area of the country, Germany (DE) is in the lead. Since the various indicators in Fig. 2 have different measurement units, they can only be summarized after normalization. The normalized ratio of the number of private cars to the country's area, population and gross domestic product is shown in Fig. 3.

In Fig. 3 it is seen the same proportion as in Fig. 1, but here the quantities do not have units of measurement because they are normalized. Normalized quantities were summarized, regardless of what their measurement units were before normalization.

The normalized ratio of railway length to the country's area, population and gross domestic product is shown in Fig. 4.

In Fig. 3 and 4 are shown the normalized data of Fig. 1 and 2. Therefore, the regularities in Fig. 3 and 4 are the same as in Fig. 1 and 2. It was necessary to normalize the data in order to generalize different indicators (measured in different dimensions). It is possible to summarize normalized indicators by calculating the arithmetic or geometric mean. The results of summarizing the indicators are shown in Fig. 5 and 6.

The data in Fig. 5 show that private car transport is most developed in Poland, Italy and Germany.

The data in Fig. 6 show that railway transport is most developed in Poland, Lithuania and Germany. A summary of the numbers in Fig. 5 and 6 show that in those countries with

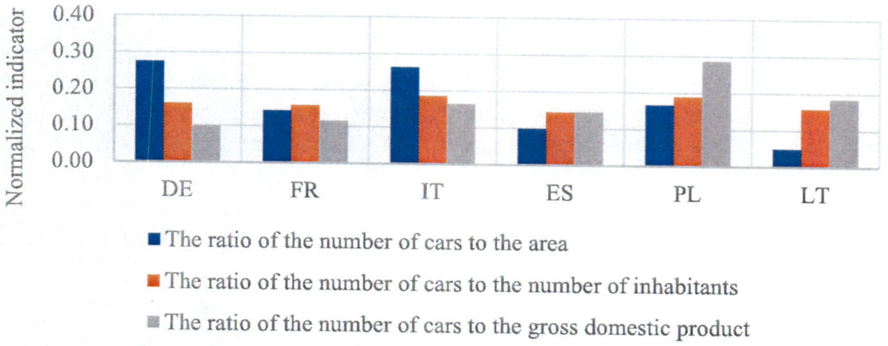

Fig. 3. The normalized ratio of the number of private cars to the country's area, population and gross domestic product.

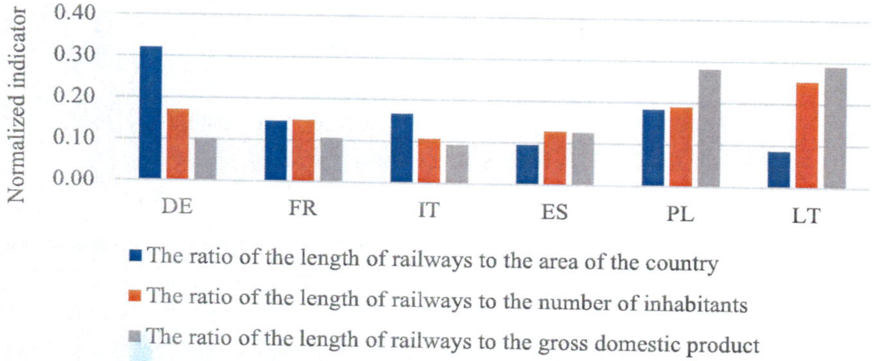

Fig. 4. The normalized ratio of railway length to the country's area, population and gross domestic product.

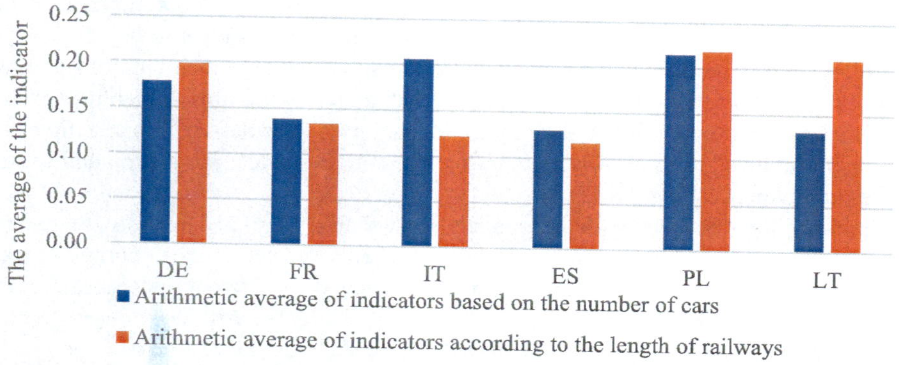

Fig. 5. Generalization of the normalized indicators of the development of automobile transport.

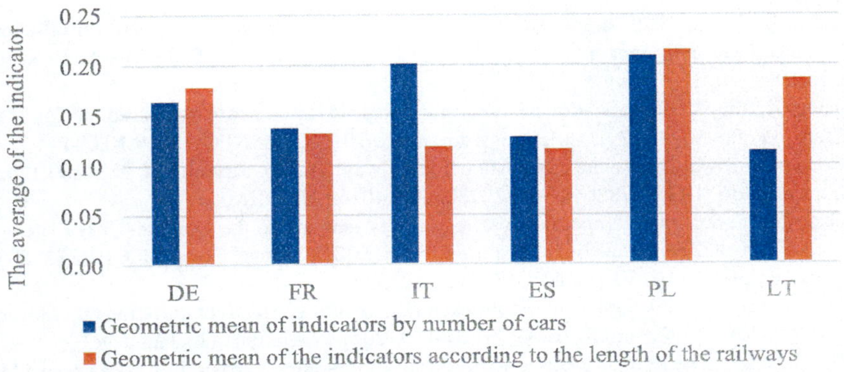

Fig. 6. Generalization of the normalized indicators of the development of railway transport.

the most developed private car transport, railway transport is also the most developed in them. It follows that in European countries the transport system (as a whole of car transport and railway transport) is developed proportionally. As a small exceptionit can be seen that the disproportionately developed private car transport is in Italy, while the significantly developed railway transport is in Lithuania.

4 Conclusions

A review of the scientific literature and an examination of practical aspects of the development of automobile transport and rail transport show that it is important that both of these modes of transport develop in a balanced way.

The author of the article suggests assessing the development of private car fleet and railways, respectively, in the ratio of the number of private cars or the length of railways to the area of the country, to the population and to gross domestic product, respectively.

The core of the proposed method is the normalization and generalization of the relative indicators, which are measured in different measurement units, by the principle of arithmetic or geometric mean.

Summarizing the results of the study, it can be seen that the disproportionately developed car transport is in Italy, while the significantly developed railway transport is in Lithuania. Though overall in European countries the transport system (as a whole of private car transport and rail transport) is developed proportionally.

References

1. Arnz, M., Krumm, A.: Sufficiency in passenger transport and its potential for lowering energy demand. Environ. Res. Lett. **18**(9), 094008 (2023). https://doi.org/10.1088/1748-9326/ACEA98
2. Bešinović, N.: Resilience in railway transport systems: a literature review and research agenda. Transp. Rev. **40**(4), 457–478 (2020). https://doi.org/10.1080/01441647.2020.1728419

3. Çelik, G., Yılmaz, Z.: Comparison of traditional and green public transportation vehicles in terms of CO_2 emissions, pp. 203–222 (2023). https://doi.org/10.4018/978-1-6684-5996-6.CH010
4. Chen, X., Liu, Y.: Visualization analysis of high-speed railway research based on CiteSpace. Transp. Policy **85**, 1–17 (2020). https://doi.org/10.1016/J.TRANPOL.2019.10.004
5. Cowie, J.: Towards a financially sustainable railway. Transp. Sustain. **14**, 281–294 (2022). https://doi.org/10.1108/S2044-994120220000014016/FULL/XML
6. Gao, D., Li, S.: Spatiotemporal impact of railway network in the Qinghai-Tibet Plateau on accessibility and economic linkages during 1984–2030. J. Transp. Geogr. **100**, 103332 (2022). https://doi.org/10.1016/J.JTRANGEO.2022.103332
7. Gedgaudas, A., et al.: Taršių automobilių kaita Lietuvoje. Verslas Technol. Biomed.: Inovacijų Įžvalgos 2021: Straipsnių Rinkinys **1**, 120–130 (2021). [In Lithuanian language]
8. Jin, S., et al.: The influence of high-speed rail on ice–snow tourism in northeastern China. Tour. Manage. **78**, 104070 (2020). https://doi.org/10.1016/J.TOURMAN.2019.104070
9. Liu, K., et al.: Study on dissipative structure of mega railway infrastructure project management system. Eng. Constr. Archit. Manage. (2023). https://doi.org/10.1108/ECAM-10-2022-1021/FULL/XML
10. Moločka, V.: Veiksnių, turinčių didžiausią įtaką saugiam pavojingųjų krovinių vežimui geležinkelių ir kelių transportu, tyrimas. [In Lithuanian language]. https://talpykla.elaba.lt/elaba-fedora/objects/elaba:48306594/datastreams/MAIN/content
11. Toh, C.K., Sanguesa, J.A., Cano, J.C., Martinez, F.J.: Advances in smart roads for future smart cities. Proc. Roy. Soc. A **476**, 2233 (2020). https://doi.org/10.1098/RSPA.2019.0439.12
12. Vaičiūnas, G.: Comparison of railway development in the countries of the world. In: Prentkovskis, O., Yatskiv, I., Skačkauskas, P., Maruschak, P., Karpenko, M. (eds.) TRANSBALTICA 2022. LNITI, pp. 706–715. Springer, Cham (2023). https://doi.org/10.1007/978-3-031-25863-3_68
13. Vaičiūnas, G., Bureika, G.; Steišūnas, S.: Rail vehicle axle-box bearing damage detection considering the intensity of heating alteration. Eksploatacja niezawodnosc – Maint. Reliabil. **22**(4), 724–729 (2020). https://doi.org/10.17531/EIN.2020.4.16
14. Vaičiūnas, G.: Rating of railway development in Europe country's. In: Prentkovskis, O., Yatskiv, I., Skačkauskas, P., Junevičius, R., Maruschak, P. (eds.) TRANSBALTICA 2019. LNITI, pp. 561–567. Springer, Cham (2022). https://doi.org/10.1007/978-3-030-94774-3_55
15. Van, P.M., Nhur, L.Q., Trung, V.T.: Public transport development orientation with transit and land-use integration toward a livable environment. In: AIP Conference Proceedings, vol. 2560, p. 1 (2023). https://doi.org/10.1063/5.0125564
16. Yu, F., Lin, F., Tang, Y., Zhong, C.: High-speed railway to success? The effects of high-speed rail connection on regional economic development in China. J. Reg. Sci. **59**(4), 723–742 (2019). https://doi.org/10.1111/JORS.12420
17. Zhang, Z., Jis, L., Qin, Y.: RAMS analysis of railway network: model development and a case study in China. Smart Resil. Transp. **3**(1), 2–11 (2020). https://doi.org/10.1108/SRT-10-2020-0013

Perceiving the Resilience of Land Transport Critical Entities

David Rehak(✉) and Heidi Janeckova

Faculty of Safety Engineering, VSB - Technical University of Ostrava, Lumirova 13, 700 30 Ostrava – Vyskovice, Czech Republic
{david.rehak,heidi.janeckova}@vsb.cz

Abstract. Land transport sector is one of the most important critical infrastructure sectors, essential for the proper functioning of almost every country and the continuity of their essential services. The issue of critical infrastructure protection has undergone a major transformation since its inception. Although in the early days this issue was addressed mainly in the context of the protection of a specific element of critical infrastructure, currently the issue of critical infrastructure protection is based on the principle of strengthening the resilience of critical entities. Based on these facts, the paper presents general and specific criteria for designating critical entities in the area of land transport. According to these criteria, critical land transport entities are recommended. In the case of these entities, attention is subsequently paid to their resilience, especially to the factors determining individual types of resilience, namely organizational, technical, social, and economic.

Keywords: Critical infrastructure · Critical entity · Transport · Resilience

1 Introduction

Land transport is critical infrastructure for most countries in terms of social, economic, and territorial cohesion. Due to the significant importance of this sector and the possibility of cascading impacts of disruptions to dependent sectors (healthcare and energy) [1, 2], it can be concluded that this is a sector that may be described as critical [3]. In certain cases, owners and operators of services in this sector are referred to as critical entities defined by a set of criteria [2]. It is therefore essential to take steps to build and maintain the quality of their elements in advance of potential incidents, i.e. events that can significantly disrupt the delivery of essential services [4].

The issue of defining critical entities is not currently addressed in any professional literature. Modest references are presented in the contributions of Pursiainen and Kytömaa [5], Ibraimova and Lysik [6] or Mazzeo et al. [7], who are merely reacting to the adoption of Directive (EU) [4]. The authors agree that the Directive does not sufficiently define the issue of resilience and does not establish a uniform framework for its assessment, although there is a sufficient body of literature on which to draw from.

One possible concept that can be used to maintain the most important societal functions, economic activities, public health and safety, and the environment is the resilience

O. Prentkovskis et al. (Eds.): TRANSBALTICA 2023, LNITI, pp. 553–561, 2024.
https://doi.org/10.1007/978-3-031-52652-7_55

of critical entities. These entities are responsible for the protection and performance of the functions. In the context of critical infrastructure, resilience is defined as *"the ability of a critical entity to prevent, protect against, respond to, resist, mitigate, absorb, adapt to, and recover from incidents"* [4].

The issue of resilience within the general concept of critical infrastructure has been addressed in recent years, e.g. by Rød et al. [8], Cantelmi et al. [9], Guo et al. [10] or Liu et al. [11]. A more specific viewpoint in the context of the transport critical infrastructure sector has subsequently been addressed, for example, by Belokas et al. [12], Ashrafi [13], Huang [14] and Ientile et al. [15]. Although these are important publications, none of them explicitly focuses on the critical entities' resilience, both in the field of land transport and in general terms.

For this reason, the aim of the paper is to define criteria for the provision of essential services in the land transport sector, based on which critical entities can be identified. For these entities, the variables determining the resilience of critical land transport entities are subsequently defined in a framework.

2 Criteria for Designating Critical Entities

The definition of criteria for the identification of critical entities is the first legislative step for each EU Member State at national level following the adoption of the Directive (EU) [4]. Based on these criteria, it is possible to identify critical entities that are owners or operators of critical infrastructures.

In the context of this paper, the following section focuses on entities owning or operating critical infrastructure in the area of land transport, i.e. road and rail transport. Critical entities in this area and others are designated based on general criteria defined in the Directive (EU) [4] and specific criteria that are defined by EU member states at the national level. The general criteria are defined as follows:

a) the entity provides one or more essential services,
b) the entity operates, and its critical infrastructure is located, on the territory of that Member State,
c) an incident would have significant disruptive effects, as determined in accordance with Article 7(1), on the provision by the entity of one or more essential services or on the provision of other essential services in the sectors set out in the Annex that depend on that or those essential services.

In order to assess the fulfilment of the first criterion, it is necessary to define the essential services provided in the Member State concerned. For this purpose, it is necessary to rely on the definition of an essential service, which is defined by the Directive (EU) [4] as *"a service essential for the continuity of essential social functions, economic activities, public health and safety or the environment"*. Based on this definition, it is possible to formulate the basic services provided in the field of land transport which are considered the specific national criteria:

1. the operation of Trans-European Transport Network (TEN-T) infrastructure;
2. the operation of services related to the transport of passengers and freight in the TEN-T;

3. the operation of intelligent transport systems in the TEN-T.

The 2nd criterion can in effect be considered binary. Either the entity and its critical infrastructure are in the territory of one of the Member States or they are not located there and then there is no need to deal with them. However, in this context, it is worth noting that if an entity provides an essential service in or to six or more Member States, then it is designated as a critical entity of European importance.

The last criterion is the assessment of the level of so-called significant disturbance. In determining the significance of a disturbance, Member States must further consider the following circumstances [4]:

1. the number of users relying on the essential service provided by the entity concerned,
2. the extent to which other sectors and subsectors as set out in the Annex depend on the essential service in question,
3. the impact that incidents could have, in terms of degree and duration, on economic and societal activities, the environment, public safety and security, or the health of the population,
4. the entity's market share in the market for the essential service or essential services concerned,
5. the geographic area that could be affected by an incident, including any cross-border impact, considering the vulnerability associated with the degree of isolation of certain types of geographic areas, such as insular regions, remote regions or mountainous areas,
6. the importance of the entity in maintaining a sufficient level of the essential service, considering the availability of alternative means for the provision of that essential service.

Based on the above criteria, the following section of the paper describes potential critical entities in the land transport sector.

3 Critical Land Transport Entities

In order to be designated as critical entities, owners or operators in the land transportation sector must meet two limiting conditions. The first condition is the fulfilment of all three criteria presented in the previous part of this paper. The second condition is that they must fall within the category of entities defined in the Annex to (EU) Directive [4]. In the case of land transport, these are the categories of rail and road transport entities.

Categories of rail transport entities:

I. Infrastructure operators within the meaning of Article 3(2) of Directive 2012/34/EU of the European Parliament and of the Council [16]. Such an operator means *"any entity or enterprise entrusted with the establishment, management and maintenance of railway infrastructure, including traffic management and signalling".*

II. Railway enterprises within the meaning of Article 3(1) of Directive 2012/34/EU [16]. Such an enterprise is defined as *"any public or private enterprise licensed in accordance with this Directive whose principal activity is the carriage of goods or passengers by rail and which provides traction; enterprises which merely provide traction are also included".*

III. Operators of service facilities within the meaning of Article 3(12) of Directive 2012/34/EU [16]. Such an operator means *"any public or private entity responsible for managing one or more service facilities or for providing one or more services to railway enterprises"*.

Categories of road transport entities:

A. Road authorities within the meaning of Article 2(12) of Commission Delegated Regulation (EU) 2015/962 [17]. Such an authority is defined as *"a public authority responsible for the planning, control or management of roads within its territorial jurisdiction"*.
B. Intelligent Transport Systems Operators within the meaning of Article 4(1) of Directive 2010/40/EU of the European Parliament and of the Council [18]. These are *"systems using information and communication technologies in the field of road transport, including infrastructure, vehicles and users, and in the field of traffic and mobility management and interfaces with other modes of transport"*.

Based on the above conditions, potential critical entities in the land transport sector can be identified in general terms (see Fig. 1).

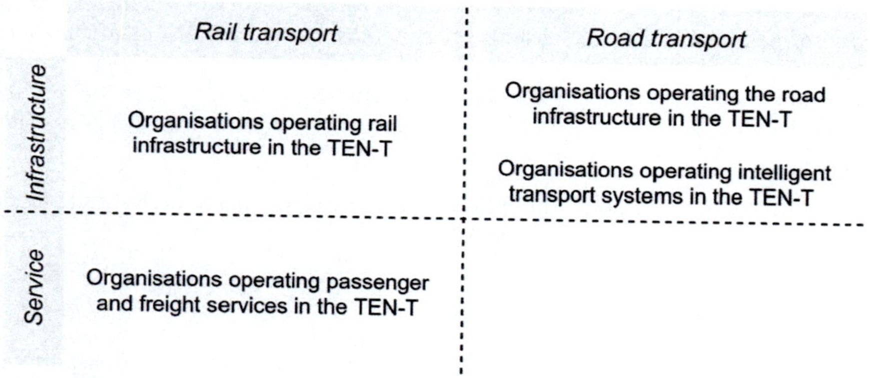

Fig. 1. Critical land transport entities.

In the case of rail transport, the critical entities are expected to be identified as organisations operating rail infrastructure in the TEN-T (i.e. organisations operating high-speed corridors, strategic transport terminals or signalling and signalling equipment) and organisations operating passenger and freight services in the TEN-T (i.e. passenger rail operators, freight rail operators and organisations providing traction).

In the case of road transport, the critical entities are expected to be identified as organisations operating the road infrastructure in the TEN-T (i.e. organisations operating motorways and dual carriageways) and organisations operating intelligent transport systems in the TEN-T (i.e. processing and providing traffic information, traffic and freight management, road safety and security against external threats, and linking vehicles to the transport infrastructure).

4 Resilience of Critical Land Transport Entities

Since the beginning of the research, the issue of critical infrastructure resilience has undergone a significant transformation. Although resilience research was initially mainly associated with the issue of technical resilience of critical infrastructure elements [19], the focus is currently on the resilience of the entity itself [4], i.e. the ability of an organisation to absorb and adapt to a changing environment [20].

On the basis of the above, a number of publications have been produced in which individual authors determine the individual factors and variables of the above mentioned areas. In the field of technical resilience research, for example, Bertocchi et al. [21], Kozine et al. [22] and Rehak et al. [23]. And within the second area, i.e. organizational resilience, the authors Denyer [24], Patriarca et al. [25] and Tasic et al. [26].

However, in the context of the Directive (EU) [4], the resilience of critical entities cannot be viewed solely based on the above areas. Attention must also be paid to other areas and variables that shape resilience. It is thus primarily an area of economic resilience, which denotes the ability of an organisation to face costs arising not only from a crisis [27–30]. Since the resilience of a given system is also largely influenced by stakeholder engagement, it is essential to pay attention to the area of social resilience [27, 31–33]. It can be understood as a society's ability to adapt not only to challenges before a disaster, but also after it [34].

Based on the above, the same determining components of the individual fields of resilience, which are resistance, robustness, recoverability, and adaptability, can be identified. In the context of the problem at hand, resilience can be defined as the ability of a critical entity to prevent incidents from occurring, whereas robustness as the ability to absorb the impact of an incident that has already occurred [34]. Another component that determines the resilience of a given subsystem is the so-called recoverability, or the ability of the subsystem to restore its operation to its original or desired level of performance. The last component that resilience is shaped by is adaptability, which is the ability of the critical entity to prepare the element for a possible re-exposure to an adverse event that has already occurred in the past [23].

Based on the above, Fig. 2 provides a framework to illustrate the variables determining the resilience of critical land transport entities.

In addition to the individual components of resilience, Fig. 2 illustrated above also shows the specific variables of the resilience domains. In the context of organisational resilience, it is thus mainly about crisis preparedness, which represents a set of measures to increase the preparedness of an entity for possible incidents. These measures include, for example, failure analyses [35]. The second variable is the ability to predict. This variable denotes the ability of the system to predict a possible incident in time, where the resulting time interval can be used to implement further preventive measures [34]. Depending on that variable, the time required to activate protective measures to prevent the spread of the consequences caused by the incident, i.e. the ability to react, follows [36]. Another variable in that field are the so-called recovery processes, which are defined as processes that support the rapid recovery of the required element performance [23]. Examples include business continuity plans or contingency plans [37]. The final variables in the area of organisational resilience are risk management and innovation processes. Both of these variables represent important internal processes of the

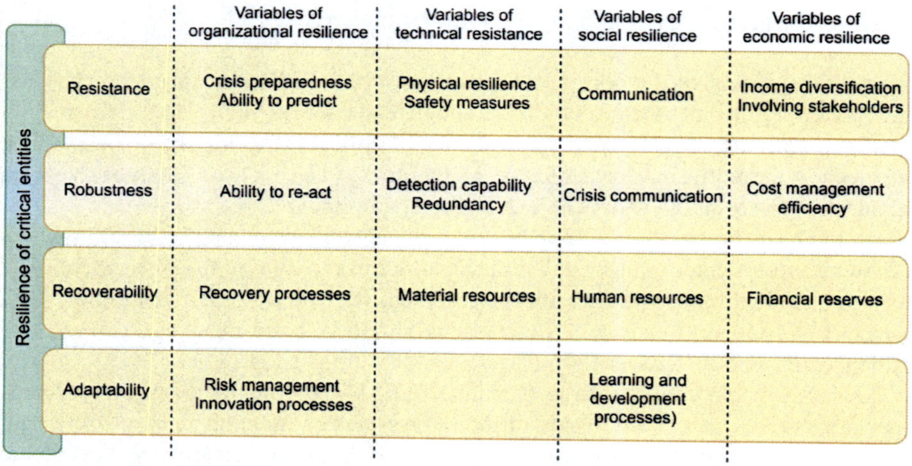

Fig. 2. Variables determining the critical entities resilience.

organisation, where risk management consists mainly of the coordination of activities and innovation processes in the reliability and external security of the technologies used [36].

In terms of the structural characteristics of buildings and the technologies used, physical resilience, together with safety measures (i.e. a set of organisational and regime measures and technical means), represent some of the fundamental variables in the field of technical resilience [36]. Apart from the above, the area of technical resilience is also determined by the detection capability, i.e. the probability and/or time of incident detection, or redundancy, which indicates the ability to immediately substitute the performance of the disrupted part of the element or to enhance its capabilities [38]. Last but not least, it is also about the availability of the necessary components that are needed to carry out repairs or replacements of damaged parts of the elements (i.e. material resources) [36].

The following area, i.e. the area of social resilience, is determined by a total of five variables. Thus, it is primarily a communication variable that expresses the ability to reliably exchange information, which must be ensured both during the normal operation of the entity and during its possible disruption (crisis communication) [39, 40]. Furthermore, the smooth flow of information must also be ensured not only between individuals but also between groups, both within the entity and in relation to the parties concerned, i.e. the public or stakeholders [41]. The area of social resilience is further determined by the availability of human resources and their knowledge, skills and attitudes (i.e. learning and development processes) [42].

One of the most important areas where the critical entities resilience is shaped is in the area of economic resilience. Within the context of this area, income diversification expresses the ability of a critical entity to maintain financial stability in the event of a failure of an income source [43]. One of the ways in which financial stability of a critical entity can be achieved (spreading the financial costs) is by involving stakeholders who contribute to the financing of the repair of the critical infrastructure [44]. Cost management efficiency is also an essential variable for economic resilience. This variable thus reflects the ability of the critical entity to limit/reduce expenditure that is not necessary to overcome the incident which threatens the entity. Finally, financial resources, or their availability (financial reserves), also play an important role in this area, which enable the rapid restoration of a specific element of critical infrastructure [36].

The variables presented above were recommended as key factors determining the resilience of critical entities. These factors are used to assess the level of resilience. Based on the level of achievement of individual variables, critical entities should propose measures to strengthen resilience in all four areas, i. e. organizational, technical, social, and economic.

5 Conclusion

Land transport represents one of the key sectors of the critical infrastructure system through which efficient transport of people and freight is undertaken. However, the importance of this sector is even more fundamental, due to its significant importance to dependent sectors that are negatively affected in the event of its disruption or failure. For this reason, the concept of resilience is used in the context of critical infrastructure, though until recently it consisted of resilience of critical infrastructure elements.

However, the adoption of the new Directive (EU) has changed the existing perception of critical infrastructure resilience. Attention is thus currently focused on the resilience of critical entities that own or operate the elements in question. However, for an entity to be designated as critical, it needs to meet not only the general criteria set out in the Directive (EU), but also specific criteria that are defined by the EU member states at the national level.

The main contribution of this study is therefore the definition of specific (national) criteria for designating critical entities in the land transport sector and the definition of factors (variables) determining the resilience of these entities. As the Directive (EU) does not sufficiently address the issue of resilience, the Authors also define the variables of the different resilience domains (i.e. technical, organisational, social and economic resilience) that determine the resilience of critical land transport entities.

Based on the above, the presented findings are significant in the context of the issue of identifying critical land transport entities. At the same time, the basic aspects for future research on resilience building of critical entities are provided.

Funding. This work was supported by the Ministry of the Interior of the Czech Republic [grant number VK01030014] and by the VSB – Technical University in Ostrava [Grant number SP2023/086].

References

1. Rehak, D., Patrman, D., Foltin, P., Dvorak, Z., Skrickij, V.: Negative impacts from disruption of road infrastructure elements performance to dependent subsystems: methodological framework. Transport **36**(6), 510–524 (2021)

2. Rehak, D., Novotny, P.: Bases for modelling the impacts of the critical infrastructure failure. Chem. Eng. Trans. **53**, 91–96 (2016)
3. Theoharidou, M., Kandias, M., Gritzalis, D.: Securing transportation critical infrastructures: trends and perspectives. In: Georgiadis, C.K., Jahankhani, H., Pimenidis, E., Bashroush, R., Al-Nemrat, A. (eds.) E-Democracy ICGS3 2011 2011. LNICST, vol. 99, pp. 171–178. Springer, Heidelberg (2012). https://doi.org/10.1007/978-3-642-33448-1_24
4. Directive (EU) 2022/2557 of the European Parliament and of the Council of 14 December 2022 on the resilience of critical entities and repealing Council Directive 2008/114/EC
5. Pursiainen, Ch., Kytömaa, E.: From European critical infrastructure protection to the resilience of European critical entities: what does it mean? Sustain. Resil. Infrastruct. **8**, 85–101 (2023)
6. Ibraimova, A., Lysik, A.: Critical entities resilience directive (CER) – broader scope and more stringent obligations. https://www.technologylawdispatch.com/2023/04/data-cyber-security/critical-entities-resilience-directive-cer-broader-scope-and-more-stringent-obligations/. Accessed 15 May 2023
7. Mazzeo, O., Longo, A., Zappatore, M.: Cloud computing and critical infrastructure resilience. In: Barolli, L. (ed.) AINA 2023. LNNS, vol. 655, pp. 115–126. Springer, Cham (2023). https://doi.org/10.1007/978-3-031-28694-0_11
8. Rød, B., Lange, D., Theocharidou, M., Pursiainen, C.: From risk management to resilience management in critical infrastructure. J. Manage. Eng. **36**(4) (2020)
9. Cantelmi, R., Di Gravio, G., Patriarca, R.: Reviewing qualitative research approaches in the context of critical infrastructure resilience. Environ. Syst. Decis. **41**(3), 341–376 (2021)
10. Guo, D., Ming, S., Kingsford, O.: Resilience assessment frameworks of critical infrastructures: State-of-the-art review. Buildings **11**(10), 464 (2021)
11. Liu, X., Fang, Y., Zio, E.: A hierarchical resilience enhancement framework for interdependent critical infrastructures. Reliabil. Eng. Syst. Saf. **215** (2021)
12. Belokas, G. A., Saroglou, C., Moschovou, T., Vlahogianni, E.: Transport infrastructure resilience: from geotechnical asset to transport network. In: Proceedings of the International Conference on Natural Hazards and Infrastructure, ICONHIC, Athens (2022)
13. Ashrafi, B., Naseri, M., Barabady, J.: Resilience of a transportation network: importance of vulnerable nodes. In: Probabilistic Safety Assessment and Management, PSAM (2022)
14. Huang, Z., et al.: Resilience assessment of tunnels: Framework and application for tunnels in alluvial deposits exposed to seismic hazard. Soil Dyn. Earthq. Eng. **162** (2022)
15. Ientile, S., et al.: Sensor based monitoring systems for resilient road infrastructures. In: Proceedings of the International Conference on Natural Hazards and Infrastructure (2022)
16. Directive 2012/34/EU of the European Parliament and of the Council of 21 November 2012 establishing a single European railway area
17. Commission Delegated Regulation (EU) 2015/962 of 18 December 2014 supplementing Directive 2010/40/EU of the European Parliament and of the Council with regard to the provision of EU-wide real-time traffic information services
18. Directive 2010/40/EU of the European Parliament and of the Council of 7 July 2010 on the framework for the deployment of Intelligent Transport Systems in the field of road transport and for interfaces with other modes of transport Text with EEA relevance
19. Council Directive 2008/114/EC of 8 December 2008 on the identification and designation of European critical infrastructures and the assessment of the need to improve their protection
20. ISO 22316. 2017. Security and Resilience – Organizational Resilience – Principles and Attributes. International Organization for Standardization, Geneva
21. Bertocchi, G., et al.: Guidelines for critical infrastructure resilience evaluation. Italian Association of Critical Infrastructures' Experts, Roma (2016)
22. Kozine, I., Petrenj, B., Trucco, P.: Resilience capacities assessment for critical infrastructures disruption: the READ framework. Int. J. Crit. Infrastruct. **14**(3), 199–220 (2018)

23. Rehak, D., Senovsky, P., Slivkova, S.: Resilience of critical infrastructure elements and its main factors. Systems **6**, 21 (2018)
24. Denyer, D.: Organizational resilience: a summary of academic evidence, business insights and new thinking. BSI and Cranfield School of Management, Cranfield (2017)
25. Patriarca, R., Di Gravio, G., Costantino, F., Falegnami, A., Bilotta, F.: An analytic framework to assess organizational resilience. Saf. Health Work **9**(3), 265–276 (2018)
26. Tasic, J., Amir, S., Tan, J., Khader, M.: A multilevel framework to enhance organizational resilience. J. Risk Res. **23**(6), 713–738 (2020)
27. Labaka, L., Comes, T., Hernantes, J., Sarriegi, J.M., Gonzalez, J.J.: Implementation methodology of the resilience framework. In: Proceedings of the 47th Hawaii International Conference on System Science, vol. 47 (2014)
28. Tonn, G., Erwann, M.K., Howard, K.: Insurance, economic incentives and other policy tools for strengthening critical infrastructure resilience: 20 proposals for action. Wharton School of the University of Pennsylvania, Pennsylvania (2016)
29. Fisher, M.K., Gamper, K.: Policy evaluation framework on the governance of critical infrastructure resilience in Latin America. IDB Improving lives (2017)
30. Resilience for sustainable, inclusive growth. https://www.mckinsey.com/capabilities/risk-and-resilience/our-insights/resilience-for-sustainable-inclusive-growth. Accessed 20 May 2023
31. OECD. Good Governance for Critical Infrastructure Resilience; OECD Publishing, Paris (2019)
32. Bucci, S., et al.: After hurricane sandy: time to learn and implement the lessons in preparedness, response, and resilience; The Heritage Foundation Emergency Preparedness Working Group, Washington, DC, USA (2013)
33. Kwok, A., Doyle, E., Becker, J., Johnston, D., Paton, D.: What is 'social resilience'? Perspectives of disaster researchers, emergency management practitioners, and policymakers in New Zealand. Int. J. Disast. Risk Reduct. **19**, 197–211 (2016)
34. Rehak, D., Flynnova, L., Slivkova, S.: Concept of Resistance in the Railway Infrastructure Elements Protection. In: Prentkovskis, O., Yatskiv (Jackiva), I., Skačkauskas, P., Junevičius, R., Maruschak, P. (eds.) TRANSBALTICA 2021. LNITI, pp. 419–428. Springer, Cham (2022). https://doi.org/10.1007/978-3-030-94774-3_41
35. EN 13306. 2017. Maintenance - Maintenance terminology
36. Rehak, D., Senovsky, P., Hromada, M., Lovecek, T.: Complex approach to assessing resilience of critical infrastructure elements. Int. J. Crit. Infrastruct. Prot. **25**, 125–138 (2019)
37. Hatton, T., Grimshaw, E., Vargo, J., Seville, E.: Lessons from disaster: creating a business continuity plan that really works. J. Bus. Contin. Emer. Plan. **10**(1), 84–92 (2016)
38. NIAC (National Infrastructure Advisory Council). Critical Infrastructure Resilience: Final Report and Recommendations. U.S. Department of Homeland Security, Washington, DC (2009)
39. Organizational Resilience Framework. https://www.build-resilience.org/organizational-resilience-framework.php. Accessed 29 May 2023
40. Vymetal, S.: Crisis communication and risk communication. Grada, Prague (2009)
41. Clair, J.A., Waddock, S.: A "total" responsibility management approach to crisis management and signal detection in organizations". Int. Handb. Organ. Crisis Manage. 299–313 (2007)
42. Armstrong, M.: Armstrong's Handbook of Human Resource Management Practice, 3rd edn. Kogan Page, London (2014)
43. Diversification: An essential part of the strategy, without which the investor cannot do. https://finex.cz/diverzifikace-portfolia/. Accessed 29 May 2023
44. Disaster Recovery Guidance Series - Private Sector Participation in Disaster Recovery and Mitigation. https://www.gfdrr.org/sites/default/files/publication/Private_Sector_Guidance_Note_DRAFT%206_LOWRES.pdf. Accessed 30 May 2023

Prevention of Crisis Situations During the Operation of the Critical Infrastructure of Railway Transport

Valerii Samsonkin[1], Oksana Yurchenko[1](✉), Iuliia Bulgakova[1], Oleksandra Soloviova[1], and Assem Akbaeva[2]

[1] State University of Infrastructure and Technologies, 9, Kyrylivska St., Kyiv, Ukraine
yurchenko_og@gsuite.duit.edu.ua
[2] Academy of Logistics and Transport, Shevchenko str., 97, Almaty, Republic of Kazakhstan

Abstract. Railway transport is a complex technogenic system, which aims to ensure high-quality and fast transportation of goods and passengers, as well as the protection of its economic performance from the influence of negative factors and emergency situations sourced by railway transport operation, in unity with the environment and interaction of the users with the services. This complicates the process of establishing the railway risks management system and requires the formation of an effective mechanism for identifying, analysing, and preventing. Risks reduction and emergency prevention activities in railway transport should be based both on the operation and continual improvement of the system of technical and economic indicators of traffic safety and on the mechanisms of risk redistribution and management. The insufficient number of theoretical developments on improving the level of railway transport operations safety requires researching the causes and consequences of emergency situations that occur during railway transportation. The result of experimental studies of the statistical database of technical failures and failures in the signaling and communication division of the railway is presented in the paper. The principles and directions of crisis emergency prevention are considered. The main problems in finding an effective way to control the process of keeping the alarm systems in the required technical condition are defined. It is proposed to apply the method of preventing crisis situations (in the example of railway transport) for each systematization parameter in the developed classifier.

Keywords: Risk management · Safety management system · Experimental research · Crisis prevention · Failure statistics · Operation and maintenance of railway transport

1 Introduction

For the last five years, the world has been experiencing a period of global emergencies, starting with the COVID-19 pandemic, and ending with war in Ukraine and related to the war humanitarian disasters.

Two directions of effective emergencies treatment can be defined:

O. Prentkovskis et al. (Eds.): TRANSBALTICA 2023, LNITI, pp. 562–573, 2024.
https://doi.org/10.1007/978-3-031-52652-7_56

(a) to be well prepared mean developing scenarios, training staff (public authorities' staff of all levels from national to local, staff of companies and special services), to outreach among the people, etc.;
(b) to forecast events on time, place, and nature, affected branches of the economy, and thus try to prevent it. It's better to prevent a crisis than to respond to it effectively.

The Resilience science deals with issues of direction (a) [1]. Forecasting methods, statistical methods, management theory, and risk theory are applied for direction (b). The current research is focused on the direction (b). It proposes a method of preventing crisis situations (MPS), which is based on two theoretical approaches:

- the method of identifying hidden statistical patterns. The method in its modern state has been developed during the implementation of the Erasmus+ Project "Crisis and Risks Engineering for Transport Services" (CRENG) No 598218218-EPP-1-EPP EPPKA2-CBH-JP [2].
- the process of risk management [3] and risk prevention for rail transport [4]. To clearly understand and identify the concepts of crisis and an emergency the classification [5] was applied.

The object of the MPS application is railway transport, and the area of application is the signalling and communication division.

The study aims to develop the method of preventing crises in railway transport, and its empirical confirmation. The method will provide an opportunity to identify the causes of crises and develop a management solution to prevent them.

The novelty of the work consists in the following: the method of preventing crisis situations (MPS) is based on two theoretical approaches: the authors' method of identifying hidden statistical patterns and the theory of risk management. In the same way, sources of information are real statistics on the failures of trains in the safe movement of trains in the railway network.

In the practice of the signalling and communication division of the Regional Branch "Southern Railway" of JSC "Ukrainian Railway" the proposed method will allow to avoid precarious situations by preventing on the organizational level of influence possible crisis points considering its geographical localization, stage of the technological process, time, and possibly other characteristics. Therefore, research in the field of development and implementation of management approaches to the business units of the company, which are based on real data of the units functioning, is of great importance. It is important to note the great value of scientific novelty and practical significance of the paper because of its pure authoring.

2 Literature Review

Any phenomena that threaten the natural environment, people's lives and health, functioning of economic facilities are called emergency situations.

Emergency situations, or transport accidents, are disasters, accidents, serious incidents, and incidents that occurred during the railway rolling stock movement and led to the death or injury of people, damage to technical equipment, cargo, railway transport facilities, and traffic disruption.

Railway transport is responsible for all threats sourced by the rolling stock operation which caused damage to other persons (injury or death of passengers, causing damage to the life, health, or property of company employees and third parties). Besides that, railway transport bears substantial losses in case of damage to technical means and railway infrastructure [6].

Railway tracks, passenger platforms, places of loading and unloading of goods, railway stations in general and areas where rolling stock and trains move are zones of high danger by themselves, even under normal operating conditions. This danger increases many times when the operation conditions and traffic safety are violated. That can lead to emergency situations with human casualties, significant material losses, environment damage, and long interruptions in trains movement.

Scientific works, which considered the principles of railway companies management, reveal the problem of identifying and managing risks. However, the problem-solution approaches are at an early stage of its development.

In particular, the authors [7] claim that the human factor is the most significant for the safety of railway traffic. The statistical analysis of railway accidents showed that the human factor is the predominant element. The events affected traffic safety was analysed in the study. Such an analysis made it possible to propose measures of improving the safety management system in terms of safety culture regulation. But the study did not specify the data to be used to identify risks.

A risk-based approach to rail transport management has been successfully implemented in the railway transport system of Great Britain. In fact, the developed approach to the railway activity regulation focuses on setting priorities of identified risks. The priorities should be updated and revised regularly. This approach relies on two factors:

(1) Regulatory supervising, inspection, control of execution and compliance mode.
(2) Requirement for all railway operators to have an effective safety management system, to make it a regulatory requirement that is subject to review. The success of this approach has been proved by the fact that Great Britain has the lowest mortality rate per billion train-km in Western Europe [8]. However, this idea hasn't widely used till the moment due to significant differences of the railway transport management system of Great Britain from other systems.

For example, in Poland, a complex threat management system is used in the railway in frame of separate modules. The work [9] described a complex of decision-making systems of Polish railway. The presented system offered extensive support to railway companies. The database of cases of potentially dangerous situation was created due to the specific coding system. In addition, the system makes it possible to exchange the data between interested users. Thus, risk management and assessment approaches described in [9] is an advisory tool of supporting the railway traffic safety in frame of general safety methods in the conditions of Polish railway.

Jan Prochazka, Dana Prochazkova [10] made a critical analysis of accidents and failures in railway automation systems. The purpose of the analysis is to identify the causes of failures and their minimization. The presented results of the study of the causes of failures refer to previous studies [11, 12, 14]. As a result, integral risk was assessed [12, 13]. The study is aimed at implementing the cyber security process in the railway transport safety management system.

Sylwia Werbińska-Wojciechowskaa, Klaudia Winiarskab, and Honorata Poturaj [15] presented the concept of risk assessment in the transport system according to three parameters: 1) probability of failure, 2) consequences of failure, 3) availability of service resources. The authors investigate the problem of distribution of transport system maintenance resources in the context of operational risk assessment. At the same time, only the vehicle maintenance system was considered.

Authors A. Di Graziano, V. Marchetta in [16] developed a risk management methodology to support decision-making processes in the process of controlling factors that affect the level of safety of railway systems with a limited resource. This methodology is characterized by sufficient uniformity. However, the proposed concept is characterized by significant exclusivity, as it is applied to a local railway in the southern part of Italy.

Therefore, the state-of-the-art analysis has shown that the problem of prevention and minimization of losses from emergency and crisis events is topical for modern Ukraine and is a necessary condition for the effective implementation of the safety management system in railway transport.

Existing of a great variety of emergency situations, different of mechanisms of its occurrence, the peculiarities of its development and the scale of its consequences produce a need for a systematic approach to them. This is achieved by a clear classification of emergency situations [6].

According to Ukrainian legislation, an emergency situation is defined as "a violation of the normal conditions of life and activity of people in a separate territory or object on its territory, or on a water body, caused by an accident, disaster, natural disaster or other dangerous event, in particular epidemic, epizootic, epiphytotic, a fire that led (may lead) to the occurrence of a large number of victims, threats to life and health of people, their deaths, significant material losses, as well as to the impossibility of living in the territory or object, conducting economic activities there" [4].

Terms "Accident" and "Serious Accident" are used in the European railway legislation. Definitions of terms are given in Directive 2004/49/EC of the European Parliament and of the Council from April 29, 2004, in Article 3 point (k) and (l).

According to this document, an "Accident" is an unwanted or unintended sudden event or a specific chain of such events which have harmful consequences; accidents are divided into the following categories: collisions, derailments, level-crossing accidents, accidents to persons caused by rolling stock in motion, fires, and others.

"Serious accident/occurrence" means any train collision or derailment of trains, resulting in the death of at least one person or serious injuries to five or more persons or extensive damage to rolling stock, the infrastructure or the environment, and any other similar accident with an obvious impact on railway safety regulation or the management of safety; "extensive damage" means damage that can immediately be assessed by the investigating body to cost at least EUR 2 million in total [2].

Railway legislation is determined by the Federal Railroad Administration in the USA. According to paragraph 225 (definition) of chapter 49, part 225 of the Code of Federal Regulations, the term accident/ incident is used in relation to railway events [6]:

Accident/Incident on the railway is:

1. Any event with a "road user" at a road crossing with railway tracks at the same level. The term "road user" includes cars, buses, trucks, motorcycles, bicycles, agricultural

machines, pedestrians, and all other types of ground motorized and non-motorized transport.

2. Any collision, derailment, fire, explosion, natural disaster, or other event involving railway rolling stock (which was in motion or standing), which resulted in losses, the value of which exceeds the current threshold for reporting the loss of railway equipment, signaling system, rolling stock, road surface. The reporting threshold (since 2008) is 8.5 thousand USD.
3. Each death, injury or occupational disease that is a new case and meets the general reporting criteria.
4. Occupational disease.
5. Thus, in the US, "accident/occurrence/incident" is a collective term used to describe the entire spectrum of railway accidents, including medical ones. These include collisions, accidents and other events that cause death or injury to any person, or occupational diseases of railway staff Fig. 1 [6].

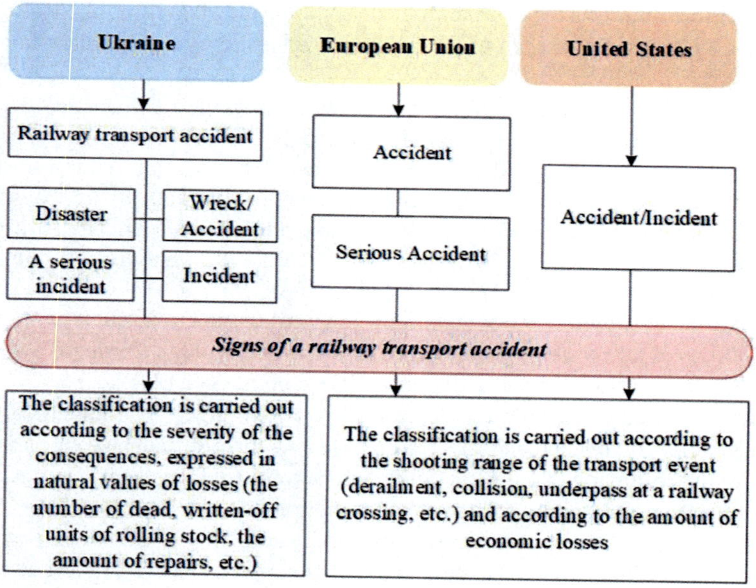

Fig. 1. Types and characteristics of railway transport platforms.

Emergency situations prevention and elimination of its consequences are an important task of the sectoral policy of railway transport. The effectiveness of such a policy is reflected in whether a disaster was averted and how quickly the consequences were eliminated in the event of one [19]. The dependence of choosing the methods of dealing with emergency situation of high probability and the consequences of a possible emergency on the given detailed classifications require further research, especially for cases with sanitary and irreversible losses.

Normative acts regulating this activity cover a wide range of diverse knowledge and actions, in particular, determine the possibility of involving the forces and means of railway and the State Emergency Service of Ukraine to eliminate the consequences of railway events.

3 Methods of Preventing Crisis Situations

The theory of self-organizing systems based on the single principles of self-organization is presented in detail in the work of Prof. Samsonkina V. M. [17]. In particular the MPS for railway transport is based on the principles set out in [20]. The *main principles* topical for current research are the following:

(1) The initial data for methodology is the registered statistics database (DB) on transport events (disasters, accidents, incidents), halting and failure of technical means, personnel errors that can lead to crisis situations. It is obvious that the use of statistics of crisis situations is problematic, because of its insufficient amount or absence. However, the statistics of traffic safety violations can be used in full on railway transport. Cases of safety violation differ only by its consequences. Thus, technological chains that lead to an emergency are often repeated and this can be prevented by breaking this dangerous chain.
(2) The DB records are systematized by nine parameters provided in Table 1.

Table 1. Parameters of systematization.

Parameters marking	Relevant information in the DB
WHAT	A failure event that occurred and classified according to the event classifier
WHEN	Event time: hour, day of the week, working/ non-working hours
WHERE	Localization or place of events occurrence
WHY	The cause of the events according to the cause classifier
HOW	Prerequisites for events, or basic reasons of the event
FOR WHAT	Motivation of the violator: intentionally/ unintentionally
CONSEQUENCES	What violations of traffic safety led to?
WHO	The violator identification data: division/ position/ surname
TO WHOM	To whom losses are granted: company/individual/railway

For further system analysis it is necessary to develop a classifier for each systematization parameter:
(1) Classifier'what,
(2) Classifier'when,
(3) Classifier'where etc.

This task is not difficult, because such classifiers have already been existing for the most parameters in the necessary form or with different names. Further, the output real DB is converted (marked) by systematization parameters.

(3) The chain of cause-effect relations looks as:

$$\text{HOW} \rightarrow \text{WHY} \rightarrow \text{WHAT} \rightarrow \text{CONSEQUENCES}$$

That is, the parameter HOW - prerequisites for events, or basic reasons of the event is added to the traditional chain which is commonly used in the control of traffic safety. This is explained by the fact that the causes of events objectively lie in the technology space and appear randomly, situationally. The real cause lies in the space of service organization that describes the parameter HOW.

(4) Dangerous places that may become a crisis or emergency trigger are determined by identifying patterns in statistics that describe the behavior of the transport system and processes in time flow. Such patterns are identified by analysis of graphics built from systematization parameters data: one-dimensional variant or number series, two-dimensional (in the two-parameters space), a three-dimensional version (in the three-parameter space). There are two signs of patterns (Table 2). As a result of successive system analysis at first the critical events, then the critical reasons, and afterwards based on the critical reasons the critical prerequisites are identified. The critical prerequisites are named as critical risks of possible crisis situations. Critical prerequisites should be sought in the space HOW if the purpose of the system analysis is to identify shortcomings in the process organization, not to punish the violators. This principle is based on the principle of "bottleneck" in self organization theory.

It is obvious, the "extension" of the bottlenecks is the way to increase efficiency and security in an economically optimal way.

(5) Management decision – critical risk management roadmap.

Table 2. Visual features of dangerous patterns.

<table>
<tr><th>Feature</th><th>Representation of the feature on the graphics of the analysis of systemizing parameters changes over the DB time frame</th></tr>
<tr><td>1</td><td>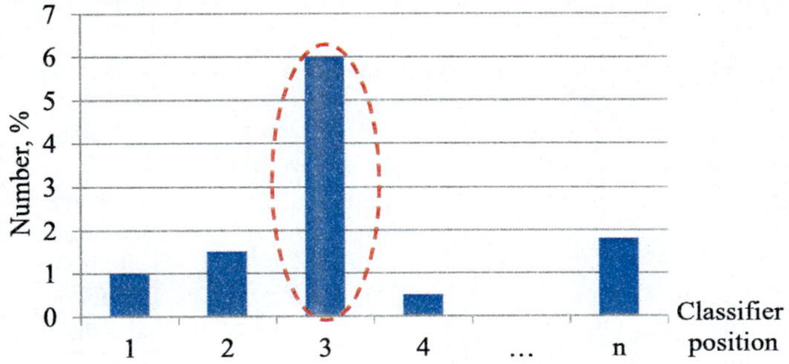
</td></tr>
<tr><td>2</td><td>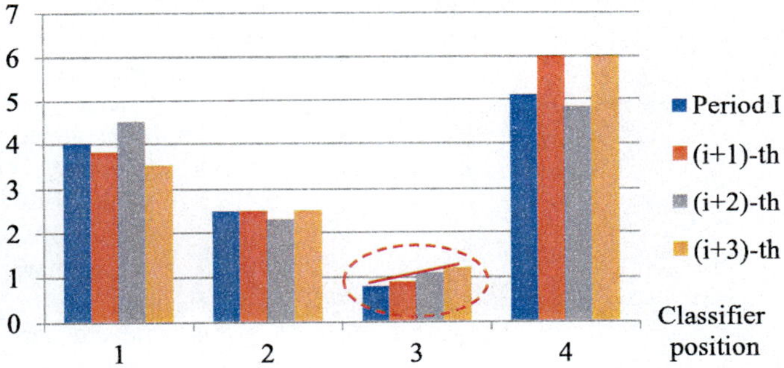
</td></tr>
</table>

4 Research Results

The results of experimental research are presented below.

Stage 1 is the monitoring (continuous control) of technical means condition in accordance with the current regulations and the identification of various types of deviations from the regulations (failures, errors, crashes, breakdowns, transport accidents). All deviations are recorded in paper or electronic journals. These recorded events are statistics of violations. Afterwards all statistics are systematized (according to principle 2) by all nine or fewer systematization parameters as allowed by the available statistics.

The experimental research presented in the paper is based on statistics DB of technical failures and failures in the signaling and communication division of the Regional Branch "Southern Railway" of JSC "Ukrainian Railway" from 13.06.2018 till 29.12.2019 [18], i.e., for 1.5 years. This data is real. The DB Excel form is presented in Table 3. DB contains 335 events of failures described by the way summarized in Excel form.

Stage 2. Based on the analysis of the manifestation of equipment failure events, it was established that three types of failures are more than 80%. These failures are rail circles,

Table 3. A fragment of the failure DB

№	Title of a failure (WHAT)	When did the failure happen? (WHEN)		Place of failure (WHERE)		The cause of the failure (WHY)	
		Date	Time	Station	Track section	Name	Run
1	Fake engagement of track circuits	13.06.2018	04:00		Lozova – Kharkiv passengers	Track circuits (track elements of track circuits) Jumper shorting	Merefa – Pokotylivka
2	Fake engagement of track circuits	15.06.2018	03:11	Lozovz	Lozova-Kharkiv passengers	Tripods, relay cabinets, track, transformer, shooting, cable boxes Failure of the arrester and transformer during a lightning discharge	Track post 939 km – Lozova
…	…	…	…	…	…	…	…

traffic lights and rail switches. Statistics confirm the existing problem of maintenance of technical means in the signaling division.

A detailed analysis manifestation of equipment failure events indicates the absence of any pronounced patterns.

Stage 3. The statistics of technical failures in the "WHAT-WHY" space, that is, events and causes, was analyzed. As a result, it was defined that three causes belong to the zone of unacceptable or critical risks. These causes are: 1. ALE (failure of auto-blocking elements), 3. Burn. L (burnout of the lamp) and 4. EC (failure of relay elements of electrical centralization). All these risks require constant attention and control from the top managers of the service and signaling and communication distance.

Stage 4. The following risk management solution is proposed:

I. for acceptable risks the existing regular control procedure is functioning taking into account mitigation measures.

II. for unacceptable (critical) risks a procedure of additional measures is proposed and presented in the form of a road map. Such measures can be special meetings, involvement of experts, analysis of the dynamics of failure statistics and causes, and others. The criterion for determining the time of a special measure is set as no later than one month from the moment of increase in negative statistics of risks occurrence.

5 Discussions of the Results

Statistical methods are the most effective ones, especially when managing complex systems, such as railway signalling and communications.

Thus, the presented example of the practical application of the developed methodology can act as an algorithm of risk management in the distances and railway signalling and communication services. The developed methodology can be effectively applied practically in the conditions of the Ukrainian railway system, in particular if the appropriate software is developed. The disadvantage of the method is the need to accumulate a great data arrays of failures statistics and to monitor constantly the implementation of the management decision program. However, this shortcoming can be avoided if a computer program is used, which will ensure the automatic accumulation of statistics.

6 Conclusions

Any disruption of railway traffic resulted financial costs for lamination of its consequences. Therefore, to categorize emergency event, indicators of economic losses are often considered. Such an approach is more objective and universal. Its advantage is possibility of general assessment of the state of traffic safety on road sections.

The proposed Crisis Prevention Methodology is based on the theory of self-organization and the theory of risk management. This gives hope for the effectiveness of the technique.

The technique refers to the statistical approach, in which the characteristic of the system state is statistical data on its parameters. The most significant has been selected from all system parameters. Statistical management methods are efficient under the following conditions: information reliability, a long history, and promptly update the analysis provided by digital technologies.

The proposed methodology allows to avoid many crisis-dangerous situations by preventive organizational influence on possible crisis points from the point of view of geographical location, technological process location, time, and possibly other characteristics.

The performance and efficiency of the methodology is shown in the railway signaling and communication division operation. This provided an opportunity to identify critical causes and develop a management solution to prevent them.

References

1. Disaster Risk Management in the Transport Sector. A Review of Concepts and International Case Studies, 157 p. The World Bank (2015)
2. Samsonkin, V., Nikolaienko, I., Bulhakova, Yu.: ta In: Inzhynirynh kryz ta ryzykiv transportnykh posluh. Kolektyvna monohrafiia, 312 p. Talkom, Kyiv (2021). ISBN 978-617-8016-10-4. [in Ukrainian]
3. Keruvannia ryzykom. Metody zahalnoho otsiniuvannia ryzyku. DSTU IES/ISO 31010 (2013). https://khoda.gov.ua/image/catalog/files/dstu%2031010.pdf. [in Ukrainian]
4. Pro zatverdzhennia Polozhennia pro systemu upravlinnia bezpekoiu rukhu na zaliznychnomu transporti, Nakaz Ministerstva infrastruktury Ukrainy № 842 (2020) (Ukraina). https://zakon.rada.gov.ua/laws/show/z0351-21#Text. [in Ukrainian]
5. Volianskyi, P.: Metodolohichni pidkhody do upravlinnia ryzykamy v protsesi likvidatsii naslidkiv nadzvychainykh sytuatsii. Invest.: praktyka dosvid (13), 134–136 (2013). [in Ukrainian]
6. Dymkovych, N.: Ohliad kryzovykh ta nadzvychainykh sytuatsii na zaliznychnomu transporti: Kvalifikatsiina robota, 60 p. Kyiv (2023). https://files.duit.edu.ua/uploads/fuzt/certification/2023/bachelors/dymkovich.pdf
7. Kalem, A., Lindov, O., Šimić, E.: Safety culture in the function of optimization of railway safety management system. In: Karabegović, I. (ed.) NT 2021. LNNS, vol. 233, pp. 700–708. Springer, Cham. (2021). https://doi.org/10.1007/978-3-030-75275-0_77
8. Lefsrud, M., Macciotta Pulisci, R., Nkoro, A.: Railway association of canada, role of safety management systems (SMS) in the railway industry and potential for enhancement of the railway safety act (RSA) (2017). https://tc.canada.ca/sites/default/files/migrated/appendix_e.pdf
9. Michalak, M., et al.: Central threat register – a complex system for risk analysis and decision support in railway transport. IET Intel. Transp. Syst. **14**(8), 970–981 (2020). https://doi.org/10.1049/iet-its.2019.0695
10. Prochazka, J., Prochazkova, D.: Risk management of selected elements of critical transport infrastructure. In: Proceedings of the 32nd European Safety and Reliability Conference (ESREL 2022), pp. 234–241 (2022). https://doi.org/10.3850/978-981-18-5183-4_R07-02-100-cd
11. Procházková, D.: Principles of risk management of complex technological facilities, 364 p. ČVUT, Praha (2017). https://doi.org/10.14311/BK.9788001061824
12. Procházka, J., Procházková, D.: Risk management of railway transport systems. ČVUT, Praha (2022). https://doi.org/10.14311/BK.978800101069950
13. Prochazkova, D., Prochazka, J.: Tools for risk management of technical facilities operation. Eur. J. Eng. Res. Sci. **5**(4), 494–500 (2020). https://doi.org/10.24018/ejers.2020.5.4.1854
14. Kertis, T., Procházková, D.: Information power supporting the rail systems safety. In: Safety and Reliability – Safe Societies in a Changing World. Ed. 1, pp.2939–2947 (2018). ISBN 978-0-8153-8682-7
15. Werbińska-Wojciechowska, S., Winiarskab, K., Poturaj, H.: Risk of maintenance resource sharing in transport systems. In: Probabilistic Safety Assessment and Management PSAM 16, Honolulu, Hawaii, vol. 3. pp. 1653–1662 (2022). ISBN 978-1-7138-6375-5
16. Di Graziano, A., Marchetta, V.: A risk-based decision support system in local railways management. J. Rail Transp. Plann. Manage. **20**, 475–482 (2021). https://doi.org/10.1016/j.jrtpm.2021.100284
17. Druz, V., Samsonkin, V.: Edynaia teoryia samoorhanyzuiushchykhsia system, 117 p. Talkom (2022)

18. Statystyka tekhnichnykh vidmov u sluzhbi syhnalizatsii ta zviazku Rehionalnoi filii «Pivdenna zaliznytsia» AT «Ukrainska zaliznytsia» za period z 13.06.2018 po 29.12.2019 roky. https://kart.edu.ua/wp-content/uploads/2020/09/statistika-vidmov.xlsx. [in Ukrainian]
19. Akbayeva, A., Muratbekova, G., Altayeva, Z., et al.: Development of safety methods on artifical structures of railway lines. Eastern-Eur. J. Enterp. Technol. **6/1**(120), 43–52 (2022). https://doi.org/10.15587/1729-4061.2022.255275
20. Samsonkin, V., Sotnyk, V., Yurchenko, O., et al.: Devising a methodology to manage the performance of technical tools of rail transport signaling systems based on the risks of their functioning. Eastern-Eur. J. Enterp. Technol. **6/3**(120), 32–43 (2022). https://doi.org/10.15587/1729-4061.2022.268715

Determination Modelling of Ukraine's High-Speed Railways with Shared Use for Passengers and Cargo Service

Andrii Pozdniakov(✉), Viktor Myronenko, and Olga Pozdniakova

State University of Infrastructure and Technologies, 9, Kyrylivska St., Kyiv 04071, Ukraine
andrei.pozdniakov.mail@gmail.com

Abstract. The possibilities of applying the principle of joint use of future high-speed railways for both passenger and individual cargo transportation in the transport system of Ukraine are considered. Methods of substantiating the rational proportion of passenger and freight trains on the line are proposed, as well as determining the technological and economic parameters of the functioning of high-speed railways as part of a multimodal transportation system. An economic-mathematical model is described, which allows to optimize design and technological solutions in the construction and operation of high-speed railways depending on the economic and other conditions of their functioning.

Keywords: High-speed railways · Passenger and cargo transportation · Shared use · Transport networks

1 Introduction

Considering the goals of the National Transport Strategy of Ukraine for the period until 2030 (NTS-2030) to form the transport market and increase the share of non-state-owned transport operators in railway transport to 25% by 2025 and to 40% by 2030 [1], the need for consistency of train schedules, terminal operations, and passenger convenience and mobility, taking into account urban transport, becomes even more urgent, especially after the appearance of a separate operator of the new high-speed railway (HSR) infrastructure.

In addition, NTS-2030 envisages the phased introduction of high-speed rail connections (up to 400 km/h) between the main centres of Ukraine on separate 1435 mm tracks and their use for combined passenger and freight traffic (for accelerated delivery of goods with high added value), as well as joining the national HSR network to the Trans-European network TEN-T.

Even in China, with its powerful passenger flows and long distances, not all high-speed railway lines are profitable, therefore, they also carry high-value cargo that requires urgent delivery [2, 3]. In Ukraine, using very expensive HSR lines only for passenger service, will be obviously inefficient and unprofitable. Our preliminary estimates, made before the full-scale Russian war against Ukraine, show that the payback period for investments in the infrastructure and rolling stock of HSR, if it is used only for passenger

O. Prentkovskis et al. (Eds.): TRANSBALTICA 2023, LNITI, pp. 574–585, 2024.
https://doi.org/10.1007/978-3-031-52652-7_57

transportation, will be hundreds of years, even with maximum passenger flows and ticket prices. On the other hand, with relatively small passenger traffic in Ukraine, HSR will have sufficient free capacity for freight transportation.

Ukraine has a favourable transit geographical location, and that is why the NTS-2030 provides for the development of an efficient competitive multimodal national transport system and strengthening of the transit potential of Ukraine's transport sector: - ensuring interoperability (operational compatibility) of the national transport network with the world multimodal transport network, including with a 1435 mm gauge railway track network; - passenger, freight and logistics infrastructure complexes established as integral parts of multimodal clusters of rail, road, air and waterborne transport [1].

Therefore, there is no alternative to the shared use of our future HSR for the passenger service and cargo transportation, as a part of international multimodal transport systems. The study of the balance of cargo and passenger transportation on the HSR according to the criteria of maximum profitability for the infrastructure operator, considering the interests of other market participants, is an important scientific and applied problem and a prerequisite for the development of high-speed multimodal transportation in Ukraine.

The object of the study presented in this work is a set of factors and processes affecting profitability and other performance indicators of the HSR infrastructure operator.

The aim of the study is the technological and economic substantiation of the business model of the future HSR of Ukraine with shared use for transportation of both passengers and cargo, considering the balance of interests of infrastructure and transportation operators. This involves solving the following problems as the main objectives of the study:

1. development of a model for the optimal functioning of multimodal passenger and cargo transportation in common corridors of high-speed railways, terminals and motorways;
2. determination of the rational proportion of passenger and cargo traffic on a single infrastructure, considering the balance of interests of infrastructure and transportation operators;
3. determination of the optimal number of stations on the HSR line, which ensures its profitability, reliability, convenience for users and a balance of interests.

For these purposes, the methods of analytic modeling are used to identify the main relationships of the parameters of the proposed models in order to assess their impact on the performance of the HSR infrastructure operation.

2 Analysis of Recent Research

According to the analysis of the work in the field of high-speed multimodal transportation, certain helpful provisions and ideas were identified, which was used as a basis for the technological justification of the business model of Ukraine's future HSR network operator. In the works of Hammadi [4], the general foundations and optimal models of the construction and operation of multimodal transport are investigated. Researcher Janić [5], undertakes a multidimensional examination of the infrastructural, technical/technological, operational, economic, social, and environmental performances of high-speed rail (HSR) systems, including their overview, analysis of some real-life cases, and limited (analytical) modeling. In the works of Brunello [6], Mahboob [7] methods of optimization and availability of high-speed lines and adjacent transport networks are investigated. Maximizing the use of the HSR network for passenger and freight transportation is one of the main issues in the research of Ghilas et al. [10] and Li et al. [11]. The authors emphasize that due to the unevenness of the passenger flow during the day, the use of the capacity of HSR in off-peak periods may be low. A potential approach to solving these problems is the integration of a joint freight and passenger transport system [12]. The fixed schedule of passenger trains makes it possible to fully use the remaining capacity of the existing network for freight transportation and obtain economic benefits [13] from the maximum use of the capacity of HSR. The shared use approach of the HSR network involves the sharing of at least three resources: stations, tracks and trains [14]. The choice of common resources determines the amount of capacity intended for passenger and freight transportation, which is related to the efficiency of the use of the capacity of the HSR. This study focuses on the reserve capacity of the existing train timetable and describes an economic-mathematical model that allows to optimize the design and technological decisions in the construction and operation of high-speed railways to increase the flexibility of the use of capacity. Myronenko et al. [8, 9] studied the problems of passenger and cargo multimodal transportation in Ukraine, including the combined use of railway lines by trains of different categories. These and other works are a sufficient basis for the transition to the level of substantiation of the business model of the future HSR system, which makes it not only convenient for users, but also as an attractive project for investors.

3 Comprehensive Model of Passenger and Freight Transportation in Shared Use of High-Speed Railway Mainlines

The future transport system, which includes high-speed railways, being an extremely high-cost element of infrastructure (compared to conventional railways), is not efficient if it does not provide for the integrated use of infrastructure. Firstly, for mass passenger service, as well as for freight transportation (in certain niches of the market). Secondly, as a system-forming component in other types of economic activity (such as tourism, hotel business or residential construction in the area of attraction of the HSR). The future functioning of such a transport system can only be predicted using appropriate complex mathematical models.

It should be noted that the balance of cargo and passenger transportation on the HSR is possible taking into account two restrictions arising from the need to satisfy the demand for transportation:

1. The need to transport all willing passengers.
2. The need to transport the maximum possible amount of cargo within the limits of the available transport capacity.

One of the problems of organizing the shared use of the HSR infrastructure for freight and passenger traffic is the different speed, as well as the different acceleration/deceleration dynamics of freight and passenger trains. This leads to the well-known in the practice and theory of transportation phenomenon of mutual "removal" of trains of different speed categories and loss of capacity of lines. Some analytical dependencies reflecting this phenomenon are presented in this paper.

In order to realize the advantages of high-speed traffic, it is necessary that the distances between the stations are maximum, and the time of acceleration and deceleration of the train is minimum. The calculations show that when the distance between the stations is less than 15 km and the acceleration/deceleration time is more than 3 min, then high-speed traffic has no effect at all, because when the maximum speed increases, the average speed of trains does not increase, but decreases. This should be taken into account during technical/economic substantiation and design of HSR.

When trains with different speeds are running on the line, the phenomenon of "removal" of trains of one speed category by trains of another speed category occurs, which leads to loss of line capacity [4]. The mathematical model of this phenomenon is as follows. The proposed basic capacity model reflects the influence of the following factors on the capacity of the line, on which trains of two speed categories run:

$$N_q = \frac{24 - t_{reg}}{j} - \left[1 + \frac{2}{j}\left(\frac{1}{v_q} - \frac{1}{v_a}\right)l\right]N_a; \tag{1}$$

where N_q is the total number of freight trains per day, train; t_{reg} is the regulated time of interruptions in traffic during the day (for example, "windows" for carrying out scheduled works in the track and power facilities), hours; j is the time interval between trains, hours; l is the length of the limiting run of the line (distance between stations, where average train travel time is the longest), km; v_a is the average speed of passenger trains, km/h; v_q is the average speed of freight trains, km/h; N_a is the total number of passenger trains per day, train.

$$v_a = v_a^{max}\left(1 - \frac{v_a^{max}}{2l}t_{ab}\right), \tag{2}$$

$$v_q = v_q^{max}\left(1 - \frac{v_q^{max}}{2l}t_{qb}\right), \tag{3}$$

where v_a^{max} is the maximum speed of passenger trains, km/h; v_q^{max} is the maximum speed of freight trains, km/h; t_{ab} is the acceleration/deceleration (braking) time of passenger trains, hours; t_{qb} is the acceleration/deceleration (braking) time of freight trains, hours.

The total railway line capacity is defined as:

$$N = N_a + N_q; \tag{4}$$

where N is the total number of trains per day:

$$N = \left[24 - t_{reg} - 2\left(\frac{1}{v_q} - \frac{1}{v_a}\right)N_a l\right]. \tag{5}$$

The application of the model is illustrated by the following example (at the minimum and maximum number of passenger trains on the line, provided that the rest of its capacity is used for freight trains).

It should be noted that the estimates of train speeds and line capacity obtained here do not contradict the results obtained in work for operating speeds of trains and "ultimate"/"practical" capacities, but they are calculated using a simpler method.

4 Rational Proportion of Passenger and Freight Traffic on a Single Infrastructure

From the practice of the world's railways, including Ukrainian railways, it is known that passenger transportation is often loss-making. In this case, subsidies are required. In modern domestic practice, this is the so-called "cross-subsidization" of losses of passenger transportation at the expense of revenues from freight transportation and other sources. In the future, this is unacceptable, both from the point of view of the rail transport legislation of the European Union and the laws of the market economy.

Therefore, an analytical tool is needed to substantiate the rational ratio of passenger and freight traffic on a single infrastructure, considering the balance of interests of infrastructure operators and transportation undertakings, taking into account that the infrastructure operator cannot bear losses from providing it for socially necessary passenger transportation.

One of the criteria for the shared use of the HSR infrastructure for freight and passenger transportation can be the aggregate income from running along the line of freight and passenger trains, which have different sources and indicators of profitability (the cost of a ticket in a passenger train and the tariff for freight transportation of the corresponding category of cargo).

The mathematical model for calculating revenue at the maximum speed of passenger trains v_a^{max} has the following general form:

$$\begin{cases} d_a N_a + d_q N_q \Rightarrow max \\ d_a > 0, N_a > N_{amin}, N_a < N_{amax}, \\ d_q > 0, \end{cases} \tag{6}$$

where d_a is the train revenue rate for passenger transportation, (for example, euro/train-km); d_q is the train revenue rate for container transportation, euro/train-km.

In the analysis, the following restrictions were adopted, reflected below in the system of equations and inequalities (7):

$$\begin{cases} d_a N_a + d_q N_q \Rightarrow max \\ N_q = \frac{24 - t_{reg}}{J_t} - \left[1 + \frac{2}{J_t}\left(\frac{1}{v_q} - \frac{1}{v_a}\right) l\right] N_a \\ T_{REG} = 3 \\ 0,03 \le J_t \le 0,3; \\ 250 \le v_a \le 350; \\ 150 \le v_q \le 290 \\ 30 \le l \le 70 \\ d_a > 0, \\ 2 < N_a < 18, \\ d_q > 0, \\ 70 < N_q < 102 \end{cases} \tag{7}$$

In order to ensure the correctness, efficiency and adequacy of the economic-mathematical model of transportation on the same infrastructure in trains of different categories, which give different incomes to the infrastructure operator, forward working hypothesizes are put:

1. the fee for using the railway infrastructure (infrastructure charges per train-kilometer when passing a train of a certain category on a certain route) must compensate the infrastructure operator for the costs of its construction and operation;
2. costs for the construction and operation of the railway infrastructure directly depend on the maximum train speed for which it is designed and inversely on the average distance between stations on the railway line, and the influence of the maximum speed on these costs is stronger than the distance between the stations.

These assumptions are confirmed by the current systems of infrastructure charge rates for the use of railway infrastructure, in particular, the following system operates at Deutsche Bahn AG [15]:

Table 1. Infrastructure charge rates.

Rates, euro/train-km	d_q	d_a
min	2.98	2.65
max	4.22	12.17

The data of Table 1 show that passenger trains have the highest rates for using the infrastructure, which is due to their higher speed compared to freight trains that run on the same infrastructure. Taking into account the above, the income of the infrastructure operator can be represented by the Eq. (8), which is given without intermediate statements (relating to the accepted weighting factors of influence on costs and revenues from transportation of such factors of the model as the speed of trains v_q, v_a and the distance between intermediate points l):

$$\sum D = d_q\left[2\frac{v_q}{l}(24 - t_{reg}) - (5 - 4\frac{v_q}{v_a})N_a\right] + d_a N_a. \quad (8)$$

Formula (1) reflects the "ideal" situation, when the necessary number of passenger trains N_a is passed through the infrastructure element of length l, and the rest of the capacity of the infrastructure is used to pass freight trains in the number N_q. The results of calculations according to Formula (8) are shown in Fig. 2.

It can be seen from the Fig. 1 graphs that even under "ideal" conditions, when there is the necessary freight flow to use all the capacity of freight trains in the amount of N_q after passing the required number of passenger trains N_a, the total revenue from transportation $\sum D$ decreases with the increase in the number of passenger trains. That is, unprofitability and the need to subsidize passenger transport have another confirmation here. Passenger transport is a "social order" (public obligations), so in all countries it is subsidized in one way or another by the customers of these transports, if they are unprofitable for the carrier.

However, this approach did not encourage the carrier to reduce its own operating costs. It was considered another approach to be more productive, when the amount

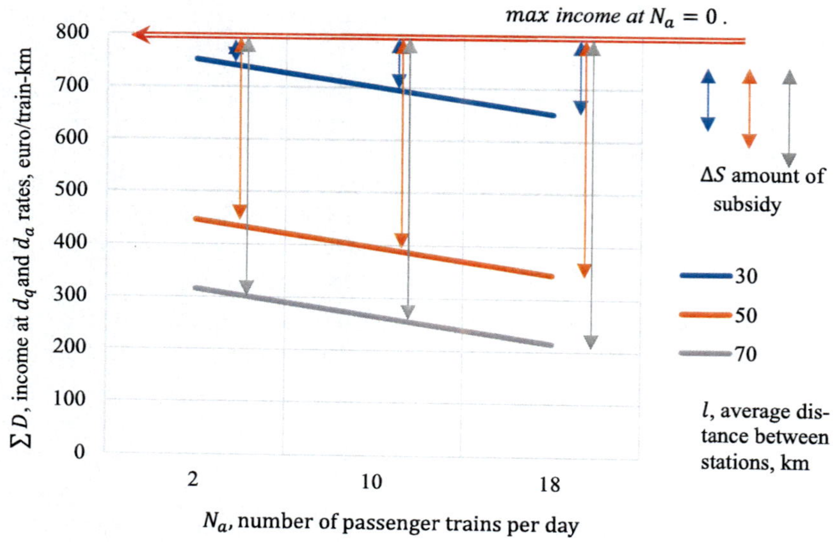

Fig. 1. Dependence of income and subsidy volume at different rates of charges for the use of infrastructure and the number of passenger trains.

of subsidies based not on the amount of losses, but on the amount of subsidising the under-received income.

$$\Delta \sum D_S = \sum D_{N_a=0} - \sum D, \tag{9}$$

where: $\sum D_{N_a=0}$ is the revenue from infrastructure use only by freight trains, in the absence of passenger trains ($N_a = 0$).

An example of the dependence displayed by Formula (9) is shown in Fig. 2.

Analysis of the results of calculations according to the model shows that the profitability of the infrastructure operator decreases when:

1. increasing the length of the limiting run of the line;
2. increasing the number of passenger trains;
3. reducing the speed of freight trains;
4. increasing the speed of passenger trains (the latter is explained by the "removal" of more profitable cargo trains by passenger trains).

On the other hand, profitability is greatest with the minimum length of the limiting run of the line, the minimum number of passenger trains and the minimum difference between the speeds of passenger and freight trains.

The results of the infrastructure operator's revenue analysis show that the greatest dependence is related to the minimum length of the limiting run and the speed, especially of freight trains, and more precisely the minimum difference between the speeds of passenger and freight trains. Taking into account these conclusions, the weighting factor of influence on the profitability of the infrastructure operator from distance at the level of $\alpha_l = 0.2$ Also, the weighting factor of influence on the profitability of the infrastructure operator from speed at the level of $\alpha_v = 0.8$ were taken.

Based on these limitations, the weighted average revenue rate of the railway infrastructure usage fee was calculated:

$$d_q^{'} = \alpha_v d_q^v + \alpha_l d_q^l. \tag{10}$$

$$d_a^{'} = \alpha_v d_a^v + \alpha_l d_a^l, \tag{11}$$

where $d_q^{'}$ is the weighted average income rate of the charge for infrastructure use by freight trains; $d_a^{'}$ is the weighted average income rate of the charge for infrastructure use by passenger trains; α_l is the speed influence weight factor; α_v is the distance influence weight factor; d_q^v is the weighted average income rate of the charge for infrastructure use by freight trains depending on the speed; d_q^l is the weighted average income rate of the charge for infrastructure use by freight trains depending on the distance; d_a^v is the weighted average income rate of the charge for infrastructure use by passenger trains depending on the speed; d_a^l is the weighted average income rate of the charge for infrastructure use by passenger trains depending on the distance.

The weighted average income rate of the charge for infrastructure use by passenger and freight trains is defined as:

$$d_{aq} = \frac{d_a^{'} N_a + d_q^{'} N_q}{N_a + N_q}. \tag{12}$$

Formulas (10), (11) and (12) are used to substantiate the weighted average value of the rates, which must be used in the formation and assessment of the adequacy of the model of the optimization of infrastructure management processes, as well as in the design of the parameters of the HSR infrastructure, in particular, the optimal number and location stations and stops on the line, which affects both the required capital investments, their payback period, and the HSR operation conditions.

5 Optimal Number of Stations on a HSR Line

The location and number of the nodes of corresponding HSR network along particular lines is one of the main aspects of its design, construction and operation [a]. In other words, more or less stations on the line means shorter or longer runs between them, that impacts on construction and operation costs, economic performance and other aspects of HSR life cycle. In our opinion, when determining the optimal number of stations on the line, one should consider such an economic factor as subsidizing passenger transportation in that part of the infrastructure operator's costs that are not covered by revenues from passenger and cargo transportation, when the line is operated on the shared-use principle.

As it was mentioned above, the amount of subsidising the under-received income in case of increasing the number of passenger trains which may be less profitable for infrastructure operator than freight trains is an important factor in decision-making.

The amount of subsidy during the operation of HSR, S_{OP}, can be determined as:

$$\begin{cases} S_{OP} = D_{max} - D_{act} \\ D_{max} = \frac{24-t_{REG}}{J_t} d_q' \\ D_{act} = \left\{ \frac{24-t_{REG}}{J_t} - \left[1 + \frac{2}{J_t}\left(\frac{1}{v_q} - \frac{1}{v_a}\right) l\right] N_a \right\} d_q' + N_a d_a' \\ S_{OP} = \frac{24-t_{REG}}{J_t} d_q' - \left\{ \frac{24-t_{REG}}{J_t} - \left[1 + \frac{2}{J_t}\left(\frac{1}{v_q} - \frac{1}{v_a}\right) l\right] N_a \right\} d_q' + N_a d_a' \end{cases} \tag{13}$$

where D_{max} is the maximum income on HSR line under "ideal" conditions (see above); D_{act} is the actual income on HSR line, taking into account the "removal" of profit-making freight trains by less profitable passenger trains taking more capacity of the line.

As can be seen from (13), the amount of subsidies is affected by the distance between stations l, both directly and in the form of more complex dependencies, for example, the impact on train speed. The average value of this distance, for a certain length of the line, affects the required number of stations that need to be built on the line, and therefore, the amount of capital investments, their payback period, and the efficiency of using the HSR infrastructure.

The Formula (14) for calculating the total capital expenditure of E_C for infrastructure construction per one station is given below:

$$E_C = C_{1km} \sum l_{tr} + \sum_{i=1}^{n_{tr}} n_{toi} C_{toi} + C_B, \tag{14}$$

where C_{1km} is the construction cost of 1 km of station track, euro; l_{tr} is the overall train length, m; $\sum l_{tr}$ is the total length of all tracks at the station, m; n_{tr} is the number of tracks at the station; n_{toi} is the number of rail turnouts of a certain type; C_{toi} is the cost of the rail turnouts of a certain type, euro; C_B is the cost of all the building at the station, euro.

For further calculations, it was assumed that an average of 3 tracks is be enough for one station, and the distance between the axes of the tracks is 5 m in average.

The total capital costs $\sum E_C$ for the construction of the infrastructure of all stations are determined by this equation:

$$\sum E_C = C_G + K_{st}\left[C_{1km} \sum l_{tr} + \sum_{i=1}^{n_{tr}} n_{toi} C_{toi} + C_B\right]; \tag{15}$$

where C_G is the capital costs for the construction of the infrastructure of a single control centre of HSR, independent of the number of stations, euro; K_{st} is the number of stations on the line.

In order to calculate the optimal number of stations, the objective function of the optimization problem was formulated. This function reflects the search for the minimum total costs of subsidizing losses from the passage of passenger trains on HSR lines $\sum S_{OP}$ instead of more profitable freight and costs for the construction and maintenance of the infrastructure of HSR lines and stations $\sum E_C$ which depend on the number of station K_{st}. After entering the notation $x = K_{st}$ and substituting x into Formulas (13) and (15), that reflect the two components of the objective function described above, the first derivative $\frac{dY(x)}{dx}$ of the objective function by differentiating was founded it with respect to the argument x. Omitting the intermediate transformations, the final Formula (17) is presented, according to which the optimal number of stations K_{st} on the HSR line of length L is calculated, at which the total costs reflected by the objective function (Formula (16)) is minimal.

$$Y(x) = \sum S_{op} + \sum E_C => min, \tag{16}$$

$$x = \sqrt{\frac{2N_a d'_q\left(\frac{1}{V_q} - \frac{1}{V_a}\right)L}{J_t\left\{C_B + (2n_{tr} + 1)\overline{C}_{to} + 2\left[\frac{n_{tr}}{2}(m_B l_B + l_{loc}) + e(42n_{tr} - 37)\right]\frac{C_{1km}}{1000}\right\}}}; \tag{17}$$

where: $x = K_{st}$ is the optimal number of stations on the HSR line of length L.

The results of calculations based on the mathematical model described here are presented in Fig. 2.

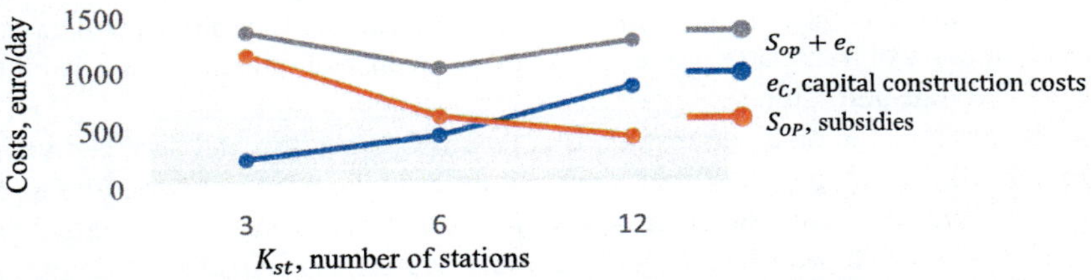

Fig. 2. Amount of subsidies and amount of capital expenditures for different number of stations.

As can be seen from Fig. 2, the proposed economic-mathematical model makes it possible to reflect the reduction in the need for subsidies S_{OP} when the number of stations K_{st} on the line increases, as its operation becomes more flexible when the size and structure of the train flow changes, the throughput capacity is better used. On the other hand, the construction of a larger number of stations requires higher capital costs e_C, which is also evident from the results of the calculation based on the model. Instead, the objective function, which reflects the total costs $S_{OP} + e_C$, has a clear minimum (in our case at $K_{st} = 6$), which corresponds to the optimal number of stations on the line under the given conditions of its construction and operation.

6 Conclusions

For the technological and economic substantiation of the business model of the future high-speed railway system in Ukraine, it is advisable to use the economic-mathematical model of the rational proportion of passenger and freight traffic on a single line.

The model is based on considering the costs for railway construction and operation, charges for using the railway infrastructure, the maximum and average operational speed of trains, other operational variables and, as proposed by the authors, the factor of subsidizing the non-profitable passenger transportation.

The results of the calculations show that the profitability of the rail infrastructure operator depends on various factors, such as the average length of the limiting run of the line, the number and speed of freight and passenger trains. The highest profitability is observed with the minimum length of the limiting run of the line, the minimum number of passenger trains and the minimum difference between the speeds of passenger and freight trains.

A methodology for determining the rational proportion of passenger and cargo traffic on railways is proposed, considering the balance of interests of infrastructure and transportation operators. The authors emphasize the need to change the approach to subsidizing passenger transportation, in particular, focusing on unearned revenues from freight transportation in the case of shared use of high-speed railways for passenger and freight transportation.

The parameters of the business model were identified, including indicators of technology and economics of passenger and cargo transportation. These parameters are used to analyse the technological aspects of the business model, such as the optimal distance between stations, the required number of trains (carriages/wagons/containers in trains) for passenger and freight transportation. This enables the calculation of transportation costs and the price of transport services to achieve a balance of interests of infrastructure operators and transportation.

On the basis of the obtained results, it is possible to analyse and simulate various scenarios for the development of the infrastructure of high-speed railways. The market strategy of the infrastructure operator is determined in order to ensure the optimal technological base and achieve a balance of interests of various participants in the system at the required level of quality of transport service for users.

References

1. National transport strategy of Ukraine for the period until 2030/Національна транспортна стратегія України на період до 2030 року | Кабінет Міністрів України (2018). https://www.kmu.gov.ua/ua/npas/pro-shvalennya-nacionalnoyi-transportnoyi-strategiyi-ukrayini-na-period-do-2030-roku. Accessed June 2023
2. Mewati, S.: Is the Chinese High-Speed Rail operating at a loss? – Quora (2016). https://www.quora.com/Is-the-Chinese-High-Speed-Rail-operating-at-a-loss. Accessed June 2023
3. Sevillano, E.G.: Spain's high-speed rail: one in four AVE stations used by fewer than 100 passengers a day | In English | EL PAÍS (2016). https://english.elpais.com/elpais/2016/05/23/inenglish/1464003478_520592.html. Accessed June 2023
4. Hammadi, S., Ksouri, M.: Multimodal Transport Systems, p.256. ISTE/Wiley, London/Hoboken (2014)
5. Milan, J.: A multidimensional examination of performances of HSR (high-speed rail) systems. J. Mod. Transp. **24**, 1–21 (2016)
6. Brunello, L.R.: High Speed Rail and Access Transit Networks, 163 p. Springer, Cham (2018). https://doi.org/10.1007/978-3-319-61415-1
7. Mahboob, Q., Zio, E.: Handbook of RAMS in Railway Systems: Theory and Practice, 765 p. CRC Press (2018)
8. Myronenko, V., Samsonkin, V., Rudkovskyic, S., Rostislav, D., Feldman, A.: Ukraine in system of transport links. Int. J. Eng. Res. Dev. (IJERD) **13**(9), 1–6 (2017)

9. Myronenko, V., Hrushevska, T.: Problems of passenger and freight trains combined traffic on high-speed railway lines. ETiL **76**, 101–106 (2018)
10. Ghilas, V., Demir, E., Woensel, T.: Integrating passenger and freight transportation: model formulation and insights, vol. 441, 23 p. Technische Universiteit Eindhoven, Eindhoven (2013)
11. Li, Z., Shalaby, A., Roorda, M.J., Mao, B.: Urban rail service design for collaborative passenger and freight transport. Transp. Res. Part E: Logist. Transp. Rev. **147**, 102205 (2021)
12. Fatnassi, E., Chaouachi, J., Klibi, W.: Planning and operating a shared goods and passengers on-demand rapid transit system for sustainable city-logistics. Transp. Res. Part B: Methodol. **81**(Part 2), 440–460 (2015)
13. Li, S., Zhu, X., Shang, P., Li, T., Liu, W.: Optimizing a shared freight and passenger high-speed railway system: a multi-commodity flow formulation with Benders decomposition solution approach. Transp. Res. Part B: Methodol. **172**, 1–31 (2023)
14. Behiri, W., Belmokhtar-Berraf, S., Chu, C.: Urban freight transport using passenger rail network: scientific issues and quantitative analysis. Transp. Res. Part E: Logist. Transp. Rev. **115**, 227–245 (2018)
15. Standard Rates and other Provisions of DB Cargo AG. https://www.dbcargo.com/resource/blob/9916364/fa3ea49b53069565bcfd0ba4dcc6f376/Standard-Rates-and-other-Provisions-of-DB-Cargo-AG-2023_EN-data.pdf. Accessed June 2023

The Purpose of the Research Agenda for Rail Wagon Predictive Maintenance

Mateusz Zajac(✉), Artur Kierzkowski, Agnieszka Tubis, Tomasz Kisiel, Franciszek Restel, Ewa Mardeusz, and Jacek Ryczynski

Wroclaw University of Science and Technology, Wyb. Wyspianksiego 27, 50-370 Wroclaw, Poland

mateusz.zajac@pwr.edu.pl

Abstract. Monitoring the technical condition of rail wagons remains a significant challenge, especially concerning freight wagons. The article outlines the purpose of research agenda "Innovative Predictive Operation System for Wagon-Environment Interaction". The research aimed to develop a dedicated solution for predicting failures and monitoring component wear in railway rolling stock. The system relies on a sensor infrastructure and predictive algorithms. This enables the detection of abnormal operating conditions before failures occur by analyzing patterns of abnormal behavior. The result of the research is creation of the effective monitoring and predictive system, which enables early detection of failures and optimization of maintenance and operation of the railway rolling stock.

Keywords: Railway transport · Condition based maintenance · Freight wagon maintenance · Digital twin

1 Introduction

The decreasing costs of sensors have led to the increasing prevalence of data-driven monitoring systems. Combined with advancements in technology and Big Data handling methods, this development provides the basis for individual and systemic benefits during associated transformative processes (such as Digital Transformation) – a paradigm observed in various industrial sectors, including manufacturing, supply, logistics, construction, policy-making, and education.

The aim of the research is a creation of the agenda entitled "Innovative Predictive Operation System for Wagon-Environment Interaction", implemented by Tankwagon Sp. z o.o. and Wroclaw University of Science and Technology. The objective of the agenda is to develop an application that assists in the early detection of potential failures, primarily in freight wagons. The detection of abnormalities occurs by comparing the patterns of correct operation of rail wagons on the railway infrastructure with the current results of individual parameters obtained from sensors placed on the wagons.

Tankwagon Sp. z o. o. is an experienced partner in the rental and maintenance of various types of freight wagons. The company maintains approximately 1,500 rail wagons, mainly consisting of Eanos, Eaos, Fac, Fcs, Fas, Res, Uagpps, and Z wagons

O. Prentkovskis et al. (Eds.): TRANSBALTICA 2023, LNITI, pp. 586–593, 2024.
https://doi.org/10.1007/978-3-031-52652-7_58

used for transporting food, chemicals, light and heavy petroleum products, and gases. Tankwagon is a company with over 25 years of railway experience. Since 2013, as the Entity in Charge of Maintenance (ECM), they have ensured the highest quality of technical and organizational services for the safety of railway rolling stock operation.

At the current stage of the research, 80 freight wagons are to be equipped with telemetry units containing various sensors, enabling the monitoring of current technical parameters of the vehicle components. The availability of data from the sensors paves the way for the use case of "acceleration-based infrastructure monitoring", which allows for monitoring both the railway infrastructure and rolling stock without disrupting daily operations. The article presents research agenda related to implementation, adopted research methodology, planned benefits, and the novelty of the results of the wagon remote diagnostic system.

2 Literature Review

The Condition Based Maintenance (CBM) method can be applied to the maintenance of rail wagons, including freight wagons. CBM is an approach to maintenance where decisions regarding maintenance and repairs are based on the actual condition of the equipment rather than a predetermined schedule. Effective implementation of CBM in the maintenance of rail wagons requires the appropriate installation of sensors, monitoring systems, and data analysis. This may require investments in technological infrastructure but can lead to long-term optimization of the railway rolling stock maintenance process.

An example of a model for maintaining the technical condition of rail vehicles based on an integrated sensor system is presented in [1] or [2]. In the case of rail wagons, CBM relies on monitoring various parameters and indicators that can indicate their technical condition. These can include temperature measurements, vibrations, pressure, speed, component wear, or other indicators related to specific subsystems. When such data is collected and analyzed, threshold values can be established to indicate the need for inspection or repair. The principles of this type of maintenance are described in more detail in [3]. They do not necessarily apply to the entire rail wagon. In articles [4, 5], and [6], the focus is limited to specific components or subsystems and determining their suitability for fulfilling transport missions.

Articles [7] and [8] highlight elements that have a critical impact on the technical condition of rail wagons and specify the conditions under which they should undergo repairs. Remote diagnostics of the technical condition also enable better planning of spare parts inventory, as shown in [9]. Nowadays, remote data collection and transmission have become much simpler, often utilizing large datasets. An example of big data-based analysis is presented in [10]. One significant challenge in sensor development has been achieving autonomy, and solutions to this problem are presented in [11]. In a way, CBM represents an innovative approach to the maintenance and reliability of both vehicles and infrastructure [12].

The application of CBM in the maintenance of rail wagons, especially freight wagons, can bring numerous benefits. It allows for minimizing wagon downtime since inspections and repairs are only performed when necessary. Additionally, by monitoring the technical condition of wagons, potential faults can be detected earlier, avoiding more serious breakdowns and enhancing safety in railway operations.

Data collection can also enable the creation of a digital twin of the railway rolling stock, facilitating advanced simulations related to the degradation of individual wagon components. A digital twin is a virtual model that reflects the real parameters, properties, and technical condition of the wagons. It is used to simulate the behavior of wagons in various operating conditions. Data collected from accelerometers allow for monitoring actual vibrations and oscillations recorded during wagon operations. This data is then utilized to create a digital twin that accurately replicates the real behavior of the wagon.

Using a digital twin, advanced simulations can be conducted to forecast the degradation of individual wagon components based on various factors such as load, speed, and weather conditions. These simulations enable the identification of potential issues related to specific wagon components like bearings, brakes, or structural elements before actual failures occur. Conducting simulations based on a digital twin allows for a better understanding of degradation processes and the impact of various factors on the technical condition of wagons. This, in turn, facilitates more informed decisions regarding maintenance, repair planning, and the overall optimization of the railway rolling stock maintenance process. The extension of CBM to include simulations is discussed in [13], while the issues of remote diagnostics and creating operational models are further described in [14].

In addition to monitoring the technical condition of railway vehicles, remote diagnostics of signals from sensors mounted on the wagons also enable the assessment of the condition of railway tracks. In this case, accelerometers are employed, as described in [15]. Mounting accelerometers on freight wagons allows for collecting data on vibrations and oscillations that occur during movement on the tracks. This data can then be analyzed to assess the technical condition of the railway tracks or specific sections of the rail network.

There are several benefits to monitoring track vibrations using accelerometers on freight wagons. One benefit is the early detection of faults. Monitoring vibrations can help identify early signs of infrastructure damage, such as gaps, cracks, misalignments, or track instability. This allows for prompt maintenance actions to be taken, preventing more severe failures that could disrupt railway operations or pose safety hazards.

Furthermore, optimization of maintenance is possible. Data from accelerometers enables a more precise determination of problematic areas on the tracks, allowing maintenance efforts to be focused on specific areas. This contributes to more efficient resource utilization and cost reduction. Monitoring changes over time is also possible. Systematically collecting data from accelerometers during multiple journeys on the tracks enables trend analysis and identification of areas that require special attention and regular monitoring.

In summary, the CMB approach, supported by data-driven monitoring systems and advanced simulation techniques, can be applied to the maintenance of rail wagons, including freight wagons. It offers benefits such as minimized downtime, early fault detection, and the optimization of maintenance processes. Additionally, by utilizing data collected from accelerometers, digital twins of the rolling stock can be created, facilitating advanced simulations for component degradation analysis. Moreover, monitoring track vibrations using accelerometers on wagons enables early detection of infrastructure faults and maintenance that is more efficient planning.

3 Discussion on Adopted Research Methodology

The development of a solution aimed at supporting Tank wagon employees in the proactive diagnosis of wagon and track infrastructure conditions in real-time, along with reporting actions to ensure uninterrupted operations, has been achieved. A "digital twin" has been established to enhance operational processes related to rolling stock. The comprehensive and integrated solution has led to increased work efficiency and cost reduction.

The research agenda focused on the creation of a dedicated solution for the prediction of failures and the monitoring of component wear in railway rolling stock. This solution was built upon a sensor infrastructure and prediction algorithms. The designed system was capable of identifying and reporting abnormal operational conditions of the rolling stock using a predictive model before signs of failure manifested. Different patterns of abnormal behavior corresponded to pre-failure states (with symptoms of failure), failures, and degraded performance. The research challenge involved the formulation of appropriate error patterns (abnormal behavior) that enabled the early detection of impending failure or degradation, leading to wear and tear estimation. It was necessary to monitor specific components for deviations from normal operating patterns, utilizing various measurement signals from the sensors. Selecting relevant data from a substantial dataset, characteristic of the operational state of a specific segment of the rolling stock and encompassing a range of sub-patterns distinctive of subpar performance, posed another research challenge. Defining acceptable tolerance limits that could trigger failure states also presented difficulties.

Patterns of parameter variations in abnormal operating states resulting from failure were developed. Failure states could arise from the malfunction of one or multiple elements simultaneously (parallel failure). Such failure states deviated from the standard operating pattern. Enhancing operational processes and maintenance activities faced the additional challenge of simultaneously representing the static and dynamic facets of the system. Equally vital was the monitoring of rolling stock element parameters. Striking a balance between achieving the highest quality control over rolling stock operations and minimizing data acquisition requirements proved to be pivotal. The research centered on formulating appropriate models that could ascertain the states of individual vehicles. A crucial aspect of fleet diagnostics lay in the potential transferability of the model from one instance to others with a similar profile. Initiating the construction of a model for each type of rolling stock from scratch was deemed impractical and costly.

The second research concern was the establishment of a "digital twin" to enhance vehicle operational processes and maintenance systems. The primary objective of leveraging augmented reality technology was to aid specialists by providing virtual information, with the goal of offering enhanced support during tasks. The culmination of this study was the creation of a demonstrator showcasing the outcomes of the development efforts.

4 Expected Outcomes from Research Agenda

The research hypothesis has been validated, affirming the feasibility of determining the condition of rolling stock components based on signatures extracted from data collected by acceleration, velocity, pressure, and temperature sensors embedded in individual rolling stock modules. The validation was achieved through the application of a model designed to identify impending failure states, enabling reliable assessments of rolling stock reliability. During this study, a diagnostic system was developed for various wagon types, spanning diverse operational conditions and configurations (measured signal range and location). Operational research was also conducted, involving the creation of reliability models from historical data and the preparation of infrastructure characteristics.

Subsequent stages of work involved validating the developed tool, assessing vehicle conditions during technical inspections, refining algorithms and models, and further calibration. Recorded signal characteristics provided insights into a well-functioning system, initial or advanced wear, and malfunction. Diagnostic signal recording encompassed key rolling stock components, selected based on service history, failure records, and railway operator expertise. Measurement channel parameters, sensor sensitivity, and measurement point distribution were determined after preliminary tests to optimize the measurement system while controlling costs and minimizing excessive data recording and transmission to the data server.

In railway transportation, rolling stock failures can have profound consequences, impacting safety and finances significantly, often resulting in multimillion Polish zloty losses due to infrastructure damage or other vehicle-related issues. In extreme cases, such failures can lead to railway accidents causing loss of life. Presently, most railway carriers and wagon owners rely on preventive maintenance strategies involving manual measurements and entries into documentation. Human involvement and manual data input introduce the potential for errors at each stage of the information flow.

Although some predictive strategies are in use, they tend to focus on individual components and prove less effective in modern transport systems due to heightened safety and reliability expectations from customers. Consequently, the ongoing development of solutions is expected to enable dependable railway maintenance by leveraging intelligent transport systems and integrating proactive maintenance concepts with safety tools. This approach aims to tackle crucial railway system issues, such as safety and disruptions, including delays stemming from the aftermath of failures or adjustments to work plans. The anticipated outcomes would manifest as new service technology, aligned with the evolution toward "intelligent railways". This aligns with Maintenance 4.0 concepts and leverages innovative solutions based on Big Data, information integration, and digital twins. The results align with global trends in railway vehicle operation and maintenance strategies, underscoring the innovative nature of the proposed solution.

4.1 Implementation of Maintenance 4.0 Strategy

The implementation of the Maintenance 4.0 strategy has enabled proactive and real-time diagnostics of wagon and track infrastructure conditions where sensor-equipped wagons are operational, alongside the reporting of actions to ensure uninterrupted operations.

Presently, the majority of rolling stock managers adhere to preventive maintenance strategies primarily grounded in systematic servicing and scheduled inspections and repairs derived from the Vehicle Maintenance Documentation (VMD). It is assumed that inspection intervals are appropriately and optimally determined, striking a balance between minimizing technical maintenance costs and maximizing the potential for detecting early-stage failures. Predictive solutions in railway transportation are currently confined mainly to complex electrical systems within locomotive equipment, with a notable absence of such solutions for wagons and their components.

Conversely, a wide array of available sensors allows for the collection of extensive operational data on wagon parameters and characteristics. However, a critical innovation lies in the integrated real-time analysis of this data to foresee potential issues before they materialize. The key lies in a comprehensive approach that amalgamates diagnostic data from various wagon components to assess the rolling stock's real-time condition. This approach hinges on the implementation of comprehensive predictive solutions ensuring continuous monitoring of vehicle equipment elements through the measurement of crucial variables like temperature, vibrations, brake performance, and more. Concurrently, continuous recording of the vehicle's operational state, coupled with the uninterrupted transmission of this data to data servers, facilitates the introduction of a virtual black box service. This represents an additional innovation supporting the Maintenance 4.0 strategy. The collected data is timestamped, enabling accurate reconstruction of the device's status until a malfunction occurs.

However, the primary purpose of the conducted diagnostics revolves around the accumulation of operational data, forming the foundation for determining optimal timing for maintenance, replacements, or repair activities. Measurement data from sensors monitoring the wagon's state undergo continuous analysis through a diagnostic algorithm, delivering real-time feedback on potential failure risks. This process generates large datasets, fueling a proactive system for identifying and rectifying faults, signaling necessary service interventions. The developed damage analysis, rooted in a virtual reality subsystem, guides service personnel regarding elements necessitating service operations and outlining the required scope of actions.

4.2 Digital Twin Creation of Rail Wagon

A "digital twin" was generated to enhance wagon operational processes and maintenance systems. The concept of a digital twin comprises three fundamental elements: a) the physical wagon existing in the real world, b) the virtual wagon existing in the digital realm, c) two-way connections facilitating data and information interchange between the virtual and real counterparts.

A pivotal facet of the devised solution aimed at enhancing operational procedures and maintenance activities is its capacity to concurrently depict both static and dynamic facets of the designated system. To the best of current knowledge, no operators in Poland currently employ "digital twin" solutions within the sphere of wagon operation and maintenance. While digital twins are a core component of the Industry 4.0 paradigm, their integration is predominantly pronounced in maintenance strategies concerning production machinery facilities. Nonetheless, the application of this concept can also be extended to the domain of railway fleet management. Through the utilization of a digital twin, it becomes viable to replicate the genuine behaviors of individual wagon components and elements, investigating the influence of usage processes and present operational conditions on the vehicle's performance. Moreover, the twin can be harnessed to conduct virtual experiments pertaining to alterations in wagon operation and maintenance protocols. This opens doors to identifying solutions to issues by refining the product or altering operational procedures prior to the occurrence of catastrophic failures.

4.3 Support for Decision-Making Processes Related to the Operation of Rail Wagons

Enhancing decision-making processes pertaining to wagon task allocation is achieved by amalgamating data regarding track infrastructure conditions and the vehicle's operational readiness and technical specifications. A pivotal innovation arising from the intended outcomes is the fusion of track infrastructure data with real-time diagnostic information concerning the wagon's state. Through specialized measurement apparatuses for railway infrastructure, decision-makers access up-to-the-minute insights into infrastructure conditions. Concurrently, monitoring systems for individual wagon components and elements ascertain its preparedness to undertake a specific transport task within given operational contexts.

The analytical integration of external parameters and diagnostic data supports the decision-making process regarding task progression, potential adjustments, or task completion. As the period between inspections lengthens due to the predictive capabilities, wagon operators/managers can make well-informed choices to allocate a particular transport task to a rail vehicle. This facilitates operating the wagon in proximity to a potential servicing location. In terms of wagon fleet management, such action significantly curtails expenses linked to transporting a malfunctioning wagon from a distant location to a service center for repair.

5 Conclusions

Throughout the outlined research, the predictive approach, entailing the constant monitoring of rolling stock conditions to swiftly identify any concerning indicators, was expanded. The research extended the notion of sustaining operational fitness by perpetually assessing technical conditions, thereby eliminating superfluous efforts and concentrating solely on current necessities. This distinction represents a significant advantage over previous maintenance models, which were either reactive or adhered to inspection schedules. The outcome of predictive maintenance implementation aimed to ensure optimal utilization of technical service resources and the complete technical readiness of rolling stock.

The proposed maintenance strategy ensured an enhanced understanding of managed wagon conditions, leading to a reduction in unplanned downtimes and the minimization of substantial costs related to unplanned repairs and the transportation/handling of wagons scattered across Europe. These costs were formerly borne by operators due to the prevailing approaches.

Acknowledgements. The presented scientific work is a part of the project "Innovative Predictive Operation System for Wagon-Environment Interaction", POIR.01.01.01-00-0196/22-00. The project financed by The National Centre for Research and Development.

References

1. Pfaff, R., Shahidi, P., Enning, M.: Connected freight rail rolling stock: a modular approach integrating sensors, actors and cyber physical systems for operational advantages and condition based maintenance. In: Proceedings of the Asia Pacific Conference of the Prognostics and Health Management Society (2017)
2. Bernal, E., Spiryagin, M., Cole, C.: Onboard condition monitoring sensors, systems and techniques for freight railway vehicles: a review. IEEE Sens. J. **19**(1), 4–24 (2018)
3. Ma, Q.: Condition-based maintenance applied to rail freight car components: the case of rail car trucks (Doctoral dissertation, Massachusetts Institute of Technology) (1997)
4. Osman, H., Yacout, S.: Condition-based monitoring of the rail wheel using logical analysis of data and ant colony optimization. J. Qual. Maint. Eng. **29**(2), 377–400 (2023)
5. Bernal, E., Spiryagin, M., Cole, C.: Wheel flat detectability for Y25 railway freight wagon using vehicle component acceleration signals. Veh. Syst. Dyn. **58**(12), 1893–1913 (2020)
6. Sobaś, M.: The causes and prevention of rolling bearing overheating of the axle box in the freight wagons while driving. Pojazdy Szynowe (2021)
7. Sobaś, M.: Wheel wear during the operation of rail vehicles: causes and effects. Pojazdy Szynowe (2021)
8. Gawlak, K.: Safety critical components (SCC) in the maintenance management system for railway vehicle. Diagnostyka **23** (2022)
9. Yao, N., Jia, Y., Tao, K.: Rail weld defect prediction and related condition-based maintenance. IEEE Access **8**, 103746–103758 (2020)
10. Fumeo, E., Oneto, L., Anguita, D.: Condition based maintenance in railway transportation systems based on big data streaming analysis. Procedia Comput. Sci. **53**, 437–446 (2015)
11. Schiavo, A.L.: Fully autonomous wireless sensor network for freight wagon monitoring. IEEE Sens. J. **16**(24), 9053–9063 (2016)
12. Milković, V., Lisjak, D., Kolar, D.: New reliability-based model of stock optimisation for railroad passenger wagon maintenance. FME Trans. **48**(4), 914–921 (2020)
13. Gruber, A., Yanovski, S., Ben-Gal, I.: Condition-based maintenance via simulation and a targeted Bayesian network metamodel. Qual. Eng. **25**(4), 370–384 (2013)
14. Bragança, C., et al.: Calibration and validation of a freight wagon dynamic model in operating conditions based on limited experimental data. Veh. Syst. Dyn. **60**(9), 3024–3050 (2022)
15. Otte, T., et al.: Condition monitoring of rail infrastructure and rolling stock using acceleration sensor data of on-rail freight wagons. In: ICPRAM, pp. 432–439 (2022)

Estimation of Running Smoothness and Derailment Stability Considering the Parameters of Passenger Car Suspension

Gintautas Bureika(✉)

Vilnius Gediminas Technical University, Saulėtekio al.11, 10223 Vilnius, Lithuania
gintautas.bureika@vilniustech.lt

Abstract. The main features of passenger car dynamics such as running smoothness and a derailment stability are estimated in this publication. The running smoothness is evaluated considered the parameters of carbody vibrations and by oscillation of bogies. The oscillation of body, suspension and other running gear elements occur while train moving by track irregularities. The four-axle passenger car movement on tangent track with given vertical and lateral rail irregularities is simulated by using software package „Universal Mechanism" (UM). The dependence of oscillation amplitudes of passenger car body on track irregularities considering the suspension vertical and lateral stiffness were defined. The variation of derailment stability taking into account changing the stiffness of car suspension elements analysed in this work.

Keywords: Railway · Tangent track · Dynamics of passenger car · Derailment stability · Nadal criteria · Universal Mechanism

1 Introduction

Worldwide railway transport is one of the main types of transport, and the operation of passenger car and multi-units fleet is one of the most important area of economic activity of railway undertakings. In addition, railway transport is one of the main types of land transport. An enterprise "LTG Link" of State Company "Lithuanian Railways" (hereinafter – LTG) managers a passenger car fleet which consists of passenger locomotives, coaches, railcars and self-propelled units (electric and diesel multi-units). A passenger car is a railway vehicle intended for carrying passengers. Other types of coaches are restaurant, luggage, mail, service and special cars. The comfort and safety of passenger transportation depends on the rationality of vehicle fleet structure and their technical characteristics [2, 4]. The possibility of widely implementing complex mechanization and automation in the production and repair of wagons, as well as their operation (intelligence train traffic management, monitoring of technical state, automated formation of trains, etc.) are of great importance.

With the improvement of traction vehicles, increasing movement speed, at the same time, increasing traffic safety requirements are being raised. So, more and more attention

O. Prentkovskis et al. (Eds.): TRANSBALTICA 2023, LNITI, pp. 594–603, 2024.
https://doi.org/10.1007/978-3-031-52652-7_59

is paid to the study of dynamic processes of rail vehicles and the study of the rail/vehicle interaction forces [10]. This makes possible to determine the most important structural parameters of the track and rolling-stock, to select the right materials, to assess deformations, to ensure the durability, reliability and durability of the mechanisms in order to ensure the greatest possible smoothness and comfort of movement [5, 6].

This article examines one of the main characteristics of wagon dynamics – smoothness of movement and derailment stability. The efficiency of the suspension of the passenger car is described, the characteristics of the oscillation processes and vibrations of the elements are analysed. Oscillations of the wagon body and bogies arise when moving over the unevenness of the railway track. Vertical and lateral and longitudinal, i.e. dynamic fluctuations of wagon masses have a significant influence on the smoothness of displacements and should not exceed critical values. Thus the ability of the vehicle elastic suspension to suppress body vibrations should be confirmed. One of the main parameters that determine the smoothness of the vehicle ride are the amplitudes and accelerations of the wagon element masses. The maximum permissible quasi-static acceleration of the passenger car body is 0.8 m/s^2.

The study of rolling-stock dynamics under real conditions (i.e. field conditions) is very time-consuming or too complicated due to the railway line in-service occupation during the tests. Computer simulation is used to facilitate this process. For simulation studies, the Author used the software package "Universal Mechanism" (hereinafter – UM), which was developed to study the dynamics of various mechanical objects moving in space. The objects are connected to multifunctional systems and analysed.

The Nadal criterion is widely used to assess rolling-stock derailment conditions. The values assessed by the Nadal criterion are shown in Fig. 1.

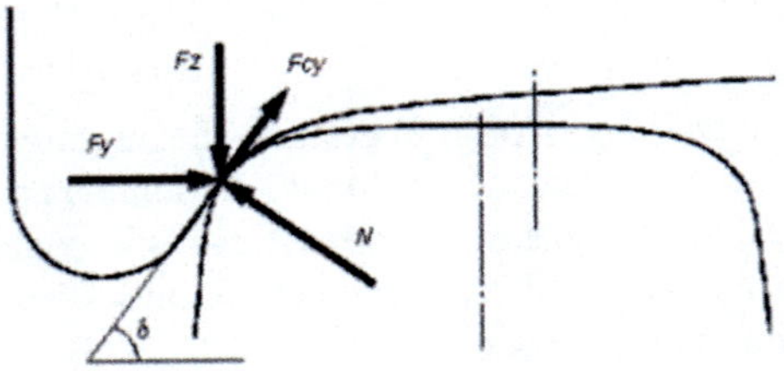

Fig. 1. Interaction of rail and wheel: Fy – lateral force; Fz – vertical force; N – normal force; F_{cy} – tangent force in rail/ wheel contact; δ – angle of rail/wheel contact surface.

Nadal criterion K_{Nad} is calculated:

$$K_{Nad} = \frac{F_y}{F_z} = \frac{tg\,\delta - \mu_y}{1 + \mu_y \cdot tg\,\delta} < 0.85; \quad (1)$$

where μ_{y} – friction coefficient in lateral direction.

Nadal's criterion K_{Nad} takes into account the shape of the wheel rolling surface, the contact angle and the coefficient of friction, but does not take into account the longitudinal forces and the angle of the curve on which wheel-set moves. A derailment occurs when Nadal's criterion reaches a value greater than 0.85. According to International Union of

Railways (UIC), vehicle running on rails is safe as long as this criterion does not exceed the value of 0.80 for a curve with a radius greater than 250 m. The value of this criterion should be higher than allowed for an empty wagon due to too little force in the vertical Z axis. In other words, with less vertical force, it is easier for the wheel to climb the rail head, i.e. a loaded wagon is more stable than an empty one. Nadal's criterion is often used in field tests due to the difficulty of measuring the angle of wheel rolling attack.

When traveling in a car, besides safety passengers concerned also with comfort. The main factor affecting comfort is vibrations, which can occur due to natural and forced oscillations of the wagon, which occur due to unevenness of the track [2, 8]. The resulting vibrations of rolling stock can be undesirable, especially if unevenness occurs frequently on the track in high speed. Due to vibration, the activity of some human organs and systems is disrupted. The nervous, cardiovascular, bone and muscle systems are most affected.

Driving smoothness and derailment stability problematic solutions for rolling-stock depend on an adjustable parameter. The dynamics of the rolling-stock depends on some nonlinear parameters, such as wheel/track running profile and creepage forces [3, 9]. Also, the coefficient of friction affects the amplitude of the lateral movement of the wheel when the wheel acquires a sinusoidal lateral yaw, which is shown in Fig. 2.

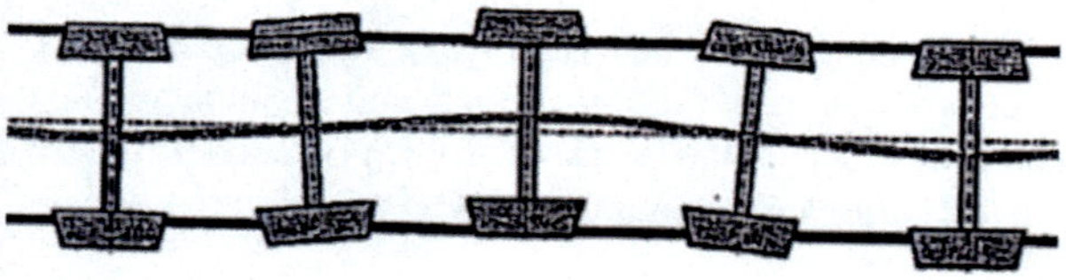

Fig. 2. Oscillation of rail vehicle wheel-sets causing sinusoidal movement.

The oscillation of the wheel-sets (Fig. 2) grounds the hunting of whole carbody and causes discomfort for the passengers, i.e. an obnoxious sound and vibrations can be felt in the passenger car [11]. At the same time, due to the intense contact of the wheels with the rails, it accelerates the wear of the wheels and the rails. These wear processes are caused by lateral and longitudinal creepage forces.

One of the most important factors limiting the maximum permissible speed of vehicles is the stability of running on railway track. The purpose of this study is to determine the suspension parameters that ensure the maximum driving stability of the passenger car. The dependence of these parameters on track irregularities and vertical and lateral suspension parameters is examined. In order to ensure the derailment stability of passenger cars on the track, the most rational combinations of wagon suspension parameters and stability criteria were estimated.

2 Modelling of Rail Track and Passenger Car

Railway rolling stock moves along uneven tracks. When the train is moving, due to the geometric irregularities of the track and operational damage, vibrations of the rolling stock occur, especially if the irregularities are felt and moved at high speed [7, 12].

When creating the track profile, allowable track irregularities are taken into account, which would not limit the maximum permissible speed of the wagon, i.e. 120 km/h. The modelling of the track profile was based on the track safety requirements used in the USA (the UM program is focused on the standards applied in EU), according to which the track is divided into 9 categories depending on the irregularities and the speed of the wagon movement. The UM program simulates a 200-m tangent track. It includes lateral track irregularities and sinusoidal vertical ones up to 16 mm (Fig. 3). Roughnesses are assigned separately for each rail. In order to obtain the quite big vibrations of the wagon, the track is modelled with frequent irregularities.

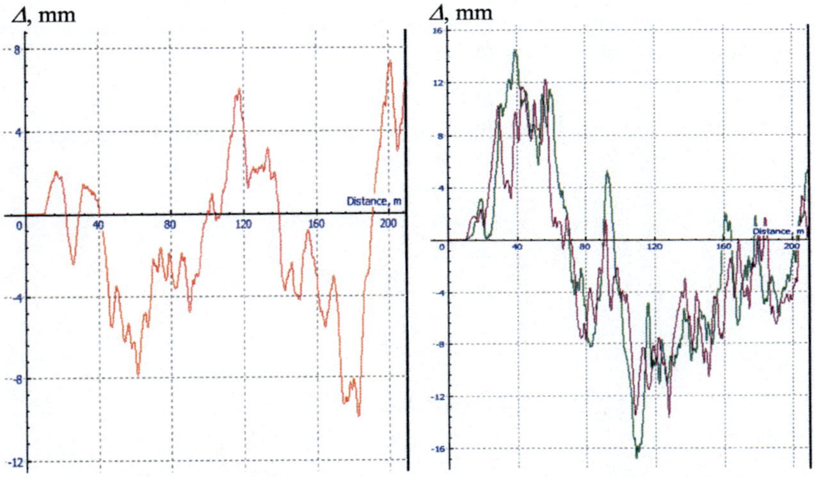

Fig. 3. Irregularities of rail track: left) – vertical; right) – lateral.

Various elastic elements are used in the suspensions of passenger cars. When examining vertical dynamics, the characteristics of forces influencing vertical dynamics, i.e. dependence of the element's vertical force P on its static deformation f_{st}. Optional variables – vertical stiffness of the suspension springs were taken from 323 to 534 kN/m and transverse – from 120 to 200 kN/m, respectively.

UM studies the dependence of vertical and transverse displacements of the wagon body on suspension parameters: vertical and lateral stiffness. The wagon body and suspension are assumed to be symmetrical about the centre line [1]. The suspension calculation model is presented in Fig. 4.

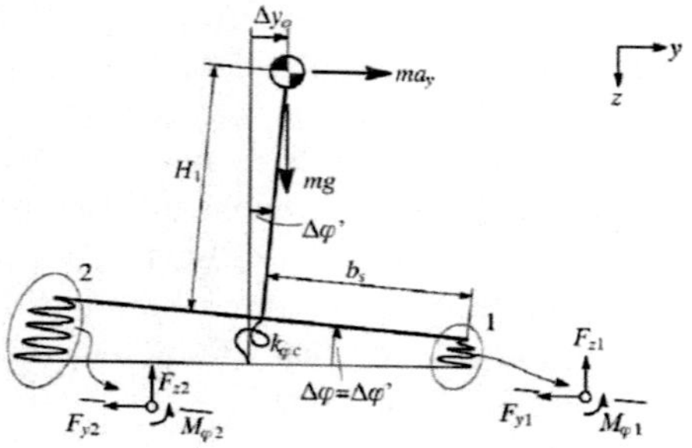

Fig. 4. Model of calculation of suspension oscillation: a_y – lateral acceleration, m/s^2; $\Delta\varphi' = \Delta\varphi$ – angle of forced spring oscillation, rad; g – gravitation, m/s^2; H_1 – distance from mass centre, m; b_s – distance from suspension fulcrum, m.

The diagram of forces and deformations of the springs is presented in Fig. 5.

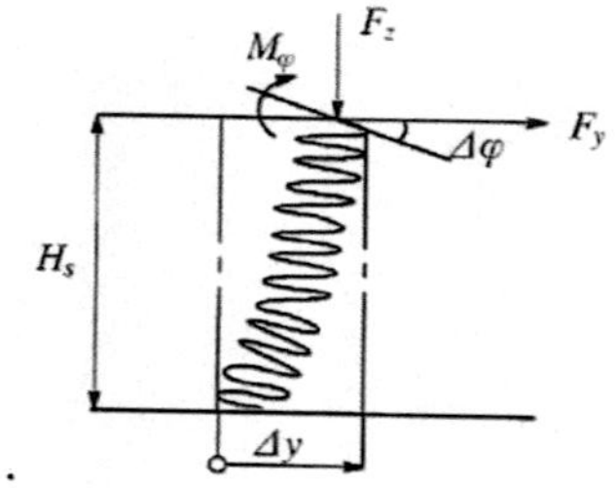

Fig. 5. Scheme of exerted forces, moments and deformations of spring.

The calculation matrix of the suspension springs in the coordinate plane was written as (Andersson *et al.* 2007):

$$\begin{pmatrix} F_y \\ M_\varphi \end{pmatrix} = \begin{pmatrix} k_{yy} & k_{y\varphi} \\ k_{\varphi y} & k_{\varphi\varphi} \end{pmatrix} \begin{pmatrix} \Delta y \\ \Delta\varphi \end{pmatrix}; \quad \text{where } k_{y\varphi} = k_{\varphi y} < 0. \tag{2}$$

where F_y – lateral force, N; $M\varphi$ – bending torque, Nm; k_{yy} – transversal stiffness, N/m; $k_{y\varphi} = k_{\varphi y}$ – angular stiffness, *N/m;* $k_{\varphi\varphi}$ – bending stiffness, *N/m;* $\Delta y'$ – lateral displacement, *m;* $\Delta\varphi$ – yaw angle, rad.

Suspension lateral displacement $\Delta y'$ (Andersson *et al.* 2007):

$$\Delta y' = \frac{ma_y}{2k_{yy}} \left[1 - 2k_{yy} \cdot \frac{H_1 - \frac{k_{y\varphi}^2}{k_{yy}}}{2b_s^2 k_{zz} + k_{\varphi\varphi} - mgH_1 - 2\frac{k_{y\varphi}^2}{k_{yy}}} \right]; \tag{3}$$

where m – suspend mass of car, kg; a_y – lateral acceleration, m/s^2; k_i – suspension stiffness in corresponding i-plane, N/m; g – gravitation, m/s^2; H_1 – distance between center of car mass and suspension (Z axis), m; b_s – distance between center of car mass and suspension (Y axis), m.

It can be seen from Formulas (1–3) that the vertical deformation of the suspension f_{st} and the transverse displacement $\Delta y'$ depend on the static load of the suspension, i.e. from the gross weight of the wagon. When examining the vertical and transverse displacements of the carbody, three points were selected. It can be seen from Formula (3) that the transverse displacement of the wagon body $\Delta y'$ increases as the distance H_1 from the point of connection of the body to the suspension increases. The points located at the level of the seated and standing passenger (person) were selected. The coordinates of the wagon points are shown in Fig. 6.

Fig. 6. The overall layout of wagon body coordinates.

The dependence of the wagon rolling criterion and body accelerations on the type of wheels and vertical and lateral suspension parameters were determined. Diagrams of carriage body and bogie displacements are shown in Fig. 7.

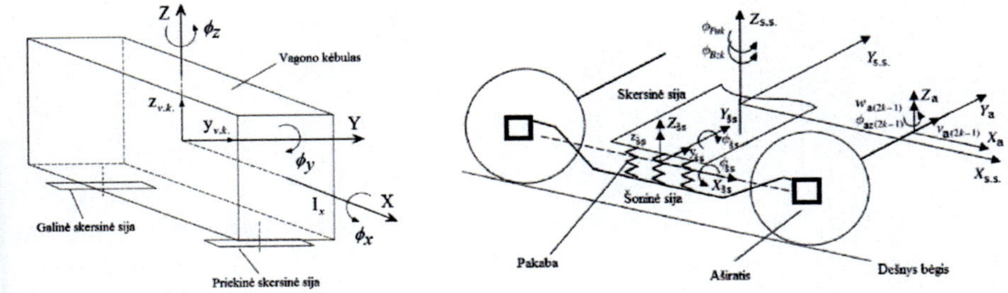

Fig. 7. The general schemes of displacements: carbody and bogie.

All elements of the wagon, except the spring suspension, are modelled as rigid bodies and can move in all directions of the Cartesian coordinate system. The suspension consists of springs and friction wedges that allow lateral and vertical displacements. The connection between the wheels and the side beams allows the side beams to rotate relative to the wheels. During this deflection, the bogie frame gets the shape of a parallelogram. When calculating the degrees of freedom of the wagon, this parallel deflection of the side beams is denoted by Φ_{pa}. The results of calculating the degrees of freedom (D.O.F.) of the wagon model are presented in Table 1.

Table 1. Calculation results of degree of freedom

Car element	Degree of freedom							Number	D.O.F
	x	y	z	Φx	Φ_y	Φ_z	Φ_{pa}		
Body	–	+	+	+	+	+	–	1	5
Bogie	–	+	+	–	+	+	+	2	10
Total degree of freedom									15

Thus, the computer model of the passenger car has 15 D.O.F. The linear displacement of the wagon body and bogie movement in the X direction and the angular displacement of the bogie around the axis X are not evaluated.

3 Examination of Passenger Car Running Stability

The simulation is performed on a tangent track section with selected vertical and transverse rail irregularities. The results of this studies are presented in graphic form below. The running speed is 120 km/h.

The graphs of the carbody vertical displacements with suspension vertical stiffness of 323 kN/m and lateral stiffness of 120 kN/m are presented in Fig. 8.

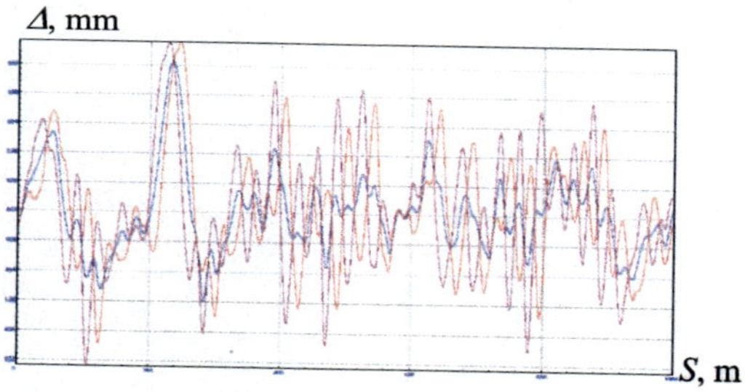

Fig. 8. The vertical displacement of carbody of 4-axle passenger car.

The results of the carbody lateral displacements with suspension vertical stiffness of 534 kN/m and lateral stiffness of 200 kN/m are presented in Fig. 9.

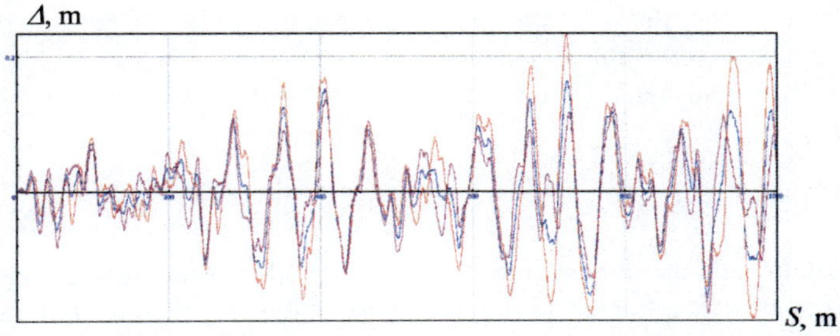

Fig. 9. The lateral displacement of carbody of 4-axle passenger car body.

The diagrams of the carbody lateral accelerations with stiffness vertical stiffness of 534 kN/m and lateral stiffness of 200 kN/m are presented in Fig. 10.

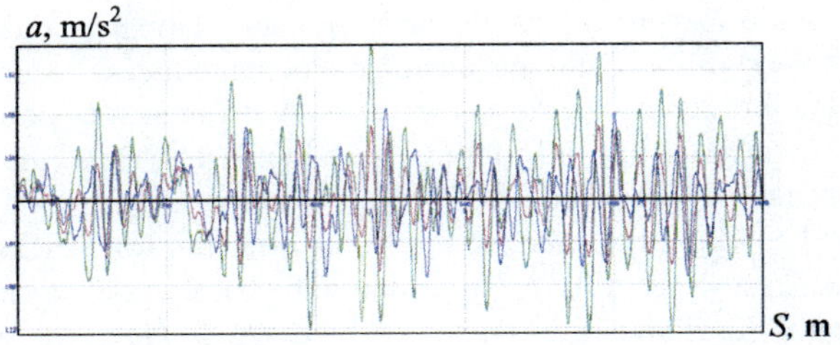

Fig. 10. The lateral accelerations of carbody of 4-axle passenger car body.

The variation of Nadal criteria K_{nad} of the first bogie of 4-axle passenger car is presented in Fig. 11.

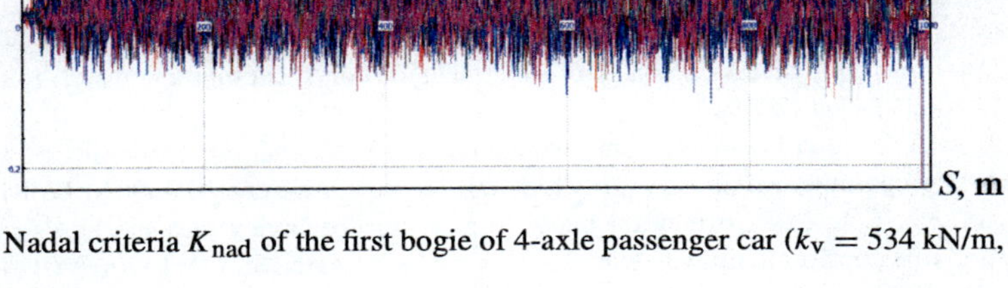

Fig. 11. Nadal criteria K_{nad} of the first bogie of 4-axle passenger car ($k_{\text{v}} = 534$ kN/m, $k_{\text{s}} = 200$ kN/m).

During simulation tests with increased stiffness of the wagon suspension parameters, the value of the Nadal criterion K_{nad} increased from 0.2 to 0.27. It should be noted that with higher stiffness suspension elements, the probability of derailment is higher.

4 Conclusions

It has been established that the parameters of the carbody lateral movements are mainly influenced by the stiffness of the elastic elements of the suspension. By increasing the vertical stiffness of the passenger car suspension springs from 323 kN/m to 534 kN/m and the lateral stiffness of the springs from 120 kN/m to 200 kN/m, the amplitude of the vertical swing of the passenger car body increases by 6 mm, and the amplitude of the lateral swing increases by 11 mm. With an increase in the vertical and lateral stiffness of the suspension springs of a passenger car by 20% at a speed of 120 m/h, the value of the Nadal criterion increases by 18%. Thus, the derailment stability of the passenger is correspondingly reduced.

The stiffness of the suspension affects the lateral and vertical vibrations of the body, i.e. an increase in the rigidity of the suspension elements reduces the amplitude of body movements, while vibration acceleration increases. A stiffer suspension is called "stiff" because the impact of track on the running gear increases. Thus, the comfort of the passenger car is principally reduced.

At a speed of 120 km/h, the value of the Nadal criterion for increased rigidity of a passenger car reaches a value of 0.27. By reducing the hardness of suspension elements by 20%, the value of the Nadal criterion decreased by 26%. Thus, while the stiffness of the suspension of a passenger car decreases, the derailment stability increases. When the speed of a passenger car decreases from 120 km/h to 140 km/h, the amplitudes of lateral and vertical movements of the body and their accelerations decrease by 35%. Thus, with an increase in the speed of a passenger car, its resistance to derailment decreases.

Summarizing the results presented above, it can be concluded that for passenger cars, the suspension with adjustable stiffness depending on the running speed is best justified, i.e. semi-active or active suspensions.

References

1. Andersson, E., Berg, M., Stichel, S.: Rail Vehicle Dynamics, p. 224. Railway Group KTH, Stockholm (2007)
2. Ballew, B., Chan, B.J., Sandu, C.: Multibody dynamics modelling of the freight train bogie system. Veh. Syst. Dyn. **49**(4), 501–526 (2011)
3. Bureika, G., Levinzon, M., Dailydka, S., Steišūnas, S., Žygienė, R.: Evaluation criteria of wheel/rail interaction measurement results by trackside control equipment. Int. J. Heavy Veh. Syst. **26**(6), 747–764 (2019)
4. Dižo, J., et al.: Evaluation of ride comfort in a railway passenger car depending on a change of suspension parameters. Sensors **21**, 8138 (2021). https://doi.org/10.3390/s21238138
5. Dumitriu, M., Leu, M.: Correlation between ride comfort index and sperling's index for evaluation ride comfort in railway vehicles. Appl. Mech. Mater. **880**, 201–206 (2018). https://doi.org/10.4028/www.scientific.net/AMM.880.201

6. EN 12299:2009. Railway applications – Ride comfort for passengers – Measurement and evaluation
7. Keršys, R., Bazaras, Ž., Griškevičius, P.: Vagono vertikalios dinamikos modeliavimas. [Modelling of wagon vertical dynamics]. Transportas **15**(5), 220–227 (1999). (in Lithuanian)
8. Kulkarni, R.A., Qazizadeh, Q., Berg, M.: Unsupervised rail vehicle running instability detection algorithm for passenger trains (iVRIDA). Measurement **216**, 112894 (2023)
9. Martinez-Casas, J., Di Gialleonardo, E., Bruni, S., Baeza, L.: A comprehensive model of the railway wheelset–track interaction in curves. J. Sound Vib. **333**(18), 4152–4169 (2014). https://doi.org/10.1016/j.jsv.2014.03.032
10. Polach, O.: On non-linear methods of bogie stability assessment using computer simulations. Proc. Inst. Mech. Eng. Part F J. Rail Rapid Transit. **2020**(1), 13–27 (2006). https://doi.org/10.1243/095440905X33251
11. Steišūnas, S., Bureika, G., Liudvinavičius, L.: Survey of assessment methods of rolling-stock chassis hunting and derailment processes. In: TRANSBALTICA 2013: The 8th International Conference, Vilnius, Lithuania, 9–10 May 2013, pp. 218–224 (2013)
12. Wheel-rail interface book. Edited by Lewis, R. and Olofsson, U.: Wood-Head Publishing Limited, 841 p. (2009)

Approaches to Improving the Locomotive Maintenance Organization System Through the Introduction of Reliability Centered Maintenance

Oleksandr Ochkasov[1], Maksym Ocheretniuk[1], and Viačeslav Petrenko[2](✉)

[1] Ukrainian State University of Science and Technology, Lazaryana 2, Dnipro 49010, Ukraine
[2] Vilnius Gediminas Technical University, Plytinės st. 27, 10105 Vilnius, Lithuania
viaceslav.petrenko@vilniustech.lt

Abstract. The efficiency of the locomotives depends on the rational use of technical resources intended for the smooth operation of railway transport. The organization of maintenance of locomotives is one of the most important areas in the railway sector and is constantly under development and improvement. One of the directions for improving the locomotive maintenance system is the use of the RCM methodology – Reliability Centered Maintenance – to optimize the planning of processes related to the maintenance of locomotives. In order to improve the maintenance system of locomotives, the authors made calculations concerning the predicted wear of wheelsets in locomotives based on a critical analysis of data obtained from the locomotive depot. The resulting calculations make it possible to predict the setting of locomotives for maintenance, considering the previous analytical data obtained during maintenance. The aim of improving locomotive maintenance is the reduction in downtime for locomotives that are idle waiting for maintenance. The paper presents the statement of the problem with the use of input data, analysis of the data, and their application for further prediction of equipment wear. The research shows that the application of the RCM methodology on the example of a locomotive depot allows for obtaining predictive data for subsequent use in the management of the maintenance system.

Keywords: Railway · Rolling stock · Locomotives · Reliability Centered Maintenance (RCM) · Maintenance organization

1 Introduction

Improving the organization of maintenance for railway rolling stock is an urgent issue for the railway industry of Ukraine, which is determined by the increase in the number of failures in the operation of rolling stock. As a result, there is an increase in the cost of maintenance and repair of wagons and locomotives, a decrease in profitability due to unscheduled repairs and lengthy downtime of rolling stock on maintenance [1].

When managing the technical condition of locomotives, the following types of maintenance systems are used: maintenance system to failure (reactive maintenance), routine-preventive maintenance system (periodic maintenance), maintenance system by condition (predictive maintenance), the system of integrated management of the technical

O. Prentkovskis et al. (Eds.): TRANSBALTICA 2023, LNITI, pp. 604–613, 2024.
https://doi.org/10.1007/978-3-031-52652-7_60

condition (prescriptive maintenance) [2]. Given the complexity of the locomotive as a technical object, as well as the technical possibilities of introducing diagnostic systems on locomotives, in most cases a combination of the systems listed above is used. A detailed analysis of the feasibility of using one or another locomotive maintenance system is described in [3–5]. In [3], the authors considered the features of choosing a maintenance system for locomotives equipped with on-board diagnostic systems. The paper [4] is devoted to a comparative analysis of the use of various approaches in the organization of the locomotive maintenance system of Ukrainian Railways. Relationships of the stages in the transition from a routine-preventive system to a maintenance system for locomotives according to technical conditions, as well as a comparative analysis of locomotive maintenance strategies, are given in the paper [5].

Most of the techniques for choosing the parameters of the maintenance system are based on the use of statistical data on the reliability of locomotives in operation and are focused on optimization criteria. Optimization criteria, as a rule, are the adjustment of overhaul periods and repair volumes, taking into account operating conditions, repair organization, reliability level, minimizing the maintenance costs, minimizing the number of unscheduled repairs, and minimizing the life cycle costs. In addition to using the optimization approach, methodologies of RCM (Reliability Centered Maintenance), RBM (Risk Based Maintenance), and RAMS (Reliability, Availability, Maintainability and Safety) have become widespread recently [3].

Currently, when building maintenance systems for technical facilities, the risk-based maintenance methodology (RBM) for traction rolling stock is widely used. The RBM combines both optimization approaches to building maintenance systems and elements of RCM using preventive maintenance approaches.

The basic concept of RBM is the concept of "risk" and its minimization or support at a given level. The concept of "risk" is widely used in the economic and technical spheres. The usage examples of RBM approaches in railway transport are given in [6–8]. The complexity of using RBM is the need to collect and analyze large amounts of statistical data. Moreover, the RBM approach is more appropriate to use when managing a rolling stock fleet, since the construction of an individual maintenance system is complicated by the duration of data collection.

One of the directions for improving the rolling stock maintenance system is the use of the RCM methodology to ensure the continuous functioning of railway enterprises and transport companies. The RCM is maintenance aimed at ensuring the reliability of equipment. It is a methodology that enables an enterprise to optimize an asset maintenance and repair program [9–11].

RCM is a systematic approach to maintaining, analysing, and accounting for possible failures of all systems of a research object or an entire company. The major priority is given to safety and efficiency from the viewpoint of the profitability of the asset being serviced, which makes it possible to identify and classify effective and possible for applying tasks in the preventive maintenance for the fleet [9, 12]. Therefore, RCM is about reducing maintenance costs, focusing on critical system functions, and excluding maintenance activities that are not strictly necessary. The methodological foundations of approaches in various fields of technology are considered in [9, 12, 13]. Examples of using the RCM methodology in the development of rolling stock maintenance systems are given in [14–16].

The main objective of the RCM methodology is to extend the life of the equipment and reduce downtime in the most efficient way. The aim of this methodology is to identify and prevent known or predictable types of functional failures of equipment, system, or other technical facility. When applying this methodology in the maintenance of rolling stock, a systematic approach to maintenance, analysis, and accounting of possible failures of all rolling stock systems is noted [14]. Top priority is given to safety and economy in terms of the cost-effectiveness of the serviced nodes, which allows you to identify and classify effective and possible applying tasks in the preventive maintenance for the fleet of locomotives.

The RCM methodology is used to build a maintenance system focused on ensuring the operability of the locomotive, in this case, the requirement for a fully serviceable condition of the locomotive is not always ensured. This approach is relevant in the context of the need to reduce the cost of maintenance performance and repair of locomotives. At the same time, a key factor in the RCM methodology is the performance of engineering analysis for the functions of locomotive nodes and systems in order to define how each type of failure can be detected and prevented, as well as the consequences of a failure occurrence. One of the main tools in the RCM methodology is a system for monitoring operational parameters, which for locomotives is implemented through a maintenance system, the use of monitoring, and diagnostic systems in operation.

From the viewpoint of approaches to the management of the technical condition of the locomotive, the RCM methodology is focused on implementing a combination of approaches: reactive, periodic, and predictive maintenance. For each major node or system of the locomotive, the most appropriate approach is selected. When choosing an approach, the criticality of the failure, the rate of failure development, the nature of the development of degradation processes, and the engineered capability in monitoring and diagnosing operational parameters are taken into consideration. This ensures high efficiency in monitoring the technical condition of the locomotive with a reduction in maintenance costs.

The aim of the approaches in the RCM methodology is to concentrate efforts on maintaining the locomotive in working condition, reducing the consequences of failures, and introducing individual planning of the frequency and scope of maintenance and repair work for each specific locomotive.

The Reliability Centered Maintenance methodology makes it possible to implement an individual maintenance system for rolling stock, requires the collection of a large amount of data for each individual rolling stock unit, and also requires the fastest possible processing and analysis. When considering the implementation of the Reliability Centered Maintenance methodology, one can choose groups of equipment or individual technical nodes of the rolling stock.

The results of solving the problem of building an individual locomotive maintenance system using RCM approaches are further used both when planning setting for locomotive maintenance and for planning the load level of locomotive repair shops, and the volume of required resources. An example of building a load model for a locomotive repair shop using data on the reliability of locomotive nodes and resources required for repair work is given in [17].

2 Research Methods Used for the Investigation

In this paper, as an example of the use of RCM approaches, the authors conducted research concerning the possibility of defining an individual remaining resource for locomotive nodes and further planning work related to the maintenance of locomotives.

As a node for the research, the wheelsets of the locomotive were chosen. The choice of wheelsets is explained by the simplicity of obtaining statistical data on wear, as it does not require sophisticated measuring equipment; on the other hand, the wheelset is an important part, the condition of which directly affects traffic safety. During the research, we collected statistical data on the wear of wheelsets of eight-axle passenger electric locomotives ChS7 in one of the locomotive depots of the Ukrainian Railways.

The purpose of the research was to analyze the wear values of the wheelsets of electric locomotives, on the basis of the analysis, calculate the expected moment of the next maintenance for wheelset turning, and also make a predictable calculation of the ultimate wearing of the wheelsets rim. The data obtained from the locomotive depot allowed us to analyze and define the predicted residual life for each wheelset and for the locomotive as a whole.

According to the technical documentation [18], the limiting parameters for the use of a wheelset on an electric locomotive are the boundary values of the rim thickness and the flange of the wheelsets. For the investigated electric locomotives, the boundary values of the rim thickness range from 95 mm (h_{max}) and 45 mm (h_{min}). The boundary values of the flange thickness t fact of the wheelset are 30 mm – t_{max} and 25 mm – t_{min}, respectively.

Based on the analysis of the process in the operation of electric locomotives, it was found that in 95% of cases, the setting of the locomotive for wheelsets turning occurs due to a decrease in the thickness of the wheelset flange to a critical value $t_{min} = 25$ mm. This happens for a number of technical reasons, which include the natural wear of V_{wn} (metal-on-metal friction), the mode and areas of operation of the locomotive, and the quality of the metal from which the locomotive rims are made is of great importance. The t_{min} parameter limits the locomotive's being in operation. W_{ith} a value of $t_{min} <$ 25 mm, the operation of the locomotive is prohibited. In order to restore the allowable value of the flange thickness t fact to allowable values within $t_{min} - t_{max}$, maintenance of the locomotive is carried out for wheelsets turning. While performing maintenance on wheelsets turning in order to eliminate the rejection parameter for the wheelset flange thickness t fact in the process of wheelsets turning, along with a change in the flange thickness t fact, the value of the rims thickness of the wheelset h fact also changes. From the above, it follows that during the maintenance of wheelsets turning, due to the critical value of the locomotive flange parameters, both the value of the locomotive flange parameter and the value of the wheelset rims thickness change.

The main task of the research is to analyze the statistical data collected at the depot and perform a predictive calculation concerning the setting of a locomotive for maintenance on wheelsets turning, as well as perform a predictive calculation of the residual period in the operation of the locomotive wheelsets, which (a period) depends on the thickness of the wheelset rim.

Definition of the predicted residual period (mileage) in the operation P_{expl} of the wheelset until the next maintenance for wheelset turning is defined:

$$P_{expl} = \frac{t_{res} \cdot 10^4}{V_g}, \tag{1}$$

where t_{res} – the residual thickness of the wheelset flange, mm; V_g – the rate of natural wear of the wheelset flange thickness, mm/km.

The Eq. (2) is used to define the residual thickness h_{res} of the wheelset rim:

$$h_{res} = h_{fact} - h_{min}, \tag{2}$$

where h_{res} – the residual value of the wheelset rim thickness, mm; h_{fact} – the actual value of the wheelset rim thickness, mm; h_{min} – the minimum value of the wheelset rim thickness, mm.

To define the natural wear rate V_{wn} of the wheelset rim thickness, equation can be used:

$$V_{wn} = \frac{h_{fact1} - h_{fact2}}{d_1 - d_2} \cdot 10^4, \tag{3}$$

where V_{wn} – the rate of natural wear of the wheelset rim thickness, mm; h_{fact1} – the actual value of the wheelset rim thickness at the mileage of d_1; h_{fact2} – the actual value of the wheelset thickness at the mileage of d_2; d_1 – the value of the locomotive (wheelset) mileage when measuring the value of the wheelset parameters h_{fact1}; d_2 – the value of the locomotive (wheelset) mileage when measuring the value of the wheelset parameters h_{fact2}.

The turning factor has a significant impact on the wear rate of the rim thickness. To define the predicted residual period of operation P_{expl} for the wheelset, we use equation:

$$P_{expl} = \frac{h_{res} \cdot 10^4}{V_{wn} + V_m}, \tag{4}$$

where h_{res} – the residual value of the wheelset rim thickness, mm; V_{wn} – the rate of natural wear of the wheelset rim thickness, mm/km; V_m – the wear rate of the wheelset rim thickness, taking into account turning, mm/km.

Setting a locomotive (on the example of an eight-axle locomotive) for maintenance on wheelset turning is carried out from the condition of choosing the minimum residual mileage for all wheelsets in the locomotive:

$$P^{locom}_{expl} = \min\{P^1_{expl}, P^2_{expl}, P^3_{expl}, P^4_{expl}, P^5_{expl}, P^6_{expl}, P^7_{expl}, P^8_{expl}\}. \tag{5}$$

Thus, in order to implement individual prediction of the residual life for locomotive wheelsets using the approaches of the RCM methodology, it is necessary to organize a system for collecting operational indicators, such as the fringe thickness and the rim thickness. Having a sufficient amount of statistical data, it is possible to predict the mileage of the locomotive at which it is necessary to perform the wheelset turning, as well as the resource remaining for the wheelset change. Let us consider a practical example of using this approach.

3 Results

To conduct research on predicting the wear of wheelsets in locomotives, four main-line electric passenger locomotives of the ChS7 series were chosen as research objects. All selected electric locomotives were operated in the same conditions: the same railway line, similar trains combinations, all locomotives from one locomotive depot. When performing the research, the results of monitoring the parameters of wheelsets in electric locomotives from the moment in the overhaul of electric locomotives were collected. When performing overhauls, new wheelsets are installed. At the time of data collection, the mileage of electric locomotives after overhauls ranged from 250 thousand km to 400 thousand km. The average mileage of a wheelset after a major overhaul before the change is 400–500 thousand km. For the operation area under examination. For two electric locomotives, statistics were collected on the operating interval of 350 and 400 thousand km of mileage. For the other two, it was in the range of up to 250 thousand km of mileage. Thus, it became possible to further predict the wear of the wheelset rim in the electric locomotives, and compare the predicted resource with the actual wear data of the wheelsets under similar operating conditions.

For clarity in displaying the results of the analysis, the paper presents data on the analysis of the wear for the first wheelset in each electric locomotive. As noted earlier, one of the most important parameters that limit the setting of the locomotive for maintenance for wheelset turning is the wheelset flange thickness. In this regard, an analysis was made of the wear rate of the wheelset flange. When performing the analysis, the rate of change in the wheelset flange thickness per unit of the mileage was defined. An example of a graph showing the wear rate of the wheelset flange is shown in Fig. 1. As can be seen from the graph, there is uneven wear of the flange during the operation of electric locomotives. Similar uneven wear of the flange was observed in all electric locomotives. Predicting the residual thickness of the flange without an analysis of additional conditions that characterize the operating conditions is not advisable because of the low predictive accuracy. The reasons for uneven wear can be seasonal factors, the condition of the railway track, etc.

The parameter on which the service life of the wheelset depends is the rim thickness. During the operation of the locomotive, this parameter varies from 90 mm to 45 mm. The reasons for the rim thickness reduction are two factors: natural wear and the performance of maintenance for wheelset turning due to wear of the wheelset flange (artificial wear). Based on the analysis of statistics, graphs of wear for the wheelset rim in the electric locomotive are constructed. As one can see from the graph, the main reason for changing the rim thickness is the rim returning. The steps on the wear graph correspond to the moments when there is a wheelset turning. Similar to the flange wear, the rim wear, the frequency of turning, and the reduction in the rim thickness at wheelset turning are not uneven. Thus, the long-term planning in the remaining life of the wheelset must be performed after each wheelset turning. Refinement of the value of the residual resource of the wheelset must be performed after each control of the parameters of the wheelset in operation. A linear function can be used as a predictive function (Fig. 2). Despite the approximate nature of the prediction, an important parameter is the predicted rate of reduction in the rim thickness. Using the capabilities of modern information systems based on the regular collection and analysis of statistics on the geometric dimensions for each wheelset, it becomes possible to obtain an individual prediction of the resource before the wheelset changing. At each step of turning and checking the geometry of the wheelset, the remaining resource will be specified.

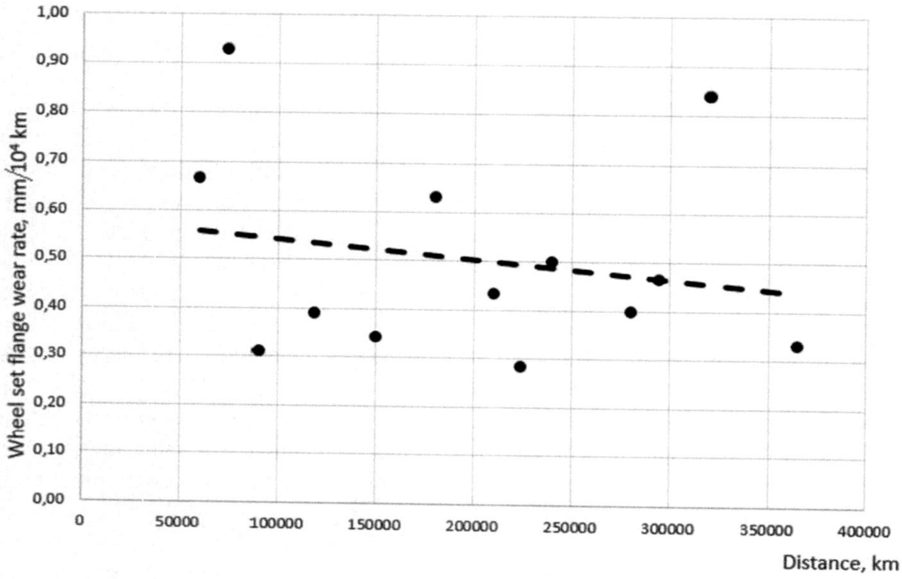

Fig. 1. Wheelset flange wear rates, where "points" is the wear data, "line" is the linear trend.

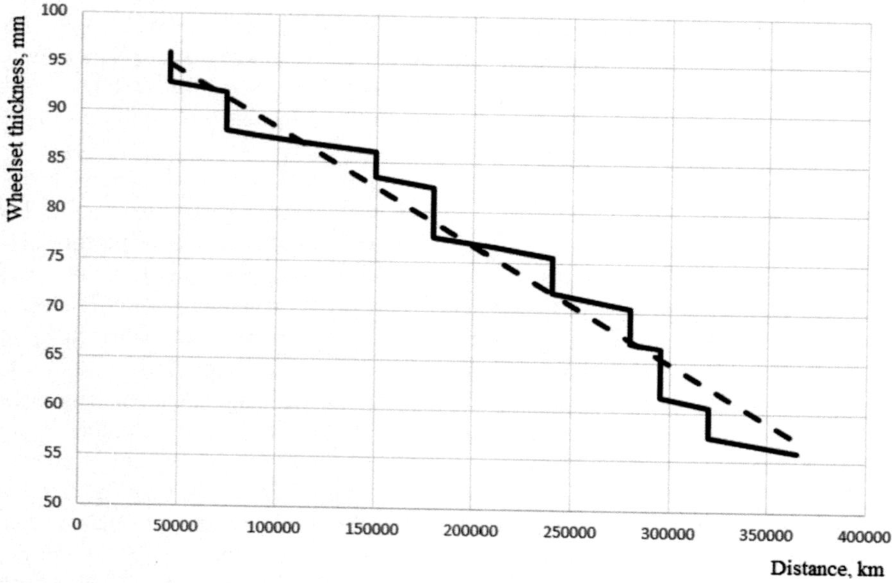

Fig. 2. Wear of the wheelset rim, because of turning, where "solid line" is measurement results, "dotted line" is the linear trend.

An example of the analysis of the collected data on rim wear and the results of defining the average wear rate of the rim for each wheelset for four electric locomotives is shown in Fig. 3.

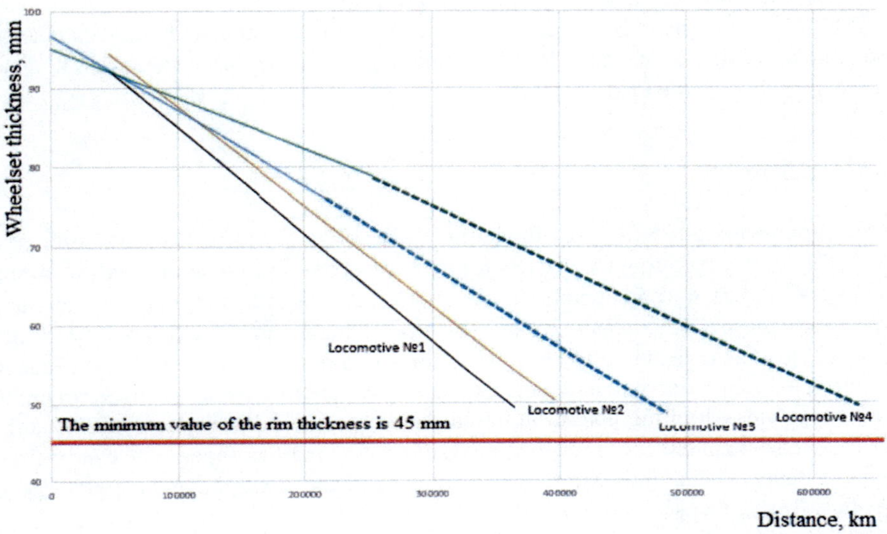

Fig. 3. The graph of the predicted wear of wheelset rims, where "solid line" is real wear, "dotted line" is predicted wear.

The graph (Fig. 3) shows the reduction in the wheelset rim thickness of the locomotive, relative to the mileage. Lines 1 and 2 are marked with a solid line, built based on real data from locomotives No. 1 and No. 2. Lines 3 and 4 for locomotive No. 3 and No. 4, respectively, consist of two parts. The solid line is real data, and the dashed line is predicted data, which is built on the basis of the collected turning and wear data. The predicted line is not constant, so each time the system receives new parameters, the predicted line will change due to the automatic recalculation of the system. The mileage of the electric locomotive, corresponding to the boundary value of the rim, is the point of intersection of the graph of the average wear rate with the horizontal line of the minimum rim thickness.

As one can see from the graph, during operation, the rim wear rate increases significantly, which is explained by an increase in the number of turnings, each turning significantly reduces the resource of the wheelset before changing.

Having analyzed the data, we can conclude that the information collected for each locomotive needs to be analyzed regarding to its individual parameters. Due to the regularity of data collection and automated analysis, it becomes possible to predict the setting of rolling stock for maintenance, therefore, to plan rationally the time of maintenance and the load on the enterprises' infrastructure. It should be noted that timely monitoring and predicting the parameters of wheelsets can significantly affect the total amount of time required for the locomotive to be in operation. This in turn will lead to a decrease in operating costs for maintenance and an increase in the profit of the enterprise. An analysis of the data collected on the downtime of electric locomotives made it possible to define that during the operation of one locomotive, the idle time in waiting only for maintenance for wheelset turning is 10 days a year, which corresponds to the potential shortfall in profits of the enterprise.

The given example of predicting the residual life of a wheelset before its change shows the feasibility of using the RCM methodology for nodes whose wear has a linear or predetermined character.

4 Conclusions

When implementing the RCM methodology, it is obligatory to implement systematic monitoring of the operating parameters for the locomotive. Based on the analysis of the dynamics of the changes in operational parameters, the intervals for conducting control inspections, the periodicity period, and the size of maintenance and repair work are adjusted. RCM is considered to be optimal for a combination of several maintenance strategies, namely reactive maintenance, planned maintenance (based on operating time or time intervals), limiting condition-based maintenance, and preventive maintenance service. These maintenance strategies, instead of being applied separately, can be used simultaneously. This allows you to take advantage of the strengths of each strategy to minimize lifecycle costs.

It is most expedient to implement the RCM methodology for nodes whose failures are caused by wear and the law of failure development is known. Such nodes include wheelsets, brake pads, brushes for traction motors, etc.

The paper gives an example of an individual predicting the remaining resource (mileage before changing) of wheelsets for passenger electric locomotives ChS7. When performing the research, statistics were collected on the wear of wheelsets from the moment of the overhaul to the mileage of 350 and 400 thousand km. For two locomotives. It is also and for two other electric locomotives, in the range of up to 250 thousand km of mileage.

The analysis of the wear rate for flanges and rims of wheelsets in electric locomotives is carried out. The paper presents the results of the analysis of the wear of the first wheelset for each electric locomotive since the wear of the first wheelset defined the moment for setting the electric locomotive for maintenance with wheelset turning. The analysis showed uneven wear and different wear rates of wheelsets for different electric locomotives, despite the same operating area. It is shown that the main factor limiting the service life of the wheelset is the number of wheelsets turning, since in the process of turning the wheelset rim thickness in the electric locomotive is significantly reduced.

Using the proposed approach does not allow predicting for a long interval. However, it allows for ensuring safety, due to regular monitoring and predicting the remaining resource, efficiency, by reducing downtime in waiting for maintenance and increasing the completeness in the use of the wheelsets resource.

References

1. Ochkasov, O., Ocheretniuk, M., Skvireckas, R.: Approaches to the improving the locomotive fleet management system. In: Transport Means 2021: Proceedings of the 25th International Scientific Conference, pp. 1054–1058 (2021). https://transportmeans.ktu.edu/wp-content/uploads/sites/307/2018/02/Transport-Means-2021-Part-III.pdf
2. Lee, J., Ni, J., Singh, J., Jiang, B., Azamfar, M., Feng, J.: Intelligent maintenance systems and predictive manufacturing. ASME. J. Manuf. Sci. Eng. **142**(11), 110805 (2020). https://doi.org/10.1115/1.4047856
3. Bodnar, B., Ochkasov, O.: System choice of the technical maintenance of locomotives equipped with on-board diagnostic systems. In: Transport Means 2017: Proceedings of 21st International Conference, Juodkrante, Lithuania, 20–22 September 2017, Part I - pp. 43–47. Kaunas University of Technology, Kaunas (2017)
4. Bondar, B., Ockasov, O., Petrenko, V., Martishevskij, M.: Implementing intelligent monitoring of the technical condition of locomotive hydraulic transmissions. In: Prentkovskis, O., Yatskiv, I., Skačkauskas, P., Maruschak, P., Karpenko, M. (eds.) TRANSBALTICA 2022. LNITI, pp. 726–736. Springer, Cham (2023). https://doi.org/10.1007/978-3-031-25863-3_70
5. Tartakovskyi, E., Ustenko, O., Puzyr, V., Datsun, Y.: Systems approach to the organization of locomotive maintenance on Ukraine railways. In: Sładkowski, A. (ed.) Rail Transport—Systems Approach. SSDC, vol. 87, pp. 217–236. Springer, Cham (2017). https://doi.org/10.1007/978-3-319-51502-1_5
6. Grencik, J., Poprocký, R., Galliková, J., Volna, P.: Use of risk assessment methods in maintenance for more reliable rolling stock operation. In: MATEC Web of Conferences (2018)
7. Xu, S., Chen, C., Lin, Z., Zhang, X., Dai, J., Liu, L.: Review and prospect of maintenance technology for traction system of high-speed train. Transp. Saf. Environ. **3**(3) (2021)
8. Bulakh, M., Okorokov, A., Baranovskyi, D.: Risk system and railway safety. Paper Presented at the IOP Conference Series: Earth and Environmental Science (2021)
9. Afefy, I.H.: Reliability-centered maintenance methodology and application: a case study. Engineering **2**(11), 863–873 (2010)
10. Carretero, J., et al.: Applying RCM in large scale systems: a case study with railway networks. Reliab. Eng. Syst. Saf. **82**(3), 257–273 (2003)
11. Rezvanizaniani, S., Valibeigloo, M., Asghari, M., Barabady, J., Kumar, U.: Reliability centered maintenance for rolling stock: a case study in coaches' wheel sets of passenger trains of Iranian railway. In: 2008 IEEE International Conference on Industrial Engineering and Engineering Management, pp. 516–520 (2008)
12. Hussain, M.: Enhancing Operational Efficiency with Reliability Centered Maintenance (RCM) (2023)
13. Cotaina, N., et al.: Study of Existing Reliability Centered Maintenance (RCM) Approaches Used in Different Industries (2000)
14. Cheng, H., Cao, Y., Wang, J., Zhang, W., Zeng, H.: A preventive, opportunistic maintenance strategy for the catenary system of high-speed railways based on reliability. Proc. Inst. Mech. Eng. Part F J. Rail Rapid Transit **234**(10), 1149–1155 (2020)
15. Wei, D.I.: Reliability maintainability and maintenance information management of locomotive and car. J. Beijing Jiaotong Univ. **31**(6), 15–17 (2007)
16. Singh, S.B., Suresha, R., Sachidananda, K.H.: Reliability centered maintenance used in metro railways. J. Eur. des Syst. Automatises **53**(1), 11–19 (2020)
17. Bodnar, B.E., Ochkasov, O.B., Bodnar, E.B., Hryshechkina, T.S., Ocheretnyuk, M.V.: Simulation of locomotive repair organization by the methods of queue systems theory. Наука та прогрес транспорту (5), 28–40 (2018)
18. Rules of technical operation of railways of Ukraine, 256 p. Transport of Ukraine, Kyiv (1995)

Customer Center as a Tool for Increasing Competitiveness in Rail Freight Transport

Eva Nedeliakova(✉)

University of Žilina, Univerzitná str. 1, 010 26 Žilina, Slovakia
eva.nedeliakova@uniza.sk

Abstract. Competitiveness in the transport market is perceived as a feature of a company to be successful in economic competition with sustainable growth. The subject of this contribution is a proposal to improve the services of the customer center of a rail freight company. The goal is to emphasize the importance of automating the processes of communication and exchange of information between the carrier and the customer. Based on the results of a research in the form of the Delphi method, where professionals from rail freight companies were addressed, the author proposed basic principles of customer center operation and communication applications that represent a fast, safe and effective tool for information exchange, data collection and their evaluation within the rail company. New applications strengthen business relations between the carrier and the customer using modern technologies. This leads to an improvement in the company's position on the market and an increase in the level of the company's competitiveness.

Keywords: Rail freight transport · Customer center · Competitiveness

1 Introduction

Competitiveness can be understood as the art of a company to produce goods or services while maintaining sufficient efficiency, while sustainable business growth is maintained [1]. Another definition is based on the assumption that the competitiveness of a company is a certain potential, with the help of which a company can be successful in the field of economic competition [2].

Definitions of the competition vary, but in general it can be said that it is a competition or battle between two or more entities that appear on the same market and offer the same or similar products [3].

Rail freight transport has an irreplaceable place in the economic growth of the country. The latest EU legislation documents seek to support the growth of the competitiveness of rail freight transport on the market [3, 4]. The goal is to relieve the burden on road freight transport and transfer services to railways. The reason is the overloaded road network and the associated high costs of maintaining the road infrastructure, a high proportion of greenhouse gases resulting from road freight transport, extensive congestion and a high number of accidents [4, 5]. Rail freight transport represents one of the ecological types of transport, it does not burden the road network and offers the transport of goods in large volumes over longer distances in a relatively short period of time.

O. Prentkovskis et al. (Eds.): TRANSBALTICA 2023, LNITI, pp. 614–623, 2024.
https://doi.org/10.1007/978-3-031-52652-7_61

The European Union approaches rail freight transport as a potential that can sufficiently provide high-quality, safe, ecological and efficient movement of goods over long distances, even in cooperation with other modes of transport [5]. The railway package documents promote awareness of the adequate use of rail freight transport capacity, the transfer of performance from road transport to rail transport, which helps in the competitiveness of rail freight transport [6].

In addition to state-owned enterprises, private ones also entered the rail freight transport market, the market is liberalized. Elements of interoperability are introduced in practice, which primarily help international transport. Despite these efforts, the share of road and rail freight transport in terms of transported tons or transport performance has been different for a long time [3, 7]. Road transport has a significant predominance, which, according to the White Paper, needs to be relieved and transferred to rail transport [5].

The issue of competitiveness is closely related to the company's ability to properly communicate with customers [8, 9]. Although railway freight companies have a certain system of communication with customers, complex services concentrated in one place, in the form of a customer center, are often missing [10, 11]. This reduces the competitiveness of companies [12]. Customer centers are better supported in rail passenger transport [6].

For the above reasons, it is very important to address this issue. The problem of competition in rail freight transport is one of the top priorities because it affects a sustainable transport system [13]. This contribution emphasizes the importance of the existence of a customer center that will simplify communication, make processes transparent and improve customer service [14]. As evidenced by the author's research in companies, the customer center is the basis for higher utilization of rail transport services [15]. This paper is a result of the research KEGA 014ŽU-4/2020 Six Sigma and progressive education of quality management in the study program railway transport under the requirements of transport undertakings and VEGA 1/0379/20 Socio-economics aspects of rail transport market services deregulation in the context of EU legal framework and sustainable mobility and KEGA 010ŽU-4/2023 Innovative approaches in teaching in the field of transport focused on railway traffic management, with the support of risk and crisis management.

2 Research Background and Methodology

The author's research focused on addressing 258 professionals in Europe who work in the marketing and customer communication sections of rail freight companies. The Delphi method was used from February to May 2023, which approached experts in three rounds. The response rate was 67%. All rounds were completed by 105 respondents who were willing to contribute to the improvement of proposals for an efficiently functioning customer center in rail freight transport. The respondents were from the Czech Republic, Poland, Hungary, Germany, Italy, Spain, and Slovakia. This success was due to the fact that the author of the research actively cooperates with railway companies on several international researches. Nevertheless, some information remains confidential at the request of companies. However, what can be presented is concentrated in the following parts of this contribution.

The aim of the research was to find out the experience of experts in the field of communication and services for customers, their opinions, suggestions for the proper functioning of customer centers. At the same time, it was necessary to find out to what extent customer centers in railway freight companies function as centers that provide complex services in one place.

The research methodology consisted in the preparation of expert questionnaires, which included 12 questions in 3 areas. The areas were divided into: Forms of communication with customers, Principles of customer center and other ways of customer care, Design and specification of innovative customer center activities.

The way a company communicates with its customer, otherwise known as the communication channel, can be diverse [10]. The rail freight carrier chooses this communication channel in order to best satisfy his customers while maintaining the efficiency of his work [11]. The customer center is one of the ways in which the carrier can transmit and receive information from customers [16].

A customer center is a communication channel, which represents complex communication, including the entire transport process from planning to invoicing [14]. This mean of communication has a very broad scope in the field of transport covering:

1. the transport planning process,
2. provide transport documents,
3. help with the selection of railway wagons,
4. inform the customer about the arrival of railway wagons for loading/unloading,
5. convey information from the operation to the customer, for example about delays, obstacles during transportation,
6. necessary to combine the means of impersonal and personal communication [17].

This definition of the customer center was interpreted by the respondents and served as the basis for the research.

3 Research Results

3.1 Forms of Communication with Customers in Rail Freight Companies

Interesting results of the research were shown in the first part of the questions, where finally, out of 105 respondents who were willing to answer, 4 ways of how railway freight companies communicate with customers were found.

As can be seen from Fig. 1, the largest share of responses (51%) was chosen by companies that use a department that primarily has another role to communicate with customers. Communication with customers, their requests, informing and processing requests is a secondary task of the given business entity. It can therefore be said that this option is economically efficient due to the lower personnel costs of the company. The company does not need to hire staff who are only dedicated to customer communication, but use the potential of employees who have this role as a secondary one. This model can be used in the case that the employees of the given department really have time and functionality to provide information to customers. However, companies have admitted that this is often at the expense of service quality. The existence of a customer center is in the second place according to the research results with 25%.

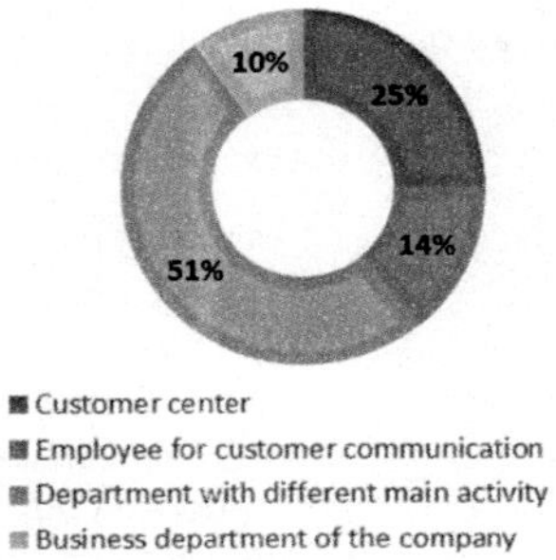

Fig. 1. Entities communicating with the customer.

The oldest form is telephone communication, it offers a quick exchange of information, the possibility to respond to information and ask questions. It is therefore understandable that this communication channel is the most common and widespread among companies operating in transport, as stated by 75% of the respondents. Of course, it is supplemented by communication via email. But according to respondents, both of these systems are not separate enough and their use is tedious.

A separate information system have 73% of companies, but individual forms of communication complement each other. It is the same with communication via e-mail, which represents a common form of communication in 84% of companies and offers a backward search for information and data that are stored on the server. Personal contact is the least represented 8%, as most carriers are legal entities and therefore communicate with customers in the above-mentioned forms.

An interesting result of the research is that companies that have implemented a unified customer center system have expressed that they have positive results in relation to customer satisfaction. They would welcome innovations in service automation that would simplify communication.

A customer center is a tool by which a company can communicate with its customers. Therefore, it is important to know not only how communication is carried out, but also who is carrying out the communication and exchange of information on the part of the given company.

3.2 Competitiveness and Priority Principles in Customer Centers

One of the results of this research is the establishment of basic principles that, according to the experts in the field, should be met by well-established customer centers in rail freight transport. Respondents were asked to indicate which principles of customer care are a priority for them and which answers they would assign to them. As shown in Table 1, they are arranged from number 1 to 4 according to the priority assigned to them by experts from rail freight companies. Table 1 documents the correctly set processes of the customer center.

The respondents expressed themselves in the order of priorities as shown in the table. It is necessary to provide customers with information not only to the necessary

Table 1. The principles of the customer center arranged by priority.

Questions	Answers of respondents according the priority
1. Who are the carrier's customers	It is important that the carrier realizes who the addressee of his information is. Each carrier keeps records of its transports, its customers and also which customer used the service / which one and to what extent. It must therefore formulate the provided information and submit it in a sufficient form and quantity so that it is understandable and comprehensible to the customer
2. What information need to be conveyed to customers?	The carrier must also respond to other changing customer requirements, such as when the wagon will be ready for loading/unloading, where the shipment is currently located, active assistance in filling out transport documents, information on the specifics of transport
3. What means must be used to transmit information so that the information is not violated?	The information that the customer receives should be understood, logical, complete. Therefore, it is necessary that the chosen communication channel and the means that move the message to the customer do not influence and change the given message
4. How to implement it?	In order for the customer center to be effective and fulfill its function to a sufficient extent, a certain objective of the customer center must be established, which will be measurable. In this way, it will be possible to establish efficiency, benefit, coordination and changes in the process of the customer center, so that its functioning will be positive and flexible in the future. In the next line, the carrier can choose whether to solve the issue himself or leave this task to an external company. He must choose the right methods, who and how will coordinate and control the activity of the customer center

extent, for example reports on the bill of lading, conclusion of the transport contract, familiarization with the transport schedule.

A properly configured customer center system provides information that the customer can find on the carrier's online platform. The carrier does not have to share with each customer separately, for example, the transport schedule, the sample bill of lading, manuals for filling in, other regulations necessary for the conclusion of the transport contract, the catalog of freight wagons and the like. The customer center represents a certain platform that combines documents and information, forms a point where the customer, operations and management of the company meet.

Carriers currently offer their customers opportunities to communicate and submit information on their website. The form of the customer center is implemented on various communication platforms such as MS Teams, the activity of the centers takes place online using tables in the Microsoft Excel application. In these shared online spreadsheets, employees can see which request is handled, which is waiting to be handled, and which is in the process of being resolved. With the help of tables, various data are also monitored, such as the number of fulfilled requests, the time of fulfilling the request, the type of demand and transportation, and others, with the help of which the efficiency is determined and the planning of the activity of the customer center is set for the future.

Most functions are only available to registered users who have a contract with the carriers. It also includes the scope of access rights within the use of the customer portal. The portal is connected to other carriers' systems, from which they draw current data from the field of operation. To monitor the current location of the wagons, the carriers use technology based on GPS signals, the so-called Global Navigation Satellite Systems (GNSS) technology. Carriers have implemented measuring units on the wagons that are most used for transportation. These are primarily high capacity sliding wall wagon series E, for example Eas or Eamos, Shimmns or Habbillns wagons. The sensor, which is placed using magnets on the underside of the wagon structure, has dimensions of approximately 16x8x6 centimeters and includes a battery, integrated memory, sensors for sensing the direction, speed or impact force of the wagon, and the sensor also includes a SIM card.

From the point of view of the software solutions, various programs are used, for example Wagon Monitoring System (WMS). The current position of the wagon, passage through checkpoints, speed of the wagon, position of the wagon, history of the wagon position and others are displayed. The obtained information is primarily used only for own purposes of companies, for the business, operation or maintenance departments. According to the research results, the respondents state that the basic benefits of train tracking should include more benefits also for customers:

a) reduction of the total operating costs of freight wagons,
b) efficient planning of wagon maintenance,
c) optimal planning of wagon cycles,
d) identification of violent damage to wagons,
e) overall, more efficient use of freight wagons.

3.3 Design and Specification of Innovative Customer Center Activities

Proposals were defined based on the results of the Delphi method and knowledge of the functioning of communication between customers and carriers in the countries that participated in the research. The aim of the research was to simplify customer access to free freight train routes and increase the automation of the communication between customer and carrier. A practical guide for customers was created. The customer can enter the transport parameters himself, and the system will subsequently generate trains that meet the specified parameters and have free capacity.

The result of the research is the proposal of four software applications, which are aimed at increasing the automation of the communication between the carrier and the customer, to increase the comfort, speed, relevance and efficiency in the exchange of information for customers. Applications should be a permanent part of customer centers, which would improve the quality of services for customers. They make it possible to expand the portfolio of services where it is important to receive and process information regarding the planning, implementation and progress of transportation. These are the following applications:

1. Timetables of freight trains.
2. Selection of a rail wagon.
3. Intelligent calculation of price for transportation.
4. Current position of the wagon.

The mentioned applications are an extension of the customer portal for customer centers and their trial operation is currently in selected companies in the V4 countries. The applications were created in cooperation with software engineers and the University of Zilina (Slovakia).

Freight Train Timetables. The freight train timetable application helps the customer in the initial stages of planning and carrying out the transport. Thanks to this application, the user can find out some useful information that he needs when planning the transport, namely:

1. numerical designation of the station of departure and the station of destination,
2. narrative and technical authorization of selected stations,
3. search for free train connections that the carrier has,
4. information about the transport route: length of the transport route, route category, load category, graphical representation of the transport route,
5. information on transported goods according to the Harmonized Nomenclature of Goods, NHM code,
6. the dispatch authorization of individual stations offers information on what operations can be carried out in a given station in connection with freight transport, for example connection to a combined transport terminal.
7. technical equipment of stations, for example ramps, track scales, cranes, etc.,
8. information about sidings provides an overview of the name and number of sidings that fall under a given station.

The application shows the customer the trains that the carrier has available. Train connections are displayed: type of freight train, train number, station of departure and destination, estimated time of departure and arrival of the train, date of departure and date of arrival of the train. In the event that there is no direct train connection between the departure and destination stations, a combination of train connections is displayed that allows transportation.

Furthermore, the application provides information about the transport route: transport distance, basic information about the track (track category, load category, electrification, number of tracks on the track). In the case of a combination of several types of tracks, the application will display those that represent the limiting parameters.

Selection of a Rail Wagon. This application is an automated guide in choosing a rail wagon. In this application, the user can choose it according to the criteria that are important to him. The user can choose a rail wagon according to two basic criteria: according to the type of transported commodity and according to the type of rail wagon suitable for transporting the given commodity.

Intelligent Calculation of Price for Transportation. The application serves as an aid in determining the approximate amount of fees. All current prices, rates and calculation methodologies are included in this app, so users always have access to current prices. The application consists of two sub-systems, namely the system for calculating freight charges and the system for calculating additional fees. By subsequently adding these prices, the resulting price is available. In addition, for the calculation, it is necessary to indicate the ownership of the rail wagon, which can be private, customer's or owned by the carrier.

When using all the applications in this user guide, the system will search for the necessary data so that the user does not have to re-enter them.

Current Position of the Wagon. The application offers the user to track the current location of rail wagons that are used for transportation. The user has access to information about the current position of the wagon, but also about the past movement of the wagon along the network. This application also offers a graphical representation of the current position of the wagon and its past movement (see Fig. 2), while information on speed and delay is available.

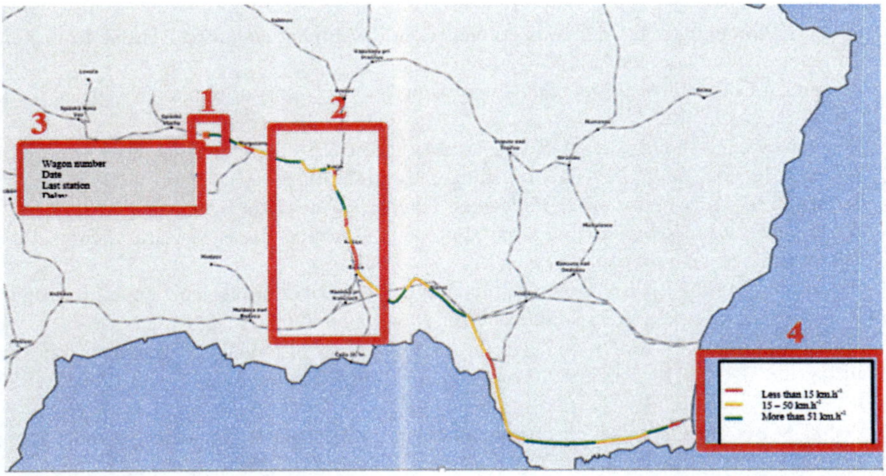

Fig. 2. Current position of the wagon as an information for customer: 1 – the current position of the wagon is represented by an orange square; 2 – the history of the movement of a wagon separated by color according to the speed it had on the given section; 3 – information about the train and the transportation (wagon number, date of filing, last recorded station, current speed, and delay); 4 – legend that contains information about speed intervals, which are divided into individual colors displayed on the map during the last movement of the wagon.

4 Conclusion

It is very important that the company chooses and uses such communication channels and ways of presenting information so that the customer understands them as well as possible, knows how to work with them and can return to them in time. On the other hand, according to the research results, companies that have been in the transport market for a long time and have regular customers have established systems and methods of communication that work favorably for all parties involved.

This research was based on the investigation of forms of communication and exchange of information in various companies operating on the rail freight transport market. Of course, this issue is extensive and it is always important to compare it not only within rail transport but also with other modes of transport, which was also part of the research described above.

The creation of applications, as a result of this research, which the customer can use at any time without the need to involve an employee of the carrier increases the comfort of obtaining information, reduces the personnel burden of the carrier, and represents a significant saving in terms of time. However, it represents difficulty in introducing applications, their design, and software solutions, connecting individual systems, regular updates and maintaining them in an operational state.

References

1. Lall, S.: Promoting industrial competitiveness in developing countries: lessons from Asia. Commonwealth Econ. Pap. **39**, 65–70 (2009)
2. Blažek, L.: Competitive ability of enterprises: analysis of factors of economic success, p. 211 (2008)
3. Nedeliaková, E., Štefancova, V., Kulka, A.: Methods of Evaluating the Quality of Transport Services, 1st edn. Edis, University of Zilina, Slovakia (2013)
4. Parmenter, D.: Key Performance Indicators, 1st edn. Wiley, Great Britain (2018)
5. White paper 2011. Roadmap to a Single European Transport Area - Towards a competitive and resource efficient transport system
6. Pálková, A., Mašek, J., Pribula, D.: Co-modality as a tool for improvement of public transport. Transp. Means: Sustain. Res. Solutions Proc. **1**, 646–650. (2022). ISSN 1822-296X
7. Černá, L., Pribula, D., Bulková, Z., Abramović, B.: Draft of public rail passenger transport during the COVID-19 pandemic. Logi: Sci. J. Transp. Logist. **1**(1), 77–88 (2023). ISSN 1804-3216
8. Dedik, M., et al.: Optimization of timetables on the Bratislava-Zilina-Kosice route in the period after the end of the COVID-19 pandemic. Sustainability **14**(9) (2022)
9. Dolinayova, A., Domeny, I.: Competition on the railway market in a segment of public service obligations in terms of effectiveness: study in V4 countries. J. Competitiveness **14**(3), 41–58 (2022)
10. Jančár, A.: Correct distribution of the selected type of goods on the loading area of the semi-trailer. PHD Progress **10**(2), 1–7 (2022)
11. Konečný, V., Zuzaniak, M., Brídziková, M., Jaśkiewicz, M.: Regional differences in the impact of the COVID-19 pandemic on the demand for bus transport in the Slovak Republic. Logi: Sci. J. Transp. Logist. **14**(1), 146–157 (2023). ISSN 1804-3216

12. Dedik, M., Valla, M., Bulková, Z., Paďourová, A.: Aplikácia progresívnych logistických technológií a dopravných systémov na prepravný reťazec v osobnej deprave v kontexte pandémia COVID-19. In: Horizonts of Railway Transport 2022. Vlak "70" Slovak Republic, pp. 64–71 (2022)
13. Mašek, J., et al.: Proposal for using IT solutions in public passenger transport in Slovak Republic to reduce the spread of COVID-19. Logi: Sci. J. Transp. Logist. **14**(1), 181–191 (2023). ISSN 1804-3216
14. Čulík, K., Harantová, V., Kalašová, A.: Traffic modelling of the circular junction in the city of Žilina. Adv. Sci. Technol. Res. J. **13**(4), 162–169 (2019)
15. Harantová, V., Mazanec, J., Štefancova, V., Mašek, J., Foltynova, H.B.: Two-step cluster analysis of passenger mobility segmentation during the COVID-19 pandemic. Mathematics **11**(3), 1–17 (2023)
16. Domeny, I., Dolinayova, A.: Possibilities for introducing a new night train connection in Middle Europe. Transp. Tech. Technol. **17**(1), 6–12 (2021)
17. Blecharz, P.: Quality and Xustomer, 1st edn. Ekopress, Praha (2015)

Author Index

O. Prentkovskis et al. (Eds.): TRANSBALTICA 2023, LNITI, pp. 625–627, 2024.
https://doi.org/10.1007/978-3-031-52652-7